Lecture Notes in Artificial Intelligence 3398

Edited by J. G. Carbonell and J. Siekmann

Subseries of Lecture Notes in Computer Science

Doo-Kwon Baik (Ed.)

Systems Modeling and Simulation: Theory and Applications

Third Asian Simulation Conference, AsiaSim 2004
Jeju Island, Korea, October 4-6, 2004
Revised Selected Papers

 Springer

Series Editors

Jaime G. Carbonell, Carnegie Mellon University, Pittsburgh, PA, USA
Jörg Siekmann, University of Saarland, Saarbrücken, Germany

Volume Editor

Doo-Kwon Baik
Korea University
Department of Computer Science and Engineering
1,5 Ga, Anam-dong, Sungbook-gu, Seoul, Korea 136-701
E-mail: baikdk@korea.ac.kr

Library of Congress Control Number: 2004118195

CR Subject Classification (1998): I.6, I.2, C.2, J.1, J.2, J.3

ISSN 0302-9743
ISBN 3-540-24477-8 Springer Berlin Heidelberg New York

Springer is a part of Springer Science+Business Media

springeronline.com

© Springer-Verlag Berlin Heidelberg 2005
Printed in Germany

Typesetting: Camera-ready by author, data conversion by Olgun Computergrafik
Printed on acid-free paper   SPIN: 11382416   06/3142   5 4 3 2 1 0

Preface

The Asia Simulation Conference (AsiaSim) 2004 was held on Jeju Island, Korea, October 4–6, 2004. AsiaSim 2004 was the third annual conference on simulation in the Asia-Pacific region. The conference provided the major forum for researchers, scientists and engineers to present the state-of-the-art research results in the theory and applications of systems modeling and simulation. We were pleased that the conference attracted a large number of high-quality research papers that were of benefit to the communities of interest.

This volume is the proceedings of AsiaSim 2004. For the conference full-length versions of all submitted papers were refereed by the respective international program committee, each paper receiving at least two independent reviews. Careful reviews from the committee selected 81 papers out of 178 submissions for oral presentation. This volume includes the invited speakers' papers, along with the papers presented at the conference. As a result, we publish 80 papers for the conference in this volume.

In addition to the scientific tracks presented, the conference featured keynote talks by two invited speakers: Yukia Kagawa (Akia Prefectural University, Janpan) and Bo Hu Li (Beijing University of Aeronautics and Astronautics, China). We are grateful to them for accepting our invitation and for their talks. We also would like to express our gratitude to all contributors, reviewers, program committee and organizing committee members who made the conference very successful. Special thanks are due to Yun Bae Kim, the Program Committee Chair of AsiaSim 2004 for his hard work in the various aspects of conference organization.

Finally, we would like to acknowledge partial financial support for the conference by the Korea Research Foundation. We also would like to acknowledge the publication support from Springer.

November 2004   Doo-Kwon Baik

Conference Officials

Committee Chairs

Honorary Chair   Shoji Shinoda
     (Chuo University, Japan)
     Bo Hu Li
     (Beijing University of Aeronautics
     and Astronautics)

General Chair   Doo-Kwon Baik
     (Korea University, Korea)

Program Chair   Yun-Bae Kim
     (Sungkyunkwan University, Korea)

Advisory Committee

Chan-Mo Park, Pohang University of Science and Technology, Korea
Seong-Joo Park, Korea Advanced Institute of Science and Technology, Korea
Sadao Takaba, Tokyo University of Technology, Japan

Organizing Committee

Xiao Yuan Peng, Beijing University of Aeronautics and Astronautics, China
Soo-Hyun Park, KookMin University, Korea
Toshiharu Kagawa, Tokyo Institute of Technology, Japan
Kang-Sun Lee, MyongJi University, Korea
Guang Leng Xiong, Vice President of CASS, Tsinghua University, China

Program Committee

Keebom Kang, NPS, USA
Sang Hyun Ahn, University of Seoul, Korea
Xu Dong Chai, Beijing Simulation Center, China
Voratas Kachitvichyanukul, Asia Institute of Technology, Thailand
Shinsuke Tamura, Fukui University, Japan
Ning Li, Beijing University of Aeronautics and Astronautics, China
Chi-Myung Kwon, Donga University, Korea
Kyung-Seob Kim, Yonsei University, Korea
Hyung-Gon Moon, Korea Institute for Defense Analysis, Korea
Jin-Woo Park, Seoul National University, Korea
Zi Cai Wang, Harbin Institute of Technology, China
Seok-Jun Yoon, Sejong University, Korea

Table of Contents

Military Simulation

Medical Simulation – I

General Applications – I

Network – I

e-Business

Numerical Simulation

Modeling and Simulation Methodology – II

Traffic Simulation

Network – II

Aerospace – II

Network – III

General Applications – II

Transportation

Virtual Reality

Network – IV

Medical Simulation – II

Engineering Applications

Discrete Huygens' Modelling and Simulation for Wave Propagation

Yukio Kagawa

Department of Electronics and Information Systems, Akita Prefectural University,
84-4, Tsuchiya-Ebinokuchi, Honjo, Akita 015-0055, Japan
Y.Kagawa@akita-pu.ac.jp

Abstract. With the advent of digital computers, wave field problems
are best solved by computers. Present paper introduces a physical model
for wave propagation, which is traced on the computer based on the
mechanism equivalent to the Huygens' principle. The approach does not
require the numerical solution of the differential equation or wave equation which is the mathematical model of the wave propagation phenomena. It only requires the manipulation tracing the sequence of the impulse
scattering at the discretized nodes due to the Huygens' principle. The
paper describes the discrete Huygens' modelling for the acoustic field
and demonstrates the some applied examples.

1 Introduction [1, 2]

Sound and vibration problems belong to a classical physics and their theory
has been well established. The acoustic fields are governed by wave equation of
Helmholtz equation of scalar or vectorial type, which can analytically be solved
only for the linear fields of simple geometry under simple boundary conditions.
For most practical problems, however, numerical approaches must be devised,
which essentially provide approximate procedures but more accurate solutions
than the analytical approaches. The numerical approaches can accommodate the
domain shape and boundary conditions closer to those of the actual problems of
interest. The approaches also provide numerical experiment or computer simulation capability, with which the computer-aided design (CAD) leading to optimal
design is possible.

Finite difference method is one of the oldest among others, which produce a
discretized alternative to the differential equation. A set of simultaneous equation is derived, which is then numerically solved. As the mesh for discretization
are normally fixed, the domain shape and boundary conditions are not always
properly prescribed. The approximation due to this discretization does not cause
much problem, if the meshes are chosen to be much smaller than the wavelength
of interest. The finite difference method now revives and is frequently in use
for the transient analysis, referred as finite difference-time domain (FD-TD)
method. Finite element method is one of the most powerful and versatile numerical techniques, which is capable of accommodating the inhomogeneous media
with general boundary shape and boundary conditions, but unbounded domain

D.-K. Baik (Ed.): AsiaSim 2004, LNAI 3398, pp. 1–11, 2005.

must be considered for most sound radiation and scattering problem. For the last case, the boundary element method is inherently suitable. These methods are all said to be based on the mathematical or numerical models which are numerically solved.

Finite element and boundary element methods are particularly suitable for the stationary or harmonic problems. When the time-domain (transient) solution is required, in many cases, the time-domain is discretized in terms of the finite difference scheme. On the other hand, the discrete Huygens' model is a physical model alternative to the FD-TD, which is particularly suitable for the time-domain solution.

The propagation and scattering of waves are simulated as the sequence of impulse scattering as Huygens' principle states. This is a time-domain analysis method and its algorithm is very simple, as the solution of the differential equation is not required.

The present paper introduces a discrete counterpart to the Huygens' principle and demonstrates some applied examples of the acoustical problems. The simulation also includes sound source identification and tomography in time reversal procedure.

The discrete Huygens' modelling is a synonym of the transmission-line matrix model, which is best explained by the equivalent transmission-line theory as it was originally developed by P. B. Johns for the electromagnetic field problem analysis [3, 4].

What is particular of the method is that the solution is sought in an equivalent transmission-line network in time domain, while the network has traditionally been solved in frequency domain. The solution procedure therefore results in the scattering of the impulses at the nodes taken at the equal-spaced intervals in the field. As computers become increasingly powerful both in execution time and memory size, engineers favor the time-domain approach, which provides the full wave solution.

2 Propagation

2.1 Discrete Huygens' Model

We consider here the sound wave propagation radiated from a point source as shown in Fig. 1, in which the sound wave radiates spherically. According to Huygens, a wave front consists of a number of secondary point sources which give rise to spherical waves, and the envelope of these waves forms a new spherical wavefront which again gives rise to a new generation of spherical waves.

Now we consider the Huygens' principle in the discrete sense in order to implement this process of sequences on a digital computer. The two-dimensional sound space is considered in which the above mentioned process occurs adhering to the directions of the Cartesian coordinates. An array of nodes and mesh which are separated by the distance Δl are shown in Fig. 2. The propagation takes place between isolated nodes of the matrix. When a sound-impulse of unit amplitude or delta function is incident to one of the nodes, the sound-impulses scatter in

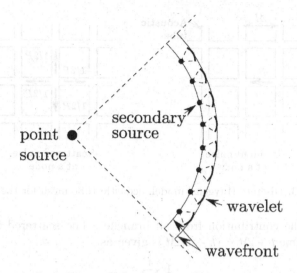

Fig. 1. Huygens' principle.

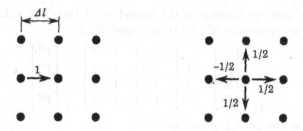

Fig. 2. Discrete Huygens' model.

four directions as shown in the right figure of Fig. 2 like a collision of elastic balls (molecules in the gaseous medium). Each scattered pulse has one-quarter of the incident energy, and the magnitude of the scattered pulse is 1/2, in which the coefficient of the reflected one to the incident direction is negative.

This is equivalent to the travelling and scattering of the pulses over an orthogonal mesh made of transmission lines, or tubes as shown in Fig. 3. The orthogonally connected transmission lines forms impedance discontinuity at the connecting node because the characteristic impedance Z ($= \rho c$, where ρ density and c sound speed) is the same for each branch and three branches are connected to one at the node, so that the impedance looked from one node is one-third of that of the incidence branch. When an impulse of amplitude P is incident on one of the branches, the scattering takes place as shown in the right figure of Fig. 3 due to impedance discontinuity at the node.

We then proceed to the more general case in which four impulses $P^1 \sim P^4$ are incident to the four branches at the same time $t = k\Delta t$ (where $\Delta t = \Delta l/c$ is the time delay required for a pulse to travel the distance Δl, c is propagation speed along the branch and k is the integer). The response can be obtained by

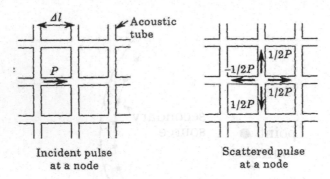

Fig. 3. Discrete Huygens' model, acoustic tube mesh for the field.

superposing the contribution from all branches. The scattered impulse S^n at branch n at time $t + \Delta t = (k+1)\Delta t$ is given as

$$_{k+1}S^n = \frac{1}{2} \sum_{m=1}^{4} {}_kP^m - {}_k P^n \tag{1}$$

where $_kP^n$ is incident impulse at the branch n at time $t = k\Delta t$. This can be rewritten in the expression of a scattering matrix as

$$\begin{bmatrix} S^1 \\ S^2 \\ S^3 \\ S^4 \end{bmatrix}_{k+1} = \frac{1}{2} \begin{bmatrix} -1 & 1 & 1 & 1 \\ 1 & -1 & 1 & 1 \\ 1 & 1 & -1 & 1 \\ 1 & 1 & 1 & -1 \end{bmatrix} \begin{bmatrix} P^1 \\ P^2 \\ P^3 \\ P^4 \end{bmatrix}_k \tag{2}$$

Pressure $P_{i,j}$ at the node is given by

$$_kP_{i,j} = \frac{1}{2} \sum_{n=1}^{4} {}_kP^n \tag{3}$$

where subscripts i, j represent the node position $(x, y) = (i\Delta l, j\Delta l)$.

The scattered pulses travel along the branches in the reversal directions. When the field is divided into square meshes, the scattered pulses become incident pulses to the adjacent elements at whose node scattering again takes place. The incident pulses on a node at position (i, j) are represented by the scattered pulses at the adjacent nodes as

$$_{k+1}P^1_{i,j} = {}_{k+1}S^3_{i-1,j}, \quad _{k+1}P^3_{i,j} = {}_{k+1}S^1_{i+1,j},$$
$$_{k+1}P^2_{i,j} = {}_{k+1}S^4_{i,j+1}, \quad _{k+1}P^4_{i,j} = {}_{k+1}S^2_{i,j-1}. \tag{4}$$

Repeating the operation of equations (2)–(4) on all nodes, the impulse response in the field can be traced at successive time intervals. The method is inherently a time domain solution method which is quite suitable for the simulation and visualization of a wave propagation behavior on the computer. It is possible to show that the method is equivalent to the FD–TD method. However, the present model provides a physical model, which does not require the solution of the wave equation.

2.2 Propagation [5, 6]

From the scattering algorithm it is possible to establish a wave equation. By comparing it with the wave equation in free field, one finds that the propagation speed in the network is $1/\sqrt{2}$ of that of the free field. This factor is $1/\sqrt{3}$ for the three-dimensional field. Fig. 4 illustrates the wave propagation from the center as the scattering. A sinusoidal impulse train is excited at the center. It resembles the wave propagating radially in the water surface when a stone is thrown into the water.

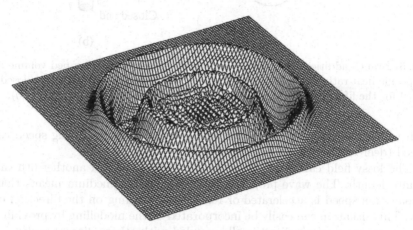

Fig. 4. Cylindrical wave generation for a point excitation (Sinusoidal train).

2.3 Non-reflective Boundary [5, 8]

In many application, an infinite field domain must be simulated. To realize this condition is to provide the non-reflective boundary surrounding the field of interest. The simplest and the most practical way is to terminate the boundary by the characteristic impedance Z_0 ($= \rho c_0$, c_0 is the propagation speed in the free field) of the free field. Considering the characteristic impedance of our network field whose traveling speed is lower by the factor $1/\sqrt{2}$, the reflection coefficient at the boundary is

$$\Gamma = \frac{Z_0/\sqrt{2} - Z_0}{Z_0/\sqrt{2} + Z_0} = -0.17157 \tag{5}$$

The reflection from the element adjacent to the boundary, with this relation present to the the branch connected to the boundary, is as small as -50 dB for the normal incidence. Many other modellings are attempted to reduce the reflection. For example, one assumes that traveling is only possible [7], and another is to provide absorbing multi-layers [8].

2.4 Inhomogeneous, Lossy and Flowing Media [9]

The medium can be inhomogeneous and lossy which can be realized by providing a proper volume or a tube with closed end connected to the node as the 5th tube,

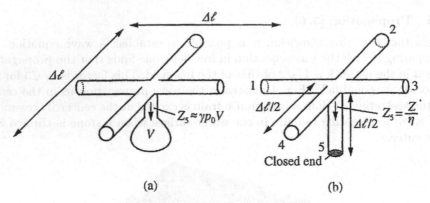

Fig. 5. Branch addition for propagation velocity variation; (a) lumped volume model γ: specific heat ratio $p_0 \approx p$: ambient pressure V: a volume; (b) equivalent closed tube model for the fifth branch (velocity is controlled by the choice of parameter η).

with which the field characteristic impedance or the propagating speed can be varied (decreased). This is illustrated in Fig. 5.

The lossy field can be modelled by the introduction of another 6th tube of infinite length. The wave propagation in the flowing medium means that the propagation speed is accelerated or retarded depending on the direction of the flow. This situation can easily be incorporated in the modelling by providing the extra arms at one node directionally coupled with other adjacent nodes.

2.5   Examples [10, 11]

The acoustic wave is scalar so that it is straightforward to extend to the axisymmetric case. The first example is the sound focusing by a concave lens submerged in water.

Figs. 6 and 7 show the lens and the focusing of the wave.

Figs. 8 and 9 demonstrate an acoustic duct phenomena under the sea surface due to the temperature and saline gradient distribution in the ocean.

3   Back Propagation [11]

3.1   Process in Time Reversal

The particular feature of the present model is that because it is a physical model, the back propagation in time-reversal is easily implemented. Our scattering matrix was expressed by

$$_{k+1}\{S\} = \frac{1}{2}[A]_k\{P\} \tag{6}$$

Premultiplying $[A]^{-1}$ on both side, as $[A]^{-1} = [A]$, one has

$$_k\{P\} = [A]\,_{k+1}\{S\} \tag{7}$$

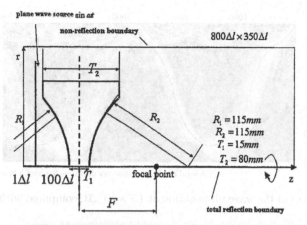

Fig. 6. Cross-sectional view of a concave lens.

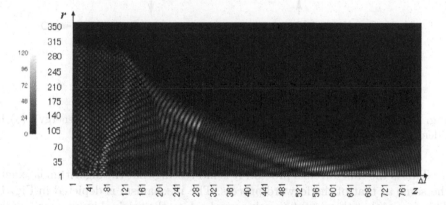

Fig. 7. The power distribution on the analytical plane.

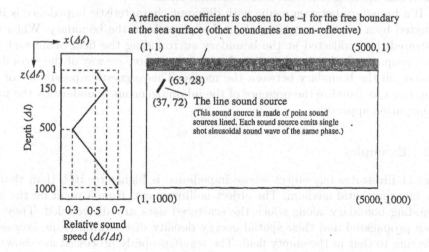

Fig. 8. Sound speed variation in the convergence zone propagation.

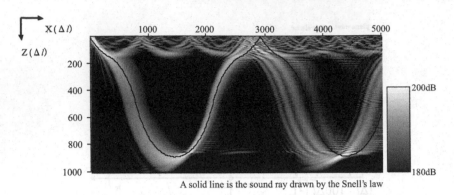

A solid line is the sound ray drawn by the Snell's law

Fig. 9. The traces of the wave propagation at $1.7 \times 10^4 \, \Delta t$, compared with Snell's sound rays.

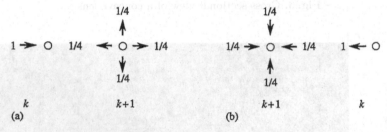

Fig. 10. Scattering and concentration in a two-dimensional element (energy). (a) Incident and radiated impulses; (b) time-reversed impulses.

This is the time reversal process. It is interesting that the algorithm is exactly the same as that of the forward process. The mechanism is depicted in Fig. 10. This means that the output signals come back to the original input source point, which can be utilized for the source identification.

If a boundary between media with different characteristic impedance is illuminated by a source, scattering also takes place from the boundary. When the scattered data collected at the boundary surrounding the field of interest are back-propagated, the signals will go back to the source by way of the secondary sources on the boundary between the media. This suggests a possibility of tomography to visualize the presence of the inhomogeneous boundary by the back propagation approach.

3.2   Examples

Fig. 11 illustrates our object whose impedance is higher by 10% than that of the environmental medium. The object is illuminated from a point on the surrounding boundary, along which the scattered data are all recorded. They are back propagated and their spatial energy density distribution (time-averaged) referring to that in the empty field. The result properly processed are shown in Fig. 12.

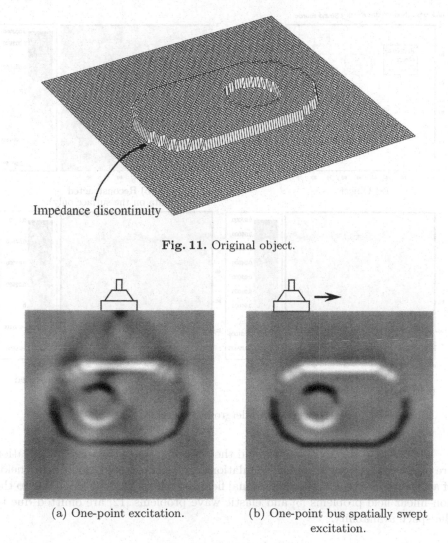

Fig. 11. Original object.

Impedance discontinuity

(a) One-point excitation.

(b) One-point bus spatially swept excitation.

Fig. 12. Reconstructed.

Fig. 13 demonstrates another example, a model for underground prospecting. A circular object with the characteristic impedance by 10 % higher is buried beneath the ground surface (Fig. 13(a)). Impulsive sound of a unit intensity is emitted at a point on the surface and the scattered data are collected over the surface. After the similar process to the previous case, one obtains the picture as shown in Fig. 13(b). When two boreholes are provided on both sides to collect the scattered data in addition to those on the ground surface, the results improve as shown in Fig. 13(c), in which a circular boundary is clearly visualized. Fig. 13(d) is the case when the data are all available on the boundary surrounding the object.

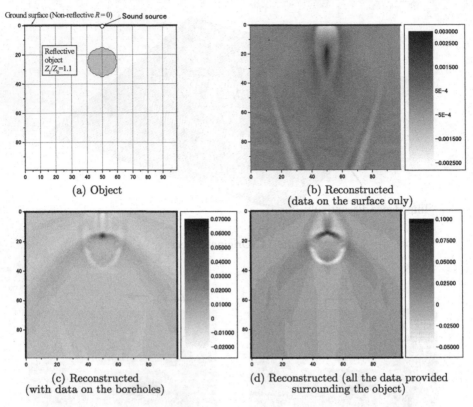

(a) Object

(b) Reconstructed
(data on the surface only)

(c) Reconstructed
(with data on the boreholes)

(d) Reconstructed (all the data provided
surrounding the object)

Fig. 13. Underground sounding.

The discrete Huygens' models and their mechanism of the wave propagation are presented, and some applied simulations are demonstrated. Some other fields of applications to the three-dimensional field modelling [11], the extension to the non-linear field problems [5] and elastic wave problems [12] are omitted due to the space availability.

References

1. Y. Kagawa: Computational Acoustics – Theories of Numerical Analysis in Acoustics with Emphasis of Transmission-line Matrix Modelling. Int. Symp. on Simulation, Visualization and Auralization for Acoustic Research and Education, Tokyo, 2–4 April, (1997) 19–26
2. Y. Kagawa: Discrete Huygens' model and its application to acoustics. 10th Int. Congress on Sound & Vibration, Stockholm, 7–10, July, (2003) P147 (Specialized keynote speech)
3. P. B. Johns and R. L. Beurle: Numerical solution of 2-dimensional scattering problems using a transmission-line matrix. Procs. Inst. Elec. Eng., **118** (1971) 1203–1208

4. P. B. Johns: The solution of inhomogeneous waveguide problems using a transmission-line matrix. IEEE Trans. Microwave Theory and Techniques, **MTT–22** (1974) 209–215

5. Y. Kagawa, T. Tsuchiya, B. Fujii and K. Fujioka: Discrete Huygens' model approach to sound wave propagation. J. Sound & Vib., **218**(3) (1998) 419–444

6. Y. Kagawa, N. Yoshida, T. Tsuchiya and M. Sato: Introduction to Equivalent Circuit Network Modeling. Morikita-Shuppan, Tokyo (2000) (in Japanese)

7. J. A. Morente, J. A. Porti and M. Khalladi: Absorbing boundary conditions for the TLM method. IEEE Trans. Microwave Theory and Techniques, **MTT–40**, 11 (1992), 2095–2099

8. L. Chai, S. Nogi, N. Wakatsuki and Y. Kagawa: Absorbing boundary conditions in Transmission-line Matrix(TLM) modeling. To be published on J. Japan Society for Simulation Technology.

9. Y. Kagawa, T. Tsuchiya, T. Hara and T. Tsuji: Discrete Huygens' modelling simulation of sound wave propagation in velocity varying environments. J. Sound & Vib., **246**, 1 (2001) 419–436

10. L. Chai, S. Nogi, N. Wakatsuki and Y. Kagawa: Focusing characteristics simulation of lenses in time-domain with axis-symmetric TLM modelling. J. Japan Society for Simulation Technology, **23**, 2 (2004) 142–150

11. Y. Kagawa, T. Tsuchiya, K. Fujioka and M. Takeuchi: Discrete Huygens' model approach to sound wave propagation – Reverberation in a room, sound source identification and tomography in time reversal. J. Sound & Vib., **225**, 1 (1999) 61–78

12. Y. Kagawa, T. Fujitani, Y. Fujita, L. Chai, N. Wakatsuki and T. Tsuchiya: Discrete Huygens' modelling approach to wave propagations in a homogeneous elastic field. J. Sound & Vib., **255**, 2 (2002) 323–335

Some Focusing Points in Development
of Modern Modeling and Simulation Technology

Bo Hu Li

Beijing University of Aeronautics and Astronautics,
P.O.Box 3905, Beijing 100854, P.R. of China
bohuli@moon.bjnet.edu.cn

Abstract. Today, M&S combined with High Performance Computing is becoming the third important means for recognizing and rebuilding the objective
world besides the theory research and experiment research. At present, the tendency of system modeling and simulation technology is developing towards
networkitization, virtualization, intelligentization, collaboratization and pervasivization. Based on the research fruits of paper1 and author's recent research
projects, this paper further discusses the meanings of modern M&S and its architecture. Furthermore, some focusing points in recent research and application of M&S are also discussed and prospected, including networkitized M&S
technology based on modern network techniques, M&S technology on synthe-
tized nature environment, Intelligent system modeling and intelligent simulation system, M&S technology of complex system and open, complex, huge system, virtual prototyping engineering technology, high performance computer,
pervasive simulation technology etc.. At last, some conclusions are given.

1 Introduction

Through more than half century, driven by various application's requirement and
related technology, modern modeling & simulation have formed into more complete
technology system, and developed rapidly as an general and strategic technology. At
present, modeling and simulation technology is developing towards the modern direction with the feature of networkitization, virtualization, Intelligentization, collaboratization and pervasivization. It has been successfully applied in many high-tech fields,
such as aeronautics & astronautics, information, biology, material, energy, advanced
manufacturing, etc., and other various application fields, including industry, agricul-
ture, business, education, military affairs, traffic, society, economy, medicine, entertainment, daily life service etc. Along with High Performance Computing, M&S is
becoming the third important means for recognizing and rebuilding the objective
world besides the theory research and experiment research.

Based on paper [1] and author's recent research, the connotation, technology system, and some aspects for M&S technology that should be paid more attention to in
recent research and application are further discussed. At last, some conclusions are
given.

D.-K. Baik (Ed.): AsiaSim 2004, LNAI 3398, pp. 12–22, 2005.

2 The Connotation and Technology System for Modern M&S [1]

2.1 The Connotation for Modern M&S

M&S technology is a synthesized multidisciplinary technology based on analog theory, model theory, system theory, information technology and speciality technology related to simulation application domain. Based on models, M&S technology is used to participate in activities including study, analysis, design, manufacture, experiment, running, evaluation, maintenance and disposal (all lifecycle) of existing or tentative systems by taking advantage of computer system, physical effect device related with application and simulator.

2.2 The Technology System for Modern M&S

With half century's development, the technology system of M&S has formed progressively. Generally, the technology system of M&S mainly includes modeling technology, infrastructure technology for M&S and simulation application technology. It is shown in Fig.1. The more detailed illustration is shown in paper 1.

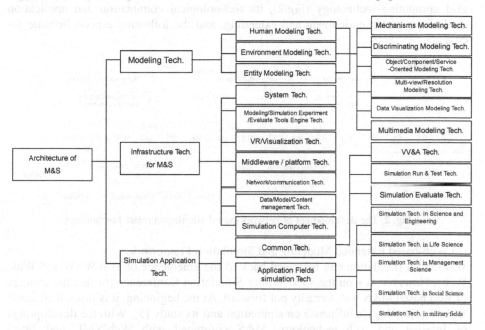

Fig. 1. The technology system of M&S

3 Some Focuses of Modern M&S and Its Application

3.1 Networkitized Modeling and Simulation Technology

The networkitized modeling & simulation generally refers to the technology to realize system modeling, simulation experiment and evaluation on the basis of modern network technology. Modern network-based modeling and simulation began with Distributed Interactive Simulation technology. With the military application and the development of Internet technology, distributed simulation system appeared at the beginning of 1980's, which has many limitation in scalability, performance and its application, limited by network technology then. In order to satisfy the demand for higher degree of fidelity, US DoD put forward the concept of ADS (Advanced Distributed Simulation) in mid 1980's, and its DARPA and Land Forces established together the SIMNET project. From then on, an epoch in ADS was marked, and went through several typical periods such as DIS, ALSP and HLA. At present, HLA has been applied successfully in military area, i.e., STOW project, JADS project and JWARS project, and so on. At the same time, HLA has also gotten broadly attention for civil use, and is becoming a standard for next generation distributed simulation system. HLA is just the beginning of modern network-based simulation and has some limitations on some aspects, e.g. dynamic management and scheduling for simulation resources, capability for self-organization and fault tolerant, security mechanisms, support for management of project etc. [2], but it will make more progress with simulation requirement, simulation technology and other supporting technology. In recent years, with the developing of Internet technology, web/web service technology and grid computing technology (Fig.2), its technological connotation and application mode have gotten enrichment and extending, and the following aspects become focuses:

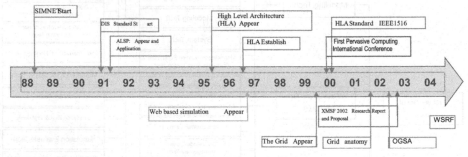

Fig. 2. The development of network-based simulation related technology

(1) Web-based Extended Modeling and Simulation Framework.
Web-based simulation can be traced back to the emergence of WWW (World Wide Web). However, it's on the 1996 Winter Simulation Conference [6] that the concept of web based M&S was formally put forward. At the beginning, it is just talked about WWW technology's influence on simulation and its study [3]. With the development of Internet and web technology, M&S combined with Web/XML and Internet/Network promoted the XMSF [7] founded in 2000. The nature of XMSF is to enable the interoperability and reusability of modeling & simulation in more scalable range (especially including C4I system). At present, XMSF is still in the state of study. However, some demonstration systems also appeared. Besides US DoD, many other country or association carried out the technology research, e.g. SCSI (the Society for Modeling &Simulation International), The Swedish Defense Research Agency (FOI) etc.. In China, the similar things have been put forward and related technolo-

gies are being studied. Current emphases of web-based modeling & simulation are to establish and improve XMSF related standards, technology framework and application mode to emerge HLA SOM/FOM and BOM (Base Object Model) and to combine HLA application's development with web/web service, which enable the execution of federation under the condition of wide area. Web-based simulation and XMSF have much meaning to push distributed modeling and simulation to develop towards the direction of standardization, componentization and the embedded simulation which promotes the merging of simulation and real system.

(2) Grid computing based simulation grid technology.

Grid technology [5] is an emerging technology, which is getting more and more attentions. Its essence is to solve the dynamic share and collaborative operation of various resources on network, e.g. computing resource, storage resource, software resource, data resource, etc.. The combination of simulation and grid will provide huge space for simulation application to access, use and manage simulation resource, at the same time, it also will gives solution for many simulation challenges, e.g. collaborative development of simulation application, the coordination, security and fault tolerance of simulation execution, simulation model and service discovery mechanism, new management mechanism of simulation resource, resource monitoring and load balance, etc. At present, many research projects of simulation combined with grid were carried out to systematically study the simulation oriented grid application, e.g. SF-Express, CrossGrid, NessGrid, DSGrid and FederationX Grid etc.. In China, we also have begun this kind of research, e.g. the simulation grid project CoSim-Grid in Beijing Simulation Center [2], and the grid-based RTI development in Beijing University of Aeronautics and Astronautics, which also have got some achievements.

Through the study and practice, authors put forward the connotation of Simulation Grid: "it is a synthesized multidisciplinary technology and important tool, with the background of application field, synthetically applying complex system modeling technology, advanced distributed simulation technology, VR (virtual reality) technology, grid technology, management technology, system engineering technology and application field related specialized technology, to realize the share and reuse of various resource in grid/federation safely (including model/computing/storage/ network/data/information/knowledge/software resource, and application related physical effect device and simulator, etc.), collaboratively cooperation, dynamic optimization of scheduling executions, etc., to support the various activities in full life cycle (from argumentation, design to maintenance and disposal) for existing or tentative complex system/project in engineering or non-engineering field". A suggested Simulation Grid architecture is illustrated as fig.3.

Currently, Simulation Grid research related key technologies include: simulation grid architecture and system technology, simulation grid service resource dynamically synthetic management technology, simulation field resource gridization (transform to grid service) technology, the cooperation technology in collaborative modeling & simulation, QoS of simulation grid and real-time realization technology, simulation execution monitoring and optimized scheduling technology, simulation grid application development technology, simulation grid visualization technology, simulation grid security technology and simulation grid portal technology, etc.

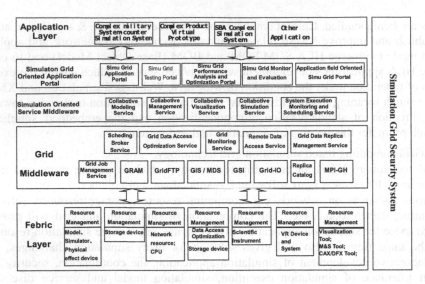

Fig. 3. Architecture of Simulation Grid

Simulation Grid is a common technology for both military application and civil application, which will bring huge reformation and innovation for modeling & simulation application mode, management mode and infrastructure technology, also take great social effects and economic benefit.

3.2 The Prospects of Modeling and Simulation Technology on Synthesized Nature Environment [1]

The behavior and its decision-making of entities can be acted and influenced by nature environment in real world. The effect of nature environment must be considered in simulation system in order to get more realistic simulation outcome true to nature. Some simulations furthermore aim at obtaining either inside transformation of nature environment or environmental affection and influence. The modeling and simulation technology on synthesized nature environment involves the modeling and simulation of terrain/physiognomy/geology/ocean/atmosphere/space/electromagnetic environment, etc.. Especially, the modeling and simulation for electromagnetic environment are considered more, which include modeling and simulation for EMC, electromagnetic transient, frequency spectrum, high frequency environment and microwave environment, etc.. The purpose of synthesized nature environment modeling, with variable intentions, is to build up the model of virtual environment in relation to real multidimensional data throughout real environment. Simulation applications on the basis of environment models convey the influence of nature environment on entity functions and decision-making activities. The static and dynamic multidimensional field methods can be applied in modeling technology of synthesized nature environment. Meanwhile, the interaction between virtual environment and entities will also be tested. The modeling and simulation technology on synthesized nature environment have widely applied to battlefield simulation, military training, space research, industrial inspection and evaluation, multi-knowledge research etc..

Currently, the modeling and simulation technology on synthesized nature environment paces in direction of standardization, multi-disciplinary fusion, real-time dynamic and distributing collaboration, etc., such as SEDRIS [8].

The key technologies to synthesized nature environment involve environment feature abstraction, dynamic modeling and data description of multi-field data, man-machine interaction and interface techniques of virtual environment, multi-resolution modeling, model compression and simplification, inside modeling of nature environment, modeling of virtual environment interface, feature identification and multidimensional navigation in virtual environment, synthesized nature environment non-sight visualization, real-time performance of virtual environment, IBR(Image-Based Rendering), volume rendering and modeling and simulation specification of synthesized nature environment etc..

3.3 The Intelligent System Modeling and Intelligent Simulation System [1]

At present, the research on human intelligent system mechanism (e.g. cerebra and nerve tissue) based on simulation technology and the research on various knowledge based simulation systems have become important fields for simulation application. The typical researches are such as the research on simulation based embedded intelligent system (e.g. simulation can be used in knowledge based system to describe the knowledge related to time or to express the reasoning process related to time.).

Currently, there are many further research focuses for intelligent system modeling, such as research on knowledge model and its expression standardization, especially the research on agent oriented model, ontology oriented model, distributed reasoning oriented network model, mobile communication oriented reasoning model, evolution model, self-organized model and fault-tolerance model, virtual human, etc.. Take the agent-oriented simulation for example, it depends on the agents' ability to be independent and autonomous, and completes the agents-based simulation of social activity and reactivity by the asynchronous message passing in agents. Obviously, the agents-bases simulation of complex social system is one hot topic while this field faces the challenge such as assuring its trustworthiness by controlling the autonomy [14].

At the same time, intelligent simulation system which introduces the technology of artificial intelligence into simulation system accelerates the development of system M&S technology, such as introduce decision support and intelligent plan (typically, expert system, dynamic programming, qualitative inference, fuzzy inference, simulated anealling, neural network, genetic algorithm, etc.) into simulation system to construct the intelligent simulation environment with intelligent front end, the knowledge based collaborative modeling and simulation environment, the simulation environment with intelligent capability of evaluation and analysis, and so on. For example, in simulation of air fight (many to many), the genetic algorithm is used in the selection of object and the study of rules on elusion of missiles, and the neural network is used in the selection of flight mobile. Undoubtedly, the knowledge based intelligent integrated modeling/simulation/evaluation environment will be the emphases of further research and development of intelligent simulation system.

3.4 Modeling and Simulation Technology in Complex System and Open, Complex, Huge System [1]

Complex system/open complex huge system, such as military counterwork system, complex manufacture system, zoology system, human brain system, human body system, complex celestial body system and social economy system, etc., have several unique features shown as follows: complex system composition, complex system behavior, complex interaction between system and its subsystems, complex interaction between system and its environment, complex energy interchange, and so on. The research on these systems has become subjects that are vital to human development. And what's more, modeling and simulation for complex system will become the most important approach to study all kinds of complex systems.

Since 1980's, complex theory and complex science put forward abroad have been viewed as the "science of the 21th century". Dr. Xuesen Qian put forward the science of system, as well as the concept of the "open complex huge system" and synthesis integration methodology from the qualitative to the quantitative. He made important contribution to the research and application of complex theory. In the study of supporting environment of modeling and simulation for complex system, one typical example is RSSE project started in China in 1995 [15].

The modeling and simulation research subjects of the complex system will include parameter optimization, qualitative/fuzzy/organum/induction/combination of the qualitative and the quantitative, adaptation and evolvement, expectancy/learning/adaptation/self-organization, multi-paradigm, parallel and qualitative, based on knowledge/intelligent agents/composition of ontology theory [1] etc. We can foresee that with the successful development of the complex system/open complex huge system M&S technology will greatly improve the modeling/simulation technology's status and effects in progressing of the human society.

3.5 Virtual Prototyping Engineering Technology

In recent years, under the pulling of the requirements of drastic competition of manufacturing and the pushing of the evolution of associated technologies, virtual prototyping technology whose core is system modeling & simulation has developed rapidly. Virtual prototyping technology for complex products is an integrated technology which combine specialistic techniques related to product and modeling and simulation techniques, information techniques, multimedia/virtual reality techniques, AI techniques, system techniques and management techniques, etc.. Virtual prototyping engineering (VPE) technology can be applied in the whole product lifecycle, from system concept design in earlier stage, engineering development, performance evaluation/training in later stage to disposal, to improve the T (Time to market), Q(Quality), C(Cost) of enterprise product development. There are many key technologies and related research fields existing in the development and implementation for VPE, such as the system technology for VPE, the modeling technology for VP, the collaborative simulation technology [4], the management technology for VPE, system concept design and system performance evaluation technology, virtual environment technology, the VV&A technology of model and the supporting platform technology etc.. A typical complicated multidisciplinary virtual prototyping engineering research is the project "Virtual Prototyping Engineering Technology of Complex Product" supported by 863/CIMS subject of China [16].

The further development tendency is to apply the M&S technology into the whole product lifecycle in the robust, collaborative and integrated mode, which includes the multi-domain, multi-dimension, multi-scale, dynamic and evolving modeling technology for complex product, optimization theory and method of multidisciplinary, the whole life cycle management technology of virtual prototyping engineering for complex products, grid platform technology for virtual manufacturing, knowledge based multidisciplinary, distributed and collaborative modeling and simulation for complex product, all aspects dynamic modeling for enterprise etc..

3.6 High Performance Computing (HPC) [17–19]

With the development of simulation application, singular processor cannot meet the need, especially in the fields where complex mathematic models are widely used such as chronometer, weather, biology, medicine and energy. Thus the employ of HPC in simulation is becoming another focus of simulation technology.

Distinct from "general computer", HPC is defined as computer that is purposefully for the optimization design of scientific computing. Vector computer appearing in 1970s could be the first generation HPC, which greatly increased speed of scientific computing by incorporating vector pipelining parts into computers. In the early years of 1980's, the dominance of vector computer was broken by the development of Very Large Scale Integration (VLSI) and microprocessor technology and Massively Parallel Processor (MPP) system became the mainstream of HPC development at the beginning of 1990s. MPP was composed of multi microprocessors through high-speed Internet and processors communicated and corresponded by exchanging message. Symmetric Multiprocessor (SMP) system appearing several years before MPP is a compute system composed by fewer microprocessors sharing physical memory and I/O bus, which is the upgrade and buildup of singular processor, which has been widely used in commercial computing. In the middle and late period of 1990s, a trend was the combination of advantage of SMP and the scalability of MPP, which was the later developed Cache Coherent-Non Uniform Memory Access (CC-NUMA) structure, i.e. distributed shared memory. While developing of CC-NUMA, the cluster system also grows quickly. Similar to MPP structure, cluster system is multiprocessor computer nodes connected through high-speed network. The nodes are often commercial computers, which can work alone. Because of sweeping economy with low cost, cluster system has the higher performance/price ratio advantage than MPP [17].

Realization technology of HPC includes computing mathematics (computing model and algorithm), computer architecture and components structure technology. Nowadays HPC technology is developed in two directions: to develop high performance computer with fast operation capability in depth and to develop high performance server with universal application foreground in width [18]. The former is mainly used in scientific engineering computing and technical design such as Cray T3E. The latter can support computing, transaction processing, database application, network application and service such as SP of IBM and Dawn 2000 of Chinese Academy of Sciences.

The development trend of HPC is mainly on networkitized, open and standardized mainstream architecture and diverse application oriented. Trend on network is of most importance. The main usage of HPC is host computer in network computing environment. Later usage will be in network and will appear millions of clients with all critical data and applications in high performance server. Client/Server style will come to the second generation of collective server style. At present, Grid has become a very important new technology of HPC. Dawn 4000A is a newly developed HPC system for Grid by Institute of Computing Technology, Chinese Academy of Sciences. It is a self-developed 64-bits HPC system with the max computing speed over 100000 billion per second. One of its research goals is to develop HPC to be used as nodes in Grid and acquire Grid enable technology to support Grid application. Dawn 4000A adopts cluster architecture for Grid.

Though application style of network computing environment is still Internet/Web in the near years, information grid style will gradually become main stream in 5~10 years. An important trend of architecture is that super sever is taking over super computer to become main architecture of high performance computing. Low-end products of HPC are mainly SMP, middle-end products are SMP, CC-NUMA and cluster, and high-end products will be cluster using SMP or CC-NUMA nodes [19].

3.7 Pervasive Simulation Technology Based on Pervasive Computing Technology

Pervasive Computing [12], a combination of computing, communication and digital media, presents a new computing pattern. It is aimed to build the smart space, which combines the information space composed of computing and communication and the physical space where people live. In this smart space, people can transparently obtain the computing and communication services everywhere at anytime. The concept of pervasive computing was first presented by Mark Weiser at the article "The Computer for the 21st Century" in Scientific American in 1991. The Ubicomp International Meeting initiated in 1999, the Pervasive Computing International Meeting initiated in 2000 and the start publication of IEEE Pervasive Computing in 2002 have indicated that the research on pervasive computing has attached much importance in the academe.

Pervasive Simulation technology is a new simulation pattern, which integrates the hardware and software of computing and communication, sensor, device and simulator. It combines the information space and the physical space. Significantly, it introduces the simulation into the real system and embeds that into our lives. At present, the focuses on Pervasive Simulation are the advanced pervasive simulation architecture based on Web-based distributed simulation technology, grid computing technology and pervasive computing technology; the collaborative management and integration technology of the information space and the physical space, the self-organization, self-adaptability and high toleration of pervasive simulation based on pervasive computing, the application technology of pervasive simulation [13].

The combination of Pervasive Computing, Grid Computing and Web/Web service technology, is expanding the application boundary of the smart space which is people-centric, sensitive to the environment and convenient to obtain the computing ability everywhere at anytime. It is the infrastructure of the coming IT technology.

The Pervasive Simulation, combined of Pervasive Computing, Grid Computing and Web/Web service, will promote the research, developing and application of modern Modeling and Simulation into a new era.

4 Summary

(1) Through more than half century, driven by various application's requirements and related technologies, modeling and simulation have formed into synthetic technology system. In the new century, the tendency of modern modeling and simulation technology is developed towards networkitization, virtualization, intelligentization, collaboratization and pervasivization.

(2) The following are the problems that should be paid more attention to: modern network technology based distributed M&S and its application, intelligent system modeling and intelligent simulation system, synthetic nature environment M&S, complex system and open, complex, huge system M&S, virtual prototyping engineering technology and its application, HPC, pervasive simulation.

(3) The application of simulation technology has made rapid progress in servicing for the system covering the whole lifecycle, the overall system and all aspects management.

(4) The development of system M&S technology should be integrated closely with its application, the development direction should be a propitious circulation which covers following stages: the requirements of application should be used to pull the development of modeling and simulation system, modeling and simulation system can bring a breakthrough in M&S technology, and this technology also accelerates the development of the system which recurs to provide services to the application.

(5) The development of simulation industry is the basis and guarantee for the development and application of M&S.

(6) Modern M&S technology has become one of the important means to recognize and rebuild the objective world. The problem of the Modern M&S subject construction and the person training have attached more and more attention [9-11]. So, establishing the discipline of M&S is the urgent demand to its further development and application.

Acknowledgments

Author should like to express the sincere thanks to my colleagues, Xudong Chai, Wenhai Zhu, Yanqiang Di, Peng Wang, Guoqiang Shi, Juan Tan, Runmin Yin, Baocun Hou, for their help and valuable contribution.

References

1. Bo Hu Li, et al, The Development of Modern Modeling & Simulation, Front of China Science (Chinese Academy of Engineering), Vol.6, Higher Education Press, Beijing, 2003
2. Bo Hu Li, Xudong Chai, Wenhai Zhu, Yanqiang Di et al.. Research and Implementation on Collaborative Simulation Grid Platform, SCSC, San Jose,USA,2004

3. Page, E.H., Buss, A., Fishwick, P.A., etc., Web-Based Simulation: Revolution or Evolution? submitted Transactions on Modeling and Computer Simulation, February, 1999
4. Yangqiang Di, Xudong Chai, Bo Hu Li, Study on Simulation Model of Virtual Type for Complex Product, ICAM, December 4-6, 2003. Beijing, China
5. Foster, I. The Grid: A New Infrastructure for 21st Century Science, Physics Today, 55 (2). 2002.
6. Fishwick, P.A.. Web-Based Simulation: Some Personal Observations, In: Proceedings of the 1996 Winter Simulation Conference, pp. 772-779, Coronado, CA, 8-11 December.
7. Donald Brutzman, etc.,: Extensible Modeling and Simulation Framework (XMSF): Challenges for Web-Based Modeling and Simulation, Findings and Recommendations Report of the XMSF Technical Challenges Workshop and Strategic Opportunities Symposium, October 2002, http://www.movesinstitute.org/xmsf/XmsfWorkshopSymposiumReportOctober2002.pdf
8. The SEDRIS web site: www.sedris.org
9. Fujimoto, Principles for M&S Education, http://www.sisostds.org/webletter/siso/iss_61/art_299.htm, Simulation and Technology Magazine, 2000.
10. Yurcik, W. and R. Silverman, The 1998 NSF Workshop on Teaching Simulation to Undergraduate Computer Science Majors, SCSC, Vancouver, Canada, 2000
11. H. Szczerbicka, J. Banks, T.I. Ören, R.V. Rogers, H.S. Sarjoughian, and B.P. Zeigler (2000), Conceptions of Curriculum for Simulation Education (panel), WSC, Orlando, Florida.
12. M. L. Dertouzos. The Future of Computing, Scientific American, 281(2):52–55, Aug. 1999
13. Robert Grimm, Janet Davis, etc., Programming for Pervasive Computing Environments, Technical Report UW-CSE-01-06-01, University of Washington
14. Tuncer Ören,, Future of Modelling and Simulation: Some Development Areas, 2002 Summer Computer Simulation Conference.
15. Bo Hu Li, Xingren Wang, et al. "Further Research on Synthetic Simulation Environment", The Proceedings of the SCSC, 2001
16. Bo Hu Li, Xudong Chai, et al." Virtual Prototyping Engineering for Complex Product", The Proceedings of SeoulSim 2001, Korea, 2001
17. Jianping Fan, et al. Present State and Perspectives of High Performance Computer Research, Information Technology Flash Report, Vol. 1 No. 5, Oct. 2003
18. Jizhong Han, et al. Present State and Trends of High Performance Computing, Chinese Engineering Science, Jan. 2001.
19. Zhiwei Xu, The Application and Trends of High Performance Computer, www.pcworld.

Research of Collaborative Modeling/Simulation Supporting Environment Based on Grid

Ni Li, Hongtao Zheng, Lijuan Xu, and Xiao-Yuan Peng

School of Automation Science and Electrical Engineering,
Beijing University of Aeronautics and Astronautics,
Post Code 100083, Beijing, China
astlini@hotmail.com

Abstract. Grid can realize the share of computing resource, storage resource, information resource, knowledge resource and expert resource through Internet. Simulation Grid, which integrates grid technology and modeling/simulation technology, will greatly promote the development of SBA (Simulation-Based Acquisition) supporting environment and become an important tool for the modeling and simulation of complex systems. GLOBUS Project is the most representative grid research project and the Simulation Grid Architecture was build on the base of GLOBUS Toolkit so that all kinds of simulation resources can be shared and reused provided as simulation grid services. Implement technique and research status of the Simulation Grid architecture are presented in this paper in detail.

1 Introduction

Grid is an integrated environment of resources and users of grid can use the resources of grid such as scientific instruments, high-performance computers, databases, experts, etc. just like using local resources so that they can solve complex problems collaboratively [1]. The current complex simulation engineering supporting environment lacks the ability to realize the secure share and reuse, collaborative interaction, optimizing schedule of all kinds of resources (including models resource, computing resource, storage resource, data resource, information resource, knowledge resource, physical devices, simulators and so on) in the simulation federation. Simulation Grid, which integrates grid technology and modeling/simulation technology, will greatly promote the development of SBA (Simulation-Based Acquisition) supporting environment and become the important tool for the modeling and simulation of complex systems [2].

A collaborative modeling/simulation environment was built on the base of current grid technology, which is Simulation Grid Prototype system (Fig. 1 shows the architecture). Simulation Grid architecture includes simulation grid portal layer, simulation grid services management layer, simulation grid services layer, basic grid platform layer and grid resources layer. In detail,

Grid resources layer provides all kinds of resources including models, toolkits, network, computing, storage, data, information, knowledge resources, physical devices and simulators.

D.-K. Baik (Ed.): AsiaSim 2004, LNAI 3398, pp. 23–30, 2005.

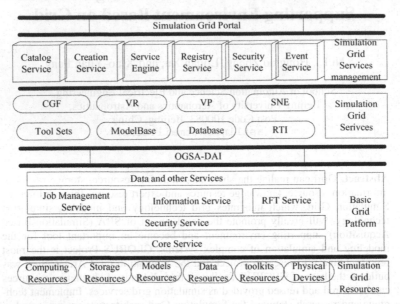

Fig. 1. Shows the five-layer architecture of Simulation Grid based on which simulation resources can be shared and reused

Basic grid platform layer is an open grid services environment developed on the basis of GLOBUS Toolkit 3.x to provide the core services for Simulation Grid middleware, including dynamic/ flexible resource sharing and schedule technology, high throughput resource management technology, data collection/analysis and security technology, and so on. OGSA-DAI (Data Access and Integration) Project aims to develop grid services to realize data access and integration. Through the OGSA-DAI interfaces, disparate, heterogeneous simulation data resources in the grid can be accessed and controlled as though they were a single logical resource [3].

Simulation grid services layer provides core services for collaborative modeling/simulation/evaluation/visualization/management application. Simulation resources are shared in Simulation Grid by developing and deploying simulation grid services according to the OGSA specification.

Simulation grid services management layer is built over the simulation services and provides uniform management function for the simulation resources. Grid platform supporting function can be further extended in this layer by Catalog Service, Creation Service, Service Engine and so on.

Simulation grid portal layer faces simulation grid users to provide the graphical interactive portal, safely and conveniently supporting collaborative modeling/simulation research of the distributed users in a virtual organization.

The implement technique and research status of the Simulation Grid architecture are presented as follows in detail.

2 Basic Grid Platform

At present, grid technology has been developed rapidly. Especially, GLOBUS 3.x version adopted OGSI (Open Grid Services Infrastructure), which integrated Web Service mechanism, specified the creation, management and information exchange means of grid services to provide an extensible platform for different grid applications [4]. GLOBUS 3.x includes core service module, security service module, resource management service module, information service module and data service module [5]. And Java COG (Commodity Grid Toolkits) provides a cross-platform client tool-kit to access the remote Globus services [6]. The function modules of Globus and Java COG were analyzed and encapsulated, based on which AST-GRID software package and services (shown in Fig.2) were developed and integrated to form the basic grid services platform and realize the basic grid function such as security authentication, resource query, job scheduling, file transfer and information management. The above work provided a foundation for the realization of simulation grid services and simulation grid services management.

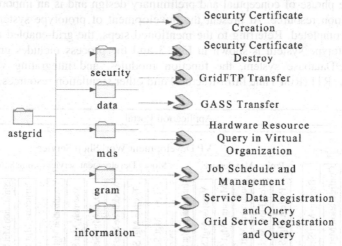

Fig. 2. Shows the function framework of astgrid software package to form the basic grid services platform

3 Simulation Grid Services and Grid-Enabling Simulation Resources

To build Simulation Grid, the emphasis is put on how to make the existing simulation resources grid-enabled and implement the simulation grid services and their portals. Grid-enabling simulation resources means according to OGSA specification, uniformly encapsulating the resources in the simulation grid as grid services and providing service accessing interfaces to share all kinds of simulation resources. Different simulation resources have different function and application modes so that different methods should be adopted to make them grid-enabled. However, there exist common approaches and the steps as follows can be taken:

Analyzing the function of simulation resources, and separating the background logic modules from the application interfaces;

Partitioning the background logic modules into dependent function modules and determining their input and output relation;

Encapsulating the function modules to grid services based on OGSA and determining the service interfaces (including the input parameters and return type). OGSA was implemented with Java, while many existing simulation resources were developed with C/C++. JNI (Java Native Interface) technology can be adopted to realize invoking C/C++ codes, dynamic linking libraries or COM components with Java to avoid reduplicate developing work.

Modifying the original application interfaces to simulation grid service portal. The portal accepts users' input, invokes the corresponding grid services and displays the simulation result to users.

Collaborative modeling/simulation platform of complex system virtual prototype can support the collaborative modeling/simulation of complex system virtual prototype in the phrase of conceptual and preliminary design and is an important modeling/simulation resource. At present the development of prototype system has been basically completed. Referring to the mentioned steps, the grid-enabled architecture of the prototype system is shown in Fig. 3 and the process includes grid-enabling Modelbase/Database system, the function modules, and integrating visualization technology, RTI (Run Time Infrastructure) and other simulation resources.

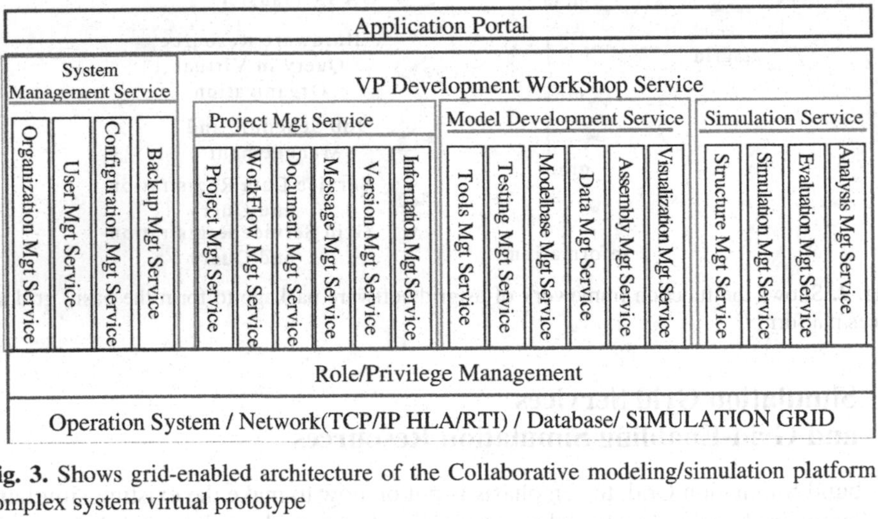

Application Portal			
System Management Service	VP Development WorkShop Service		
	Project Mgt Service	Model Development Service	Simulation Service
Organization Mgt Service · User Mgt Service · Configuration Mgt Service · Backup Mgt Service	Project Mgt Service · WorkFlow Mgt Service · Document Mgt Service · Message Mgt Service · Version Mgt Service · Information Mgt Service	Tools Mgt Service · Testing Mgt Service · Modelbase Mgt Service · Data Mgt Service · Assembly Mgt Service · Visualization Mgt Service	Structure Mgt Service · Simulation Mgt Service · Evaluation Mgt Service · Analysis Mgt Service
Role/Privilege Management			
Operation System / Network(TCP/IP HLA/RTI) / Database/ SIMULATION GRID			

Fig. 3. Shows grid-enabled architecture of the Collaborative modeling/simulation platform of complex system virtual prototype

The bottom layer is the supporting layer. Windows operation system/network (TCP/IP network protocol and High Level Architecture/RTI)/database/grid technology provide support for the platform.

The second layer is the role/privilege layer. A set of relation tables was created in the database to facilitate the role/ privilege management of the system.

The third layer is the function-driving layer, which is responsible for system management and the development, management and evaluation of the products. Each function module is encapsulated as simulation grid service according to OGSA specification.

The top layer is the friendly application portal based on Simulation Grid and Web technology is taken to provide a collaborative environment for the grid users to meet their modeling/simulation needs.

3.1 Grid-Enabling Modelbase System

Modelbase system is a main part of the model development-integration module of the platform system. The management, assembly and simulation of models are all realized based on modelbase system so that grid-enabling modelbase system is the first step for grid-enabling platform system. Modelbase is the aggregation of models that deposits models, provides model definition and holds models' characteristic information. Modelbase management system is used to access and manage modelbase. B-S architecture was adopted to realize grid-enabling modelbase system. Users with proper privilege can access modelbase grid service through its portal in the web browser without additional client software configuration. The modelbase management function was encapsulated in grid services and business logic is separated from portal display. The system is extensible and its function has been basically completed (see Fig.4). Models can be indexed, added, deleted and modified in the system and uploading, downloading models from modelbase are also supported.

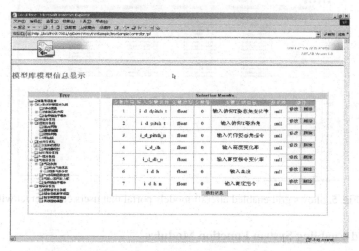

Fig. 4. Shows grid-enabled Modelbase System portal

3.2 Grid-Enabling Database System

There are a lot of modeling/simulation data resources in the collaborative modeling/simulation environment. Grid technology is used to integrate data so that its physical storage is transparent to users and disparate, heterogeneous data resources can be federated and shared. In Simulation grid, transparent data access and the manipulation of relation database (including Oracle, SQLServer, MySql) and XML database (such as Xindice) have been realized based on OGSA-DAI 4.0. Its components have been used to develop simulation grid data services that offer the capabilities of data federation and distributed query processing within a virtual organization. Users of simulation grid data services have no need to know about the database type and location and data can be accessed with the uniform method.

3.3   Grid-Enabling Virtual Prototype Models

JNI technology can implement the interaction among Java, C/C++ and Fortran languages. This makes it possible that the simulation models implemented with C/C++ and Fortran can be reused and encapsulated as simulation grid services to be published and shared. The models of an aircraft have been grid-enabled based on JNI, and the corresponding portal development for the model grid service has been completed (shown in Fig.5). Models are shared as grid services and model testing and simulation environment is also provided to users. This general technology method laid the foundation for grid-enabling other simulation resources realized with C/C++ such as CGF (Computer Generated Force) models, visualization technology, etc.

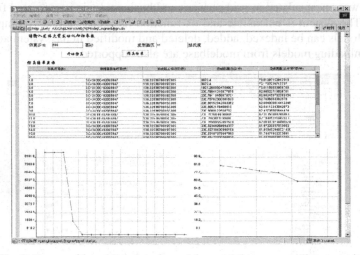

Fig. 5. Shows grid-enabled aircraft models portal that users can interact with

3.4   Grid-Enabling System Function Modules

The system function modules of the platform system ensure collaboratively applying modeling/simulation/project management technology for the full lifecycle of complex system virtual prototype. According to the steps mentioned above these function modules can be encapsulated and reused to share the modeling/simulation/project management resources in Simulation Grid.

Grid-enabling HLA (High Level Architecture) can also integrate their both advantages. RTI is responsible for data exchange and communication, while grid technol-

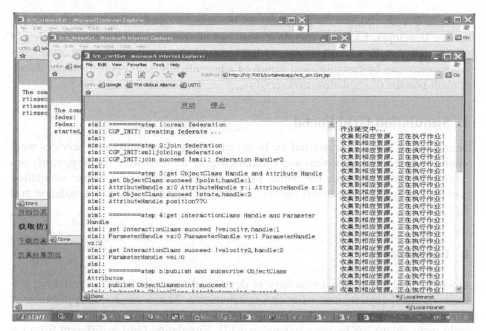

Fig. 6. Shows RTI grid service portal to support existed simulation system

ogy ensures the dynamic requirements for software and hardware in the process of simulation so that the process of distributed interactive simulation has more flexibility, fault tolerance and extensibility. At present, based on our basic grid platform, the supporting of grid-enabled RTI simulation application in Linux OS has been realized. According to the resources status in the simulation grid, concerned with the resource requirements of the simulation system the best simulation scenario can be set. The status of federates can be monitored in the simulation process and simulation results can be returned to users through web browser (Shown in Fig.6). And the multimedia workshop grid service was developed so that video meetings can be conveniently held through Internet in a virtual organization. Multimedia information such as video, audio and text information can be interacted among users through the web browser to help building a collaborative environment.

4 Simulation Grid Services Management

The function of grid platform supported by GLOBUS and Web services can be further extended so that the services and resources in Simulation Grid can be effectively organized and managed. A more friendly resource scheduling and interaction ability can be provided to users. Simulation Grid services Management layer includes six modules, which are Catalog Service, Creation Service, Service Engine, Registry Service, Security Service and Event Service. At present, Catalog Service and Security Service have been implemented based on the grid information service and security service provided by basic grid platform layer. Simulation grid services can be indexed and safely managed. And Creation Service has also been developed, which can help generating code framework according to the service information specified by the service developer. Existing function classes can be reused and the service function of service data, notification, logging and lifecycle management can be supported. Services codes can be compiled and deployed automatically to support the rapid development process of simulation grid services.

5 Simulation Grid Portal

Simulation grid portal is realized by Web technology. Simulation grid services were safely accessed and controlled and simulation grid services were friendly, easily accessible to users. Through the portal, users can query, select and invoke simulation grid resources and conveniently develop, update, maintain and destroy simulation grid services.

6 Conclusion

Modeling and simulation technology is developed towards digitize, virtualize, networkize, intelligentize, integratize and collaboratize. The research and application of modeling and simulation technology based on next generation Internet technology (grid technology) has become an emphasis [7]. Simulation Grid is different from the traditional collaborative environment. It is built on WAN and can cross operation platforms. The sharing of simulation hardware resources such as simulators, simulation devices and simulation software resources such as entities/environment/ geometry models, simulation algorithm and simulation scenario libraries is revolutionarily deep and wide. Simulation is more open, dynamic, flexible and extensible.

Simulation Grid technology can be adopted both in civil area and military area. It can bring great transform and innovation to the application mode, management mode and supporting technology of modeling/simulation to generate considerable social and economical benefits.

References

1. I. FOSTER, C. KESSELMAN, S. TUECKE: The Anatomy of the Grid: Enabling Scalable Virtual Organizations, International J. Supercomputer Applications (2001)
2. Bo Hu li, et al, Research of SBA Supporting Environment, Proceedings of System Simulation Conference (2003)1-9
3. http://www.ogsadai.org.uk/ (2004)
4. S. TUECKE, K. CZAJKOWSKI, Open Grid Services Infrastructure, http://www.ggf.org/ogsi-wg (2003)
5. http://www.globus.org/ (2004)
6. Gregor von Laszewski, Ian Foster, Jarek Gawor, Peter Lane, A Java Commodity Grid Kit, Concurrency: Experience and Practice (2001)
7. Bo Hu li, Modeling & Simulation In China-Present And Future, Asian Simulation Conf./the 5th Int. Conf. On System Simulation and Scientific Computing (2002)1-11

Using Capacity Simulation for Quality Goal Setting

Young Hae Lee, Chang Sup Han, and Jung Woo Jung

Department of Industrial Engineering, Hanyang University, 426-791, Ansan, S. Korea
yhlee@hanyang.ac.kr, cshan@ssi.samsung.com
jungjw@scm.hanyang.ac.kr

Abstract. Quality improvement continues to be a major concern in industries. Yet it often takes place without a formal evaluation of the possible expenditures and expected quality improvement target involved. Especially the inspection and rework cost structure and the inspection timing are the major factors to be decided in manufacturing strategies. This paper will propose a computer simulation model for evaluating 3 different manufacturing strategies with regard to inspection and rework cost and the inspection timing in order to implement quality improvement program cost-wise effectively. Computer simulation technique can be used as a decision support aid. Through statistical estimation of specification limits prior to run simulation, we can anticipate the results and validate simulation model as well. The objective of this research is to define and quantify the cost of quality and to demonstrate the use of simulation technique in quality improvement goal-setting with proposed the across-the-board variance reduction technique.

1 Introduction

Satisfactory product/service quality goes hand-in-hand with satisfactory product/service cost. One of the major obstacles to the stronger quality programs in earlier years was the mistaken notion that the achievement of better quality required much higher cost. A major factor in this mistaken concept of the relationship between quality and cost was the unavailability of the meaningful dataset. Indeed, in earlier years, there was a widespread belief that quality could not be practically measured in cost terms [1]. Part of the reason for this belief was that traditional cost accounting system which has not been attempted to quantify quality as a cost term. Consequently, quality cost did not easily fit into the older accounting system. Another part of the reason was that some quality-control proponents themselves were unwilling to encourage the measurement of cost of quality. They were concerned that such identification might lead to unwisely drastic reduction in these costs and consequently to reduction in quality program themselves [1]. Today, we not only recognize the measurability of quality cost but also understand its importance to the management and engineering of modern total quality control as well as to the business strategy planning for enterprises. The measurement of quality cost can provide a clear communication term to the quality-control practitioners. The use of quality cost can provide a powerful management tool for assessing the overall effectiveness of quality management [2]. Dale and Plumkett [3] emphasized their value for facilitating performance measure and improvement activities. Predating them, Groocock [4] stressed the importance of setting target for quality cost reduction and planning actions to meet targets. But there

D.-K. Baik (Ed.): AsiaSim 2004, LNAI 3398, pp. 31–39, 2005.

is no such article to deal with cost of quality and quality goal setting at the same time. Quality cost is categorized into the cost of prevention, appraisal, internal and external failure. Besterfield [2], in particular, distinguished between direct and indirect quality cost, and summarized that the latter have a subsequent impact on the organization. Because of the intermediate nature of most indirect cost, effort has been historically focused on the estimation of indirect cost. Simulation has been used in a variety of situations. The literature which proposes to quantify the effects of internal failure to inventory level and cycle time is presented by Karen et al. [5]. Using activity analysis and simulation, hidden cost in the factory to support personnel and inventory level is investigated. Flowers and Cole [6] reported on reducing the inspection cost by improving the utilization of inspection personnel. In more recent literature, Gardner et al. [7] addressed the time to inspect operations to reduce overall inspection and internal failure costs.

The opportunity taken in this research is to investigate the effect of inspection and rework to the proposed cost of quality measurement, total cost per piece (TCPP), to overcome deficiencies above mentioned. The objective of this research is to define and to quantify quality cost, and set quality goal using simulation. Such a measure will aid firms to make the investment decision on advanced manufacturing systems. We setup the simulation model for quality goal setting with regards to quality cost and the time to inspect by defining quality cost for financial evaluation and reviewing relevant simulation techniques. The output of the simulation model is then used as a target or as a guideline for quality improvement program in practice. An example is provided to justify the proposed cost of quality (COQ) measurement, and is used to generate a revised COQ according to the proposed variance reduction methods. We will describe overall research approach, and propose a cost related measurement called Total Cost per Piece (TCPP) in section 2. In section 3, the simulation output is analyzed, and the proposed variance reduction method is released in section 4. In section 5, we conclude this research.

2   The Description of Proposed Research

2.1   Overview

A simulation model through ProModel was built to quantify the proposed cost of quality measurement and to get statistics such as the number of bad parts depend on the different inspection strategies. In doing this, we can recognize the insight of the failure cost to our COQ measurement. Statistical estimation of the expected bad parts in each process is taken for evaluating the validity of the simulation result. Through statistical estimation, we can validate the output of the simulation model.

We consider material and operation cost as direct cost, inspection and rework cost as an internal failure cost, and recall cost as an external failure cost. The sum of these cost factors is represented as total cost per piece (TCPP). It represents how much cost will be spent to make one completed product in a specific manufacturing system. TCPP defined as equation (1). It will be used as a COQ measurement in this study.

$$TCPP = \frac{\sum_i \left(\begin{array}{l} Material\ Cost_i + Processing\ Cost_i + Inspection\ Cost_i \\ + Rework\ Cost_i + Recall\ Cost_i \end{array} \right)}{Total\ Number\ of\ Good\ Parts} \qquad (1)$$

The goal of quality improvement is to achieve a low TCPP in this study. In order to accomplish the goal, we have to make decisions on "how much to improve the quality" and "how to set the goal?" In this study, we considered the variance reduction method as a way of quality improvement. In the sense of comparing rework and inspection cost to recall cost for shipped bad parts, we can figure out the equilibrium point that has the same TCPP of the target system. The corresponding percentage of the variance reduction will be set as a quality improvement goal.

2.2 Model Building

In this study, the five step manufacturing process is considered. Each process has its probability distribution of processing time for each process and specification limit, which determine whether they are good or bad parts. Capability index C_P, is given. We assumed that the process capability follows the normal distribution. Manual inspection requires U(0.3, 0.1) minutes to complete per piece per operation. The inspector finds 90% of the bad parts, but 10% of bad parts will be shipped to customers. The defective parts that are found are sent to rework station. Rework is assumed to be always successful and required one-half time that the item has been in process until it is found to be defective. In order to consider external failure cost, if a bad part is shipped to customers, we incur a cost of $145 per part due to lost goods and shipping & handling cost. Inspection and rework cost are $20 and $25 per hour, respectively. The required model parameters and model descriptions are summarized in Table 1 in terms of the distribution of processing time, specification limit and mean value of each process.

There are many strategies in operating a manufacturing facility. In order to minimize the manufacturing risk on returned products, inspection is inevitable. The important decision on inspection is the time to inspect. The models are constructed according to the following strategies.

Strategy I: Inspection is performed once after all process are completed
Strategy II: Inspection is performed after each process is completed
Strategy III: No inspection is performed (baseline model)

In strategy I, the inspection will be performed at the end of the process. The inspection process measures the five different specifications of the final product. We defined a user distribution called "Type II error" which is used for representing inspector's error (detect 90% of bad parts; miss 10% of bad parts). It is used when the model generates random specification limit. After inspection, we have two alternatives. 90% of detected bad parts will be reworked at rework station with additional cost. And 10% of bad parts which incurs external failure cost ($145 value) will be shipped to the customer. In rework station, we assumed that rework is always successful and requires 1/2 time that the item has been I process until it is found to be defective.

Table 1. An information of sample manufacturing processes

	Process 1	Process 2	Process 3	Process 4	Process 5
Processing Time	U(35, 0.5)	T(2, 5, 7)	U(1.75, 1)	N(3, 1)	U(5.4, 2)
USL	49.6	19.4	122	0.005	18
LSL	30	12.8	121	0.00475	17
Mean	35	15	121.4	0.0048	17.5
C_p	0.9	1.1	0.6	0.7	0.8

3 Output Analysis

3.1 Simulation Output

We can recognize that material and operation costs are not much different between the two different inspection strategies because we have approximately the same number of entries in each system. But the other three cost factors are much different because they are dependent upon each manufacturing strategy in terms of inspection, rework, and recall. The result is summarized in Table 2.

Table 2. A summary of simulation run on each strategy

	Operation Cost	Material Cost	Inspection Cost	Rework Cost	Recall Cost	TCPP
Strategy I	10.77	114.14	0.50	6.03	6.90	139.35
Strategy II	9.25	106.93	0.51	0.277	0.322	117.29
Strategy III	9.16	64.13	N/A	N/A	104.86	178.15

As we mentioned earlier in the model overview, rework cost depends on the time to detect and the number of bad parts shipped to the customer. While recall cost depends on the number of bad parts shipped to the customers. In rework cost, strategy I has a longer time to detect because rework operation occurs at the end of process 5, while strategy II rework is done at the end of each process. Consequently, strategy II has a higher cost than that of strategy I. Recall cost only depends on the number of bad parts shipped to the customers at the end of process. Obviously strategy III has a high cost because it doesn't have an inspection and rework operation. It tells us a significant correlation between the recall cost and COQ measurement, TCPP.

From the simulation, we can quantify the cost of quality measurement. We also recognized that introducing rework and inspection is the efficient way to decrease the recall cost. The time to detect bad parts is also a critical factor to reduce quality cost. Empirically, early detection is preferable because the rework cost function is heavily relying on the detection time. The quantified measurement will be used later in quality improvement goal setting.

3.2 Estimating the Probability of Out-of Specification Limit

Before running a simulation model, a statistical estimation of out-of-specification probability is advised. Because it provides us a credibility of a simulation model itself and a validity of a simulation model. By statistically evaluating capability index prior to simulation, we can anticipate how the simulation model work and validate the fitness of the model. Two process capability measures or indices are widely used in industry: (1) the C_p index, an inherent or potential measure of capability and (2) the C_{pk} index, a realized or actual measure of capability. It should also be noted that the product specification limits are set with respect to product design(internal and external customer) needs, while "process spread" as measures by "σ", is a function of process, material, equipment, tooling, operation method, and so forth. Capability indices link the product's design-related specification to the process-related outcomes. If $C_p < 1$, we can declare that the process is potentially incapable, while $C_p > 1$, we can declare that the process is potentially capable in first. We consider C_p index because it is not realized yet. From the given data set, we can easily recognized that the proposed process has a relatively low C_p. Now we can derive each process's spread from the following equation (2).

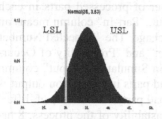

$$\sigma = \frac{USL - LSL}{6C_p}$$

$$Z_{LSL} = \frac{30-35}{3.63} = -1.337, \ Z_{USL} = \frac{46-35}{3.63} = 4.02 \qquad (2)$$

$$P(z < -1.337) \cong 0.08379, \ P(z < 4.02248) \cong 1$$

$$P(\text{within specification limits}) = 0.9158$$

Potential capability C_p measures the inherent ability of a process to meet specification and provides the process can be adjusted to target. We should note that high C_p 's does not automatically assure of a high quality product [9]. In practice, the C_p measure is useful since we can typically adjust target without great expenditure. However, variance reduction typically is both time-consuming and cost-intensive activity and requires extensive process improvement efforts. We have mean, standard deviation, and specification limits for each process. Therefore, we can anticipate the probability of a product within the specification limit using simple Z-transformation. From the cumulative standard normal distribution table, we can read corresponding probability on x-value. The difference between two numbers is the probability within specification limits. By the same token, we can estimate each process's probability within the specification limit prior to run the simulation model in Table 3.

Table 3. A calculation of expected probability

	Process 1	Process 2	Process 3	Process 4	Process 5
Mean	35	15	121.4	0.0048	17.5
Standard Deviation	3.629	1	0.2778	0.00006	0.2830
LSL	30	12.8	121	0.00475	17
USL	49.6	19.4	122.2	0.005	18
Z-value of LSL	-1.37756	-2.2	-1.44	-0.8333	-2.40038
Z-value of USL	4.00248	4.4	2.16	3.333	2.40038
Probability	0.9158	0.9861	0.9073	0.7972	0.9836

Table 4. A comparison between probabilistic vs. simulation results

	Process 1	Process 2	Process 3	Process 4	Process 5	Sum
Part In	440	432	426.4	424.8	421.6	
Probability of bad parts	0.08422	0.01391	0.09263	0.20276	0.01638	
Expected bad parts	37.05	6.01	39.5	86.13	6.90	179.59
Simulated bad parts	35	4	37	98	12	186
Difference	-2.05	-2.01	-2.50	11.87	5.1	10.41

Table 4 shows that process 1 will make 91.58% good parts while 8.42% will be bad one. For example, when we process 1,000 parts in process 1, the process will roughly produce 915 good parts and 85 bad ones. By the same token, we can anticipate an expected number of good and bad parts from process 2 to process 5, respectively. Note that the result of a previous process will be the input value of the following successor process.

Therefore, we can estimate the number of good and bad parts ahead of simulation run. Based on the simulation result in terms of number of processed parts in each process, we can pre-estimate the number of bad parts out. "Part-In" column means an actual number of entities generated from simulation package. The "Expected Number of Bad Parts" column stands for the product of "Part-In" and "Probability of Occurring Bad Parts." The figures in "Number of Bad Parts in Simulation Output" column come from the actual simulation run. The number of bad parts in simulation output is approximately 10 more parts than that of probabilistic bad parts. The causes of difference stem from the dynamic nature of ProModel and instability of the process, 8 hr-warm up period and 5 replications only. After arriving at a stable state, the difference might be squeezed. In the same manner, we got 17.8 bad parts from the statistical inference of Strategy I while we got 18.828 bad parts from simulation. But this computation tells us the inherent capability of each process to produce within specifications when the process mean is on target. We do not know where the actual mean location is. The location of mean is critical factor in quality improvement issue. It can be done by a sample Z-test which will be discussed in the following section.

3.3 Test Concerning the Population Means

ProModel generates random numbers using a built-in math function based on the user given parameters. If built-in function works properly, ProModel will return reasonable mean values. The fitness test of a simulation result such a mean value of a random variable can be verified by a sample Z-test. The Z-test tells us the location of the sample mean against the considered given distribution. A two-tailed Z-test was done to determine whether the two means are equal or not. Letting μ and σ denote a mean and a standard deviation, respectively, of the distribution, the plans to test the following null and alternative hypothesis:

$$H_0: \mu = \text{mean of given distribution}$$
$$H_1: \mu \neq \text{mean of given distribution}$$

And 50 sample data were randomly taken from the sample distribution: that is, sample size n is 50. The normal distribution is taken for generating specification random variable called measure 1. We would like to determine at 5% significance confidence level if the average value of random variable is equal to a hypothesized value of 35, given population mean of process 1. For a two-tailed test at the $100\alpha\%$ significance level, critical value of the test statistic are $100(\alpha/2)^{th}$ and $100(1-(\alpha/2))^{th}$ quantifies of the distribution of the statistic. The sample values of the specification, measure 1, are normally distributed with unknown mean and known standard deviation 3.63. Let $\mu =$ mean of given distribution. Then the test hypothesis will be $H_0: \mu = 35$. vs. $H_1: \mu \neq$ 35. Let is the sample mean. A statistic Z is chosen by the following *Central Limit Theorem*. Therefore, we decide that we can assume that the population of specification value is normally distributed. Using central limit theorem, the mean of a sample size n selected from the sample distribution is normally distributed with mean μ and standard deviation. Thus the following equation follows the standard normal distribution. Using Z as the test statistic, the critical values for two tailed-test are as follows. From actual 5 simulation replications, I got an average value of 35.173. H_O will be rejected if the observed value of Z fall in the interval $(-\infty, -1.96)$ or $(1.96, \infty)$. Detailed test results and dot-lot are shown below in detail. From the given data associated with process capability, C_p, we will derive each process's standard deviation. So the derived σ surely reflects the inherent process capability to meet the specification. We supposed, for the sake of argument, that the sample data should reflect the inherent incapability of the process. Now, we were interested in determining if the sample mean of the generated random variable is equal to the mean of a given distribution. And also we interested in the normality of the sample data themselves. If the null hypothesis were not rejected, the considered process has approximately the same mean value under the considered test condition. After 5 simulation runs, we have corresponding 50 sample values of each process. A sample data set is omitted due to space limitation. It can be available upon request.

For example, in inspection process 1, we assumed that the specification random variable was normally distributed, N(35, 3.63). Using these data set, we took Z-test with α = 0.05. The rejection criteria is $|Z| > Z_{0.025} = 1.96$.

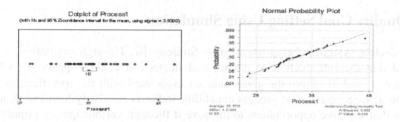

Fig. 1. Dot & Normal plot for Process 1

From the result, we can conclude that null hypothesis is not rejected. If the process is repeated many times when the null hypothesis is true, we could conclude that μ = 35 with α = 0.05. Through analyzing the results of sample Z-test and normal probability plot, we can conclude that ProModel properly generates random numbers under the given condition. Therefore we can validate the use of ProModel as a simulation tool. In addition we can also verify that the sample gas the approximately same out of specification probability that coincide with the value of Chapter 5.3. For example, when we applied specification limits to the normal probability plot, we can count the number of sample data that are out of specification limits. For example, we have 4 data points out of 50 samples I process(see fig. 1). Then we can easily compute the probability of out-of-specification as follows: The result also coincides with the result of the probability of out-of-specification which considered in Chapter 2.

Potential capability, C_P, measures the inherent ability of a process to meet specification limits. By the way, we derive σ from the given C_P. So a low C_P, process obviously has the low inherent ability to meet specification limit in terms of mean and standard deviation. Therefore these test results coincide with our expectation mentioned earlier prior to test. From the Z-test results, we can verify the location of the sample mean and conclude that ProModel properly generates specification random variables according to the given condition. From the normality probability plot, we can verify the normality of the sample data and compute the probability of out-of-specification limit without many efforts. Consequently, we validate the reliability of the adapted simulation package, ProModel. Here are the additional Z-test results of each remaining process in Table 5.

Table 5. A result of Z-test for process 2 to process 5, respectively

	Process 2	Process 3	Process 4	Process 5
Test	H_0: μ=15 H_1: $\mu \neq 15$ σ=1	H_0: μ=121.4 H_1: $\mu \neq 121.4$ σ=0.278	H_0: μ=0.0048 H_1:$\mu \neq 0.0048$ σ=0.00006	H_0: μ=17.5 H_1: $\mu \neq 17.5$ σ=0.2083
Mean	14.900	121.387	0.004804	17.4850
Standard Deviation	1.114	0.296	0.000061	0.2380
Mean of S·E	0.141	0.039	0.000008	0.0295
95% C·I	(14.623 ,15.177)	(121.310 ,121.464)	(0.004787 ,0.004821)	(17.4273 ,17.5428)
Z-value	-0.71	-0.33	0.49	-0.51
Probability	0.479	0.743	0.627	0.611

4 Quality Goal Setting Using Simulation

Our baseline system is being operated by Strategy III. Through simulation, we confirmed that the introduction of rework and inspection could improve the COQ in strategy I and II. From the given data set associated with the specification limits, we can also recognize that process capability index is relatively low(less than 1). It means that we have opportunities to improve it through various quality improvement activities such as training, equipment calibration and tight inspection on a raw materials in advance, etc. In the sense of comparing the cost of quality measurement, we can figure out the quality improvement target using simulation. So the next issue is: how much variance reduction is needed to improve actual process capability from present situation to Strategy II and III. In doing so, we can determine how much across the board reduction is needed from the as-is condition to the to-be situation. It indicates that we have to improve process capability through quality improvement activity. Through across-the-board variance reduction, we can achieve improved COQ index, that is, low TCPP. In this research, 10% across-the-board variance reduction is considered. Note that k^{th} 10% variance reduction is 10% reduction of $(k-1)^{th}$ variance, not the 1st variance. For example, process 1 follows normal distribution, N(35, 3.63). The variance, σ^2, is 13.17690. 10% reduction of it is 11.85921. Therefore a revised new σ is 3.4437 instead of 3.63. Using run time interface module in ProModel, we can instantly simulate its effect on TCPP. After each variance reduction trial, we can get a corresponding TCPP as follows: As a result of simulation, our baseline model(Strategy III) need a 7 times of 10% across-the-board variance reduction to reach the same effect of Strategy I, while our baseline model(Strategy III) need a 15 times of 10% across-the-board variance reduction to achieve Strategy II(see Fig. 2). Using these results, we can set our quality improvement goal. It can be used as a guideline during quality improvement plan implementation.

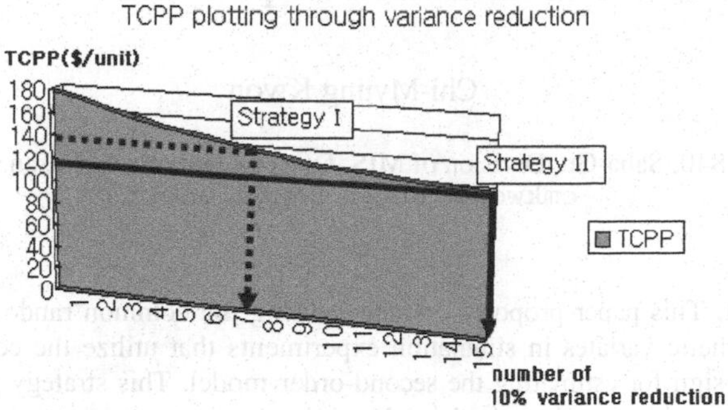

TCPP plotting through variance reduction

Fig. 2. TCPP Plotting through variance reduction

5 Conclusion

We defined and quantified the proposed cost of quality measurement using simulation according to the different inspection strategy. It can demonstrated that the significance of rework and inspection in reducing TCPP. Then we validated the simulation package itself through comparing the number of bad parts between statistically estimated value and actual simulated one. Finally, we set the quality improvement goal using simulation by the variance reduction method. In order to get the same effect of Strategy I from baseline model strategy III, strategy III, needs a 7 times of 10% across-the-board variance reduction. At the same token, 15 times of 10% across-the-board -variance reduction is needed to reach Strategy II from the baseline strategy III. In doing this, we can set the quality goal, how much efforts to be done, using simulation result.

References

1. Feigenbaum, A.V. (1983) "Total Quality Control", New York, Mc Graw-Hill
2. D.H. Besterfield (1979), "Quality Control", New Jersey, Prentice-Hall
3. B.G. Dale & J.J. Plumkett(1991), "Quality Costing", London, Chapmam and Hall
4. J.M. Groocock(1991), "The cost of quality, London, " Pitman Publishing
5. Karen E. Schmahp, Yasser Dessouky and David Rucker(1997), "Measure the cost of quality: case stidy," Production and inventory management journal-fourth quarter
6. Flowers, A.D., and J.R. Cole(1985), "An application of computer simulation to quality control in manufacturing," IEE Transactions 17, pp. 277-283
7. Gardner, L.L., M.E. Grant, and L.J. Roston(1992), "Converting simulation data to comparative income statement," Proceedings of the Winter Simulation Conference
8. Gardner, L.L., M.E. Grant, and L.J. Roston(1995) , "Using simulation to access cost of quality," Proceedings of the Winter Simulation Conference
9. William J. Koralik(1995), "Creating quality concepts, systems, and tools", Mc Graw-Hill

Application of Central Composite Design to Simulation Experiment

Chi-Myung Kwon

Hadan-Dong 840, Saha-Gu, Division of MIS, Dong-A University, Pusan, Korea 604-714
cmkwon@daunet.donga.ac.kr

Abstract. This paper proposes a strategy using the common random numbers and antithetic variates in simulation experiments that utilize the central composite design for estimating the second-order model. This strategy assigns the common random numbers to the cubic points and a center, and the antithetic variates to the axial points and one center. An appropriate selection of the axial points allows this strategy to achieve variance reduction in estimation of first and second order coefficients of a given model. Under certain conditions, the proposed method yields better estimates for the model parameters than independent streams. In estimation of the intercept, independent random number assignment strategy shows better results than the proposed strategy. The simulation experiments on the (s, S) inventory system supports such results.

1 Introduction

In simulation experiments, the analyst is often concerned with estimating a second-order linear model of the response of interest and the selected level of design variables. For instance, an average inventory cost per period in (s, S) inventory system shows a quadratic type of response surface in the space of s and S [2]. One of the popular and useful designs for fitting such a model is the central composite design (CCD) [4, 7]. Even though many research suggested that a second-order model offers a better approximation to the true underlying relationship between the response and the selected levels of input factors, as Tew [9] noted, the CCD has received little attention from the simulation community. We consider that development of an efficient strategy utilizing the CCD in simulation experiment would help this design gain broad application for estimating a second-order linear model.

In this paper, our main interest is to obtain more precise model estimators in a simulation experiment utilizing the CCD. To estimate a first-order linear model more accurately, Schruben and Margolin [8] proposed the variance reduction technique of common random numbers (CRN) and antithetic variates (AV) to one simulation experiment and showed that their method yields superior estimates of model parameters to the strategy of either CRN alone or independent random number streams under condition that the design admits orthogonal blocking into two blocks. However, the CCD and other designs for estimating second-order linear model do not accommodate Schruben and Marolin's random number assignment rule directly since such designs do not allow orthogonal blocking into two blocks. Tew [9] suggested a correlation induction strategy of the CRN and AV in simulation experiments utilizing the CCD without the restriction of orthogonal blocking. He used the CRN to induce posi-

D.-K. Baik (Ed.): AsiaSim 2004, LNAI 3398, pp. 40–49, 2005.
© Springer-Verlag Berlin Heidelberg 2005

tive correlations between responses across all design points and the AV to induce negative correlations across replicates within the simulation experiments. This method yields better estimates for model parameters than the independent random number streams.

In this paper, different from Tew's random number assignment strategy, we propose a method using the CRN and the AV across the design points in one simulation experiment designed to estimate the second order model. Compared to Tew's strategy, our rule of random number assignment is much simpler in the view of application.

Although the general second order design does not admit orthogonal blocking into two blocks, by selecting appropriate axial points, adding center points, and correcting the pure quadratic terms for their means in the CCD, we may have the design which accommodates the Schruben-Marolin's correlation induction rule. We conjecture that such efforts to take the advantage of Schruben and Margolin's method in the CCD may be effective in reducing variability of parameter estimators under certain conditions. We investigate the simulation efficiency of the proposed method in estimating model parameters analytically and apply it to the (s, S) inventory system of a single commodity to evaluate its performance.

2 Simulation Experiments Using Central Component Design

In this section, we present the statistical framework of simulation experiments necessary to estimate the coefficients of a second-order linear model. For this purpose, we define a second-order linear model, and identify the associated CCD for estimation of model parameters.

2.1 Second-Order Linear Model

Consider a simulation experiment consisting of m deign points where each design point is defined by p design variables. We let y_i denote the response at design point i. We assume that the relationship between the response of y_i and the design variables is linear in the unknown parameters and second-order. If we select three evenly spaced levels for each design variable, then each variable can be coded to -1, 0, and 1. We let x_{ij} be the coded level of the jth design variable at the ith design point. Then we can write

$$y_i = \beta_0 + \sum_{j=1}^{p} \beta_j x_{ij} + \sum_{j=1}^{p} \beta_{jj} x_{ij}^2 + \sum_{j<k} \beta_{jk} x_{ij} x_{ik} + \varepsilon_i \qquad (1)$$

for $i = 1, 2, \ldots, m$ and $k = 1, 2, \ldots, p$; where β_0 is the constant coefficient, β_j ($j = 1, 2, \ldots, p$) are the first-order coefficient, β_{jj} ($j = 1, 2, \ldots, p$) are the pure second-order coefficient, β_{jk} ($j, k = 1, 2, \ldots, p$; $j < k$) are the mixed second-order coefficient, and ε_i is the error term.

In order to represent the equation (1) in matrix form, we let $\mathbf{y} = (y_1,..., y_m)'$ be the response vector of m design points. We also let $\boldsymbol{\beta}$ and $\boldsymbol{\varepsilon}$ denote the vector of coefficients and that of error terms, respectively, in (1). We denote that $\mathbf{X}$ is a design matrix whose ith row vector is equal to $(1, x_{i1} \cdots x_{ip}, x_{i1}^2 \cdots x_{ip}^2, x_{i1}x_{i2} \cdots x_{ip-1}x_{ip})$ ($i = 1, 2 \ldots m$). Then we can express the equation (1) as

$$\mathbf{y} = \mathbf{X}\boldsymbol{\beta} + \boldsymbol{\varepsilon}. \tag{2}$$

Many simulation studies assume that the simulation outputs of responses $\mathbf{y}$ have the multivariate normal distribution $\mathbf{y} \sim N_m(\mathbf{X}\boldsymbol{\beta}, \boldsymbol{\Sigma})$. Throughout the remainder of this paper, for the correlation induction strategy considered, we will use the ordinary least squares (OLS) estimate and the weighted least squares (WLS) estimate for $\boldsymbol{\beta}$.

2.2 Central Composite Design

We shortly review the central composite design in this section. For a more detailed discussion, see the references [1, 4]. We consider a second-order CCD used to fit the second-order model given in (1). Such a design is the 2^p factorial (two levels of each variable coded to -1 and 1) augmented by the axial $2p$ points positioned at α and $-\alpha$ along the axes of the design variables, and c center points positioned at the center $(0,0,...,0)$. Thus, the number of total design points is equal to $m = 2^p + 2p + c$.

For instance, in a case of $p = 2$ and $c = 2$, the CCD has 10 design points. This design is as given in (3), where the first 4 upper points are the cubic points, the points 5 and 10 are the center points and the other points are the axial points.

$$\mathbf{X} = \begin{matrix} \beta_0 & \beta_1 & \beta_2 & \beta_{11} & \beta_{22} & \beta_{12} \\ \begin{bmatrix} 1 & -1 & 1 & 1 & 1 & -1 \\ 1 & -1 & -1 & 1 & 1 & 1 \\ 1 & 1 & 1 & 1 & 1 & 1 \\ 1 & 1 & -1 & 1 & 1 & -1 \\ 1 & 0 & 0 & 0 & 0 & 0 \\ 1 & \alpha & 0 & \alpha^2 & 0 & 0 \\ 1 & -\alpha & 0 & \alpha^2 & 0 & 0 \\ 1 & 0 & \alpha & 0 & \alpha^2 & 0 \\ 1 & 0 & -\alpha & 0 & \alpha^2 & 0 \\ 1 & 0 & 0 & 0 & 0 & 0 \end{bmatrix} \end{matrix}. \tag{3}$$

We choose the factorial portion of the design so as to estimate the first-order and two factor interaction terms. The augmented axial points and observations at the center allow for the estimation of the second-order terms and error term, respectively, in the model. Although the design in (3) includes two centers, we may add even number of centers to the design so that a proposed random number assignment strategy is applicable to the design. We will discuss this in next section.

3 Correlation Induction Strategy

In computer simulation, random number streams that derive a simulation model are under the control of experimenter and completely determine the simulation output. We let $R = (r_1, r_2, ..., r_g)$ denote the complete set of random number streams used for the simulation run at a given design point. The independent random number streams, which involve randomly selecting R for each design point, do not induce correlations between responses across the design points. Thus, it usually serves as the baseline strategy to which other correlation induction strategies are compared. When we apply the CRN, $R = (r_1, r_2, ..., r_g)$ randomly selected, to two design points, it is known that a positive correlation is induced between two responses. Given a set of streams R, the AV, denoted by $\overline{R} = (1 - r_1, 1 - r_2, ..., 1 - r_g)$ induces a negative correlation between two responses [8].

In addition to the normality assumption on the responses, much of analysis of alternative strategies for estimating the coefficients of the linear model in (2) assume that the simulation outputs of responses across all design points have equal variances, and the amounts of induced correlations between two responses are equal across the design points if the CRN (or AV) is used. Also empirical simulation experiments have shown that an absolute value of induced correlation by the CRN is greater than that obtained by the AV [8]. We establish the following assumptions on the responses based on the above discussion:

Assumption 1: homogeneity of response variances across the design points

$$Var(y_i) = \sigma^2 \text{ for } i = 1, 2, ..., m.$$

Assumption 2: homogeneity of induced correlation between two responses across the design points $(i \neq j)$

$$Cov(y_i, y_j) = \begin{cases} 0 & \text{if independent random number streams is used.} \\ \rho_1 \sigma^2 (\geq 0) & \text{if the CRN is used.} \\ \rho_2 \sigma^2 (\leq 0) & \text{if the AV is used.} \end{cases}$$

Assumption 3: relationship of induced correlation by the CRN and AV.

$$0 \leq -\rho_2 \leq \rho_1.$$

For applying the correlation induction strategies of the CRN and AV to the second-order CCD, we divide the design points into two blocks: the first block includes all cubic points and one center, and the second block includes all axial points and one center. Then the first block consists of the $(2^p + 1)$ design points and the second block consists of $(2p + 1)$ design points. We use the CRN for the design points of the first block, and the AV for those of the second block. For instance, the CCD in (3) has 10 design points. The CRN is assigned to the upper 5 points that correspond to the 4 cubic points and a center, and the AV is assigned to the lower 5 points corresponding to the 4 axial points and one center. Then realized responses at the m design points according to assigned random number strategy have the following type of response covariance matrix from the assumptions 1 and 2:

$$Cov\ (\mathbf{y}) = \sigma^2 \begin{bmatrix} 1 & \rho_1 & \cdots & \rho_1 & \rho_2 & \cdots & \cdots & \rho_2 \\ \rho_1 & 1 & \vdots & r & \vdots & \vdots & \vdots & \vdots \\ \rho_1 & \rho_1 & \cdots & 1 & \rho_2 & \cdots & \cdots & \rho_2 \\ \rho_2 & \rho_2 & \cdots & \rho_2 & 1 & \rho_1 & \cdots & \rho_1 \\ \vdots & \vdots & \vdots & \vdots & \rho_1 & 1 & \vdots & \rho_1 \\ \vdots & \vdots & \vdots & \vdots & \vdots & \vdots & \ddots & \vdots \\ \rho_2 & \rho_2 & \cdots & \rho_2 & \rho_1 & \rho_1 & \cdots & 1 \end{bmatrix} \tag{4}$$

4 Parameter Estimation

In this section, through statistical analysis, we provide the estimators for parameters and their variances, and investigate what conditions are necessary for proposed strategy to ensure an improvement in variance reduction.

4.1 Parameter Estimators and Their Covariance

To represent the equation (4) as a matrix form, we let $\mathbf{G} = (1,0,...,0)'(1,0,...,0)$ be the square matrix whose dimension corresponds to the number of columns of $\mathbf{X}$. We also define the m-dimensional vector $\mathbf{z}$ that determines the random number assignment strategy applied to each design point. The ith component of $\mathbf{z}$ has 1 if the CRN is assigned to the ith design point. Also if the AV is assigned to this point, the ith element of $\mathbf{z}$ is -1. Then we can write the response covariance matrix in (4) as

$$Cov(\mathbf{y}) = \sigma^2 [\frac{1}{2}(\rho_1 - \rho_2)\mathbf{X}\mathbf{G}\mathbf{X} + \frac{1}{2}(\rho_1 + \rho_2)\mathbf{z}'\mathbf{z} + (1 - \rho_1)\mathbf{I}]. \tag{5}$$

Rao [6] and Watson [10] showed that a sufficient condition for the equivalence of OLS estimator and WLS estimator for $\boldsymbol{\beta}$ in (2) is that the covariance matrix of responses has the following representation:

$$Cov(\mathbf{y}) = \mathbf{X}\mathbf{H}\mathbf{X} + \mathbf{w}'\theta\mathbf{w} + \sigma^2\mathbf{I}. \tag{6}$$

In equation (6), the $\mathbf{w}$ is a vector such that $\mathbf{w}'\mathbf{X} = \mathbf{0}$, and $\mathbf{H}$, θ and σ^2 are arbitrary. We recall the assignment vector $\mathbf{z}$ whose first $(2^p + 1)$ elements are 1's and whose remaining $(2p+1)$ elements are -1's. Multiplying $\mathbf{z}'$ by $\mathbf{X}$ gives

$$\mathbf{z}'\mathbf{X} = (2^p - 2p, 0,...,0, 2^p - 2\alpha^2,...,2^p - 2\alpha^2, 0,...,0). \tag{7}$$

Selection of α such that $2^p - 2\alpha^2 = 0$ makes $\mathbf{z}$ to be orthogonal to all column vectors of $\mathbf{X}$ except its 1st column vector. In a case of $p = 2$ and $\alpha = \sqrt{2}$, the $Cov(\mathbf{y})$ in

(5) is of the form in (6) and $z'X = 0$. Therefore the WLS estimator for β is equivalent to the OLS estimator for β.

We next consider a more complicate situation of $p \geq 3$. If we correct the pure quadratic terms in X for their means, and partition the design matrix as $X = (1 | X^*)$, then $1'X^* = 0$. Also, as the case of $p = 2$, appropriate selection of α makes $z'X^* = 0$. Thus, such a design matrix X admits orthogonal blocking in two blocks, and all OLS estimator for β other than β_0 in (2) are unaffected by the proposed random number assignment strategy [8]. We now compute the covariance of the OLS estimates of parameters in (2). Substitution of (5) into covariance formula of the OLS estimator for β yields

$$Cov(\hat{\beta}_{OLS}) = (X'X)^{-1} X' Cov(y) X (X'X)^{-1}$$
$$= \sigma^2 (1 - \rho_1)(X'X)^{-1} + \frac{\sigma^2}{2} (\rho_1 - \rho_2) \begin{bmatrix} 1 & 0' \\ 0 & O \end{bmatrix}. \tag{8}$$

4.2 Performance of Parameter Estimator

As we noted earlier, inclusion of independent random number streams strategy provides us an opportunity to compare the simulation performances achieved by the combination of the CRN and AV. For this purpose, we use notations $\hat{\beta}_{inp}$ and $\hat{\beta}_{com}$ to represent the OLS estimators obtained from independent strategy and combined strategy of the CRN and AV, respectively. If we use independent random number streams strategy across m design points, then from the assumptions 1 and 2, we have $Cov(y) = \sigma^2 I$. Therefore we obtain the covariance of the OLS estimator for β in (2) as follows:

$$Cov(\hat{\beta}_{ind}) = \sigma^2 (X'X)^{-1}. \tag{9}$$

Many criteria have been proposed for comparing design coefficients in estimating β in (2) [1]. The widely used criteria include the estimators' variances; the determinant and trace of $Cov(\hat{\beta})$. We let $\hat{\beta} = (\hat{\beta}_0, \hat{\beta}_1, ...)$. Comparison of equation (8) to (9) gives

$$Var(\hat{\beta}_l)_{com} \leq Var(\hat{\beta}_l)_{ind} \text{ for } l \neq 0. \tag{10}$$

Therefore in estimating $\hat{\beta}_l (l \neq 0)$, the combined strategy is superior to independent streams. If we let $(X'X)_{11}^{-1}$ be the first–row and first–column element of $(X'X)^{-1}$, then

$$Var(\hat{\beta}_0)_{ind} = \sigma^2 (X'X)_{11}^{-1} \tag{11}$$
$$Var(\hat{\beta}_0)_{com} = (1 - \rho_1)\sigma^2 (X'X)_{11}^{-1} + (\rho_1 - \rho_2)\sigma^2 / 2. \tag{12}$$

Correcting the pure quadratic terms in (1) by their column means, we find that $(\mathbf{X'X})_{11}^{-1} = 1/m$ [4]. In equation (12), we see that a variance reduction of first term may not compensate a variance increase of second term as the number of design points increases. Due to this reason, combined strategy may be worse in estimation of β_0 than independent strategy. Although we may identify the determinants and traces for two estimator's dispersion matrices in (8) and (9), the computation procedures obtaining them are not simple. Instead of comparing them analytically, we utilize the simulation results to analyze the efficiency of the estimators with respect to the determinant and trace.

5 Simulation Experiment

We conducted simulation experiments on the (s, S) inventory system in order to evaluate the performance of method of the CRN and AV presented in Section 3. We briefly describe the (s, S) inventory system and illustrate an implementation of correlation induction strategy. We also present a summary of the simulation results.

5.1 (s, S) Inventory System

In the (s, S) inventory system, an order is placed when the level of inventory on hand falls below the level s, and the amount of the order is placed up to S. We consider the standard infinite horizon, single product, and periodic review inventory model with full backlogging and independent demands. The times between demands are IID exponential random variables with a mean of 0.1 and the sizes of the demands are also IID random variables with probability in Tables 1.

Table 1. Distribution of Demand Size

demand	1	2	3	4
Probability	0.2	0.3	0.3	0.2

At the beginning of each period, we review the inventory level and decide how many items to order. When a demand occurs, it is satisfied immediately if the inventory level is greater than or equal to the demand size. If the demand size is greater than the inventory level, the excess of demand over supply is backlogged and satisfied by future deliveries. Total inventory control cost consists of the ordering, holding and shortage costs. The ordering cost includes the set-up cost of $32 and the item cost of $3 times order amount. When an order is placed, its delivery lag is a random variable that is uniformly distributed on the interval (0.5, 1). A holding and a shortage costs per item per period are $1 and $5, respectively (see the example in Chapter 1 of [3]).

To evaluate inventory-ordering policies of determining s and S, we assume that the average inventory cost per period (response) would be approximated by the second-order linear model in (1), and we estimate it by use of the CCD.

5.2 Assignment of Random Number Streams

Consider the two factors (s and S) second-order CCD with two centers. Table 2 presents applied strategy of random number streams and the levels of s and S at each design point used for simulation experiment.

The inventory model includes three random components: (a) demand sizes, (b) the times between demands, and (c) item delivery time. We assign the CRN, $R = (r_1, r_2, r_3)$, to the design points 1-5, and its AV to the design points 6-10.

Table 2. Second Order CCD and Random Numbers Assignment Strategy

Design point i	Factor combination $S(x_{1i})$	$S(x_{2i})$	assignment Strategy
1	20(-1)	80(+1)	R
2	20(-1)	60(-1)	R
3	40(+1)	80(+1)	R
4	40(+1)	60(-1)	R
5	30(0)	70(0)	R
6	34($\sqrt{2}$)	70(0)	$\overline{R}$
7	16(-$\sqrt{2}$)	70(+1)	$\overline{R}$
8	30(0)	84($\sqrt{2}$)	$\overline{R}$
9	30(0)	56(-$\sqrt{2}$)	$\overline{R}$
10	30(0)	70(0)	$\overline{R}$

5.3 Numerical Results

We simulated this system for 360 periods at each design point by use of the AweSim simulation language [5]. We set the initial inventory level to be 60. Table 3 shows the estimates for model parameters obtained under the independent random number strategy and the combined strategy of the CRN and AV.

Table 3. Parameter Estimates for β

Estimate	Independent Run	CRN and AV Run
$\hat{\beta}_0$	123.014	122.956
$\hat{\beta}_1$	2.619	2.575
$\hat{\beta}_2$	0.511	0.656
$\hat{\beta}_{11}$	2.893	2.605
$\hat{\beta}_{22}$	0.948	0.903
$\hat{\beta}_{12}$	-0.565	-0.561

Table 4 presents the performances of estimates for $Var(\hat{\beta}_i)$, the determinants and traces of $Cov(\hat{\beta})$ obtained from independent run and combination run of CRN and AV as well as percentage reduction in variances of estimates achieved by strategy utilizing the CRN and AV. From this Table, we note that combination method of the CRN and AV is better than independent strategy in estimating the first and second order parameters. The percentages in reducing variance for $(\hat{\beta}_0, \hat{\beta}_1, ..., \hat{\beta}_{12})$ are around 90%. However, in estimation of the β_0, we observe that combination strategy of CRN and AV inflates variance of estimate over than 30% compared to independent strategy. Thus, in estimating the β_0, the combination method of the CRN and AV is worse than independent assignment strategy. These simulation results seem to be consistent with the analytical results in equation (10)-(12). With respect to the design criteria of determinant and trace of $Cov(\hat{\beta})$, strategy using the CRN and AV across the design points is better than independent random number assignment strategy.

Table 4. Performance Measure for estimtor

Performance Measure	Independent Run	CRN and AV Run	Variance Reduction (percent)
$Var(\hat{\beta}_0)$	.249	.330	32.5(inflation)
$Var(\hat{\beta}_1)$	.340	.036	89.4
$Var(\hat{\beta}_2)$	.534	.036	93.3
$Var(\hat{\beta}_{11})$	.755	.094	87.5
$Var(\hat{\beta}_{22})$	.933	.048	94.9
$Var(\hat{\beta}_{12})$	.626	.069	89.0
det $Cov(\hat{\beta})$	5.646×10^{-3}	5.730×10^{-9}	-
Trace $Cov(\hat{\beta})$	3.437	0.612	-

6 Summary and Conclusions

In this paper, we proposed a combined method of the CRN and AV applicable to the CCD in one simulation experiment. We divide the design points into two blocks. The first block includes the cubic points and one center, and the second one contains the axial points and one center. Then we assign the CRN to the design points in the same block and use the AV across two blocks. This strategy is shown to be superior to the independent random number strategy in estimating the first and second order parameters. However, in estimating the model intercept, independent strategy is better than the proposed method. We perform simulation experiments utilizing the proposed method on the (s, S) inventory system. Simulation study shows similar results to analytical ones.

We consider that a combination of the CRN and AV is a preferred strategy to independent strategy if our interests focus on the effects of factors on the response other than the model intercept. Even though our simulation results are from the CCD with only two factors, we hope that this work will stimulate more application of the combination of the CRN and AV in a simulation area of the CCD.

References

1. Box, G.E.P., Draper, N.R.: Empirical Model Building and Response Surface. John Wiley & Sons, New York (1987)
2. Kwon, C.M.: Application of Stochastic Optimization Method to (s, S) Inventory System. J. Of the Korean Society for Simulation, Vol. 12. No.2. (2003) 1-11
3. Law, A.M., Kelton, W.D.: Simulation Modeling and Analysis. 3rd. Ed. McGraw-Hill, New York (2000)
4. Myers, R.H.: Response Surface Methodology. Edward Brothers, Ann Arbor (1976)
5. Pritsker, A.A.B., O'Reilly, J.J.: Simulation with Visual SLAM and AweSim. John Wiley & Sons, New York (1999)
6. Rao, C.R.: Least Squares Theory Using an Estimated Dispersion Matrix and Its Application to Measurement of Signals. Proceedings of the 5th Symposium on Mathematical Statistics and Probability (1967) 355-372
7. Sanchez, S.M.: Robust Design: Seeking the Best of All Possible Worlds. Proceedings of the WSC (2000) 69-76
8. Schruben, L.W., Margolin, B.H.: Pseudorandom Number Assignment in Statistically Designed Simulation and Distribution Sampling Experiments. JASA, Vol. 73. (1978) 504-525
9. Tew, J.D.: Using Composite Designs in Simulation Experiments. Proceedings of the WSC (1996) 529-537
10. Watson, G.S.: Linear Least Squares Regression. Annals of Mathematical Statistics. Vol. 38. (1967) 1679-1699

A Study on the Machine Loading Problem Considering Machine-Breakdown in Flexible Manufacturing Systems

Seong Yong Jang[1], Doosuk Kim[2], and Roger Kerr[2]

[1] Department of Industrial and Information System Eng.,
Seoul National University of Technology, Korea
[2] School of Mechanical and Manufacturing Eng., in Uni. of New South Wales, Australia
z2246477@student.unsw.edu.au

Abstract. In this paper, the machine loading problem (MLP) as an example of combinatorial problems is presented within a Flexible Manufacturing Cell (FMC) environment. In general, MLP can be modeled as a general assignment problem and hence formulated as 0-1 mixed integer programming problem which have been effectively solved by a variety of heuristics methodologies. Recently, Neural Networks have been applied to various problems with a view of optimization. Especially, Hopfield Networks to solve MLP having objective of balancing workload of machines is discussed in this paper. While heuristic methods may take a long time to obtain the solutions when the size of problems is increased, Hopfield Networks are able to find the optimal or near-optimal solutions quickly through massive and parallel computation. By such capabilities of Hopfield Networks, it is also possible to approach real-time optimization problems. The focus in this paper is on the quality of solutions of Hopfield Networks by means of comparison with other heuristic methods through simulation, considering machine-breakdown. The simulation results show that the productivity of parts is highly improved and that utilization of each machines is equalized by achieving a balanced workload on machines.

1 Introduction

A Flexible Manufacturing System (FMS) is a production system consisting of a set of identical and/or complementary numerically controlled machines which are connected through an automated transportation system. Each process in an FMS is controlled by a dedicated computer. Operational problems of an FMS are concerned with optimal utilization of specific FMS components and are effectively separated into 4 subproblems: planning, grouping, machine loading and scheduling[1]. The group of machines is called Flexible Manufacturing Cells (FMC) which can be defined as a system of two or more machines, controlled by an integrated hierarchical computer network, with or without automated materials handling into and within the system. Loading problem is raised after solving the planning and grouping problem of an FMS. Machine loading problem (MLP) is to deal with the allocation of part operations amongst available machines for a given product mix so that some system performance criterion is optimized.

Stecke [2] has described six objectives for the loading problem.

1. Balance the assigned machine processing times
2. Minimize the number of movements from machine to machine, or equivalently, maximize the number of consecutive operations on each machine

D.-K. Baik (Ed.): AsiaSim 2004, LNAI 3398, pp. 50–58, 2005.
© Springer-Verlag Berlin Heidelberg 2005

3. Balance the workload per machine for a system of group of pooled machines of equal sizes
4. Unbalance the workload per machine for a system of group of pooled machines of unequal sizes
5. Fill the tool magazines as densely as possible
6. Maximize the sum of operation priorities

A number of authors have dealt with loading problem, presented the methods for solving the problem and proposed future work for improving the performance. The first mathematical formulation for MLP is given by Stecke[2].. The grouping and loading problems are formulated as 0-1 mixed integer programming problems and they are solved by the methods of linearizing the nonlinear terms. Kusiak[3] presented the loading problem as general assignment model or general transport model. Branch and Bound algorithm[4, 5] and heuristic methods[6, 7] are popularly applied to MLP. In addition, metaheuristic methods such as fuzzy-based heuristics[8], simulated annealing [9] (SA) and genetic algorithms [10]have recently been used to solve MLP.

However, these methodologies are time-consuming and solutions to large scaling problems are rarely obtained. Hence, real-time control is not possible in FMS environment.

Since Hopfield and Tank[11] solved the traveling salesman problem (TSP), Hopfield Networks have been very popular for solving combinatorial optimization problems including general assignment problem[12] and job shop problem[13]. Unfortunately, it is difficult to generate valid solutions, which are dependent on the values of the weighting parameters in the energy function. Even though valid solutions may be obtained, the quality of the solutions are poor, as reported by Wilson and Pawley[14].

Nevertheless, many researchers have used Hopfield Networks approach for solving optimization problems through variations of Hopfield Networks which can be broadly categorized as either deterministic or stochastic. The stochastic approaches are perhaps more successfully able to improve solution quality. The ability to massive, parallel and quick computation of Hopfield Networks can be suitable for dynamic scheduling problems necessary to find the quick solution in response to unexpected events.

This paper is organized as follows: In section 2, firstly the mathematical formulation of the MLP is modeled as zero-one integer programming. In addition, the mapping steps and the computing procedure of Hopfield Networks are introduced. The transformation of the formulation of MLP into energy function of Hopfield Network is explained and then energy function of Hopfield Networks considering machine-breakdown is presented in section 3. The modeling method for experiment work of MLP and results of simulation are showed with evaluations in section 4. Section 5 contains a brief summary and conclusion.

2 Machine Loading Problem (MLP) and Hopfield Networks

2.1 Statements of MLP

The model of MLP can be constructed using elements below. It is assumed as given a set of n single-operating part types $j_1, j_2, \bullet \bullet \bullet, j_n$ that are to be processed on a set of m parallel identical machines $M_1, M_2, \bullet \bullet \bullet, M_m$, m not being a very lager number.

All machines are capable of processing any part and every operation requires one arbitrary machine. Job j_n should be granted one machine and the required resources during the non-interrupted processing time of length p_j.

MLP is formulated as the following zero-one integer programming problem:

$$\min \sum_{i=1}^{m} \sum_{j=1}^{n} p_j \bullet X_{ij} \tag{1}$$

$$\text{s. t.} \sum_{j=1}^{n} p_j \bullet X_{ij} \le b_i \tag{2}$$

$$\sum_{i=1}^{m} X_{ij} = 1, \ j = 1, 2, \bullet \bullet \bullet n \tag{3}$$

$$X_{ij} = 0 \text{ or } 1 \ i = 1, 2, \bullet\bullet\bullet, m, j = 1, 2, \bullet\bullet\bullet, n \tag{4}$$

where, j : job index
           i : machine index
          p_j : processing time of job j
          b_i : processing time available on machine i
          X_{ij} : 1 if job j is processed on machine i, 0 otherwise

Equation (1) is objective function minimized.. Inequality (2) says that the time available on each machine cannot be exceeded. Equation (3) confine that each part should be processed by only one machine and equation (4) indicates that X_{ij} is zero-one decision variable which is 1 if the j part is processed on the i machine and 0 otherwise.

2.2 Hopfield Network

Hopfield and Tank[11] introduced a deterministic neural network model with a symmetrically interconnected network. Hopfield Networks can present as two functions and two equations.

- Energy Function : $E(x) = -\dfrac{1}{2} \sum\limits_{i=1}^{N} \sum\limits_{j=1}^{N} T_{ij} V_i V_j - \sum\limits_{i=1}^{N} I_i V_i$

- Input-output Activation Function : $V_i = g(u_i) = \dfrac{1}{2}(1 + \tanh \lambda u_i)$

- Differential Motion Equation : $\dfrac{du_i}{dt} = -\dfrac{u_i}{\tau_i} + \sum\limits_{j=1}^{N} T_{ij} V_j + I_i$

- Update Equation of Neuron : $u_i^{t+1} = u_i^{t} + \dfrac{du_i^{t}}{dt} \Delta t$

Where, N is the number of neurons. V and u are output neuron and input neuron respectively. T_{ij} is weighting parameter and I is external input neuron. λ and τ are gain factor and time constant without loss of generality respectively. Δt is time-step.

Rammanujam and Sadayappan[15] prescribe the following steps in mapping a problem onto Hopfield Networks which use an energy function:

1. Choose a representation scheme which allows the outputs of the neurons to be decoded into a solution to the problem.
2. Derive an Energy Function whose minimum value corresponds to best solutions of the problem to be mapped.
3. Derive connectivity(T_{ij})and input bias currents(I_i) from the Energy Function.
4. Set up initial values for the input to the neurons which completely determine the stable output of the neurons in the network.

Steps 1, 2 and 3 are very important in mapping the problem onto the network and will play a critical role in deciding whether a specific problem can be solved easily using neural networks or not. They stated that there is no direct method for mapping constrained optimization problems on to a neural network except through addition of terms in the energy function which penalize violation of the constraints.

Generally, three fundamental steps are adopted to accomplish the schema of solving neural optimization problems. The first step in the solution method is to construct an energy function whose minimum corresponds to the solution of the problem. The second step is to set up threshold principles for judging and updating the state of the networks to lead the energy function towards a minimum value. The modification of weights does not need to be considered because the weights are fixed. The third step is to repeatedly perform the comparison of the energy function values calculated by the two successive iterations of updating the states of the networks. The updating continues until the value of the energy function is unchanged several times.

3 Application of Hopfield Networks for Solving MLP

3.1 Energy Function of MLP

The approach of Hopfield Networks for the optimization problem consists of three main steps. First, the constrained optimization problem should be equivalently converted into an unconstrained optimization problem. Second based on the unconstrained optimization problem a dynamic system of differential equations is derived. Third according to dynamic system the VLSI implementation is constructed.

The constrained optimization problem of MLP is converted into an equivalent unconstrained optimization problem by using penalty function method. The energy function of the machine assignment is as follow

$$E = \frac{D}{2} \sum_i^m \left(\sum_j^n p_j \bullet X_{ij} \right)^2 + \frac{C}{2} \sum_i^m \left(\sum_j^n p_j \bullet X_{ij} - APT \right)^2$$
$$+ \frac{A}{2} \left(\sum_i^m \sum_j^n X_{ij}(1 - X_{ij}) \right) + \frac{B}{2} \sum_j^n \left(\sum_i^m X_{ij} - 1 \right)^2$$

(5)

where APT is average processing time. A, B, C and D are weighting parameters and are positive. The first term is to minimize objective function and the second term is constraint to equalize the utilization of each of m machines. b_i in equation (2) is replaced with APT. The third term, which moves values of neurons toward zero or one, i. e., to a vertex of the hypercube, is called a binary value constraint because it takes minimum value when all the neurons take the value zero or one. Last term indicates that only one part is processed on one machine.

3.2 Energy Function Considering Machine-Breakdown

If one machine breaks down, energy function can be described as follows:

$$E = \frac{D}{2}\sum_{i}^{m-1}\left(\sum_{j}^{n}p_j\bullet X_{ij}\right)^2 + \frac{C}{2}\sum_{i}^{m-1}\left(\sum_{j}^{n}p_j\bullet X_{ij}-APT\right)^2$$
$$+\frac{A}{2}\left(\sum_{i}^{m-1}\sum_{j}^{n}X_{ij}(1-X_{ij})\right)+\frac{B}{2}\sum_{j}^{n}\left(\sum_{i}^{m-1}X_{ij}-1\right)^2 \tag{6}$$

According to this unconstrained optimization objective function, the dynamic system of differential motion equation of the energy function (5) is derived as follow:

$$\frac{\partial E}{\partial X_{st}}(\vec{X}) = -D\bullet p_t\sum_{j}^{n}P_j\bullet X_{sj}$$
$$-C\bullet p_t\left(\sum_{j}^{n}p_j\bullet X_{sj}-APT\right)-A\left(\frac{1}{2}-X_{st}\right)-B\left(\sum_{i}^{m}X_{it}-1\right)$$

The behavior of a network is influenced by initial conditions as there are domains of attractions associated with each stable state of a network. Wilson and Pawley[14] experimented the following techniques in their work on the TSP. The initial condition of neurons is set by

$$X_{init} = -\frac{X_0}{2}\ln(m\times n-1)\pm Noise \text{ } \textbf{\textit{where Noise}} = 0.1\times X_{init}$$

Activation function for output of neuron used is $X_{ij}=\dfrac{1}{1+\exp\left(-2X_{init}\big/X_0\right)}$.

For the digital simulation, discrete time steps(Δt) are used. The update equation is

$$X_{ij}^{t+1} = X_{ij}^{t} + \frac{dX_{ij}}{dt}\Delta t$$

4   Results and Evaluation for MLP

4 1 Experimental Description

For modeling the simulation of MLP, the assumptions are needed as follows:

1. Jobs are assigned to the system at predetermined intervals
2. Set-up time is included in processing time
3. Material handling time is negligible
4. No preempting of jobs
5. The processing time for each operation is uniformly distributed between 1 and 20
6. Running of 100 replications with 10000 time horizon

Figure 1 illustrates an FMC with three machines via ARENA software.

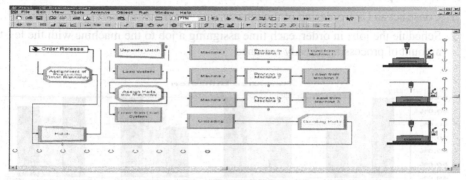

Fig. 1. ARENA simulation of MLP

ARENA[16] software as a tool for solving MLP and for modeling real-time control in the manufacturing environment is used. Hopfield Networks is implemented in VBA (Visual Basic for Applications) integrated into ARENA software.

The algorithms of Hopfield Networks to solve MLP are as follows

Step 1. To initialize the parameters A, B, D and increment step time.

Step 2. To compute $E(t_0)$

Step 3. Select one nuron randomly X_{ij}

Step 4. To compute dX_{ij}/dt according to differential motion equation

Step 5. To compute new $X_{ij}(t + \Delta t)$ accroding to activation function and upate equation

Step 6. To defive $E(t_0 + \Delta t)$

Step 7. To derive $\Delta E = E(t_0 + \Delta t) - E(t_0)$

Step 8. Let $t_0 \leftarrow (t_0 + \Delta t)$

Step 9. To check $\Delta E < \text{threshold}$, else go to Step 3, then output solution.

4.2 Results of Experiment

To assess the performance, the heuristic methods of Dogramaci[17] and Baker[18] are added in the experiment.

The steps of the Dogramaci's method (Ali heuristic) are

1. List the n jobs in an order using a certain priority rule
2. Among the jobs that have not yet been assigned to a processor choose the first one in the ordered list, assign it to the earliest available processor, and remove it from the ordered list Repeat this step until all the jobs are assigned
3. Next consider separately each of the m processors and the jobs assigned to it in Step 2

Also, Baker presents the following algorithm

1. Construct an LPT ordering of the jobs
2. Schedule the jobs in order, each time assigning a job to the machine with the least amount of processing already assigned

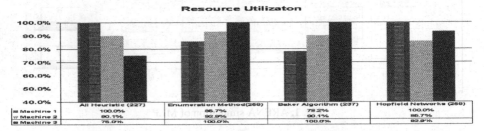

Fig. 2. Result of Simulation for MLP with 9 jobs and 3 machines

The results of simulation in figure 2 indicate that machine utilization and throughput of Hopfield Networks are superior to those of the heuristic methods of Dogramaci and Baker. Hopfield Networks have nearly found optimal solutions as compared with enumeration method which can find the optimal solution through fully search, in small size problem. As shown in figure 3, the solutions of Hopfield Networks are nearly coincident with those of enumeration method. The performance measure is:

$$\textbf{unbalance} = \sum_{i=1}^{m} \sum_{j}^{n} \left(p_j X_{i,j} - APT\right)^2$$

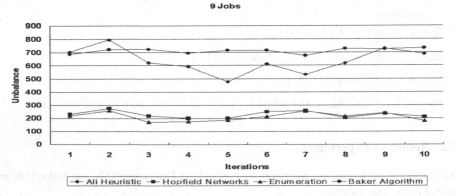

Fig. 3. The comparison among methods and result of simulation with 9 jobs and 3 machines

4.3 Evaluation and Analysis

The sensitive analysis is performed with comparison between Hopfield Networks and enumeration method. Figure 4 shows the results of the analysis, which Hopfield Networks are possible to find optimal or good solutions.

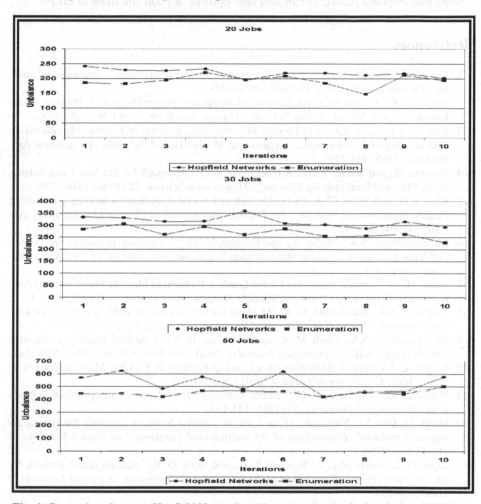

Fig. 4. Comparison between Hopfield Networks and enumeration method as the number of jobs

5 Conclusion

Hopfield Networks is presented as a method for solving the machine loading problem and the mathematical formulation as energy function of Hopfield Networks is introduced in this paper. To date, while the various methodologies have been reported and presented for solving MLP with the formulation of zero-one integer programming, the approach of real-time environment have been reported little. However, Hopfield Networks can make use of such problems as real-time scheduling problem by quickly obtaining the solutions owing to the computing capability with massive and parallel. In this paper, computing time is not considered as the performance measure, because a lot of literature has proven that heuristic methods and enumeration method are time-consuming and are not suitable to larger scaling problems. The simulation experiment shows that Hopfield Networks can find near-optimal or good solutions to MLP.

References

1. Kusiak, A. "Flexible Manufacturing Systems: A Structural Approach". *International Journal of Production Research*. 23 (1985) 1057-1073.
2. Stecke, K.E. "Formulation and Solution of Nonlinear Integer Production Planning Problems for Flexible Manufacturing Systems". *Management Science*. 29 (1983) 273-288.
3. Kusiak, A. *Loading Models in Flexible Manufacturing Systems*, in *Flexible Manufacturing Systems: Recent Developments*, A.Raouf and M. Ben-Daya(Eds.) Elsevier Science: Amsterdam. (1995) 141-156.
4. Berrada, M., and Stecke, K. E. "A Branch and Bound Approach for Machine Load Balancing in Flexible Manufacturing Systems".. *Management Science*. 32 (1986) 1316-1335.
5. Kim, Y.D., and Yano, C. A. "A new branch and bound algorithm for loading problems in flexible manufacturing systems". *The International Journal of Flexible Manufacturing Systems*. 6 (1994) 361-382.
6. Mukhopadhyay, S.K., Midha, S., and Krishna, V. M. "A heuristic procedure for loading problems in flexible manufacturing systems". *International Journal of Production Research*. 30 (1992) 2213-2228.
7. Kuhn, H. "A Heuristic Algorithm for the Loading Problem in Flexible Manufacturing Systems". *The International Journal of Flexible Manufacturing Systems*. 7 (1995) 229-254.
8. Vidyarthi, N.K., and Tiwari, M. K. "Machine loading problem of FMS: a fuzzy-based heuristic approach". *International Journal of Production Research*. 39 (2001) 953-979.
9. Mukhopadhyay, S.K., Singh, M. K., and Srivastava, R. "FMS machine loading: a simulated annealing approach".. *International Journal of Production Research*. 36 (1998) 1529-1547.".
10. Basnet, C. "A Genetic Algorithm for a Loading Problem in Flexible Manufacturing Systems". *MS/IS Conference in Korea*. 4 (2001), pp. 284-287.
11. Hopfield, J.J., and Tank, D. W. "Neural computation of decisions in optimization problems". *Biological Cybernetics*. 52 (1985) 141-152.
12. Gong, D., Gen, M., Yamazaki, G., and Xu, W. "Neural Network approach for general assignment problem". *Proceedings of the International Conference on Neural Networks*. 4 (1995) 1861-1866.
13. Zhou, D.N., Cherkassky, V., Baldwin, T. R., and Hong, D. W. "Scaling neural network for job-shop scheduling". *Proceedings of the International Conference on Neural Networks*. 3 (1990) 889-894.
14. Wilson, G.V., and Pawley, G. S. "On the stability of the traveling salesman problem algorithm of Hopfield and Tank". *Biological Cybernetics*. 58 (1988) 63-70.
15. Ramanujan, J., and Sadayappan, P. "Optimization by neural networks". *Proceedings of IEEE International Conference on Neural Network*. 2 (1988) 325-332.
16. Kelton, D.W., Sadowski, P. R., and Sadowski, A. D., *"Simulation with ARENA"*. (2002), Sydney: McGraw-Hill.
17. Dogramaci, A., and Surkis, J. "Evaluation of a heuristic for scheduling independent jobs on parallel identical processors". *Management Science*. 25 (1979) 1208-1216.
18. Baker, K.R., *"Introduction to Sequencing and Scheduling"* (1974) John Wiley & Sons.

Performance Analysis of Alternative Designs for a Vehicle Disassembly System Using Simulation Modeling

Eoksu Sim[1], Haejoong Kim[1], Chankwon Park[2], and Jinwoo Park[1]

[1] Department of Industrial Engineering, Seoul National University,
Shillim-dong, Kwanak-gu, Seoul, 151-742, Korea
{ses,ieguru}@ultra.snu.ac.kr, autofact@snu.ac.kr
[2] Department of e-Business, Hanyang Cyber University,
17 Haengdang-dong, Seongdong-gu, Seoul, Korea
chankwon@hycu.ac.kr

Abstract. Recently some regulations such as EPR (Extended Producer Responsibilities) have forced companies to develop an interest in recycling systems. In particular, these regulations are focused on the reuse of product materials to resolve environmental disruption or waste disposal. Unfortunately, a large amount of research has been conducted only in the field of mass production, rather than in the automotive disassembly systems that are more likely to have significant opportunities for reuse and recycling. Therefore, it has become necessary to conduct analytic research on systems with the capacity to deal with the annually increasing number of vehicles for disassembly. In this paper, we evaluate four existing vehicle disassembly systems and propose a new and improved alternative disassembly system. We also conduct performance analysis for each system using simulation modeling. Furthermore, we propose 4 alternatives and test their efficiencies in order to improve their performances.

1 Introduction

Vehicles have become an essential necessity in people's lives and in the modern economic system. In this regard, mass production of vehicles has evolved in order to meet the large demand. Therefore, a lot of research on the automotive production process has been conducted at the commencement point of the vehicles' lifecycle.

Each year, over a million cars are scrapped in Korea and approximately 400,000 scrap cars are processed by the Dutch car disassembly industry [1]. Although there are many requirements for cars to be disassembled, a majority of the research has focused on the disassembly process of general electronic equipment such as PCs, rather than that of cars.

Companies, governments and many researchers consider the disassembly and recycling of cars at the end point of their lifecycle, however, as an important research topic. To efficiently utilize the limited, global natural resources and to resolve waste disposal and environmental disruption issues, it is increasingly important to reuse and recycle resources and products. This is interrelated with the new emerging concept of PLM (Product Lifecycle Management) that is centered on the management of the data on products and resources at all stages from design to disposal and recycling.

The EU regulation on the recycling of cars requires that 80% of cars be recycled, 85% be recovered by 2006 and 85% recycled and 95% recovered by 2015. To meet these requirements, it is necessary to establish a car disassembly system that has vari-

D.-K. Baik (Ed.): AsiaSim 2004, LNAI 3398, pp. 59–67, 2005.
© Springer-Verlag Berlin Heidelberg 2005

ous technological equipments satisfying environmental regulations. In many developed countries, in order to reuse waste from scrapped cars, laws related to recycling have been enacted and promulgated. In Germany, there is an environmentally friendly system that supports free collection of scrapped cars, removal of all operating fluids, part reuse, and dismantlement/separation/collection of plastic parts. In Japan, H motor company has formed a new firm in conjunction with 6 other companies in the recycling, disassembly, and reuse industry. The main objective of this firm is to systematically reduce the waste by reusing the scrapped plastic parts from scrapped cars [3].

Many companies all over the world have set up new ventures to handle the disassembly of cars. Car recycling system (CRS) using rail operates in the Netherlands [4]. N company in Japan operates a disassembly system using a cart-on-track conveyor and 2 post jack type lifts. West-Japan Auto Recycle Co., Ltd operates a system utilizing hangers [5]. MATEC uses an unmanned fork lift to return the cart-on-track [6]. Likewise, in many developed countries research on the installation and operation of various disassembly systems has been conducted. To deal with the increasing number of scrapped cars, we need to evaluate the existing disassembly plants and to design a new disassembly system.

2 Analysis of the Existing Disassembly Systems

We have found that some disassembly systems are not systematically organized in their processes. In this paper, we present 4 disassembly systems that are operated in some nations. We have analyzed the strengths and weaknesses of the existing disassembly systems and, based on that analysis, we propose a new and improved disassembly system. This new system has been designed by facility experts, field workers and managers who will be directly involved in the construction and operation of the new system.

Through simulations, we compare the performance of the new system with the existing systems. Based on the simulation results, we propose 4 alternatives to improve the performance of the new system and present simulation models reflecting those alternatives.

2.1 Existing Disassembly Systems

The existing systems all show the following similar sequence of disassembly operations.

- Checking operation of the components for disassembly
- Disposal of explosive components like air bags
- Setup for the removal of all operating fluids
- Removal of all operating fluids: oil, fuel, refrigerant in the air conditioning, coolant, and brake fluid
- Disassembly operation of exteriors like bumpers
- Disassembly operation of interiors like seats
- Disassembly operation of engine/transmission
- Compression of car body.

Each operation is performed in each work center. For example, the fifth operation is performed in work center 5.

2.1.1 Two Post Jack Type Disassembly System (2PTDS)

This system is currently being used by N company in Japan. Through this system, cars are moved by a cart-on-track conveyor that consists of a track and 4 powered carts which shuttle between work centers (WC). Cart 1 moves from WC1 to WC4 and returns after the car is lifted in WC4. When operation 4 is finished, the cars are then moved to WC5 by cart 2, which returns to WC4 after the car is lifted in WC5. When operation 5 is finished, the cars are then moved to WC6 by cart 3. After operation 6 is finished, cart 4 moves the car to WC7 and cart 3 returns to WC6. When operation 7 is finished, the hoist holds the car, moves it to WC8 and dumps it into the compressor.

This automated system does not require any man power to move the carts because it uses a cart-on-track conveyor. Carts move between the WCs automatically. Thus, we can expect minimization of the delays that can occur in manual operations. However, this mechanical system does not guarantee minimal costs because the powered carts are more expensive than manned carts.

2.1.2 Unmanned Forklift Type Disassembly System (UFTDS)

This system uses an unmanned forklift to move non-powered carts from WC8 to WC1. The car is moved by the same cart through all of the WCs. When operation 7 is finished, the car is held by a hoist, then dumped into the compressor in WC8 and the cart is returned by the unmanned forklift.

The unmanned forklift solves the problem of withdrawing the carts which need to be returned to WC1. However, the downside is that it takes up a large space to move, requires a sensitive control technique, and incurs installation and operation costs.

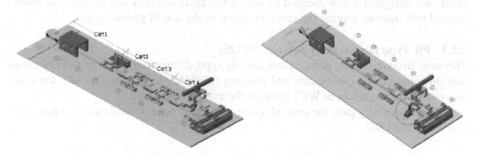

Fig. 1. 2PTDS layout   Fig. 2. UFTDS layout

2.1.3 Hanger Type Disassembly System (HTDS)

Carts move the cars from WC1 to WC4 and hangers move the cars through the other WCs.

Hangers save time because they move and lift cars without wasting loading and lifting time. A hanger can also replace the hoist; however, it is dangerous for a hanger to hold cars because problems can occur during transportation of the cars.

2.1.4   Unit Cell Type Disassembly System (UCTDS)

This system has been experimentally operated in Germany. Each WC exists as a work cell. The car is moved by cart from WC1 to WC2 and by forklift through the other WCs. If there are buffers for cars at each WC, it is possible to process in a batch operation. However, this requires additional space for the buffers. Cars are moved by forklift, which allows the possibility of congestion to occur in the plants.

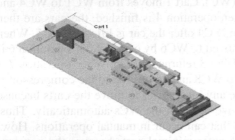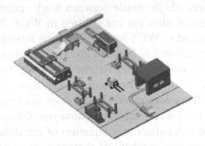

Fig. 3. HTDS layout   **Fig. 4.** UCTDS layout

2.2   The New Disassembly System

Based on the above summary of the existing disassembly systems, we propose a new type of disassembly system. Hangers can cause danger in the handling of cars whereas forklifts need space to move around. Therefore, we selected carts as transporters for moving the cars. Because the powered cart is too expensive to install in a small car-recycling company, we use a non-powered cart.

Loading the car on the cart is time-consuming so it is more efficient to use the same cart to move the car from one WC to the next. To return the cart from WC8 to WC1, we designed a new method in which the carts are returned through an underground rail, thereby solving the space problem in the use of ground space.

2.2.1   Pit Type Disassembly System (PTDS)

The new disassembly system, called the Pit type disassembly system (PTDS), uses non-powered carts to move cars and pushing machines to put the cars into the compressor. Carts can return to WC1 through the underground rail.

PTDS effectively uses the ground space and saves time to load cars on carts compared with 2PTDS.

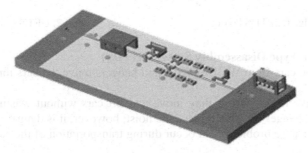

Fig. 5. PTDS layout

3 Simulation Model and Analysis

3.1 The Data for Simulation

To construct a simulation model, we need data such as operation sequences, operational job details, number of workers and operation times [7]. Since we cannot obtain operation time data, we collected operation time by observing the workers at work. We assumed that it would take the same amount of time to do the same operation in all disassembly systems.

Operation	Operation 1	Operation 2	Operation 3	Operation 4	Operation 5	Operation 6	Operation 7	Operation 8
# of workers	1		2		2		1	1
Operation time	5min	3min	10.2min	15min	9.7min	7.9min	20.8min	6min

Fig. 6. The number of workers and operation time

3.2 Performance Measures

We consider 4 criteria as performance measures.

- C1 : the time to disassemble the first car
- C2 : the number of cars disassembled in 4 hours
- C3 : the time to disassemble the 10 cars
- C4 : the average time to dissemble each car (C4).

C1 means the time to disassemble the very first car in a new batch. C2 means the number of cars dissembled in 4 hours from the arrival of a new batch. Because the manager, who will also be involved in the construction and operation of the system, expects to disassemble the cars with the 7 workers within 4 hours, it is important to know how many cars can be disassembled within 4 hours in the current system. C3 tells us how much time is required to disassemble 10 cars.

3.3 Simulation Modeling

Based on the data and criteria, we developed detailed simulation models for one of the existing systems, 2PTDS, and the newly proposed system, PTDS, using the ARENA 7.0, Rockwell software [8]. Table 1 presents some additional simulation data.

Table 1. Additional data for simulation modeling

Time to load car on cart by forklift	5 min.
Cart speed	10 meters/min.
Time to load car on cart after operation	2 min.
Time to lift/unlift cars	1 min. per each
Time to put car into compressor by hoist	2 min.
Time to push car on the cart into compressor	1 min

The following figure 7 shows the ARENA simulation model.

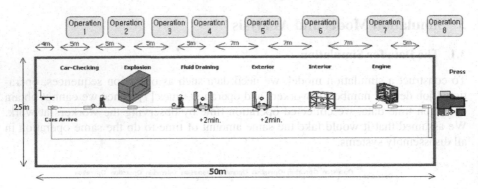

Fig. 7. ARENA animation model

3.4 Simulation Results

As we can see in Table 2, PTDS gives better results than 2PTDS for the 4 criteria. This is because PTDS saves loading time on the cart and pushing time to the compressor.

Table 2. Simulation result for PTDS

Criteria	2PTDS	PTDS
C1	102.7 min.	92.5 min.
C2	6	8
C3	348.1 min.	279.7 min.
C4	34.4 min.	27.9 min.
# of used carts	4 powered.	5 non-powered

Table 3. Changes in performance according to the # of carts

# of Carts	3	4	5	6
C2	6	7	8	8
C3	370.6	298.7	279.7	297.7

Table 3 shows that PTDS needs 5 carts, which is more than the number of carts in 2PTDS. Nevertheless, in terms of cost, the powered carts used in 2PTDS are about 5 times more expensive than the non-powered carts used in PTDS, which confirms that PTDS is more efficient and economical than 2PTDS.

4 Four Alternatives for Performance Improvement

We propose the following 4 alternatives to improve the performance of the disassembly system:

- Alternative 1(A1): relax the bottleneck operation by sharing works
- Alternative 2(A2): increase the capacity of the bottleneck operation
- Alternative 3(A3): balance the workloads
- Alternative 4(A4): A2 + A3

4.1 A1: Relax the Bottleneck Operation by Sharing Works

Table 4 shows that in the current disassembly operations, operation 7 is the bottleneck operation. Therefore, if we can quicken this operation we can improve the work efficiency by spreading the bottleneck work among the other workers.

In Table 4, the utilization of the bottleneck is 3~4 times higher than that of the compression operation. Hence, if there is no work in WC8, that worker can help with the work in WC7. In that case, the working time can be reduced from the original 20.8 minutes to about 13 minutes.

Table 4. Operation Utilization

Operations	2PTDS	PTDS	Operations	2PTDS	PTDS
Operation 1	0.1892	0.2145	Operation 5	0.3257	0.5020
Operation 2	0.3809	0.4376	Operation 6	0.2529	0.3346
Operation 3	0.1135	0.1289	Operation 7	0.6150	0.7705
Operation 4	0.5271	0.7294	Operation 8	0.1746	0.2145

4.2 A2: Increase the Capacity of the Bottleneck Operation

We can install an additional WC to increase the capacity of the bottleneck. Moreover, two workers can share the jobs in WC7 and WC8.

Fig. 8. Operation flow in Alternative 2

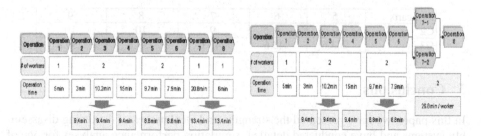

Fig. 9. Operation flow in A3   **Fig. 10.** Operation flow in A4

4.3 A3: Balance the Workloads

We can balance the whole operations in addition to the bottleneck. We consider the operations that can be shared by the same workers

4.4   A4: A2 + A3

In this alternative, we consider both A2 and A3. We increase the capacity of the bottleneck and balance the workloads in the two groups, with A2 and A3 being composed of operations 2,3,4 and operations 5,6, respectively.

4.5   Simulation Results

Table 5 shows the results of 2PTDS. We can see that A2 is the best alternative because of the workload relaxation of the bottleneck. In A3, line balancing cannot improve the performance measures because the number of carts, 4, restricts the smooth flow of the work.

Table 5. Simulation results of 2PTDS

Criteria	Basic	A1	A2	A3
C1(min.)	102.7	94.4	102.7	102.7
C2(unit)	6	6	6	4
C3	348.1	335.8	330.2	425
C4	34.8	33.6	33	42.5.

Table 6. Simulation results of PTDS

Criteria	Basic	A1	A2	A3	A4
C1	92.5	84.7	84.7	92.5	84.7
C2	8	9	9	9	10
C3	279.7	256.54	250.1	253.2	239.8
C4	27.9	25.6	25	25.3	24

In table 6, we can see that A4 is the best alternative in PTDS because after relaxing of the bottleneck, 8 carts [Table 7] can be used to streamline the overall flow in the line.

Table 7. Simulation results of PTDS

# of carts	5	6	7	8	9
C2	9	9	9	10	10
C3	243.6	243.6	243.6	239.8	239.8

5   Conclusions

In this paper, we have presented the strengths and weaknesses of 4 existing disassembly systems and have conducted detailed, simulation performance analyses for one of them, 2PTDS, and for the newly proposed system, PTDS. Based on these results, we have proposed 4 alternatives to improve the performance of the new system and presented the efficiencies of these 4.

Computational experiments demonstrated that PTDS outperforms 2PTDS in all of the performance measures. Among the 4 alternatives to improve performance, A4 showed the best results in simulation analysis.

Using the results of this study, we can install and operate the improved disassembly system described. Furthermore, after more detailed research on work analysis and design, we will be able to improve the working environment and work performance.

References

1. Marcel, A.F.D., Jan, H.M.S.: Information Technology and Innovation in Small and Medium-Sized Enterprises. Technological Forecasting and Social Change 60 (1999) 149-166
2. F. Torres, P. Gil, S.T. Puente, J. Pomares and R. Aracil. : Automatic PC disassembly for component recovery. Int. J. Adv. Manuf. Technology 23 (2004) 39-46
3. Y.S. Lim, et. al.: A Study on the development of car parts disassembly system to reduce wastes. Hyundai Motor Company (1999)
4. CRS(Car Recycling System), http://www.crs-europe.com
5. West-Japan Auto Recycle Co., Ltd(WARC), http://www.bigwave.ne.jp/~warc/html/top.htm
6. MATEC, http://www.matec-inc.co.jp/elv
7. Averill M. Law, W. David Kelton.: Simulation modeling and analysis. McGraw-Hill (1999)
8. W. David Kelton, Randall P. Sadowski, Deborah A. Sadowski. : Simulation with ARENA, 2E. McGraw-Hill (2002)

Layout of Virtual Flexible Manufacturing System and Machining Simulation*

Guofu Ding, Kaiyin Yan, Yong He, and Dan Li

School of Mechanical Engineering, Southwest Jiaotong University, P.R. China
dingguofu@163.com, Ding_guofu@yahoo.com.cn

Abstract. Virtual Manufacturing (VM) is higher computer application in product design and manufacturing and management. Verification of process production planning and control is a basic part in estimation of manufacturing, so layout of virtual factory and simulation will play an important role in VM. Here we consider typical Flexible Manufacturing System(FMS) as studied object and give a detail layout of virtual FMS, whose 3-d geometrical models are originated from a made-in-China 3-D solid CAD software, CAXA. All of the virtual objects of FMS are built according to their functions, propertied and geometrical exterior and assembly them in right position of virtual factory. In order to simulating the dynamical production planning, the consequential control procedure which is on the basis of machining process of each equipment and overall control codes are programmed, which considers transmitted belts as key connection among all of the mounted virtual machining tools and robots, etc. given results show that we can deal with some typical layout and production simulation of virtual FMS.

1 Introduction

Because of the complexity in product planning, computer simulation and modeling are rapidly becoming important and valuable aid in decision-making processes of the manufacturing factory. Simulation can generates reports and detailed statistics describing the behavior of the system under study. Based on these reports, the physical layouts, equipment selection, operation procedures, resource allocation and utilization, etc can be effectively implemented. So it can be utilized to study and optimize the layout and capacity, material handling systems and warehousing and logistics planning. To make decision in production planning of virtual manufacturing, an effective visualized simulating environment is convenient for maker.

Virtual Manufacturing (VM) is a new concept in the recent ten year's manufacturing fields. Different scholars have different definition of VM, Onosato and Iwata think that: VM is a key concept that summarized computerized manufacturing activities dealing with models and simulations instead of objects and their operations in the real world, and classifies it as two kinds: Virtual Physical System(VPS) +Real Information System(RIS) and VPS + Virtual Information System(VIS) [1]. Hitchcock thinks that: VM is an integrated, synthetic manufacturing environment exercised to enhance all levels of decision and control in a manufacturing enterprise [2]. Nahavandi and Preece described VM as a simulated model of the actual manufacturing

* This paper is supported by Scientific Development Funds of Southwest Jiaotong University, China.

D.-K. Baik (Ed.): AsiaSim 2004, LNAI 3398, pp. 68–76, 2005.

setup which may or may not exist. It holds all the information relation to the process, the process control and management and product specific data. It is also possible to have part of the manufacturing plant be real and the other part virtual [2]. Other scholars, Lin [2], Kimura [3], Xiao [2], Glonra.J. Wiens [2], etc. have their particular understanding to VM. From all of these, it is obvious that VM is abstract of realistic manufacturing, and realistic manufacturing is an instance of VM. Here we study layout of virtual FMS and its simulation of production procedure planning, wish to make an estimation to product manufacturing.

2 Building of Virtual Equipment

FMS is made up of NC equipment, material handling facilities and computer control system, etc.. to study FMS by using virtual manufacturing, two aspects need to be done, one is simulating the functions of equipment distributing in this system in a right form, the other is how to plan the production procedure and obtain the exact evaluation of this FMS. For a typical flexible manufacturing system, in which warehouse, fork truck, stock, conveyor, all kinds of CNC or DNC, etc., are planned according to the product line. A complicated process network is also constructed where nodes are named as the units composed of machining equipment and other machinery. To map it into the computer, some abstracts must be done, the geometrical model of these nodes can be simplified. However, their functions and actions should be rightly modeled.

For certain production task, to ensure the reliability of process evaluation, product plan in virtual manufacturing must be consistent with that of realistic manufacturing.

2.1 Object-Oriented FMS

To modeling these objects distributing in FMS, some methods are considered, one is object-oriented analyses. In FMS, many components can be considered as so-called objects, for example, machining stations (turning center, mill-turn center, milling machine modules, etc.), load/unload stations S/R machine, material handling and storage system, etc. A typical object can be described as following equation:

Object=Action + Property + Method + Message (communication)

All of these objects in FMS can be abstracted as a set of programmed ones and be managed by using the data structure of line list.

2.2 Geometrical Modeling of Objects

The geometrical exterior of virtual equipment is the same as real one, so, we must firstly model the geometrical part of equipment. To simulate the functions of equipment, the equipment can be thought the assembly of several essential components, for example, a robot is made up of base, arm 1, arm2, arm 3, palm and fingers, which is like Fig. 1.

In Fig. 1, parts of robot are designed in 3-D solid software as well as CAXA, Solidworks, then assembly them into a functional robot, and send these models into 3-D virtual environment, so do other object such as turning center, milling center transmitted belt, which can be represented in Fig. 2.

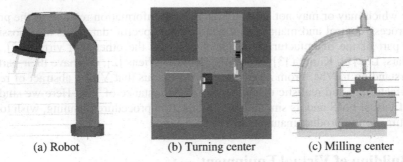

(a) Robot   (b) Turning center   (c) Milling center

Fig. 1. Components of objects

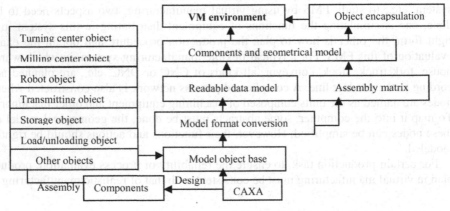

Fig. 2. Geometrical model of object

Models designed by CAXA can be exported in several data formats, one is *.3ds, the other is *.igs or *.wrl, by using the middle software, they will be converted into readable display lists that can be dealt with by OpenGL API. At the same time, relative position of each component of object is picked up in the form of coordinate transformation matrix, each part of object has own Local Coordinate System(LCS), it is positioned by the World Coordinate System(WCS) or parent coordinate system. In Fig. 3, O is origin of WCS, the others are origins of LCS, O_1 is the parent of O_2, so does the other, maybe O_j has parent or child, it is only connected with WCS. Each component has own transform matrix T, which has the following form:

$T = \begin{bmatrix} R_{3\times3} & P_{3\times1} \\ f_{1\times3} & 1\times1 \end{bmatrix}$, here R is the rotating sub-matrix described by Euler angles, P is

the position sub-matrix which represents the position of the origin of component coordinate system in the reference system of WCS or LCS, f is scalar factor sub-matrix of component. According to the parent-child relationship among components, sets of transforming matrix will be executed to locating the position of component in the object, which is:

$$T_o = T_i T_j T_k \ldots T_n \text{, and } T_n = A_r A_t A_s \tag{1}$$

where A is the product of translating, rotating or scaling matrix. The completed object is listed in Fig. 1 (a), (b), (c).

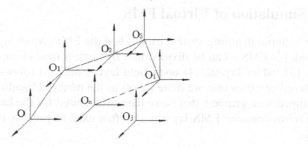

Fig. 3. Coordinate system of components in the object

2.3 Function and Action Modeling of Object

Each object has its own function and action. For example, a turning machine tool can cut turning-type part and it has own machining part scope; a robot can be in charge of loading part from transmitting belt to work piece clamping position of machine tool or unloading part from chuck to tray transmitting work piece, and moving along the belt. Actions of all of these objects should be encapsulated into a software object according to the rule of object-oriented design. In FMS, machine tool is one of the most key equipment, it can read NC codes and translate them into pulse-values which are used to drive servo-control system to make cutting tool come to a right position. Other encapsulated functions include: decoding of NC program such as G, M, S, T, N, X, Y, Z; control of multi-axis co-moving, calculating of cutting force, parameters, etc.. For a robot, its kinematical function is much more complicated, because parent-child coordinate transformations will be multiplied along arms and hands according to Fig.3, the kinematical path of robot in space is also planned, so gesture of each component of robot will be accurately defined. All of these will be briefly described in Fig. 4, which is dynamic object relative to that of static object in Fig. 3.

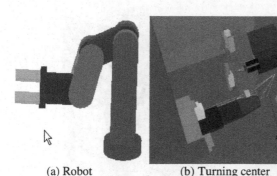

| (a) Robot | (b) Turning center | (c) Milling center |

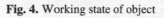

Fig. 4. Working state of object

In Fig. 4 (a), robots put work piece to another position like workspace of milling center, in (b), turning center is machining a turning work piece with cone, and in (c), milling center is machining an inner contour of part with arcs. These equipment have corresponding instruction to control them completing given function.

3 Simulation of Virtual FMS

The material handling system established the FMS. Most layout configurations found in today's FMS's can be divided into five categories: (1) in-line layout, (2) loop layout, (3) ladder layout, (4) open field layout and (5) robot-centered cell. The robot-centered cell uses one ore more robots as the material handling system. Robots can be equipped with grippers that make them well suited for the handling of rotational parts, and robot-centered FMS layouts are often used to process cylindrical or disk-shaped parts.

3.1 Layout of Virtual FMS

In order to satisfy the need of industry, several layout types can be synthetically utilized. But several problems need to be resolved, one is the right position of virtual equipment. In WCS, each equipment has a parent Coordinate Reference System(CRS) O like Fig. 4, O_1, O_2... O_n are its children, so matrix transformation will be executed among WCS and parent CRS, operation will be the same as formula (1).

Another problem is how to plan and practice process procedure in existed layout of FMS. Here we present a novel idea, it is on the basis of transmitting belt. As a tool transmitting part of work piece, it plays an important role in information interaction of material handling equipment, right plan and layout of transmitting belt gives the work piece transmitting routine and process routine and the equipment setting routine.

Such objects as robots, machine tools, are put in virtual layout space, they are arranged on the basis of transmitting belt. In virtual layout environment, the data structure of linear pointer list is used, all of the modeled objects can be added, deleted, modified, enabled or disabled action. The scene can be rotated, panned, zoomed in or out. The contents can also be saved for backup. Optimized schemes are used as guide to practice. A typical static layout of FMS can be displayed in Fig. 5.

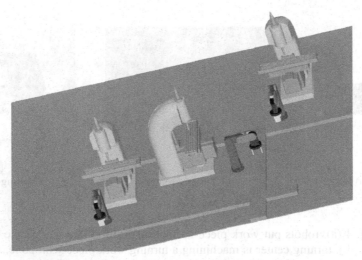

Fig. 5. Layout of three milling machine tools +three robots

In Fig. 5, three milling machine tools face different side of work space, each robot serves a milling machine tool, in charge of loading/unloading parts between machine tool and pallet.

3.2   Simulation of Production Procedure

After layout finished, process planning can be executed, on the basis of process of machined parts, it is divided into two steps, one is to deal with each sub-process of parts, the second is to plan the overall production of parts. Layout data are also used to program control codes. Actions of those equipment is based on the process procedure of parts and sequence of equipment sequence, which are dealt with according to Fig. 6.

In Fig. 6, the process and production plan are firstly organized into the control codes, they are decoded into control instructions. When control codes notify some virtual equipment to start working, it is triggered and auto-operate. All of these instructions can be described in Table. 1 as a simple language.

As an example, here gives a segment of control codes for turning center.

N01 H0100000;//No. 01 robot is ready, 0 for supplement
N02 G01HX-100;// No.01 part moves to lathe about -100cm along X
N03 L01A.txt;// No. 01 lathe starts machining, and load NC file of L01a.txt
N04 G01HH-200; // No. 01 part moves -200cm along horizontal direction

Table 1. Notation explain nation control codes

No.	Code Name	Function	No.	Code Name	Function
1	MI	Milling machine tool	10	GVR	Milled part rotates an angle around itself
2	L	Lathe	11	G	Functional code
3	V	Vertical	12	N	Flag
4	H	Horizontal	13	X,Y,Z	Coordinate values
5	GHX/GHY	Lathed Part moves to given position that robot grasps along X or Y direction	14	T	Tool no
6	GHH/GHV	Lathed part moves to machined position along vertical or horizontal direction	15	F	Rate of feed
7	GHR	Lathed part rotates an angle around itself	16	S	Rotate rate of Spindle
8	GVX/GVY	Milled Part moves to given position that robot grasps along X or Y direction	17	M	Auxiliary function
9	GVH/GVV	Milled part moves to machined position along vertical or horizontal direction			

3.3   A Case of Virtual FMS

In virtual FMS, transmitting belts should be firstly laid out, on the base of the dimension of selected belts and characteristics of modeled robots, the position of robot can be located, which also locates the position of machine tools. So layout of transmitting belts is an important part in virtual FMS. Several steps can be divided to lay out the virtual equipment.

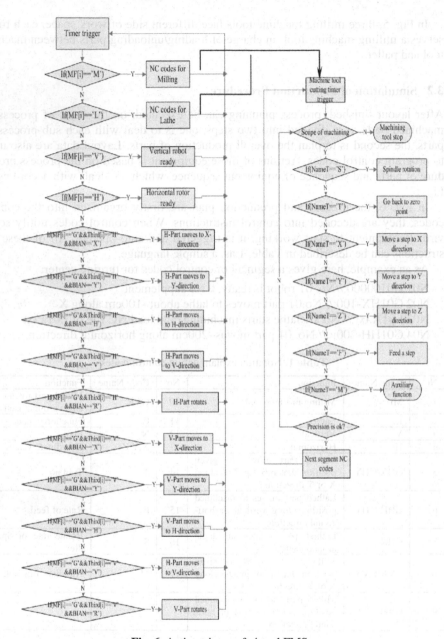

Fig. 6. Action trigger of virtual FMS

3.3.1 Ready for Layout

Supposed that the dimension of virtual plant, machine tools, transmitting belts, robots, etc., are given, which are displayed in Fig. 7. From Fig. 7, we can know that dimension of machine tools, robots, transmitting belts and their locating points, so the layout that three dimension space maps into two dimensional planar area, is presented.

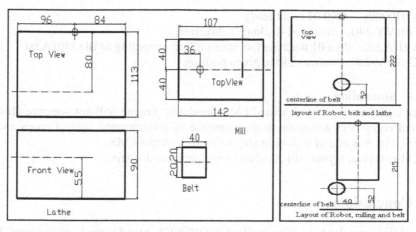

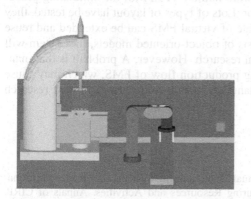

Fig. 7. A case of layout

Fig. 8. Result of layout

Fig. 9. Robot is loading the part to milling machine tool

3.3.2 Layout

Based on the data in Fig. 7, by using the developed software VPLFMS, we can lay out a virtual FMS, which is described as Fig. 8. It includes a milling machine tool, cross transmitting belts, a robot, parts.

3.3.3 Control Codes and NC Program for Milling Machine Tool

Write NC program as following:

N00 S200;
N01 G00 X85 Y140;
N09 G00 X0 Y0;
......
N10 M02;

And save it as M01A.txt, then write control codes as following:

G01VH-100;   //move -100 along horizontal direction
G01VR90;   //rotate 90 degree at cross belt position

V010000;   //N0 01 robot ready
G01VY-140; //move -140 along Y direction
M01A.txt;   //No 01 machine tool start milling according to file M01A.txt
G01VV-240; //move -240 to given position

3.3.4 Simulation

Once codes are generated, virtual FMS simulating engine will auto-operate all of the virtual equipment according to the sequence of programmed control codes and NC codes. Fig. 9 is one of actions in simulating this virtual FMS.

Other typical layout will attached to this paper as an affix.

4 Conclusions

All of these results have been testified in OPENGL-based virtual environment, based on which we develop a simulating software named VPLFMS, the simulating procedure is triggered by multi-timer simulator. Lots of types of layout have be tested, they show that the idea is reliable and the scale of virtual FMS can be extended and reuse because all of the models are on the basis of object-oriented models, the system will be very useful to future process decision research. However, A problem is that man-hour has not been consider in simulating production flow of FMS, which may cause to half-baked decision in production planning, it will be discussed in next research plan.

References

1. Iwata K, Onosato, Teramoto K, et al. Virtual Manufacturing Systems as Advanced Information Structure for Integrating Manufacturing Resources and Activities. Annals of CIRP, 1997, 46(1):335~338.
2. Mingquan ZHU, Shusheng ZAHNG. Virtual Manufacturing System and practice. China: NWPU Publish House, 2001
3. Kimura F. Product and Process Modeling as a Kernel for Virtual Manufacturing Environment. Annals of the CIRP, 1993, 42(1):147~150
4. Shuka C., Nazquez M. Chen F. F.. Virtual Manufacturing :an Overview. Computers & Industrial Engineering, Vol.31, pp.79~82, January, 1996
5. Yuan Qingke, Zhao Rujia. Virtual Manufacturing System. China Mechanical Engineering, Vol.6, pp.10~13, April, 1995
6. Junqi Yan, Xiumin Fan, Jian Yao. The Architecture and Key Techniques for VM System. China Mechanical Engineering, Vol.9, pp.60~64, 1998 (in Chinese)
7. Xiangli Han, Gang Yang, Tianyuan Xiao. Virtual Manufacturing and Its Application in Cims Engineering Research Center. High Technique Letters. pp.1~6, January,1999 (in Chinese)1. Baldonado, M., Chang, C.-C.K., Gravano, L., Paepcke, A.: The Stanford Digital Library Metadata Architecture. Int. J. Digit. Libr. 1 (1997) 108–121

A Simulation-Based Investment Justification for an Advanced Manufacturing Technology with Real Options

Gyutai Kim[1] and Yun Bae Kim[2,*]

[1] Department of Industrial Engineering, Chosun University, Kwangju, Korea
gtkim@mail.chosun.ac.kr
[2] School of Systems Management Engineering, Sungkyunkwan University, Suwon, Korea
kimyb@skku.edu

Abstract. Strategic investment projects such as advanced manufacturing technology (AMT) projects are considered to provide a company benefits beyond the immediate cash flows as well as a variety of real options that are conceived to increase their value. Traditionally, discounted cash flow (DCF) techniques have been widely used to evaluate the value of the investment projects. However, investment analysts have reduced their dependence on the use of the DCF techniques mainly because they believe that the DCF techniques lack in ability to value the real options embedded in the investment projects. As required, the real options approaches (ROA) have been proposed as the complementary for the DCF techniques. In this paper, taking a real case we present how to evaluate the value of the investment projects for the AMT involving a deferral real option with a ROA. For this purpose we first obtained the processing-related data by simulating the underlying manufacturing system. And product cost information was collected under the context of an activity-based costing (ABC) system. Based on the result of the case study we performed we suggest that the investment analysts should be careful in choosing the investment justification technique proper for the projects of interest because there may exist a great amount of difference in the value of investment projects depending on which technique is used.

1 Introduction

A highly competitive market place has urged manufacturing companies of today to invest in AMT projects and thereby they may satisfactorily meet changing customer needs. Such AMT projects are characterized to be more expensive, complex to analyze and evaluate than the projects for traditional manufacturing technologies. And much of the values of the AMT projects required to justify the adoption is derived from the benefits that are intangible, contingent, and hard to estimate. These characteristics make a relevant investment decision environment more dynamic and uncertain. Thus, what managers of a company are now concerned about is to figure out a desirable way to deal with uncertainty inherent in the AMT investment project.

An uncertain decision environment in most cases creates options for the managers to take to increase the value of the projects under consideration and thus to steer the direction of the preceding investment decisions. The examples of the options embedded in investment projects are to delay, abandon, expand, and contract the investment

---

* To whom correspondence should be addressed.

D.-K. Baik (Ed.): AsiaSim 2004, LNAI 3398, pp. 77–85, 2005.

project. To date, the DCF techniques like a net present value (NPV) have been widely used to capture the financial performance of the AMT investment projects. They implicitly view the decision environment as static and assume that once the investment project is initiated, its basic components will remain unchanged. Therefore, they do not explicitly value options and result in undervaluing the AMT investment projects [1], [2], [3]. Copeland notes that the NPV always undervalues the potential of the investment projects, often by several hundred percent [4].

Complying with the requirement, the ROA has recently gained significant attention as it is said to capture the value of the options [4], [5], [6], [7], [8]. Under the real option decision framework, any company's decision to invest or divest in real investment projects is regarded as an option. A company has a right to gather more information as events unfold before making a decision to invest in the AMT projects. By doing so, there is no downside risk beyond the initial investment in AMT. From this aspect, it can be said that investment in the AMT projects is akin to investment in a call option.

What really makes the ROA so attractive as an investment valuation tool in the business environment of today lies in its ability to clearly recognize that an investment decision maker can incorporate new information obtained over the planning horizon of an investment project to change the decision already made at the start or the previous stage. Such benefit of the ROA has been exploited to evaluate and justify investment in AMT for the place of the DCF techniques. In a recent survey, Graham and Harvey address that the ROA has been applied to addressing the capital budgeting decisions in 27% of the firms surveyed [5].

There exist some research papers presenting the application of the ROA to valuing AMT investment projects. For example, Kumar [2] considers investment decisions relating to expansion-flexible manufacturing systems, and evaluates its value using a valuation model for option proposed by Margrabe [11]. And Bengtsson [1] investigates manufacturing flexibility in terms of real options. However, MacDougall and Pike [3] suggest that care should be exercised when employing the ROA to evaluate investment projects for the AMS. They argue for the reason that the strategic value of the investment projects varies during the critical and complicated implementation phase, and thus it is difficult to incorporate such varying value into the ROA at the early stage. However, they also acknowledge that the ROA is a more effective investment valuation tool than the DCF techniques.

This paper is organized as follows. Section 2 is devoted to the brief review of the ROA. In Section 3, we will describe the investment decision problem under consideration, and how to implement the ROA for the justification of the AMT investment project. In the final section, we will make short concluding remarks, including the relevant future research directions.

2   ROA and an Economic Evaluation of Investment Projects

Recently considerable debate has been concerned with how well the DCF techniques serve real investment projects under an uncertain decision environment. The major issue of the debate is that the DCF techniques take account of nothing in the calculation of capturing the value of flexibility to make future decisions resolving uncertainty. The DCF technique is typically performed by forecasting the annual cash flows occurring during the planning horizon of an investment project, discounting them back to the present time at a risk-adjusted discount rate, then subtracting the initial investment cost from the sum of the discounted cash flows.

To be in accordance with requirement, the ROA has been recently developed at a pace, even though there does not exist much empirical evidence in a real world. Boriss and Janos report that 76% of the firms accepted the projects that failed to pass the NPV criteria [11]. 94% of these cases took strategic concerns priority over the quantitative analysis. This survey result implies that the firms do not want to miss the investment opportunity, and want to take the flexibility of the investment options by intuition.

The term "real", often used as a contrast to "financial", refers to fixed assets. The word, "option", can be regarded to represent freedom of choice after watching the external events unfold. Put them together, a real option can be defined to be the right, but not the obligation, to take an action on the underlying real asset in the future at a predetermined price (exercise price=initial investment cost), for a predetermined time, (time until opportunity disappears). Taking an action in the future is performed after revelation of additional information that increases or decreases the value of the underlying investment project. The action taken entails completely or partially to eliminate the downside risk of the investment project, causing the cash flow distribution of the investment project to become asymmetric and skewed toward a higher rate of return.

There are two major groups allowing us to calculate a real options value (ROV): discrete- and continuous-time approach. The discrete-time approach is mainly comprised of lattice models, while the continuous-time approach consists of analytical and numerical methods, and simulation. All of the approaches are well documented in the literature [6], [13], [14]. In addition to those approaches, merging a decision tree analysis or game theory with a real options pricing theory has been considered to be one of the promising research area [15], [16].

The lattice model assumes that the underlying asset follows a discrete, multiplicative stochastic process throughout time for a certain form of tree. The ROV is then calculated recursively from the end nodes of the tree in a similar way as done for a dynamic programming technique. Among various types of the lattice models, the binomial lattice model originally developed by Cox, Ross, and Rubinstein has been most popularly used. Analytical valuation methods are used to value options by solving stochastic differential equations (SDE) with boundary conditions, which are set up for a given set of assumptions. The representative examples for these methods are the Black-Scholes [9], Margrable [10], and Geske [17] models. However, the solution to the SDE does not exist in many cases and thus it must be solved numerically, either using finite-difference methods, or the Monte Carlo simulation. These are called numerical methods. Despite the difficulty in solving the SDE, it is still considered to be a valuable tool to calculate the ROV. Since it is complex to develop SDE and laborious to solve lattice model, simulation methods are regarded as very useful tools to calculate the ROV. Nevertheless, there exist very few papers in which the simulation methods were used to value the real options value [7].

Since the approaches currently in use to value the ROV are benchmarked from the financial option pricing theory, there is no perfect approach to value real investment projects. Many researchers maintain that the blind use of financial option pricing approaches to derive the ROV may significantly distort the value of real options. Sick first generally discussed the deficiency of the ROA [18]. More recently, Miller and Park seriously discussed the problems in a well-organized way [6]. They explained why exercise should be taken when applying the ROA to valuing the ROV, focusing on each of six parameters affecting an option value: the underlying asset, striking price, expiration date, interest rate, volatility, and dividends.

With the DCF techniques, the value of investment projects consists of the NPV only. However, it consists of two terms: the NPV and option premium. And it is now called a 'strategic net present value (SNPV). In this paper, we will analyze the AMT project by the NPV technique and ROA and compare two results.

3   A Problem Description

As for the case study for this research, we took one of the Korean Heavy Industry Companies and called it Chosun Heavy Industry Company that is a disguised name. This company was founded in 1962 and currently produces 42 different parts for automobile, truck, machining tools, and so on. Parts include turret slide, saddle, column, spindle box, cross rail, spindle gearbox, motor bracket, and so on. The company would plan to replace a part of its manufacturing system with an advanced manufacturing system, which consisted of 21 machines. They were lathe, milling m/c, boring m/c, and so on plus an automated material handling system.

The company launched this project with the partial financial support from the Korean Government. It required around $6 million for constructing the additional building, installing the new machines and developing the relevant computer software, and $0.35 million for purchasing the additional land for a factory site. These costs were cited in the monetary value as of year 2004 considering an exchange rate of 1,000 Won/$.

To implement an investment decision analysis and provide a logical basis for the economic analysis, we made the following assumptions:

- The planning horizon is a 10-year period, i.e., T=10.
- All the costs and selling prices will not change over the planning horizon of 10 years.
- A general inflation rate of 5% is applied to revenue, cost of goods sold, general administrative and marketing & sales expenses, and the land value.
- The salvage value of the building and machine at the end of the project life is 10% of the initial investment cost.
- The working capital is to be 20% of the total annual revenue.
- The half of the total initial investment cost are financed at an annual interest rate of 5%, and is repaid in an annual equal amount of money, including the principal and interest over 10 years.
- The marginal tax rate is determined to remain constant at 28%.
- The company's after-tax return on investment is assumed to be 10%.

3.1   Calculating a Product Cost

Figure 1 shows the conceptual structure of generating a product cost information. We first simulated the underlying manufacturing system with ARENA and obtained the

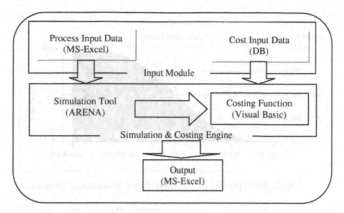

Fig. 1. The conceptual structure of the simulation model

processing-related data. The input data for simulation saved in MS-Excel was sent to the simulation program. We wrote a computer program in Visual Basic and interfaced it with the simulation model to get a product cost information. The program received data from the simulation program and the DB containing cost input data. The final result was saved in MS-Excel.

In this paper, we employed the concept of an activity-based costing (ABC) system to calculate a product cost instead of a direct labor-based costing system. It is now widely accepted that the ABC system captures what and how much resources committed for a production of a variety of products are used up by each product more properly than the direct labor-based costing system [19]. The activities identified for this case study included a processing, setup, material handling, quality-related, tooling, and other production-related activities.

3.2 Performing a Net Present Value Analysis

The NPV analysis of the underlying investment project was performed assuming that a minimum attractive rate of return (MARR) was 10%. It was determined based on the fact that most of the medium-sized heavy industry companies in Korea usually used an MARR of between 10% and 15%.

The NPV of the project was calculated with the most likely estimate of each component of the cash flow statement. It came out to be a negative of $10,254. Therefore, it would not be recommended for the company to implement the project. Since the managers of the company wanted to gain more useful information on the project, we performed a Monte Carlo simulation with Crystal Ball considering that the annual revenue, and a general administrative, marketing & selling expenses contained some amount of uncertainty. For the simulation, it was assumed that two components followed a uniform probability distribution with a range of ±20% about the respective mean and there was no correlation in the cash flows generated between years.

Figure 2 shows the results of 10,000 simulation runs. The mean turned out to be a negative of $111,521 and a standard deviation was $86,920. However, there is great uncertainty in the outcome as ascertained by the minimum and maximum NPV estimate of -$239,357 and $220,106. And there exited a probability of 44.62% that the

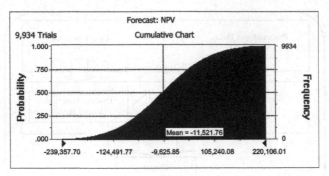

Fig. 2. NPV profile of the underlying investment project

underlying project would yield a positive NPV. Due to this desirable probability, Chosun Company would want to analyze the investment project through a less risky investment decision-making approach of the ROA.

3.3 Evaluating the Value of the Deferral Option

The underlying investment project was rejected based on the NPV criteria even though there existed a relatively high likelihood that the project would generate a sufficient amount of profit for the company. Noting this fact, the company would like to wait and see what would happen to the project in the following two years. That is, the company was willing to delay making the investment decision on the underlying project by two years. The value of such investment timing flexibility can be properly captured by an ROA, but not by the NPV technique.

Two issues in relation with the application of the ROA must be addressed. Financial option analysis requires a liquid and traded security that is used to maintain a riskless hedge in the replicating portfolio. However, real options does not have such a security. Therefore, in this paper we will take Marketed Asset Disclaimer (MAD) proposed by Copeland and Antikarov [4]. MAD assumes that the present value of the real investment project itself can be selected as the underlying asset, thus it is not necessary to find a twin security.

The second issue is concerned with the type of risk. Risk is generally classified into two groups: technical risk and market risk. Technical risk is unique, internal, and dependent on the project's ability to meet the desired requirements. This kind of risk can be resolved by making investment, and reduces the value of the real options. It can be generally analyzed with the use of probabilities. On the contrary, market risk is associated with the economic factors of market. It typically requires the passage of time to get the answer that would resolve it, and increases the value of the real options due to the perfect hedge by a portfolio in a market. The Black-Scholes model, a binomial lattice model, and so on, can be employed to analyze this kind of market risk. In this paper, we will not discriminate between technical or market risk, mainly because any investment decision that could potentially affect the market value of a company should be considered as market risk. Thus, an ROA would be proper to take for dealing with such investment decisions.

Deferral Option. It is an option to delay making an investment decision for a period of time. In this paper, it is assumed that Chosun Heavy Industry Company would like to delay making the investment decision for two years. However, they can make the decision in either year 1 or 2; i.e., this deferral option is an American call option instead of a European call option.

Table 1. Summary of data relevant to calculate the deferral option value

Items	Symbol	Amount
PV of expected cash flow	PV	$3,853,352
initial investment	X	$5,885,000
expiration date	T	2 years
volatility	σ	49%
risk-free interest rate	r_f	5.5%

To perform the ROA, we needed to have the following information: for now, it was assumed that the company invested in the project with similar risk profiles as its publicly traded stock. Therefore, the volatility of the underlying project was proxied from the weekly closing prices of Chosun Heavy Industry Company stock price over from January, 2001 to June, 2004. The risk-free interest rate of 5.5% used for an economic evaluation was taken from the rate of return of the 3-year bond issued by the Korean Government because we failed to find out the 2-year bond. The interest rate was cited as of June30, 2004. It was also assumed that the initial investment equivalent to an exercise, (striking), price grew at the risk-free interest rate of 5.5% each year.

Table 1 shows data required to calculate the deferral option value. In a financial option price problem, all the data except for volatility in the table can be obtained in the market, while all the data in a real options price problem should be estimated or determined in a way.

With the data presented in Table 1, we constructed the present value event tree for the underlying project as shown in Figure 3. The present value for time =0 generally comes from the cash flow statement. The other present values must be calculated. For this calculation, we first need to determine the up and downward ratios using Equation (2) and (3). Next, we simply multiply the present value of the previous year by the up or downward ratio to obtain the present value of the current year.

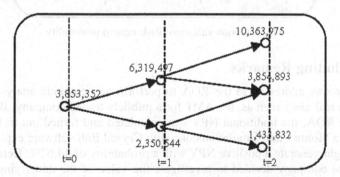

Fig. 3. Present value event tree for the underlying investment project

Figure 4 shows the option value for each period of time, which was calculated using a binomial lattice approach. The variable of "p" represents a risk-neutral probability and can be calculated using Equation (1). The upward and downward movement ratios are expressed in Equations (2) and (3), respectively.

$$p = \frac{e^{r_f} - d}{u - d}. \tag{1}$$

$$u = e^{\sigma\sqrt{\Delta t}} \tag{2}$$

$$d = \frac{1}{u} \tag{3}$$

where,

r_f : a risk-free interest rate

σ: the volatility of a project value

Δt : a time interval

According to the resulting option values shown in Figure 4, it is economically desirable for Chosun Heavy Industry Company to hold a deferral option in year 1 no matter what would happen for the economic situation of the year. However, the company had to exercise the deferral option held for one year for the underlying project in year 2 if and only if the economic situation for year 1 and 2 turned out to be favorable in a row. Otherwise, it had to stop considering the underlying project any more. By so doing, the company would earn an SNPV of $655,537. Comparing this value with the traditional NPV of -$10,254, the SNPV was increased by $665,791, which is called an option premium.

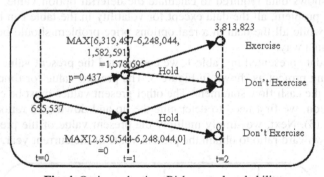

Fig. 4. Option valuation: Risk-neutral probability

4   Concluding Remarks

This paper was addressed to the ROA to perform an economic analysis of investments in a real asset such as, the AMT for a publicly traded company. Before counting on the ROA, the traditional NPV was calculated and turned out to be -$10,254. However, a Monte Carlo simulation done with Crystal Ball software explored that the project might generate a positive NPV with a probability of 44.62%. Referring to this finding, the company decided to investigate the value of the underlying investment project using the ROA. The ROA provided a strategic net present value of $655,537, and based on the result it was recommended to implement the underlying investment project for the AMT in year 2 in case which the economic situation for year 1 and 2 had been consecutively favorable.

In a real world, many companies would take a sequential investment approach to minimize risk inherent in strategic investment projects, instead of investing in the project on a full scale at one time. In this case, we can employ a real learning options approach to evaluate its economic value. It is considered to be an approach that more realistically reflects what happens in the real world. Valuing real investment projects with the real learning options approach will be one of the promising future research topics.

Acknowledgement

This research was partially supported by 2002 Chosun University Research Fund.

References

1. Bengtsson, J.: Manufacturing Flexibility and Real Options: A Review. Int. J. Prod. Econ., 74 (2001) 213-224
2. Kumar, R. L.: An Option View of Investments in Expansion-Flexible Manufacturing Systems. Int. J. Prod. Eco., 38 (1995) 281-291
3. MacDougall, S. L., Pike, R. H.: Consider Your Options: Changes to Strategic Value during Implementation of Advanced Manufacturing Technology, Omega 31(2003) 1-15
4. 4.Copeland, T.: The Real-Options Approach to Capital Allocation. Strategic Finance, Oct. (2001) 33-37
5. Graham, J. and Harvery, C.: The Theory and Practice of Corporate Finance: Evidence from the Field, J. of Fin. Eco. 60 (2001) 187-243
6. Miller, L. and Park, C.S: Decision Making under Uncertainty-Real Options to the Rescue, The Eng. Economist 47(2) (2002) 105-150
7. Rose, S.: Valuation o Interacting Real Options in a Tollroad Infrastructure Project, The Quart. Rev. of Eco. And Fin. 38 (1998) 711-723
8. Trigeorgis, L.: Real Options and Interactions with Financial Flexibility, Fin. Manag. 22(3) (1993) 202-224
9. Black, F., and Scholes, M.: The Pricing of Options and Corporate Liabilities, J. of Pol Eco. 81(1973) 637-659
10. Margrabe, W.: The Value of an Option to Exchange One Asset for Another, The J. of Fin. 33(1) (1978) 177-186
11. Borrissiouk, O. and Peli, J.: Real Option Approach to R&D Project Valuation: Case Study at Serono International S.A. Master Thesis, University of Lausanne, (2002)
12. Brach, M.A.: Real Options in Practice, John Wiley & Sons, Hoboken NJ, (2003)
13. Cox, J.C., Ross, S.A., and Rubinstein, M.: Option Pricing: A Simplified Approach, J. of Fin. Eco. 7 (1979) 229-263
14. Hull, J.C.: Options, Futures, and Other Derivatives, 5th Edn. Prentice Hall, Upper Saddle River NJ (2003)
15. Smit, H. T. J., and Trigeorgis, L., Strategic Investment: Real Options and Games, Princeton University Press, Princeton and Oxford, NJ (2004)
16. Herath, H. S. B., and Park, C. S., Real Options Valuation and Its Relationship to Bayesian Decision-Making Methods, The Engineering Economist, 36(2001) 1-32
17. Geske, R.: The Valuation of Compound Options, J. of Fin. Eco. 7(1) (1979) 63-81
18. Sick, G.: Capital Budgeting with Real Options, Monograph Series in Finance and Economics, Monograph 1989-3, New York University, Salomon Brothers Center (1989)
19. Cooper, R., Kaplan, R. S., Maisel, L. S., Morrissey, E., and Oehm, R. M., Implementing Activity-Based Cost Management: Moving from Analysis to Action, Institute of Mang. Account. Montvale NJ (1992)

Space Missile Simulation on Distributed System[*]

Jong Sik Lee

School of Computer Science and Engineering
Inha University, Incheon 402-751, South Korea
jslee@inha.ac.kr

Abstract. Demand of space missile simulation on distributed system is increasing to provide cost-effective, execution-effective, and reliable simulation of mission-effective missile systems. This paper presents modeling and simulation of space missile systems on distributed system and discusses kinetics of a missile on the earth and movement of attack/interceptor missiles. This paper focuses high performance simulation for space missile simulation with distributed system construction and execution concepts and provides simulation effectiveness with reasonable computation and communication resources. This paper analyzes simulation performance and scalability of space missile simulations in terms of simulation cost and sensitivity. Analytical and empirical results prove feasibility of space missile simulations on distributed system through reduction of network bandwidth requirement and transmission data bits within tolerable error.

1 Introduction

There is an increasing demand of missile simulation [1, 2] which is working on distributed system, collaboratively. Many ballistic missile-related organizations, including NASA and DoD, have progressed research to accomplish missile simulation objectives. The main objective of missile simulation provides cost-effective, execution-effective, and reliable simulation of ballistic missile systems. For space missile simulation, paradigms of kinetics of a missile on the earth [3, 4] and attack/interceptor missile movement have been noticed. This paper proposes high performance modeling and simulation for these missile simulations with distributed system construction and execution concepts. This paper focuses on an execution of a complex and large-scale system simulation with reasonable computation and communication resources on a distributed simulation. Especially, this paper uses a HLA [5, 6]-compliant distributed simulation which supports a flexible and high performance modeling and simulation. HLA is a technical architecture for DoD simulations and defines functional elements, interfaces, and design rules needed to achieve a proper interaction of simulations in a federation or among multiple federations.

Distributed simulation for ballistic missile system is complex and large in their size. In fact, in order to execute these complex and large-scale distributed simulations within reasonable communication and computing resources, development of a large-scale distributed simulation environment has drawn attention of many distributed simulation researchers. A complex and large-scale distributed simulation is character-

---

[*] This work was supported (in part) by the Ministry of Information & Communications, Korea, under the Information Technology Research Center (ITRC) Support Program.

D.-K. Baik (Ed.): AsiaSim 2004, LNAI 3398, pp. 86–94, 2005.

ized by numerous interactive data exchanges among components distributed between computers networked together. This paper presents simulation cost reduction through transmission-required data reduction and simulation error reduction with smoother modeling. Transmission-required data reduction decreases system execution cost by reducing an amount of communication data among distributed components. In addition, it reduces local computation load of each distributed component, increases modeling flexibility, and improves simulation performance through computation synchronization and scalability.

This paper is organized as follows: Section 2 introduces two space missile systems: kinetics of a missile on the earth and attack/interceptor missiles. And it provides modeling of the two systems. Section 3 discusses improvement of simulation performance. It considers simulation cost and error reduction. Section 4 discusses performance analysis of an attack/interceptor missile system. Section 5 illustrates an experiment and evaluates simulation performance. The conclusion is in Section 6.

2 Space Missile System

This section introduces a space missile system that is focused on movement of missile on space. This paper presents two missile systems: kinetics of a missile on the earth and attack/interceptor missiles. And it specifies modeling for the two system simulations.

2.1 Kinetics of a Missile on the Earth

This section presents a kinetics part of missile. Circulation of missile on an ideal circle is a part of the kinetics as shown Fig. 1. For missile to maintain an ideal circular orbit with radius D and speed v around a massive body, it is also required that a centripetal force, mv^2/d equals to the force of gravity. The force of gravity pulls along the line joining the two centers and has magnitude $F = GMm/d^2$, where G is the gravitational constant, M and m are the masses. The distance of a ship with center at (x,y) to the center of gravity of a massive body (x_0,y_0) is $d = ((x - x_0)^2 + (y - y_0)^2)^{1/2}$. The force is projected in the x and y directions in proportions, $p_x = x/d$ and $p_y = y/d$, respectively. In a ideal orbit with d = D (constant), the coordinate dynamics separate into two independent 2nd order linear oscillators.

Basically, frequency $w = (GM/d^3)^{1/2}$ would be to maintain to circulate, however we use a gain value instead of frequency. The gain controls a degree of movement of missile. As the gain changes, the performance will be measured and compared because gain can decide the stability, simulation time, and accuracy of system. For missile traveling, one of strong influences is gravity. Gravity is a force exerted on a body by all other bodies in relation to their and their distance away. The center of gravity allows us to aggregate particles in a rigid body into a single point that represents their gravitational interaction with any other body to consider the forces acting at the centers of gravity of interacting bodies.

Modeling for Kinetics of a Missile
This paper develops a missile model and constructs an abstraction that is for maintaining an accounting of where ships are and predicting their future destinations.

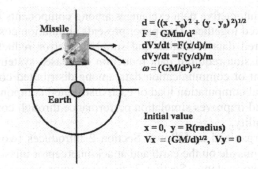

$$d = ((x - x_0)^2 + (y - y_0)^2)^{1/2}$$
$$F = GMm/d^2$$
$$dVx/dt = F(x/d)/m$$
$$dVy/dt = F(y/d)/m$$
$$\omega = (GM/d^3)^{1/2}$$

Initial value
$$x = 0, \ y = R(radius)$$
$$Vx = (GM/d)^{1/2}, \ Vy = 0$$

Fig. 1. Circulation of Missile

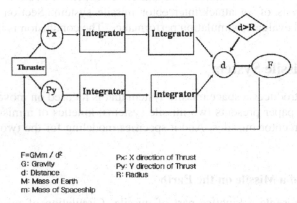

F=GMm / d²
G: Gravity
d: Distance
M: Mass of Earth
m: Mass of Spaceship

Px: X direction of Thrust
Py: Y direction of Thrust
R: Radius

Fig. 2. Modeling for Kinetics of a Missile

Overall modeling objectives are to construct a space travel scheduling and test it. The modeling is based on the differential equations which are based on Newtonian mechanics [3, 4]. Fig. 2 gives us information about the movement of missile and about how to know where ships are and predict their future destinations with several parameters. The center of gravity coordinates, mass, M, and escape radius, R are parameters that characterize earth component. When the distance from the center of gravity exceeds the threshold, R, a state event is generated which sends the ship to another model component.

2.2 Attack/Interceptor Missile System

Attack/interceptor missile system uses a geocentric-equatorial coordinate system [3, 4]. An attack missile is a ballistic flight and accounts for gravitational effects, drag, and a motion of a rotation of the earth relative to it. An interceptor missile is assigned an attack missile, and it follows its attack missile until it hits its attack missile. In modeling an attack/interceptor missile system, there are two main models: attack missile and interceptor missile. The attack missile model is the model of a sphere of uniform density following a ballistic trajectory. This model begins at an initial position with an initial velocity, moves, and stops until it meets an interceptor missile. An interceptor missile model is a model of the same sphere of an attack missile, and it begins at a certain initial position and a certain initial velocity, which are different from those of an attack missile model. An interceptor missile model follows an attack missile model assigned to it. When the interceptor missile model is close to the assigned attack missile model within a certain distance, it stops and we consider the interceptor missile hits its attack missile.

Modeling of Attack/Interceptor Missile System
An attack missile model includes three sub-models: acceleration model, velocity model, and position model. An acceleration model uses parameters (e.g. gravity, atmosphere velocity, atmosphere density, etc.) to generate its acceleration values. A ground station model calculates these parameters using position values of an attack missile model. For a real implementation of a velocity and a position models, an integrator model is developed. The integrator model has two types: a basic integrator and a predictive integrator. A velocity model receives an acceleration input from an acceleration model, and a position model receives a velocity input from a velocity model. Finally, a position model generates three dimensional position values of an attack missile and sends them to both a ground station model and a missile model. A missile model includes two sub-models: a velocity generator model and a position model. A velocity generator model receives a position update message from an attack missile model, and it generates velocity values and sends them to a position model. A position model (an integrator model) receives a velocity values and generates three dimensional missile position values. As simulation time is advanced, a position of a missile model gradually becomes closer to a position of an attack missile model.

To realize an attack/interceptor missile system presented in this paper, this paper models two systems. The first system is a basic system, which is not applied by any quantization scheme. The second system employs a quantization prediction approach.

In a basic system, there are two federates; an attack missile federate and an interceptor missile federate. An attack missile federate includes attack missile models, and an interceptor missile federate includes interceptor missile models. In a basic system, basic integrators are used for a velocity and a position models in an attack missile model, as well as for a position model in an interceptor missile model. A basic integrator works with time scheduling based on discrete time concept. A position model of an attack missile model in an attack missile federate sends three dimensional position double values (x, y, z) to an interceptor missile federate in a fixed step time due to the fact that the basic integrators are used. The results from the basic system are standard results for performance evaluation. Fig. 3 illustrates a component diagram of an attack missile and of an interceptor missile models in the basic system.

3 Simulation Performance Improvement

3.1 Simulation Cost Reduction

To reduce tremendous data communicated between attack missile model and interceptor missile model with only reasonable error, this paper utilizes a quantization prediction approach which is extended from data management approaches [7, 8, 9]. For this approach, predictive integrators are used for velocity and position models in an attack missile and an interceptor missile models. The quantization prediction system includes an attack missile and an interceptor missile federates which are the same

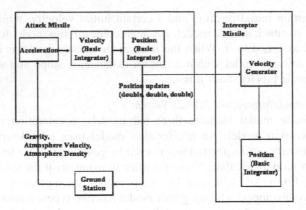

Fig. 3. Component diagram of the basic system

as the basic system. A position model in an attack missile model sends three dimensional position integer values, such as (-1, 0, 1), to a velocity generator model in an interceptor missile model. To generate velocity of an interceptor missile model, a velocity generator model needs current position values of an attack missile model; thus, it calculates current position values of an attack missile model by multiplying input integer values and current quantum size and by adding the multiplied result to old position values.

To avoid error of an attack missile position values in an interceptor missile model, a current quantum size in an interceptor missile model should be the same as the current quantum size in an attack missile model., Inter-federate communication data are tremendously reduced by sending integer values (not the double values) from an attack missile federate to an interceptor missile federate. To save more of data bits, three dimensional position integer values, such as (-1, 0, 1), are not transferred directly, and only five $(5 > \log_2 3^3)$ data bits representing three dimensional integer values are sent. Therefore, an encoder, which changes three dimensional integer values to five data bits, is needed. In order to change received five data bits to exact three dimensional integer values in an interceptor missile federate, a decoder is needed. A decoder decodes five data bits to exact three dimensional integer values and generates position values of an attack missile model as a velocity generator in an interceptor missile model does. Fig. 4 illustrates a component diagram of an attack missile and an interceptor missile models in the quantization prediction system.

3.2 Simulation Error Reduction

Quantization predictive approach causes error due to using a predictive integrator. To reduce error, this paper suggests a method. The method is to use a smoother model, which can reduce error from multi-dimensional output values of a predictive integrator with each different unfixed time advance. A velocity model and a position model outputs with variable time advance in an attack missile model since these models are developed with predictive integrators. Position values of a predictive integrator are outputted with each variable time advance for each dimension since positions of an attack missile and an interceptor missile are three dimensional. A ground station

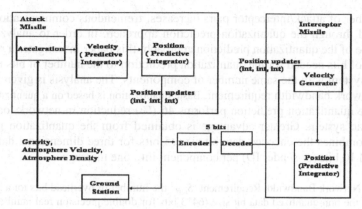

Fig. 4. Component diagram with predictive integrators

model generates gravity, atmosphere density, etc., with all three dimensional position values at a same time. The position values, outputted with each different variable time advance, include old values and current values at a given event time. Use of old values of an attack missile position in a ground station model causes error.

A smoother model receives, keeps, and updates three dimensional position values outputted from a predictive integrator with each different variable time advance for each dimension until a fixed time step of a basic integrator is advanced. A smoother model outputs three dimensional position values to a ground station model when a fixed time step advance of a basic integrator comes. By using a smoother model, the error caused by use of old values can be reduced. Fig. 5 illustrates a component diagram of an attack missile model using a smoother model.

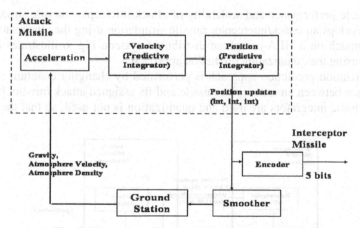

Fig. 5. Component diagram with a smoother model

4 Performance Analysis

This section analyzes performance and scalability of attack/interceptor missile simulation. As the number of attack missile and interceptor missile pairs increases, the number of messages communicated among federates increases also significantly. As the number of attack/interceptor pairs increases, tremendous communication bits can be saved through the quantization prediction approach. In order to analyze the performance of the quantization prediction approach, this paper investigates a ratio of the number of bits needed for the quantization prediction to the number of bits needed for a basic system with the same number of components. The analysis is given in Table 1 with network bandwidth requirement. Basic quantization is based on a quantization theory [11]. The quantization prediction performs 46 (%) reduction in network load relative to a basic system. Greater advantage is obtained from the quantization prediction, which combines the encoded data bit size (5 bits for three dimensional data of message and 10 bits for sender ID) per component into one message.

Table 1. Network Bandwidth Requirement (S_{OH}: the number of overhead bits for a packet (160 bits); S_D: the non-quantized data bit size (64*3 bits for double precision real numbers for three dimensions); S_Q: the quantized and encoded data bit size (5 bits for three dimensions ($\log_2 3^3 < 5 = S_Q$)); S_L: the encoded data bit size for sender ID (10 bits for 1000 N_{pair} ($\log_2 1000 < 10 = S_Q$)); N_{pair}: the number of pair components (1000), a: the ratio of active components)

Approach	Bandwidth Required for N_{pair}	Ratio for large N_{pair}	Ratio for N_{pair} =1000
Basic Quantization	$aN_{pair}(S_{OH} + S_D)$	1	1
Quantization Prediction	$aN_{pair}(S_{OH} + S_Q)$	$(S_{OH} + S_Q)$ $/(S_{OH} + S_D)$	0.46

5 Experiment and Performance Evaluation

5.1 Experiment

To evaluate performance and scalability of attack/interceptor missile simulation, this paper develops an attack/interceptor missile simulation using the quantization prediction approach on a HLA-compliant distributed system. Fig. 6 illustrates data transmission using the quantization prediction approach.

Quantization prediction approach is performed by changing quantum size related to distance between an interceptor missile and its assigned attack missile. In the basic system, basic integrators are used and quantization is not used, so that the basic sys-

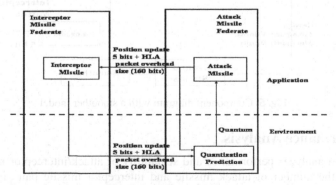

Fig. 6. Data Transmission

tem is considered a standard system in which no error occurs. This paper develops two federates: attack missile and interceptor missile. An attack missile in the attack missile federate sends position update messages, which includes encoded bits (5 bits for three dimensions) and HLA packet overhead (160 bits), to an interceptor missile. This experiment compares transmission data bits and error. Here, error is defined as difference among an attack missile positions after simulation. Transmission data bits indicate data bits that interceptor missiles receive from attack missiles.

5.2 Performance Evaluation

This paper measures error between non-quantization and quantization prediction. Ratio of error is calculated by dividing position values in quantization prediction by position in non-quantization. Table 2 shows error trajectory of quantization prediction in varying range of multiplying factors of standard quantum sizes. Larger multiplying factor indicates longer distance between an attacker missile and an interceptor missile. As simulation time increases, error decreases between non-quantization and quantization prediction. Error decreases significantly while multiplying factor varies from 40 to 1.

Table 2. Error Trajectory

Simulation Time	Range of Multiplying Factor		
	10 ~ 1.0	20 ~ 1.0	40 ~ 1.0
1.0	0.16	0.36	1.87
3.0	0.08	0.05	0.40
5.0	0.04	0.11	0.12
7.0	0.05	0.07	0.13
9.0	0.02	0.10	0.10

Table 3 illustrates ratio trajectory of Transmission Data Bits (TDB) in resulting from different ranges of multiplying factors of standard quantum sizes. The ratio of TDB is calculated by dividing TDB in quantization prediction by TDB in non-quantization. As Table 3 shows all ratios are less than 0.1. That means TDB in quantization prediction is less than 10 % of TDB in non-quantization. As simulation time increases, ratio of TDB increases since multiplying factor of standard quantum sizes decreases. As multiplying factors increases, TDB decreases significantly when using quantization prediction approach.

Table 3. Ratio Trajectory of Transmission Data Bits (TDB)

Simulation Time	Range of Multiplying Factor		
	10 ~ 1.0	20 ~ 1.0	40 ~ 1.0
1.0	0.054	0.033	0.017
3.0	0.056	0.034	0.017
5.0	0.060	0.036	0.018
7.0	0.066	0.039	0.019
9.0	0.082	0.048	0.023

6   Conclusion

This paper presents modeling and simulation of space missile systems on distributed system. Space missile systems are focused on kinetics of a missile on the earth and movement of attack/interceptor missiles. This paper proposes high performance simulation for these space missile simulations with distributed system construction and execution concepts. This paper focuses on distributed execution of space missile simulations with reasonable computation and communication resources and tolerable error.

This paper suggests simulation cost reduction through transmission-required data reduction and simulation error reduction with smoother modeling. Transmission-required data indicates minimum transmission data to achieve satisfactory simulation results though tolerable error occurs. Transmission-required data reduction decreases system simulation time and cost by reducing tremendous communication data among distributed components. In addition, reduction transmission-required data causes reduction of local computation load of each distributed component and improves simulation performance through computation synchronization and simulation flexibility and scalability. In experiment, this paper provides a HLA-compliant distributed simulation which uses a middleware for DoD simulations to achieve a proper interaction of simulation entities in a federation or among multiple federations. Use of HLA middleware supports flexible and high performance modeling and simulation of space missile systems. This paper analyzes simulation performance and scalability of space missile simulations. The analytical and empirical results show favorable reduction of network bandwidth requirement and transmission data bits and prove feasibility of space missile simulations on distributed system.

References

1. McManus, John W.: A Parallel Distributed System for Aircraft Tactical Decision Generation, Proceedings of the 9th Digital Avionics Systems Conference (1990) 505 –512.
2. Hall, S.B. and B.P. Zeigler. Joint Measure: Distributed Simulation Issues in a Mission Effectiveness Analytic Simulator. in SIW, Orlando FL (1999)
3. Roger R. Bate, Donald D. Mueller, Jerry E. White: Fundamentals of Astrodynamics. Dover Publications, New York (1971)
4. Erwin Kreyszig: Advanced Engineering Mathematics: Seventh Edition. John Wiley& Sons Inc, New York (1993)
5. High Level Architecture Run-Time Infrastructure Programmer's Guide 1.3 Version 3, DMSO (1998)
6. Defense, D.o., Draft Standard For Modeling and Simulation (M&S) High Level Architecture(HLA) - Federate Interface Specification, Draft 1 (1998)
7. Nico Kuijpers, et al.: Applying Data Distribution Management and Ownership Management Services of the HLA Interface Specification. in SIW,. Orlando FL (1999)
8. Gary Tan et. al.: A Hybrid Approach to Data Distribution Management, 4th IEEE Distributed Simulation and Real Time Application (2000)
9. Zeigler, B.P., et al. Predictive Contract Methodology and Federation Performance. in SIW, Orlando FL (1999)

A Portability Study on Implementation Technologies by Comparing a Railway Simulator and an Aircraft FTD

Sugjoon Yoon[1], Moon-Sang Kim[2], and Jun-Seok Lim[3]

[1] Department of Aerospace Engineering, Sejong University 98 Gunja-Dong,
Gwangjin-Gu, Seoul, 143-747 Republic of Korea
[2] School of Aerospace and Mechanical Engineering, Sejong University 98 Gunja-Dong,
Gwangjin-Gu, Seoul, 143-747 Republic of Korea
[3] Department of Electronics Engineering, Hankuk Aviation University 200-1,
Whajon-dong, Koyang-city, Kyungki-do, 412-791 Republic of Korea

Abstract. The paper introduces major technical specifications of the Line II railway simulators of Pusan City in Korea. Comparing design specifications of the railway simulators with those of the light aircraft Flight Training Device (FTD), the paper reveals commonality of implementation technologies applied to both simulators: Overall configurations and design philosophies are basically the same. In both programs VMEbus computing systems with UNIX are adopted as backbones of the simulators. It is found that the railway simulators are less stringent in real-time requirements than the aircraft FTD and the railway simulators are designed to be more event-driven and object-oriented. The experiences show that models may be diverse depending on the objects but implementation technologies are about the same. Maximizing portability of implementation technologies is a matter of an organization's strategy of adopting standardized processes and modular technologies available and most economic to them.

1 Introduction

Flight Simulator Industry has lead overall simulation technologies since the appearance of Link's first Trainer. Most people in simulation communities today understand or at least have some knowledge of, general requirements and technical specifications of airplane simulators. On the other hand, railway simulators have begun to attract customers' interests and formed a niche market since the beginning of 1980's. Information regarding railway simulators such as requirements, technical specifications, modeling, system components, etc is relatively unknown to simulation communities, and belongs to manufacturers of railway simulators or special research institutes. This paper reveals technical specifications of railway simulators in relative detail, which are being tested for practical use in Korea, and studies portability of simulation technologies by comparing the technical specifications with those of flight simulators. For practical comparison, technical specifications of CG-91 Flight Training Device [1], [2], [3] and PUTA (Pusan Urban Train Association) Line II Metro Simulators [4] are used as examples. Both of the simulators were developed by Korean Air between 1991 and 1998.

D.-K. Baik (Ed.): AsiaSim 2004, LNAI 3398, pp. 95–106, 2005.

2 Railway Simulators

PUTA Line II Metro Simulators provide trainees with realistic driving environment of the Pusan City metro system and enable effective and safe training. The railway simulators are composed of a pair of cabs, which means two metro drivers, one at each cab, can be trained independently at the same time by an instructor at a common control station.

Training through railway simulators is categorised into basic and application courses, which is a little more elaborated in Table 1. The basic training courses encompass manipulation of on-board equipment and familiarisation of track environment including signals on route. Their goals are familiarisation with operation environment. The application courses include abnormal situations in addition to the basic training so that trainees can be trained and tested for their responses. Relevant training results and their evaluations are provided to trainees and an instructor either in printed report forms or on screen. A CCTV camera is installed in each cab and records whole training processes for later use during performance evaluation of a trainee.

Interior and exterior of PUTA Line II Metro Simulators are shown in Figs. 1 and 2. H/W and S/W configurations of a set of railway simulators are illustrated in Figs. 3 and 4. As shown in the figures, the system hardware is composed of a pair of cabs and relevant electric signal and power lines, which are almost symmetric and connected to an Instructor Operation Station (IOS).

Table 1. Training courses of a railway simulator

Basic Training	Application Training
Initial starting procedures	Basic Training
Start, operation, stop	Emergency procedures
Manipulation of on-board equipment	Malfunctions
Response to signals on route	Operation and stop on emergency conditions
Route familiarisation by visual display of forward track views	

Fig. 1. Interior of PUTA line II metro simulators   **Fig. 2.** Exterior of PUTA line II metro simulators

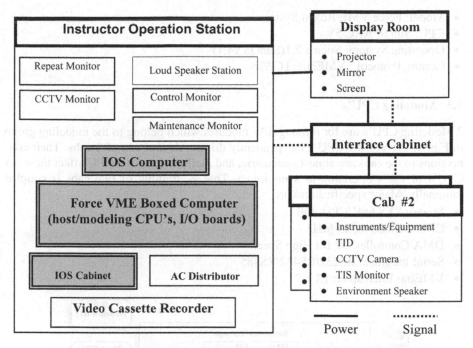

Fig. 3. H/W configuration of PUTA line II metro simulators

The followings summarize configurations of simulator components and their functions shown in Fig. 3.

2.1 IOS Computer

An IOS computer is used for a common terminal to the host computer and the operation environment of Graphical User Interface (GUI) group S/W. The monitor of the IOS computer is used as a control monitor to show IOS menus, and as a terminal of the host computer for maintenance purpose. Its technical specifications are as follows:

- Model: COMPAQ DESK PRO 2000
- Operating System: Window 95 or Window NT
- Comm. Protocol : TCP/IP

2.2 Host CPU

A host CPU or a host computer is for system maintenance and operation environment of S/W modules belonging to the host group in Fig. 4. Each of execution files and data files is controlled and monitored by the host CPU, which is a part of a Force VMEbus computer system. The system comprises 1 host CPU board, 2 modelling CPU boards, and several I/O boards. The boards in the VME rack communicate with each other on VMEbus. Relevant system specifications are as follows:

- Model: Force VME Boxed System
- CPU: Sun SPARC 5V
- Operating System: Solaris 2.1(SunOS v4.1)
- Comm. Protocol : VMEbus, TCP/IP

2.3 Modeling CPU's

2 Modeling CPU's are for running S/W modules which belong to the modeling group in Fig. 4. Each of the CPU's independently drives relevant one of 2 cabs. Their connections to the cabs are almost symmetric, and malfunctions of a CPU affect the other CPU's or cab's operation to a minimum. That is, training of two cabs is coupled minimally. Major specifications are

- Model: SYS 68K/CPU-40B
- CPU: MC68040, 25 MHz
- DMA Controller: 32 Bit High Speed
- Serial Interface: RS232/RS422/RS485
- VMEbus Interrupter: IR 1-7

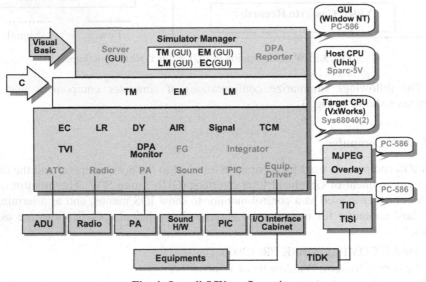

Fig. 4. Overall S/W configuration

2.4 Visual Computers

A pair of visual computers are located in the IOS cabinet. A visual computer for a cab receives position data of a train via Ethernet from the TVI(Track View Information) module on the relevant modeling CPU, and retrieves a visual image, front track-view, which is the most appropriate for the position among the filmed visual database compressed on computer hard disks. Motion JPEG is used for the purpose. A track-view image in the form of Motion JPEG is transmitted to an overlay card, and superimposed by light signals generated with computer graphics. The resultant visual image is finally transmitted to the repeat monitor in the IOS and the visual projector for a trainee in a cab. Major specifications are as follows:

- Model: COMPAQ DESK PRO 2000
- Overlay Card: Digi-Mix
- M-JPEG board: Digi-Motion (9 GB hard disks included)
- Operating System: Windows 95 or Windows NT

2.5 TIDK-TID Driving Computer

A pair of TIDK-TID driving computers is also located in the IOS cabinet as the visual computers. One computer per cab is for generating signals to TID (Track Information Display) and TIDK (Train Information Display and Keyboard). The specifications are

- Model: COMPAQ DESK PRO 2000
- FIP control box

All the S/W modules running on the above computers are programmed in either C or Visual Basic depending on their functions and hardware platforms. Table 2 summarizes major S/W modules along with their relevant operating environments.

Table 2. Programming languages and operating environments 2.6 database

	S/W module	Compiler	Operating System
Modeling CPU	Master Driver #1	GNU .68KC compiler	VxWorks 5.3
Host CPU	Master Driver #2	GNU C compiler	Solaris
PC-586	IOS S/W	Visual Basic 5.0	Windows NT/Windows 95
PC-586	Visual S/W	Visual Basic 5.0	Windows 95

2.6 Database

Major data files used by execution programs of PUTA Line II simulators are categorised into

- track data
- performance data of the trains
- electric circuit data of relevant on-board equipments
- pneumatic data of relevant on-board equipments

2.7 GUI Group (IOS) S/W

Modules in this GUI (Graphical User Interface) S/W group are programmed in Visual Basic 5.0 of Microsoft in order to decrease development process as well as its time and increase maintainability of the modules. With on-screen menus representing the GUI group S/W an instructor controls and monitors the whole system including 2 cabs. The GUI group S/W is a window to relevant S/W modules in host and target CPU's. An example of on-screen menus is shown in Fig. 5 and a part of the IOS menu hierarchy in Fig. 6. The GUI group S/W is composed of following sub-components:

- TM(GUI): Track Manager GUI invokes TM in the host CPU.
- EM(GUI): Exercise Manager GUI invokes EM in the host CPU.
- LM(GUI): Lesson Manager GUI invokes LM in the host CPU.
- EC(GUI): Exercise Controller GUI invokes EC in the target CPU.
- DPA RPT (Driver's Performance Assessment Reporter): The DPA Reporter generates either a trip-log of events or an assessment of the events in a lesson.
- Server(GUI): Server GUI invokes Server.

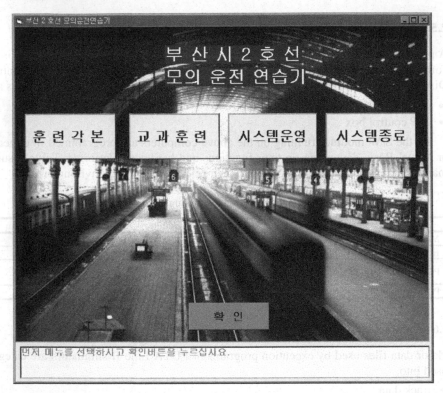

Fig. 5. An example of IOS menu on screen

2.8 Host Group S/W

The host group S/W in Fig. 4 is composed of TM (Track Manager), EM (Exercise Manager), and LM (Lesson Manager). These components are common to a pair of simulator cabs and do not communicate frequently with modeling group S/W once training lessons begin. They take important roles as lessons begin or are completed. The S/W components and their functions are listed below:

- TM (Track Manager): TM is comprised of 3 interlocking tasks, manage track sections, manage media and manage routing, which combine to handle the building and amendment of track sections, media information, and route validation and selection.

- EM (Exercise Manager): EM manages lessons and comprises Exercise Builder and Exercise Management Facilities. Exercise Builder generates a table with the route information TM provides, in order for the other modules to refer. Exercise Management Facilities changes exercises depending on selected conditions such as faults and signals.
- LM (Lesson Manager): LM supplies exercise signal information during initialization of lesson. Exercise signals consist of id, distance, type, state, fog repeater indicator, auto signal indicator and distance/speed/state that the auto signal is checked against.

2.9 Modeling Group S/W

Pairs of S/W modules in the Modeling group operate independently on two symmetrical modeling CPU boards. S/W modules belonging to this group are as follows:

- EC(Exercise Controller): EC synchronises tasks in a lesson and logs events generated during the lesson and controls a snapshot.
- LR(Lesson Reviewer) : LR reviews lessons performed.
- DY (Dynamics): DY supplies train position and speed. The DY model calculations can be split into four main areas such as longitudinal, carriage weight, cab distance, and train motion.
- AIR (Air System): The air system model provides six basic functions such as initialization control, air pressure model, response to opening and closing valves, generation of outputs on pressure thresholds, and calculation of air braking force in response to requests of dynamics.
- SIG (Signal): All signals are initialized to a default state of having the topmost bulb set. If this is not the required state then it is changed from the graphical user interface.
- DPA Mnt (Driver's Performance Assessment Monitor): DPA Mnt provides events that are used when assessing the drivers performance. The DPA monitor is provided with the speed limits, stop zones, their positions, and obstacles on the track.
- TVI (Track View Interface): TVI provides an interface on Ethernet between IOS and visual computers. It converts data/control signals of Master Driver into the format required by the visual computers.
- TCM (Train Circuit Modeler): TCM models the circuit diagrams in such areas as door control, round train & safety, traction & braking, driving & auxiliary control, train lines & car wires, and a compressor system.
- FG(Fault Generator): FG generates faults or malfunctions of on-board systems or signals, which can be selected or deselected either interactively or at a specific distance selected during definition of an exercise.
- Integ (Integrity test): Integ diagnoses system problems.
- ATC: ATC(Automatic Train Controller) emulator.
- Radio: radio driver.
- PA: PA(Passenger Address) driver.
- Sound: This simulates sounds normally generated by equipment in the cab or the environment. The audio system is controlled by the Exercise Controller.

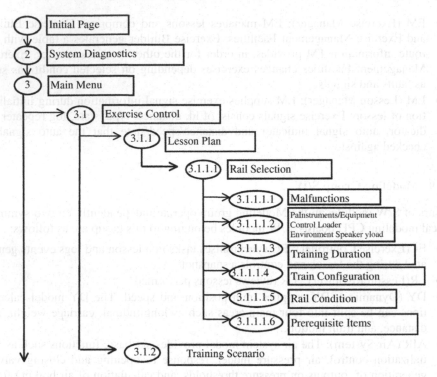

Fig. 6. An example of IOS menu hierarchy

- PIC (Passenger Information Controller): PIC driver.
- Equip Drv.: On-board equipment drivers.

2.10 Cab #1, #2

In simulation of switches, indicators, and equipment in the cabs an applied principle is to minimise their modification. If major modification is necessary, the actual system is replaced by a digitally controllable simulated system. For some equipment such as Passenger Information Controller and TIDK (Train Information Display and Keyboard) stimulation is rather adopted. That is, signal inputs to the equipment are simulated. Table 3 is a list of switches, indicators, and equipment simulated or stimulated, where AI (analog input), AO (analog output), DI (digital input), and DO (digital output) mean interfacing ports of VME I/O boards which drive the switches, indicators and instrument.

3 Airplane Simulators

In order to study portability of S/W and H/W in simulators technical specifications of PUTA Line II metro simulators are introduced above. For comparison technical sepcifications of airplane simulators are briefly described here. The airplance simulator selected as an example is a flight training device developed by Korean Air between

Table 3. A list of on-board equipment

No	Switches/Indicator/Equipment	Interface
1	Master Controller	AI
2	TRCP(Train Radio Control Panel)	RS485
3	window wiper switch	DI
4	MDH(Mode Direction Handle)	DI
5	ADU (Automatic Display Unit)	AO
6	dimmer switch	AI
7	push button switch (3EA)	DI
8	push button switch (11EA)	DI
9	indicating lamp (4EA)	DO
10	duplex pressure gauge	AO
11	DC volt meter (3KV)	AO
12	DC volt meter (150KV)	AO
13	PIC (Passenger Information Controller)	RS485
14	TIS (Train Information System)	RS485
15	NFB	DI & DO
16	head lamp control switch	DI
17	whistle valve with foot pedal	DI
18	lamp push button switch	DI
19	destination indicator	DI
20	train number indicator	DI
21	emergency brake cut out switch	DI
22	rescue operating switch	DI
23	emergency brake switch	DI

1991 qand 1995 to train pilots of a light aircraft, ChangGong-91. The aircraft was also developed by Korean Air in 1991. The flight training device is shown in Fig. 7. Its S/W and H/W configurations are illustrated in Figs. 8 and 9.

4 Portability Study of S/W and H/W

In this paper parts of technical specifications of a railway simulator and a flight training device are briefly introduced. The portability of H/W and S/W concentrated in the paper is about computer H/W, S/W, and IOS, which are actually major components of simulators. They really determine performance and value of a simulator.

Let us investigate computer H/W of PUTA metro simulators. The host and modeling computers in the simulator are of VMEbus type. VMEbus computers are also used for the ChangGong-91 FTD, and frequently chosen for computer systems of airplane, ship, and automotive simulators. Complexity of VMEbus cards' combination depends on application. As higher fidelity simulation is required, or as the simulated system is more complicated, higher performance CPU boards and more I/O channels may be included in the VMEbus computer. That is, complexity and fidelity of simulation determine configuration of a VMEbus computer. That is why different configurations of VMEbus computers are applied to PUTA metro simulators and ChangGong-91 FTD. However, computer-related technologies during implementation such as HILS (Hardware-In-the-Loop-Simulation) technology and relevant equipment, are basically the same either in railway or airplane simulators.

Fig. 7. ChangGong-91 flight training device

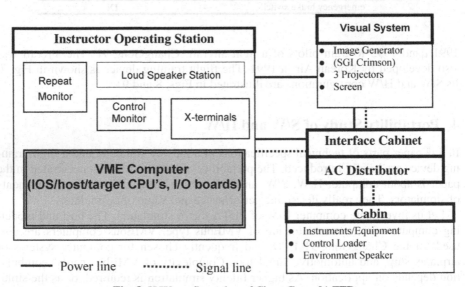

Fig. 8. H/W configuration of ChangGong-91 FTD

S/W such as real-time OS, I/O drivers, and communication S/W is closely related with computer H/W, and its portability depends on the computer H/W. Such S/W may be amended partially in order to meet interface requirements of modeling S/W and simulated/stimulated equipment. However, accumulated technologies through experience of system developments allow easy alternation of the S/W. Let us com-

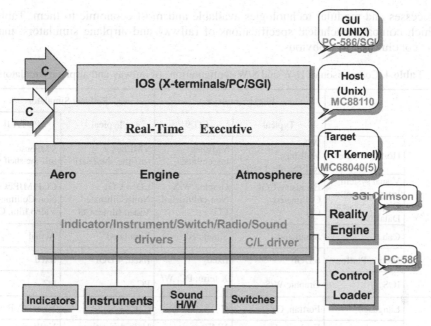

Fig. 9. S/W configuration of ChangGong-91 FTD

pare Figs. 3, 4, 8, and 9. Similarity of H/W and S/W configurations of the railway simulator and the FTD is revealed. S/W of an airplane simulator is stricter to real-time requirements and rapider response time is expected. In case of a railway simulator these constraints are less stringent. That is a rational of applying an executive-type time controller to an airplane simulator and an event-driven scheduler to a railway simulator. However, it is considered reasonable to adopt executive-type scheduler even in a railway simulator if simulation model is stricter and more sophisticated, and if exact response characteristics is required.

In case of IOS menu designs similar observation is obtained. Menu contents can be very diverse depending on simulated systems and customer's requirements. However, S/W tools, programming languages, and menu hierarchies to realize IOS menus are about the same. Standardized procedures can be set and applied independent of simulated systems based on implementation experiences. Apparent features and operating procedures of IOS' may look quite different from one another. But striking similarities in training procedures and IOS-inherent features, which enable effective training, are observed between railway and airplane simulators. For example, configuration, concept and methodology behind training lessons, normal and emergency procedures are very close to each other.

5 Conclusions

The experiences show that models may be diverse depending on the objects but implementation technologies are about the same. Maximizing portability of implementation technologies is a matter of an organization's strategy of adopting standardized processes and modular technologies available and most economic to them. Table 4, which compares technical specifications of railway and airplane simulators, makes this conclusion more obvious.

Table 4. Comparison of H/W and S/W configurations of railway and airplane simulators

		Flight Simulator		Railway Simulator	
		Typical	CG-91	Typical	PUTA II
H/W	Host Computer	W/S VMEbus (off-the-shelf set)	VMEbus (assembled)	VMEbus (off-the-shelf set)	VMEbus (off-the-shelf set)
	Visual system Image generator Display System Database	Exclusive CGI Collimated CG	Graphic W/S Non-collimated CG	LD⇒ CGI Non-Collimated Video film⇒CG	CGI+MJPEG Non-Collimated Video film, CG
	Cab	Simulated	Simulated	Simulated	Actual
	Motion Platform	6 DOF	fixed	fixed, 4 DOF	fixed
	IOS	Graphic W/S	X-term, PC, W/S	PC	PC
S/W	Language	Fortran, C	C	C, C++	Visual C, Basic
	OS	Exclus.⇒Unix POSIX comp. Executive type	VMEexec Non-POSIX Executive type	Unix derivatives POSIX comp. Event-driven	VxWorks POSIX comp. Event-driven
	Comm. Protocol	Ethernet, bus	Ethernet, bus	Ethernet, FTP	Ethernet, FTP
	System Model	Accurate Test data	Accurate Test data	Simplified Actual+Generic	Simplified Actual+Generic
	IOS Menu	GUI	GUI	GUI	GUI
Relevant Regulations		FAA, CAA, etc.	FAA, CAA, etc.	none	none

Acknowledgement

This research(paper) was performed for the Smart UAV Development Program, one of the 21st Century Frontier R&D Programs funded by the Ministry of Science and Technology of Korea.

References

1. Yoon S. et al. : Final Report for Development of a Flight Training Device, Korean Ministry of Commerce and Industry, June (1995)
2. Yoon S., Kim W., and Lee J. : Flight Simulation Efforts in ChangGong-91 Flight Training Device, Proceedings of AIAA Flight Simulation Conference, Aug. (1995)
3. Yoon S., Bai M., Choi C., and Kim S. : Design of Real-Time Scheduler for ChangGong-91 FTD, Proceedings of 'Making it Real' CEAS Symposium on Simulation technologies, Delft Univ., Nov. (1995)
4. PUTA II Simulator Requirements Specification, Doc #. SIM-001-002, Korean Air, May (1997)

Forward-Backward Time Varying Forgetting Factor Kalman Filter Based Landing Angle Estimation Algorithm for UAV (Unmanned Aerial Vehicle) Autolanding Modelling

Seokrim Choi[1], Jun-Seok Lim[1,*], and Sugjoon Yoon[2]

[1] Dept. of Electronics Eng., Sejong University,
98 Kwangjin Kunja, Seoul Korea 143-747
{schoi,jslim}@sejong.ac.kr

[2] Dept. of Aerospace Eng., Sejong University,
98 Kwangjin Kunja, Seoul Korea 143-747
sjyoon@sejong.ac.kr

Abstract. This paper discusses a landing angle estimation algorithm in UAV (Unmanned Aerial Vehicle) autolanding simulation. In UAV autolanding with radar system, the ground multipath effect and the time varying landing angle make it difficult to estimate landing angle information of UAV. This paper proposed a new algorithm based on forward-backward Kalman filter with time varying forgetting factor. This algorithm effectively handles the multipath effect and time varying landing angle environments to estimate highly accurate landing angle.

1 Introduction

Recently many people pay attention to military and commercial application of UAV. In military applications, recovery is important issue because UAV carries many sensitive sensor systems and a heap of valuable information with it. It necessitates safer landing systems by landing gears. Radar guide autolanding system is one of the candidates. Many radar guide autolanding systems equip a repeater on the UAV to produce clear echo for radar. UAV in military application should meet various landing conditions from rough ground to pitching shipboard. Under the environments, radar often has trouble in tracking low flying targets. The difficulties are due to the reflection from the surface of the earth or the sea and that of the pitching shipboard. The reflected ray consists mainly of the specular component for low grazing angles. The specular component is highly correlated with the direct component. In addition, UAV landing period is short so that the landing angle is relatively fast time varying. Therefore, the radar in autolanding system requires high-resolution landing angle tracking capability for UAV to land safely. In the UAV autolanding simulation assuming

---

* Jun-Seok Lim is the corresponding author.

D.-K. Baik (Ed.): AsiaSim 2004, LNAI 3398, pp. 107–112, 2005.

such environments, it is necessary to build the estimation function for the landing angle estimation as well as to model the landing environment modeling with respect to RF in radar.

In the paper, we develop a recursive high-resolution scheme for tracking landing angles using Kalman filter and time varying forgetting factor for the UAV autolanding simulation. Kalman based high-resolution algorithm was proposed in [1]. It estimates noise subspace by Kalman filter and applies the subspace to MUSIC method to result in high-resolution angle. It requires no a prior knowledge of the number of signals. It, however, cannot be applied to the time varying and coherent landing angle estimation. We modify this algorithm to estimate the time varying and coherent landing angle. To handle the coherent signal, we propose forward/backward Kalman filtering scheme. It has a space smoothing effect to ease the coherent specular signal. We also propose the time varying forgetting factor to estimate nonstationary subspace. It makes radar to track the time varying landing angle more accurately.

2   Forward-Backward Kalman Based Noise Subspace Estimator

Chen and et al. proposed Kalman based noise subspace estimator with the linear array in [1]. This method, however, doesn't work in coherent signal environments like UAV autolanding. It is because the estimated covariance matrix becomes singular in coherent signal case. To tackle it, this paper modifies the object function as eqn. 1 and the measurement equation as eqn. 2 with forward and backward sensor input vectors.

$$\underset{\hat{W}_i}{Min}\, J_i = \hat{W}_i^T R_{fb} \hat{W}_i, \quad \text{subject to} \quad \hat{W}_i^T \hat{W}_i = 1 \tag{1}$$

$$v_{fi} = \gamma_f(k) - X_f^T(k)\hat{W}_i^*(k), \quad i = 1, 2, \cdots, V \tag{2}$$

$$v_{bi} = \gamma_b(k) - X_b^T(k)\hat{W}_i^*(k), \quad i = 1, 2, \cdots, V, \tag{3}$$

where $\hat{W}_i(k) = [\hat{w}_{1i}(k), \cdots, \hat{w}_{Mi}(k)]^T$, $X_f(k) = [x_1(k), \cdots, x_M(k)]^T$, $X_b(k) = [x_M(k), \cdots, x_1(k)]^T$ and $R_{fb} = E(X_f X_f^H + X_b X_b^H)$. If we set $\gamma_f(k) = 0$ and $\gamma_b(k) = 0$, we can obtain the mean square error as

$$\mathbf{E}\left[\left|0 - C^T(k)\hat{W}_i^*(k)\right|^2\right] = \hat{W}_i^H R_{fb} \hat{W}, \quad \text{where} \quad C = [X_f\, X_b]. \tag{4}$$

Therefore, from eqn. 1 to eqn. 3, the measurement equations of Kalman base estimator can be modeled as

$$0 = C^T(k)\hat{W}_{id}^*(k) + v_{oi}(k), i = 1, 2, \cdots, V, \tag{5}$$

$$W_i^*(k) = \frac{\hat{W}_{id}^*(k)}{\left\|\hat{W}_{id}(k)\right\|}, i = 1, 2, \cdots, V, \tag{6}$$

Table 1. Forward-Backward Kalman MUSIC.

$G(k) = \frac{P(k-1)C(k)}{C^T(k)P(k-1)C(k)+Q(k)}$,
where $Q(k)$ is noise covariance matrix.
$P(k) = P(k-1) - G(k)C(k)^T P(k-1)$
$\hat{W}_{id}(k) = \hat{W}_i(k-1) - G(k)C^T(k)\hat{W}_i(k-1), i = 1, 2, \cdots, V$
$W_i(k) = \frac{\hat{W}_{id}(k)}{\|\hat{W}_{id}(k)\|}, i = 1, 2, \cdots, V$
DOA estimation by MUSIC with the estimated noise subspace
$\hat{W} = [W_1, \cdots, W_V]$.

where $v_{oi}(k) = [v_{fi}(k)\, v_{bi}(k)]^T$ and $\| \bullet \|$ denotes the norm operator. This modification introduces the spatial smoothing effect to the covariance matrix and prevents the degradation in coherent signal. Although this algorithm can estimate one direct signal and one correlated reflected signal, it is enough in autolanding in which the specular reflection is dominant. In Table 1, we summarize the forward-backward Kalman based noise subspace estimation algorithm.

3   Time-Varying Forgetting Factor in Kalman Based Noise Subspace Estimator

For better performance in the time-varying signal condition, it is necessary to introduce the time-varying forgetting factor (TVFF) [2] and then we propose TVFF to minimize the following forward-backward residue energy sum.

$$J = \sum_{j=1}^{k} \left[\prod_{t=j+1}^{k} \beta(t) \right] (\varepsilon_f(j)\varepsilon_f^H(j) + \varepsilon_b(j)\varepsilon_b^H(j)), \qquad (7)$$

where $0 \le \beta(t) \le 1$, $\omega(k,k) = 1$, $\omega(k-1,k) = \beta(k)$, $\varepsilon_f(k) = -X_f^T(k)\hat{W}^*(k)$ and $\varepsilon_b(k) = -X_b^T(k)\hat{W}^*(k)$. To find the forgetting factor, we minimize eqn. 7 by Newton method as follows.

$$\beta(k+1) = \beta(k) - \frac{1}{2}\frac{\partial J}{\partial \beta} = \beta(k) - \alpha Re\left[\varepsilon^H(k)\varepsilon'(k)\right], \qquad (8)$$

where $\varepsilon(k) = -C(k)\hat{W}^*(k-1)$ and $\varepsilon'(k) = \frac{\partial \varepsilon(k)}{\partial \beta} = -C(k)\frac{\partial \hat{W}^*(k-1)}{\partial \beta}$
$= -C(k)\Psi^*(k-1)$. In eqn. 8, the term of $\varepsilon^H(k)\varepsilon'(k)$ is noise sensitive. Therefore, we smooth it as follows.

$$J_\varepsilon = \sum_{j=1}^{k} \left[\prod_{t=j+1}^{k} \beta(t) \right] \varepsilon^H(j)\varepsilon'(j) = \beta(k)J_\varepsilon(k-1) + \varepsilon^H(k-1)\varepsilon'(k-1). \qquad (9)$$

Then eqn. 8 becomes as eqn. 10.

$$\beta(k+1) = \beta(k) - \alpha Re\left[J_\varepsilon(k)\right]. \qquad (10)$$

Table 2. Forward-Backward Time Varying Forgetting Factor Kalman MUSIC.

$G(k) = \frac{P(k-1)C(k)}{C^T(k)P(k-1)C(k)+Q(k)}$, where $Q(k)$ is noise covariance matrix.

$P(k) = \frac{1}{\beta(k)}\left[P(k-1) - G(k)C(k)^H P(k-1)\right]$

$J_\varepsilon = \beta(k)J_\varepsilon(k-1) + \varepsilon^H(k-1)\varepsilon'(k-1)$

$\beta(k+1) = \beta(k) - \alpha Re\left[J_\varepsilon(k)\right]$

$S(k) = \frac{1}{\beta(k)}\{(I - P(k)C^H(k))^H S(k-1)(I - P(k)C^H(k))$
$\quad + G(k)G^H(k) - P(k)\}$

$\Psi(k) = (1 - G(k)C^H(k))\Psi(k-1) + S(k)C(k)\varepsilon(k)$

$\hat{W}_{id}(k) = \hat{W}_i(k-1) - G(k)C^T(k)\hat{W}_i(k-1),$

$i = 1, 2, \cdots, V$

$W_i(k) = \frac{\hat{W}_{id}(k)}{\|\hat{W}_{id}(k)\|}, i = 1, \cdots, V$

DOA estimation by MUSIC with the estimated noise subspace
$\hat{W} = [W_1, \cdots, W_V]$.

In addition, the derivatives of $\hat{W}(k)$ and $P(k)$ are derived respectively as eqn. 11 and eqn. 12 in the same way in [3].

$$\Psi(k) = \frac{\partial \hat{W}(k)}{\partial \beta} = (1 - G(k)C^H(k))\Psi(k-1) + S(k)C(k)\varepsilon(k) \quad (11)$$

$$S(k) = \frac{1}{\beta(k)}\{(I - P(k)C^H(k))^H S(k-1)(I - P(k)C^H(k)) \atop + G(k)G^H(k) - P(k)\} \quad (12)$$

In Table 2, we summarize the TVFF forward-backward Kalman based noise subspace estimation for DOA estimation.

4   Simulation Results

We provide some simulation results to demonstrate the applicability to UAV autolanding. We track the UAV under the scenario of Fig. 1 by using a linear uniform array with four sensors and use the direct signal and the reflect signal with correlation factor of 0.8 in the coherent case as well as with correlation factor of 0 in the non-coherent case. We take a snap-shot in every 30msec. Fig. 2 shows the results of the proposed algorithm and two fixed forgetting factor conventional ones in noncoherent signal environments. The results say that the proposed method estimates the accurate landing angle under the uncorrelated signals. Fig. 3 shows the results in coherent signal environments. Especially the proposed forward-backward kalman filter effectively tackles the correlation effect between the direct signal and the reflect signal, although, without the forward-backward kalman filter, we can expect the result to be blown up because of the correlation effect between the two signals. Fig. 2 and Fig. 3 also show that the variable forgetting factor works well to keep good track of landing angle. Table 3 summarizes the estimation performance in mean and standard deviation.

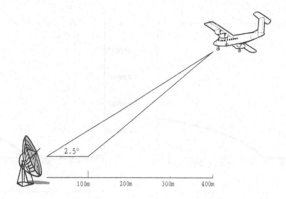

Fig. 1. Simulation Scenario of UAV Autolanding.

Table 3. Summary of estimation performance.

	Noncoherent case		Coherent case	
forget factor	mean	std	mean	std
0.98	0.015	0.128	0.084	0.075
0.8	0.016	0.021	0.028	0.039
variable	0.015	0.025	0.026	0.037

The table explains the performance more detail. When we focus on the row of the forgetting factor 0.8 and the row of variable, which look like in Fig. 2 and Fig. 3, the proposed algorithm is almost the same or better than the conventional kalman filter with the forgetting factor 0.8. To compare with the row of the forgetting factor 0.98 and the row of variable, the proposed one is superior to the conventional one.

From these simulation results, we can confirm the proposed algorithm works well in the coherent and non-coherent signal environments, which we frequently meet in the radar guided' UAV autolanding. In addition, the proposed method needs no *a priori* information on the variation rate in the landing angle and then we can say that the proposed algorithm overcomes the conventional method in landing angle estimation as well as that it is more suitable for its system performance simulation.

5   Conclusion

In the paper, we have developed a recursive landing angle tracking scheme using the forward-backward Kalman filter and time-varying forgetting factor for the UAV autolanding simulation. The algorithm could estimate the time-varying and coherent landing angle. The results have shown that the proposed algorithm can estimate the time-varying and coherent landing angle more accurately than conventional method without *a priori* landing condition and that it is more suitable for the radar guided UAV autolanding simulation.

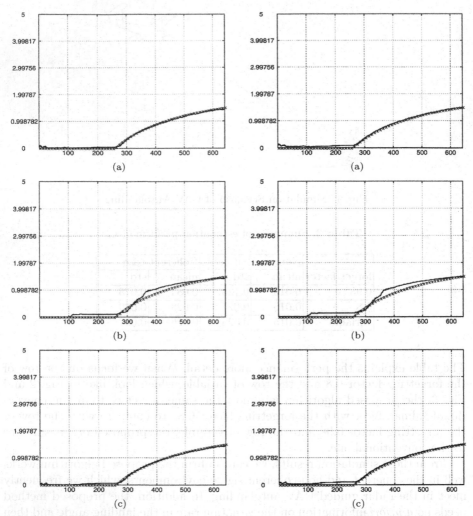

Fig. 2. DOA estimation results in noncoherent signal case. (a) fixed forgetting factor (0.8) (b) fixed forgetting factor (0,98) (c) proposed algorithm (-x-:true angle).

Fig. 3. DOA estimation results in coherent signal case. (a) fixed forgetting factor (0.8) (b) fixed forgetting factor (0,98) (c) proposed algorithm (-x-:true angle).

References

1. P. A. Thompson, "An adaptive spectral analysis technique for unbiased frequency estimation in the presence of white noise." P*roc. 13th Asilomar Conf. Circuits, Syst. Comput.* (Pacific Grove, CA), 1980, pp. 529-533.
2. Yuan-Hwang Chen and Ching-Tai Chiang, "Kalman-based estimator for DOA estimations," *IEEE Trans. on Signal Processing*, vol. 42, no. 12, pp. 3543-3547, DEC. 1994.
3. S. W. Lee et al., "Time-varying signal frequency estimation Kalman filtering," *Signal Processing*, vol. 77, pp. 343~347, 1999.

A Computational Method for the Performance Modeling and Design of a Ducted Fan System

Jon Ahn and KyungTae Lee

Sejong University, Department of Aerospace Engineering,
Gwangjin-gu, Gunja-dong 98, 143-747 Seoul, Korea
{kntlee,jonahn}@sejong.ac.kr
http://dasan.sejong.ac.kr/~aeroprop

Abstract. A computational model been developed for an axisymmetric ducted fan system to identify its design parameters and their influence on the system performance. The performance of the system has been investigated as its geometrical design parameters, its fan model parameters, and the thrust requirement are changed parametrically. The fan in the system is modeled as an actuator disk, which provides jumps in stagnation enthalpy and stagnation density across it. Current results show that increasing diffuser angle of the duct improves the propulsion efficiency of the ducted fan system, while the inlet geometry has little effect on it. Also, non-uniform fan strength distribution models have been proposed and investigated, which shows that the strength distribution of the fan increasing in the direction of the fan blade tip facilitates wake expansion and requires less power than the fan strength distribution which increases in the direction of the blade hub. The current results are to be validated in near future, when the experimental and computational results are available from parallel research works.

1 Introduction

Ducted fan has been known to improve propulsion efficiency and static thrust, along with noise reduction [6], but has seen limited applications in aerospace industry due to several reasons, such as the increased weight and complexity of system, the marginal gain under off-design conditions, and the lack of understanding in related physical phenomena, to list a few. Recently, the ducted fan system has received recurring attention, as its application is more widely accepted, especially in helicopter of gross weight range in 10,000-20,000 lbs class. Also, several UAV designs propose to utilize the ducted fan system for all their lift, thrust, and control [1], such as the recent work by Guerrero et al. [8], in which they propose a deflector-controlled micro UAV.

Earlier studies on the ducted fan system have relied mostly on wind tunnel experiments, and parametric studies have been limited to a few, due to the cost and the complication of the experimental methods. One example of those a few parametric studies is by Black and Wainauski [4], in which they identified the duct exit area ratio as the most important design parameter in the ducted fan system. More recently, Abrego and Bulaga [2] experimentally investigated the thrust, power, and control characteristics of a ducted fan system with exit vanes. In an effort to provide a theo-

D.-K. Baik (Ed.): AsiaSim 2004, LNAI 3398, pp. 113–121, 2005.
© Springer-Verlag Berlin Heidelberg 2005

retical method for the prediction of aerodynamic control parameters, Guerrero *et al.* [8] utilized the momentum method and a uniform actuator disk model, which showed drawbacks when the fan-duct interaction became prominent.

This work is focused on the provision of a practical computational tool, which can show a design direction from parametric studies, as well as on the contribution to more advanced physical understanding of the ducted fan system. To overcome the drawback of the uniform actuator disk model as mentioned above, the fan is modeled as an axisymmetric actuator disk with non-uniform radial strength distribution and its influence on the system performance is investigated, along with geometric design parameters. The primary intention of this research is to provide an efficient modeling and design method of a ducted fan system, which can be integrated into simulations of helicopters and vertical take-off aircrafts powered by ducted fans.

2 Computational Method

This study utilizes the axisymmetric transonic viscous flow prediction and design method developed by the first author Ahn and Drela [3], which is an extension of the stream-surface based Newton method proposed originally by Drela and Giles [5]. It couples Euler equations with axisymmetric integral boundary layer equations that employ two-dimensional closure and transition models, which results in an efficient and precise modeling for viscous axisymmetric flows. However, the present research excludes the boundary layer coupling, since the complicated blade-tip flow phenomena can't be modeled properly by an axisymmetric integral boundary layer model without proven modifications to closure and transition models.

A finite-volume conservation cell (dotted lines) sandwiched by two stream-surfaces is shown in Fig. 1. Since there is no mass flow across the stream surfaces, the continuity equation and the energy conservation become algebraic identities, and the number of flow variables to be determined is reduced from 4 to 2 for axisymmetric or two-dimensional problems. These two algebraic identities provide computational efficiency and facilitate the computational implementation of the changes in the mass flow and the energy in the actual flow.

Application of the momentum conservation law in two directions, one parallel to the mean flow direction in the finite volume cell and the other normal to the mean flow direction, results in two momentum equations, which are used to determine the density of the finite-volume cell and the stream-surface movement in the direction normal to the stream-surface. For the iteration process of the non-linear flow equations, Newton method is adopted for rapid convergence, typically within 7 iterations. Since a Newton solver needs an initial grid system close to the solution for the proper convergence of a non-linear system, a potential based axisymmetric stream-surface grid generator has been also developed. The geometry of the ducted fan and its initial grids are shown in Fig.2. Also, Table 1 shows the geometry and the operation parameters.

The current computational method converges rapidly within 5-6 iterations, when the flow is subsonic and flow conditions or geometries are not much different from the starting conditions. However, a typical ducted fan problem with specified thrust takes 10 iterations for a full convergence, which corresponds to 10 seconds on a

Linux based Pentium 4 computer running at 1.6 GHz. The explicit specification of thrust requirement redundantly constrains the system of equations, in a similar manner as a Lagrangian multiplier, which degenerates the quadratic convergence nature of the Newton method.

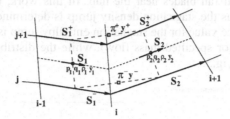

Fig. 1. Finite volume conservation cell (*dotted quadrilateral*) is defined between two stream surfaces (*bold arrow pointed lines*)

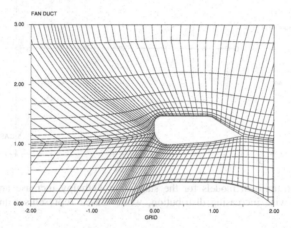

Fig. 2. A potential-based axisymmetric grid generation program is used to create the initial stream-surface grid system

2.1 Actuator Disk Modeling of the Fan

In the actual fan flow, the unsteady motion of the fan blades increases the stagnation enthalpy and the stagnation density (or stagnation pressure) in each stream-tube as it crosses the fan blades. When there is no loss, the increase in the stagnation density and the increase in the stagnation enthalpy follow isentropic relations.

In this work, the fan is modeled as an actuator disk, which provides jumps in stagnation enthalpy and the stagnation density across it, as presented in Equation (1). Note that subscripts 1, 2 denote flow properties before and after the actuator disk. The specification of the jump in stagnation enthalpy replaces the energy equation in a conservation cell at the fan location, while the jump in the stagnation density replaces the momentum equation in the mean flow direction.

$$ h_o + \Delta h_o = \frac{\gamma}{\gamma - 1} \frac{p_2}{\rho_2} + \frac{1}{2} q_2^2, \quad \rho_{o2} = \rho_{o1} + \Delta \rho_o \tag{1} $$

Non-uniform thrust and loss distributions in a fan can be modeled by assigning different jump values for each stream-tube. In Fig. 3., three models for the stagnation enthalpy distribution is proposed. Case 1 approximately represents fan blades that have less twist angle distribution near the tip than the ideal one, while Case 2 roughly describes less twisted fan blades near the hub. In this work, the fan is assumed to have no loss, and thus the stagnation density jump is determined from the isentropic relation. The reference value for the stagnation enthalpy jump is determined to satisfy the specified thrust (or specified mass flow), while the distribution models are prescribed as shown in Fig.3.

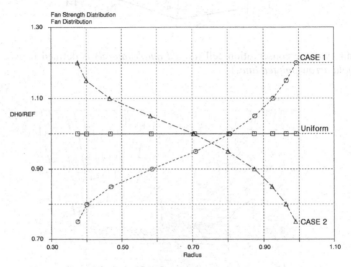

Fig. 3. Two non-uniform models for the stagnation enthalpy increase are proposed. Case1 represents a tip-wise increasing distribution, while Case2 represents a hub-wise increasing distribution

2.2 Power Requirement of the Fan

The power requirement of the fan can be calculated by applying the energy conservation for each stream-tube. Power supplied by the fan should equate to the sum of energy rates carried by each stream-tube, as shown in Equation (2).

$$P = \sum_{j=1}^{n} \dot{m}_j \Delta h_o \tag{2}$$

2.3 Thrust Calculation

In this work, the axial momentum fluxes are calculated for each stream-tube at the inlet and exit of the computational domain, whose momentum balance is the thrust generated by the stream-tube. When summed up for all the stream-tubes, the thrust generated by the fan duct system can be found.

$$T = \sum_{j=1}^{n} \dot{m}_j [(\rho u)_2 - (\rho u)_1] \tag{3}$$

2.4 Ducted Fan Geometry and Operating Conditions

The ducted fan system investigated in this research work is a computational model for the ducted tail rotor of a research helicopter, currently under design study phase in Korea. Physical geometric data and operating conditions are summarized in Table 1. A boat-tail is added on the downstream side of the computational model of the duct as can be seen in Fig. 2, since the stream-surface based Newton method can not handle such a massive separated flow behind a bluff body. A computational research parallel to this work is being performed at this point, which will provide proper shape of the boat-tail in the future. Until then, the main concern of this research is the identification of the design parameters and their influences on the system performance, not the exact prediction of the fan power requirement.

Table 1. Physical dimensions and operating conditions of the ducted fan system are presented

Fan radius	0.550 m	Center-body length	1.29 m
Duct outer radius	0.831 m	Maximum thrust	7169 N
Duct length	0.830 m	Mach No. (side flight)	0.05294
Diffuser angle	4 - 6 °	Design rotation speed	3125 rpm

3 Computational Results

The variations of power requirement of the fan duct is investigated in this section, as the geometric parameters and fan models and changed parametrically. For the geometric parameters, 3 diffuser angles and 9 duct -inlet radii are selected upon request from the funding party. For each geometric parameter change, a separate grid system is generated, and the thrust requirement is gradually increased to the maximum thrust requirement 7169 N. It usually takes 20-30 steps to meet the maximum thrust requirement, and each gradual step needs 10 iterations for convergence to engineering accuracy, which is considered to be enough for the starting solution for the next thrust step.

3.1 Diffuser Angle Changes (Uniform Fan)

The computation results for uniform fan with no loss model are summarized in Table 2, for the maximum thrust condition. Every degree increase in diffuser angle dilates the exit area of the duct by approximately 1% (See Fig. 4), which is found to give about 1% power reduction. The power saving comes from the increased mass flow into the fan duct, as can be seen in the table. The authors have found that there is no thrust generated in stream-tubes that encompass the outer surface of the fan duct as expected, which is also confirmed in the uniform stagnation density of Fig. 5. The power coefficient and the propulsion efficiency are plotted against the thrust coefficient in Fig. 5, which shows increase in diffuser angle provides more efficiency gain at higher thrust coefficients.

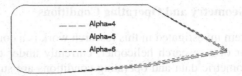

Fig. 4. Diffuser angle changes are plotted. Each degree increase results in approximately 0.5% dilation in the diameter of the fan duct exit

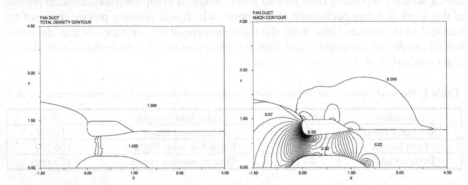

Fig. 5. Contour plot of stagnation density *(left)* shows the uniform entropy distribution in the outer flow. The inner flow possesses uniformly increased entropy distribution. Contour plot of Mach number *(right)* shows the rapid acceleration of the flow on the fan inlet surface

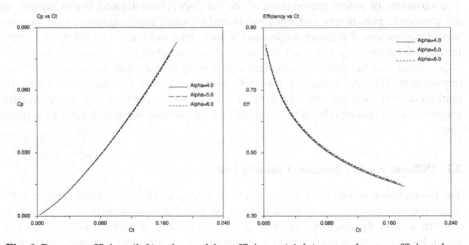

Fig. 6. Power coefficient *(left)* and propulsion efficiency *(right)* versus thrust coefficient shows power savings as the diffuser angle is increased

Table 2. Power requirement results for uniform fan model are presented below. Configuration marked in green color denotes the design selection of the funding party

Diffuser Angle	4.0	5.0	6.0
Flow (Kg/sec)	127.20	128.66	130.10
P (W)	334732	332027	330126
P (HP)	446.31	442.70	440.17

3.2 Inlet Radii Changes

As mentioned in the preceding sub-section, all the thrust is generated on the fan duct surface and on the fan surface that lie under the stagnation stream-surface. The funding party has placed much weight on the parametric study of inlet radii $r1$, $r4$, $r5$ (see Fig.7), because a small inlet radius change could move the position of the stagnation streamline drastically, which would affect the thrust. However, as can be seen in the results shown in Table 3, the inlet radius $r1$ hardly affects (less than 0.1%) the fan duct performance, which also applies to radii $r4$ and $r5$. Results for $r4$, $r5$ are omitted here, since they don't carry much physical meaning. The authors believe that the rapid acceleration induced by the suction prevents the stagnation flow from being affected by the radii change. This point is corroborated by the constant Mach number on the ellipsoidal inlet surface (about 0.5 for all radii cases, as shown in Figure 7), and by the observation that the stagnation point remains virtually same, even though the inlet radii are changed by 30%. That is, the inlet flow is dominated by suction pressure from the fan, not much by the geometry of the inlet.

Table 3. Power prediction results for the change of parameter $r1$ are presented. Configuration marked in green color denote the design selection of the funding party (maximum thrust condition)

R1	0.15	0.115	0.095
Flow (Kg/sec)	127.20	127.22	127.22
P (W)	334732	334432	334426
P (HP)	446.31	445.91	445.90

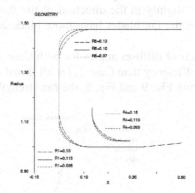

Fig. 7. Geometry changes of the inlet surface due to the change in each parameter are plotted

3.3 Fan Strength Distribution Models

Since this study assumes an ideal fan with no loss, the uniform fan strength distribution is expected to provide the best power requirement for a given thrust level. This is analogous to the uniform downwash (or elliptic lift distribution) condition for a finite wing. The fan strength distribution model Case1 requires 1% more power than the uniform fan, while Case2 needs almost 5% more power than the uniform fan, as shown in Table 4 (maximum thrust condition). For Case 1, the fan duct flow has higher stagnation pressure in the direction of the fan-blade tip, and interacts with the outer flow, which has lower stagnation pressure, facilitating the expansion of the fan

Table 4. Power prediction results for fan models shown in Fig.3. are presented

Fan Model	Case 1	Uniform	Case 2
Mass Flow (Kg/sec)	127.82	127.20	125.97
P (Watts)	338177	334732	350502
P (HP)	450.90	446.31	467.34
	(+1 %)		(+4.7 %)

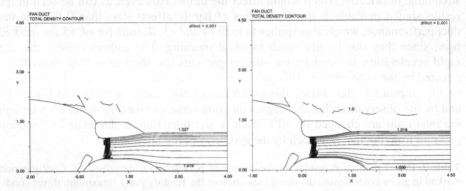

Fig. 8. Contour plot of stagnation density shows the increasing stagnation enthalpy in the direction of the fan blade tip for the fan model *Case1 (left)* shown in Fig. 3. *Case2* represents the fan strength distribution increasing in the other direction, towards the fan blade hub

duct flow. This can be seen in the stagnation density plots in Fig. 8, where the contour lines are expanding slightly in the direction of the fan-blade tip. In real world, Case 1 can happen when the fan blades have more blade angle distribution near the tip than the ideal twist.

Fig. 9 shows the effects of diffuser angles for both Case 1 and Case 2. Case 1 provides better propulsion efficiency than Case 2, for all thrust coefficient ranges studied in this work. By comparing Fig. 9 and Fig. 6, the fan strength distribution model Case

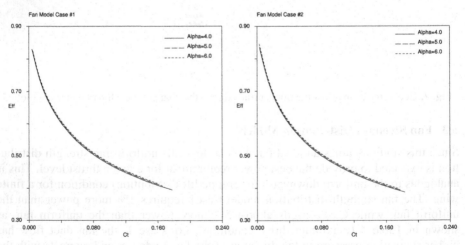

Fig. 9. Propulsion efficiency for *Case1 (left)* and *Case2 (right)* are plotted against the thrust coefficient

1 is less affected by the increase in diffuser angle than both the uniform fan and Case 2. This can be explained from the observation that Case 1 already has more expansion in the outer part of the fan-duct flow than Case 1 and the uniform fan, thus leaving less margin to be exploited by the increase in the diffuser angle.

4 Conclusion and Future Works

- Parametric studies confirm that diffuser angle is a dominant fan duct parameter.
- Inlet geometry does not affect fan duct performance at high thrust coefficients.
- Inlet stagnation flow is strongly affected by the suction pressure in front of the fan.
- Higher fan strength distribution in the direction of the blade tip facilitates the expansion of the fan-duct flow.
- Current results need to be validated in the future, possibly using experimental data.

References

1. Abd-El-Malek, M., Kroetsch, D., Lai, G., Marchetti, S., Wang, D., and Zlotnikov, D., "Component Architecture for a Combined Indoor and Outdoor Aerial Reconnaissance System," *2002 International Aerial Robotics Competition*, Competition Paper, Calgary (2002)
2. Abrego, A. I. and Bulaga, R.W., "Performance Study of a Ducted Fan System," *Proceedings of AHS Aerodynamics, Acoustics, and Test and Evaluation Technical Specialist Meeting*, San Fransisco, CA (2002)
3. Ahn, J. and Drela, M., "Newton Method on Axisymmetric Transonic Flow and Linearized 3D Flow Prediction," *38th AIAA Aerospace Sciences Meeting and Exhibit*, AIAA-98-0928 (1998)
4. Black, D.M. and Wainauski, H.S., "Shrouded Propellers – A Comprehensive Performance Study," *AIAA 5th Annual Meeting and Technical Display*, Philadelphia, PA (1968)
5. Drela, M. and Giles, M.B., "ISES: A Two-Dimensional Viscous Aerodynamic Design and Analysis Code," AIAA-87-0424 (1987)
6. Dunn, M. H., Tweed, J., and Farassat, F., "The Prediction of Ducted Fan Engine Noise Via a Boundary Integral Equation Method," *2nd AIAA/CEAS Aeroacoustical Conference*, State College, PA, AIAA-96-1770 (1996)
7. Fletcher, H. S., "Experimental Investigation of Lift, Drag, and Pitching Moment of Five Annular Airfoils," NASA TN-4117 (1957)
8. Guerrero, I., Londenberg, K., Gelhausen, P., and Myklebust, A "A Powered Lift Aerodynamic Analysis for the Design of Ducted Fan UAVs" *2nd AIAA Unmanned Unlimited Systems, Technologies, and Operations Conference*, SanDiego, CA, AIAA 2003-6567 (2003)
9. Yu, J., Jadbabaie, A., James, P., and Huang, Y., "Comparison of Nonlinear Control Design Techniques on a Model of the Caltech Ducted Fan." *Automatica Journal*, No. 37 (2001)

HLA-Based Object-Oriented
Modeling/Simulation for Military System*

Tae-Dong Lee, Seung-Hun Yoo, and Chang-Sung Jeong**

School of Electrical Engneering in Korea University,
1-5ka, Anam-Dong, Sungbuk-Ku, Seoul 136-701, Korea
{lyadlove,friendyu}@snoopy.korea.ac.kr, csjeong@charlie.korea.ac.kr

Abstract. This paper presents an HLA-based Object-oriented modeling/Simulation for Tank-helicopter combat system (HOST) which can be efficiently used for evaluation of tactics and weapon system for tank/helicopter combat. HOST is a distributed combat (tank-to-tank, tank-to-helicopter) simulation based on High Level Architecture (HLA). We design and implement an object-oriented simulation engine which can provide an unified interface to a set of interfaces for entities in HOST by using facade pattern. We design a domain specific object model (DOM) for tank-helicopter combat system based on CMMS (Conceptual Model of the Mission Space) and UML (Unified Modeling Language). We shall show how CMMS can be used to aid the design of use case diagram in UML, and presents various diagrams such as use case, class, sequence, state chart diagram in UML, which help users develop the relevant models in HLA. We shall also show how to adapt the domain specific object model to HLA by the efficient design of FOM (Federation Object Model) and TMM (Time Management Model). TMM provides time-stepped, event-driven and optimistic time models in graphic and text modes. In a graphic mode, inner federate time management (Inner-FTM) and inter federate time management (Inter-FTM) are used. Inner-FTM achieves synchronization by coordinating local objects and visualization, and Inter-FTM by coordinating federates and visualization respectively. The object-oriented design of our domain specific object model in HOST provides users with modification, extensibility, flexibility through abstraction, encapsulation and inheritance. Moreover, HLA-based model in HOST allows our system to be plugged in other HLA-compliant system, and extended into more complex system by providing reusability and interoperability

1 Introduction

High Level Architecture (HLA) [1] has a wide applicability across a full range of simulation areas, including education, training, analysis, engineering. These widely differing applications indicate the variety of requirements considered in

---

* This work has been supported by KOSEF and KIPA-Information Technology Research Center, University research program by Ministry of Information & Communication, and Brain Korea 21 projects in 2004.
** Corresponding author.

D.-K. Baik (Ed.): AsiaSim 2004, LNAI 3398, pp. 122–130, 2005.

the development and evolution of HLA. HLA does not prescribe a specific implementation, nor does it mandate the use of any particular software or programming language. Over time, as technology advances, new and different implementations will be possible within the framework of HLA. The motivation behind HLA is a common architecture to meet new and changing user needs. Further, by standardizing only key elements of the architecture and not implementation, supporting software developments can be tailored to the performance needs of applications.

Time management components of HLA consist of logical time, local time and wallclock time. The time advance services provide a protocol for the federate and RTI to jointly control the advancement of logical time. The RTI can only advance the time constrained federate's logical time to T when it can guarantee that all TSO events with time stamp less than or equal to T have been delivered to the federate. At the same time, conservative federates must delay processing any local event until their logical time has advanced to the time of that event, while optimistic federates will aggressively process events and rollback when it receives a TSO event in its past and use T as an estimate of Global Virtual Time (GVT) for fossil collection [2]. Among time constrained federates, there are three subclasses of federates: (1) conservative time-stepped, (2) conservative event-driven, and (3) optimistic. For each of these federate subclasses, three time advance service of Time Advance Request(TAR), Next Event Request (NER) and Flush Queue Request (FQR) exist. A time management cycle consists of three steps. First, the federate invokes a time management service to request its logical time to advance. Next, the RTI delivers some number of messages to the federate. A message is delivered to the federate by the RTI invoking a federate defined procedure. The RTI completes the cycle by invoking a federate defined procedure called Time Advance Grant to indicate the federate's logical time has been advanced [3].

This paper presents an HLA-based Object-oriented modeling/Simulation for Tank-helicopter combat system (HOST) which can be efficiently used for evaluation of tactics and weapon system for tank/helicopter combat. HOST is a distributed combat (tank-to-tank, tank-to-helicopter) simulation based on High Level Architecture (HLA). We design and implement an object-oriented simulation engine which can provide an unified interface to a set of interfaces for entities in HOST by using facade pattern. Entities can be accessed only through access to the unified interface of facade pattern. We design a domain specific object model (DOM) for tank-helicopter combat simulation based on CMMS(Conceptual Model of the Mission Space) and UML (Unified Modeling Language). We shall show how CMMS can be used to aid the design of use case diagram in UML, and presents various diagrams such as use case, class, sequence, state chart diagram in UML, which help users develop the relevant models in HLA. We shall also show how to adapt the domain specific model to HLA by the efficient design of FOM(Federation Object Model) and TMM(Time Management Model). FOM defines the data model shared among the federates for data exchange. TMM provides time-stepped, event-driven and optimistic time models. In a graphic mode, inner federate time management (Inner-FTM) and inter federate time management (Inter-FTM) are used. Inner-FTM achieves synchronization by coordinating local objects and visualization, and Inter-FTM by coordinating federates and visualization respectively. We provides time-stepped, event-driven and optimistic algorithms in HOST.

The remainder of the paper is organized as follows: Section 2 explains system architecture of tank-helicopter combat system. Section 3 presents a domain specific object model using CMMS and UML as well as HLA-based model including FOM and TMM. Section 4 gives a conclusion.

2 System Architecture

The design objectives of HOST are to simulate the combat between tanks and/or helicopters for the analysis of weapon and combat efficiency resulting from the experiment of a mock tactics and the analysis of the tank efficiency by increasing the number of shell and varying a tank gun aperture.

HOST system architecture consists of main processor, user interface, preprocessor, and postprocessor as shown in Figure 1(a). The main processor at the core of the design consists of system object container and simulation engine. The system object container is composed of objects representing combatants such as tank, aircraft, and the simulation engine schedules events exchanged among objects. The preprocessor inputs scenario data, field data, entity data, and simulation data. The postprocessor stores data on the simulation history and produces the game results. The system operates in both graphical and text mode. While in graphic mode, it visualizes the combatants on the terrain in real-time, in text mode, it does not visualizes simulation situations. Four types of objects exist: two types of tank (CommandVehicle, FightVehicle) and two types of helicopter (ScoutHelicopter, AttackHelicopter). As time goes on, each object changes state, and simulation time advances.

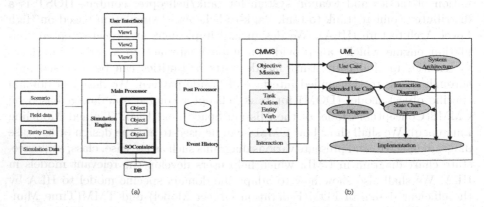

Fig. 1. (a) System architecture (b) CMMS transforming into UML.

3   Modeling

The model development process of a simulation package over HLA describes many important issues. The development experience suggests the method of development and can improve the sense of simulation. First, we suggest the domain specific object model using Conceptual Models of Mission Space (CMMS) and Unified Modeling Language (UML) [4]. Second, we provide HLA-based model using Federation Object Model (FOM) included in Federation Development and Execution Process (FEDEP) and Time Management Model (TMM). Simulation designers must have a clear understanding of the domain to be simulated in order to produce a model or simulation that is valid and sufficient for intended purpose or use. CMMS is a first abstraction of the real world and serves as an authoritative knowledge source for simulation development capturing the basic information about real entities involved in any mission and their key actions and interactions. We construct the use case through CMMS because we can acquire authoritative informations in military simulations. While the FEDEP provides guidance and assistance in federation design, federate design, and documentation for federations and federates, the FEDEP is not complete and its processes are not clear to the majority of users who are not fluent in the HLA development process. Because the FEDEP is not enough for simulation developers, analysts, military operator and user to understand and comprehend operations and activities in simulation, UML is complementary for visualizing, specifying, constructing, and documenting the artifacts of a system-intensive process.

3.1   Domain-Specific Object Modeling (DOM)

This section presents a domain-specific object model (DOM) using CMMS and UML. We shall show how to develop use cases based on CMMS.

Conceptual Models of the Mission Space (CMMS). Conceptual Models of the Mission Space (CMMS) are simulation implementation-independent functional descriptions of the real world processes, entities, and environment associated with a particular set of missions. It provides a common starting point for constructing consistent and authoritative Modeling and Simulation (M&S) representations, and to facilitate interoperability and reuse of simulation components. It also serves as a first abstraction of the real world and as a frame of reference for simulation development by capturing the features of the problem space or subject domain. Figure 1(b) shows how CMMS is transformed into UML. We first construct use case through objective and mission in CMMS. The use case is extended into extended use case by task / action / entity / verb / interaction in CMMS. Based on extended use case, we construct variant diagrams like class diagram, sequence diagram and state chart diagram. Finally, we implement classes, and design software architecture based on those diagrams.

Unified Modeling Language (UML). The UML is a modeling language for specifying, visualizing, constructing, and documenting the artifacts of a system. We shall show how to develop UML in HOST.

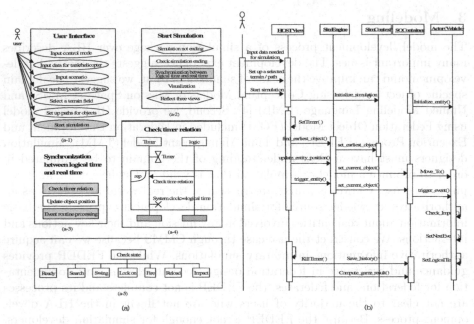

Fig. 2. (a) Usecase diagram (b) Sequence diagram.

Use Cases. Figure 2(a) shows a graphical representation for use cases in HOST [8]. Figure 2(a-1) shows user interface. User inputs control mode, entity data, scenario, number and position of objects and selects a terrain field, finally sets up a path. Figure 2(a-2) shows start simulation use case included in Figure 2(a-1) more in detail. When user starts a simulation, start simulation use case will iterate sequence: check simulation ending, synchronization between logical time and real time and visualization. Figure 2(a-3) shows synchronization between real time and logical time use case including check timer relation, update object position and event routine processing. Figure 2(a-4) shows a check timer relation use case in which rep represents repetition. After Timer starts, logic checks time relation. Figure 2(a-5) shows an event routine processing use case which processes a event after checking the object state. If wallclock time value is larger or equal to logical time, repetition ends [5].

Class Diagram. Figure 3(a) shows a class diagram which uses a facade pattern [9] which provides a unified interface to a set of interfaces in a subsystem. The facade pattern offers the benefits to shield clients from subsystem components, promote weak coupling between the subsystem and its clients. In Figure 3(a), CSOContainer is a facade. Each view visualizes information through CSOContainer. Also, the combat entity can be extended through aggregation to CSOContainer. This object-oriented architecture provides extensibility of combat objects. For example, if you want to add a missile, you only need to include the missile class with aggregation to CSOContainer. CSimEngine operates the simulation, and advances time by processing the earliest event using information

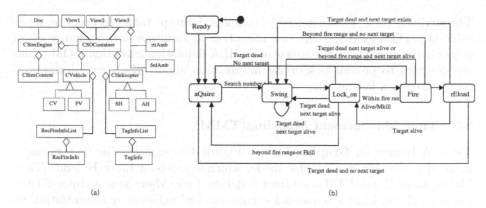

Fig. 3. (a) Class diagram (b) Statechart diagram.

stored in CSimContext. CVehicle and CHelicopter objects use two data structures: TagInfoList and RecFireInfoList storing enemy and impact information respectively.

Sequence Diagram. A sequence diagram describes dynamic behavior between objects by specifying the sequence of message generation. In a sequence diagram, objects are drawn up in order in vertical axis, and time concept is included in horizontal axis. Figure 2(b) shows a sequence diagram in HOST. User inputs data, selects a terrain and sets up a path, finally starts a simulation to HOSTView. SimEngine processes the earliest event, updates object positions, and advances time. SimEngine sends update/event messages to SOContainer. After SOContainer sends messages, the receiving object processes the event needed to the messages, and advances time. After ending a simulation, HOST saves data and computes game results.

State Chart Diagram. Figure 3(b) shows a state chart diagram which describes the dynamic behavior of object states. This diagram expresses all the states which an object can have, and the change of state when an event is processed. When the simulation starts, object is in R (Ready) state. R state is changed into Q (aQuire) state. Q state is changed into S (Swing) state when more than an enemy are detected. S state is changed into L (Lock_on) state when a target enemy lives. L state is changed into F (Fire) state within fire range. After F state, E (rEload) state is acquired. S/L/F/E states are changed into Q state when target enemy is dead or there is no additional target. The explanation about other state changes is written in Figure 3(b).

3.2   HLA-Based Modelling

Federation Object Model (FOM). According to U.S. DOD, the HLA object model template (OMT) prescribes the format and syntax for recording information in HLA object models, in which it includes objects, attributes, interactions, and parameters. We follow the HLA OMT to define our object models in HOST.

The attribute table, parameter table, complex datatype table, and routing space table of a FOM in HOST are also identified to facilitate data exchange of all object attributes during the course of an HLA federation execution. FOM tables are designed to provide descriptive information about all object informations represented in a federation.

3.3   Time Management Modeling(TMM)

Time Advance in Graphic Mode. Figure 4(a) explains the time management of a federate in a graphic mode, which consists of Inner Federate Time Management (Inner-FTM) and Inter Federate Time Management (Inter-FTM). Inner-FTM considers a sequential simulation and achieves synchronization by coordinating local objects and visualization in each federate. There may exist many entities in one federate. Each entity has its local time. Inner-FTM advances time by checking each entity's local time. In Inner-FTM, real time is synchronized with local time of objects. Inter-FTM achieves synchronization by coordinating federates using RTI TM API. RunTime Infrastructure (RTI) [6] TM API is to establish or associate events with federate time, interactions, attribute updates, object reflections or object deletion by federate time scheme, support causal behavior within a federation, support interaction among federates using different timing schemes.

Figure 4(b) shows the time-stepped algorithm in a graphic mode, where Timer( ) is called at regular interval in real time to perform visualization while processing local and remote events. It is composed of Inner-FTM, Inter-FTM and visualization. Inner-FTM processes the events of local objects and advances local object's time. Inter-FTM updates information of local objects and advances time synchronized with other federates and reflects information of remote objects. Visualization is performed in synchronization with Inner-FTM and Inter-FTM. find_earliest_object( ) is to find an object having earliest local time and trigger_event( ) is to process an event and advance the object's local time [7].

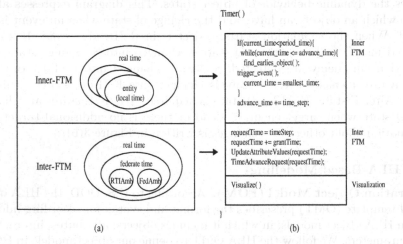

(a)   (b)

Fig. 4. Time management model and Time advance mechanism in graphic mode.

```
Time_Stepped( )
{
while(current_time<period_time){
    if(current_time - previous_time <=
            timeStep){
        previous_time = current_time;
    find_earlies_object( );
    trigger_event( );
        current_time = smallest_time;
    }
    else
      previous_time += timeStep;
    requestTime = timeStep;
    requestTime += grantTime;
    UpdateAttributeValues(requestTime);
    TimeAdvanceRequest(requestTime);
}
```
(a)

```
Event_Driven( )
{
while(current_time<period_time){
    if(Request_Flag == true){
    find_earlies_object( );
    trigger_event( );
    current_time = smallest_time;
    requestTime = current_time;
    UpdateAttributeValues(requestTime);
    NextEventRequest(requestTime);
    }
    else
      NextEventRequest(requestTime);
    If(requestTime > grantTime)
      Request_Flag = false;
    else
      Request_Flag = true;
    }
}
```
(b)

```
RollBack_Flag = fasle;
Federate::FlushQueueRequest( )
{
    while(current_time<period_time) {
        if(RollBack_Flag == true){
            Event_History_RollBack(rollback_time);
            Callback_History_RollBack(rollback_tme);
        }
        else {
        find_earlies_object( );
        trigger_event( );
        current_time = smallest_time;
        requestTime = current_time;
        Retract_Handle = UpdateAttributeValues(requestTime);
        Save_Event_History(current_time, state, Retract_Handle);
        FlushQueueRequest(requestTime);
        }
        queryLBTS(GVT);
        Delete_Before_GVT_In_Event_History( );
        Delete_Before_GVT_In_Callback_History( );

        If(requestTime > grantTime)
            Request_Flag = false;
        else
            Request_Flag = true;
        }
}
FedAmb::Reflect_and_Receive(RTIfedTime
callback_time, EventRetractHandle retract_handle,
char state)
{
Check_Flag = Check_Callback_Time( );
If(Check_Flag == false)
    Rollback_Flag = true;
else {
    Rollback_Flag = false;
Save_Callback_History(callback_time,
retract_handle, state);
    }
}
```
(c)

Fig. 5. Time management model: time-stepped, event-driven and optimistic time advance algorithms.

Time Advance in Text Mode. In text mode, three time advance schemes can be used: time stepped, event-driven and optimistic. Time-stepped time advance algorithm is shown in Figure 5(a). The variable timeStep is the value needed to advance time-stepped method. Event-driven time advance algorithm is shown in Figure 5(b). In Figure 5(b), Request_Flag is used to meet the case that the logical time of the federate may be advanced to the time stamp of the next TSO message that will be delivered to the federate. Request_Flag is set FALSE when request time is greater than grant time. In such case, the next TSO message less than request time exists. After Request_Flag is set FALSE, again NER is called.

Figure 5(c) shows an optimistic algorithm in HOST. We use two data structures: Event_History and Callback_History which stores local event time and retract handle and state respectively. Event_History stores local event informations, and Callback_History received informations from other federates. If time value in ReflectAttribute( ) and ReceiveInteraction( ) is less than the previous time stored in Callback_History data structure, Retract_Flag is set true. If Retract_Flag is true, rollback happens in both Callback_History and Event_History. If time advance is granted, all data in both Event_History and Callback_History less than GVT(LBTS) are removed.

4   Conclusion

This paper has presented the design and implementation of HOST using objected oriented and HLA concept: We have provided a DOM model using CMMS and UML, and a HLA-based model using FOM and TMM model respectively. DOM makes information exchanged easily among users, simulation developer, designers, and military operators. We have shown how CMMS can be used to aid the design of use case diagram in UML, and have presented various diagrams such as use case, class, sequence, state chart diagrams in UML, which help users develop the relevant models in HLA. We have shown how to adapt the domain specific object model to HLA by the efficient design of FOM. HOST can be used for training and analytic simulations, and operated in graphic and text modes. For each mode, TMM provides time-stepped, event-driven and optimistic time advance schemes. We have shown how Inner-FTM and Inter-FTM can be coordinated with visualization in graphic mode. The object-oriented design of our domain specific object model in HOST provides users with modification, extensibility, flexibility through abstraction, encapsulation and inheritance. Moreover, HLA-based model in HOST allows our system to be plugged in other HLA-compliant system, and extended into more complex system by providing reusability and interoperability. Moreover, we have presented two design patterns: the façade pattern which is used by object-oriented simulation engine to provide an unified interface to a set of entities in HOST, and the Model/View/Control pattern which is used for the implementation of MDI.

We believe that the design concept of HOST using CMMS, UML and design patterns as well as TMM for graphic and text modes in various time advance schemes can be efficiently exploited in the development of other HLA-based applications.

References

1. IEEE Standard for Modeling and Simulation (M&S), "High Level Architecture (HLA) Federate Interface Specification," IEEE Std 1516.1-2000
2. Richard M.Fugimoto: "Computing Global Virtual Time in Shared Memory Multiprocessors", ACM Transactions on Modeling and Computer Simulation, 1997.
3. Fujimoto, R.M: "Time Management in the High Level Architecture." Simulation Vol. 71, No.6, pp. 388-400, December 1998.
4. Larman, "Applying UML and Patterns - An Introduction to Object-Oriented Analysis and Design," Prentice Hall, 1997.
5. Bjorn Regnell: "A Hierarchical Use Case Model with Graphical Representation", IEEE, 1996.
6. T.B. Stephen, J.R. Noseworthy, J.H. Frank, "Implementation of the Next Generation RTI," 1999 Spring Simulation Interoperability Workshop.
7. T. McLean, R. Fujimoto, "Predictable time management for real-time distributed simulation," Parallel and Distributed Simulation, 2003. (PADS 2003). Proceedings. Seventeenth Workshop on , 10-13 June 2003 Pages:89 - 96

Research on Construction and Interoperability of Complex Distributed Simulation System

Ning Li, Zhou Wang, Wei-Hua Liu, Xiao-Yuan Peng, and Guang-Hong Gong

Advanced Simulation Technology Aviation Science and Technology Key Laboratory,
BUAA, 100083 Beijing, P.R. China
lining_buaa@hotmail.com

Abstract. This paper presents an assumed interoperability framework of Complex Distributed Simulation System (CDSS). On the basis of it, interoperability conceptual model of C4I simulation system, distributed weapon platform Modeling and Simulation (M&S) system and Synthetic Natural Environment (SNE) is proposed. Key technologies such as C4I modeling and simulation, SNE modeling to construct CDSS and achieve interoperability are discussed. The interoperability Data Interchange Format (DIF) based on HLA OMT, called CDSS FOM, is given and introduced, which is applied in the Attack-Defense Simulation System.

1 Introduction

The battle system is a typical complex system, which is occurred at a certain natural, geographical, political and historical environment. The factors of weapons, equipments, units, strategic purposes and tactics ideas restrict and affect each other. A great deal of information flows and physical flows are interwoven. Effect relationships among them are complex. The battle environment simulation system is a typical complex distributed virtual battlefield simulation environment, which needs to integrate interoperable and reusable simulation applications among different communities, such as C4ISR community and M&S community, functions, such as share ability between battle and acquisition, simulation types, such as virtual simulation, construction simulation, live simulation, and multi-resolution simulation based on standard support platform and advanced simulation technology to realize simulation of battlefield environment and battle process under a distributed network environment.

As the military strength multiplier, C4ISR can increase operational effectiveness on the whole. Natural environment has an important impact on efficiency of weapons and human decision-making behavior. A large-scale virtual battlefield simulation should include real C4ISR system or C4ISR simulation system and natural environment simulation system. At present, as the development of advanced simulation technology, the research on the interoperability among C4I, Synthetic Natural Environment (SNE) and weapon platform modeling and simulation (M&S) system spreads deeply. US DoD C4ISR community has developed Defense Information Infrastructure Common Operating Environment (DII COE), M&S community has published M&S Master Plan and established HLA simulation standards, C4ISR/Sim community has presents Technical Reference Model (TRM), which is intended to serve as a framework for the interoperability of C4I and M&S system, to promote development of requirement analysis, design, realization and test between heterogeneous systems.

D.-K. Baik (Ed.): AsiaSim 2004, LNAI 3398, pp. 131–140, 2005.
© Springer-Verlag Berlin Heidelberg 2005

In recent years, DMSO has a remarkable development on natural environment modeling and simulation. SNE Conceptual Reference Model has been presented, Synthetic Environment Data Representation and Interchange Specification (SEDRIS) project has been developed. The above efforts make foundation for interoperability of these information fields.

2 Research on Interoperability of Complex Distributed Simulation System

2.1 Interoperability Framework of Complex Distributed Simulation System

The TRM is very similar to the SNE Conceptual Reference Model developed by Birkel at 1998. SNE Conceptual Reference Model is more focused on functionality and environmental effects but still is oriented towards interfacing to weapons and C4I systems. TRM is more focused on information exchange. The SNE Conceptual Reference Model authoritatively extends the description of those classes in the TRM that deal with the environment. SEDRIS provide a method to express environment data, but neither the simulations nor the C4ISR systems use the SEDRIS format.

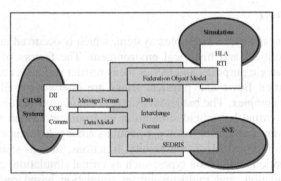

Fig. 1. Is a primary assumption framework to realize interoperability among three heterogeneous information systems i.e. C4ISR system, platform M&S system and SNE. The simulation standards architecture is the HLA and its standard for representation is the FOM. The C4I standards architecture uses the DII COE and its standard for representation is either message formats (VMF, USMTF, etc.) or data models. SNE adopts SEDRIS specification. The differences of formats are unfavorable to exchange data and reuse among heterogeneous systems. A standard and uniform Data Interchange Format (DIF) can solve this problem

As shown in Fig.1, the complex system is composed of distributed systems based on different architecture frameworks and standards, including weapon platform entities/systems, C4ISR system, natural environment, human and their interactions, having a large amount of elements and giving priority to simulation applications. In this paper, the integration of information systems with these characteristics is called Complex Distributed Simulation System (CDSS).

In a CDSS, DIF can be standardized and described as the style of HLA OMT (FOM) when HLA application is the main application. HLA OMT DIF uses keywords and nested parenthesis to describe Federation Object Model (FOM), Simulation Object Model (SOM) and the interchange of data via RTI when a federation is executed.

When other systems are integrated to HLA simulation, the exchange data can be unified and described as HLA OMT.

XML is a standardized language. XML and HLA OMT DIF can be translated into each other. We can also use XML to describe FED, OMT DIF, battle document, scenario, and HLA Object etc.

2.2 Interoperability Conceptual Model of CDSS

A typical C4I simulation system is composed of Exercise Control Module, C4I oriented Visualization Module, Knowledge/Rules Database, Scenario Database and Multi-level C4I Simulation & Models Module. C4I Module includes Physical Models, Sensor/Detector Models, Human Decision-making Behavior Models, Command & Control Models and Communication Models.

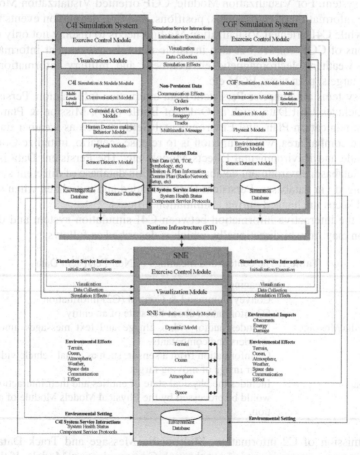

Fig. 2. On the basis of the above framework, real C4I system is modeled in this paper. According to C4ISR/Sim TRM and SNE Conceptual Reference Model, an Interoperability Conceptual Model of integration of C4I simulation system, distributed simulation system and SNE simulation system is presented, which is shown in Fig. 2. The Interoperability Conceptual Model is also valuable as a reference model for real C4I, M&S and SNE systems

M&S system model takes example for a typical Computer Generated Forces (CGF) simulation, which includes Exercise Control Module, CGF oriented Visualization Module, Simulation Database and Multi-resolution CGF Simulation & Models Module.

SNE simulation system is composed of Exercise Control Module, Natural Environment Visualization Module, Environment Database and SNE Simulation & Models Module, which includes Dynamic Model and four SNE fields: Terrain, Ocean, Atmosphere and Space.

The developing environments of three simulation systems are different. The running support environment of CDSS is HLA RTI. As shown in Fig.2, the common interactions among simulation systems are Simulation Service Interactions, which includes Initialization and Control of Simulation Execution, Visualization, Data Collection and Simulation Effects. Subscribing and publishing information differs in different system. For Visualization Module, CGF oriented Visualization Module only subscribe information of entity states, positions, and fire, detonation events in its own system, while C4I oriented Visualization Module needs to display not only states and interactions of CGF entities but also invisible natural environment information such as wind, weather, electromagnetic environment and military information such as minatory targets, plan view, ECM etc.

Three systems all comprise Persistent Data interaction information. Persistent Data usually includes Unit Data, such as Order-of-Battle (OB), Mission & Plan Information, Communication Plan, and Environmental Setting, such as weather data, terrain data in the combat area, which function is to release scenario, initialize Communication Module and Environmental Effects Models etc. The Persistent Data is saved in scenario database, environment database, simulation database of different systems and be exchanged via databases. Persistent Data is loaded and initialized when simulation starts and keeps unchanged during a simulation process.

All of the interactive information between C4I simulation system and distributed simulation may flow bi-directionally, which is described as follows.

Table 1. The content and meanings of Non-Persistent Data

Content	Meanings
Orders	Convey Command & Control (C2) information
Reports	C2 information about the state of an entity
Multimedia Message	Includes audio, video, image and text messages among commanders and combatants.
Imagery	C2 information from a sensor, such as aerial vehicle video or radar image of moving target
Track Data	Includes the physical state of entities and their interactions which would be processed by the Physical Models Module of a simulation.

Transmission of C2 information, Multimedia Message and Track Data between and within two systems is carried out through Communication Models. If the simulation offers a high degree of fidelity in C2, it may associate these interactions with Communication Effects. Orders, Reports and Imagery from C4I simulation system would interface to Behavior Models of a CGF simulation, affecting the decision making of simulated battle units.

The interoperability of distributed simulation system and SNE simulation system embodies the following two aspects:

Environmental Effect: A direct influence of the environment on sensors, weapons, and/or units/platforms (e.g., ducting of acoustic energy by ocean's vertical density structure; impaired tank movement due to rain-soaked soil; influence of dynamics performance of aerial vehicle due to wind field).

Environmental Impact: A direct influence on the environment caused by sensors, weapons, and/or units/platforms (e.g., CGF bomb crater on runway, cooling of atmosphere due to smoke from burning tanks, Obscurants).

Besides, there are interactions among the four fields of SNE (e.g., influence of ocean wave height by surface wind change).

The interoperability of C4I simulation system and SNE simulation system is similar. We mainly consider Environmental Effect: A direct influence of the environment on sensors (active and passive), ECM, units and commanders (e.g., ducting of acoustic energy by ocean's vertical density structure; influence of sensors performance by atmosphere, ocean, terrain and electromagnetic environment; weather environment affects behavior of commanders).

3 Key Technologies to Construct CDSS

3.1 C4I Modeling and Simulation

There are two ways for C4I modeling and simulation, one way is from bottom to top, starting with microcosmic segments modeling, then integrating, running and analyzing; another way is from top to bottom, describing macro-system based on widely used Lanchester equation.

We use the two methods for C4I system modeling and simulation. The two methods are applied to different C4I simulation levels: battle unit level and battle command center level.

No matter C3I, C4I or C4ISR, the kernel is Command & Control (C2), so we discuss C4I modeling and simulation of battle unit level taking example for C2 modeling.

3.1.1 Conceptual Model of C2 Object

On referring to Joel S. Lawson's C2 Process Conceptual Model presented in the mid-1970s, we present a universal Conceptual Model of C2 Object as shown in Fig.3. C2 model is an independent and reusable object component module for multi-levels of battle units, which has a strong interoperability with battlefield, natural environment and other superior/inferior C2 objects.

Command and Control: includes several independent but interrelated modules which are Information fusion, Object recognition, Threat assessment and Target assignment; receives external information which is subscribed from self and enemy CGF system, objects in C4I simulation system, other external intelligence, superior assignments/orders, natural environment information; releases corresponding C2 orders and reports.

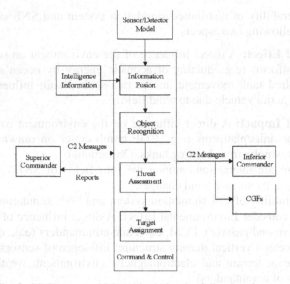

Fig. 3. C2 Conceptual Model includes the following three parts

Interface to superior C2 objects
Interface to inferior C2 objects and CGF battle units

Communication system and Data Link modeling will not be discussed in this paper owing to the restriction of pages.

3.1.2 Battle Command Center Level C4I Modeling

For battle command center level C4I modeling of CDSS, we mainly consider influence of C4ISR, information fusion and natural environment on Lanchester equation.

Influence of C4ISR on Lanchester Equation: Typical Lanchester equation considers that military strength of each side is equal to its operational effectiveness and does not consider "multiplier" function of C4ISR system. Scholars such as Taylor and Dockery etc. extend Lanchester equation, modify its coefficients and add influence of different weapon platforms, C2, communication and reconnaissance on it. It is formulated as

$$\frac{dx}{dt} = -F_y \left[\prod_{k=1}^{n} \frac{R_{ky}^2}{\pi (k_{3y} R_{ey})^2} y \cdot \frac{k_{1y}}{k_{2y}} \right] \tag{1}$$

$$\frac{dy}{dt} = -F_x \left[\prod_{k=1}^{m} \frac{R_{kx}^2}{\pi (k_{3x} R_{ex})^2} x \cdot \frac{k_{1x}}{k_{2x}} \right] \tag{2}$$

where F_y is the shoot rate of y; R_{kx} is the kill radius of weapon k, which belongs to military strength x; R_{ky} is the kill radius of weapon k, which belongs to military strength y; R_{ex} is the surveillance radius of military strength x; R_{ey} is the surveillance radius of military strength y; K_{1x} is the amendatory coefficient of reconnaissance ability improvement of military strength x; K_{1y} is the amendatory coefficient of reconnaissance ability improvement of military strength y; K_2 and K_3 is the loss coefficient of reconnaissance network latency.

Influence of C4ISR Multi-sensor Information Fusion on Lanchester Equation:
Yin Hang presents an improved Lanchester equation which reflects the influence of
C4ISR multi-sensor information fusion system efficiency. Suppose that multi-sensor
information fusion system directly increase self field system efficiency or decrease
that of the opposing site, and then it is given by

$$\frac{dx}{dt} = -a\left[k_{Cx}k_{Bx}\left(1-e^{-P_{dx}}\right)\left(1-e^{-P_{tx}}\right)\left(1-e^{-P_{fx}}\right)P_{jx}\left(e^{-\sigma x} + P_{ix}\right)e^{\frac{1}{tx}}\right]y \tag{3}$$

$$\frac{dy}{dt} = -b\left[k_{Cy}k_{By}\left(1-e^{-P_{dy}}\right)\left(1-e^{-P_{ty}}\right)\left(1-e^{-P_{fy}}\right)P_{jy}\left(e^{-\sigma y} + P_{iy}\right)e^{\frac{1}{ty}}\right]x \tag{4}$$

where K_c is the communication condition coefficient; K_B is the counter/anti-counter
coefficient; P_d, P_t, P_f is the space, time and frequency domain cover rate; P_i is the
probability of object recognition, P_j is the probability of object detection, σ is aim
precision (variance of detecting value and real value), t is system response time.

Influence of Natural Environment on Lanchester Equation: Concerning influence
of natural environment on Lanchester equation, we map the influences of natural
environment on weapon platforms/entities, units into the corresponding coefficients
instead of adding new modification coefficients. (e.g. natural environment has an
influence on detecting distance of sensor/detector and broadcasting of communication
signal, so R_e, K_c will be affected. The above two parameters can be calculated in CGF
models respectively, together with other coefficients, applying to solving Lanchester
equation.)

3.2   SNE Modeling and Simulation

Considering characteristics of SNE data model, we use a method which combines the
Object Oriented technology and the data field concept to construct Environment Data
Model (EDM). Natural environment can be divided into two types according to its
distributed characteristic in space dimensions. One type of environment object such as
atmosphere, ocean and space, which attributes (e.g. atmosphere temperature/humidity,
seawater salinity, wave height) distributes continuously in time/space, is considered
as variable data field in substance. Another type of environment object which has
clear boundary, integrated function and distributes discretely in space, mainly exists
in terrain such as tree, building, ground vehicle etc. This type of object is independent
individual and we use Object Oriented method to model it.

Terrain environment modeling is realized by expanding OpenFlight format terrain
database to meet the needs of visualization, sensors and CGFs. Using OpenFlight
API, users can define own data types and save/load them to/from OpenFlight database. Generally, mathematic model of atmosphere and ocean environment is higher-
order differential equation set. It is impossible to be calculated in real time using the
current micro-computer. It is expressed as 2D, 3D or multi-dimensional gridded data
and saved in database instead. The gridded data is generated off-line via environmental mathematic model or comes from observational data and is loaded when running. We can use numerical value atmosphere model (e.g. NCAR MM5 model) to
generate forecasting data and simulate atmosphere environment.

SEDRIS Environment Data Coding Specification (EDCS) defines the sort, attribute, data allowable value of EDM objects. EDCS provides methods to classify environment objects and clarify their attributes. The steps to achieve interoperability between SNE and other systems are: first, convert EDCS into HLA OMT (FOM); and then design FOM/SOM for the SNE.

3.3 Some Problems on CDSS FOM Design

To realize interoperability of CDSS, a unified DIF should be mapped out. CDSS FOM is defined on the basis of interoperability framework and interoperability conceptual model. The synthetic FOM is composed of the following three parts: platform distributed simulation system FOM, C4I simulation system FOM and synthetic natural environment system FOM. To sum up, certain key areas are identified, including platform entity, aggregate entity, C4I systems, communications device, communications network, detecting information of sensors, C2 orders, multimedia messages like secret audio, video, imagery and text, intelligence information and natural environment information. These areas formed the basis for deriving the object classes, attributes, and interactions of CDSS FOM.

The design of CDSS FOM embodies the idea of component and reuse. A series of Base Object Models (BOM) are designed and integrated, some authoritative and widely used FOMs are adopted for reference at the same time. These BOMs/FOMs are: Scenario Description BOM, C2 BOM, Data Link BOM, Environment Federation BOM, Multimedia Communication BOM, RPR-FOM and SISO C4I FOM, ISR FOM etc.

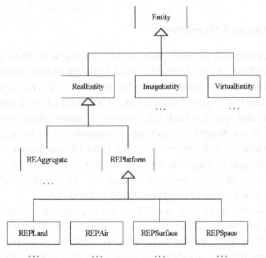

Fig. 4. CDSS FOM object class has a hierarchical structure, which is shown in Fig.4. *Entity* is the base class of entity object; its child classes include *RealEntity* (real entity), *ImageEntity* (mirror of real entity), *VirtualEntity* (virtual entity of a simulation e.g. virtual observer). *RealEntity* is classified as *REAggregate* (aggregate entity) and *REPlatform* (platform entity). *REPlatform* is classified as *REPLand* (land-based entity), *REPAir* (air-based entity), *REPSurface* (surface-based entity), *REPSpace* (space-based entity). These classes are then classified in more detail

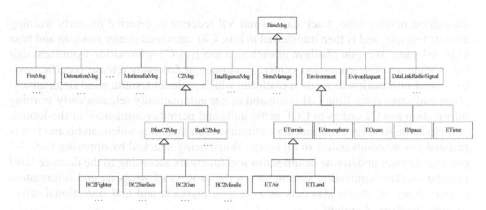

Fig. 5. C2 information of non-persistent data such as orders, reports, imagery and other information such as intelligence information, reconnaissance information, multimedia message, natural environment information is modeled as Interaction BOMs. The structure of CDSS interaction class is shown in Fig.5, which includes *BaseMsg* (base class of interaction class), *FireMsg* (interaction class for fire event), *DetonationMsg* (interaction class for detonation event), *MultimediaMsg* (multimedia message), *C2Msg* (C2 information), *Environment/EnvironRequest* (natural environment interaction classes), *DataLinkRadioSignal* (interaction class for data link radio signal) etc.

In CDSS FOM, we design the following two schemes to express detecting target object class in C4I simulation system and CGF simulation system.

Use *ImageEntity* Class to Describe Detecting Target Entity: Use *RealEntity* class to describe CGF entity; use *ImageEntity* class to describe detecting image of real CGF entity, which is likely to be an illusion. This scheme has a clear entity object class structure; however, it makes the structure of FOM huge.

Adhere to Real Entity Object Class; Use Certain Attribute Value of Object Class to Describe Multi-source Detecting Target: Attributes of *Entity* class are expressed as *Entity{ForceId, EntityType, EntityID, Position, Velocity, Orientation, Appearance, EntityMarking, CurrentTimeStamp}*. Object class of multi-source detecting target adheres to real entity object class and it is distinguished and classified as different sites / detecting target via different sensor types / detecting target of the same sensor type but via different detectors by attribute *ForceId* (e.g. blue, red, neutral, radar detecting, early warning aircraft detecting, satellite detecting, radar local fusion, early warning aircraft local fusion, satellite local fusion, fusion center result etc.). The definition of *ForceId* can be extended according to the scale of simulation system. Because detecting target object is wrapped in CGF entity object class, this scheme has a compact structure and can be better reused compared with the above scheme.

4 An Example of CDSS FOM Application

CDSS FOM is applied in the Attack-Defense Simulation System. As shown in Fig.6, the simulation system is a HLA-compliant system developed to validate the performance of red weapon platform virtual prototyping (VP) integration system. During simulation running time, track data of red VP federate is detected by early warning aircraft federate, and is then transferred to blue C4I command center federate and blue CGF federates. Weapon platform of each side receives C2 information from local unit C2 module and C4I command center. Environment federates release specific environmental information to weapon platform, which includes terrain, wind field, atmosphere and space data. Blue C4I command center automatically releases early warning information and C2 orders to CGF battle units and permits commander-in-the-loop to participate in. Real time multimedia communication including video, audio and text is realized among commanders of all levels. When being attacked by opposing site, C4I command center updates its health status and functions according to the damage level calculation. This simulation system shows the process of weapons and information countermeasures, which also acts as a weapon platform and C4I operational effectiveness assessment system.

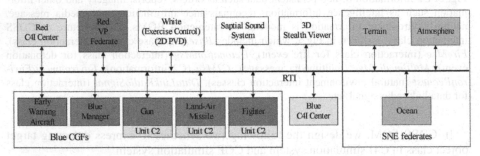

Fig. 6. Architecture for Attack-Defense Simulation System

5 Conclusion

On the basis of CDSS interoperability conceptual model, we accomplish interoperability DIF: CDSS FOM. It is a reference FOM, which research is likely to be a helpful foundation for interoperability and integration of real or simulated C4ISR system, weapon platform M&S system and SNE. CDSS FOM itself needs to be developed, extended and certificated by standard department.

References

1. Birkel, P.: SNE Conceptual Reference Model. Fall Simulation Interoperability Workshop, SISO Paper 98F-SIW-018 (1998)
2. Donald H. Timian, etc.: Report Out of the C4I Study Group. Fall Simulation Interoperability Workshop, SISO Paper 00F-SIW-005 (2000)
3. Li Ning Gong Guanghong, Peng Xiaoyuan: Research on C4I Simulation in Distributed Simulation System. The Fifth Symposium of Instrumentation and Control Technology SPIE Proceeding Vol.5253 (2003) 848-852
4. Yin Hang, Li Shaohong: The model of Multisensor Data Fusion effect in base of Lanchester. Journal of System Simulation,Vol. 16 No. 3 (2004) 411-412
5. John D. Roberts, Verlynda Dobbs. Application of a C4I Reference or Starter FOM to an Existing Simulation Environment. Fall Simulation Interoperability Workshop, SISO Paper 00F-SIW-067, 2000.

A Flexible and Efficient Scheme
for Interest Management in HLA

Changhoon Park[1], Koichi Hirota[1], Michitaka Hirose[1], and Heedong Ko[2]

[1] The University of Tokyo, 4-6-1 Komaba, Meguro-ku, Tokyo 153-8904, Japan
{wawworld,hirota,hirose}@cyber.rcast.u-tokyo.ac.jp
[2] KIST, 39-1 Hawolgok-dong, Seongbuk-gu, Seoul 136-791, Korea
ko@imrc.kist.re.kr

Abstract. In this paper, we present a new scheme to improve both the flexibility and efficiency of interest management for scaling up distributed interactive simulation. Our approach is explained by describing Area of Interest (AoI) update and Degree of Interest (DoI) estimation. We introduce the concept of resolution into the existing spatial model and propose a new relevance model named dynamic resolution spatial model (DRSM). And, a 2-tiered process is presented which based on the proposed model and the existing grid model. Our scheme will enable an interest manager to support various applications and sophisticate strategy for large-scale DIS.

1 Introduction

With the recent increases in network bandwidth and graphics performance for personal computers, there is a growing interest in distributed interactive simulation (DIS) that allow multiple simultaneous participants to interact in a synthetic battle-space [1,2,3]. This environment will enable geographically dispersed participants to interact not only virtual space itself but also other participants just as if they were in the same physical space. DIS has the potential to offer an attractive infrastructure for variety of collaboration applications including, military and industrial team training, collaborative design and engineering, virtual meetings, and multiplayer games [1].

In order to achieve the impression of a single shared virtual world with all participants, DIS should maintain the consistency of the virtual world database managed by participating hosts. So, every virtual world is updated not only its local player but also other remote participants. Therefore, each host transmits state update messages to other hosts through network whenever it's virtual space changes. Moreover, a specific applications area poses a large-scale DIS compensating hundreds of thou-sands of participants. As the number of participants in a DIS increases, the state update exchanged among participants' hosts increases to an enormous level. The increased computation overhead occurs at each host for the processing the incoming state updates. Furthermore, the communication networks (WANs and LANs) sup-porting DIS is overwhelmed by the resulting traffic load. The limitation of network bandwidth makes the scale up effort more difficult. For the scalability, it is widely believed that DIS will have to employ bandwidth reduction techniques such as relevance filtering, dead reckoning, packet compression and aggregation and so on [2,3,4].

The rest of the paper is organized as follows. The next section contains a summary of the concept of interest management and relevance model. Section 3 de-scribes our approach of proposed scheme. The proposed model and 2-tiered process are described

D.-K. Baik (Ed.): AsiaSim 2004, LNAI 3398, pp. 141–149, 2005.
© Springer-Verlag Berlin Heidelberg 2005

in Section 4. Section 5 describes experimental results our scheme. Finally, a brief summary and conclusion appear in Section 6.

2 Related Work

This section describes interest management and relevance model. The role and function of interest management is explained by introducing the concept of neighborhood. And, we define AoI and DoI to introduce relevance model. Then, the spatial model and the grid model are presented as examples of existing relevance model.

2.1 Interest Management

For the scalability of DIS, the concept of interest management was developed to address the explosion of state update messages. In traditional DIS such as SIMNET, each host broadcasts state update messages to the rest player so that the overall consistency of the shared virtual world is maintained. This means that the amount of state update increases polynomial to the number of participants. But, in some experiments [2], as much as 90% of state update is useless for the receiver. Namely, this approach wastes a significant amount of network bandwidth for unnecessary consistency among virtual worlds of distributed hosts. To overcome this problem, interest management reduces the destination of state update messages to a smaller relevance set of hosts.

Fig. 1. DIS can restrict the scope of the consistency by using the neighborhood determined by an interest manager. Let N(pi) denote the neighborhood of player pi. For an arbitrary player, its state update messages are transmitted to not the rest player but its neighborhood. In the worst case, every player's neighborhood is the rest of players. This is the same result when interest management is not used. On the contrary, the best case is that every player's neighborhood is empty set. Therefore, the size of neighborhood is very important factor for the scalability of DISs.

In this paper, neighborhood is defined as a set of participants whose state update is necessary for an arbitrary player. The request of state update is dependent on DoI between arbitrary participants. DoI quantifies the relevance and is defined by the relevance model which is described in the next section. This approach may be similar to the human perceptual limitation in real world. Therefore, the introduction of interest management makes the exchange of state update message for the consistency depends on not the overall number of player but the size of neighborhood. We believe that interest management is a key teahouse to reduce the amount of state update in a DIS gracefully. But, the load of interest management itself should be considered at the same time because this also can reduce the scalability irrelevant to the original intension.

2.2 Relevance Model

Relevance model defines AoI and DoI for interest management. AoI consists of focus and nimbus to represents a player's interest. Focus describes the observer's allocation of attention, while nimbus describes the observed object's manifestation or observability. And, DoI quantify the relevance that one player has of another and can be defined as a function in terms of each player's focus and nimbus.

At present, there are two popular relevance models:the grid model and the spatial model. The spatial model was previously proposed and developed by [5], which aims to provide key abstraction for mediating distributed interaction among participants in DISs [5]. This model was demonstrated in various CSCW systems such as MASSIVE, DIVE, CyCo and so on [1]. And, the grid model has investigated as an approximation of the spatial model. Most of all, military systems such as SIMNET, ModSAF and NPSNET are based on the grid model [4,6,7.8]. For example, below figure depicts the focus and nimbus of each model. The spatial model is an entity-based and the grid model region-based.

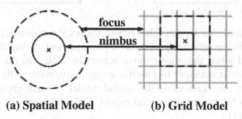

(a) Spatial Model   (b) Grid Model

Fig. 2. Focus and nimbus for Spatial Model and Grid Model:In the spatial model, focus or nimbus can be represented as a circle whose center is the same as the location of an avatar. And, DoI is represented by means of intersection calculation between focus and nimbus. On the other hand, the gird model partitions the virtual space into grid cells so that a player must be included in an arbitrary cell. and the value of DoI is determined according to whether nimbus cells include focus cell or not. If focus is the same as nimbus, DoI is symmetric.

3 Overview

In this section, we introduce our approach of the proposed interest management scheme by means of AoI (Area of Interest) monitor and DoI (Degree of Interest) estimation. The goal of the proposed scheme is to enhance the efficiency as well as the flexibility of interest management.

3.1 AOI Monitor

The AoI is represented as a sub-region in an arbitrary domain for the player's inter-est and can be updated with the interaction between the player and the virtual space. Therefore, an interest manager should keep track of every player's AoI in order to reflect updated AoI whenever the neighborhood is decided. At this time, the flexibil-ity of interest management is closely related to whether the size and location of AoI can be changed dynamically or not. And, the efficiency of interest management can be measured in terms of the total number of AoI Update messages transmitted to an interest manager.

In the case of using the grid model, the AoI is represented in terms of the grid cell that is pre-defined to partition the virtual world. And, the spatial model allows defining not only the location of AoI but also its size individually. Therefore, the spatial model is more flexible than the grid model from this point of view. We believe that the flexibility is important to introduce sophisticated strategies for interest management.

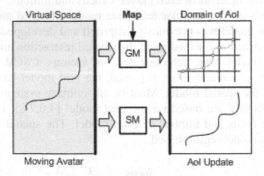

Fig. 3. AoI Update and State Update:This figure shows the transmission of AoI Update and State Update to explain the efficiency of interest management. Using the grid model, AoI Update only is generated when the player move across its cells. On the other hand, the spatial model requires an interest manager to monitor every movement of the player. Therefore, the number of AoI Update messages using the spatial model is much greater than that of the grid model. So, this makes the use of the spatial model difficult practically for large-scale DISs in spice of its flexibility [1].

This paper will propose a new relevance model named dynamic resolution spatial model (DRSM) by introducing the concept of resolution in order to overcome the overload of spatial model for monitoring every movement of the player. The concept of resolution provides an interest manager with the control of the fidelity of AoI monitoring. Namely, it can be possible to make the occurrence of DRSM's AoI updates similar with that of grid model's AoI updates. We believe that a new relevance model will improve the efficiency with keeping the flexibility of the spatial model.

3.2 DOI Estimation

The role of interest management is to decide the neighborhood in order to reduce the amount of state update message among participating participants. To achieve this, an interest manager not only keeps track of AoI update of every player but also estimates DoI between participating participants. Whenever a player transmits an AoI update, an interest manager estimates every DoI between a corresponding player and the rest of participants.

As the grid model is a kind of region-based model, it calculates the DoI by means of a cell of a grid logically overlaid on the virtual environment as mentioned before. It means that DoI calculation depends not on a player but on the cell in which a player locates and the relationship among cells. On the other hand, the spatial model is a kind of entity-based model which requires one-to-one DoI calculation for every two participating participants.

Therefore, the load of DoI estimation of the spatial model is more expensive than that of the grid model. Although, the spatial model enables to estimate accurate DoI for complex strategies, the spatial model requires the load of DoI computation which increases according to the number of participating participants. On the other hand, the grid model requires relatively inexpensive load of DoI computation with simple comparison and constant operations. But as the value of DoI of the gird model is 0 or 1, this means that there are limitations of application.

This paper proposes 2-tiered process in order to support both the accurate and efficient DoI computation by using both a region-based model and an entity-based model. In the first phase an interest manager is based on the grid model and decides the candidate with low load. And, the second phase decides the neighborhood by applying the proposed dynamic resolution spatial model (DRSM) into the candidate. Therefore, the high-cost and accurate DoI computation depends on the number of not overall participants but the candidates.

4 Proposed Scheme

This paper presents a new scheme for both flexible and efficient interest management. This scheme will enable to wide applications area and facilitate to adopt sophisticated strategies for large-scale DISs. In first, dynamic resolution spatial model (DRSM) is proposed. Then, we present 2-tiered process based on the grid model and the proposed DRSM at the same time.

4.1 Dynamic Resolution Spatial Model

This paper proposes a new relevance model named Dynamic Resolution Spatial Model (DRSM) by introducing the concept of resolution. The spatial model requires the player's host to generate AoI Update whenever an avatar moves. This means that an interest manager should keep track of every movement of the player for interest management as described before. To resolve this inefficiency, the resolution is introduced into the spatial model in order to limit the generation of AoI Update against the moving avatar.

The DRSM defines the resolution as a parameter of a function to map the moving avatar within a virtual space into the AOI within an arbitrary domain. In this mapping function, resolution allows to control the generation of AoI Update against the movement of an avatar. Therefore, the resolution enables an interest manager to control the fidelity and cost of AoI monitor.

In this paper, the resolution is applied to the distance of a moving avatar. So, every player generates AoI Update when the distance between an avatar's current location and the location at the previous AoI Update is greater than the resolution. Below figure shows the concept of resolution.

But, the introduction of resolution raises an error by using the existing DoI estimation of the spatial model, as it should make uses of the exact location of avatar. When a player generates AoI Update, an interest manager estimates DoI with the rest participants. At this time, an interest manager cannot know exact location of the rest participants. So, the resolution of the rest of the player should be considered to estimate DoI as following expression. Although this expression may include unnecessary

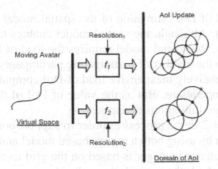

Fig. 4. AoI Update and moving avatar (Resolution1 < Resolution2):AoI Update is generated at filled circles against a moving avatar. Within the domain of AoI, AoI Update is generated when an avatar locates on the boundary of a circle whose radius means the resolution. Therefore if the resolution decreases, the loads of AoI monitor increases with the number of filled circle for the same moving avatar. After all, the resolution is closely related to AoI monitor.

participants, neighbor participants should not be excluded. And, they can be considered as a kind of candidate.

$$DoI_{DRSM}(X, Y) = 1 - \left[\frac{dist(X, Y) - Y.resolution}{X.focus + Y.nimbus} \right] \qquad (1)$$

It should be noted that there may be popping situation that the elements of neighbor change immediately if the resolution increases. This is the same reason with the boundary of the grid model. But, we can control this by the focus and resolution.

4.2 2-tiered Process

Although DRSM enable to control the fidelity of AoI monitor by means of the resolution, the load estimating DoI is remained yet. Whenever a player's host transits the AoI Update, an interest manager should estimate DoI with the rest player. This means that one-to-one DoI estimation is required to decide the neighborhood for each player because DRSM is a kind of entity-based model. And, this load is more computationally expensive than that of the grid model which is a kind of the region-based model. For the efficient the DoI estimation of our scheme, this paper presents a 2-tiered process for interest management as illustrated in figure.

After all, this paper proposed a new scheme which makes the interest management efficient and flexible with taking advantages of entity based model and region based model. In other words, our proposed scheme solves the inefficiency of spatial model and the inflexibility of the grid model.

5 Experimental Results

In this section we compare the cost of spatial model, grid model and our scheme in terms of AoI Update and DoI estimation. Before describing experiments, we define a simple model of avatar's behavior within DISs. It is necessary to simulate the movement of the avatar in order to evaluate interest management for large-scale DISs which compensate thousand numbers of participants. This model is defined in terms

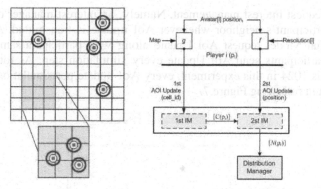

Fig. 5. 2-tiered process for Interest Management:For the efficient the DoI estimation of entity-based model, the first phase is to decide a set of candidates from all participants by using a grid model which is a kind of region-based model. Then, a set of neighbor is decided from candidates in the second phase by using an alternative of spatial model which is a kind of an entity-based model. Our scheme aims to support various areas of applications with a flexible control of AoI (Area of Interest) and avoid excessive overload of interest management itself by supporting a different level of QoS (Quality of Service) for monitoring AoI.

of the direction and distance as following figure 6. Although this movement does not mean that of real player's movement, we can simulate concrete and uniform movement by controlling the average and diversion of possibility distribution function. The movement is generated by means of applying two random number generators which are based on the standard normal distribution and gamma distribution into the direction and distance of the model respectively.

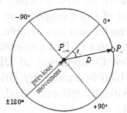

Fig. 6. Our behavior model is defined in terms of direction and distance. The direction depends on the previous movement. If the value of direction is 0 then an avatar moves to the same direction to previous movement. And, if the value is -90 or +90, then an avatar moves left or right. Distance is the same meaning with the name.

After presenting behavior model, spatial model, grid model and our scheme should be normalized to obtain meaningful experimental results. The output of inter-est management is neighbor. So, it is possible to make the result of the participants' neighbor similar by adjusting each model's AoI. Then, we will compare their cost for interest management. Finally, we make the movement of simulated participants same for each model. This is achieved by controlling the seed number of random number generator for behavior model.

First, we compare the number of AoI Update which is closely related to the cost of interest management. Each participating host sends AoI Update to an interest manager in order to request interest management. Namely, an interest manager decides corresponding participant's neighbor whenever AoI update is transmitted. As mentioned before, Spatial Model request AoI Update along with participant's movement. So, almost all participants send AoI Update every simulation step. As total number of participants is 1024 in this experiment, every AoI update per simulation step is 1024 as illustrated in following Figure 7.

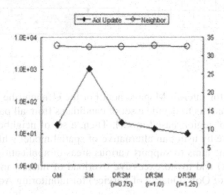

Fig. 7. AoI Update of Spatial Model is very high comparing with others. Although DRSM is alternative of Spatial Model, our proposed DRSM is similar with GM with respect to the cost of interest management. In result, DRSM has both flexibility of Spatial Model and low cost of Grid Model.

But, DRSM need to solve the amount of DoI computation that depends on the total number of participants because this model is a kind of entity based model. To solve this problem our scheme presents 2-tiered process. In the case of Grid Model as a region-based model, the amount of DoI Computation is not depends on the total number of participants. As Grid model requires simple comparison in order to decide corresponding neighbor. So, the cost of DoI computation is almost constant. By applying 2-tiered process, the DoI Computation of DRSM is depends on not overall participant but its candidate which is decided by Grid Model. Therefore, the cost of DoI Computation can be dramatically reduced. And, new cost for Gird Model is acceptable in view of overall cost.

Table 1. This table shows the result of experiments in order to compare the cost of DoI Computation. Grid Model requires the lowest cost for the computation of DoI as mentioned before. On the other hand, Spatial Model requires the highest cost. In this experiment, the total number of participants is 1024. So, the amount of DoI is 1024¡¿1023. Although proposed DRSM reduce the cost of AoI Update, the amount of DoI computation is critical factor. In our 2-tiered process consisting of Gird Model and DRSM, the cost of DoI is reduced as dependent on the size of candidate. At this time, the size of candidate depends on the subdivision of Gird Model.

Base Model	Size		Cost	
	Input	Output	AoI Update	DoI Computation
GM	1024	32.7	19.0	18.0*c
SM	1024	32.3	1024.0	1024*1023
DRSM	1024	32.4	14.2	14.2*1023
GM+DRSM	1024+96.6	96.9+32.3	10.1+10.2	10.1*c+14.2*96.6

6 Conclusion

For large-scale DISs, this paper has described a new scheme which allows efficient and flexible interest management with taking advantages of both entity based model and region based model. There are many researches on interest management to improve the scalability of DISs but it is hard to find comments about the cost of interest management itself. Although an interest management can reduce the amount of state update among participating hosts on the network gracefully, interest management will not be used actually if its efficiency and flexibility are poor.

Our scheme consists of a new relevance model named the dynamic resolution spatial model (DRSM) and 2-tiered process. DRSM is proposed by introducing the concept of resolution into the spatial model. This means that the rate of AoI update is dependent on the value of resolution to improve the efficiency of the spatial model. Therefore, the proposed model enables to control not only AoI but also the fidelity of AoI monitor individually in run-time through resolution. For the efficient DoI computation, 2-tiered process is presented. As DRSM provides one-to-one DoI as an entity based model, the load to compute DoI increases according to the number of all participants whenever AoI update is received. To overcome this, the first step determines candidates by using the grid model with acceptable costs. Then, the second step requires DoI calculation with not the rest players but candidates to determine the neighborhood by using the DRSM.

References

1. Singhal, S., Zyda, M.:Networked Virtual Environments Design and Implementation. Addison Wesley. (1999)
2. Morse, K. Bic, L.: Interest Management in Large-Scale Distributed Simulations. Presence, Vol. 9, No. 1, (2000)
3. Bassiouni, M.A., Chiu, M.H., Loper, M., Garnsey, M., Williams, J.:Performance and Reliability Analysis of Relevance Filtering for Scalable Distributed Interactive Simulation. ACM Transactions on Modeling and Computer Simulation. Vol. 7, No. 3, (1997) 293–331
4. Van Hook, D. J., Calvin, J., Rak, S.:Approaches for relevance filtering. In Proceedings of the 11th DIS Workshop on Standards for the Interoperability of Distributed Simulation. (1994) 367–369
5. Benford, Fahlen, L.:A Spatial Model of Interaction in Virtual Environments. inProc. Third European Conference on Computer Supported Cooperative Work. (1993)
6. Steinman, J. S., Wieland, F.: Parallel Proximity Detection and the Distribution List Algorithm. In Proceedings of the 1994 Workshop on Parallel and Distributed Simulation. (1994) 3–11
7. Morse, K., Steinman, J.: Data Distribution Management in the HLA:Multidimensional Regions and Physically Correct Filtering. In Proceedings of the 1997 Spring Software Interoperability Workshop. (1997)
8. Abrams, H., Watsen, K.:Three-Tiered Interest Management for Large-Scale Virtual Environments. ACM Symposium on Virtual Reality Software and Technology. (1998) 125–129

Advanced M&S Framework Based on Autonomous Web Service in Korea: Intelligent-XMSF Approach

Mye M. Sohn

The School of Systems Management Engineering, Sungkyunkwan University
300, Chunchun-dong, Jangan-Gu, Suwon, Kyunggi-do, 440-746, Korea
myesohn@skku.edu

Abstract. The ROK Armed Forces have developed its own division-level modeling and simulation (M&S) in the quarter century since it started to operate M&S, and achieved success in compliance testing to the international standard High Level Architecture/Run-time Infrastructure (HLA/RTI). This laid the foundation for using the simulation developed by the ROK Armed Forces in combined exercises with the United States. However, it is time to review the M&S standards adopted by the ROK Ministry of National Defense (MND), while various kinds of studies are performed, to enable easy interoperability and reusability between the M&Ss that HLA/RTI aims to develop. Ongoing studies include component-based Model-Driven Architecture (MDA), an Extensible Modeling and Simulation Framework (XMSF) supported by the Defense Modeling and Simulation Office (DMSO), and Intelligent XMSF (I-XMSF), the advanced framework of XMSF. I-XMSF is an architecture in which not only the human operator but also the software agent can play a role as components in the simulation. This paper introduces eXtensible Rule Markup Language for Web Service (XRML-S), the core element for implementation of I-XMSF, and shows how it is integrated into I-XMSF. In addition, the paper presents a way to further develop the M&S of the ROK Armed Forces, which has just begun to really flourish.

1 Introduction

It's been about 30 years since military simulation called wargame was introduced into the ROK Armed Forces. In the early 1970's, the ROK Armed Forces began to operate A Tactical Logistical and Air Simulation (ATLAS) which is theater-level analysis model. A variety of M&Ss have been utilized over each Service, the Korea Institute for Defense Analyses (KIDA), the Joint Chiefs of Staff (JCS), and the Combined Forces Command (CFC) for training, analysis and acquisition [12]. The ROK Armed Forces had focused on operation of M&S until the Chang Jo 21 (CJ 21) model was developed successfully by ROK Army in 1999 [10]. While the ROK Armed Forces had accelerated the development of stand-alone M&S, the U.S. DoD Defense Modeling and Simulation Office (DMSO) announced High Level Architecture/Runtime Infrastructure (HLA/RTI) to achieve interoperability and reusability between distributed M&Ss, which enacted an IEEE 1516, the international standard. The ROK Ministry of National Defense (MND) also adopts HLA/RTI as the domain military standard. It means that that HLA/RTI should be applied with in all M&Ss of ROK Armed

D.-K. Baik (Ed.): AsiaSim 2004, LNAI 3398, pp. 150–158, 2005.
© Springer-Verlag Berlin Heidelberg 2005

Forces to be developed in the future. Because it is mandatory that interoperability with the M&S of ROK-U.S. in conducting combined and joint simulation like Ulchi Focus Lens (UFL) [14]. However, rapid evolution of the World Wide Web (WWW) and Internet technology brings changes to distributed simulation architecture, which is represented by HLA/RTI. That is, various attempts have been made to link M&S with the Internet and the WWW technologies. DMSO is also sponsoring some studies. In this regard, this paper aims to examine the changes in the distributed simulation area and to propose an architecture Intelligent-Extensible Modeling and Simulation Framework (I-XMSF) that can be the leader of these changes. I-XMSF is the addition of autonomy to XMSF and can be utilized by software agents such as computer generated forces (CGF) or semi-autonomous forces (SAF).

The following sections describe the above issue. Section 2 reviews the literature on distributed simulation architecture. Section 3 proposes advanced M&S framework based on intelligent semantic web. Section 4 proposes M&S policy to acquire technical sovereignty of M&S in Korea. Finally, Section 5 concludes with a summary and the potential contributions.

2 Literature Review

2.1 From SIMNET to HLA and Beyond

We well understand that M&S provides better insight into weapons acquisition, reduces training costs, and develops force mixes of weapon quantities and types. Starting with the SIMulator NETwork (SIMNET) project in the early 1980's, various efforts have continued to integrate distributed simulations and/or simulators for military purposes. HLA/RTI was proposed in 1996. HLA/RTI aims at cost effectiveness as well as achievement of effective military training, analysis and acquisition tasks through interoperation and reuse of Command, Control, Communications, Computers and Intelligence (C4I) Systems, support utilities and war-fighters as well as simulation and simulators. The U.S. has clarified in its M&S master plan (MSMP) that it would not allocate a budget to the development of an M&S system that did not comply with HLA/RTI, and has adopted a policy of totally banning the operation of such systems in the future.

Recently, there have been ongoing studies replacing HLA/RTI using web-based commercial open standards as concern has been raised that HLA make the same mistake as ADA [3, 10]. These studies examine plans to settle issues of the interoperability and reusability of M&Ss, which HLA tried to solve through a unique middleware called RTI. The Object Management Group (OMG) proposes Model Driven Architecture (MDA), which is an integration of various industry standards such as CORBA, XML/XMI, JAVA, Web Services, and .NET [10]. In 2002, the Extensible Modeling and Simulation Framework (XMSF) project was launched by the Modeling, Virtual Environment and Simulation (MOVES) Institute of the Naval Postgraduate School, and several research projects are being undertaken on web-based modeling and simulation [17, 18]. At the October 2003 WebSim Symposium, DMSO and the Defense Information Systems Agency (DISA) declared publicly that Web Services, C4I and M&S interoperability are essential [15]. This validates the fundamental importance of XMSF work.

2.2 Comparison of HLA/RTI and Web Services

Web service is a buzzword on the Internet and WWW. One of main factors prompting the appearance of web services is Business-to-Business Electronic Commerce (B2B EC). For a successful B2B EC, heterogeneous applications used by customers, vendors and/or business partners have to interact and integrate seamlessly. To achieve this based on de-facto standards such as Universal Description, Discovery, and Integration (UDDI), Web Service Description Language (WSDL), and Simple Object Access Protocol (SOAP), web service has an architecture in which three software agents - the service provider, service requester, and service registry – interact. This architecture is known as Service Oriented Architecture (SOA).

We can recognize an interesting fact when comparing the interaction process of software agents comprising SOA with the interaction process of federates comprising a federation of HLA/RTI. A federate is an HLA-compliant simulation application that interacts with other federates through RTI. For this purpose, each federate publishes their service ability in the form of a Simulation Object Model (SOM), subscribes the contents to receive from other federates, and interacts through RTI. Figure 1 modifies some of interaction process presented by [4]. This paper will use this example in explanation.

Fig. 1. Interaction procedure through RTI

Let's take a look at SOA. Each simulation application is service provider and service requester at the same time. As a service provider, air traffic simulation notifies UDDI that it can perform aircraft simulation. In addition, as a service requester, air traffic simulation searches for UDDI to find a simulation application that can provide the guidance simulation it requires. If Radar & ATC Simulation publishes its guidance ability as a service provider to UDDI, interaction between the two simulation applications can be accomplished through SOAP. This means that the interoperability and reusability between simulation applications, that HLA aims to accomplish through sophisticated middleware, RTI, can be also accomplished using a commercial standard. Furthermore, simulation can be implemented by web service orchestration or choreography because it has to be integrated with long-life, multi-step interaction between different models.

2.3 The Next Generation of M&S Based on Web Services

Web service is more attractive than HLA/RTI in that it can implement autonomous simulation application through ontology. When ontology-based web service is implemented, software agents such as SAF and CGF as well as human can play a role as an entity of simulation. [1, 16]. The military area that uses a military terms is regarded as the most suitable area for implementing an autonomous web service by the application of ontology. The DARPA Agent Markup Language for Service (DAML-

S) is a markup language that enables intelligent agents to perform tasks and transactions automatically through the implementation of ontology. Extensible Rule Markup Language (XRML) is the markup language proposed for the implementation of autonomous Business-to-Business Electronic Commerce (B2B EC), and there are ongoing studies to expand this language into intelligent web service [7, 14]. The above study started from the fact that the use of the web is expanding from human agents to software agents, and that such expansion would enable implementation of an autonomous web. As studies are performed on Battle Management Language (BML) which enable the agent to perform automated processing of Command and Control (C2) information, it is expected that utilization of agents such as CGF and SAF will increase in future exercise and training [5]. Intelligent XMSF connecting an autonomous web service and XRML in this study is also a framework that can automate the operation of M&S through utilization of an agent.

3   Advanced M&S Framework
Based on Autonomous Web Service – Intelligent XMSF

3.1   XRML for Autonomous Business Processing

XRML is concerned with identification, and interchange with structured rule format, and is accessible by various applications of the implicit rules embedded in Web pages. Ultimately, it achieves knowledge-sharing between human and software agents in such a way as to allow software agents to browse and process the original Web page for human comprehension [7]. To do this, XRML has three components: Rule Identification Markup Language (RIML), RSML, and Rule Triggering Markup Language (RTML). Figure 2 depicts their role simply.

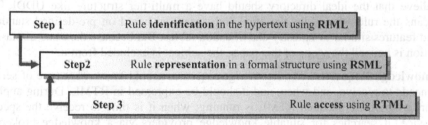

Fig. 2. Three components of XRML

However, in reality it is impossible for an RTML-embedded application to comprehend all rules for interaction. In addition to that, if interaction partners are limited to RSML-type rules, XRML's application range cannot avoid being restricted as well. To overcome these restraints, we have turned our attention to the effectiveness of linking XRML and web services together, with continuing researches in this area.

3.2   XRML for Web Service: XRML-S

XRML-S architecture has a triangular structure similar to SOA. In contrast to SOA, XRML-S implies XRML components like RIML, RSML and RTML, as some of the agents to achieve its goals as depicted in Figure 3.

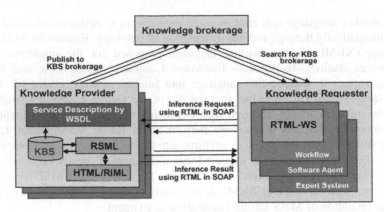

Fig. 3. Components of XRML-S

There are three types of agent in XRML-S. Each function is detailed as follows.

Knowledge Provider. The knowledge provider performs knowledge processing after receiving a knowledge processing request embedded in RTML, and returning the result to the knowledge requester. The knowledge provider which is selected for knowledge processing derives inferential results through the process of identifying and structuring pieces of knowledge from its own web page. Of course, before deriving such results, a registering step to knowledge brokerage first is taken. Knowledge provider acts as knowledge requester, vice versa.

Knowledge Brokerage. The knowledge brokerage is a registered distributed directory showing what type of service the knowledge provider can be provided with. I believe that the ideal directory should have a multi-tier structure like UDDI. This means the rules should be classified and structured based on pre-defined standards and features similar to industry classification tables. But because the issue of classification is beyond the scope of this paper, the subject is excluded from it.

Knowledge Requester. The knowledge requester should know what type of service it needs to receive, and when, and it should be expressed in RTML. During application programs, embedded RTML is running; when it is time to receive the specific service, it searches for suitable knowledge providers via a knowledge brokerage. Search results are displayed to the human decision-maker in the web browser. The RTML is modified by downloading its contents from the knowledge provider's WSDL after an individual has made a decision. The modified RTML is encoded into a SOAP message and transferred to the knowledge provider. Inferential results from knowledge provider processes are also returned in the SOAP message.

3.3 M&S Based on XRML-S: Intelligent XMSF

This paper will use federation of aircraft traffic simulation and radar & ATC simulation as explained in section 2.2, to connect M&S and XRML-S. Two simulations, which are both knowledge provider and requester at the same time, express the aircraft and guidance service that they can provide as WSDL and register as a knowledge brokerage. The RTML in XRML-S architecture is composed of a triggering condition, and a list of relevant rules and inference results to be returned. If XRML and web services are integrated with each other, changes in the relevant rule decision process occur. In other words, the relevant rule or knowledge provider is not pre-determined, but determined through 'search-decision-interaction' processes.

The first markup set of RTML, <WhenTrigger> and </WhenTrigger> expresses the rule trigger condition. If aircraft traffic simulation conforms to the condition <WhenTrigger>, and </WhenTrigger> specifies, for example, when the speed of the aircraft reaches 10,000km/h, location becomes 127 degrees longitude, 37 degrees latitude, and altitude 12,000km, and the simulation agent starts searching for knowledge brokerage. However, from this condition on its own, it cannot tell what should be searched. Therefore RTML requires an additional tag, which can associate the service RTML should receive with the knowledge providers registered in the knowledge brokerage. Herewith the tag set <ServiceName > and </ServiceName> are added.

```
<RTML Version="0.5">
<ServiceName>GuidanceService</ServiceName>
<WhenTrigger>
    <AND>
        <PeriodicAttribute>
            <Velocity>10,000</Velocity>
            <Position>
                <longtude>127degrees</longitude>
                <lat>37degrees</lat>
                <height>12000</height>
            </position>
        </PeriodicAttribute>
    </AND>
</WhenTrigger>
</RTML>
```

When the specified condition of <WhenTrigger> is met, the aircraft traffic simulation embedded RTML finds the knowledge providers which process the service <ServiceName> specifies by searching for the WSDL root element <definitions> that is registered in the knowledge brokerage. The search process is accomplished by comparing keywords between the value enveloped by the tag set <ServiceName> and the web service term specified in WSDL <definitions>. In reality, it is difficult to assume both values are exactly the same. Therefore, even in the case that fragment of the values enveloped with the tag set <ServiceName> are included in the WSDL file, it is selected as a knowledge provider candidate. Because the final decision on the selected knowledge providers is made by a decision-maker or commander, the somewhat inaccurate keyword search process can be compensated for.

It is widely understood that XMSF aims to accomplish interactionability, shareability between humans and agents, compatibility, composability, and reusability between M&S components by integrating web technology and XML, Internet and Networking, and M&S. Shareability of XMSF can be accomplished more or less by utilizing XML. However, existing XMSF will be face with restrictions if it tries to expand the role of agent into the function of substituting for the decision-making duties of a commander, beyond simple data processing. This paper attempts to connect XMSF and XRML-S as the framework for enabling up-to-rule processing by a software agent, and calls this Intelligent XMSF (I-XMSF). The biggest advantage of I-XMSF is that it can contribute to the implementation of autonomous XMSF, since operational command or other information required for an M&S operation can be fully shared by the commander and the software agent that can replace the commander.

4 Suggestion of M&S Development Policy in Korea

To deter war and enhance combat readiness, ROK-US conducting combined exercises such as the UFL, Reception, Staging, Onward Movement and Integration (RSOI) and Foal Eagle (FE) in the Korean Peninsula [2, 8]. At the same time, the ROK Armed Forces conduct its own Apnok-kang and Hokuk Exercises to perfect combat readiness [8]. It is true that exercises were used to M&Ss whether it is aggregate-level or the services-level such as air defense, navy operation, and ground fighting simulations. In UFL 02, three models such as Corps Battle Simulation (CBS), Research, Evaluation, and System Analysis (RESA), and Air Warfare Simulation Model (AWSIM) were used to portray combat and associated activity [6]. Given that most M&Ss used in exercises have been developed in or provided from the U.S., it can be readily understood that the ROK Armed Forces would have to pay a considerable amount to purchase or use such an M&S. To make matters worse, it becomes more difficult for the ROK Armed Forces to obtain the source code, even if the M&S were purchased, since Intellectual Property Rights would be enforced, raising questions as to whether it would properly reflect the battlefield environment, operation plan and weapons system of the ROK Armed Forces. As a result, it is a question of fitting the body to the clothes, and not the clothes to the body.

Such a change of environment has caused the ROK Armed Forces to have an interest in the development of M&S. The ROK Army finally succeeded in developing the CJ 21 model, in the quarter century since it started to utilize M&S. This served as an opportunity to confirm the M&S development ability of the ROK Armed Forces, and it passed the HLA compliance testing administered by the DMSO in 2003. The ROK Navy has also completed HLA compliance testing for the Navy war exercise simulation prototype model under development [12]. Now it is possible to integrate CJ 21 with existing U.S. simulations. Furthermore, the technical foundation has been laid for utilizing the model developed by the ROK Armed Forces in ROK-US combined exercises. However, it is unreasonable to expect that the resolution of technical problems alone will result in or be generally connected to utilization of the ROK Armed Forces M&S in a combined exercise. This is because both the political and the economic aspects of combined exercise need to be considered.

It is certain that the ROK Armed Forces have been dependent on the U.S. – or free acquisition or purchase of simulation developed by the U.S. – in the M&S area thus far. We can take it that the role of the ROK Armed Forces may not exceed its personnel support and share of exercise cost in ROK-U.S. Armed Forces while the ROK Armed Forces do not possess their own simulations, capable of reflecting its methods of operation or weapons systems. In the end, the U.S. has obtained the advantages of securing economic benefit, by dividing the exercise cost between themselves and the ROK Armed Forces, as well as by utilizing it as a test bed for M&S in which new technologies have been applied. In this environment, the role of the ROK Armed Forces is only very marginal. Raising the role of the ROK Armed Forces in combined exercise from 'marginality' to 'equality' means that the U.S. would relinquish part of the benefit it obtains from the exercise. In addition, it can be imagined that the U.S. would expect to obtain other types of benefit in exchange for this loss.

It is useless to discuss the technical differences between the ROK and the U.S. in terms of maturity of development ability or level of M&S related technology. In this respect, it might be difficult to provide the 'other types of benefit' that the U.S. would expect if we establish and pursue the U.S.-dependent policy, in which we follow the M&S policy of the U.S. only because ROK-U.S. combined exercise are necessary in the Korean peninsular, and we accept the standards established by the U.S. as our standards. So what shall we do?

First, we need to acquire a thorough knowledge of the M&S policy pursued by the U.S. Department of Defense, since combined exercise with the U.S. are necessary in the Korean peninsula to deter war. This is because we should be in a position to determine a reasonable level of sharing, through thorough analysis and understanding of the policies of the U.S. DoD, if it is unavoidable for the ROK Armed Forces to share the economic burden. At the same time, we need to request the transfer of M&S development technology, under the justification of burden sharing. Whilst we cannot expect that the U.S. will conduct exercises in the Korean peninsular in place of the ROK Armed Forces, technology transfer must be accomplished for the ROK Armed Forces to take the initiative in combined exercise.

Second, it is necessary to take a careful look at the trend of changes in the U.S. as well as current policies, and to prepare for such changes in advance. The distributed simulation policy of the U.S. has recently been changed. DMSO supports funding for XMSF study, and there are ongoing efforts to solve the interoperability and reusability issue between M&Ss that HLA/RTI aims to accomplish. As SIMNET, DIS and ALSP gave way to HLA/RTI, current distributed simulation, IEEE 1516, has been threatened by more universal and accessible commercial standards. It is necessary to obtain 'technical sovereignty' in the M&S area in the future, by aggressively accepting such changes and intensively conducting studies in related areas. The proposal of Intelligent XMSF, an advancement of the XMSF study in this paper, is meaningful in this context.

5   Conclusion and Further Research

Everyone agrees with the need for M&S efficiency in military training, analysis and acquisition. The change in the defense environment, such as the ending of the cold war requires effective execution of the national defense budget although the existence of the threat from North Korea. Complexity of weapons systems and changes in the battlefield environment require the execution of various kinds of operation. The rapid development and relative cheapness of information technology serves as a foundation to enable the development of an M&S that can deal with such questions, and HLA/RTI was presented as a technical framework.

Accelerating the development of M&S since the late 1990's, the ROK Armed Forces have currently developed an HLA/RTI compliant simulation for interoperability in the ROK-US combined exercises in the future, and have already achieved some desirable results. The ROK Armed Forces need to advance one step further from just accepting the decisions of the U.S., to pioneer in a sector in which it can take the lead in development, if we want to have 'technical sovereignty' in the M&S area. I-XMSF is very relevant in this context.

This paper has presented the architecture of Intelligent XMSF in brief, and aims to demonstrate not only that Intelligent XMSF can accomplish interoperability and reusability between M&Ss but also that a software agent can execute a role as an entity of M&S through future implementation. For this purpose, we have recently conducted studies for the implementation of I-XMSF.

References

1. Akimoto H., Kawano K. and Yoshida M.: Consideration to Computer Generated Force for Defense Systems. IEEE (2001).
2. Boose W., Hwang Y. Morgan P. and Scobell A. (Editors): Recalibration the U.S.-REPUBLIC OF KOREA alliance, http://www.carlisle.army.mil/ssi/pdffiles/00050.pdf (2003).
3. Brutzman D. and Tolk A.: JSB Composability and Web Services Interoperability via Extensible Modeling and Simulation Framework (XMSF), Model Driven Architecture (MDA), Component Repositories, and Web-based Visualization. http://www.MovesInstitutes.org/xmsf (2003).
4. Canazzi, D.: yaRTI, a Ada 95 HLA Run-Time Infrastructure, Lecture Notes in Computer Science, Publisher: Springer-Verlag Heidelberg (1999).
5. Carey A., Heib R. and Brown R.: Standardizing Battle Management Language – Facilitating Coalition Interoperability. Proceedings of the 2001 Fall Simulation Interoperability Workshop (2001).
6. Donlon, J. J.: Simulation Interoperability Between the U.S. and Korea. Proceedings of the 2003 Spring Simulation Interoperability (2003).
7. Lee, J. and Sohn M.: eXtensible Rule Markup Language. Communications of the ACM. Vol. 45, No. 5. ACM Press, (2003) 59-64.
8. Ministry of Defense: Military Exercises, http://www.mnd.go.kr (2004).
9. Park Y. and Kim Y.: Development of Division and Corps-level wargame model (Chang jo 21). Proceedings of the 1999 Fall MORS-K Conference (1999).
10. Parr S. and Keith-Magee R.: The Nest Step – Applying the Model Driven Architecture to HLA. Proceedings of the 2003 Spring Simulation Interoperability Workshop (2003).
11. ROK Army. History of ROK Army wargame. http://www.army.go.kr (2004).
12. ROK Navy: Certification Acquisition of Navy Training Model. http://www.navyclan.com/roknavy (2003).
13. ROK MND. STAndard Management Information System (STAMIS). http://ditams.mnd.go.kr:8080/trm1/index.php (1998).
14. Sohn, M. and J. Lee: XRML with Web Services. Proceedings of Korea Management Information System, pp. 1069-1076. Spring (2003) (in Korean).
15. Tolk, A.: A Common Framework for Military M&S and C4I Systems. Proceedings of the 2003 Spring Simulation Interoperability Workshop (2003).
16. Tolk, A: Computer Generated Forces - Integration into the Operational Environment. RTO SAS Lecture Series on 'Simulation of and for Military Decision making (2002).
17. Ulriksson J, Moradi F., and Svenson O. A Web-based Environment for building Distributed Simulations. European Simulation Interoperability Workshop (2002).
18. Veith L.: World Wide Web-based Simulation. International Journal of Engineering. Vol. 14, No. 5 (1998).

ECG Based Patient Recognition Model
for Smart Healthcare Systems

Tae Seon Kim and Chul Hong Min

School of Information, Communications and Electronics Engineering
Catholic University of Korea, Bucheon, Korea
tkim@catholic.ac.kr
http://isl.catholic.ac.kr

Abstract. Patient adaptable ECG diagnosis algorithm is required and ECG based user recognition method is essential for patient adaptable smart healthcare system. For this, we developed ECG lead III signal based patient recognition model using artificial neural networks. To extract patient's feature from ECG signal, we used three level noise removing method. After noise cancellation, number of vertices, signal intervals and detailed signal shapes were extracted from ECG signal as patient features. To show the validity of proposed model, we modeled recognition models for seven adults and we tested them under the artificial stress conditions including running, drinking and smoking and proposed model showed 92% of recognition accuracy rate.

1   Introduction

Recently, every people and their environment are fully networked and we can communicate at any time and at any place in the "ubiquitous environment". At the same time, due to the increment of elderly people, there are tremendous needs for automated and customized smart healthcare system on ubiquitous environment. To realize smart healthcare system, various technologies including smart sensor technology, customized diagnosis technology, and network technology are required. As a preliminary step to realize smart healthcare system, we proposed ECG based patient recognition method. Practically, it is nearly impossible to construct standardized diagnosis rule for heart disease since electrocardiogram (ECG) signal is different from patient to patient. And even, it is varying on measured time and one's condition for same patient. Therefore, patient adaptable ECG diagnosis algorithm is required and CCG based user recognition method is essential for patient adaptable smart healthcare system. For this, we developed ECG lead III signal based patient recognition model using artificial neural networks.

There are steady efforts of researchers to develop automated ECG diagnosis system [1][2]. Earlier researches focused on development of standardized diagnosis engine using expert knowledgebase. Togawa showed more advanced diagnosis research results. He considered patient's age and gender for diagnosis [3]. Szilagyi developed customized to an individual subject or specific waveforms based ECG waveforms detection method [4]. However, all of those researches focused only on patient customized diagnosis methods and not considered automated patient recognition approach. To realize customized smart healthcare system on ubiquitous environ-

D.-K. Baik (Ed.): AsiaSim 2004, LNAI 3398, pp. 159–166, 2005.
© Springer-Verlag Berlin Heidelberg 2005

ment, minimization of user interference is desirable. Therefore, we want to monitor and diagnosis peoples health conditions with minimal interference. For this, we propose the ECG based automated people recognition model without any user interference to implement it on patient recognition for customized healthcare in ubiquitous environment. Figure 1 shows the basic recognition system flow. Three level noise cancellation schemes including dynamic baseline method, noise cancellation using signal base type and vertex point analysis method were used to remove noise components. After noise cancellation, number of vertices, signal intervals and detailed signal shapes were extracted from ECG signal as patient features and artificial neural networks were used for recognition engine.

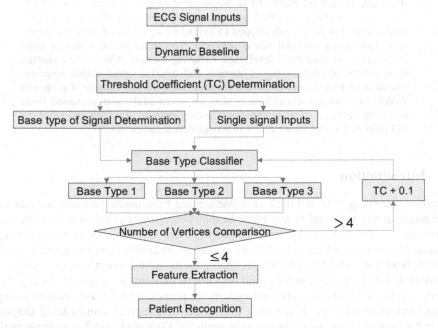

Fig. 1. Basic flow of noise detection and cancellation module and feature extraction module for patient recognition model

2   Noise Detection and Cancellation Methods

Before the extraction patient's features from ECG signal, it is necessary to remove noise parts. In this paper, we used three level noise cancellation schemes including dynamic baseline method, noise cancellation using signal base type and vertex point analysis method.

2.1   Dynamic Baseline Method

Conventionally, 1~50 Hz IIR band pass filter was widely used to remove baseline noise. The band pass filter is very effective to control the fluctuations of baseline as shown in Figure 2. However, as shown in Figure 3, it can make distortion and shape changes to original signal and make it made changes on important features including signal intervals and vertices. Therefore, to keep the important feature information, we introduced dynamic baseline method. Based on peak point, dynamic baseline algorithm divided ECG signal into series of single signal set and average value of each single signal was used as new baseline. In this case, drift of baseline can be dynamically adjusted for each signal and we can keep the shapes of original signal.

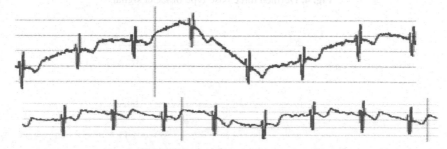

Fig. 2. Noise removal result using IIR band pass filter. Upper plot shows noised signal and lower plot shows the filtered signal output

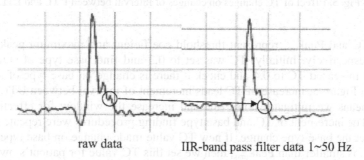

raw data   IIR-band pass filter data 1~50 Hz

Fig. 3. Distortion of signal after IIR band pass filtering

2.2 Base Type Definition for Noise Cancellation

After removing baseline noise, we defined base type for each single signal as shown in Figure 4. In general, defined base type of signal does not change for normal conditions; they could be changed by excessive noise or significant changes of patient condition. During the measurement step, we can recognize that the single signal which has different base type from others is not appropriate to describe patient's normal condition. Therefore, we skip them for recognition since they have abnormal information. Based on base types of signal, different feature extraction methods are applied at feature extraction step. For example, we can measure the interval from baseline point to another baseline point return from negative regions only at base type 2. Other types always make increasing of signal from baseline.

To classify the base type of signal, we defined upper threshold line (UTL) and lower threshold line (LTL) as shown in below.

$$UTL = Baseline + TC \times (Peak_{max} + Baseline) \tag{1}$$

$$LTL = Baseline - |\ UTL - Baseline\ | \tag{2}$$

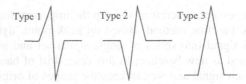

Fig. 4. Defined three base type of ECG signal

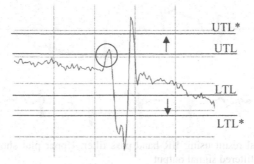

Fig. 5. Effect of TC changes on changes of interval between UTL and LTL

Where TC and $Peak_{max}$ represent threshold coefficient and maximum peak value of signal, respectively. Initially, TC was set to 0.3 and find base type of signal. After then, we increased TC to 0.4 and check it there is change on base type of signal. As shown in Figure 5, increase of TC leads increment of interval between UTL and LTL which means we mitigates noise rule and increase the window for effective signal ranges. The increment of TC and base type finding procedures were repeated until we can detect the base type change. If new TC value made change on base type of signal and UTL is smaller than $Peak_{max}$, then we set this TC value for patient's own characteristic. After determination of base type using customized TC value, this base type can be compared to every single signal of future measurement. After comparison, type matched single signals are selected for recognition and mismatched single signals are rejected since it is possible that noise factors made base type changes.

2.3 Vertex Point Analysis

After determination of base type of signal, we analyze vertices points of signal to remove noised signal. To extract the vertex point, we sampled out of threshold level signal and take differentiation to find sign changes. For each single signal, we count the number of vertices and compare it to average number of vertices for previous signal patterns. If the difference is bigger than 4, 0.1 is added to TC and return to base type classifier as shown in Figure 1. At the type classifier step, base type of returned signal was classified and compared it to normal signal to discard noised signal. At normal heart rate, new single signal was fed to system at every 0.75~0.85 second. In other words, elimination of few number of single signal set does not make problems such as significant delay or lack of information. Therefore, for occasional noise component, it is much effective to find noised single signal and discard it rather than removing them from original signal.

3 Feature Extraction

After removing noise, seventeen features are extracted from ECG signal for patient recognition. The several kinds of signal interval information, detail shape of signal and several peak values are selected for feature of signal.

3.1 Signal Interval

According to base type of signal, different approach is required to extract signal interval. Figure 6 shows one example to find signal internal for base type 1 signal. In this case, we can extract three kinds of signal interval as shown in Figure 6 (a), (b), and (c). In (a), interval information is extracted based on UTL. For (b) and (c), LTL and baseline is used respectively. However, for base type 3 signals, only two types (interval (a) and (c)) of interval information is available since they don't have sign change at the baseline level.

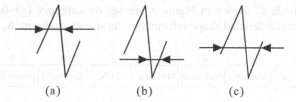

(a)   (b)   (c)

Fig. 6. Three kinds of signal internal

3.2 Detailed Shape Definition

Based on different rules from base type signal definition, another signal shape classification scheme, "detailed shape definition method" was applied for feature extraction. For detailed shape definition, shapes of upper baseline and lower baseline were characterized and used as one of recognition feature from ECG signal. In this work, detailed shapes are classified to eight shapes as shown in Figure 7. In figure 7, shape (a), (b), and (c) are set for upper part of baseline and shape (d), (e), and (f) are designed for lower part of baseline for ECG signal. Shape (g) and (h) give information if the position of minimum value was located before the maximum peak point (shape (g)) or after the maximum peak point (shape (h)). They are especially effective to

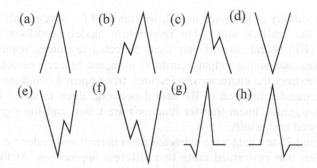

Fig. 7. Defined eight kinds of detailed shape of ECG signal

extract the signal features for patients who have base type 3 signal. For example, ECG signal shown in Figure 3 can be classified to base type 3 signal and detailed shape (c) in Figure 7. Also, it is classified to detailed shape (g) since the minimum point is located earlier than maximum peak point.

3.3 Feature Representation

In this paper, total 17 features are extracted using signal interval extraction and signal type and shape classification methods. Figure 8 shows one example to represent 17 features for ECG signal shown in Figure 3. The average value of base type classification results is set on category (a) and integer number representation of base type classification results are set on category (b). The category (c) is used to represent number of vertices and category (d) and (e) represents the amplitude from baseline to maximum peak point and minimum peak point, respectively. Category (f) represents ratio between value of (d) and (e). Measured values from three kinds of signal interval extraction methods as shown in Figure 8 are set on category (g)~(i). The category (j)~(q) represent the detailed shape information as shown in Figure 8.

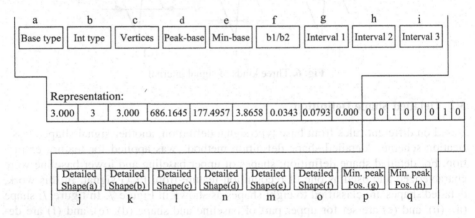

Fig. 8. Feature representation example

4 Recognition Results

To show the validity of proposed model, we modeled for seven adults and we tested them under the artificial stress. For recognition model, feed-forward error back-propagation (BP) neural networks are used. Selected seventeen features are feed to network inputs and single output neuron is assigned to each person. For network training, we extract the characteristic features from normal condition measurement. To optimize transfer function of BP neural network, three kinds of functions (log-sigmoid, tan-sigmoid, linear-transfer function) are tested and log-sigmoidal transfer function showed best results.

For test, total 25 test data were extracted from normal and under the artificial stress condition. Test was performed using two different approaches. At the first level, 5 minute of ECG signal is stored and some of them were used for training some of them were used for testing. In this case, 100% of recognition results are obtained since it has very limited changes on environment and patient. For second level, some of measured signal at normal condition was used for training and test signals were measured after artificial stress conditions including drinking coffee and alcoholic drink, smoking, and running up and down of stairs. Figure 9 shows the typical signal shapes of seven candidates and Figure 10 shows signal shapes of tested examples.

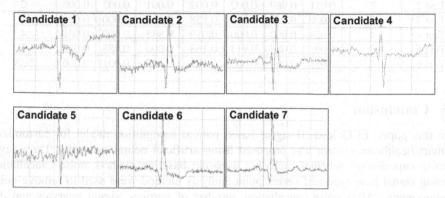

Fig. 9. Typical signal shapes of seven candidates

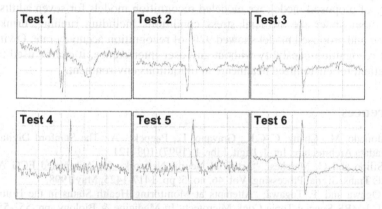

Fig. 10. Example of tested signal shapes

Table 1 showed network outputs for test samples shown in Figure 10. For test sample 1 to 4, patient recognition model accurately find among seven candidates. Among 25 test patterns, test pattern 5 and 6 are the only signal to fail to recognize. For test signal 5, model recognized it for candidate 3 even though it was came from candidate 2. In this case, feature values (signal intervals and number of vertices) of candidate 2 are very similar to those of candidate 3 and detailed shapes are repeatedly changed during measurement. Therefore, it is expected to improve detailed shape definition step. Test sample 6 was measured after running up and down of stairs. In this case, model recognized that this signal was not belong to any candidates. In other words, measurement after physical exercise made significant changes on ECG signal and it can be considered as future works.

Table 1. Network outputs for six test samples

Test Signal		Network Outputs							Decision
Signal number	Actual candidate number	1st	2nd	3rd	4th	5th	6th	7th	(candidate number)
Test 1	1	0.986	0.056	0.001	0.004	0.0000	0.000	0.000	1
Test 2	2	0.011	0.969	0.002	0.002	0.001	0.002	0.000	2
Test 3	3	0.000	0.046	0.000	0.858	0.254	0.000	0.000	3
Test 4	4	0.028	0.002	0.000	0.059	0.831	0.000	0.001	4
Test 5	2	0.001	0.083	0.914	0.163	0.000	0.088	0.083	3
Test 6	4	0.018	0.001	0.000	0.185	0.448	0.001	0.000	N/A

5 Conclusion

In this paper, ECG lead III signal based patient recognition model for customized smart healthcare system was proposed using artificial neural networks. Three level noise cancellation schemes including dynamic baseline method, noise cancellation using signal base type and vertex point analysis method were used to remove noise components. After noise cancellation, number of vertices, signal intervals and detailed signal shapes were extracted from ECG signal as patient features. To show the validity of proposed model, we modeled recognition models for seven adults and we tested them under the artificial stress conditions including running, drinking and smoking and proposed model showed 92% of recognition accuracy rate. Owing to the patient recognition capability without any user interference, it can be used to recognize patient for customized healthcare in ubiquitous environment.

References

1. Baldonado, M., Chang, C.-C.K., Gravano, L., Paepcke, A.: The Stanford Digital Library Metadata Architecture. Int. J. Digit. Libr. 1 (1997) 108–121
2. R. Silipo and C. Marchesi, "Artificial Neural Networks for Automated ECG Analysis," IEEE Trans. Signal Processing, Vol. 46, no. 5, pp. 1417-1425, May 1998.
3. M. Ogawa and T. Togawa, "Attempts at Monitoring Health Status in the Home," Proc. IEEE-EMBS Special Topic Conf. Microtech. In Medicine & Biology, pp. 552-556, Lyon, France, Oct. 2000.
4. S. Szilagyi, Z. Benyo, L. Szilagyi and L. David, "Adaptive wavelet-transform-based ECG waveforms detection," Proc. IEEE Int'l Conf. Engineering in Medicine and Biology Society, vol. 3, pp. 2412-2415, Cancun, Mexico, Sep. 2003.

Three-Dimensional Brain CT-DSA
Using Rigid Registration and Bone Masking
for Early Diagnosis and Treatment Planning

Helen Hong[1], Ho Lee[2], Yeong Gil Shin[2,3], and Yeol Hun Seong[4]

[1] School of Electrical Engineering and Computer Science BK21: Information Technology,
Seoul National University, San 56-1 Shinlim-dong Kwanak-gu, Seoul 151-742, Korea
hlhong@cse.snu.ac.kr
[2] School of Electrical Engineering and Computer Science, Seoul National University
{holee,yshin}@cglab.snu.ac.kr
[3] INFINITT Co., Ltd., Taesuk Bld., 275-5 Yangjae-dong Seocho-gu, Seoul 137-934, Korea
[4] Dept. of Radiology, Seoul National University Bundang Hospital, 300, Gumi-dong,
Sungnam-si, Kyunggi-do, Korea
radimage@snubh.org

Abstract. This paper proposes an accurate and robust three-dimensional brain CT-DSA using rigid registration and bone masking for early diagnosis and treatment planning of intracranial aneurysms. Our method is composed of the following four steps. First, a set of feature points within skull base are selected using a 3D edge detection technique. Second, a locally weighted 3D distance map is constructed for leading our similarity measure to robust convergence on the maximum value. Third, the similarity measure between feature points is evaluated repeatedly by selective cross-correlation. Fourth, bone masking is performed to completely remove bones. Experimental results show that the accuracy and robustness of our method are much better than conventional methods. In particular, our method can be useful for the early diagnosis and treatment planning of intracranial aneurysms.

1 Introduction

Accurate intracranial vascular imaging is necessary to detect cerebral aneurysms, arterial stenosis, and other vascular anomalies as well as to perform safe and proper surgical clipping or endovascular treatment in a patient with ruptured or nonruptured aneurysms. Recently, the usefulness of computed tomographic (CT) angiography with intravenously administered contrast material for the early diagnosis and treatment planning has been advocated [1-3]. Contrast material-enhanced CT angiography may allow better delineation of the aneurysm neck, shape, orientation, its relationship to the parent artery, and important adjacent bone structures. However, there is a limitation to get a clear vascular anatomy where arteries are obscured by bone, as in the petrous or carvenous portion with their intricate mingling of bone and arteries.

A conventional approach of delineating vessels from bone is to subtract nonenhanced images from contrast material-enhanced images [4-5]. This subtraction technique assumes that tissues surrounding vessels do not change in position or density during exposure. However, even minor patient movement results in severe distortion

D.-K. Baik (Ed.): AsiaSim 2004, LNAI 3398, pp. 167–176, 2005.

or artifacts since the contrast between vessels and surrounding tissues is significantly smaller than that between bone and surrounding tissues. Without the proper handling of patient motion artifacts, the clinical application of a subtraction method is very limited. Even though the motion of a patient can be minimized by taking special precautions of either patient, acquisition system, or both [6], artifacts cannot be completely avoided. To reduce patient motion artifacts, we have to use a retrospective image processing techniques.

Several registration techniques have been proposed to reduce patient motion artifacts in brain CT angiography. Yeung et al. [7] proposed a 3D feature detector and a 3D image flow computation algorithm for matching feature points between enhanced and nonenhanced images of brain CT angiography. The processing time for finding 3D feature points takes too much time since the interest index of each voxel position is found by comparing variance values in all 13 directions. Venema et al. [8] developed a global matching method using gray value correlation of skull bases in two time interval brain CT angiography. The alignment by minimizing the difference of intensity values of the both skull bases may not be accurate because pixels at the same position of individual CT scans may have different density. For the same purpose, Kwon et al. [9] proposed a mutual information-based registration. However, the similarity measure which uses limited information coming from the area of skull base does not guarantee the exact alignment.

The current registration methods can minimize patient motion artifact, however, they still need some progress in computational efficiency, accuracy and robustness for getting a clear vascular anatomy. In this paper, we propose an accurate and robust three-dimensional brain CT-DSA (digitally subtraction angiography) using rigid registration and bone masking. Our method has the following four steps. First, a set of feature points within the skull base is selected using a 3D edge detection technique. Second, a locally weighted 3D distance map is constructed for leading our similarity measure to robust convergence on the maximum value. Third, the similarity measure between feature points is evaluated repeatedly by selective cross-correlation. Fourth, the bone masking is performed for completely removing bone. Experimental results show that our method is more accurate and robust than the conventional methods. In particular, our method can be useful for the early diagnosis and treatment planning of intracranial aneurysms.

The organization of the paper is as follows. In Section 2, we discuss how to extract and select feature points in an angiographic images. Then we propose a similarity measure and an optimization process to find exact geometrical relationship between feature points in the enhanced and nonenhanced images. Finally, a masking technique for removing bone is described. In Section 3, experimental results show how the method accurately and robustly extracts the intracranial aneurysms from the CT angiography. This paper is concluded with a brief discussion of results in Section 4.

2 Three-Dimensional CT-DSA

Fig. 1 shows the pipeline of our method for three-dimensional CT-DSA using rigid registration and bone masking. To find exact geometrical relationship between enhanced and nonenhanced images of CT angiography, one dataset is fixed as mask volume whereas the other dataset is defined as contrast volume which taken after injecting a contrast material to mask volume. The contrast volume is moved during the iterative alignment procedure. Interpolating the contrast volume at grid positions of the mask volume is required for each iteration depending on the transformation. Since rigid transformation is enough to align the skull base, we use three translations and three rotations about the x-, y-, z-axis. After transformation, the mask is slightly widened by means of dilation with 1-pixel to allow partial-volume effects and acceptable small mismatch. Finally, extracted vessels are displayed by a conventional volume rendering technique.

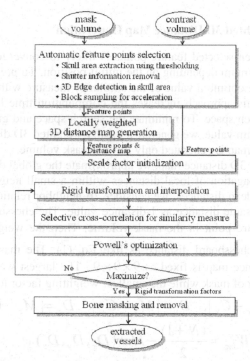

Fig. 1. The pipeline of three-dimensional CT-DSA using rigid registration and bone masking

2.1 Automatic Feature Points Selection

A traditional approach of finding the correspondence between mask and contrast volume of brain CT angiography requires voxel by voxel correspondence test for entire volume. This is computationally expensive and cannot be a clinically accepted technique. Since most artifacts appear in the region where strong edges are present in the subtracted image, we can accelerate the registration procedure by processing voxels belonging to image edge area only instead of all voxels in the volume.

Our feature identification uses a 3D operator for utilizing spatial relations in volume data. At first, in order to align skull base which represents rigid object boundaries in brain CT angiography, we use pixels with a CT number above a chosen threshold. Shutter information of CT angiography is removed since it leads to misalignment by detecting the boundary of the shutter as a feature. Then 3D edge detection technique [10] is applied to the skull base in mask and contrast volume, respectively. Just like their two-dimensional counterparts, 3D edges are usually defined as discontinuities of image intensity caused by transition from one homogeneous 3D region to another 3D region of different mean intensity. The location of rigid object boundaries can be computed by detecting the local maxima of gray-level gradient magnitude. Since rigid object boundaries are scale-dependent image features, they can be only detected by using derivatives which allow finer tuning for detecting the required edge scale. Among detected features, we select a predefined number feature points enough to ensure even distribution of feature points in the skull base.

2.2 Locally Weighted 3D Distance Map Generation

Registration between selected feature points is likely to converge to local maximum near to global maximum depending on the initial position. To prevent this occurrence we need to find the optimum value of the similarity measure within the global search space. However, this approach increases the computation time by the proportion to the size of the search space. To minimize the search space and guarantee the convergence to the optimum value, we generate a locally weighted 3D distance map per each feature point. This map is generated only for the mask volume.

For generating a 3D distance map, we approximate the global distance computation with repeated propagation of local distances within a small neighborhood mask. To approximate Euclidean distances, we consider 26-neighbor relations for a chessboard distance to be the same distance as 1-distance value. The chessboard distance is applied to each feature point in the mask volume. Then the weighting factor W_{xyz} is multiplied to the chessboard distance as like Eq. (1). The mask size N of locally weighted 3D distance map is fixed as $9 \times 9 \times 9$. The largest weighting factor is assigned to the center of mask while the smallest weighting factor to boundary of mask.

$$D_x = |M_x - C_x|, \quad D_y = |M_y - C_y|, \quad D_z = |M_z - C_z|$$
$$W_{xyz} = \frac{(N+1)}{2} - Max(D_x, D_y, D_z) \tag{1}$$

where M and C is the current and the center position of a locally weighted-3D distance mask. D_x, D_y, D_z is the difference of x-, y-, and z-axis between the current and the center position in the locally weighted 3D distance mask.

Fig. 2(a) shows a locally weighted 3D distance map when N is fixed as $5 \times 5 \times 5$. Fig. 2(c) shows the cut plane of the volume of the locally weighted 3D distance map when the distance map of Fig. 2(a) is applied to feature points as shown in Fig. 2(b). Each color shown in Fig. 2(c) represents the distance value from a selected feature point.

2.3 Selective Cross-Correlation and Optimization

The similarity measure is used to determine the degree of resemblance of windows in successive frames. Several similarity measures have been devised and applied to angiographic images – the sum of squared intensity differences, cross-correlation, the entropy of the difference image and mutual information. However, most of these

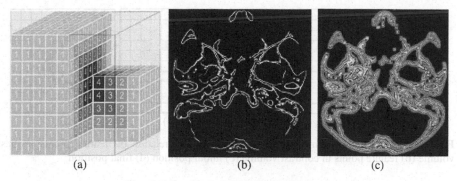

Fig. 2. A locally weighted 3D distance map

similarity measures are sensitive to mean gray-level offset and local dissimilarities caused by contrasted vessels. We propose the selective cross-correlation as a similarity measure which only uses feature points near to skull base. Our approach reduces sensitivity to mean gray-level offset or local dissimilarities by incorporating distance information of a locally weighted distance map into the similarity measure.

As can be seen in Eq. (2), the local weighting factor of a 3D distance map in mask volume is multiplied to the distance in contrast volume. We assume that distances of feature points in contrast volume, $D_C(i)$, are all set to 1. Then the selective cross-correlation SCC reaches maximum when feature points of mask and contrast volume are aligned correctly.

$$SCC = \frac{1}{N_C} \sum_{i=0}^{N_C-1} D_C(P_C(i))W_M(T(P_C(i))) \qquad (2)$$

where N_C is total number of feature points in contrast volume, $P_C(i)$ is the position of i-th feature point in contrast volume. The weighting factor of the current feature point of contrast volume, $W_M(i)$, is obtained from the corresponding locally weighted 3D distance map in mask volume.

To evaluate the selective cross-correlation of large samples from volume dataset, we use the Powell's method. Since the search space of our similarity measure is limited to the surrounding skull base, we do not need a more powerful optimization algorithm such as simulated annealing. Fig. 3 shows the process of optimizing the selective cross-correlation. Fig. 3(a) and (b) are feature points selected from mask and contrast volume, respectively. In the initial position as Fig. 3(c), the number of matching feature points is 6 pixels. Finally, the number of matching feature points becomes 12 pixels when the selective cross-correlation reaches the maximum value.

2.4 Bone Masking and Removal

A traditional approach for enhancing vessels after matching is to subtract registered contrast volume to mask volume. However it is very difficult to remove bone completely using a traditional subtraction technique since densities between mask and registered contrast volume can be different even in the same pixel position. In addi-

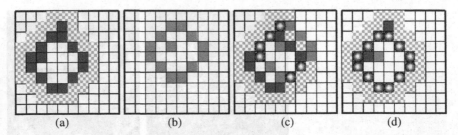

Fig. 3. The process of optimizing the selective cross-correlation (a) feature points in mask volume (b) feature points in contrast volume (c) initial position (d) final position

tion, partial-volume effects near to bone area and slight amounts of mismatch make it possible to generate artifacts in the subtraction image. For more complete bone removal, we propose a bone masking and removal technique instead of a traditional subtraction technique.

Our bone masking and removal process first identifies bone pixels in mask volume by applying a threshold value of 1500HU. These identified bone pixels, as a mask, is then slightly widened by means of dilation with 1-pixel to allow for partial-volume effects and slight amount of mismatch. Finally, pixels in contrast volume which are corresponding to pixels in mask volume are set to an arbitrarily low value. This results in the removal of bones in the volume rendering image of contrast volume.

3 Experimental Results

All our implementation and test were performed on an Intel Pentium IV PC containing 3.0 GHz CPU and 1.0 GBytes of main memory. Our method has been applied to five different patient's datasets with intracranial aneurysm, as described in Table 1, obtained from MDCT (Multi-Detector Computed Tomography). We assume that image and pixel sizes are the same in enhanced and nonenhanced images. The average processing time including volume rendering of five different patient's datasets is less than 60 seconds.

Table 1. Image conditions of experimental datasets

Case #	Image size	Slice #	Pixel size	Slice thickness	FOV
1	512 x 512	220	0.32 x 0.32	0.3	163 x 163
2	512 x 512	220	0.34 x 0.34	0.3	176 x 176
3	512 x 512	110	0.35 x 0.35	1.0	180 x 180
4	512 x 512	130	0.44 x 0.44	1.0	227 x 227
5	512 x 512	220	0.31 x 0.31	0.3	160 x 160

The performance of our method is evaluated with the aspects of accuracy and robustness. For accuracy assessment, we show results of visual inspection in Fig. 4 and 5. Fig. 4 shows two-dimensional comparison of a regular subtraction method and the proposed method. Note that although enhanced images in Fig. 4(b) and (d) appear nearly identical, subtraction images in Fig. 4(c) and (e) are very different. Fig. 4(c) is of poor quality whereas Fig. 4(e) is of relatively good quality. This difference underlines the necessity of registration. However registered subtraction has still artifacts near to skull base as shown in Fig. 4(e). The final result of our method in Fig. 4(f) is obtained by masking the registered enhanced image in Fig. 4(d) with the mask image which has already been obtained by applying our bone masking to Fig. 4(a). All pixels in Fig. 4(f) corresponding to the mask image are set to a tissue-equivalent value (40HU).

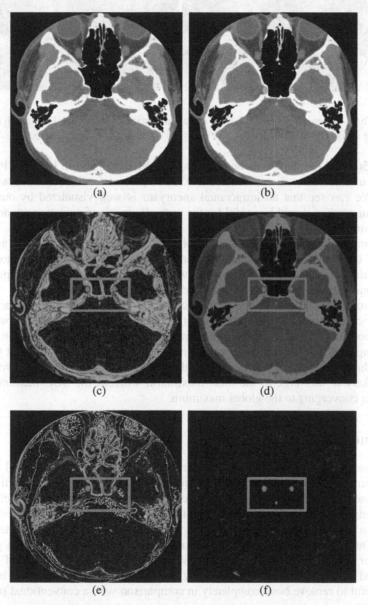

(a)   (b)

(c)   (d)

(e)   (f)

Fig. 4. Comparison of regular subtraction and the proposed method in patient 1 (a) nonenhanced image (b) enhanced image (c) regular subtraction image ((b) minus (a)) (d) registered enhanced image (e) registered subtraction ((d) minus (a)) (f) final result of the proposed method

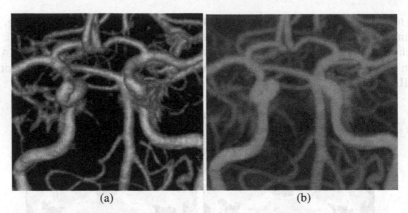

(a)   (b)

Fig. 5. Three-dimensional brain CT-DSA of patient 5 with an intracranial aneurysm (a) volume rendering image (b) MIP image

Fig. 5(a) and (b) show volume rendering and MIP (Maximum Intensity Projection) images of three-dimensional brain CT-DSA of patient 5 with an intracranial aneurysm. We can see that an intracranial aneurysm is well visualized by our method. Thus, our method would be useful for the early diagnosis of patients with an intracranial aneurysm.

Our method can be also used for treatment planning to perform safe and proper surgical clipping or endovascular treatment in a patient with ruptured or nonruptured aneurysms as Fig. 6. In Fig. 6(b) and (d), we can clearly see aneurysms of petrous and carvenous portions where arteries are obscured by bone.

The robustness of the selective cross-correlation (SCC) criterion has been evaluated by comparing SCC measure traces (represented by circle dot line) with cross-correlation (CC) measure traces (represented by square dot line). As shown in Fig. 7, the changes of SCC measure are smooth near to the maximal position, but CC measure is changed rapidly. This means that the CC measure for similarity evaluation is more likely to converge to the local maximum, whereas the SCC measure is more robust in converging to the global maximum.

4   Conclusion

In this paper, we have developed an accurate and robust three-dimensional brain CT-DSA using rigid registration and bone masking. The feature point selection within skull base makes our registration more accelerate by processing voxels belonging to image edge area only. The locally weighted 3D distance map per each feature point allows to minimize the search space and to guarantee the convergence to the optimum value. Our selective cross-correlation measure reduces sensitivity to mean gray-level offset or local dissimilarities by incorporating distance information of a locally weighted distance map into the similarity measure. In addition, the bone masking is also useful to remove bone completely in comparison with a conventional subtraction technique.

Our method has been successfully applied to the enhanced and nonenhanced brain CT images of five different patients with intracranial aneurysms. Experimental results

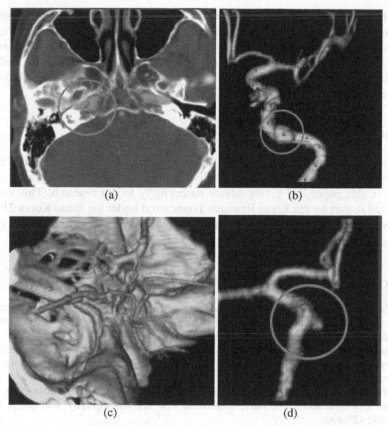

Fig. 6. Three-dimensional brain CT-DSA of patients 3 and 4 with intracranial aneurysms (a) axial image with an aneurysm at petrous portion (patient 3) (b) the result of the proposed method (patient 3) (c) volume rendering image with an aneurysm at carvenous portion (patient 4) (d) the result of the proposed method (patient 4)

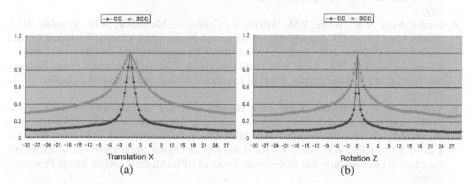

Fig. 7. Comparison of SCC and CC traces in patient 5 (a) translation in the x-direction (b) rotation in the z-direction

show that our method is clinically promising by the fact that it is very little influenced by image degradation occurred in bone-vessel interface. For all experimental datasets, we can clearly see intracranial aneurysms as well as arteries on the volumetric images. Our method could be well applied to detect cerebral aneurysms for early diagnosis and to help surgeons perform safe and proper surgical clipping or endovascular treatment.

Acknowledgements

The authors are grateful to Prof. Sung-Hyun Kim from the Seoul National University Hospital of Bundang, Korea for providing brain CT and CT angiography datasets shown in this paper and giving advice unsparingly to our research. This work was supported in part by the Korea Research Foundation under the Brain Korea 21 Project and INFINITT Co., Ltd. The ICT at Seoul National University provides research facilities for this study.

References

1. Alberico, R.A., Patel, M., Casey S., Jacobs, B., Maguire, M., Decker, R., Evaluation of the Circle of Willis with Three-Dimensional CT Angiography in Patients with Suspected Intracranial Aneurysms, AJNR Am J Neuroradiol, Vol. 16 (1995) 1571-1578.
2. Zouaoui, A., Sahel, A., Marro, B., et al., Three-Dimensional Computed Tomographic Angiography in Detection of Cerebral Aneurysms in Acute Subarachnoid Hemorrhage, Neurosurgery, Vol. 41 (1997) 125-130.
3. Velthuis, B.K., Rinkel, G.J.E., Ramos, L.M.P., et al., Subarachnoid Hemorrhage: Aneurysm Detection and Preoperative Evaluation with CT Angiography, Radiology, Vol. 208 (1997) 423-430.
4. Meijering, E.H.W., Niessen, W.J., Viergever, M.A., Retrospective Motion Correction in Digital Subtraction Angiography: A Review, IEEE Trans. on Medical Imaging, Vol. 18, No. 1 (1999) 2-21.
5. Buzug, T.M., Weese, J., Strasters, K.C., Motion Detection and Motion Compensation for Digital Subtraction Angiography Image Enhancement, Philips J. Res., Vol. 51 (1998) 203-209.
6. Jayakrishnan, V.K., White, P.M., Aitken, D., Crane, P., McMahon, A.D., Teasdale, E.M., Subtraction Helical CT Angiography of Intra- and Extracranial Vessels: Technical Considerations and Preliminary Experience, AJNR Am J Neuroradiol, Vol. 24 (2003) 451-455.
7. Yeung, M.M., Yeo, B.L., Liou, S.P., Banihashemi, A., Three-Dimensional Image Registration for Spiral CT Angiography, Proc. of Computer Vision and Pattern Recognition (1994) 423-429.
8. Venema, H.W., Hulsmans, F.J.H., den Heeten, G.J., CT Angiography of the Circle of Willis and Intracranial Internal Carotid Arteries: Maximum Intensity Projection with Matched Mask Bone Elimination – Feasibility Study, Radiology, Vol. 218 (2001) 893-898.
9. Kwon, S.M., Kim, Y.S., Kim, T., Ra, J.B., Novel Digital Subtraction CT Angiography based on 3D Registration and Refinement, Proc. of SPIE Medical Image: Image Processing (2004).
10. Nikolaidis, N., Pitas, I., 3D Image Processing Algorithm, Wiley Inter-Science Pulication (2001). Frederik Maes, Andre Collignon, Dirk Vandermelen, Guy Marchal, Paul Suetens

Embedded Healthcare System for Senior Residents Using Internet Multimedia HomeServer and Telephone

Sung-Ho Ahn[1], Vinod C. Joseph[1], and Doo-Hyun Kim[2,*]

[1] Embedded GUI Research Team, ETRI, Daejeon, Korea
{ahnsh,vinod}@etri.re.kr
[2] School of Internet and Multimedia Eng., Konkuk University, 1 Hwayang-Dong,
Kwangjin-Gu, Seoul, 143-701, Seoul, Korea
doohyun@konkuk.ac.kr

Abstract. Healthcare has been primarily concentrated in the hospital domain. With the current technology trends, some e-healthcare services have evolved. However, healthcare has not been accessible for the common man. Generic healthcare services available today focus on technology marvel. This paper illustrates our implementation of the healthcare service that uses the conventional telephone. The usage of telephone for facilitating Voice and Video over IP (V2oIP) communication to the home generates a new lease of life to the senior residents of the society. This embedded healthcare system uses the telephone connected to the HomeServer that connects to the hospital network over the Internet. The telephone uses a special sequence character to invoke the healthcare application. The application connects to the hospital network using Session Initiation Protocol (SIP) to facilitate multimedia communication. The output can be configured for display on the TV connected to the HomeServer and/or any conventional display.

1 Introduction

This paper describes the implementation and design of our embedded healthcare system. This healthcare system is embedded in a HomeServer and connected with home network devices and healthcare peripherals through the HomeServer. The healthcare system connects to the hospital in the external world over Ethernet. The healthcare module is loaded to the HomeServer and facilitates V2oIP communication over SIP. SIP is a signaling protocol used on top of UDP.

The HomeServer is an embedded system that has Qplus OS – our flavor of embedded linux. HomeServer is developed to play the central role of managing all home network devices and healthcare services in the household. HomeServer can be used as a digital television, surveillance equipment with a camera attached to the Home-Server, messaging server as well as a streaming server, a Data Mining Server or a Healthcare/Telecare device based on the users needs. Our paper discusses the design of the healthcare service that is one of the important services provided to the home environment. We named the embedded healthcare system Talk2TV.

This paper is organized as follows. Section 2 illustrates the necessity of this design to meet the needs of our senior residents and denotes the operation and block diagram

---

* Corresponding author.

D.-K. Baik (Ed.): AsiaSim 2004, LNAI 3398, pp. 177–186, 2005.

architecture of our embedded healthcare solution based on SIP. Section 3 describes the process scenarios of several embedded healthcare solutions and denotes the message sequence charts of the operation scenarios. Finally, section 4 concludes the paper summarizing the needs to develop a better society.

2   Embedded Healthcare System

Figure 1 shows the total system environment of the healthcare solution. This healthcare system can support a user from within a home or outside the home by connecting to the HomeServer. The following devices are critical to the operation of the embedded healthcare system:

- Motion Detector: Motion detector generates the event signal. And this signal is sent to the HomeServer that transcodes the video and sends it to the doctor.
- IM (Instant Messaging): The transcoded SIP message is transmitted to the user's PDA
- Video Surveillance Signal Transmit: Activates the video surveillance signal.
- Healthcare Peripherals like ECG, BP, SpO2 and respiration measurement devices.

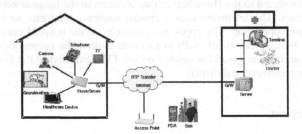

Fig. 1. Environment of Embedded Healthcare System

We can control the surveillance camera in home through the HomeServer and track the movement of the elderly person. The external user can be alerted by some messages generated by the HomeServer and motion detector and directed for display on a remote user device such as PDA or SmartPhone through the Internet. Pressing some buttons on a generic telephone and other series of pre-defined control mechanisms can control the Healthcare System. The healthcare peripherals connected to the HomeServer include Blood pressure (BP) device, diabetes detector, weight measurement device, temperature/respiration measurement and/or Electrocardiogram (ECG) device. These devices may be connected to the HomeServer over the serial or USB ports. The users input details are sent to the remote hospital network over the Internet. The doctor at his remote location monitors all records at his Hospital Terminal and sends the output back to the user for display. The doctor also watches the user movements shown through the video surveillance system attached to the HomeServer. On detecting critical condition, the doctor alerts the system to message/call a series of stored family contacts and/or ambulance.

This system has the Qplus OS and a traditional network layer over its OS. It has the device driver of USB camera and that of VoIP card. It also has a video codec-H.261/MJPEG/MPEG-4, RTP stack, SIP/SDP module for session management, VoIP card stack and Healthcare stack. To facilitate video surveillance, it has the V2oIP module, video surveillance module, and a healthcare module to co-operate the different tasks. Finally, it includes the healthcare application and its GUI.

The hardware architecture of the embedded healthcare system is connected to the camera through the USB/serial port for video communication and/or video surveillance. The embedded healthcare system is connected to the conventional telephone through the SLIC of VoIP card in HomeServer. The DTV for display is connected to this system through the S-Video port. Healthcare devices are connected to the system through the serial port and/or USB port. Finally, the embedded healthcare system on HomeServer is connected to the Internet over Ethernet.

Considering the implementation point of view; the healthcare application program, video streaming software, VoIP stack and the video codec are loaded on an embedded healthcare system. The video signals are displayed on DTV through the video receiver. The received video signals from the surveillance camera are sent to the remote healthcare application. In that case, the received signals are encoded by the video codec and can be sent to the remote healthcare device or the access point of PDA/SmartPhone over the Internet. The received encoded video signals are regenerated and displayed by the decoder. The Healthcare System can be used to facilitate the senior residents and home-alone children to be monitored and provided healthcare services. The patient/doctor/remote monitoring person chooses the selected camera and sends the healthcare surveillance camera output to the remote doctor on his device. Thereby, the elderly person/child can be monitored constantly based on input from doctor, family members or the monitored person himself. The following devices are critical to the operation of the embedded healthcare system:

- Motion Detector: Motion detector generates the event signal. And this signal is sent to the HomeServer that transcodes the video and sends it to the doctor.
- IM (Instant Messaging): The transcoded SIP message is transmitted to the user's PDA
- Video Surveillance Signal Transmit: Activates the video surveillance signal.
- Healthcare Peripherals like ECG, BP, SpO2 and respiration measurement devices.

The hardware architecture of the embedded healthcare system is connected to the camera through the USB/serial port for video communication and/or video surveillance. The embedded healthcare system is connected to the conventional telephone through the SLIC of VoIP card in HomeServer. The DTV for display is connected to this system through the S-Video port. Healthcare devices are connected to the system through the serial port and/or USB port. Finally, the embedded healthcare system on HomeServer is connected to the Internet over Ethernet.

3 Process Scenarios and Message Flows

The operation of the healthcare system is illustrated on the basis of the process scenarios shown in this section. The message sequence charts shown in this section illustrate the overall operation of the system according to the scenarios. The embedded healthcare system has four process scenarios:

3.1 User Operated Normal Mode

Figure 2 shows the process scenario in case of the user operated normal mode. The operational details for this mode are described below:

1. User sends a special sequence character using the conventional telephone to activate the Healthcare device connection to connect to the doctor
2. User receives request from doctor for concerned measurement(s)
3. He uses the concerned healthcare devices and reports measurement.
4. Doctor monitors measurements and reports variations to patient

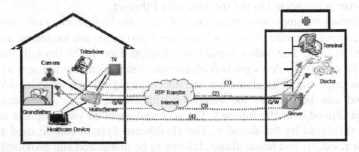

Fig. 2. Scenario of User Operated Normal Mode

Figure 3 shows the message sequence chart in case of the user operated normal mode. In the normal operation mode, the user (usually an elderly person) who wishes to connect to the Hospital Network just uses his conventional telephone and dials a special sequence number. The special sequence numbers have been devised for elderly users to facilitate easiness of not adapting to the complex IP address scenario of actual usage. The special sequence number, for example #*0~9 is the default code for the Healthcare System to facilitate a connection to the users default Hospital Net-

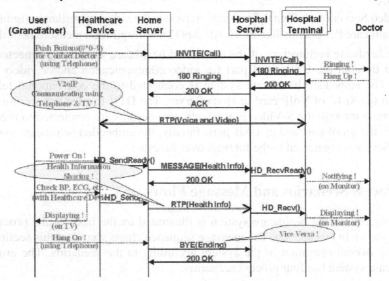

Fig. 3. Message Sequence Chart of User Operated Normal Mode

work; his family doctor in a practical scenario. The HomeServer recognizes special sequence characters beginning with #* as commands to the Healthcare System. The HomeServer then creates a SIP connection using a SIP INVITE method to the Hospital Server that in turn connects to the corresponding Hospital Terminal. The SIP leg is established between the HomeServer and the Hospital Terminal on the hospital network. The RTP transaction begins on successful SIP transaction and enables voice and video communication between the doctor at the concerned Hospital Terminal and the user at home.

The doctor instructs the user to deliver the required measurements. The user activates the healthcare devices connected to the HomeServer, which in turn activates the HD_SendReady() function in the healthcare application. The HomeServer wraps the following measurement message in a SIP MESSAGE method. The receipt of the following message at the Hospital Server invokes HD_RecvReady() call to the Hospital Terminal to facilitate actual measurement data transfer. Subsequently, the user sends the HD_Send() call to the hospital server using the RTP payload. The following measurement information is displayed on both the user terminal and the hospital monitor. The elderly home user and the doctor discuss on the measurements using voice and video.

The user or the doctor can then disconnect the voice and video communication. The SIP BYE method invokes the termination of the communication session between the user and the doctor as shown in Figure 3.

3.2 User Operated Emergency Mode

Figure 4 shows the process scenario in case of the user operated emergency mode. The operational details for this mode are described below:

1. Patient sends a special sequence character(emergency code, #*9) to activate the Healthcare System connection to alert the doctor's Hospital Terminal.
2. Simultaneously, the Healthcare System connects to the family.
3. Doctor performs surveillance on patient and takes appropriate action. (Doctor may report to ambulance and/or send a nurse and/or decide to treat the patient on his own).

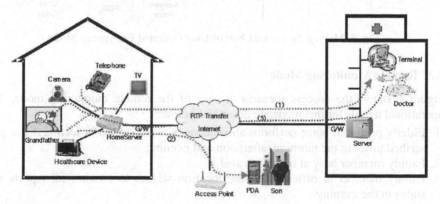

Fig. 4. Scenario of User Operated Emergency Mode

Figure 5 shows the message sequence chart in case of the user operated emergency mode. The operation of the Healthcare System in emergency mode involves the elderly user pressing an emergency sequence key or a single-key mapping customized to the user needs. The telephone attached to the HomeServer is used for key messaging the Healthcare System. Other devices that could be used include remote controller and PDA/SmartPhone. The DTMF tone processor at the HomeServer obtains the special sequence characters and forwards the message to the Healthcare System on obtaining emergency healthcare contact sequence numbers. The Healthcare System instructs the HomeServer device to initiate a SIP transaction with the hospital network and/or other configured emergency monitoring services. The SIP transaction begins with the SIP MESSAGE method for emergency notification, which invokes a SIP leg from the doctor and/or the monitoring authority. Several cameras configured at his home and connected to the HomeServer then monitor the user. The doctor and/or monitoring authority take appropriate action based on user condition. The actions include direct visit by doctor/nurse to patient at his residence, ambulance visit or other non-critical actions based on user condition.

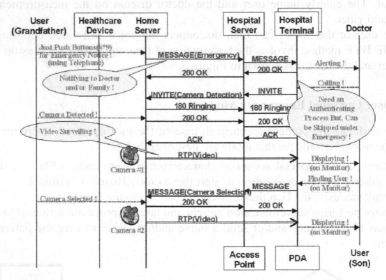

Fig. 5. Message Sequence Chart of User Operated Emergency Mode

3.3 Remote Monitoring Mode

Figure 6 shows the process scenario in case of the remote monitoring mode. The operational details for this mode are described below:

1. Elderly person at home performs all sequences as illustrated in section 3.1 at prescribed times in the morning, afternoon and evening
2. Family member busy at meeting/travel
3. Family member at office or remote location wishes to monitor all reports and status in the evening
4. Family member connects to the HomeServer and monitors stored information.
5. He takes appropriate action based on reports and/or discusses with doctor.

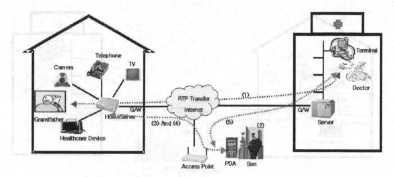

Fig. 6. Scenario of Remote Monitoring Mode

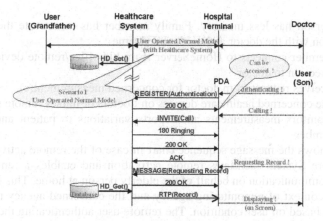

Fig. 7. Message Sequence Chart of Remote Monitoring Mode

Figure 7 shows the message sequence chart in case of the remote monitoring mode. The Healthcare System facilitates a family member or doctor to monitor the measurements of the home user remotely. This usage scenario is extremely helpful when a concerned family member or doctor is not available for real-time contact due to emergency. The elderly person initiates a transaction to the hospital network and delivers the measurements at the prescribed time and schedule. The remote user can then authenticate the Healthcare System using the SIP REGISTER method and monitor the measurement records. The remote user then takes the appropriate action based on the needs of the context. The remote user is generally considered to be a concerned family member who has to know about the status of the elderly person at home. He may initiate a transaction with the doctor's Hospital Terminal to discuss on the measurement reports stored in the users Healthcare System.

3.4 Remote Activation Mode

Figure 8 shows the process scenario in case of the remote activation of the Healthcare System. The operational details for this mode are described below:

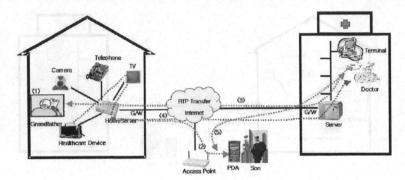

Fig. 8. Scenario of Remote Activation Mode

1. Elderly person has less memory. Family member has to activate the concerned consultation with the doctor at the scheduled time
2. Family member connects to HomeServer from his office/remote device and activates connection to doctor
3. Patient receives request from doctor for concerned measurement(s)
4. He uses the concerned healthcare devices on doctors or nurses remote supervision
5. Doctor monitors measurements and reports variations to patient and/or remote family member

Figure 9 shows the message sequence chart in case of the remote activation mode. The Healthcare System facilitates remote activation and enables a remote user to activate the communication on behalf of the elderly person at home. This is extremely useful in the context of a monitoring agency and the concerned agency takes appropriate actions based on user condition. The remote user authenticating the Healthcare System to make a call to the Hospital Terminal initiates this procedure. The remote

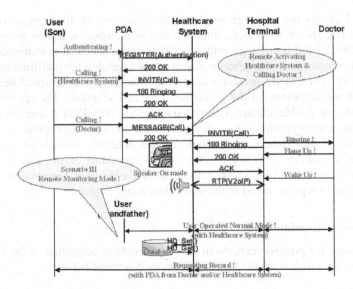

Fig. 9. Message Sequence Chart of Remote Activating Mode

user initiates a SIP transaction to the Healthcare System. The user invokes a SIP MESSAGE method to facilitate the Healthcare System to initiate a SIP transaction to the Hospital Terminal. The doctor at the Hospital Terminal uses the V2oIP channel to instruct the elderly user to begin the user operated normal mode.

3.5 Snapshots of Embedded Healthcare System

Figure 10 demonstrates the screen shot on the user TV containing the measurement factors, doctor's picture, user's picture. The healthcare information chart demonstrates the user's measurements for ECG, BP, temperature, etc. The doctor and the user discuss on the measurements when they are seeing each other.

Figure 11 shows the exhibition of embedded healthcare system in SoftExpo2003, which was held at COEX in Seoul on December 2003.

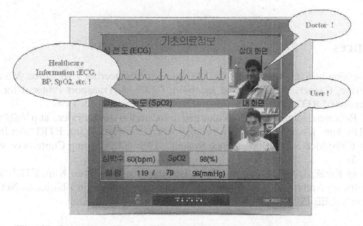

Fig. 10. Snapshot of Embedded Healthcare System Screen Structure

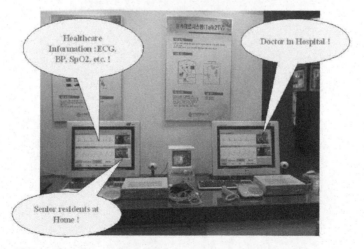

Fig. 11. Snapshot of Embedded Healthcare System In SoftExpo2003

4 Conclusion

Current technology trends enable e-Healthcare services to be available dynamically on user demand. However, the evolution of these precious and much needed services to the senior residents and needed persons is still at its infancy. We presented a novel approach to facilitate a common man to utilize these much-needed services using the conventional telephone. This paper also presents the feature of the Healthcare System that enables a remote user/doctor to activate and monitor the elderly person and facilitate e-Healthcare into every household. HomeServer is a common home device of the future and integrating the healthcare service to the HomeServer provided a new evolution in e-Healthcare. The system design is integrated to the HomeServer to enhance the existing features of the HomeServer, create added potential to a new healthcare market and primarily to provide these valued services to the needed persons of our society.

References

1. "SIP: session initiation protocol", rfc2543, Internet Engineering Task Force, March 1999.
2. Schulzrine, Casner, Frederick, and Jacobson, "RTP: A Transport Protocol for Real-Time Applications," RFC 1889, Internet Engineering Task Force, Feb. 1996
3. ITU-T Recommendation H.261 - Video codec for audiovisual services at p*64kbits/s, 1993
4. Sung-Ho Ahn, Kyung-Hee Lee, Ji-Young Kwak, Doo-Hyun Kim, ETRI "An Implementation of Embedded Video Surveillance System", 19th KIPS Spring Conference Vol.10 No.1 (2003. 5)
5. Ji-Young Kwak, Dong-Myong Sul, Sung-Ho Ahn and Doo-Hyun Kim, ETRI "An Embedded Software Architecture for Connected Multimedia Services in Ubiquitous Network Environment", IEEE ISORC2003/WSTFES2003, Hakodate, Japan

self-CD: Interactive Self-collision Detection for Deformable Body Simulation Using GPUs

Yoo-Joo Choi[1], Young J. Kim[1], and Myoung-Hee Kim[1,2,*]

[1] Department of Computer Science and Engineering
Ewha Womans University, Seoul, Korea
{choirina,kimy}@ewha.ac.kr
[2] Center for Computer Graphics and Virtual Reality
Ewha Womans Univesity, Seoul, Korea
Tel: +82-2-3277-4418, Fax: +82-2-3277-4409
mhkim@ewha.ac.kr.

Abstract. This paper presents an efficient self-collision detection algorithm for deformable body simulation using programmable graphics processing units (GPUs). The proposed approach stores a triangular mesh representation of a deformable model as 1D textures and rapidly detects self-collisions between all pairs of triangular primitives using the programmable SIMD capability of GPUs [1]. Since pre-computed spatial structure such as bounding volume hierarchy is not used in our algorithm, our algorithm does not require expensive runtime updates to such complex structure as the underlying model deforms. Moreover, in order to overcome a potential bottleneck between CPU and GPU, we propose a hierarchical encoding/decoding scheme using multiple off-screen buffers and multi-pass rendering techniques, which reads only a region of interests in the resulting off-screen buffer.

1 Introduction

Interactive simulation of deformable objects in real-time is crucial for many applications such as virtual reality, surgical simulation and 3D interactive games [2, 3]. Typical examples of objects under deformation include soft tissues, human organs, articulated human characters with clothes on, biological structure, elastic material, etc. In these applications, collision detection is considered as a major computational bottleneck, which makes real-time deformable body simulation difficult. The problem becomes even more complicated when simulated objects are non-convex and deform severely over time, because self-collisions inside an object may happen during the object deformation. In this case, we need to correctly handle both self- and inter- object collisions in order to realistically and robustly simulate object deformation.

Traditionally, many researchers have advocated the use of spatial structure such as bounding volume hierarchies (BVH) to efficiently check for collisions between generic polygonal models. However, since the shape of a deformable object is constantly changing at runtime, pre-computed BVHs may be inefficient for collision detection of severely deforming objects. Moreover, BVHs must be dynamically updated at runtime,

---

* Corresponding author.

D.-K. Baik (Ed.): AsiaSim 2004, LNAI 3398, pp. 187–196, 2005.

and this imposes substantial computational costs. Worse yet, in order to detect self-collisions inside a deforming object, one should be able to discern potentially colliding parts of the object from non-colliding parts so that he or she can focus on checking collisions only between colliding parts. However, this task is non-trivial since there is no easy way to exactly determine which part of a model is collision-free and which part is not. Moreover, any primitive (e.g., triangle) consisting of a deformable model can be a potentially colliding part. As a result, it is quite challenging to report all pairwise self-collisions inside a deformable model. It is no coincidence that many researchers in the computer-aided geometric design community also have recognized that detecting and resolving self-colliding surfaces is one of the difficult problems appearing in applications like swept volume computations [4], Minkowski sums [5], arrangement problems [6], etc.

Main Results: In this paper, we present an effective approach to detect self-collisions inside a deformable object using programmable GPUs. The proposed approach is attractive in the following senses:

- **Robust Self-collision Detection Using GPUs:** All pairwise self-collisions between primitives in a deformable model are rapidly detected using the programmable SIMD capability of GPU.
- **Rapid Hierarchical Readback:** Using a hierarchical readback scheme, reading the collision results from GPU back to CPU is considerably accelerated.
- **No Precomputation:** Expensive runtime updates of complex spatial structure such as BVHs are not necessary.
- **Generality:** No assumption about input primitives is required as long as the primitives are polygonized.
- **Easy Extension to Inter-object Collisions:** Self- and inter-object collision can be simultaneously handled with a minimal modification.

Organization: The rest of the paper is organized in the following manner. We give a brief summary of the related work in Sec. 2 and present an overview of our approach in Sec. 3. Sec. 4 describes our self-collision detection algorithm in more detail. We present acceleration methods to improve the overall performance in Sec. 5. Sec. 6 presents the implementation results of our algorithm. We conclude the paper in Sec. 7 and discuss future work.

2   Related Work

In this section, we briefly review previous work related to collision detection of deformable objects.

Van den Bergen [7] have suggested a collision detection scheme for deformable objects using axis-aligned bounding box (AABB) trees. He describes a way to speed up an overlap test between AABBs and also shows how to quickly update AABB trees as the underlying model is deformed. However, since his method is based on AABB trees, the computational cost of updating AABB trees is still high as the model is severely deformed. Moreover, he does not consider self-collisions inside a model. Another previous work using BVH by Larsson and Moeller [8] also imposes a substantial computational cost and storage overhead for complex deformable objects. Volino et al. [9], Hughes et al. [10], and Lau and Chan [11] have suggested self-collision detection algorithms for deformable objects. However, these methods are too conservative or restricted to curved surface models.

Most of the earlier approaches [3, 12–14] that employ GPUs for collision detection have been restricted to convex objects or focused on inter-object collisions only. Recently, the GPU-based approach by Govindaraju et al. [12] has been extended to handling self-collisions [15]. However, the efficiency of the method is governed by the underlying image space resolution and relative visibility of the self-intersecting areas.

3 Background and Overview

In this section, we give a brief introduction to using GPUs as a general-purpose processor and also provide an overview of our self-collision detection algorithm.

3.1 General-Purpose Computing on GPUs

Recently, many research work [16, 17] has shown how non-graphical, generic computation can be performed on GPUs. The reasons why one uses GPUs as general-purpose computing processors can be summarized as follows [18]:

- **Performance:** A 3 GHz Pentium 4 processor can perform six giga-flops, whereas a pixel shader running on modern GPUs such as NVIDIA GeForce FX 5900 Ultra, can perform 20 giga-flops, roughly equivalent to a 10 GHz Pentium 4.
- **Load Balancing:** We can obtain the overall performance speedup of a CPU-intensive application by balancing its work load on CPU and GPU.
- **Performance Growth:** The development cycle of GPU shows more steep growth rates than that of CPU for the past decade, and this trend is expected to continue.

GPU is not a serial processor, but a stream processor. A serial processor executes an instruction and updates memory sequentially. However, a stream processor works differently by executing a function (e.g., pixel shader in GPU) on a set of input data in parallel and producing a set of output data (e.g., shaded pixels in the frame buffer). Each data element passed to the stream processor is processed independently without any inter-dependency. In GPUs, arbitrary data type can be stored at floating point textures or off-screen buffers and a pixel shader can perform generic calculations on the stored data. This process is actually realized by rendering a simple geometry (e.g., quadrilateral) onto a viewing window whose dimension is equal to the number of data elements necessary for the calculation.

3.2 Overview of Our Approach

The main idea of our self-collision detection algorithm is to store each vertex of triangle primitives at 1D textures, and employ pixel shaders to detect collisions between all possible pairwise combinations of triangles in the model. The pairwise checking is realized by interweaving these 1D triangle textures.

The overall pipeline of our algorithm can be subdivided into four major steps as follows:

1. **Collision Geometry Setup:** The position of each triangle is stored at three 1D textures. Then, a quadrilateral is texture-mapped using the 1D triangle textures while the textures are periodically duplicated across the quadrilateral both in horizontal and vertical directions
2. **Topology Culling:** A pair of triangles is symmetrically represented on the quadrilateral; i.e., the (i, j) and (j, i) pairs represent the same triangle pairs. We need to avoid examining these duplicate triangle pairs. Moreover, pairs of adjacent triangles sharing edges or vertices are excluded in the pairwise collision detection since they are reported as being collided all the time, even though in practice such collision between neighboring primitives can be avoided. We use the stencil test supported by GPUs to efficiently implement the topological culling step.
3. **Pairwise Collision Detection:** The triangle pairs that survive the stencil test are examined for collisions using the separating axis testing (SAT) [7, 19]. We first perform three SATs based on the AABBs of triangles, followed by eleven SATs based on the vertex/face and edge/edge combinations of a pair of triangles.
4. **Hierarchical Readback of Colliding Pairs:** In order to effectively read collision results back from GPU, we render the collision results onto off-screen buffers using a multi-pass rendering technique, where each pass hierarchically encodes lower-level collisions results. By reading backwards from the highest level buffer to the lowest level buffer, we can selectively obtain needed collision reports without reading the entire contents of the lowest level buffer.

4   Pairwise Collision Detection Using Texturing

In this section, we describe an internal representation of all pairs of primitives on GPUs and how to detect collisions between them using texturing and programmable shaders.

4.1   Collision Geometry Setup

Initially, we store the position of each triangle at 1D textures. Then, we attempt to detect collisions between all pairs of triangles by using pixel shaders. The coordinate of a vertex in a triangle is stored at texels in multiple 1D textures. Since typical texturing hardware supports only four channels (RGBA) per texel, three 1D textures are required to represent a single triangle. These three textures are periodically texture-mapped to a quadrilateral both in horizontal and vertical directions. As a result, all pairwise combinations of triangles are represented on the texture-mapped quadrilateral. Fig. 1 illustrates how to construct six 1D textures and how to map them to a quadrilateral.

4.2   Pairwise Collision Detection

Rendering a textured quadrilateral invokes pixel shaders to compute certain operations on each pixel in the frame buffer. In our case, pixel shaders perform a collision checking between all pairs of triangles by referencing six 1D textures.

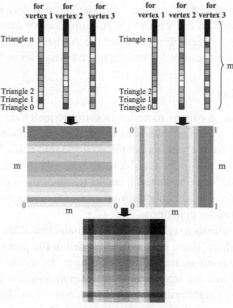

Quad. with six textures (All pairs of triangles)

Fig. 1. Representation of All Pairs of Triangle Primitives as a Set of 1D Textures. Here, n is the total number of triangle primitives and m is the closest number to n in powers of two ($m \geq n$).

We use a separating axis test (SAT) similar to the one proposed by Gottschalk et. al [19] to check for collisions between triangle pairs. However, instead of fifteen SATs for OBBs, we use only eleven separating axes that are calculated based on two vertex/face (VF) and nine edge/edge (EE) combinations from a pair of triangles. In the VF case, we use the face normal of F as a separating axis and, in the EE case, we use the cross product of two edge (EE) directions.

Once the rendering is complete, the frame buffer contains final results and by reading the buffer from GPU back to CPU, we can obtain the final results.

5 Performance Acceleration Techniques

In order to improve the performance of our collision detection algorithm, we employ various techniques to accelerate both collision detection and readback routines. This section introduces such performance acceleration techniques in detail.

5.1 AABB Culling

The most time-consuming part of our algorithm is per-pixel operation which realizes a pairwise triangle/triangle collision checking. We can accelerate this process by culling away distant triangles. We employ an AABB-based overlap test, which has been favored by many earlier collision detection algorithms. In our case, we simply considers additional three face normal directions of an AABB as separating axes in the SAT. These face normals correspond to the directions of principal axes. As a result, before applying eleven SAT tests derived by the VF and EE combinations of triangles, we first perform the AABB overlap test to determine whether the AABBs of a triangle pair overlap. If the AABBs overlap, we continue to perform more costly eleven SAT tests.

5.2   Topology Culling

In the basic algorithm, a pair of triangles is symmetrically represented on a texture-mapped quadrilateral. Therefore, we need to invoke pixel shaders selectively in order to prevent the shaders from examining duplicated triangle pairs. We use HW-supported stencil tests to achieve this objective. More specifically, when we render a textured quadrilateral, we mask out duplicated triangle textures by setting the corresponding stencil bits to zeros. As a result, only relevant areas are rendered; i.e., the same triangle pairs are not considered for collisions.

Notice that topologically neighboring triangle pairs that share vertices and edges are always reported colliding. These should be excluded in the pairwise collision detection by clearing the corresponding stencil bits, since they are collided from the beginning. With extra computations, one may be able to distinguish a real triangle collision from the shared topology of neighboring triangles. This type of a collision problem between adjacent triangles are beyond the scope of this paper and we will attempt to address these issues in our up-coming companion paper. Fig. 2 depicts the result of the stencil masking.

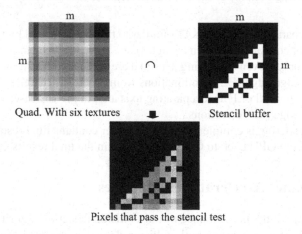

Fig. 2. Topology Culling Using Stencil Tests. Texels corresponding to duplicated and adjacent triangle pairs are masked out by the stencil test.

5.3   Hierarchical Readback

There are several limiting factors to GPU-based geometric computations. The readback time between CPU and GPU is one of them, imposed by the current bus architecture between CPU and GPU. In order to reduce the readback overhead, the number of readback operations and the size of a readback buffer must be minimized as much as possible. We explain how we can improve the readback performance by devising a selective readback scheme based on a hierarchical encoding/decoding of frame buffer results.

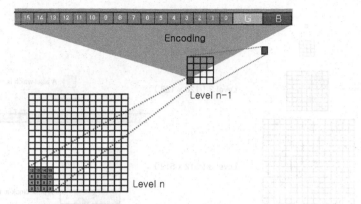

Fig. 3. Hierarchical Encoding of Off-screen Buffers. Each pixel in the higher level buffer encodes the contents of 16 pixels in the lower level buffer.

In order to realize the hierarchical encoding, first of all, we do not directly render the textured quadrilateral onto the frame buffer; instead, we render it onto off-screen buffers. More specifically, if m is the closest number to the total number n of triangle primitives in powers of two ($m \geq n$), the result of collision detection is stored at an $m \times m$ off-screen buffer. Then, we consecutively render the off-screen buffer to another off-screen buffer by reducing the dimension of the off-screen buffer by a factor of 4×4. By repeating this process, we construct hierarchical readback structure. In this structure, each pixel in the higher level buffer encodes the contents of 16 pixels in the lower level buffer. The contents of the lowest level buffer contains only Boolean values, which denote the results of collision detection; i.e., TRUE means a collision. Notice that we use only one channel (e.g., Red) among the four channels (RGBA) available in the off-screen buffer since we encode only 16 pixels of a lower level buffer. Fig. 3 illustrates the procedure of a hierarchical encoding of off-screen buffers.

For example, if the size of the lowest level off-screen buffer is 512^2, additional three buffers in sizes of 128^2, 32^2, and 8^2 are created. In order to analyze the content of the lowest level buffer (level 3, 512^2), we first read the highest level buffer (level 0, 8^2). By examining each pixel in the level 0 buffer, we select only non-zero pixels and, for each non-zero pixel, we extract the corresponding 16 pixels in the level 1 buffer. We recursively repeat this process until we reach the level 3 buffer. In Fig. 4-(a), the non-empty pixels denote those pixels that need to be read from GPU. Pixels in level 0 and level 2 are locally read whenever necessary. Fig. 4-(b) explains the relationship between a pixel in a higher level buffer and its corresponding block of pixels in a lower level buffer. The non-empty pixels in Fig. 4-(b) denote the pixels representing the colliding result of a triangle pair.

6   Results and Analysis

We tested the performance of the proposed approach on five different benchmarking models as shown in Table 1. Fig.5 shows deformed can, cloth and pipe benchmarking models. Our experiments were carried out on a 3.4 GHz Pentium 4 PC equipped with a GeForce 6800 GPU (NV40).

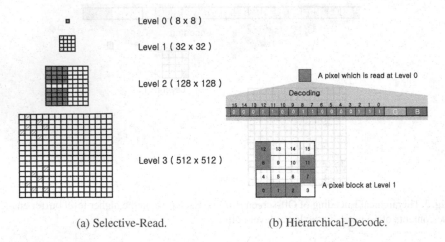

(a) Selective-Read.   (b) Hierarchical-Decode.

Fig. 4. Selective Readback and Hierarchical Encoding of an Off-screen Buffer. (a) Pixels in level 0 and level 2 are selectively read. (b) Relationship between a pixel in a higher level and a block of pixels in a lower level.

Table 1. Complexity of Benchmarking Models.

Model Complexity	Stomach	Liver	Pipe	Table Cloth	Can
Number of Vertex	1116	139	467	323	899
Number of Face	372	274	930	588	966

Table 2. Algorithm Performance. Our algorithm can detect self-collisions for the benchmarking models at 16~67 FPS.

Model	Updating Texture	Culling	Collision Detection	H-Encoding	Selective Readback	Total
Stomach	0.0002	0.0024	0.0123	0.0007	0.0004	0.0163
Liver	0.0002	0.0024	0.0109	0.0007	0.0004	0.0149
Pipe	0.0004	0.0084	0.0488	0.0021	0.0008	0.0608
Cloth	0.0004	0.0084	0.0488	0.0021	0.0007	0.0606
Can	0.0004	0.0084	0.0488	0.0021	0.0009	0.0609

For moderately complex deformable models, we were able to detect all self-collisions at 16~67 frames per second as shown in Table 2. Moreover, in order to prove the effectiveness of our hierarchical readback scheme, we compared the performance of our algorithm with and without the hierarchical readback scheme as shown in Table 3; The performance was improved by about 73.6 % using the hierarchical readback scheme.

7   Conclusions

In this paper, we have proposed an efficient self-collision detection algorithm for a deformable model using programmable GPUs. Using the programmable SIMD capability of GPUs, we rapidly detect collisions between all pairs of triangle primitives in a de-

Table 3. Comparison of Readback Performance. Notice that, using the hierarchical readback scheme, the readback performance is improved by about 73.6%.

Readback Type	Stomach	Liver	Pipe	Table Cloth	Can
Full Readback	0.0035	0.0036	0.0142	0.0132	0.0135
Encoding/Selective Readback	0.0011	0.0011	0.0030	0.0029	0.0030

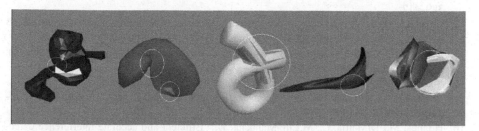

Fig. 5. Deformed Models with Self-Collisions. Self-colliding areas are colored red.

formable model without relying on costly spatial structure such as BVH. Moreover, using the hierarchical encoding of collision results stored at off-screen buffers, we improve the performance of reading the collision results back from GPU. For a moderately complex deformable model, we are able to report all self-collisions at interactive rates.

There are a few limitations of our work. First of all, unfortunately, we have been unable to fully optimize the implementation of our collision detection algorithm on the experimental GPU (GeForce 6800) that we use throughout our implementation. As a result, we limit our experiments to the benchmarking models of moderately complex sizes (i.e., about 1K triangles). However, we expect to apply our approach to more complex models by using more optimized implementation of our algorithm as well as by combining classical SW-based approaches like BVH. We do not consider checking for self-collisions between neighboring triangle primitives since most of physically-based applications prevent inter-penetrations between neighboring primitives.

Finally, we will like to extend our framework to other proximity calculations requiring pairwise computations, such as distance calculation or penetration depth computations.

Acknowledgements

This work was partially supported by the Korean Ministry of Science and Technology under the NRL Program, the Korean Ministry of Information and Communication under the ITRC Program and the grant No. R08-2004-000-10406-0 from the Korean Ministry of Science and Technology.

References

1. Rost, R.J.: OpenGL Shading Language. Addison Wesley (2004)
2. Cotin, S., Delingette, H., Ayache, N.: A hybrid elastic model for real-time cutting, deformations, and force feedback for surgery training and simulation. In: The Visual Computer. Volume 16. (2000) 437–452

3. Lombardo, J., Cani, M., Neyret, F.: Real-time collision detection for virtual surgery. In: Proceedings of Computer Animation 99. (1999) 33–39

4. Elber, G., Kim, M.S.: Offsets, sweeps, and Minkowski sums. Computer-Aided Design **31** (1999) 163

5. Ramkumar, G.D.: Tracings and Their Convolutions: Theory and Application. PhD thesis, Standford (1998)

6. Halperin, D.: Arrangements. Handbook of Discrete and Computational Geometry (1997) 389–412

7. van den Bergen, G.: Efficient collision detection of complex deformable models using AABB trees. In: Journal of Graphics Tools. Volume 2. (1997) 1–13

8. Larsson, T., Akenine-Moeller, T.: Collision detection for continuously deforming bodies. In: Proceedings of Eurographics. (2001) 325–333

9. Volino, P., Magnenat-Thalmannhen., N.: Efficient self-collision detection on smoothly discretized surface animations using geometrical shape regularity. In: Proceedings of Eurographics 94. (1994) 155–166

10. Hughes, M., DiMattia, C., M.Lin, Manocha, D.: Efficient and accurate interference detection for polynomial deformation and soft object animation. In: Proceedings of Computer Animation Conference. (1996) 155–166

11. Lau, R.W., Chan, O.: A collision detection framework for deformable objects. In: Proceedings of the ACM Symposium on Virtual Reality and Technology. (2002) 113–120

12. Govindaraju, N.K., Redon, S., Lin, M.C., Manocha, D.: CULLIDE:interactive collision detection between complex models in large environments using graphics hardware. In: ACM SIGGRAPH/EUROGRAPHICS Graphics Hardware. (2003) 25–32

13. Hoff, K., Zaferakis, A., Lin, M., Manocha, D.: Fast and simple 2D geometric proximity queries using graphics hardware. In: Proceedings of Symposium on Interactive 3D Graphics. (2001) 145–148

14. Baciu, G., Wong, W., Sun, H.: Recode: an image based collision detection algorithm. In: The Journal of Visualization and Computer Animation. (1999) 181–192

15. Govindaraju, N.K., Lin, M.C., Manocha, D.: Fast self-collision culling in general environment using graphics processors. In: Technical Report TR03-044 of University of North Carolina at Chapel Hill. (2003)

16. Krueger, J., Westermann, R.: Linear algebra operators for gpu implementation of numerical algorithms. In: Proceedings of SIGGRAPH. (2003) 908–916

17. Bolz, J., Farmer, I., Grinspun, E., Schroeder, P.: Sparse matrix solvers on the gpu: Conjugate gradients and multigrid. In: Proceedings of SIGGRAPH. (2003) 917–924

18. Pellacini, F., Vidimce, K.: Chapter 37. a toolkit for computation on gpus. In: GPU Gems , Addison Wesley. (2004) 621–636

19. Gottschalk, S., Lin, M.C., Manocha, D.: OBBTree: A hierarchical structure for rapid interference detection. In: Proceedings of SIGGRAPH 96. (1996) 171–180

On the Passive Vibration Damping
by Piezoelectric Transducers
with Inductive Loading

Takao Tsuchiya[1] and Yukio Kagawa[2]

[1] Doshisha University,
1-3 Miyakodani, Tatara, Kyotanabe City, Kyoto 610-0321, Japan,
ttsuchiy@mail.doshisha.ac.jp
[2] Akita Prefectural University,
84-4 Tsuchiya-Ebinokuchi, Honjo City, Akita 015-0055, Japan

Abstract. The increase of the effectiveness of the passive damping in mechanical vibrations is discussed with piezoelectric transducers. An electrical resistor with shunt inductor is connected between the electrodes of the piezoelectric transducers, which transformed the vibrating energy into the electrical energy by means of the electromechanical coupling. The electrical equivalent circuit modeling has been developed for an electromechanical coupling system. The circuit parameters are determined by the modal analysis based on the finite element model. Some examinations are demonstrated for a thin plate. It is found that the introduction of the electrical shunt inductor is more effective than the simple electrical resistor termination because the electrical inductance cancels the equivalent shunt capacitance or damped capacitance between the electrodes.

1   Introduction

The damping in plates and structures are often required to reduce or control their vibrations. The simple way to control the vibrations is to apply a viscoelastic damping layer over the plates and structures. Extensive investigations have been made since Oberst's pioneering work [1]. To control the damping, the use of the piezoelectric layer transducers on the surface of the plates and structures has been proposed by many investigators [2, 3]. The vibrating energy is expected to be transformed into the electrical energy by means of the electromechanical coupling and dissipated in terms of the electrical energy. In our previous paper [4, 5], we have discussed the effective range of the passive damping of the mechanical vibrations with piezoelectric transducers.

The present paper discusses the improvement of effectiveness of the passive mechanical damping with piezoelectric transducers, to which the electrical resistor with shunt inductor is connected to cancel the effect of the shunt or damped capacitance. To evaluate the damping effect, the electrical equivalent circuit modeling developed for the piezoelectric vibrations is used [4–6]. The circuit parameters of the equivalent circuit are determined by the modal analysis

D.-K. Baik (Ed.): AsiaSim 2004, LNAI 3398, pp. 197–204, 2005.
© Springer-Verlag Berlin Heidelberg 2005

based on the finite element model. Examination is demonstrated for a thin plate in which a one-dimensional thickness vibration plays a main role. The damping effectiveness is evaluated in terms of the loss factor of the total system.

2   Equivalent Circuit Modelling

The discretized finite element expression for the steady-state vibration of a piezo-electric body partially electroded has the following form:

$$
\begin{bmatrix} (1+j\eta)\boldsymbol{K} - \omega^2\boldsymbol{M} & \boldsymbol{P}_p & \boldsymbol{P}_q \\ \boldsymbol{P}_p^* & -\boldsymbol{G}_{pp} & -\boldsymbol{G}_{pq} \\ \boldsymbol{P}_q^* & -\boldsymbol{G}_{pq}^* & -\boldsymbol{G}_{qq} \end{bmatrix} \begin{bmatrix} \boldsymbol{\xi} \\ \boldsymbol{\phi}_p \\ \boldsymbol{\phi}_q \end{bmatrix} = \begin{bmatrix} \boldsymbol{f} \\ 0 \\ 0 \end{bmatrix}
\tag{1}
$$

where $\boldsymbol{K}, \boldsymbol{M}, \boldsymbol{G}$ and $\boldsymbol{P}$ are stiffness, mass, capacitance and electromechanical coupling matrices, $\boldsymbol{\xi}$ and $\boldsymbol{\phi}$ are displacement and electric potential vectors, $\boldsymbol{\phi}_p$ and $\boldsymbol{f}$ are applied voltage to the electrodes and external force. Subscript p refers to the nodes associated with electrode and subscript q to the nodes of the non-electroded region, ω is vibratory angular frequency, η is mechanical loss factor, and $(\ )^*$ denotes the transposition of a matrix or a vector. In eq. (1), the structural damping is assumed without electrical damping.

Eliminating the electric potential in eq. (1), a standard equation of motion is given:

$$
\left(\widetilde{\boldsymbol{K}} + \widetilde{\boldsymbol{A}}/C_d + j\omega\widetilde{\boldsymbol{R}} - \omega^2\widetilde{\boldsymbol{M}} \right) \boldsymbol{\xi} = \boldsymbol{f}
\tag{2}
$$

where

$$
\begin{aligned}
&\widetilde{\boldsymbol{K}} = \boldsymbol{K} + \boldsymbol{P}_q\boldsymbol{G}_{qq}^{-1}\boldsymbol{P}_q^*, \ \ \widetilde{\boldsymbol{A}} = \boldsymbol{A}_p\boldsymbol{A}_p^*, \ \ \widetilde{\boldsymbol{R}} = \eta\boldsymbol{K}/\omega, \\
&\widetilde{\boldsymbol{M}} = \boldsymbol{M}, \ \ \boldsymbol{A}_p = \boldsymbol{P}_q\boldsymbol{G}_{qq}^{-1}\boldsymbol{G}_{pq} - \boldsymbol{P}_p, \\
&C_d = \boldsymbol{G}_{pp} - \boldsymbol{G}_{pq}\boldsymbol{G}_{pp}^{-1}\boldsymbol{G}_{pq}^*
\end{aligned}
\tag{3}
$$

and C_d is shunt or damped capacitance.

Now the normalized coordinate system $\overline{\boldsymbol{X}}$ is introduced so that

$$
\boldsymbol{\xi} = \overline{\boldsymbol{\Xi}}\,\overline{\boldsymbol{X}} = \sum_{m=1}^{N} \overline{\boldsymbol{\xi}}_m\overline{X}_m
\tag{4}
$$

where $\overline{\boldsymbol{\Xi}}$ is the modal matrix defined as

$$
\overline{\boldsymbol{\Xi}} = \left[\overline{\boldsymbol{\xi}}_1\, \overline{\boldsymbol{\xi}}_2 \cdots \overline{\boldsymbol{\xi}}_N \right]
\tag{5}
$$

where $\overline{\boldsymbol{\xi}}_m$ is the eigenvectors corresponding to the m th mode which is the solution of the characteristic equation for the undamped free vibration. N is the degree of freedom of the system. Substituting eq. (4) into the eq. (2) and pre-multiplying $\overline{\Xi}^*$ on both side of the equation, the equation of motion can be expressed as

$$(\overline{K}^{\text{diag}} + \overline{A}^{\text{diag}}/C_d + j\omega\overline{R}^{\text{diag}} - \omega^2\overline{M}^{\text{diag}})\overline{X} = \overline{F} \tag{6}$$

where the superscript ^{diag} denotes the diagonal matrix. The diagonal components of $\overline{K}^{\text{diag}}$, $\overline{M}^{\text{diag}}$ and $\overline{R}^{\text{diag}}$ are respectively expressed as follows

$$\overline{K}_m = \overline{\xi}_m^*\widetilde{K}\overline{\xi}_m, \quad \overline{M}_m = \overline{\xi}_m^*\widetilde{M}\overline{\xi}_m,$$
$$\overline{R}_m = \overline{\xi}_m^*\widetilde{R}\overline{\xi}_m, \quad \overline{A}_m = \overline{\xi}_m^*\widetilde{A}\overline{\xi}_m \tag{7}$$

These parameters correspond to the modal stiffness, modal mass, modal damping and modal force factor for the m th mode, respectively. $\overline{F}$ in eq. (6) is expressed as

$$\overline{F} = \overline{\Xi}^* f \tag{8}$$

The modes are thus all decoupled and the following N independent single mode equations are derived

$$\left(\overline{R}_m + j\omega\overline{M}_m + \frac{1}{j\omega\overline{C}_m} + \frac{\overline{A}_m^2}{j\omega C_d}\right)\overline{V}_m = \overline{F}_m \tag{9}$$

where $\overline{V}_m = j\omega\overline{X}_m$ is the displacement velocity, and $\overline{C}_m = 1/\overline{K}_m$. Making equivalence of $\overline{V}_m$ to the electric current and $\overline{F}_m$ to the electromotive force, the electrical equivalent circuit is derived as shown in Fig.+1 for the electrical input terminal. In the figure, $\overline{\xi}_m$ and $\overline{A}_m$ are the transformation ratio (winding turn number) of the ideal transformer, respectively. The equivalent circuit is valid for each mode. Their input terminals are connected in parallel for the multi-mode system.

3 Passive Vibration Damping with Piezoelectric Transducers

3.1 Electrical Resistive Loading

We now consider the piezoelectric transducer with the simple electrical resistor connected to the electrical output terminals. The mechanical energy is transformed by means of the electromechanical coupling and the dissipation is taken

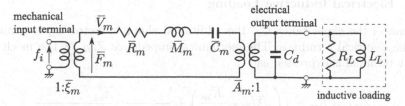

Fig. 1. Electrical equivalent circuit for the m th mode with the inductive loading connected to the electrical output terminals.

place in terms of the electrical energy in the resistor. When the resistor R_L is only connected to the electrical output terminals in Fig. 1, the displacement velocity at the resonant angular frequency ω_m is expressed as

$$\overline{V}_m = \frac{1 + (\omega_m C_d R_L)^2}{\overline{R}_m\{1 + (\omega_m C_d R_L)^2\} + R_L \overline{A}_m^2} \overline{F}_m \tag{10}$$

For minimizing the displacement velocity, one derives the impedance matching relation as follows.

$$R_{L0} = \frac{1}{\omega_m C_d} \tag{11}$$

The resonant angular frequency ω_m at the matching condition is expressed as

$$\omega_m = \sqrt{\frac{\overline{K}_m}{\overline{M}_m}\left(1 + \frac{1}{2\overline{\gamma}_m}\right)} \tag{12}$$

where $\overline{\gamma}_m$ is the capacitance ratio expressed as

$$\overline{\gamma}_m = \frac{C_d}{\overline{A}_m^2 \overline{C}_m} = \frac{\overline{K}_m C_d}{\overline{A}_m^2} \tag{13}$$

The modal damping $\overline{r}_m$ at the resonant frequency is therefore expressed as

$$\overline{r}_m = \overline{R}_m + \frac{\overline{A}_m^2 R_{L0}}{2} = \overline{R}_m\left(1 + \frac{1}{2\overline{\eta}_m \overline{\gamma}_m}\right) \tag{14}$$

where $\overline{\eta}_m = \omega_m \overline{C}_m \overline{R}_m$ is the loss factor of the total system when the electrical terminals are short circuited. When the resistor is connected to the electrical terminals, the loss factor is increased to

$$\overline{\eta}'_m = \omega_m \overline{C}_m \overline{r}_m = \overline{\eta}_m + \frac{1}{2\overline{\gamma}_m} \tag{15}$$

It is expected that the mechanical loss or the damping increases when the capacitance C_d is made smaller.

3.2 Electrical Inductive Loading

To cancel the capacitance C_d, the additional shunt inductor L_L is connected to the electrical terminals. The mechanical impedance Z_{mL} at the mechanical input terminal is expressed as

$$\overline{Z}_{mL} = \overline{R}_m + j\left(\omega \overline{M}_m - \frac{\overline{K}_m}{\omega}\right) + \frac{\overline{A}_m^2}{\dfrac{1}{R_L} + j\left(\omega C_d - \dfrac{1}{\omega L_L}\right)} \tag{16}$$

When the electrical anti-resonant angular frequency $\omega_e = 1/\sqrt{L_L C_d}$ matches the mechanical resonant angular frequency $\omega_m = \sqrt{K_m/\overline{M}_m}$, the modal damping $\overline{r}_m$ is maximized to be

$$\overline{r}_m = R_m + A_m^2 R_L = \overline{R}_m \left(1 + \frac{1}{\overline{\eta}_m \overline{\gamma}_m} \right) \tag{17}$$

In this matched condition, the loss factor is

$$\overline{\eta}'_m = \overline{\eta}_m + \frac{1}{\overline{\gamma}_m} \tag{18}$$

which is about twice as large as the loss factor of equation (15) when the loss factor of the electromechanical system $\overline{\eta}_m$ is small enough. The inductance L_{Lm} at the resonance ω_m is given by

$$L_{L0} = \frac{1}{\omega_m^2 C_d} \tag{19}$$

4 Numerical Demonstrations

We consider a thickness vibration of a thin plate as shown in Fig. 2. A piezo-electric ceramics of a plate $10 \times 10 \times 5$mm is considered for the vibrator. The material properties of the piezoelectric ceramics are tabulated in Table 1. The vibrator is divided into the cubic elements in which the division is $(x : y : z) = (1 : 1 : 40)$. The direction of the displacement and the polarization are parallel to the z axis. One end of the vibrator is fixed at $z = 0$ and the another end is driven by the external force of $f = 1$ (N). Both surfaces of the transducer are fully electroded.

4.1 Modal Parameters of the Vibrator

The natural frequencies and the modal parameters of the vibrator without electrical loading are calculated up to the fifth mode as tabulated in Table 2. All parameters depend on the mode. The shunt or damped capacitance calculated is 131pF.

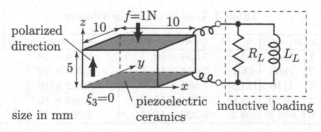

Fig. 2. A thickness vibrator and electrical inductive loading.

Table 1. Material properties of the piezoelectric ceramics (NEPEC6; Tokin Corp., Japan).

$c_{33}^E = 1.10 \times 10^{11}$ (N/m^2)	$e_{33} = 15.5$ (C/m^2)
$\varepsilon_{33}^S = 6.552 \times 10^{-9}$ (F/m)	$\rho = 7730$ (kg/m^3)
$\eta = 0.001$	$k_{33} = 0.5$

Table 2. Modal parameters of the vibrator.

m	f_m (Hz)	$\overline{K}_m$ (GN/m)	$\overline{M}_m$ (g)	$\overline{R}_m$ (N/m/s)
1	193.187	2.572	1.745	2.042
2	646.317	31.52	1.911	5.956
3	1086.274	89.09	1.912	9.871
4	1526.181	175.2	1.905	13.765
5	1967.879	289.1	1.891	17.583

4.2   Damping with Simple Resistor

The damping with the simple electrical resistor R_{L0}, which is connected between the electrodes of the transducer, is considered. The loss factors $\overline{\eta}_m$ calculated from the modal parameters are tabulated in Table 3, which shows the increase in connecting the electrical resistor.

4.3   Damping with Additional Shunt Inductor

The effect of the introduction of the additional shunt inductor is now considered. The frequency response of the displacement at the force driving point for the various inductances L_L for the fundamental mode ($m = 1$) is shown in Fig. 3. In this case, no resistor is connected ($R_L = \infty$). The resonant frequency is shifted or split into two frequencies but no considerable damping is achieved by connecting the electrical inductor as expected because no energy is dissipated in the inductor.

The case when both electrical resistor and inductor are connected in parallel between the electrode of the vibrator is then considered. From eq. (19) the matched inductor L_{L0} for the fundamental mode is calculated to be 5.181mH. Fig. 4 shows the frequency response of the displacement at the force driving point

Table 3. Loss factor of the vibrator.

m	$R_{L0}(\Omega)$	$\overline{\gamma}_m$	$\overline{\eta}_m$ (open circuited)	$\overline{\eta}_m$ (resistive loading)
1	5881	3.51	9.638×10^{-4}	1.436×10^{-1}
2	1868	43.0	7.674×10^{-4}	1.263×10^{-2}
3	1116	121.6	7.562×10^{-4}	5.112×10^{-3}
4	795.0	238.9	7.534×10^{-4}	3.093×10^{-3}
5	616.8	394.2	7.520×10^{-4}	2.268×10^{-3}

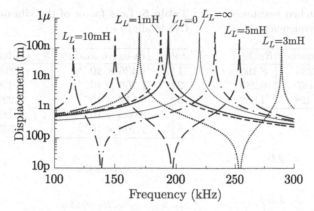

Fig. 3. Frequency response of the displacement for various inductances ($m = 1$, $R_L = \infty$).

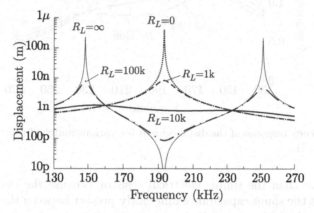

Fig. 4. Frequency response of the displacement for various resistances ($m = 1$, $L_{L0} = 5.181\text{mH}$).

for the various resistances R_L. The displacement at the natural resonant frequency ($f_1 = 193.2\text{kHz}$) decreases as the value of R_L increases. However the resonant frequency splits into two frequencies ($f_L = 148.4\text{kHz}$ and $f_H = 251.5\text{kHz}$). There are three matched resistances to be considered which correspond to three resonant frequencies (f_L, f_1 and f_H) as tabulated in Table 4. Fig. 5 shows the frequency response of the displacement for the matched resistances calculated in Table 4. The maximum damping is achieved for $R_L = 8.196\text{k}\Omega$ which is calculated for the lower resonant frequency f_L. The loss factors $\overline{\eta}'_m$ calculated from the modal parameters are tabulated in Table 5. They increase in connecting the electrical shunt inductor.

5   Conclusions

The passive damping of a simple piezoelectric vibrator is numerically examined. The numerical results show that the introduction of electrical inductor is

Table 4. Matched resistance at resonant frequencies.

Resonant frequency	R_L(kΩ)
$f_L = 148.235$	8.196
$f_1 = 193.187$	6.289
$f_H = 251.428$	4.832

Table 5. Loss factor of the vibrator with inductive loading.

m	$\overline{\eta}_m$ (resistive loading)	$\overline{\eta}_m$ (inductive loading)
1	1.436×10^{-1}	2.862×10^{-1}
2	1.263×10^{-2}	2.426×10^{-2}
3	5.112×10^{-3}	9.224×10^{-3}
4	3.093×10^{-3}	5.186×10^{-3}
5	2.268×10^{-3}	3.537×10^{-3}

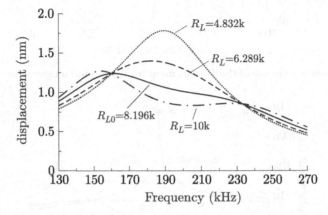

Fig. 5. Frequency response of the displacement for various matched resistances ($m = 1$, $L_{L0} = 5.181$mH).

more effective than the simple electrical resistor because the electrical inductance cancels the shunt capacitance inherently present between the electrodes of the piezoelectric devices.

References

1. H. Oberst, "Material of high inner damping," Acoustica, 6(1), pp. 144-153, 1956.
2. N. W. Hagood and A. Flotow, "Damping of structural vibrations with piezoelectric materials and passive electrical networks," J. Sound and Vib., 146(2), pp. 243-268, 1991.
3. J. J. Hollkamp, "Multimodal passive vibration suppression with piezoelectric materials and resonant shunts," J. Intelligent Mater. Sys. and Struct., 5, pp. 49-57, 1994.
4. T. Tsuchiya and Y. Kagawa, "Finite element based equivalent circuit modelling for piezoelectric vibratory systems," Proc. 5th Asian Simul. Conf., pp. 270-273, 2002.
5. T. Tsuchiya and Y. Kagawa, "On the passive vibration damping with electromechanical transducers," Proc. Int. Conf. Acoustics 2004, pp. IV3247-IV3250, 2004.
6. Y. Kagawa, T. Tsuchiya and N. Wakatsuki, "Equivalent circuit representation of a vibrating structure with piezoelectric transducers and the stability consideration in the active damping control," Smart Mater. and Struct., 10, pp. 389-394, 2001.

Path Sensitization and Sub-circuit Partition of CUT Using *t*-Distribution for Pseudo-exhaustive Testing

Jin Soo Noh, Chang Gyun Park, and Kang-Hyeon Rhee

Dept. of Electronic Eng., Multimedia & Biometrics Lab.,
Chosun University, Gwangju Metropolitan city, (Daehanminkook)501-759, Korea
njinsoo@vlsi.chosun.ac.kr, {cgpark,multimedia}@chosun.ac.kr

Abstract. This paper presents a new pseudo-exhaustive testing algorithm that is composed of the path sensitization and sub-circuit partitioning using *t*-distribution. In the proposed testing algorithm, the paths, for the path sensitization the, between PIs and POs based on the high *TMY*(test-mainstay) nodes of CUT(circuit under test) are sensitized and the boundary nodes, for the partitioned sub-circuits, are defined on the level of significance α on *t*-distribution respectively. As a consequence, when $(1-\alpha)$ is 0.2368, the most suitable of the performance to operate the singular cover and consistency operation in the path sensitization. And when α is 0.5217, the most suitable of the performance to partition the sub-circuit in sub-circuit partitioning.

1 Introduction

Advances in semiconductor manufacturing technology continue to increase the complexity of integrated circuit. According to the increasing of circuit complexity, the testing is to be importance problem more than design. Because exhaustive testing is impractical for ASIC design, partitioning which allows exhaustive test of sub-circuits offers an engaging alternative. Goel[1] observed that with each twice of the number of gates in a circuit, the test cost increases as the square of the previous test cost.

As the input pins of a circuit grow with the technique, the requirement of the test patterns for a circuit, no matter combinational or sequential circuits, is much more than before in the test stage of ASIC design process. It usually consider to generate exhaustive test patterns for CUT. However, exhaustive testing uses the generated all the combination of the test patterns, the test time might be intolerable if the number of inputs is large. Since the test pattern requirement of pseudo exhaustive testing[2, 3] is much fewer than exhaustive testing, there are many methods and architectures proposed[4-8] to solve this problem.

There are many kinds of testing methods but the pseudo-exhaustive testing method[9-13] for CUT is very widely used because it ensures the high fault coverage[14-16]. In the pseudo-exhaustive testing methods, especially the path sensitization and the sub-circuit partitioning are confidently schemes. So, DFT(design for testability) is appeared as new subject[17] using the many testing methods.

In this paper, a new pseudo-exhaustive testing is proposed for the test pattern generation of CUT using *t*-distribution.

The proposed scheme deals with *TY*(testability) value of each nodes in CUT to the population that composed of the raw data on *t*-distribution. And *TY* is tested on the

D.-K. Baik (Ed.): AsiaSim 2004, LNAI 3398, pp. 205–213, 2005.

confidence interval. Then the paths which have the nodes of the high *TMY* is sensitized for the path sensitization and the boundary nodes of the partitioned sub-circuits is searched for the partitioning of CUT.

The organization of this paper is as follows: Sec. 2 illustrates the theoretical principles for the pseudo-exhaustive testing. Sec. 3 addresses the proposed algorithm using *t*-distribution and it is experimented for the performance evaluation in Sec. 4. Lastly, the conclusion is drawn in Sec. 5.

2   Pseudo-exhaustive Testing

There are mainly two kinds of the pseudo- exhaustive testing method in CUT. One is the path sensitization method which sensitizes the path from PIs to POs of CUT using the singular cover and the consistency operation. While the internal nodes are logically controlled, CUT is exhaustively tested. The other is the sub-circuit partitioning method which partitions CUT to the several sub-circuits block. While each sub-circuit's output is observed at PO, all sub-circuits are exhaustively tested. McCluskey [18] proposed pseudo-exhaustive testing as an alternative, where the circuit is partitioned into s subcircuits. Each partition has an upper bound c on the number of inputs, while the total number of edges between the partitions is minimized. In general, a wide variety of automated techniques for circuit partitioning and testability enhancement are available, and their impacts on CUT differ greatly. However, all efficient methods for general partitioning and test vector generation for digital circuits require significant amounts of Design for Testability(DFT) hardware[19].

These two methods are required the shortest test length more than the exhaustive testing moreover, can be detected for the single stuck-at fault as well as multiple faults due to the occurred errors of fabrication.

2.1   Path Sensitization

For the path is sensitized through a logic gate, the singular cover must be played with the truth table.

Table 1. Singular cover of logic gates

AND		NAND		OR		NOR		XOR	
In	out	in	out	in	out	in	out	in	out
1 1	1	0 X	1	1 X	1	0 0	1	X X'	1
0 X	0	X 0	1	X 1	1	1 X	0		
X 0	0	1 1	0	0 0	0	X 1	0	X X	0

The singular cover of each logic gate is illustrated in Table 1. This operation traces forward the path from PIs to POs of CUT. In somewhere internal node, if stuck-at fault is existed then the faulty node to be the output of before logic gate. The consistency operation must be applied to the logic gate and so, the input of logic gate could be defined. The consistency operation of each logic gate is illustrated in Table 2. This operation traces backward the paths from POs to PIs of CUT.

Table 2. Consistency operation of logic gates

Out	Test patterns of input				
	AND	NAND	OR	NOR	XOR
0	0 X	1 1	0 0	1 X	0 0
0	X 0	1 1	0 0	X 1	1 1
1	1 1	0 X	1 X	0 0	0 1
1	1 1	X 0	X 1	0 0	1 0

2.2 Partitioning of CUT

Figure 1 shows the partitioning method of CUT. While each sub-circuit Si is exhaustively tested, CUT is simultaneously pseudo-exhaustively tested[13].

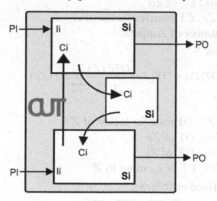

Fig. 1. Partitioning for pseudo-exhaustive testing

Sub-circuit is configured by the boundary nodes on the cutline in CUT. Then, in the partitioned sub-circuit, the number of inputs is expressed by

$$sc(Si) = Ii + Ci \qquad (1)$$

Where

Si: partitioned sub-circuit

sc: number of sub-circuits

Ii: number of inputs in Si

Ci: number of inputs on the cutline from the other sub-circuits

Then, the cost of pseudo-exhaustive testing is defined by

$$COST(Pseudo - exhaustive\ testing)$$

$$= Testlength$$

$$= \sum_{(all\ Seg.\ Si)} 2^{(sc(Si))}$$

$$= \sum_{i=1}^{SC} 2^{(sc(Si))} \qquad (2)$$

2.3 Testability

In CUT, all nodes must be assigned to the fixed logical value for pseudo-exhaustive testing. This operation is called controllability(CY)[15]. After this operation, the each node's value is observed at POs of CUT, it is called observability(OY). Then, the function that called testability(TY) is obtained from CY and OY.

CY is defined as follows

$$CY(Zo) = \frac{CTF(Z)}{n} \sum_{i=1}^{n} CY(Zi) \tag{3}$$

Where

Z : some logic gate
Zo : Z output
Zi : Z input
$CY(Zo)$: CY of Zo
$CTF(Z)$: CY transfer factor of Z
n : number of Z inputs

OY is defined as follows

$$OZ(Zi) = OY(Zo)\frac{OTF(Z)}{n-1} \sum_{k=1}^{n-1} CY(Zk) \tag{4}$$

Where

$OTF(Z)$: OY transfer factor of Z
$OY(Zo)$: OY of Zo
$OY(Zi)$: OY of Zi
$CY(Zk)$: CY of k_{th} input in Z

Then, TY can be defined as follows

$$TY = CY \times TY \tag{5}$$

3 Proposed Algorithm for Pseudo-exhaustive Testing

In this paper, the efficient algorithm of the possible path sensitization and sub-circuit partitioning is proposed. The proposed algorithm is composed of 2 parts. One scheme searches the high TMY nodes for the path sensitization and the other scheme searches the boundary nodes for the partitioned sub-circuit. The test pattern would be to generate based on this paths and sub-circuits.

After obtain TY, the high TMY nodes and the boundary nodes that existed in CI are defined, t-distribution is used as follows

$$CI = X' \pm t(n-1;\frac{a}{2})\frac{s}{\sqrt{n}} \tag{6}$$

Where

CI : confidence interval
a : level of significance
s : sample standard deviation
n : number of nodes
$n-1$: degrees of freedom
X' : sample mean

For the path sensitization, TYs of all nodes are obtained by (4) and TYs are dealing with population. From X' and s, when $(1-\alpha)$ is converged to 0, the nodes existed in CI would be the high TMY. In Figure 2, the region that existed high TMY is illustrated on t-distribution.

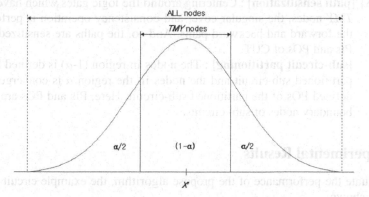

Fig. 2. The existing high TMY nodes on t-distribution

The consistency operation and the singular cover is performed to the logic gate of forward and backward to trace the path which have the high TMY. Then, the paths between from PIs to POs are sensitized and so the defined PIs are used to generate the test pattern of pseudo-exhaustive testing. Let the number of PIs be w. Thus, the number of test pattern is generated w^2.

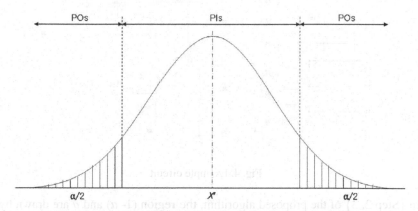

Fig. 3. The existed PIs and Pos of sub-circuits on t-distribution

In Figure 3, when CUT is partitioned to several sub-circuits, the existence of boundary nodes are shown on t-distribution. POs of sub-circuits is existed in the region that α is converged to 0 and PIs that include Ii and Ci are existed in the region $(1-\alpha)$ and the test pattern is generated by Eq. (1). In the below, there are 3 steps that the proposed algorithm of path sensitization and sub-circuit partitioning is executed.

{Step 1} The TY values of the entire nodes in CUT are treated as the population which composed of the raw data.

{Step 2} [**path sensitization**] : In the region (1-α) on t-distribution, *CI* of *TYs* is estimated and then the high *TMY* nodes are searched.

[**sub-circuit partitioning**] : The nodes that are existed in the region (1-α) and α on t-distribution are defined.

{Step 3} [**path sensitization**] : Centering around the logic gates which have the high *TMY* nodes, the singular cover and consistency operation is performed on the forward and backward paths. And so, the paths are sensitized between PIs and POs of CUT.

[**sub-circuit partitioning**] : The nodes in region (1-α) is defined PIs of the partitioned sub-circuit and the nodes in the region α is converged to 0, is defined POs of the partitioned sub-circuit. Here, PIs and POs are to be the boundary nodes of sub- circuits.

4   Experimental Results

To evaluate the performance of the propose algorithm, the example circuit in Figure 4[20] is shown.

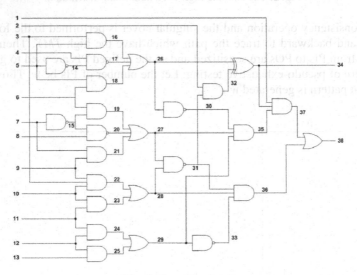

Fig. 4. Example circuit

In {Step 2, 3} of the proposed algorithm, the region (1- α) and α are drawn by $t(n-1)$;(α /2) of t-distribution for the path sensitization and sub-circuit partitioning respectively.

4.1 Path Sensitization

The existed nodes in *CI* are to be the high *TMY*. According to the region (1- α), PIs that included sensitized paths are illustrated in Figure 5.

Table 3 shows the number of generated test patterns compared with [20]. When (1-α) is 0.1579, the pseudo- exhaustive test pattern is least generated.

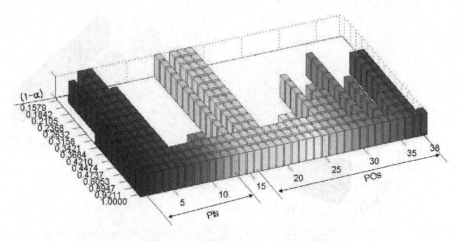

Fig. 5. Node definition by t-distribution for the path sensitization

Table 3. Experimental result of the proposed path sensitization

$1-\alpha$	(A)	(B)	(A)/(B) %
0.1579	15	143	10
0.1842	18	"	13
0.2105	25	"	17
0.2368	40	"	28
0.2632	40	"	28
0.3158	40	"	28
0.3421	40	"	28
0.3684	71	"	50
0.4210	71	"	50
0.4474	71	"	50
0.4737	71	"	50
0.6053	261	"	183
0.8947	8182	"	5729
0.9211	8192	"	5729
1.0000	8192	"	5729

(A) : Proposed algorithm
(B) : [20]

4.2 Sub-circuit Partitioning

According to the region α, POs of the partitioned sub-circuits is shown in Figure 6 and the number of sub-circuits and the generated test patterns are illustrated in Table 4 compared with the exhaustive testing. When α is 0.7826, the number of sc are 18 and then the number of test pattern is least generated to 100.

5 Conclusion

The generated pseudo-exhaustive test patterns that using the proposed testing algorithm is applied to the stuck-at faults existing on all nodes and verified functionally

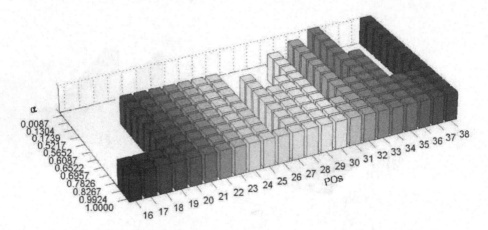

Fig. 6. Sub-circuit's PO by t-distribution

Table 4. Experimental results of the proposed sub-circuit partitioning

α	sc	(A)	(B)	(A)/(B) %
0.0087	2	1,536	8,192	18.75
0.1304	3	800	"	9.77
0.1739	4	788	"	9.62
0.5217	12	196	"	2.39
0.5652	13	156	"	1.90
0.6087	14	152	"	1.86
0.6522	15	144	"	1.76
0.6957	16	144	"	1.76
0.7826	18	100	"	1.22
0.8267	19	102	"	1.25
0.9924	22	106	"	1.29
1.0000	23	118	"	1.44

(A) : Proposed algorithm
(B) : Exhaustive test patterns

and exactly on POs of CUT. Thus, it is confirmed that the availability for the pseudo-exhaustive testing. As a consequence, in the case of path sensitization, when (1-α) is 0.2368, it is most suitable of the performance to operate the singular cover and consistency operation.

And in the case of sub-circuit partitioning, when α is 0.5217, it is the most suitable of the performance to partition the sub-circuit.

In the proposed schemes, according to the region (1-α) and α are changed in the level of significance, it could be known the prediction to sensitize the paths, to partition the sub-circuits and to generate the number of test patterns. And so, it is applicable in the design for testability and the computer aided test systems.

Acknowledgement

This research is supported by Chosun University, Korea, 2001.

References

1. Goel, P. "Test Generation Costs Analysis and Projections," The 17th Design Automation Conference, ACM/IEEE, June 1980, pp. 77-84.
2. M. Abramovici, M. A. Breuer and A. D. Friedman, Digital Systems Testing and Testable Design, Computer Science Press, New York, 1990.
3. M. L. Bushnell and V. D. Agrawal, Essentials of Electronic Testing for Digital, Memory & Mixed-Signal VLSI Circuits, Kluwer Academic Publishers, 2000
4. E. J. McCluskey, "Verification Testing A Pseudo - Exhaustive Test Technique", IEEE Trans., 1984. C-33, (6), pp.541-546
5. R. Srinivasan, S. K. Gupta, and M. A. Beruer, "Novel Test Pattern Generators for Pseudo-Exhaustive Testing". Proceedings of international test conference, pp. 1041-1050, 1993
6. E. Wu and P. W. Rutkowski, "PEST - A Tool for Implementing Pseudo-Exhaustive Self Test", Proc. of 1st European Design Automation Conf., pp. 639-643, Mar. 1990.
7. C.I. H. Chen and J. T. Yuen, "Automated Synthesis of Pseudo-Exhaustive Test Generator in VLSI BIST Design", IEEE Trans. on VLSI Systems, vol. 2, no. 3, pp. 273-291, Sep. 1994.
8. D. Kagaris, F. Makedon and S. Tragoudas, "A Method for Pseudo-Exhaustive Test Pattern Generation", IEEE Trans., 1994, CAD-13, 990, pp. 1170-1178
9. Kang Hyeon Rhee, "A Study on the Pseudo-exhaustive Test using a Net-list of Multi-level Combinational Logic Circuits," Jour. of KITE, vol. 30-B, no. 5, May 1993
10. McCluskey, E.J., "Verification testing A Pseudo- exhaustive test technique," IEEE Trans. Computers, vol. C-33, no. 6, pp. 541 546, June 1984.
11. McCluskey, E.J., "Exhaustive and Pseudo- exhaustive Test," Built-in Test Concepts and Techniques, Tutorial, ITC83.
12. J.G. Udell "Test set generation for pseudo-exhaustive BIST," in Dig. Papers, IEEE 1986 Int., Conf. Computer Aided Design (Santa Clara, CA), pp. 52-55, Nov. 11 13, 1986.
13. Itahak Shperling and Edward J. McCluskey, "Circuit Segmentation for Pseudo-exhaustive Testing," Stanford Univ., CRC Tech. Report, no. 87-2, CSL no. 87-315, May 1987.
14. Kang Hyeon Rhee, "A Study on the Development of Fault Simulator for the Pseudo-exhaustive Test of LSI/VLSI," Jour. of KITE, vol. 32-B, no. 4, Apr. 1995
15. Barry W. lohnson, Design and Analysis of Fault-Tolerant Digital System, Addison Wesley pp. 554-565, 1989.
16. D.T. Wang, "An algorithm for the detection of tests set for combinational logic networks" IEEE Trans. Computers, vol. C-25, no. 7, pp. 742 746, July 1975.
17. S. Bozorgui-Nesbat and E.J. McCluskey, "Design for Autonomous Test," IEEE Trans. Computers pp. 866-875, Nov. 1981.
18. McCluskey, E. J. and S. Bozorgui-Nesbat, "Design for Autonomous Test," IEEE Trans. on Computer, Vol. C-30, No 11, Nov. 1981, pp. 866-875.
19. Udell, J. G. Jr., and E. J. McCluskey, "Efficient Circuit Segmentation for Pseudoexhaustive Test," Proc. Int. Conf. on Computer-Aided Design, 1987, pp. 148-151.
20. Udell, "Test pattern generation for pseudo- exhaustive BIST," CRC Tech. Report, Stanford Univ.

Computer Simulation of Photoelastic Fringe Patterns for Stress Analysis

Tae Hyun Baek[1] and Myung Soo Kim[2]

[1] School of Mechanical Engineering, Kunsan National University,
#65 Miryong-dong, Gunsan City, Jeonbuk, 573-701, The Republic of Korea
thbaek@kunsan.ac.kr
[2] School of Electronic and Information Engineering, Kunsan National University,
#65 Miryong-dong, Gunsan City, Jeonbuk, 573-701, The Republic of Korea
mskim@kunsan.ac.kr

Abstract. Photoelastic fringe pattern for stress analysis is investigated by computer simulation. The first simulation is to separate isoclinics and isochromatics from photoelastic fringes of a circular disk under diametric compression by use of phase shift method. The second one is an analysis of photoelastic fringe patterns on perforated tensile plate with conformal mapping and measured data. The results of both simulations agree well with those of optical experiments.

1 Introduction

Photoelasticity is an optical measurement method for stress analysis. It can perform whole-field measurement due to two-dimensional signal processing of light. Also it can perform non-contact measurement because of transmission or reflection of light on a specimen. Photoelasticity allows one to obtain the principal stress directions and principal stress differences in a specimen [1-3]. The principal stress directions and the principal stress differences are provided by isoclinics and isochromatics, respectively. In this paper, two kinds of computer simulations are carried out. The first one is computer simulation to separate isoclinics and isochromatics from photoelastic fringes of a circular disk under diametric compression by use of phase shift method [4]. The second one is an analysis of photoelastic fringe patterns on perforated tensile plate with conformal mapping and measured data through computer simulation [5].

2 Separations of Isoclinics and Isochromatics

2.1 Method of Analysis

The optical arrangement to separate isoclinics and isochromatics from photoelastic fringes is a circular polariscope set-up in Fig. 1. In Fig. 1, P, Q, R, and A stand for linear polarizer, quarter-wave plate, retarder (stressed specimen) and analyzer, respectively. The orientation of the elements is written by a subscript that means the angle between the polarizing axis and the chosen axis. For example, P_{90} indicates a polarizer whose transmission axis is $90°$ to the x-axis. $R_{\alpha,\delta}$ stands for the stressed sample taken as a retardation δ and whose fast axis is at an angle of $(90° - \alpha)$ with the x-axis. Also, Q_ϕ is a quarter wave plate with a fast axis at ϕ. A_θ indicates an analyzer whose transmission axis is θ.

D.-K. Baik (Ed.): AsiaSim 2004, LNAI 3398, pp. 214–221, 2005.
© Springer-Verlag Berlin Heidelberg 2005

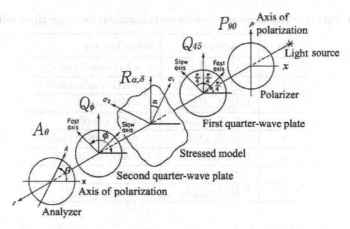

Fig. 1. Optical arrangement of a circular polariscope

In Fig. 1, the components of electric fields in light along and perpendicular to the analyzer, E_x and E_y, are obtained using Jones calculus as follows [6];

$$\begin{pmatrix} E_x \\ E_y \end{pmatrix} = \begin{bmatrix} \cos^2\theta & \sin\theta\cos\theta \\ \sin\theta\cos\theta & \sin^2\theta \end{bmatrix} \begin{bmatrix} i\cos^2\phi + \sin^2\phi & (i-1)\sin\phi\cos\phi \\ (i-1)\sin\phi\cos\phi & i\sin^2\phi + \cos^2\phi \end{bmatrix}$$

$$\begin{bmatrix} e^{i\delta}\cos^2\alpha + \sin^2\alpha & (e^{i\delta}-1)\sin\alpha\cos\alpha \\ (e^{i\delta}-1)\sin\alpha\cos\alpha & e^{i\delta}\sin^2\alpha + \cos^2\alpha \end{bmatrix} (\frac{i+1}{2}) \begin{bmatrix} 1 & i \\ i & 1 \end{bmatrix} k\,e^{i\omega t} \tag{1}$$

where $i = \sqrt{-1}$. Output light intensity from the analyzer, I, becomes

$$I = \overline{E}_x E_x + \overline{E}_y E_y \tag{2}$$

where $\overline{E}_x$ and $\overline{E}_y$ are the complex conjugates of E_x and E_y. When Eq. (1) is used in Eq. (2), Eq. (2) becomes

$$I = K[1 - \sin 2(\theta-\phi)\cos\delta - \sin 2(\phi-\alpha)\cos 2(\theta-\phi)\sin\delta] \tag{3}$$

where K is the maximum light intensity emerging from the analyzer [7]. The angle α and the relative retardation δ indicating the principal stress direction and the principal stress difference are the parameters to be obtained by phase shift method. In this paper, 8-step phase shift method is applied to get the angle α and the relative retardation δ with controlling θ and ϕ in Eq. (3). That is summarized in Table 1 [4]. The equations in Table 1 are used for calculation of isoclinics α and isochromatics δ as

$$\alpha = \frac{1}{2}\tan^{-1}\left(\frac{I_5 - I_6}{I_1 - I_2}\right) \tag{4}$$

$$\delta = \tan^{-1}\left[\frac{(I_1 - I_2)\cos 2\alpha + (I_5 - I_6)\sin 2\alpha}{1/2\{(I_4 - I_3) - (I_8 - I_7)\}}\right] \tag{5}$$

Table 1. Optical arrangements and their intensity equations for 8-step phase shift method

No.	Arrangement	Output Intensity
1	$P_{90}Q_{45}R_{\alpha,\delta}Q_{45}A_{-45}$	$I_1 = K(1 + \cos 2\alpha \sin \delta)$
2	$P_{90}Q_{45}R_{\alpha,\delta}Q_{-45}A_{45}$	$I_2 = K(1 - \cos 2\alpha \sin \delta)$
3	$P_{90}Q_{45}R_{\alpha,\delta}Q_{-45}A_0$	$I_3 = K(1 - \cos \delta)$
4	$P_{90}Q_{45}R_{\alpha,\delta}Q_{45}A_0$	$I_4 = K(1 + \cos \delta)$
5	$P_{90}Q_{45}R_{\alpha,\delta}Q_0A_0$	$I_5 = K(1 + \sin 2\alpha \sin \delta)$
6	$P_{90}Q_{45}R_{\alpha,\delta}Q_{90}A_{90}$	$I_6 = K(1 - \sin 2\alpha \sin \delta)$
7	$P_{90}Q_{45}R_{\alpha,\delta}Q_0A_{45}$	$I_7 = K(1 - \cos \delta)$
8	$P_{90}Q_{45}R_{\alpha,\delta}Q_{90}A_{45}$	$I_8 = K(1 + \cos \delta)$

2.2 Computer Simulation

The circular disk specimen with photoelastic fringe constant, $f_\sigma = 5.254$ N/cm, is used for computer simulation. Its radius (r) and thickness (t) are $r = 3.81$cm and $t = 0.476$cm. Diametrical compression load, $P = 26.7$N, is applied to the specimen. Using the given conditions of specimen, theoretical values of α and δ are

$$\delta = \frac{2\pi t}{f_\sigma}(\sigma_1 - \sigma_2) \tag{6}$$

$$\alpha = \frac{1}{2}\tan^{-1}\left[\frac{2\tau_{xy}}{(\sigma_x - \sigma_y)}\right] \tag{7}$$

where σ_1 and σ_2 are principal stress components and σ_x, σ_y, and τ_{xy} are stress components. When Eq. (6) and Eq. (7) are used in the eight equations of Table 1, eight phase-shifted photoelastic fringe patterns are obtained. After the eight fringe patterns, $I_1 \sim I_8$, are applied to Eqs. (4) and (5), isoclinics α and isochromatics δ are obtained as in Fig. 2. For comparison, isoclinics α and isochromatics δ from optical experiment [4] are presented in Fig. 3.

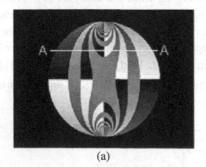
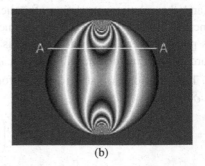

(a)   (b)

Fig. 2. Images of (a) isoclinics α and (b) isochromatics δ from computer simulation for a disk under diametric compression

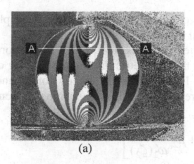 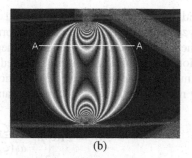

(a)   (b)

Fig. 3. Images of (a) isoclinics α and (b) isochromatics δ from optical experiment for a disk under diametric compression

The isoclinic data on the line A-A in Fig. 2(a) and Fig. 3(a) are shown in Fig. 4 for quantitative comparison. Also the isochromatic data on the line A-A in Fig. 2(b) and Fig. 3(b) are shown in Fig. 5. As shown in Figs. 4 and 5, the results of computer simulation agree well with those of optical experiment.

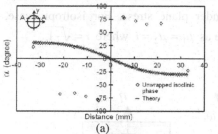 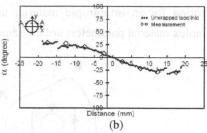

(a)   (b)

Fig. 4. Distribution of isoclinics along line A-A obtained from (a) computer simulation and (b) optical experiment

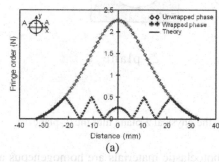 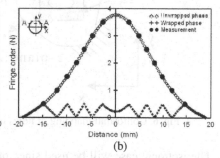

(a)   (b)

Fig. 5. Distribution of isochromatics along line A-A obtained from (a) computer simulation and (b) optical experiment

3 Photoelastic Fringe Patterns on Perforated Tensile Plate

3.1 Theoretical Formulation

From a design or analysis consideration, stress raisers associated with holes or notches continue to command appreciable attentions. Through computer simulation, hybrid full-field stress analysis around an elliptical hole of tensile loaded plates is analyzed with photoelasticity and conformal mapping in this section [5].

The present technique employs general expressions for the stress functions with traction-free conditions that are satisfied at the geometric discontinuity using conformal mapping and analytical continuation. In the absence of body forces and rigid body motion, the stresses under plane and rectilinear orthotropy can be written as [8-11]:

$$\sigma_x = 2\,\mathrm{Re}\!\left[\mu_1^2\,\frac{\Phi'(\zeta_1)}{\omega_1'(\zeta_1)} + \mu_2^2\,\frac{\Psi'(\zeta_2)}{\omega_2'(\zeta_2)} \right] \tag{8a}$$

$$\sigma_y = 2\,\mathrm{Re}\!\left[\frac{\Phi'(\zeta_1)}{\omega_1'(\zeta_1)} + \frac{\Psi'(\zeta_2)}{\omega_2'(\zeta_2)} \right] \tag{8b}$$

$$\tau_{xy} = -2\,\mathrm{Re}\!\left[\mu_1\,\frac{\Phi'(\zeta_1)}{\omega_1'(\zeta_1)} + \mu_2\,\frac{\Psi'(\zeta_2)}{\omega_2'(\zeta_2)} \right] \tag{8c}$$

Prime denotes differentiation with respect to the argument. Complex material parameters $\mu_l(l = 1, 2)$ of Eqs. (8) are the two distinct roots of the characteristic equation for an orthotropic material under plane stress. For isotropic case, two complex material parameters are the same as $\mu_1 = \mu_2 = i$ where $i = \sqrt{-1}$.

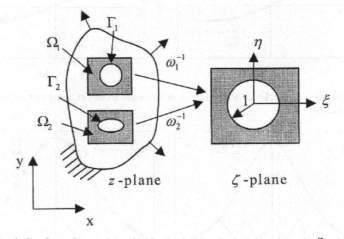

Fig. 6. Conformal mapping of holes in the physical z-plane into the ζ-plane

The isotropic case will be used since photoelastic materials are homogeneous and isotropic. The two complex stress functions, $\Phi(\zeta_1)$ and $\Psi(\zeta_2)$ are related to each other by the conformal mapping and analytic continuation. For a traction-free physical boundary, Γ, the two functions within sub-region Ω of Fig. 6 can be written as Laurent expansions, respectively,

$$\Phi(\zeta_1) = \sum_{j=-m}^{m} c_j \zeta_1^j \quad \text{and} \quad \Psi(\zeta_2) = \sum_{j=-m}^{m} (\overline{c_j} B \zeta_2^j + \overline{c_j} C \zeta_2^j) \tag{9}$$

Complex quantities B and C depend on material properties. The coefficients of Eqs. (9) are $c_j = a_j + ib_j$, where a_j and b_j are real numbers. ω^{-1} that is the inverse of the mapping function ω maps the geometry of interest from the physical z-plane into the ζ-plane as shown in Fig. 6. Conformal transformation from a simple geometry, i.e., a unit circle, in the ζ-plane to multiple geometric discontinuities can be given by Fig. 6.

Combining Eqs. (8) and (9) gives the following expressions for the stresses through regions Ω_1 and Ω_2 of Fig. 6 [9-11].

$$\{\sigma\} = [V]\{c\} \tag{10}$$

where $\{\sigma\} = \{\sigma_x, \sigma_y, \tau_{xy}\}^T$ and $\{c\} = \{a_j, b_j\}^T$. $[V]$ are rectangular coefficient matrices whose sizes depend on the number of terms m of the power series expansions of Eqs. (9). Using relation of stress components in photoelasticity, $\{c\}$ is obtained as arbitrary function G in Eq. (11).

$$G_n\{c\} = \left\{\frac{\sigma_x - \sigma_y}{2}\right\}_n^2 + \{\tau_{xy}\}_n^2 - \left\{\frac{Nf_\sigma}{2t}\right\}_n^2 \tag{11}$$

A truncated Taylor series expansion about the unknown parameters can linearize Eq. (11) and an iterative procedure is developed with

$$(G_n)_{i+1} \cong (G_n)_i + \sum_{n=-m}^{m} \left(\frac{\partial G_n}{\partial c}\right)_i \Delta c_n \tag{12}$$

Knowing $\{\sigma\}$ at various "n" locations enables one to solve for the best values of unknown coefficients $\{c\}$ in the non-linear least squares sense from Eq. (12). For measured fringe orders and predetermined m terms of Eqs. (9), the coefficients $\{c\}$ can be obtained by the non-linear least squares method [12].

3.2 Computer Simulation

For analysis of photoelastic fringe patterns on perforated tensile plate, a plate with an elliptical hole at the center is used as a specimen. Fig. 7 is the specimen. It is made of polycarbonate (PSM-1) and thickness of the specimen is 3.175mm. Material properties are $E = 2{,}482$MPa and $v = 0.38$ [13]. Photoelastic fringe constant of the plate is 7005 N/m-fringe. Tensile load is applied to the specimen in a loading fixture.

Figure 8 shows the isochromatic fringes of the specimen under tensile load P=153N when the plate is viewed in the light field polariscope of optical experiment [5].

In order to get accurate fringe data, fringes are twice multiplied, and sharpened by using the fringe multiplication algorithm and the sharpening algorithm [5, 14]. Left half of Fig. 9(a) is twice-multiplied fringe pattern of upper half of Fig. 8. Right half of Fig. 9(a) is twice-multiplied fringe pattern that is obtained by computer simulation. In the measurement of fringes, isochromatic data are excluded on the edge of the hole since fringes near the edge of the hole are too dense. Locations of the measured data

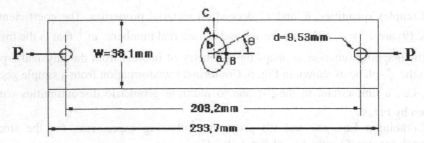

Fig. 7. Finite-width uniaxially loaded tensile plate containing a hole (Elliptical hole: a = 4.763mm, b = 9.525mm)

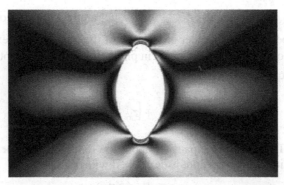

Fig. 8. Light-field isochromatic fringe pattern of a loaded tensile plate containing an elliptical hole (P =153N)

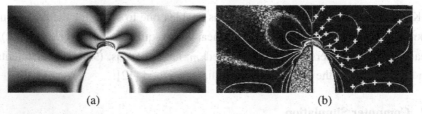

(a)   (b)

Fig. 9. Comparison between actual and simulated fringes of the specimen. (a) twice-multiplied actual (left) and calculated fringes (right). (b) Fringe-sharpened lines from actual (left) and simulated fringes (right)

are selected arbitrarily in the region as indicated "+" marks near the hole in Fig. 9(b). Numbers of data points are determined to be four to five times as many as the number of undetermined coefficients, $\{c\}$ of Eqs. (9), which is estimated by the non-linear least squares method. Left half of Fig. 9(b) is sharpened fringe pattern of upper half of Fig. 8. Right half of Fig. 9(b) is sharpened fringe pattern that is obtained by computer simulation. The "+" marks in Fig. 9(b) show the locations where data are collected. In Fig. 9, the number of terms of the power series expansions is chosen as $m = 9$ because simulated fringe patterns agree excellently with experimental fringe patterns [5]. As shown in Fig. 9, the fringe patterns from computer simulation agree well with those of optical experiment.

4 Conclusion

Photoelastic fringe pattern for stress analysis is investigated by computer simulation. The first simulation is to separate isoclinics and isochromatics from photoelastic fringes of a circular disk under diametric compression by use of phase shift method. The isoclinics and the isochromatics are made separate excellently by the simulation of phase shifting method. The isoclinic data and the isochromatic data obtained by computer simulation agree well with those by optical experiment of photoelasticity. The second one is an analysis of photoelastic fringe patterns on perforated tensile plate with conformal mapping and measured data. The results of simulation are quite comparable to those of optical experiments.

Acknowledgements

This work was supported by grant No. R05-2003-000-11112-0 from the Basic Research Program of the Korea Science and Engineering Foundation (KOSEF), and by grant 2004 from the Korea Sanhak Foundation of the Republic of Korea.

References

1. Dally, J.W., Riley, W.F.: Experimental Stress Analysis, 3rd ed., McGraw-Hill, New York (1991)
2. Burger, C.P.: Photoelasticity. In: Kobayashi, A.S. (ed.): Handbook on Experimental Stress Analysis, Prentice-Hall, Inc., Englewood Cliffs, New Jersey (1986)
3. Cloud, G.L.: Optical Methods of Engineering Analysis, Cambridge University Press (1995)
4. Baek, T.H., Kim, M.S., Morimoto, Y., Fujigaki, Y.: Separation of Isochromatics and Isoclinics from Photoelastic Fringes in a Circular Disk by Phase Measuring Technique. KSME Int. J., Vol. 16, No. 2. Korea Society for Mechanical Engineers (2002) 175-181
5. Baek, T.H., Kim, M.S., Rhee, J., Rowlands, R.E.: Hybrid Stress Analysis of Perforated Tensile Plates using Multiplied and Sharpened Photoelastic Data and Complex Variable Techniques, JSME Int. J. Series A, Vol. 43, No. 4. Japan Society for Mechanical Engineers (2000) 327-333
6. Theocaris, P.S., Gdoutos, E.E.: Matrix Theory of Photoelasticity, Springer-Verlag, Berlin Heidelberg New York (1979) 50-55
7. Quiroga, J.A., Gonzales-Cano, A.: Phase Measuring Algorithm for Extraction of Isochromatics of Photoelastic Fringe Patterns. Applied Optics. Vol. 36, No. 2 (1997) 8397-8402
8. Savin, G.N.: Stress Distribution Around Holes. NASA Technical Translation, NASA TT F-601 (1970)
9. Rhee, J., He, S., Rowlands, R.E.: Hybrid Moire-numerical Stress Analysis Around Cutouts in Loaded Composites. Experimental Mechanics. Vol. 36 (1996) 379-387
10. Baek, T.H., Rowlands, R.E.: Experimental Determination of Stress Concentrations in Orthotropic Composites. J. of Strain Analysis. Vol. 34. No. 2 (1999) 69-81
11. Baek, T.H., Rowlands, R.E.: Hybrid Stress Analysis of Perforated Composites Using Strain Gages. . Experimental Mechanics. Vol. 41. No. 2 (2001) 195-203.
12. Sanford, R.J.: Application of the Least-squares Method to Photoelastic Analysis. Experimental Mechanics. Vol. 20 (1980) 192-197
13. Photoelastic Division, Measurement Group, Inc. Raleigh. NC. USA
14. Baek, T.H., Burger, C.P.: Accuracy Improvement Technique for Measuring Stress Intensity Factors in Photoelastic Experiment. KSME Int. J. Vol. 5. No. 1 (1991) 22-27

Development of a Power Plant Simulation Tool
with Component Expansibility
Based on Object Oriented Modeling

Sang-Gyu Jeon[1] and Gihun Son[2]

[1] Graduate School, Sogang University, Shinsu-Dong Mapo-Gu Seoul 121-742, Korea
jeons1g1@sogang.ac.kr
[2] Department of Mechanical Engineering, Sogang University,
Shinsu-Dong Mapo-Gu Seoul 121-742, Korea
Gihun@sogang.ac.kr
www.sogang.ac.kr/~boil

Abstract. A power-plant simulation tool has been developed for improving component expansibility. The simulation tool is composed of a graphic editor, a component model builder and a system simulation solver. Advanced programming techniques such as object-oriented modeling and GUI(Graphical User Interface) modeling are employed in developing the simulation tool. The graphic editor is based on the OpenGL library for effective implementation of GUI while the component model builder is based on object-oriented programming for efficient generalization of component models. The developed tool has been verified through the simulation of a real power plant.

1 Introduction

A thermal power plant is a system which produces electricity with the combustion heat of fossil fuel. The thermal energy converts liquid water into steam at high temperatures and pressure, which drives turbines coupled to an electric generator. The power system has various types of components such as heat exchangers, pumps, pipes, valves and tanks. Furthermore, the component-component connections for a power system are quite complicated. For simulation of the complex system, since manual coding is very tedious and not efficient, simulation tools based on GUI(Graphical User Interface) such as MMS(Modular Modeling System), PRO_TRAX and US3(Unix simulation software system) have been developed [1~4]. However, the existing simulation tools have the limitation in component expansibility because of constructive problems and marketing strategies. They generally can not be expanded by the user who wishes to add his own component model algorithm into the simulation tool for a specific system. In this paper we describe a new simulation methodology for improving the expansibility limitation of existing tools. Advanced programming techniques such as GUI modeling and object-oriented modeling with implementation inheritance and interface inheritance are employed in developing the simulation tool.

D.-K. Baik (Ed.): AsiaSim 2004, LNAI 3398, pp. 222–229, 2005.

2   Structural Elements of Simulation Tool

The present simulation tool is composed of a graphic editor, a simulator and a model builder as shown in Fig. 1. The graphic editor includes a component editor, a component library, a system design tool and a netlist builder for component-component connection. The component editor creates and edits graphically component icons and then adds them into the component library. The system design tool provides the functions placing graphical icons selected from the component library and constructing a proposed   system. The netlist builder derives automatically the connection information of components from system design data and transfers the netlist information to the simulator. The simulator includes the system simulation solver and the input/output data processing. The system simulation solver calculates the plant dynamics with the netlist data received from the netlist builder and the model objects created from the model builder. Once component icons are placed in the system design tool, the component objects are created automatically from the model builder. The data exchange between model objects and the simulator is made through the general interface, which is explained later.

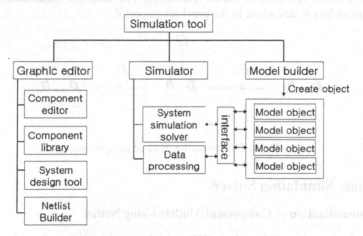

Fig. 1. The structure of simulation tool

3   Graphic Editor

Since a power plant has complex connections of various components, the graphical procedures in which various components are placed and connected with each other are essentially required for efficient system simulation. OpenGL library is employed for effective implementation of graphical function. The component editor and library are developed to enable users to add or change component icons. Users can create or change graphically component icons and save the icons in the component library. In the present tool, the system design tool does not generate a source code for simulation unlike other commercial simulation tools. This may be more convenient to the general users. Once users complete a system design and execute simulation, the simulation can be executed immediately by the simulator   sing netlist created from netlist builder.

4   Fluid-Thermal Modeling for Power System

Fig. 2. shows the connection state of power plant components such as heat exchangers, pumps, valves and tanks. The node is defined as the point connecting each component. The governing equations of fluid-thermal phenomena are mass, energy and momentum equations as follows.

$$\frac{d}{dt} M = F_i - F_o \tag{1}$$

$$\frac{d}{dt} Mh = F_i h_i - F_o h_o + Q + V \frac{dp}{dt} \tag{2}$$

$$F = a \triangle p + b , \tag{3}$$

where M, F, h, Q and p denote component mass, mass flow rate, enthalpy, heat and pressure respectively, and a and b are coefficients which determine the relationship between pressure and mass flow rate. F, h, p are the major variables calculated from Equations (1~3), and the other variables can be computed from the equation of state and heat transfer relations in Steam Tables[5]. The detailed introduction of fluid-thermal modeling is described in the previous paper[6].

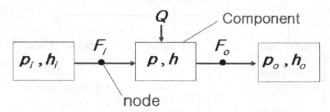

Fig. 2. The connection state of components

5   System Simulation Solver

5.1   Systematization of Component Models Using Netlist

The simulation solver is based on GUI modeling rather than tedious hand coding. When a user places components and connects them with wires for simulation, the netlist builder produces the connection data between the components from the graphical data. The connection information is required to calculate the pressure equation and energy equation. Specially, the pressure equation is solved by matrix operation as follows:

$$A\vec{p} = \vec{q} \quad \rightarrow \quad \vec{p} = A^{-1}\vec{q} \tag{4}$$

In this paper, a Gauss elimination method is modified in order that the zero parts of matrix A is not calculated for the efficiency of a calculation.

5.2   Model Interface for General Access to Model Objects

The simulation solver requires calculating a matrix to derive enthalpy from the energy conservation equations or pressure from the pressure equations. Some data used in this calculation are given from each component object, but each component object is created from the component model builders which have their own class name. Therefore, the general interface that the simulation solver can access is required. In this paper, the interface is made by an abstract class to apply the interface method of COM(Component Object Model) to model objects[7].

The Interface Class of Component Model Builders

```
class CInterface
{
public:
    virtual double GetValue(int, int)=0;
    virtual void SetValue(int, int, double)=0;
    virtual CString GetString(int)=0;
    virtual void SetString(int,CString)=0;
    virtual void Save(CString&)=0;
    virtual void Load(CString&)=0;
    virtual void Command(int)=0;
};
```

5.3 Model Builder Based on Object-Oriented Methodology

The power system simulation requires various and reliable component models to derive simulation results close to a real power system. Therefore, it is desirable that the engineers of various fields develop their specialized models and integrate them altogether. For this, C++ which is an object-oriented language is used to develop the model source codes. The model builder is constructed by class-based programming techniques such as model encapsulization, information hiding, interface inheritance and implementation inheritance. This enables users to easily add a new model builder into the simulation tool and improves component expansibility.

6 Test of Simulation Tool

The developed simulation tool has been tested through the simulation of a real power plant as shown Fig. 3. The input data for fluid-thermal model is given through a dialog box assigned to each component. In the simulation of water supply system, the pressure of a feed water tank and main steam is used for the boundary conditions of pressure equations. In Fig. 3, the digits displayed on components show calculation results of mass flow rate(F), enthalpy(H), pressure(P), level(L), the position of a valve(y) when the level and pressure of a drum is maintained to 50% and 174 kg/cm^2 respectively. As shown Fig. 3, the mass flow rate of inflow from the feed water tank is equals to the mass flow rate of outflow into the main steam in steady state. This verifies that pressure equations governing the mass conservation are calculated correctly.

7 Test of Component Expansion

We describe the component expansion procedure. As an example model, we choose a separator, which enhances the quality of main steam mechanically. The new compo-

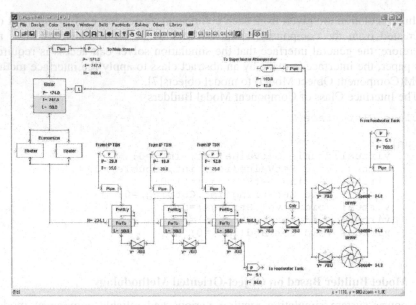

Fig. 3. The simulation result of feed water system in Honam power plant

nent model can be used immediately on equal term with the existing other components. The component expansion can be accomplished through the following three steps by users.

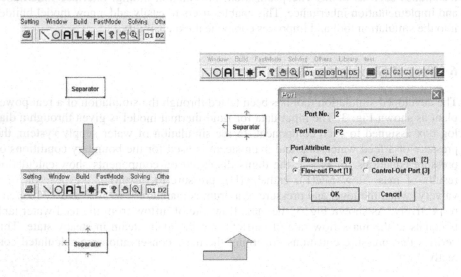

Fig. 4. Creation of a new component icon

◆ **Creation of a New Component Icon**

A new component icon can be created easily using the component editor, as shown in Fig. 4. Using graphic tools and an attribute dialog box, a user draws the shape of a icon using graphic tools, and then places the ports which the information of component models is transferred and component icons are connected with wires. Finally, a user assign attributes of the ports and saves a component icon in the component library

◆ **Development of a New Model Builder**

For developing a new model builder, a user must have an expert knowledge about power plant dynamics and C++ programming language. Using the inheritance of object-oriented programming, the model builder including an interface with the system simulation solver is constructed as follows:

The Class of Separator Model Builder

```
class CSeparator : public CFlowBody
{
// Port
    CPipeFlowInPort FInPort;
    CPipeFlowOutPort FOutPort;
    CControlInPort CtrInPort;
// variables
    SIMDATA F0, F1, P0, P1, H0, H1, AD, V, Z2, Q, U,
            v0, v1, T0, T1, A, B;
// member functions
            void Initialize();
            void SolvePressure();
            void SolveFlow();
            void SolveEnergy();
public:
            CSeparator();
};
```

The class of separator model builder inherits the CBody class, and the member functions of interface class are overridden by defined methods. The functions of the separator for computations are private members and are called through the command interface. Finally, the separator model builder is completed by creating an instance of the class. Since the CreateInstance function creates an instance of the class dynamically at runtime, the system simulation solver can access the separator object using the interface pointer of CSeparator class. The separator object is created when the icon is placed in the system design tool as in Fig. 5.

The Function Creating an Instance of the Class

```
Cinterface* CSolver::CreateInstance(int part,
int Dsn, int Elm)
{
    switch(part)
    {
        ......
        case dSEPARATOR:
            return new CSeparator;
        ......
    }
}
```

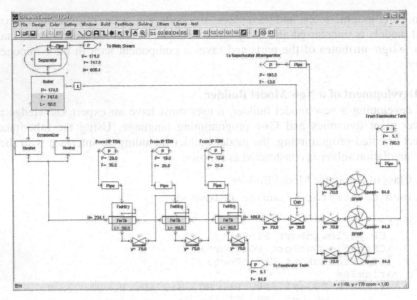

Fig. 5. Placement of a new component icon

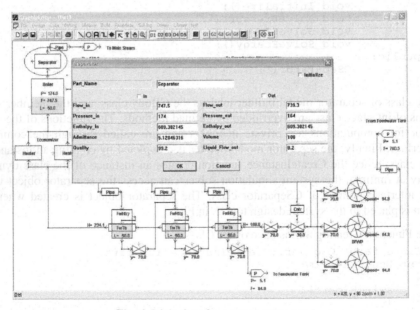

Fig. 6. Dialogbox for a new component

◆ **Creating a Definition File**

The definition file includes the information of component variables and dialog box. The model builder has the function returning the path of the definition file, and the graphic editor can use the definition file through the function of the model builder. As shown Fig. 6, the graphic editor shows the dialog box using definition file.

8 Conclusions

The simulation tool has been developed to facilitate the expansion of component models with object-oriented methodology, the general interface of component models and component editor. The functions of the simulation tool have been verified through the simulation of Honam power plant.

Acknowledgement

This work has been supported by KESRI(04516), which is funded by MOCIE (Ministry of commerce, industry and energy).

References

1. US3 : User Guide Release 5. Technologies, Columbia, Maryland (1994)
2. ProTRAX : http://www.traxcorp.com/ptmb.html
3. MCKim C. S., M. T. Matthews : Modular Modeling System Model Builder. Intersociety Energy Conversion Engineering Conference, Washington D. C. (1996)
4. C. A. Jones, N. S. Yee, and G. F. Malan : Advanced Trainingg Simulator Development Tools For Window[TM]. Society for Computer Simulation Conference v.27 no. 3, Framatome Technologies, Lynchburg, Virginia (1995)
5. C. A. MEYER, R. B. McCLINTOCK, G. J. SILVESTRI, R. C. SPENCER JR : Steam Tables, Sixth Edition, ASME (1997)
6. Gihun Son, Byung Hak Cho, Dong Wook Kim, Yong Kwan Lee : Implementation of Thermal-Hydraulic Models into a Power Plant Simulation Tool. The Korea Society for Simulation conference, Seoul (1998) 95–99
7. Eung Yeon Kim : COM Bible, SamYang Publishing Co. (1997)

Fast Reconstruction of 3D Terrain Model
from Contour Lines on 2D Maps

Byeong-Seok Shin and Hoe Sang Jung

Inha University, Department of Computer Science and Engineering
253 Yonghyeon-Dong, Nam-Gu, Inchon, 402-751, Korea
bsshin@inha.ac.kr, g2031372@inhavision.inha.ac.kr

Abstract. In order to create three-dimensional terrain models, a method that reconstructs geometric models from contour lines on two-dimensional map can be used. Previous methods divide a set of contour lines into simple matching regions and clefts [1],[2]. Since long processing time is taken for reconstructing clefts, its performance might be degraded in the case of handling complicated models containing a lot of clefts. In this paper, we propose a fast reconstruction method, which generates triangle strips with single tiling operation for simple region that does not contain branches. If there are some branches in contours, it partitions the contour lines into several sub-contours by considering the number of vertices and their spatial distribution. We implemented an automatic surface reconstruction system by using our method.

1 Introduction

Terrain modeling and rendering is an important technology in interactive computer games, geographic information system (GIS) and flight simulation. Commonly used source data for terrain modeling is height field produced by scanning a region with laser range scanners in satellite or airplane. However, since they inherently contain huge amount of sample data it is not easy to manipulate them in consumer PC's. In the case of using regularly-sampled height field, we can construct quad-tree structures [3],[4] or binary-tree structures called ROAM (Real-time Optimally Adaptive Mesh) [5], and control the number of polygons downloaded to rendering pipeline by applying real-time simplification and CLOD (Continuous Level-of-Detail) method [6-8]. Unfortunately, most terrain scanning is performed by irregular sampling such as LIDAR (light detection and ranging) scanning and reconstructed to irregular triangle meshes such as TIN (triangulated irregular network). Considerably long time is required to reconstruct TIN's and it is very hard to simplify that kind of mesh models.

Another method to reconstruct terrain model is to use contour lines on a 2D map. Since a contour line is obtained by sub-sampling the points that have the same height value with regular interval, we can make 3D terrain model by reconstructing original geometry in-between two consecutive contour lines lost in sampling step. 2D maps composed of contour lines are considerably cheaper than height field data such as DEM (digital elevation map), DTED (digital terrain elevation data) and LIDAR data. Triangles generated from two contour lines construct a set of triangle strips, which is optimal representation for graphics hardware.

D.-K. Baik (Ed.): AsiaSim 2004, LNAI 3398, pp. 230–239, 2005.

In this paper, we take into account automatic reconstruction methods for 3D terrain models from 2D contour lines. It identifies corresponding vertices on two adjacent contour lines, and reconstructs geometry by connecting the vertices with edges. A commonly used surface reconstruction method proposed by Barequet *et al.* divides a pair of corresponding contour lines into simple matching regions and clefts [1],[2]. It reconstructs triangle meshes by applying their tiling method to simple region, and triangulates cleft regions with dynamic programming method [9]. Although it has an advantage of generating considerably accurate models, it is not efficient while manipulating contour lines containing a lot of clefts.

We propose a fast reconstruction method of 3D terrain model. It generates a triangle strip with single tiling operation for a simple matching region that does not contain branches. If there are branches, our method partitions the contour line into several sub-contours by considering the number of vertices and their spatial distribution. Each sub-contour is processed by using the same method applied to matching regions.

In Section 2, we briefly review the related work. In Section 3 and 4, we explain our method and show experimental results. Lastly, we conclude our work.

2 Related Work

Surface reconstruction process is composed of three components: correspondence determination, tiling and branch processing [10] (see Fig. 1). Correspondence determination is to identify the contour lines that should be connected to a specific contour. Tiling is to produce triangle strips from a pair of contour lines. Branch processing is to determine the correspondence when a contour line faces to multiple contour lines.

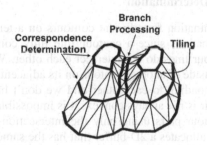

Fig. 1. Correspondence determination, tiling and branch problems occurred in adjacent slices

Assume that contour lines are defined on slices parallel to XY plane. When a contour is sufficiently close to its adjacent contour(s) in z-direction, it is easy to determine the exact correspondence using only the consecutive contours. Otherwise, we have to exploit prior knowledge or global information about overall shape of target objects [11]. Most of the reconstruction methods assume that a pair of corresponding contours should be overlapped while projecting a contour onto its corresponding one in perpendicular direction [12].

In order to cover the area between two adjacent contours with triangle strips, we define *slice chords*. A slice chord is a connecting edge between a vertex of a contour and another vertex on its corresponding one. Each triangle is composed of two slice chords and an edge of two consecutive vertices on a contour. Barequet and Sharir presented a method to exploit the partial curve matching technique for tiling matched regions of a pair of contours, and the minimum-area triangulation technique for remaining regions [7],[8]. Bajaj *et al.* defined three constrains for surface to be reconstructed and derived exact correspondence and tiling rules from those constraints [10]. Since it uses multi-pass tiling algorithm, complicated regions are reconstructed after processing all the other regions. Kline *et al.* proposed a method to generate accurate surface model using 2D distance transformation and medial axes [13].

Branches occur when a slice has N contours and its corresponding slice contains M contours ($M \neq N$, M, $N > 0$). Meyers *et al.* simplified multiple-branch problem into one-to-one correspondence by defining composition of individual contour lines [12]. They proposed a special method for canyons between two neighboring contours of the same slices. Barequet *et al.* [7],[8] and Bajaj *et al.* [10] presented methods for tiling branches that has complex joint topology and shape.

3   Terrain Model Reconstruction Method

We propose a high-speed terrain model reconstruction algorithm that does not directly handle clefts by extending the Barequet's method. A set of contour lines on a map can be separated into consecutive slices arranged with regular interval. Each slice contains contours that have the same height value, which are called neighboring contours. Let a pair of consecutive slices be $<S_n, S_{n+1}>$ and sets of contours belong the slices be C_i^n and C_j^{n+1}. A contour is composed of a set of vertices $\{v(x,y,n)\}$.

3.1   Correspondence Determination

Correspondence determination for adjacent contours on a terrain map is easier than that of the other applications since a contour line fully contains its corresponding contours. That is, contour lines do not intersect each other. When a vertex of a contour is projected onto inside of another contour on its adjacent slice, two contour lines are regarded as corresponding to each other and we don't have to check remaining vertices. Since a contour is not always convex, it is impossible to determine the exact correspondence of contour pairs using only the intersection test of bounding polygons. So our algorithm allocates a 2D-buffer that has the same size of a slice and fills up interior pixels of a C_j^{n+1} as its identifier with boundary-fill algorithm. Then it projects vertices $v(x,y,n)$ of C_i^n onto a slice S_{n+1}, and checks whether the buffer value is identical to the contour ID or not at the projected position $\{v(x,y,n+1)\}$ (see Fig. 2). When the vertices of more than one neighboring contours correspond to the same contour, the contour lines may produce branches.

3.2   Tiling

After we determine the correspondence of adjacent contours C_i^n and C_j^{n+1}, we produce triangle strips by connecting them. Vertices on the same contour maintain connectivity with neighboring ones. However, there is no topological information between two vertices on different contours. Therefore we have to determine the corresponding vertex on the adjacent contour for a specific vertex. Normally, topologically adjacent vertices have the smallest distance in comparison to the remaining vertices.

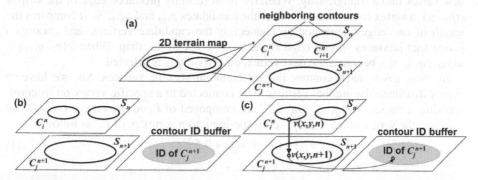

Fig. 2. Correspondence determination between two adjacent contours (a) extracts contour lines from a 2D map (b) fills up interior of C_i^{n+1} as its ID's in contour ID buffer (c) projects vertices of C_i^n onto S_{n+1} and checks the ID of corresponding points

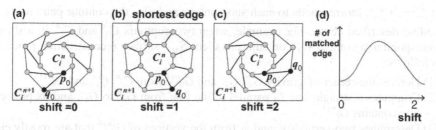

Fig. 3. An example of determining optimal shift value and the shortest edge (a) shift=0 (b) shift=1 (c) shift=2 (d) change the number of matched edge according to the shift value

Vertex lists of a pair of corresponding contours $< C_i^n, C_j^{n+1} >$ are $(p_0, \cdots, p_{l-1})$ and $(q_0, \cdots, q_{l-1})$. For simplicity, assume that the number of vertices of two contours is the same. When the length of a slice chord $e(p_k, q_k)$ is less than a pre-defined threshold ε, the slice chord is called a *matched edge*. It changes the index of the starting vertex s from 0 to l-1 for C_j^{n+1} while fixing the index of the starting vertex for C_i^n as 0. It determines the value of s that maximizes number of matched edges among slice chords $e(p_k, q_{(s+k) \bmod l})$ defined by two vertex list $(p_0, \cdots, p_{l-1})$ and $(q_s, \cdots, q_{l-1}, q_0, \cdots, q_{s-1})$. Here, s is the relative offset of two lists and called a *shift*. When applying the shift value that maximizes the number of matched edges, a slice chord that has the minimum length is regarded as the shortest edge. Fig. 3 shows an example of determining the optimal shift value. While increasing the index s from 0 to l-1, in the case of Fig. 3(a) and (c), length of each edge $e(p_k, q_{(s+k) \bmod l})$ becomes longer and the number of matched edge decreases. When the shift value is 1 (see Fig. 3(b)), the number of matched edges is maximized and $e(p_0, q_2)$ is the shortest edge. A triangle strip is generated from the shortest edge.

In order to produce visually pleased triangle strips, each triangle should be similar to an equilateral triangle. When we assume that a contour is composed of equally distant vertices, we have only to minimize the length of a slice chord while inserting a new vertex into a triangle strip. When the most recently produced edge of the strip is $e(p_k, q_k)$, a vertex to be added is one of the candidates p_{k+1} and q_{k+1}. So it compares the length of two edges generated by connecting the candidate vertices, and chooses a vertex that produces shorter edge as the next vertex of the strip. When $|e(p_k, q_{k+1})| > |e(p_{k+1}, q_k)|$, p_{k+1} becomes the next vertex, otherwise q_{k+1} is selected.

In most cases, each contour has different number of vertices. So, we have to evenly distribute the number vertices to be connected to a specific vertex on its corresponding contour. When C_i^n and C_j^{n+1} are composed of l and m vertices ($m > l$), we compute the vertex index t of C_j^{n+1} corresponding to a vertex p_k of C_i^n as follows:

$$t = \lfloor (1 + (m-l)/l)k + 0.5 \rfloor. \tag{1}$$

3.3 Branch Processing

A branch occurs when a contour C_j^{n+1} corresponds to N contours $\{ C_i^n, \cdots, C_{i+N-1}^n \}$. Our method partitions C_j^{n+1} into N sub-contours $\{ \overline{C}_j^{n+1}, \cdots, \overline{C}_{j+N-1}^{n+1} \}$, determines which of $\{ C_i^n, \cdots, C_{i+N-1}^n \}$ corresponds to each sub-contour and tiles the contour pairs using the method described above. For example, when two contours C_0^n and C_1^n on a slice S_n correspond to a contour C_0^{n+1} on adjacent slices S_{n+1}, surface reconstruction process is as follows:

(1) Derives the center-of-gravity G_0^n, G_1^n and G_0^{n+1} from C_0^n, C_1^n and C_0^{n+1}.
(2) Computes a straight line L_1 parallel to the line from G_0^n to G_1^n, and its projected line contains G_0^{n+1}.
(3) Determines two vertices v_a and v_b from the vertices of G_0^{n+1} that are mostly close to L_1 (Fig. 4(a)).
(4) Partitions the line segment $\overline{v_a v_b}$ according to ratio of the area of C_0^n and C_1^n (A_0, A_1) and denotes the point as M. Area of contours can easily be computed by counting the number of pixels while calculating the center-of-gravity (Fig. 4(b)).
(5) Computes a straight line L_2 perpendicular to the line segment $\overline{v_a v_b}$ and contains the point M.
(6) Determines two vertices v_c and v_d from the vertices of C_0^{n+1} that are mostly close to L_2 (Fig. 4(c)).
(7) Inserts shared vertices along with the line segment $\overline{v_c v_d}$. Interval of shared vertices is identical to that of ordinary vertices of original contour lines. Assume that z-values of S_n and S_{n+1} are n and $n+1$ respectively. Let the coordinates of points v_c, v_d, and M be (x_c, y_c, n), (x_d, y_d, n) and (x_M, y_M, n). Z-values of shared vertices should be interpolated along with a curve from v_c to v_d via new points M^* of which coordinates is $(x_M, y_M, n+0.5)$ (Fig. 4(d)).
(8) Partitions a contour line C_0^{n+1} into two sub-contours $\overline{C}_0^{n+1}$ and $\overline{C}_0^{n+1*}$ by duplicating the shared vertices.
(9) Reconstructs contour pairs $<C_0^n, \overline{C}_0^{n+1}>$ and $<C_1^n, \overline{C}_0^{n+1*}>$ using the tiling method described in the section 3.2 (Fig. 4(e) and (f)).

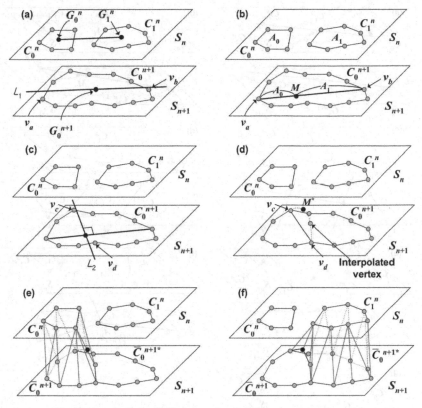

Fig. 4. An example of tiling a contour pair that contains branches

In the case of $N \geq 3$, it partitions entire contour into several sub-contours and tiles the contour pairs by recursively applying the above method until all of the sub-contours have one-to-one correspondence with adjacent contour lines. Fig. 5 shows how to reconstruct geometric models when 3-to-1 correspondence occurs. All the center-of-gravity points G_0^n, G_1^n and G_2^n of the neighboring contours can be located near a straight line (case1) or not (case2). It can be decided by performing linear regression of the center-of-gravity points, and evaluating the distance from each point to the interpolated line. In the case 1, we have to modify the above procedures as follows:

(2-1) Computes a straight line L_1 parallel to the line interpolated from G_0^n to G_2^n, and its projected line contains G_0^{n+1}.

(3-1) Determines two vertices v_a and v_b from the vertices of G_0^{n+1} that are mostly close to L_1 (Fig. 5 (a)).

(4-1) Partitions the line segment $\overline{v_a v_b}$ according to ratio of the area of C_0^n, C_1^n and C_2^n (A_0, A_1 and A_2) and denotes the points as M_1 and M_2 (Fig. 5 (b)).

(5-1) Computes straight lines L_2 and L_3 perpendicular to the line segment $\overline{v_a v_b}$ and passes by the point M_1 and M_2 (Fig. 5 (c)).

(6-1) Determines vertices v_c and v_d from the vertices of C_0^{n+1} that are mostly close to L_2 and L_3 (Fig. 5 (d)).

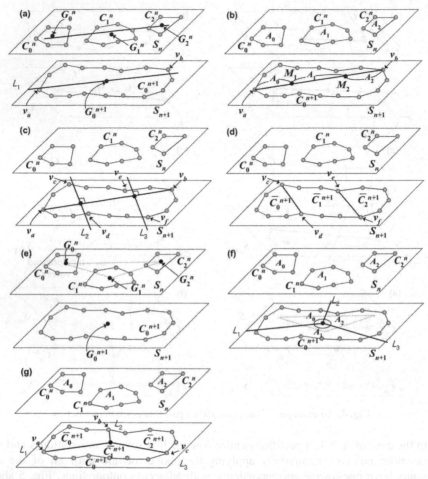

Fig. 5. An example of tiling a contour pair that contains 3-to-1 branches (case1 and case2)

In order to handle the case2, branch processing procedures should be modified as follows:

(2-2) Computes a virtual triangle T composed of G_0^n, G_1^n and G_2^n (Fig. 5(e)).

(3-2) Partitions the area of C_0^{n+1} according to ratio of the area of C_0^n, C_1^n and C_2^n (A_0, A_1 and A_2). All the partition lines L_1, L_2 and L_3 is most perpendicular to the edges of T and segment the edges based on the ratio of A_0, A_1 and A_2 (Fig. 5(f)).

(4-2) Determines vertices v_a, v_b and v_c from the vertices of C_0^{n+1} that are mostly close to L_1, L_2 and L_3 (Fig. 5(g)).

After finding the correspondence of contours of a slice and sub-contours of its consecutive slice, remaining procedure is identical to the basic reconstruction algorithm.

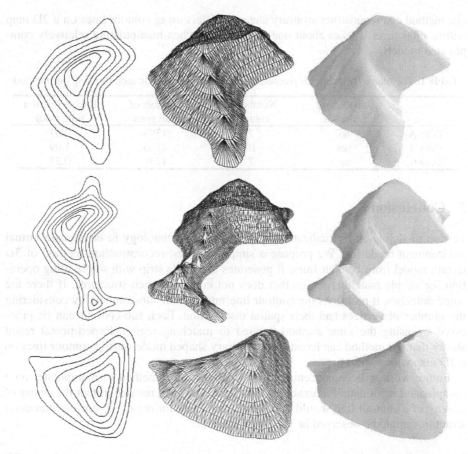

Fig. 6. Examples of reconstructed models and their shaded images with our method (left) 2D contour lines for specific regions (middle) wire frame model of reconstructed model (right) shaded image

4 Experimental Results

In order to show how efficient our method, we implement terrain model reconstruction system which extracts vertex lists for contour lines from 2D map, and produces three dimensional models by tiling corresponding contours. It is implemented on a PC equipped with Pentium IV 2.2GHz CPU, 1GB main memory, and NVIDIA GeForce4 graphics accelerator. We use a set of contours of 2D maps for several areas.

Fig. 6 shows images produced by rendering 3D terrain model with our method. It can reconstruct accurate models from contour lines even when we deal with considerably complex regions. It produces geometric models fairly well not only for simple matching area but also for branches.

Table 1 shows the time to be taken for reconstruction of geometry according to the size of data (number of slices and the number of vertices on a contour) and complexity (shape of contour lines and existence of branches). Experimental results show that our method can reconstruct arbitrary shaped models using contour lines on a 2D map within short time. It takes about one second even when manipulating relatively complicated models.

Table 1. The time to be taken for reconstructing several geometric models using our method

	Existence of branches	Number of contous	Number of polygons	Reconstruction time (sec)
Data A	no	7	1393	0.67
Data B	yes	10	2136	1.09
Data C	no	7	1241	0.55

5 Conclusion

Terrain modeling and visualization is an important technology in large-scale virtual environment rendering. We propose a simple and fast reconstruction method of 3D terrain model from contour lines. It generates a triangle strip with single tiling operation for simple matching region that does not contain branch structures. If there are some branches, it partitions the contour line into several sub-contours by considering the number of vertices and their spatial distribution. Each sub-contour can be processed by using the same method applied to matching region. Experimental result shows that our method can reconstruct arbitrary shaped models using contour lines on a 2D map within short time.

Future work will be concentrated on improving the method to manipulate more complicated input data. For example, our algorithm has a restriction that the center of gravity of a contour line should be located on its interior region. Therefore unnatural structures might be observed in 3D terrain models.

Acknowledgement

This research was supported by University IT Research Center Project.

References

1. Barequet, G., Sharir, M.: Piecewise-Linear Interpolation between Polygonal Slices. Computer Vision and Image Understanding, Vol. 63, No. 2 (1996) 251-272
2. Barequet, G., Shapiro, D., Tal, A.: Multilevel Sensitive Reconstruction of Polyhedral Surfaces from Parallel Slices. The Visual Computer, Vol. 16, No. 2 (2000) 116-133
3. Lindstrom, P., Pascucci, V.: Terrain Simplification Simplified : A General Framework for View-Dependent Out-of-core Visualization. IEEE Transactions on Visualization and Computer Graphics, Vol. 8 (2002) 239-254
4. Lindstrom, P.: Out-of-core Simplification of Large Polygonal Models. ACM SIGGRAPH (2000) 259-262
5. Duchaineau, M., Wolinsky, M., Segeti, D., Miller, M., Aldrich, C., Mineev-Weisstein, M.: ROAMing Terrain: Real-time Optimally Adaptive Meshes. ACM SIGGRAPH (1997) 81-88

6. Lindstrom, P., Koller, D., Ribarsky, W., Hodges, L., Faust, M., Turner, G.: Real-time Continuous Level of Detail Rendering of Height Fields. ACM SIGGRAPH (1996) 109-118
7. Rottger, S., Heidrich, W., Slasallek, P., Seidel, H.: Real-time Generation of Continuous Level of Detail for Height Fields. Proceedings of 6th International Conference in Central Europe on Computer Graphics and Visualization (1998) 315-322
8. Shin, B., Choi, E.: An Efficient CLOD Method for Large-scale Terrain Visualization. Proceedings of ICEC 2004 (2004) to be published
9. Klincsek, G.: Minimal Triangulations of Polygonal Domains. Annals of Discrete Mathematics, Vol. 9 (1980) 121-123
10. Bajaj, C., Coyle E., Lin, K.: Arbitrary Topology Shape Reconstruction for Planar Cross Sections. Graphical Models and Image Processing, Vol. 58, No. 6 (1996) 524-543
11. Soroka, B.: Generalized Cones from Serial Sections. Computer Graphics and Image Processing, Vol. 15, No. 2 (1981) 154-166
12. Meyers, D., Skinner, S., Sloan, K.: Surfaces from Contours. ACM Trans. on Graphics, Vol. 11, No. 3 (1992) 228-258
13. Klein, R., Schilling, A., Strasser, W.: Reconstruction and Simplification of Surface from Contours. 7th Pacific Conference on Computer Graphics and Applications (1999) 198-207

Contention Free Scheme for Asymmetrical Traffic Load in IEEE 802.11x Wireless LAN

Soo Young Shin and Soo Hyun Park

School of Business IT, Kookmin University
861-1, Chongnung-dong. Sungbuk-gu, Seoul, 136-702, Korea
{syshin,shpark21}@kookmin.ac.kr

Abstract. Every MAC (Medium Access Control) sub-layers of IEEE 802.11x, including IEEE 802.11e, defines Connection-based and CF (Contention Free)-based service functions in common. In this paper, a New-CF method is proposed. In the proposed method, conventional Round Robin method, which is used as a polling method by IEEE 802.11x PCF (Point Coordination Function) or IEEE 802.11e HCCA, is modified to give weights to channels with heavier traffic load and to provide those weighted channels with more services. Based on NS-2 simulations, it is verified the proposed method shows better throughput in general, particularly under asymmetrical traffic load conditions.

1 Introduction

With applications over 802.11 WLANs increasing, customers demand more and more new features and functions. One very important feature is the support of applications with quality of service (QoS). Thus, support of video, audio, real-time voice over IP, and other multimedia applications over 802.11 WLAN with QoS requirements is the key for 802.11 WLAN to be successful in multimedia home networking and future wireless communications.[1][4]

The performance of WLAN devices is directly linked to the speed of data transfer. MAC sub-layer, which is in charge of connections, plays an important role in the transfer speed. Taking into account the fact that IEEE 802.11a and IEEE 802.11b commonly uses MAC and IEEE 802.11e also provides with the backward capability, there is a limit to the performance improvement of WLAN devices without enhanced MAC, specially in case there are many collisions on channel access.[6] Therefore, the overall system performance is significantly affected by the collision rate in WLAN and by the error recovery speed of the crashed packets.

This paper investigates a buffer state based scheduling algorithms based on the idea of generalized processor sharing (GPS) which was proposed by Parekh and Gallager. [11][12] Recently, the idea of GPS has received a great deal of attention in the context of process scheduling and packet transmission. Our proposed scheduling algorithm, modified WGPS, is closely related to the process scheduling algorithms proposed in [3][7][9][11][12]. We modified the conventional simple polling functions by adopting WGPS (Wireless Generalized Processor Sharing) algorithm. The modified algorithm is adaptive to unbalanced and highly variational channel state. In chapter 2, the conventional 802.11 MAC contention (DCF) and CF (PCF) service,

D.-K. Baik (Ed.): AsiaSim 2004, LNAI 3398, pp. 240–249, 2005.
© Springer-Verlag Berlin Heidelberg 2005

and the enhancements of the draft 802.11e MAC over the baseline 802.11 standard have been briefly summarized. In chapter 3, a new polling technique, which refers the information of stacked data capacity by means of WGPS algorithm, has been suggested. NS-2 simulations which verifies the performance improvement of the new algorithm has been reported in chapter 4. Lastly, conclusions in chapter 5.

2 MAC of WLAN [6][8]

2.1 Contention Scheme (DCF) [8]

Before data transmission, the sender checks the state of wireless links and then is given the right for media access. At this stage, the sender uses NAV[1] to solve hidden node problems and uses backoff at the time of ending frames to avoid collision.

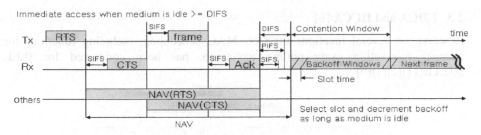

Fig. 1. An example of Contention-based Scheme (DCF)

Fig. 1 shows an example of CSMA/CA technique of Contention-based scheme (DCF). DCF method, which is a fundamental connecting method in WLAN MAC, provides with contention-based services and uses the backoff algorithm as a connection method. In the back-off algorithm, slots located in the latter part of the contention window would have lower selection probability. Consequently, the overall throughput would be dropped and the packet transmission delayed.

2.2 Contention Free Scheme (PCF) [8]

CF function is proposed for real-time transmission and provides with CF services which reserve and send time slots. It has a disadvantage, however, that CF function can not provide with adaptive services under highly variational traffic conditions. For example, in case of real-time traffics such as voices and moving images, CF function provides with uniform services only without regard to the required traffics.[4] It supports uniform access to media and contention functions are provided in turn. And during CF service period, near isochronous services are provided with. Polling procedures are conducted by PC (Point Coordinator) and every transmission process requires confirmation and reply procedures. In the beginning stage of CF period, PC

---

[1] Network Allocation Vector (NAV): An indicator, maintained by each station, of time periods when transmission onto the wireless medium (WM) will not be initiated by the station whether or not the station's clear channel assessment (CCA) function senses that the WM is busy.

(AP) transmit a beacon frame. In the beacon frame, the data of maximum duration time or CFP_Max_Duration are included and every STA[2], which are receiving this data, would setup up NAV.

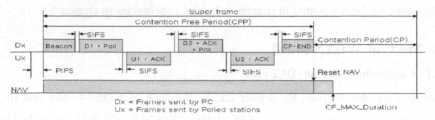

Fig. 2. An example of Contention Free Scheme (PCF)

2.3 EDCA and HCCA [6]

Research on several functions for QoS MAC scheduling, including functions for channel reservation using polling technique, has been conducted for IEEE 802.11e.[1][2][4][6]

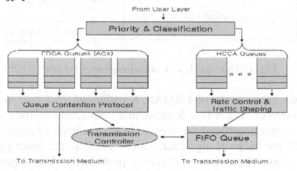

Fig. 3. IEEE 802.11e MAC architecture [1]

Figure depicts a block diagram of a MAC system implementation that facilitates the operation of the QoS channel access functions specified by the 802.11e draft. Traffic arriving at the MAC from the upper-layers is first classified as being served by the EDCA or the HCCA. Traffic classified as EDCA is mapped to one of the four queues (access categories) depending on its priority level.[6] The access of an individual queue to the channel is based on its associated parameters and the CSMA/CA algorithm that is executed by it. The MAC system requirements to implement HCCA are more complex. After admitting a stream and completing the scheduling process, the AP polls individual stations in accordance with the schedule arrived by it. These polls grant stations with a TXOP for a specified duration. As polls or TXOPs are addressed to a station and not to an individual stream, the station must ensure fairness in serving individual streams emanating from it. This is done using the Rate Control and Traffic Shaping blocks in Fig. 3.[6][1]

---

[2] Station (STA) : Any device that contains an IEEE 802.11 conformant medium access control (MAC) and physical layer (PHY) interface to the wireless medium.

3 Proposed New-CF algorithm

3.1 WGPS Algorithm [9]

To complement the disadvantage of conventional polling (RR), a study on the new polling technique, which is adaptive to the frequency and magnitude of traffic generations, is required. WGPS technique is an algorithm which is taking into account the states of the scheduler, which dominates packet scheduling in wireless network, and the balancing. In WGPS, unfairness caused by each flow's channel state is compensated by dynamically controlling its service share.[11][12]

A WGPS server with rate R and compensation index $\Delta(0<\Delta<1)$ is a fluid-flow server with following properties:

- φ_i : Each flow i is associated with its service share φ_i
- $S(t)$: Set of all flows which is backlogged and in good channel state at time t
- $C(t)$: Set of all flows which requires compensation at time t
- $\varphi_i(t)$: At any time t, each flow i is associated with its compensated service share $\varphi_i(t)$ which is given by

$$\phi_i(t) = \begin{cases} \phi_i \cdot (1+\Delta) & \text{if } i \in S(t) \text{ and } i \in C(t) \\ \phi_i & \text{if } i \in S(t) \text{ and } i \notin C(t) \\ 0 & \text{if } i \notin S(t) \end{cases} \tag{1}$$

- R : Rate of a WGPS Server
- $V(t)$: Server maintains virtual time, $V(t)$, which is given by

$$\frac{\partial V}{\partial t} = \begin{cases} \dfrac{R}{\sum_i \phi_i(t)} & \text{if } \sum_i \phi_i(t) \neq 0 \\ 0 & \text{otherwise} \end{cases} \tag{2}$$

where $V(0)$ is given by an arbitrary value.

- $W_i^c(t)$: Each flow i is associated with its compensation counter, $W_i^c(t)$, which stores the amount of service required to compensate the flow i and is given by

$$\frac{\partial W_i^c}{\partial t} = \begin{cases} \phi_i \cdot \dfrac{\partial V}{\partial t} & \text{if } i \in B(t) \text{ and } i \notin S(t) \\ -\phi_i \cdot \Delta \dfrac{\partial V}{\partial t} & \text{if } i \in S(t) \text{ and } i \in C(t) \\ 0 & \text{otherwise} \end{cases} \tag{3}$$

- $B(t)$: Set of all flows backlogged at time t and $W_i^c(t)$ is initialized by 0 when flow i becomes backlogged

Let $W_i(\tau, t)$ be the amount of flow i traffic served in an interval $(\tau, t]$, then $W_i(\tau, t)$ satisfies

$$\frac{\partial W_i(\tau,t)}{\partial t} = \phi_i(t) \cdot \frac{\partial V}{\partial t} \tag{4}$$

3.2 Modified WGPS Algorithm

WGPS algorithm has been modified by 4 contents. Firstly, the modified WGPS measures the capacity of stacked data at sender's buffer and stores the measured value. Secondly, arbitrary virtual time is continued during a certain period at the discrete moment. Service share φ_i which is constant for the life time of a flow, each flow in WGPS is associated with a time-varying compensated service share $\varphi_i(t)$. It is resulted from the fact that it is impossible to update polling lists in real time and service time. That is, WGPS is modified to generate a new polling list using the stacked data in additional polling list update intervals. Thirdly, total rate of R is rounded off because of the integer number of polling. Fourthly, let all pollable stations have at least 1 polling chance for minimum QoS guarantee. This is because it is impossible to predict the present state of the station, which is resulted from the fact that the updated polling list is based on the information of the previous interval. In contrast to the modified algorithm, conventional schedulers give uniform weighting to all pollable stations so that all the channels get a single chance in turn.

While service share φ_i determines objective service share which should be guaranteed to each flow as long as the flow is backlogged, compensated service share is a temporary service share which WGPS utilize in its instant service allocation and the update of virtual time. In order to realize the objective service allocation in the wireless networks where flows may not be servable and require compensation due to the bad channel state, each flow's compensation service share is dynamically changed according to the channel state, the buffer state, and the necessity of compensation as in eq. (1). In this paper, the major factor is the buffer states so that if there are data in transmission buffers the value of 1 is set and if no data exist then 0 is set as a flagging information of 1 bit. The 1 bit of information is received in the stage of data transmission and Coordinator (server), which has a function of WGPS, refers the information and make a service with priority considering possible.

The following equation is for 1 bit information,

$$I_i(t) = \begin{cases} 1 & if\ Q_i(t) > 0 \\ 0 & otherwise \end{cases} \tag{5}$$

where $Q_i(t)$ is the number of backlogged traffic at t on translator side and $I_i(t)$ is the partial buffer information at t.

Since a polling list update interval in WLAN contains the maximum duration time of CFP_Max_Duration value inside the beacon frames, arbitrary values can be used in the range of CFP_Max_Duration time. If the maximum rate is supposed to R, then the equation can be expressed as follows.

A WGPS server with rate R and compensation index $\Delta(0<\Delta<1)$ is a fluid-flow server

$$\phi_i(t)' = \begin{cases} \phi_i \cdot pri \cdot (1 + \Delta) & if\ I_i(t) = 1\ and\ i \in S(t)\ and\ i \in C(t \\ \phi_i \cdot (1 + \Delta) & if\ I_i(t) = 0\ and\ i \in S(t)\ and\ i \in C(t. \\ \phi_i & if\ i \in S(t)\ and\ i \notin C(t) \\ 0 & otherwise \end{cases} \tag{6}$$

$$\frac{\partial V}{\partial t} = \begin{cases} \dfrac{R}{\sum_i \phi_i(t)'} & \text{if } \sum_i \phi_i(t)' \neq 0 \\ 0 & \text{otherwise} \end{cases} \tag{7}$$

where *pri* is priority value. (*pri* is a value representing the importance or urgency of data and generally larger than 1.0)

In addition to the application of the weightings according to buffers' information $I_i(t)$, Compensation counter stores the amount of service required to compensate flow *i*. As in eq. (7), the compensation counter of a flow *i* in bad channel state is increased by a rate of φ_i which is the same as the increase rate of normalized service received by other error-free flows. If flow *i*'s channel is recovered and the flow is being served with a increased service share of $\varphi_i \cdot pri \cdot (1+\Delta)$ or $\varphi_i \cdot (1+\Delta)$, then the compensation counter is decreased by a rate of $\varphi_i \cdot pri \cdot \Delta$ or $\varphi_i \cdot \Delta$ which is same as the increase rate of normalized compensational service received by flow *i*. There remains a stage of making integer value from the service amount values. As already explained, it is done by rounding off the number of polling. The value of constant 1 is to guarantee a balancing between all stations which is pollable.

$$P_i = 1 + \text{int}\left(\frac{\partial V}{\partial t} + 1/2\right) \tag{8}$$

where P_i is Polling amount for next time interval with integer value and *int()* is function of integer. Collected set of P_i is updated regularly with the interval of pre-defined polling list update.

3.3 New-CF Scheme for IEEE 802.11x WLAN

To send one partial information bit of buffers in a packet, usable extra bits in the transmission packet can be located. There is no extra bit, however, in the packet frame which is defined by IEEE 802.11x MAC specification.[6][8] There is one way to use extra bits, which have not been used by the frame, without modification of the standard specification and another way is to attach an additional bit ahead or behind the packet. However, the last method will affect the system performance since 1 bit per packet acts as overheads. Therefore, we consider the processing speed of scheduler and balancing between STAs weightily. It is important to determine the adequate interval of the polling list modification. In this paper, the period of polling list modification is the same with the period of beacon generation. In IEEE 802.11x, the period of beacon generation is flexible since the size of super frame and of data frame is variable. It is more critical for system performance to maintain the optimum intervals under the condition of highly variational data traffics.

Fig. 4 shows an example of the New-CF processing flow.

Fig. 5 shows an example of New-CF scheme which enables adaptive CF-poll by using the modified WGPS algorithm. In the figure, at the changing time of polling list, PC changes its polling list taking into account channels' buffer information. After polling list is changed, PC tries polling to Ux (Users). At this stage, PC tries another adaptive polling to U1 (User #1) who has an amount of data in its buffer.

The proposed new technique changes its scheduling techniques repeatedly under control of AP (PC). In New-CF technique, which is operated in infrastructure based mode, STAs or Users measure the amount of packets which is stacked in transmission buffers, attach partial information (1 bit) inside the frames and transmit the data frame. PC refers the internal information of data frame received, stores it until the polling list changing time is reached, modifies the polling list using the modified WGPS algorithm in accordance with the required channel weights and lastly performs the adaptive polling.

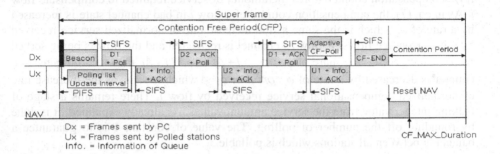

Fig. 4. An Example of New-Contention Free (New-CF) Scheme

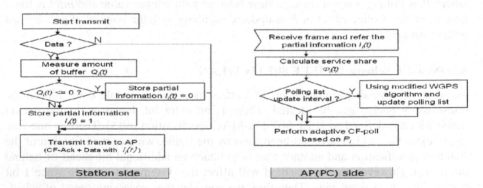

Fig. 5. New-CF scheme (Modified WGPS algorithm)

4 Simulation Result

NS-2 (Network Simulator 2) has been used for the simulations. 13 levels of traffic load are selected to generate CBR (Constant Bit Rate) traffics. Three methods of conventional DCF, PCF and the proposed New-CF were simulated for comparison of each method's performance. In addition, a case of unbalanced traffic environment, which contains video data, is also simulated by generating huge traffics (5 times of voice traffic) in a specific channel. Since the network has been operated on UDP protocols, the number of dropped packets was measured for comparisons. In the simulation model, 10 STAs and 1 AP is interconnected as an infrastructure and AP is connected to wired node. Wired traffic is excluded for throughput measurement. In case of DCF model, MAC, in which every node uses CSMA/CA, is set up. In case of

Table 1. Simulation Parameters

Parameter	Value	Parameter	Value
Time	50 sec	CBR interval	0.001~3 sec
STA	10	CBR size	210, 510
AP	1	Super Frame size	300 TU[a]
Wired Node	1	Limits	500*500 units
Bandwidth	11Mbps	DSSS_CWMin	31
Buffer size	100	DSSS_CWMax	1,023
Slot size	$20\mu s$	ShortRetryLimit	7
MRLT	512 TU	RTSThreshold	3,000 bytes

[a] TU: Time Units (1 TU = $1024\mu s$)

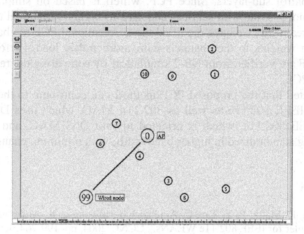

Fig. 6. Topology graph

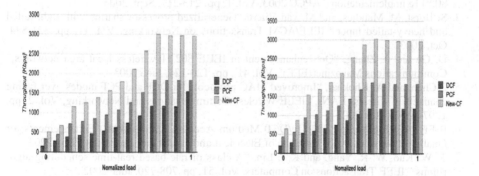

Fig. 7. Throughput (Symmetrical Traffic)   **Fig. 8.** Throughput (Asymmetrical Traffic)

PCF model, 7 STAs in PCF mode and 3 STAs in DCF mode is assumed to make regularly operating super frames. In case of New-CF model, simulation parameters are the same with PCF case except that the modified polling technique is applied in the New-CF model. Table 1 shows the simulation parameters and the topology graph of models are shown in Fig. 6.

Fig. 7 and 8 show each technique's throughput. In case of the proposed New-CF technique, the measured throughput is increased under both balanced and unbalanced conditions, especially compared with the cases of conventional DCF and PCF. Comparing with a case of PCF, for example, the proposed technique showed better throughput by up to 28 % under balanced traffic condition and by up to 34 % under unbalanced condition. The transmission failure rate rises as traffic loads increase. In relative comparison with throughput, it was inverse proportional.

5 Conclusion

IEEE 802.11b uses PCF to satisfy QoS and also uses contention-based DCF for media access control sub-layers. Since PCF, which is based on simple Round Robin method, can not provide with services adaptive to highly variational or asymmetrical data flow in real time, we have modified the conventional PCF to change the polling list by giving weights to the channels with more traffic load. Performances of the proposed PCF are verified from NS-2 simulation by comparing the results with those of DCF and PCF.

It is expected that the proposed PCF method can contribute to the improved performance of IEEE 802.11b as well as 802.11x MAC, which uses DCF and PCF in common, IEEE 802.11e, which is oriented toward QoS MAC, and IEEE 802.11n, which is being standardized in higher bandwidth, with minimum changes.

References

1. L. W. Lim, R. Malik, P. Y. Tan, C. Apichaichalermwongse, K. Ando, and Y. Harada, "A QoS scheduler for IEEE 802.11e WLANs," CCNC 2004, pp. 199-204, Jan. 2004
2. S. Rajesh, M. Sethuraman, R. Jayaparvathy, and S. Srikanth, "QoS algorithms for IEEE 802.11e implementation," APCC 2003, Vol. 1, pp. 213-217, Sept. 2003
3. S. Borst, M. Mandjes, and M. van Vitert, "Generalized processor sharing with light-tailed and heavy-tailed input," IEEE/ACM Transactions on Networking, Vol. 11, pp. 821-834, Oct. 2003
4. D. Gu and J. Zhang, "QoS enhancement in IEEE 802.11 wireless local area networks," Communications Magazine(IEEE) , Vol. 41 , pp. 120-124, June 2003
5. Z.Chen, and A. Khokhar, "Improved MAC protocols for DCF and PCF modes over fading channels in wireless LANs," IEEE Wireless Communications and Networking, Vol. 2, pp. 1297-1302, Mar. 2003
6. IEEE. "IEEE Standard. 802.11e/D6.0 Medium Access Control (MAC) Enhancements for Quality of Service (QoS)," Institute of Electrical and Electronics Engineers, Nov. 2003
7. T. W. Kuo, W. R. Yang, and K. j. Lin, "A class of rate-based real-time scheduling algorithms," IEEE Transactions on Computers, Vol. 51, pp. 708-720, June 2002
8. IEEE. "IEEE Standard for Wireless LAN Medium Access Control (MAC) and Physical Layer (PHY) specifications," Institute of Electrical and Electronics Engineers, November 1999.
9. M. R. Jeong, H. Morikawa, and T. Aoyama, "Wireless packet scheduler for fair service allocation," APCC/OECC '99. Fifth Asia-Pacific Conference on Communications, Vol. 1, pp. 794-797, 18-22 Oct. 1999
10. D. C. Stephens, J. C. R. Bennett, and H. Zhang, "Implementing scheduling algorithms in high-speed networks," IEEE Journal on Selected Areas in Communications, Vol. 17, pp. 1145-1158, June 1999

11. A. K. Parekh, and R. G. Gallager, "A Generalized Processor Sharing Approach to Flow Control in Integrated Services Networks: The Single-Node Case," IEEE/ACM Transactions on Networking, Vol. 1, pp. 344-357, Jun. 1993
12. A. K. Parekh, and R. G. Gallager, "A Generalized Processor Sharing Approach to Flow Control in Integrated Services Networks-the multiple Node Case," INFOCOM '93. , Vol. 2, pp. 521-530, Apr. 1993

Performance Analysis of the Dynamic Reservation Multiple Access Protocol in the Broadband Wireless Access System

KwangOh Cho and JongKyu Lee

Dept. of Computer Science and Engineering,
Hanyang University, Korea
{kocho,jklee}@cse.hanyang.ac.kr

Abstract. In this paper, we analyzed the DRMA(Dynamic Reservation Multiple Access) protocol with Rayleigh fading, shadowing, and capture effect for the Broadband Wireless Access System. We consider the TDMA-based protocols, since these are based on the standard for IEEE 802.16 WMAN, IEEE 802.20 MBWA, and ETSI HIPERLAN/2. The results of analytical model will be apply the development of collision resolution algorithm, scheduling algorithm, and the dynamic change of system parameters.

1 Introduction

The increased demands for mobility and flexibility in our daily life are demands that lead the development from wired LANs to wireless LANs (WLANs) or wireless MANs (WMANs). Today a wired LAN can offer users high bit rates to meet the requirements of bandwidth consuming services like video conferences, streaming video etc. With this in mind a user of a WLAN or a WMAN will have high demands on the system and will not accept too much degradation in performance to achieve mobility and flexibility. This will in turn put high demands on the design of WLANs ro WMANs of the future.

There are two competing standards for WLANs offering high bit rates, HIgh PErformance Radio Local Area Network-type 2 (HIPERLAN/2) defined by ETSI and IEEE 802.11a defined by IEEE. Also standards for WMANs are IEEE 802.16 WMAN and IEEE 802.16 Mobile Broadband Wireless Access (MBWA) defined by IEEE. These standards operate in the 5 GHz band.

The Performance Analysis of the Medium Access Control (MAC) for IEEE 802.16 WMAN, IEEE 802.20 MBWA and ETSI HIPERLAN/2 is performed in this paper by the means of queueing theory and simulation. we present an exact analytical approach for the delay of TDMA-based MAC protocol in Rayleigh fading [1].

The paper is organized as follows: In Section 2, the generalized MAC protocol and wireless capture model is described. The analysis model for the DRMA protocol is described and analyzed in Section 3, 4. In Section 5, some numerical results are reported. Finally, we give concluding remarks in Section 6.

D.-K. Baik (Ed.): AsiaSim 2004, LNAI 3398, pp. 250–259, 2005.

2   Background

2.1   Generalized MAC Protocol

In IEEE 802.16 WMAN, IEEE 802.20 MBWA and ETSI HIPERLAN/2 the
MAC protocol is based on Time Division Duplex (TDD) / Time Division Multiple Access (TDMA) using a fixed MAC frame. In the beginning of each MAC
frame the Access Point (AP) or Base Station (BS) decides and informs every
Mobile Terminal (MT) or Subscribe (SS) at what point in the MAC frame they
shall transmit and receive information. Time slots are allocated dynamically and
depend on the need for transmission. An MT (or SS) informs the AP (or BS) of
its need for transmission by means of a Resource Request (RR) control signal.

A MAC frame comprises a broadcast phase, downlink phase, uplink phase
and a random access phase. The broadcast phase comprises fields for broadcast
control, frame control and access feedback control. The uplink and downlink-
phase carries traffic from/to an MT (or SS) in bursts of packets. The MAC
frame structure is shown in Fig. 1.

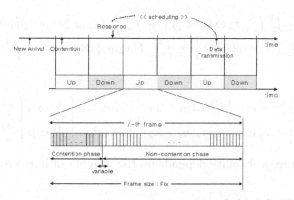

Fig. 1. General Frame Structure of Dynamic TDMA System.

WMAN, MBWA and HIPERLAN/2 uses a centralized control mechanism
and the AP (or BS) is responsible for assigning radio resources within the MAC
frame. Any MT (or SS), that wants initial access to the radio network, uses
contention slots for sending a Resource Request message to the AP (or BS). If
a collision occurs during a resource request the AP (or BS) will inform the requesting MT (or SS) in the following frame and a backoff procedure in contention
slots will take place.

2.2   Wireless Capture Model

The test packet can be received successfully-that is, it captures the receiver in
the presence of other overlapping or interfering packets-if its instantaneous power
is larger than the instantaneous joint interference power by a minimum certain
threshold factor z. This effect is the capture effect, and the threshold factor

is called the capture ratio. As is often done in the literature on the subject, instantaneous power is assumed to remain approximately constant for the time interval of packet duration. To use a convenient method of analyzing capture probabilities, we consider the weight function approach for the Rayleigh fading channels based on Laplace transforms [2].

We consider that a wireless network consists of M terminals $(s_1, s_2, \cdots, s_M)$. We define R_n as the set of terminals, which means n terminals transmit a packet at the same time. The s_1 denotes a receiver which wants to receive the packet form a certain transmitter $s_i(s_i \in R_n)$. If the s_1 receives the packet successfully from s_i, the instantaneous signal power w_s should exceed the joint interference signal power w_L from $n-1$ terminals $(R_n - \{s_i\})$ by the capture ratio z. However, the s_1 does not receive the packet successfully form s_i if only w_s is captured from w_L, since the w_s also includes the joint interference signal with multipath fading, shadowing, and near-far effect by itself. Let w_f be the joint interference signal for only s_i and w_0 denote the desired signal power of a packet, the w_0 and w_f are included in the w_s. The capture effect, denote $q(n|z)$, for n colliding packets is followings:

$$q(n|z) = \frac{2}{\sqrt{\pi}} \int_0^1 \int_{-\infty}^{\infty} r_1 \exp(-x_1^2) \left[\frac{1}{\sqrt{\pi}} \int_{-\infty}^{\infty} f(x_1, y_1) \exp(-y_1^2) dy_1 \right] dx_1 dy_1$$
$$\times \frac{2}{\sqrt{\pi}} \int_0^1 \int_{-\infty}^{\infty} r_2 \exp(-x_2^2) \left[\frac{1}{\sqrt{\pi}} \int_{-\infty}^{\infty} f(x_2, y_2) \exp(-y_1^2) dy_2 \right] dx_2 dy_2 \quad (1)$$

where

$$f(x_1, y_1) = \left[\sqrt{z} r_1^2 \exp\left\{ \frac{\sqrt{2}}{2} \sigma_s(y_1 - x_1) \right\} \right] \cdot arctan \left[\frac{1}{zr_1^4} \exp\left\{ \frac{\sqrt{2}}{2} \sigma_s(x_1 - y_1) \right\} \right]$$
$$f(x_2, y_2) = \left[\sqrt{z} r_2^2 \exp\left\{ \frac{\sqrt{2}}{2} \sigma_s(y_2 - x_2) \right\} \right] \cdot arctan \left[\frac{1}{zr_2^4} \exp\left\{ \frac{\sqrt{2}}{2} \sigma_s(x_2 - y_2) \right\} \right]$$

This capture probability depends on the number of interferers, n.

3 Analysis Model

3.1 System Model

Fig. 2 is the system model to analyze the performance for dynamic reservation access protocols in WMAN, MBWA and HIPERLAN/2. The analysis model for the number of users or terminals will be assumed a finite population model or a infinite population. Also, the arrival type of a user or a terminal will be supposed a packet or a message with several packets. If a packet or a message arrives, a user or a terminal will store arrival packets or messages to the buffer, and does not store. The contention scheme for the channel reservation of a arrival packet or message uses slotted ALOHA. And the number of contention slots will be assumed fix or variable value. If a reservation request of a arrival packet or message fail, a user or a terminal must consider the contention resolution method for re-request. The AP (or BS) can store survival packets or messages after reservation request to virtual queue. Also the size of a virtual queue will be assumed finite or infinite [3–8].

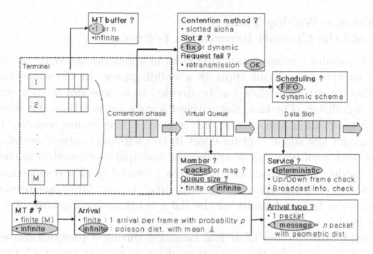

Fig. 2. System Model.

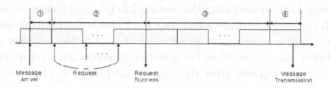

Fig. 3. The Component of Delay Time.

3.2 Delay Model

In Fig. 3, the delay time consists of following 4 parts:

1. Average waiting time before a reservation request
2. Average waiting time until the channel reservation is success
3. Average queueing time until tagged message is scheduled
4. Average transmission time of tagged message

4 Packet Delay Time Analysis

4.1 Average Waiting Time Before a Reservation Request

Assume that the length of a fixed frame, F, the arrival point of i-th message in a frame, τ_i. The τ_i is a i.i.d random variable with uniform distribution U_i, $Unif(0, F)$. If X is a number of arrival message during a previous frame $(0, F]$, a waiting time for reservation request is following

$$E\left[\sum_{i=1}^{X}(F - \tau_i)\right] / E[X] = \frac{F}{2} \qquad (2)$$

4.2   Average Waiting Time
Until the Channel Reservation Is Success

We must consider a wireless capture model in a generalized MAC protocol. The wireless capture model can apply in a uplink phase. In the contention phase of a uplink phase, time slots can be divided into time slot with a collision or without a collision. The time slot without a collision is a time slot that doesn't receive a reservation request and only receive a reservation request. This case must consider the single capture effect in the wireless capture model. And the time slot with a collision can be applied the multiple capture effect in the wireless capture model. These case exist a success probability of a reservation request in time slot a collision. Also a scheduled time slot for data transmission uplink phase exists a fail probability of the data transmission because of a wireless capture model.

We define the number of arrival messages during a previous frame, X, the number of time slots for the contention phase in current frame, C, the number of successful requests after the contention phase, S.

If s messages are success after the contention phase, s messages can be divided into s_1 with the single capture effect and $s - s_1$ with the multiple capture effect.

We must find a probability, $Pr(J = j)$ that J is the number of time slots with a collision. we assume that the number of reservation requests, n, and the number of success requests after the contention phase, s. Then we have

$$Pr(J = j) = \frac{{}_{C-s}C_j \times {}_jH_{n-s-(j\times 2)}}{\sum\limits_{j=1}^{C-s} {}_{C-s}C_j \times {}_jH_{n-s-(j\times 2)}} \tag{3}$$

And if the number of reservation requests, X is k, we obtain $Pr(S = s|X = k)$ without the wireless capture model by 'Random n object v cells' as following [9]:

$$Pr(S = s|X = k) = \frac{(-1)^s C! k!}{C^k s!} \sum\limits_{j=s}^{\min\{C,k\}} (-1)^j \frac{(C-j)^{k-j}}{(j-s)!(C-j)!(k-j)!} \tag{4}$$

Now, we can apply the wireless capture model.

$$Pr(S = s|X = k) = \sum\limits_{s_1=0}^{s} \left\{ Pr(S = s_1|X = k) \times \sum\limits_{j=s-s_1}^{\min(\lceil \frac{k-s_1}{2} \rceil, C-s_1)} \right.$$
$$\left. \times Pr(J = j) \binom{j}{s - s_1} q(n|z)^{s-s_1} [1 - q(n|z)]^{j-s+s_1} \right\} \tag{5}$$

Because that X is the number of messages during the previous frame, X follows a poisson distribution and $Pr(S = s)$ is given by

$$Pr(S = s) = \sum\limits_{i=0}^{\infty} \left[\sum\limits_{s_1=0}^{s} \left\{ Pr(S = s_1|X = k) \times \sum\limits_{j=s-s_1}^{\min(\lceil \frac{k-s_1}{2} \rceil, C-s_1)} \right. \right.$$
$$\left. \left. \times Pr(J = j) \binom{j}{s - s_1} q(n|z)^{s-s_1} [1 - q(n|z)]^{j-s+s_1} \right\} \right] \tag{6}$$

Now, we can easy solve $E[S]$.

$$E[S] = \sum_{s=0}^{\infty} s \times \Pr(S = s) = \sum_{s=0}^{C} s \times \Pr(S = s) \qquad (7)$$

Assume that the number that a user or a terminal try the reservation request until the request is a success, Z. The Z is given by

$$\Pr(Z = z) = q^{z-1}p , \quad z = 1, 2, 3\cdots , \quad q = 1 - p \qquad (8)$$

where $p = \begin{pmatrix} M-1 \\ K-1 \end{pmatrix} / \begin{pmatrix} M \\ K \end{pmatrix}$, $M = \lfloor E[X] \rfloor$, $K = \lfloor E[S] \rfloor$

We can obtain the average try number that a user or a terminal requests the reservation as following

$$E[Z] = \frac{1}{p} = \frac{\begin{pmatrix} M \\ K \end{pmatrix}}{\begin{pmatrix} M-1 \\ K-1 \end{pmatrix}} = \frac{M}{K} \qquad (9)$$

Now we find the average waiting time until the channel reservation is success as following:

$$E[Z] \times F = \frac{M}{K} \times F = \left\lceil \frac{E[X]}{E[S]} \right\rceil \times F \qquad (10)$$

4.3 Average Queueing Time Until a Tagged Message Is Scheduled

We can obtain the average queueing time until a tagged message is scheduled using $Pr(S = s)$. The number of packets for messages that a reservation request is a success, is represented by a geometric distribution as a mean, l. And If a new message consists of the sum of successful messages, this follows a negative binomial distribution.

Assume that the sum of packets in successful messages, Y.

$$\Pr(Y = y|S = s) = \begin{pmatrix} y-1 \\ s-1 \end{pmatrix} p^s q^{y-s} , \quad y = s, s+1, s+2, \cdots \qquad (11)$$

By the wireless capture model and $Pr(S = s)$, $Pr(Y = y)$ is given by

$$\Pr(Y = y) = \sum_{s=1}^{C} \left\{ \begin{pmatrix} y-1 \\ s-1 \end{pmatrix} p^s q^{y-s} \right\} $$
$$\times \sum_{i=0}^{\infty} \left[\sum_{s_1}^{s} \left\{ \times \begin{array}{l} \Pr(S = s_1 | X = k) \\ \sum_{j=s-s_1}^{\min\left(\left\lceil \frac{k-s_1}{2}\right\rceil, C-s_1\right)} \Pr(J = j) \begin{pmatrix} j \\ s - s_1 \end{pmatrix} \\ \times q(n|z)^{s-s_1} [1 - q(n|z)]^{j-s+s_1} \\ \times \frac{e^{-\lambda}\lambda^k}{k!} \end{array} \right\} \right] \qquad (12)$$

In a steady-state, the probability that the number of packets in a virtual queue is i packets, is assumed by π_i. And the transition probability from state i to state j is assumed by p_{ij}. This p_{ij} depends on the number of time slots for the data transmission, N and the total number of packets for successful messages, e_y. But we consider N_S because of the wireless capture model. The N_S is the average number of time slots for the data transmission with the wireless capture model.

$$N_s = \lfloor N \times q(w_0|z) \rfloor \tag{13}$$

Now, we can define a transition matrix as followings:

	0	1	2	$\cdots$	N_s-1	N_s	N_s+1	$\cdots$	K-N_s	K-N_s+1	$\cdots$	K-1	$\cdots$
0	e_0	e_1	e_2	$\cdots$	e_{N_s-1}	e_{N_s}	e_{N_s+1}	$\cdots$	$\cdots$	$\cdots$	$\cdots$	e_{K-1}	$\cdots$
1	e_0	e_1	e_2	$\cdots$	e_{N_s-1}	e_{N_s}	e_{N_s+1}	$\cdots$	$\cdots$	$\cdots$	$\cdots$	e_{K-1}	$\cdots$
2	e_0	e_1	e_2	$\cdots$	e_{N_s-1}	e_{N_s}	e_{N_s+1}	$\cdots$	$\cdots$	$\cdots$	$\cdots$	e_{K-1}	$\cdots$
3	e_0	e_1	e_2	$\cdots$	e_{N_s-1}	e_{N_s}	e_{N_s+1}	$\cdots$	$\cdots$	$\cdots$	$\cdots$	e_{K-1}	$\cdots$
.									.	.	.		
N_s-1	e_0	e_1	e_2	$\cdots$	e_{N_s-1}	e_{N_s}	e_{N_s+1}	$\cdots$	$\cdots$	$\cdots$	$\cdots$	e_{K-1}	$\cdots$
N_s	e_0	e_1	e_2	$\cdots$	e_{N_s-1}	e_{N_s}	e_{N_s+1}	$\cdots$	$\cdots$	$\cdots$	$\cdots$	e_{K-1}	$\cdots$
N_s+1	0	e_0	e_1	$\cdots$	$\cdots$	e_{N_s-1}	e_{N_s}	$\cdots$	$\cdots$	$\cdots$	$\cdots$	e_{K-2}	$\cdots$
N_s+2	0	0	e_0	e_3	$\cdots$	$\cdots$	e_{N_s-1}	$\cdots$	$\cdots$	$\cdots$	$\cdots$	e_{K-3}	$\cdots$
								$\cdots$	$\cdots$	$\cdots$	$\cdots$		.
K-N_s	.							e_{N_s}	e_{N_s+1}	$\cdots$	$\cdots$		.
K-N_s+1								e_{N_s-1}	e_{N_s}	$\cdots$	$\cdots$		
.	.	.	.				.	.	.				
K-1	0	0	0	$\cdots$	$\cdots$	$\cdots$	$\cdots$	$\cdots$					$\cdots$
K	0	0	0	$\cdots$	$\cdots$	$\cdots$	$\cdots$	$\cdots$	e_0	e_1	$\cdots$	e_{N_s-1}	$\cdots$
.	.	.	.				.	.			.	.	.

$$\tag{14}$$

We find π_i as followings :

$- j = 0$ case:

$$\pi_0 = \pi_0 e_0 + \pi_1 e_1 + \cdots + \pi_{N_s} e_0 = \left(\sum_{i=0}^{N_s} \pi_i\right) e_0 \Leftrightarrow \left(\sum_{i=0}^{N_s} \pi_i\right) = \frac{\pi_0}{e_0} \tag{15}$$

$- j \geq 1$ case :

$$\pi_j = \pi_0 e_j + \cdots + \pi_{N_s} e_j + \sum_{i=1}^{j} \pi_{N_s+i} e_{j-i} = \left(\frac{\pi_0}{e_0}\right) e_j + \sum_{i=1}^{j} \pi_{N_s+i} e_{j-i} \tag{16}$$

In steady-state, the sum of π_i is 1.

$$\sum_{i=0}^{\infty} \pi_i = 1 \tag{17}$$

By the computer simulation using eqn. (14)-(17), we find the probability, π_i that the virtual queue have i packets at beginning point of a frame. And the mean of π_i is $E[Q] = \sum_{i=0}^{\infty} i \cdot \pi_i$. Therefore the average queueing time until a tagged message is scheduled, is given by

$$\left\lceil \frac{E[Q]}{N_s} \right\rceil \times F \tag{18}$$

4.4   Average Transmission Time of a Tagged Message

The average processing time of a successful message after the contention phase follows a geometric distribution with a mean p_m. If $E[L]$ is a average length of a message, the average transmission time of a tagged message is given by

$$\left\lceil \frac{E[L]}{N_s} \right\rceil \times F \tag{19}$$

4.5   Average Packet Delay Time

$$Delay = \left\lceil \frac{F}{2} \right\rceil + \left\lceil \frac{E[X]}{E[S]} \right\rceil F + \left\lceil \frac{E[Q]}{N_S} \right\rceil F + \left\lceil \frac{E[L]}{N_S} \right\rceil F \tag{20}$$

4.6   Packet Waiting Time Distribution Function

A waiting time of l time slots can be expressed, using modulo F arithmetic, by:

$$l = m \times F + r , \quad 0 \leq r < F \tag{21}$$

The number of a user or a message in system before the arrival of a marked user or message is denoted by $Z^* = Z_0^*$. The amount of those users or messages remaining in the system t time slots later is denoted by Z_t^* (not taking into account arrivals during this interval). If all servers are busy and work continuously $m \cdot N$ users or messages are served in $m \cdot F$ time slots.

Assume that the marked customer has a waiting time of at most x time slots. Thus, l time slots upon the arrival of this user or message at most $N - 1$ of the users or messages in the system prior to the arrival of the marked user or message remain in the system. Taking in account continuous service we notice that r time slots upon the arrival of the marked user or message at most $Z_r^* = mN + N - 1$ of the Z_0^* users or messages in the system before the arrival remain in the system.

$$\begin{aligned}
W \leq l \quad &\Leftrightarrow Z_l \leq N - 1 \\
&\Leftrightarrow Z_r \leq mN + N - 1 \\
&\Leftrightarrow Z_F \leq (m - 1)N + N - 1 \\
&\Leftrightarrow Z_F \leq mN - 1
\end{aligned} \tag{22}$$

We define the probability $b_\nu(r)$ that r time slots upon the arrival of a user or a message at most ν of the users or messages in the system before the arrival will remain remain in the system:

$$b_\nu(F) = \Pr(Z_F \leq \nu) \tag{23}$$

The waiting time distribution function is given by:

$$\begin{aligned}
&\Pr(W \leq l) = b_{mN+N-1}(r) , \quad l = mF + r , \quad 0 \leq r < F \\
&\Leftrightarrow \Pr(W \leq l) = b_{mN-1}(F) , \quad l = mF + r , \quad 0 \leq r < F
\end{aligned} \tag{24}$$

For the computation of the waiting time distribution the probabilities $b_\nu(r)$ have to be derived. Therefore the following relation is exploited:

$P\{$at most j users or messages are in the system at time $t\}$

$$= \sum_{k=0}^{j} P\{\text{at most } (j-k) \text{ users or messages from time } (t-r)$$

$$\text{remain at time } t \text{ in the system}$$

$$\text{and } k \text{ arrivals occur in the interval } (t-r, t]\} \tag{25}$$

With $t \to \infty$ and the memoryless property of the arrival process the following equation is obtained:

$$\Pr(Z \le j) = \sum_{k=0}^{j} \pi_k = \sum_{k=0}^{j} b_{j-k}(F) \times e_k , \quad j = 0, 1, 2, \cdots \tag{26}$$

Solving this equation in a recursive manner (for $0 \le r < D$) we arrive at :

$$b_0(F) = e_0^{-1} \times \pi_0 , \quad j = 0$$
$$b_j(F) = e_0^{-1} \times \left(\sum_{k=0}^{j} \pi_k - \sum_{k=1}^{j} b_{j-k}(F) \times e_k \right) , \quad j = 1, 2, 3, \cdots \tag{27}$$

5   Numerical Results

In this section we present numerical examples for the generalized DRMA MAC protocol. The Fig. 4 compares the number of success requests and shows the similar result. The result of the Fig. 5 can apply to change the number of time slots for the contention phase in the next frame. The Fig. 6 shows the probability of π_i in the steady-state. Finally, the Fig. 7 is the average packet delay time by the OL(offered load).

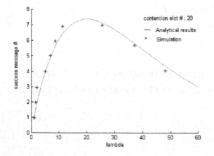

Fig. 4. Compare of analytical model and simulation model.

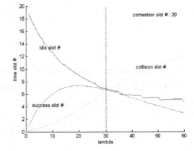

Fig. 5. Compare of time slots # in contention phase.

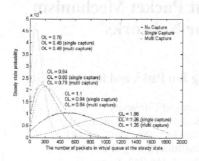

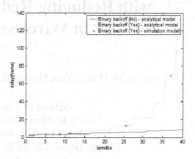

Fig. 6. Steady-steady probability of π_i.   **Fig. 7.** Compare of delay time.

6   Conclusion

In this paper we analyzed the average delay time of the TDMA-based generalized MAC protocol and derived the waiting time distribution. These results can be apply in performance studies of various modern telecommunication systems.

Acknowledgements

This work was supported by grant No. R01-2001-000-00334-0 from the Korea Science & Engineering Foundation.

References

1. A. Doufexi, S. Armour, M. Butler, A. Nix, D. Bull and J. McGeehan, "A Comparison of the HIPERLAN/2 and IEEE 802.11a Wireless LAN Standards," IEEE Communications Magazine, May 2002
2. J. H. Kim and J. K. Lee, "Capture Effects of Wireless CSMA/CA Protocols in Rayleigh and Shadow Fading Channels," IEEE Trans. on Vehicular Techonogy, 1999
3. T. Suzuki and S. Tasaka, "A contention-based reservation protocol using a TDD channel for wireless local area networks:a performance analysis," ICC '93, 23-26, May 1993
4. G. Pierobon, A. Zanella, and A. Salloum, "Contention-TDMA Protocol: Performance Evaluation," IEEE Trans. on Vehicular Technology, Jul. 2002
5. S. S. Lam, "Delay Analysis of a Time Division Multiple Access (TDMA) Channel," IEEE Trans. on Comm. Vol. COM-25, No. 12, Dec. 1977
6. I. Rubin and Z. H. TSAI, "Message Dealy Analysis of Multiclass Priority TDMA, FDMA, and Discrete-Time Queueing Systems," IEEE Trans. on Information Theory, Vol. 35, No. 3, May 1989
7. R. Jafari and K. Sohraby, "General Discrete-Time Queueing Systems with Correlated Batch Arrivals and Departures," IEEE INFOCOM 2000, Vol. 1, pp.181-188, Tel Aviv, Israel, 26-30, Mar. 2000
8. G. J. Franx, "A Simple Solution for the M/D/c Waiting Time Distribution," Operations Research Letters, 29:221-230, 2001
9. W. Szpankowski, "Packet Switching in Multiple Radio Channels: Analysis and Stability of a Random Access System," Computer Networks, No.7, pp. 17-26, 1983

Energy Efficient MAC Protocol
with Reducing Redundant Packet Mechanism
in Wireless Sensor Networks*

Jung Ahn Han[1], Yun Hyung Kim[1], Sang Jun Park[2], and Byung Gi Kim[1]

[1] School of Computing, Soongsil University
Sangdo-5dong Dongjak-gu, Seoul, 156-743, Korea
{dawndew11,yhkim602,bgkim}@archi.ssu.ac.kr
[2] Information & Media Technology institute, Soongsil University
Sangdo-5dong Dongjak-gu, Seoul, 156-743, Korea
lub@archi.ssu.ac.kr

Abstract. Wireless Sensor Networks are the technology, in which various applications such as surveillance and information gathering are possible in the uncontrollable area of human. And numerous studies are being processed for the application of ubiquitous network environment. One of major issues in sensor network is the research for prolonging the lifetime of nodes through the use of various algorithms, suggested in the mac and routing layer. In this paper, aiming at reducing energy waste, caused by redundant transmission and receipt message, we propose the mac protocol using active signal and analysis performance through simulation.

1 Introduction

Recently, the concern about micro sensor network, which can be used in the new mobile computing and various fields of communication is on the increase, and its study is actively processed in institutes and universities. Wireless sensor network is the technology, which can be applied to the various fields such as environment monitoring, medical system, and robot exploration. The number of nodes has the sensing, information processing, and wireless communication capacity in indoor or adjacent distance area and offers information with a composition of multi hop wireless network [1-3].

Currently, in the study relevant sensor networking, the development of sensor node and operating system such as MICA of UC Berkeley, WINS of Rockwell, iBagda of UCLA, u-AMPS of MIT, and Tiny OS of UC Berkeley is being progressed, and the project for satisfying the requirements of respective network layer is actively progressed in America, Europe, and the other countries.

The most sensor nodes developed currently has the small size about 1~2cm and numerous scattered nodes which consume the energy about 100μW, are designed to compose the network [4,5]. In addition, compared to the ad hoc network node, sensor network is composed of densely built up network with considerably numerous nodes, and these nodes have the high possibility of error occurrence, and topology has the

* This work was supported by Soongsil University Research Fund.

D.-K. Baik (Ed.): AsiaSim 2004, LNAI 3398, pp. 260–269, 2005.
© Springer-Verlag Berlin Heidelberg 2005

characteristics of frequent change. Besides, in contrast to the ad hoc network, each sensor nodes communicate using broadcast. Above all, in consideration of nodes, which are fixed or exposed to the external environment, it has characteristic that the nodes of sensor network should perform their function by using limited energy without being supplied energy at every requesting time. Because of this characteristics of network, many protocols suggested by the existing wireless network and wireless ad hoc network environment are not often suitable for the sensor network environment, and thereby, the protocols which fit for the new environment should be suggested. Especially, in case that some nodes cannot perform their function with the complete consumption of energy, it causes the alteration of topology and the circumstance that should ask for the re-sending as well as re-routing is occurred. Therefore, the efficient management of energy is one of major study fields, which should be settled in the sensor networks [3].

Currently, the research for efficient energy consumption efficiency in the physical, transport, and network layer are progressed. In this paper, we suggests algorithm for enhancing the efficiency of energy consumption in terms of MAC protocol and analysis its performance. The composition of this paper is as followings. In the next section, we examines the requirements of mac protocol in the sensor network, and section 3 inspects previously suggested MAC protocol. In section 4, we explain MAC protocol suggested by this paper, and section 5 evaluates its function through simulation. Lastly, in section 6 we describes future study subject, along with conclusion.

2   MAC Protocol for Sensor Networks

MAC protocol for allocating limited resources to more number of nodes, takes up the important part which can have influence on the capacity of whole network in the sensor network environment. In the design of MAC protocol, in order to guarantee high QoS for users on the whole the fair distribution of bandwidth and channel, least delay, and algorithm for enhancing the maximized process amount have been studied in the existing cellular network, Wireless LAN, and wireless ad-hoc networks environment [4, 6]. On the account that the existing network environment is not comparatively limited to the use of energy resource, the algorithm that considers the use of energy has been the secondary issue. However, in the sensor networks, because the work of nodes for a long time with a limited energy resource can influence the capacity of entire network, the existing algorithm should be corrected appropriately.

In the wireless network environment, energy is mainly consumed by the sending and receiving data, however retransmission, caused by the packet conflict, overhead of control packet, overhearing, caused by implosion and overlap, and the use of unnecessary energy, caused by ready mode can cause the great loss of energy, compared to the limited energy of sensor network nodes. In fact, in case of personal mobile communication terminal, the energy consumption rate of idle mode is about 95% of receiving mode, and occupies 71% of sending mode [7], and it appears that energy is consumed as idle: receiving: sending =1:2:2.5 in the 2Mbps Wireless LAN environment [8].

Accordingly, in case of designing MAC protocol of sensor networks, the effort which considers elements of energy consumption like this is needed in order to prolong life time of sensor nodes.

3 Related Work

Currently, research of MAC protocol for the sensor networks is mainly preformed in UC Berkeley, UCLA, and MIT and in many cases, the effort for life time prolongation of nodes which consist of network is made constantly.

The suggested methods are divided: the method to reduce the retransmission probability through avoiding conflict in the packet sending of each node, the method to decrease the energy consumption per bit by sending frame of holding the least length through regulating the length of frame, method to lower the overhearing of broadcasted information, and the method to directly control of power of each node.

The methods of medium access are largely divided into 2: A centralized method around TDMA and distributed method around CSMA [1, 9, 10].

First of all, Power Aware Multi-Access protocol with signaling for Ad Hoc Networks (PAMAS), suggested in the wireless ad-hoc, proposed algorithm, which can power off in the idle state in case of overhearing in order to prevent unnecessary energy waste, caused by overhearing of neighbor nodes when it sends data from one node to the destination node [11].

However, PAMAS has idle mode yet, and it needs the procedure to determine the time to power on. Low Power distributed MAC Protocol [12] prolonged the life time of nodes through altering the node power into ON state by using wake up signal in order to minimize energy consumption in the idle situation. However, it couldn't settle the overhearing problem fundamentally. DEANA protocol made the nodes which don't belong to the scheduling of sending and receiving mode, altered as low energy node on the basis of NAMA protocol [13].

LEACH protocol [14] is suggested to enable multiple accesses using cluster manager node in wireless ad hoc networks. S. Coleri, A. Puri and P.Varaiya made the TDMA network by using access point, which has the unlimited energy supply [9], W.Ye, j. Heidemann, D. Estrin suggested S-MAC [1], which made each node have listening mode and sleep mode periodically in order to reduce the energy consumption in the idle time. In addition, SMACS and EAR [15] in UCLA suggested the method of composing TDMA network between regional nodes without equipment like base station. In the time of sending and receiving the data, the protocol enables to communications only each nodes by using super frame structure. Like this, compared to the contention based method, centralized MAC protocol, chiefly centered on TDMA method, can reduce the conflict risk and thereby it can enhance the efficiency of energy consumption, but there are some advantages that the network situation is frequently changed, and the flexibility for network expansion like the node's addition can fall. In addition, mac protocol using the method for reducing overhead of mac header for decreasing energy consumption is proposed [16].

4 Message Redundancy

The supply of power source and battery can be given at any time, according to the user's need in wireless Internet, personal portable communication however in case of sensor nodes, the supply of power is not easy because of price, location, and characteristics of sensor. Therefore, it is inevitable that the available life time should be considered in the wireless ad hoc sensor network. So, the minimization of energy consumption of unnecessary energy in the sensor network environment is needed and the study for this in the respective hierarchy is being progressed.

The communication between nodes, which consists of sensor network, transmits information by using broadcasting. Because the obtainments of least forwarding node numbers are NP-complete, and the problem of redundant packet receiving is occurred, the effort to reduce unnecessary consumption of energy in mac layer is needed [17,18]. In case of every node's connection within the transmission area as a form of mesh, the number of node included in the area of a node, NPA is obtained by:

$$NPA = \frac{S_n \cdot \pi a^2}{\int f(x)dx} \tag{1}$$

where S_n is the number of nodes, located in the unit region and a is the range, reaching the power of node [18].

In addition, in case of supposing that each node transmits information to neighbor node only within 1 hop interval, most nodes must receive the redundant message as Fig.1. In the sensor networks, composed of NxN, if energy, required for the information sending is W_t and the energy for receiving is W_r, when data packet is sent (N,1) to (1,N), total energy consumption is given by:

$$\begin{aligned} ideal \;\; &: W_t = 2(N-1)W_{t,r} \\ redundancy \;\; &: (N-1)\{W_{t,r}(N+1)+(N-1)W_t\} \end{aligned} \tag{2}$$

5 Design of MAC

In order to prevent energy loss, caused by redundant message receiving, signal message which can control the node power is used. The nodes in the cluster area of wireless sensor network exist as a power off state in state of not performing the data sending and receiving. In case of idle mode, each node needs energy consumption with a periodic monitoring operation, however in case of off mode, least energy consumption is occurred because the node is not worked basically. In case of sending packet from a node to the neighbor node, because sending node transmits active signal to the neighbor node before sending data, it activates the node as ON state, in which the information receiving is possible from OFF state.

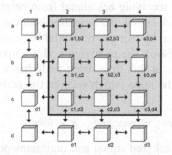

Fig. 1. Receipt of redundant message of sensor nodes

The neighbor node which receives message from a certain node transmits the packet to another neighbor node by using the active signal. The packet structure of active signal is like the Fig. 2. The part of active signal packet except control part is composed of active code which controls power as On/Off of its neighborhood nodes when the node broadcasts and Recognition Label (R-Label) for preventing the receiving redundant message and retransmission. Active code is composed of encoded k bit signal, in order to control on/off of node power, and R-label is composed of n bit random code and it judges whether the neighbor node of sending node is a previously received data or not. R-label is used when several broadcasted message fames exist in sensor network as like the case that messages are generated by several random nodes or more than 2 sink nodes broadcast message to their neighbor nodes.

Generally, when id is used as much as node numbers in the network, composed of N number of nodes, the probability of data transmission without using the redundant id is given by:

$$p(N) = 1 - \sum_{k=1}^{N} (\prod_{n=0}^{k} \frac{1}{N-n}) \qquad (3)$$

In the network of being composed of N nodes, the necessary bit numbers f(N) are obtained by (4). $<d>$ is defined to the minimum integer number more than d.

$$f(N) \approx < \log_2 N > \qquad (4)$$

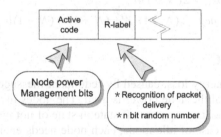

Fig. 2. The structure of active signal

The flow of sending and receiving the packet from node is like the Fig. 3. In case of sending information from a node to the neighbor node, the sending node is broadcasting active signal at first.

In this time, in case of receiving ack signal from neighbor node, packet is sent and in case of not receiving ack signal, the neighbor nodes sense the information sending, and retransmit active signal by using exponential back off algorithm of 802.11. When every transmission is finished, the receiving nodes converse the power into the OFF state automatically again. In this time, each node has m buffer, which can store R-Label of receiving, and the node which receive the active signal examines R-Label. When

$$R - label \quad in \ buffer \quad \oplus \quad received \quad R - label$$

is 0, sending node sends ack_red signal and maintains power as OFF state for broadcasting nodes to neighbor nodes except itself and abolish the received active signal packet. When it is confirmed that the receiving node is not consistent to the R-label value which is in the buffer of itself, the receiving node sends ack signal to the sending node and receives the data sending. When the energy of node is controlled by using the active signal packet, the power is altered into ON state selectively, according to the packet, entered from neighbor node through broadcasting, and it receives data. Because of this, the overhearing is minimized and the energy consumed in the idle state decreases. Therefore, the total energy consumption can be effective. The total consumed energy of an environment like Fig. 1 is as followings:

$$(N+1)(N-1)\{W_{r,t} + W_{active}\} + \delta_W \qquad (5)$$

Accordingly, the energy consumption towards existing method in the time of minimizing the redundant packet receiving by using algorithm is obtained by:

$$W_T = \frac{\{W_{r,t}(N+1) + (N-1)W_t\} + \delta_W}{(N+1)\{W_{r,t} + W_{active}\} + \delta_W} \qquad (6)$$

Where δ_W is the additional consumption amount of *active signal,* caused by retransmission.

For similar effect to R-label in active signal, 1 Byte sequence ID in Mac frame which use on IEEE 802.15.1 and IEEE 802.15.4 can use as like R-label. However, in case of existence plural sink and addition or deletion of sink or node, management is difficulty.

6 Simulation

6.1 Experiment Setup

We simulate performance using wireless sensor network packet transfer simulator to focusing evaluation of the differences energy consumption between receipt and avoidance of redundant message. The evaluation of performance compare suggested power control method by using active signal with the method using sequence ID in place of R-label and previous algorithm. In this simulation, each sensor node is disposed to random location within network area and the topology is fixed during data transmission and we assume that the broadcast message is generated single message at single sink. The length of transmitted data packet is 500~ 1500 Byte and the energy of active code used to power control is about 1.04μW and the energy consumption per bit is 0.13μJ/bit. Simulation environment is given by Table. 1.

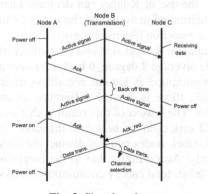

Fig. 3. Signal exchange

Table 1. Simulation environment

section	applicatoin	
Number of Nodes	10~100	
Sink	2~6	Random location
Transmisstion pakcet	500~1500 Byte	·
R-label	3~8bit	·
buffer size	1~10	packet storage
Re-transmission	random	·
Energy Consumption	0.13μJ/bit	·
Active code	1.04μW	Power control
Sequence ID	1 Byte	·

6.2 Result and Analysis

Using the simulation parameters in Table. 1, we represent result by graph. Fig. 4(a) to Fig. 4(b) shows the consumed energy and energy saving depending on the number of nodes and average degree of each node. Fig. 4(c) and Fig. 4(d) show total energy consumption depending on the number of sink and used R-label bit, and by measuring energy consumption depending on buffer size, we analysis to the optimal buffer size for stable data transmission.

Fig. 4(a) shows relative energy saving rate depending on number of nodes compared with simple power control method. In average 3 degree_4 buffer size_single sink_6 bit R-label, energy saving rate of the two methods is nearly similar. In this case, depending on receipt of redundant message by average degree, we can analysis that the difference of energy consumption occurs between previous method and the method which use R-label, sequence ID.

Fig. 4(b) shows the transition of energy consumption depending on average degree of each node which consists of sensor network comparing with the use of R-label and previous method. Under 50 node_6 bit R-label_ 4 buffer size_single sink, in using R-label, total energy consumption is uniform tendency. This results from the same transmission and receipt processing regardless of average degree of each node. On the other hand, in case of the use of simple power control method, when degree of nodes increase total energy consumption also grow due to increasing of redundant message. As the result, the use of R-label can decrease energy consumption in case that sensor nodes are distributed densely and the topology of sensor network is consisted of multi to multi connection like mesh form.

Fig. 4(c) shows the relation R-label size and the number of nodes with energy consumption under 50 node_average 4 degree_6 buffer size environments. In single sink, the level of energy consumption is similar regardless of R-label bit. On the other hand, in case of 5 sink, there is a wide difference to total energy consumption depending on R-label bit. The reason of this result is that in using 1 bit R-label, each node can distinct only 2 sink even the best case therefore, in above case that at least each node need 3 bit R-label, nodes can not distinct the message which is generated some sink node correctly. Accordingly, due to the condition of confusion message source, in using 1bit R-label, total energy consumption is less than using 8bit R-label.

We can analysis when several sink exist in sensor network, sufficient R-label bit must be secured for reliable packet transmission.

Fig. 4(d) shows the energy consumption depending on the number of sinks and buffer size. When buffer size is not sufficient, receipt of redundant message can occur for absence of message information which received at former times therefore, total energy consumption is increased remarkably. For avoidance of wasted energy due to redundant message, we can know the sufficient buffer size must be insured related with number of broadcast nodes. We obtained the buffer size for stable transmission by simulation like equation (7)

$$B(N_s) = N_s + [\frac{N_s}{2}] \qquad (7)$$

in this equation, $B()$ is buffer size, N_s is the number of broadcast nodes (in this simulation, the number of sink).

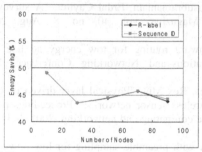

(a) Energy saving related with the number of nodes

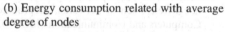

(b) Energy consumption related with average degree of nodes

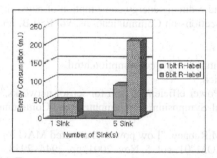

(c) Energy consumption related with R-label bit and the number of sinks

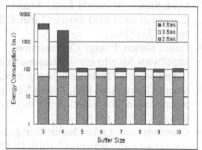

(d) The relation between buffer size and energy consumption

Fig. 4. Simulation results

7 Conclusions

In this paper, we proposed Mac protocol for efficient energy consumption, one of the important research issues and evaluated performance through wireless sensor network simulator. When power of node is controlled using suggested R-label in Active signal, each node does not change active state whenever message packet arrive but depending on R-label, each node decide whether receive or not and only in case of new arrival packet, node change power oneself and perform the process. Compared with previous method, energy consumption rate are decreased to 45~50% and in case of existing more than 2 sink, the efficiency of energy consumption is better to use of proposed mac protocol. Also the result represents that for avoidance of collision, R-label bit and buffer size used at each node must setup according to the number of broadcast node as like sink node.

References

1. 1. Wei Ye, John Heidemann, Deborah Estrin, "An energy-efficient MAC protocol for wireless sensor networks," in proceeding of the IEEE infocom, 2002, pp.1567~1576.
2. Martin Kubisch, Holger Karl, Adam Wolisz, Sizhi Charlie Zhong, Jan Rabaey, "Distributed algorithms for transmission power control in wireless sensor networks," in proceeding of IEEE Wireless Communications and Networking. Mar. 2003, vol.1, pp.558~563.
3. Ian F.Akyildiz, Weilian Su, yogesh Sankarasubramaniam, Erdal Cayirci, "A survey on sensor networks," IEEE Communications Magazine, vol. 40, no. 8, Aug. 2002, pp.102~114.
4. Rahul C. Shah, Jan M. Rabaey, "Energy aware routing for low energy ad hoc sensor networks," IEEE Wireless Communications and Networking Conference, Mar. 2002, pp.350~355.
5. Eugene Shih, Seong Hwan Cho, nathan-lckes, Rex Min, "Physical layer driven protocol and algorithm design for energy-efficient wireless sensor networks," Proceedings of the 7th annual international conference on Mobile computing and networking, 2001, Italy, pp. 272~287.
6. Amre El-Hoiydi "Spatial TDMA and CSMA with preamble sampling for low power ad-hoc wireless sensor networks," in proceeding of the Seventh International Symposium on Computers and Communications(ISCC-'02), July. 2002. pp. 685~692.
7. Mark Stemm, Randy H Katz, "Measuring and reducing energy consumption of network interfaces in hand-held devices," IEICE Transactions on Communications, vol.E80-B, no. 8, pp. 1125~1131.
8. Oliver Kasten, Energy Consumption, http://www.inf.ethz.ch/~kasten/research/bathtub/energy_consumption.html, Eldgenossische Technische hochschule, Zurich.
9. Sinem Coleri, Anuj Puri, Pravin Varaiya, "Power efficient system for sensor networks," in proceeding of the Eighth IEEE international Symposium on Computers and Communication (ISCC '03), pp. 837~842.
10. Chunlong Guo, Lizhi Charlie Zhong, Jan., M.Rabaey, "Low power distributed MAC for ad hoc sensor radio networks," IEEE GLOBECOM '01, vol. 5, Nov. 2001, pp. 2944~2948.
11. Suresh Singh, C. S. Raghavendra, "PAMAS - Power aware multi-access protocol with signaling for ad hoc networks," ACM Computer Communication Review, vol. 28, no. 3, pp. 5~26. July 1998.
12. Chunlong Guo, Lizhi Charlie Zhong, Jan. M. Rabaey, "Low power distributed MAC for ad hoc sensor radio networks," Globecom '01. IEEE, vol. 5, pp. 2944~2948. Nov 2001.
13. Venkatesh Rajendran, J.J. Garcia-Luna-Aceves, Katia Obraczka, "Energy-efficient channel access scheduling for power-constrained networks," Wireless Personal Multimedia Communications, 2002. The 5th International Symposium, vol. 2, pp. 509~513, Oct 2002.
14. Wendi Rabiner Heinzelman, Anantha Chandrakasan, Hari balakrishnan, "Energy-efficient communication protocol for wireless microsensor networks," in proceedings of the 33rd Hawaii International conference on System Sciences, pp.3005~3014. Jan 2000.

15. Katayoun Sohrabi, Jay Gao, Vishal Ailawadhi, Gregory J. Pottie, "Protocols for self-organization of a wireless sensor network," IEEE Personal Communications, vol. 7, pp. 16~27. Oct. 2000.

16. Gautam Kulkarni, Curt Schurgers, Mani Srivastava, "Dynamic link labels for energy efficient MAC headers in wireless sensor networks," Proceedings of IEEE, vol. 2, pp. 1520~1525, June 2002.

17. Wei Lou, Jie Wu, "On reducing broadcast redundancy in ad hoc wireless networks," IEEE Transactions on Mobile Computing, vol. 1, no. 2, pp. 111~122. April-June 2002.

18. Sze-Yao Ni, Yu-Chee Tseng, Yuh-Shyan Chen, Jang Ping Sheu, "The Broadcast Storm problem in a Mobile Ad Hoc Network," in Proceedings of MobiCom'99, pp. 152~162

A Performance Analysis of Binary Exponential Backoff Algorithm in the Random Access Packet Networks

Sun Hur[1], Jeong Kee Kim[1], Jae Ho Jeong[1], and Heesang Lee[2]

[1] Department of Industrial Engineering, Hanyang University, Ansan, Korea
hursun@hanyang.ac.kr
[2] School of Systems Management Engineering, Sungkyunkwan University, Suwon, Korea

Abstract. We analyze the performance of binary exponential backoff (BEB) algorithm under the slotted ALOHA protocol. When a node's message which tries to reserve a channel is involved in a collision for the n th time, it chooses one of the next 2^n frames with equal probability and attempts the reservation again. We derive the expected access delay and throughput which is defined as the expected number of messages that reserve a channel in a frame. A simulation study shows that the system is stable if the number of request slots is more than 3 times of the arrival rate.

1 Introduction

Slotted ALOHA(S-ALOHA) protocol has been widely adopted in local wireless communication systems as a random multiple access protocol [1]. In these systems, each frame is divided into small slots and each mobile terminal contends for the slot to transmit its packets at the beginning of each frame. If two or more mobile terminals contend for the same slot, then a collision occurs and none of the terminals can transmit their packets. The colliding packets are queued and retry after a random retransmission delay. The way to resolve the collision is called the collision resolution protocol, which is one of the most important issues related to S-ALOHA. In general, queued messages attempt the retransmission for the next frame with a known fixed probability. Another widely used collision resolution protocol is the binary exponential backoff (BEB) algorithm, forms of which are included in Ethernet[2] and Wireless LAN[3] standards: Whenever a node's packet is involved in a collision, it selects one of the next 2^n frames with equal probability, where n is the number of collisions that the message has ever experienced, and attempts the retransmission.

Soni and Chockalingam [4] analyzed three backoff schemes, namely, linear backoff, exponential backoff, and geometric backoff. They calculated the throughput and energy efficiency as the reward rates in a renewal process and showed that the BEB performs poorly because of the rapid growth of exponentiation in backoff delay by numerical computation. They also illustrated that the truncated BEB, which is considered in this paper, performs better since the idle length should grow only until a maximum value. More recently, Chen and Li [5] proposed the quasi-FIFO algorithm, which is another novel collision resolution scheme. They showed, by simulation results, that the proposed scheme shares the bandwidth more equally and maximizes the throughput, but no analytic result was given in [5].

D.-K. Baik (Ed.): AsiaSim 2004, LNAI 3398, pp. 270–278, 2005.

Delay distributions of slotted ALOHA and CSMA are derived in Yang and Yum [6] under three retransmission policies. They found the conditions for achieving finite delay mean and variance under the BEB. Their assumption, however, that the combination of new and retransmitted packet arrivals is a Poisson process is not valid because the stream of the retransmitted packets depends on the arrivals of new packets. This dependency makes the Poisson assumption invalid. Chatzimisios and Boucouvalas [7] adopted a two dimensional Markov chain consisted of the backoff counter and backoff stage of a given station. They presented an analytical model to compute the throughput of the IEEE 802.11 protocol for wireless LAN and examined the behavior of the exponential backoff (EB) algorithm used in 802.11. They assumed that the collision probability of a transmitted frame is independent of the number of retransmissions. As we will show later in this paper, however, this probability is a function of the number of competing stations and also depends upon the number of retransmissions that this station has ever experienced. Kwak et al. [8] gave new analytical results for the performance of the EB algorithm. Especially, they derived the analytical expression for the saturation throughput and expected access delay of a packet for a given number of nodes. Their EB model, however, is assumed that the packet can retransmit infinitely many times.

The stability is another research issue on BEB algorithm and there are many papers for this, among them are Kwak et al. [8], Aldous [9], and Goodman et al. [10]. As pointed in Kwak et al. [8], however, these studies show contradictory results because some of them do not represent the real system and they adopt different definitions of stability used in the analyses. The dispute is still going on so we do not focus on this topic but on the analytic method to evaluate the performance of the BEB algorithm. Instead, we provide the stability condition via simulation study.

In this paper, we suggest a new analytical method to find the performance measures to evaluate the system which adopts BEB algorithm, including the throughput and expected medium access delay. The throughput here is defined as the expected number of messages which can be allocated a reservation slot in a frame. The expected medium access delay is defined as the expected time from the moment that the message arrives at the system until it is successfully allocated a reservation slot. A simulation study is performed to verify our analytical computation and provide the stability condition.

This paper is organized as follows. We describe the BEB algorithm under consideration in section 2. The steady-state distribution of the number of messages at the beginning of the frame and the collision probability are given in section 3. We find mean access delay and mean number of messages which are transmitted successfully in section 4. In section 5, we give some numerical results with simulation to verify our method and provide the stability condition Finally, we conclude the paper in section 6.

2 System Description

We assume that there are infinite number of users (mobile terminals: MTs) and the total arrivals of new messages from active MTs form a Poisson process with rate λ to the system. The time is divided into slots which are grouped into frames. The length of each frame is d, a fixed value. A frame is divided into two sets of multiple slots, one of which is the group of request slots for reservation of channels and the other one is the group of data slots for transmission of the information. The number of request slots in a frame is V. When a message arrives at the system, it waits until the beginning of the next frame and randomly accesses one of the request slots to reserve a channel for transmission. If the message succeeds in the reservation, then a channel is allocated. The message is transmitted by means of the data slots. If, however, two or more messages contend for the same request slot, then a collision occurs and none of the messages can reserve the request slot.

The message which fails to get a request slot retries under the BEB algorithm: whenever a message is involved in a collision and if it was the bth ($b=0,1,\cdots,15$) collision, then it selects one of the next 2^i frames with an equal probability and attempts the reservation again, where i is given by $i = \min(b, 10)$. Therefore, if a message collides more than 10 times, it selects a frame among the next 2^{10} frames for reservation. If a message collides 16 times, then it fails to transmit and is dropped.

3 Steady-State Analysis

3.1 Steady-State Distribution of the Number of Messages in System

In this section, we obtain the steady state distribution of the number of messages at the beginning of the frame. This will be utilized to compute the collision probability, γ_c, that an arbitrary message experiences a collision when it attempts to get a request slot. Let A_n be the number of messages arrived during the n th frame, which are supposed to participate in the contention at the beginning of the $(n+1)$ th frame. N_n is the number of messages waiting in the system at the beginning of the n th frame. Also, denote J_n by the number of messages which successfully reserve a request slot at the n th frame. Then it can be shown that

$$N_{n+1} = \begin{cases} N_n - J_n + A_n, & N_n \geq 1 \\ A_n, & N_n = 0, \end{cases} \qquad (2.1)$$

$\{N_n, n \geq 1\}$ is a Markov chain. Let us denote $a_j, j = 0,1,2,\cdots$ by the steady state probability distribution of A_n, which is given by

$$a_j = \Pr(A_n = j) = \frac{e^{-\lambda d}(\lambda d)^j}{j!}, \qquad j=0,1,2,\cdots , \qquad (2.2)$$

where d is the length of a frame. We obtain the one-step transition probabilities $p_{ij} = \Pr(N_{n+1} = j | N_n = i)$, $i,j=0,1,2,\cdots$ of the Markov chain as the following:

For, $i \geq 1$

$$p_{ij} = \Pr(N_{n+1} = j | N_n = i)$$

$$= \sum_{k=\max(0,i-j)}^{\min(i,V)} a_{j-i+k} \Pr(J_n = k | N_n = i) , \qquad (2.3)$$

and $p_{0j} = a_j$. In order to compute $\Pr(J_n{=}k|N_n{=}i)$, the probability that k messages are successful in contending the request slots out of i messages, let us introduce a random variable Y_n that is the number of messages which actually participate in the contention at the n th frame. Then we have

$$\Pr(J_n = k | N_n{=}i)$$

$$= \sum_{y=0}^{i} \Pr(J_n{=}k|Y_n{=}y,N_n{=}i)\Pr(Y_n{=}y|N_n{=}i) \tag{2.4}$$

$$= \sum_{y=0}^{i} \Pr(J_n{=}k|Y_n{=}y)\Pr(Y_n{=}y|N_n{=}i).$$

Let us denote $J(y,k) = \lim_{n\to\infty} \Pr\left(J_n{=}k|Y_n{=}y\right)$ and $Y(y,k) = \lim_{n\to\infty} \Pr\left(Y_n{=}y|N_n{=}i\right)$. Then $J(y,k)$ is the probability that k messages succeed in the contention among y competing messages and $Y(i,y)$ is the probability that y messages participate in the contention among i messages in the steady state. Then $J(y,k)$ is derived in Szpankowski [11], for $0{\le}k{\le}\min(V,y)$, as the following:

$$J(y,k) = \frac{(-1)^k V! y!}{V^y k!} \sum_{m=k}^{\min(V,y)} (-1)^m \frac{(V-m)^{y-m}}{(m-k)!(V-m)!(y-m)!}. \tag{2.5}$$

Also, the probability $Y(i,y)$ is given by:

$$Y(i,y) = \binom{i}{y} r^y (1-r)^{i-y}, \qquad y=0,1,\cdots,i \quad, \tag{2.6}$$

where r is the probability that an arbitrary message participates in the contention. Since each message waiting in the system has experienced different number of collisions (let us call this number the backoff state of the message) so we derive the probability r by conditioning the backoff state. Let γ_c be the probability that an arbitrarily chosen (tagged) message experiences a collision when it contends for a request slot. We derive this unknown probability in the next subsection. Then we have

$$r = \left(\frac{1-(\gamma_c/2)^{11}}{1-(\gamma_c/2)}\right)\frac{1-\gamma_c}{1-\gamma_c^{16}} + (\gamma_c/2)^{10}\frac{\gamma_c(1-\gamma_c^5)}{1-\gamma_c^{16}}. \tag{2.7}$$

Now we can calculate the one-step transition probability $p_{ij} = \Pr\left(N_{n+1}{=}j|N_n{=}i\right)$ in equation (2.3) by plugging the equations (2.5) and (2.6) combined with (2.7) as

$$p_{ij} = \sum_{k=\max(0,i-j)}^{\min(i,V)} a_{j-i+k} \sum_{y=0}^{i} J(y,k)Y(i,y) \quad. \tag{2.8}$$

If we let π_j be the steady state probability distribution of the number of messages in system at the beginning of the frame, that is, $\pi_j \equiv \Pr(N{=}j) = \lim_{n\to\infty} \Pr\left(N_n{=}j\right)$, then it can be obtained by solving the steady state equations

$$\pi_i = \sum_{i=0}^{\infty} \pi_i p_{ij}, \qquad \sum_{i=0}^{\infty} \pi_i{=}1 \quad. \tag{2.9}$$

3.2 Collision Probability

In this subsection, we derive γ_c, the probability that a tagged message experiences a collision given that it actually participates in the contention for a request slot in a frame. Let M be the number of messages in the system at the beginning of the frame in which the tagged message is included. Then it is known that M is differently distributed from N because it contains the tagged message. The probability distribution of M is given by Kleinrock [12].

$$\Pr(M{=}m) = \frac{m\pi_m}{E(N)}, \quad where \quad E(N)=\sum_{j=0}^{\infty} j\pi_j \quad . \tag{2.10}$$

When y messages including the tagged message contend for the slot, the probability that the tagged message collides is $\sum_{i=1}^{y-1}\binom{y-1}{i}\left(\frac{1}{V}\right)^{i}\left(1-\frac{1}{V}\right)^{y-1-i}$. Therefore, we have the following:

$$\begin{aligned}
\gamma_c &= \sum_{m=2}^{\infty}\sum_{y=2}^{m}\sum_{i=1}^{y-1}\binom{y-1}{i}\left(\frac{1}{V}\right)^{i}\left(1-\frac{1}{V}\right)^{y-1-i}\cdot Y(m,y)\cdot \Pr(M{=}m)\\[2mm]
&= \sum_{m=2}^{\infty}\left\{1-\frac{V}{V-1}\left(1-\frac{r}{V}\right)^{m-1}+\frac{1}{V-1}(1-r)^m\right\}\cdot \frac{m\pi_m}{\sum_{j=0}^{\infty} j\pi_j} \quad .
\end{aligned} \tag{2.11}$$

Note that the probability that a message eventually fails to reserve a request slot and is blocked is γ_c^{16}. In order to obtain γ_c in equation (2.11), we need to know π_j but in turn γ_c should be given in order to obtain π_j from equation (2.9). So we perform a recursive computation, i.e., we initially set γ_c to be an arbitrary value between 0 and 1 and compute $\pi_j, j = 0,1,2,\cdots$ by equation (2.9). Then with this π_j, we update γ_c using the equation (2.11) and this updated γ_c is utilized to update π_j again. This recursive computation continues until both values converge.

4 Performance Measures

4.1 Expected Access Delay of Message

Now we derive the expected medium access delay of a message which is defined as the time from the moment that a message arrives at the system to the moment that it successfully reserves a request slot. It can be obtained by counting the number of frames from which a newly arrived message contends for a slot for the first time until it successfully reserves a slot. The following figure 1 illustrates the state transition of a message from its arrival at the system to be allocated or blocked. In the figure, T is the state that a message successfully reserves a request slot and B the state that it is eventually blocked. And the state (b,f) denotes that a message which has experienced $B=b$ $(b=0,1,\cdots,15)$ collisions is currently at the f th frame among its $2^{\min(b,10)}$ candidate frames to participate in the contention. Transition probabilities are shown on each arrow.

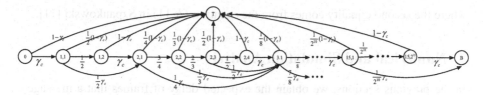

Fig. 1. State transition probability diagram

If a message reserves in the first trial with no collision (that is, $b=0$), then it experiences, on average, 3/2 frame length's delay, which is the sum of 1/2 frame length (average length from the message's arrival epoch to the beginning of the next frame) and 1 frame length. Suppose a message collides exactly $b(1\leq b\leq10)$ times until it successfully reserves a slot. This implies that the last state it stays before the reservation is one of 2^b states $(b,1),(b,2),\cdots,(b,2^b)$. Since it selects one of them with equal probability, the average number of states (or frames) it has spent in the system is given by $\frac{1}{2}+\sum_{i=0}^{b-1}2^i+\frac{1}{2^b}\sum_{j=0}^{2^b}j$. In the same manner, if $11\leq b\leq 15$, the average delay is $\frac{1}{2}+\sum_{i=0}^{10}2^i+(b-11)\cdot2^{10}+\frac{1}{2^{10}}\sum_{j=0}^{2^{10}}j$. We obtain the expected delay, $E(D)$, of frames that a message experiences from its arrival to the system until the successful reservation as the following:

$$E(D) = \sum_{b=0}^{15} E(D|B=b)P(B=b)$$

$$=\frac{1}{2}+\frac{1-\gamma_c}{1-\gamma_c^{16}}\begin{pmatrix}1+2.5\gamma_c+5.5\gamma_c^2+11.5\gamma_c^3+23.5\gamma_c^4\\+47.5\gamma_c^5+95.5\gamma_c^6+191.5\gamma_c^7+383.5\gamma_c^8\\+767.5\gamma_c^9+1535.5\gamma_c^{10}+2559.5\gamma_c^{11}+3583.5\gamma_c^{12}\\+4607.5\gamma_c^{13}+5631.5\gamma_c^{14}+6655.5\gamma_c^{15}\end{pmatrix}. \qquad (4.1)$$

4.2 Throughput

We can derive the distribution of the number, X , of messages which reserve a request slot successfully during a frame by conditioning Y and N as following:

$$\Pr(X=x)$$

$$= \sum_{n=0}^{\infty}\sum_{y=0}^{n} J(y,x)Y(n,y)\Pr(N=n). \qquad (4.2)$$

where $J(y,x),Y(n,y)$ and $\Pr(N=n)$ are given by equations (2.5),(2.6) and (2.9), respectively. The expected number $E(X)$ is given by

$$E(X) = \sum_{n=0}^{\infty} \sum_{y=0}^{n} E(X|Y=y, N=n) \Pr(Y=y|N=n) \Pr(N=n)$$

$$= \sum_{n=0}^{\infty} nr \left(1 - \frac{r}{V}\right)^{n-1} \Pr(N=n).$$

(4.3)

where the second equality comes from the equation (A.11) in Szpankowski [11].

5 Numerical and Simulation Analysis

In the previous sections, we obtain the expected delay of frames that a message experiences from its arrival to the system until the successful reservation. In this section we perform a numerical study to compare those expected delays with simulation results to verify the analytic model proposed in the previous section. The data of the parameters used for the comparison is as follows:

Table 1. Data for the parameters of numerical examples

Parameters	Data
Arrivals of new messages ~ $Poisson(\lambda)$	$\lambda = 1, 2, \cdots, 30$
Numbers of request slots (V)	$V = 15, 20, 25, 30$

For each cases, expected medium access delay ($E(D)$) and throughput ($E(X)$) are calculated by the equations (4.1) and (4.3). The computer program is written in C++. Simulation is done by Matlab 6.5 software, and 10 instances are experimented for each case. The computer simulation results given in this section are obtained by the following procedure. When a new message is generated according to the Poisson process, it is time stamped and the sojourn time in frames is measured when the message is successfully reserves a channel. Cumulative value of these sojourn times of all non-blocked messages is computed and divided by the total number of non-blocked messages (throughput) to obtain the expected medium access delay.

The fig. 2 shows the expected medium access delay as the offered load λ/V varies when $V=30$. For $V=15$, 20 and 25, we have similar results so omit those cases here. At each simulation points, 95% confidence intervals are displayed. The analytical results shown in dotted line lie within 95% confidence intervals of the simulation results shown in solid line when the offered load is smaller than, approximately, 0.3. In addition, we can see that the expected delay increases exponentially as the amount of traffics gets bigger for given number of request slots. As the λ/V increases beyond 0.3, however, the simulation results grow drastically. This implies that there are many messages waiting in the system so the system becomes unstable.

Fig. 3 shows the expected number of successful messages as the arrival rate varies when V is 30. Also, we omit the cases $V=15$, 20 and 25 since we have similar results. The 95% confidence intervals are shown for all the simulation points.

We see that, in fig. 3, while the analytic values match very well with the simulation results if λ is less than or equal to 9, the analytic model underestimates severely when λ is greater than 9, which is 0.3 times of V. As a result, we can conclude that

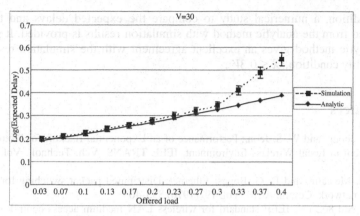

Fig. 2. Expected delay vs. offered load when $V = 30$

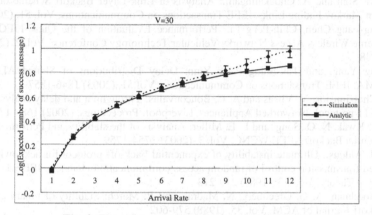

Fig. 3. Throughput vs. the arrival rate when $V = 30$

when $\lambda \leq 0.3V$ then our analytic method gives excellent results on the performance measures such as the throughput and expected medium access delay. If $\lambda > 0.3V$, however, then the system is unstable so our method is not applicable.

6 Conclusion

In this paper, we considered the performance evaluation of the BEB policy, which is a collision resolution algorithm often adopted in the random access packet networks. We obtain the performance measures by an analytic method. We obtain the steady-state distribution of the number of messages waiting in the system, which is utilized to get the probability that a tagged message experiences a collision given that it actually participates in the contention for a request slot in a frame, which has never been investigated in the literature. With these, the expected delay of frames that a message experiences from its arrival to the system until the successful reservation and the throughput are found.

In addition, a numerical study to compare the expected delays and throughput computed from the analytic method with simulation results is provided. It shows that our analytic method gives an excellent agreement with the simulation results under the stability condition, $\lambda \le 0.3V$.

References

1. D. G. Jeong, and W. S. Jeon.: Performance of an Exponential Backoff for Slotted-ALOHA Protocol in Local Wireless Environment. IEEE TRANS. Veh. Technol., Vol. 44. (1995) 470-479
2. R. M. Metcalfe, and D. G. Boggs: Ethernet: Distributed packet switching for local computer network. Comm. ACM, Vol. 19. (1976) 395-404
3. IEEE Std 802.11': IEEE standard for wireless LAN medium access control and physical layer specifications (1997)
4. P. M. Soni and A. Chockalingam.: Analysis of Link-Layer Backoff Scheme on Point-to-Point Markov Fading Links. IEEE Transactions on Communications, Vol. 51. (2003) 29-32
5. Yung-Fang Chen, Chih-Peng Li.: Performance Evaluation of the Quasi-FIFO Back-off Scheme Wireless Access Networks Vehicular Technology Conference. Vol. 2. (2003) 1344 - 1348
6. Yang Yang, and Tak-Shing Peter Yum: Delay Distributions of Slotted ALOHA and CSMA. IEEE Transactions on Communications, Vol.51. (2003) 1846-1857
7. P. Chatzimisios, V. Vitsas and A. C. Boucouvalas: Throughput and delay analysis of IEEE 802.11 protocol. Networked Appliances, Liverpool. Proceedings. (2002) 168 - 174
8. B. J. Kwak, N. O. Song, and L. E. Miller: Analysis of the stability and Performance of Exponential Backoff. IEEE WCNC, Vol. 3. (2003) 1754-1759
9. D. J. Aldous.: Ultimate instability of exponential back-off protocol for acknowledgement-based transmission control of random access communication channels. IEEE Trans. Information Theory, Vol. 33. (1987) 219-213
10. J. Goodman, A. G. Greenberg, N. Madras, and P. March: Stability of binary exponential backoff Journal of ACM, Vol. 35. (1988) 579-602
11. W. Szpankowski.: Analysis and stability considerations in a reservation multiaccess system. IEEE Trans. on Comm., Vol. com-31. (1983) 684-692
12. L. Kleinrock.: Queueing system, Vol. 1. Theory. John Wiley & Sons, Inc., New York (1975)

Modeling of the b-Cart Based Agent System in B2B EC

Gyoo Gun Lim[1], Soo Hyun Park[2], and Jinhwa Kim[3]

[1] Department of Business Administration, Sejong University,
98 Kunja-dong, Kwangjin-gu, Seoul, 143-747, Korea
gglim@sejong.ac.kr
[2] School of Business IT, Kookmin University, 862-1,
Jeongreung-dong, Sungbuk-ku, Seoul, 136-701, Korea
shpark21@kookmin.ac.kr
[3] School of Business, Sogang University, 1,
Shinsu-dong, Mapo-ku, Seoul, 121-742, Korea
jinhwakim@sogang.ac.kr

Abstract. In B2B EC area, the need for agent based system for desktop purchasing is increasing. To simulate agent based B2B e-Commerce, we model the b-cart based B2B agent framework. B-cart implies a buyer's shopping cart which a buyer carries to the seller's sites. The modeled system is designed to provide an effective architecture for employee's direct desktop purchasing from external sellers' e-marketplaces. In this paper, we propose the thirteen features of b-cart based agent system in B2B EC; *Identification, User Dialog, Collection, Trashing, Individual Purchase Decision Support, Organizational Purchase Decision Support, Negotiation, Ordering, Payment, Tracking, Recording, Record Transmission, Knowledge Maintenance.* Based on these features, we design the buying process, the system architecture, and message interfaces. We simulate the effect of the system by making a performance evaluation model. From this research we show a possibility and efficiency of the b-cart based agent system.

1 Introduction

Among EC areas, the B2B EC is being spotlighted as an interesting research area considering its size and the potential impact on the whole society. There are various B2B systems in seller-centric e-marketplaces, intermediary-centric e-marketplaces, and buyer-centric e-marketplaces. However, many B2B systems are yet operated manually by human through web browsers. Considerable issues for the future of B2B systems are: the increase of the portion of online procurement from e-marketplaces[1]; the increase of the desktop purchasing by requisitioners, which empower employees in organizational decision making for purchase[2]; the need for the adoption of intelligent agent facilities to B2B system to cope with the intelligent IT environment[3]. So far, there have been many researches about agent technologies, but they cover only the part of the emerging issues for B2B EC considering above factors. In B2B EC, it is considered that the b-cart based system will be the dominant architecture for business buyers[4]. In this paper we are going to modeling a b-cart based agent system and show the simulation results.

Section 2 describes the b-cart in B2B EC. Section 3 analyzes the role of agents in B2B EC and agent simulation environments, and section 4 defines the features and

D.-K. Baik (Ed.): AsiaSim 2004, LNAI 3398, pp. 279–288, 2005.
© Springer-Verlag Berlin Heidelberg 2005

process of the b-cart based agent system and we design the architecture for the b-cart based agent system. In section 5, we simulate the process of carts by making a performance evaluation model. The summary and conclusion are presented in section 6.

2 b-Cart

b-Cart is a new concept of buyer-oriented shopping cart[4], and defined as a buyer-cart which *resides on the buyer's site* like Fig. 1-b). *Buyer-cart is an electronic cart (e-cart) that is owned and used by the business buyers.* The buyer-cart can be classified into *s-cart, a buyer-cart that resides on the seller's site* in seller-centric e-marketplace like Fig. 1-a), *i-cart*, which is functionally similar to s-cart but *resides on the intermediary's site*, and the *b-cart*.

So far, s-cart/i-cart is popular; since the software is thoroughly developed and operated by seller/intermediary, it is easy for users to use and manage it. However, in comparison with private consumers, business buyers have to precisely keep track of the purchase progress and store records, and integrate them with the buyer's e-procurement system which might have been implemented as a part of integrated ERP (Enterprise Resource Planning) systems[3,5,6].

Therefore, s-cart (i-cart) is no longer the most effective for B2B EC because the buyer's order information is scattered in the sellers' (intermediaries') sites. However, with b-cart, a buyer can visit multiple e-marketplaces collecting items in his/her own one b-cart and make purchase orders simultaneously over multiple items inserted from different e-marketplaces. This will allow the tight integration of the information in b-cart with the e-procurement system because it is located in the buyer's site. The b-cart can be implemented by displaying an overlaid window on the buyer's PC.

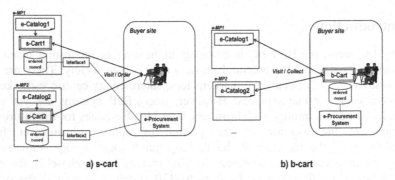

a) s-cart   b) b-cart

Fig. 1. s-cart and b-cart

3 Agent Environments in B2B EC

The need for the agent based electronic commerce is emerging as the electronic commerce evolves. Especially in B2B EC, this necessity is high because of the large amount of transactions and frequency. B2B EC needs to utilize not only buyer agents but also seller agents, and these agents should work together intelligently. In this section, we will talk about the roles of agent and the simulation environments for agent systems.

3.1 Role of Agents

According to CBB(Customer Buying Behavior) model[7], the purchase process consists of six steps: *Need Identification, Product Brokering, Merchant, Negotiation, Purchase and Delivery, Product service and Evaluation.* Among these steps, most agents are developed to assist the product brokering, merchant brokering, and negotiation process.

Blake extended this CBB model to agent-mediated B2B process model with 7 phases[8]: *Requisitioning* phase where corporate purchasing agents decide what products and services are desired, *Request for Quote* phase where businesses solicit and search for products or services from external companies/vendors. *Vendor Selection* is performed by the initiating company to determine which products/services are most compatible. *Purchase Order* is created and transmitted to the pertinent external company for the selected products/services. *Delivery* of products/services typically equates to the integration of companies' business automation mechanisms. The initiating company must compensate other collaborating companies in the *Payment Processing* phase. *Evaluation and Evolution* phase where a business must evaluate the services provided by other companies and vendors.

So we can summarize the major roles of software agents for the buyers are the collection of data from multiple e-marketplaces, filtering relevant items, scoring the products according to the customers' preference, and comparing for side-by-side examination. Buyer agents are necessary mainly in the seller-centric e-marketplaces. Buyer agents need to send out the requirement to the relevant seller agents or e-marketplaces, and interpret the received proposals.

The b-cart focused in this paper can be understood as playing the role of buyer agent. On the other hand, the seller agent substitutes for the transactional role of salesman as much as possible. In this paper, we focused on the b-cart based system, and we assume that a seller agent works in the seller's e-marketplace.

3.2 Simulation Environments for Agent System

Many agent simulation tools have been developed for the simulation of software agents or various physical agents. Agent development tools have been developed based on OAA(Open Agent Architecture) which consists of a coordination agent and several application agents in distributed environment. The objectives of OAA are supporting various mechanisms in interaction and cooperation, supporting human oriented user interface, satisfying software engineering requirement etc[9].

The known agent simulation tools are SWARM toolkit which supports hierarchical modeling developed with Objective-C in Satafe institute[10], Repast(Chicago university), Ascape(Brooking institute), Netlogo(Northwestern university), MADKIT[11], SILK[12], etc. Most recently developed tools are developed by Java based. They are mainly developed in university or institute and support agent modeling and graphic representation with object-oriented programming environment.

Agent simulation can represents the relationship between objects and the behavior mechanism by programming, so we can investigate the behavior and interaction of objects in the virtual world. Agent simulation are performed in various domains such as behavior pattern or social phenomenon simulation[13], financial market transaction simulation[14], biological/psychological modeling[15], vehicle control simulation such as airplane[16], online e-commerce simulation[17] etc.

By using agent simulation tools we can simplify and structuring complex phenomenon, make an experiment without time and space constraints, make modeling, investigate the situations by changing the conditions. In this paper, we design the simulation environment using Visual C++ with graphic user interface.

4 Design the b-Cart Based Agent System in B2B EC

4.1 Features

For this research, we define the thirteen desired features of b-cart considering intelligent buyer agent in B2B EC.

Thirteen features of the b-cart based agent system are *Identification* that identify the owner of a b-cart, *User Dialog* that dialogs to identify user's need, *Collection* that collects interesting items into the b-cart possibly from multiple e-marketplaces, and *Trashing* that trashes items that the buyer is not willing to buy from the current collection, and *Individual Purchase Decision Support* that supports the buyer's purchase decision-making process such as filtering relevant items, and scoring the products and sellers according to the customers' preference, *Organizational Purchase Decision Support* that supports the buyer organization's purchase decision-making process collaboratively connected to the organization's workflow system, *Negotiation* that negotiate with seller or seller agent, and *Ordering* that orders the selected items to sellers, and *Payment* that pays for the ordered items to sellers, and *Tracking* that tracks the progress of current purchase and history of records, *Recording* that records the collected information in a b-cart, and *Record Transmission* that transmits the information in a b-cart to the buyer's e-procurement system, and *Knowledge Maintenance* that manages the buyer's preference, ordered items, seller information, and the evaluation information in knowledge base for future decision making.

4.2 Process

The general buying process of the b-cart based agent system following the above features is like Fig. 2. This purchasing process assumes that all the desired features of b-cart are performed sequentially. Of course the process can be changed according to the business rules between sellers and buyers.

4.3 Architecture

To make a b-cart based agent system, we designed a system architecture as Fig. 3. We named it as AgentB. The arrows in the Fig. 3 mean the flow of control and data, and the dashed arrows mean the flow of messages.

4.4 Message Interfaces

Agent Communication Language (ACL) has been developed as a formal and common representation of the message for software agent [18]. The messages

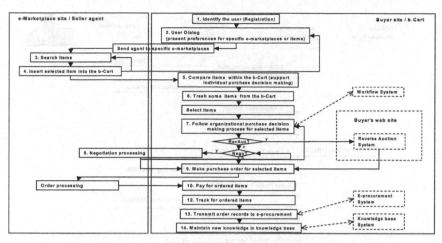

Fig. 2. Purchasing process of a b-Cart based agent system

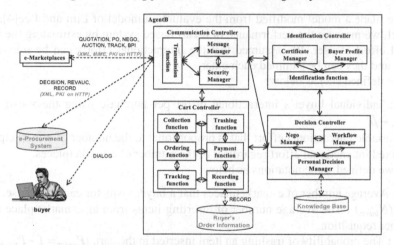

Fig. 3. The structure of the b-cart based agent system

between AgentB and e-marketplaces, and between AgentB and e-procurement system should be standardized. We designed the necessary messages to be compatible with ebXML[19] by adopting KQML (Knowledge Query and Manipulation Language) [20] as the outer language and four modified message layers derived from UNIK-AGENT[21]: DIALOG, QUOTATION, DECISION, NEGO, PO, RECORD, BPI, AUCTION, REVAUC. Fig. 4 shows the message flows for AgentB.

5 Performance Model and Simulation Result

To find the effect of the designed system comparing to previous s-cart based system, we made a performance evaluation model and made a simulation environment. We found some parameters of the model by simulating the purchasing interaction of business buyers.

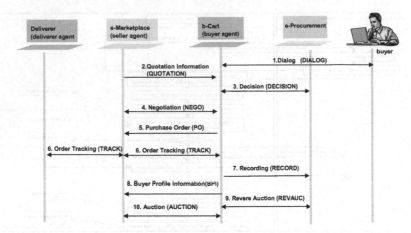

Fig. 4. Message flows

We made a model modified from the evaluation model of Lim and Lee[4]. In this model, we measure the performance of b-cart based system by estimating the interactional effort of user. We assumed that the interactional efforts can be measured by time, and they can be summed each other.

We defined e(x) as below;

e(x) : Individual buyer's interactional effort per purchase using the x-cart, **e(x)** = $\alpha N_{mp} + \beta$.

α: Variable interactional effort that is proportional to the number of e-marketplaces.
β: Fixed interactional effort regardless of the number of e-marketplaces.
And we defined some notations as below;

N_{mp} : Average number of e-marketplaces that a buyer visits for each purchase.
AVG(N_{insert}) : The average number of inserting items from an e-marketplace into the cart per requisition.
P_{trash} : The probability of trashing an item inserted in the cart. ($P_{order} = 1 - P_{trash}$).
Dialog(x) : The dialog effort for user preference using system s.
Identification(x) : The effort of identifying a buyer to enter an e-marketplace per requisition using x-cart.
Collection(x): The effort of collecting an item from an e-marketplace into x-cart.
IDS(x) : The effort for individual purchase decision supporting using x-cart.
Trashing(x): The effort of trashing an item from x-cart.
Ordering(x): The effort of making a purchase order using x-cart.
Payment(x): The effort of making a payment for the ordered items using x-cart.
Tracking(x): The effort of tracking the ordered items within x-cart.
Recording(x): The effort of recording the ordered items within x-cart.
Transmission(x): The effort of transmitting ordered records into the buyer's e-procurement system for integration from x-cart per requisition.
ODS(x): The effort for organizational purchase decision supporting using x-cart.
Negotiation(x): The effort of negotiating a purchase order using x-cart.
Knowledge(x): The effort of maintaining the knowledge within x-cart.

Table 1. Number of interactions for each step and simulation results

Simulation item	Minimum Number of Interactions		Measuring method	Simulation Results (second)	
	s-cart	b-cart		s-cart	b-cart
User Dialog	N_{mp}	1	measured by the time of the user preference input and the time of connecting a site	27.3	29.8
Identification	N_{mp}	1	measured by the ID input login time	2.1	3.3
Collection	N_{mp}	1	measured by the time from clicking an item to displaying the item in the cart	5.7	5.2
Individual Purchase Decision Support	N_{mp}	1	measured by the comparison time for the items	5.9	32.1
Trashing	$P_{trash}AVG(N_{insert})N_{mp}$	$P_{trash}AVG(N_{insert})N_{mp}$	measured by the time of deleting an item in a cart	1.5	1.8
Ordering	N_{mp}	1	measured by the time of making purchasing order with order and delivery information	23.9	23.9
Payment	N_{mp}	1	measured by the time of inputting payment information and making payment with credit card	25.7	25.7
Tracking	N_{mp}	1	measured by the time of tracking an item for delivery status	11.4	11.4
Recording	N_{mp}	1	measured by the time of recording the information in the cart; we assumed it is same to both cart	1.4	1.4
Record Transmission	N_{mp}	1	measured by the transmission time of transferring information to a prototype ERP system's specific directory	3.6	4.7
Organizational Purchase Decision Support	N_{mp}	1	measured by the duration time following the decision making workflow	23.4	24.4
Negotiation	N_{mp}	N_{mp}	measured by the negotiation duration time; we assumed it is same to both cart	22.4	22.4
Knowledge Maintenance	N_{mp}	1	measured by the time of storing the ordering knowledge to DB	3.2	3.2

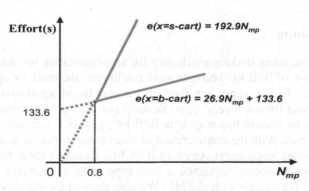

Fig. 5. e(x=b-cart) vs. e(x=s-cart)

From the characteristics and the process of b-cart, we can estimate the numbers of each step's interactions as Table 1. e(x=s-cart) and e(x=b-cart) can be calculated from the summation of the multiplications of the each step's interactional efforts in purchasing process and the number of corresponding interactions. From these assumptions we made a measurement model. The total interactional efforts for a buyer using the b-cart per purchase can be modeled as;

e(x=b-cart) = $(Negotiation(x) + AVG(N_{insert})(P_{trash} Trashing(x)))N_{mp}$ + $Dialog(x)$ + $Identification(x)$ + $Collection(x)$ + $IDS(x)$ + $Ordering(x)$ + $Payment(x)$ + $Tracking(x)$ + $Recording(x)$ + $Transmission(x)$ + $ODS(x)$ + $Knowledge(x)$

And the total interactional efforts for a buyer using the s-cart per purchase can be modeled as;

e(x=s-cart) = $(Dialog(x) + Identification(x) + Collection(x) + IDS(x) + Ordering(x)$ + $Payment(x) + Tracking(x) + Recording(x) + Transmission(x) + ODS(x) + Negotiation(x) + Knowledge(x) + AVG(N_{insert})(P_{trash} Trashing(x)))N_{mp}$

To measure the interactional time we simulated the business buyers' purchasing with each cart in Pentium PC Windows environment in T3 network in Sejong University in Seoul. We conducted the experiment for random real 3 e-marketplaces with 3 business buyers. Each buyer collects 10 items from each e-marketplace to his/her cart and selects 7 items from each collection list, and make order for the 7 items. So a buyer buys 21 items. They performed the purchasing scenario for 10 times. For *Ordering, Payment, Tracking, Recording, Negotiation, Knowledge Maintenance* we assumed that the time of using s-cart and that of b-cart are same because they are logically same process and depend only on the implementation. The simulation method and the simulation results are shown in Table 1.

Therefore, we can derive,

$$e(x=b\text{-cart}) = 26.9N_{mp} + 133.6$$
$$e(x=s\text{-cart}) = 192.9N_{mp}$$

So we can have equations as Fig. 5. From our simulation, we can know that the b-cart based agent system is efficient than the s-cart based systems with increasing e-marketplaces.

6 Conclusions

E-commerce including desktop purchasing for an organization becomes essential. As the environment of B2B EC becomes more intelligent, the need for agent based systems increases. In this paper, we showed a b-cart based agent system framework which is efficient for employees' direct desktop purchasing from e-marketplaces. We also analyzed the roles of buyer agent in B2B EC, and ACL after describing the concept of the b-cart. With the consideration of such roles, we defined the thirteen features of the b-cart based agent system in B2B EC. Based on these features, we analyzed the buying process, designed a prototype with its message interfaces and structure to be compatible with ebXML. We also showed the efficiency of the b-cart based system by simulating the designed system with proposing a performance model. Even though the b-cart based agent system has some overheads such as in development, operation, and deployment, the buyer using a b-cart can collect items from different e-marketplaces, and can order purchases simultaneously over the collected items. The b-cart based system also supports integrated services in personalized comparison on purchasing items, order tracking, financial/payment management and user's account management. It can support cost reduction by and user's account management. It can support cost reduction by providing comparative purchase and proper inventory level. This also allows a tight integration of the b-cart with e-procurement systems, supports negotiation, and provides personal decision support along with the organizational decision making process.

From our simulation we know that as the number of e-marketplaces and individual buyers in an organization increases, the need for integrated management of information and the number of interaction between e-marketplaces will increase along with the effectiveness of the b-cart based agent system in B2B EC.

Further researches will be done on the issues of making a more general environment for simulating agent based system and the deployment and standardization issues.

References

1. Lim G. G. and J. K. Lee: "Buyer-Carts for B2B EC: The b-Cart Approach", International Conference on Electronic Commerce '2000, (2000) 54-67
2. Lee J. K. and E. Turban: "Planning B2B e-Procurement Marketplaces", in Handbook of e-Business. CRC Publishing. Corp. (2001)
3. Xu, D., Wand H.: "Multi-agent collaboration for B2B workflow monitoring", Knowledge-Based Systems, vol 15, (2002) 485-491
4. Lim G. G. and J. K. Lee: "Buyer-Carts for B2B EC: The b-Cart Approach" Organizational Computing and Electronic Commerce, Vol.13, No3&4, (2003) 289-308
5. Selland C.: "Extending E-Business to ERP", e-Business Advisor, Jan. (1999) 18-23
6. Marshall M.: "ERP: Web Applications Servers Give Green Light To ERP", Information-week, Apr. (1999)
7. Guttman, R., Moukas, A., and Maes, P.: "Agent-mediated Electronic Commerce: A Survey". Knowledge Engineering Review Journal, June. (1998)
8. Blake, M.B.: "B2B Electronic Commerce: Where Do Agents Fit In?", Proc. of the AAAI-2002 Workshop on Agent Technologies for B2B E-Commerce, Edmonton, Alberta, Canada, July 28. (2002)
9. Martin, D., Cheyer, A. & Moran, D.: The Open Agent Architecture: A framework for building distributed software systems. Applied Artificial Intelligence: An International Journal. Vol. 13, No. 1-2. Jan.-Mar. (1999)
10. Minar, N., Burkhart, R., Langton C. and M. Askenazi: "The Swarm Simulation System: A Toolkit for Building Multi-agent Simulations" http://www.santafe.edu/projects/swarm/, jun,21, (1996)
11. J. Ferber, O. Gutknecht, F. Michel: MadKit Development Guide, (2002)
12. Kilgore, R.A.: SILK, JAVA and object-oriented simulation. In Proceedings of the 2000 Winter Simulation Conference, (2000) 246–252
13. Conte, R., Gilbert, N., Sichman, J.S.: MAS and Social Simulation: a suitable commitment, Lecture Notes in Computer Science, 1534 (1998) 1-9
14. O. Streltchenko, T. Finin, and Y. Yesha: Multi-agent simulation of financial markets. In S. O. Kimbrough and D. J. Wu, editors, Formal Modeling in Electronic Commerce. Springer-Verlag, (2003)
15. Kl ugl, F. and F. Puppe: The Multi-Agent Simulation Environment SeSAm. In B uning, H. Kleine (editor): Proc. of the Workshop Simulation and Knowledge-based Systems (1998)
16. Rick Kazman.: HIDRA. An Architecture for Highly Dynamic Physically Based Multi-Agent Simulations. International Journal in Computer Simulation, 1995, http://citeseer.ist.psu.edu/kazman95hidra.html (1995)

17. Janssen, M. & Verbraeck, A.: Agent-based Simulation for Evaluating Intermediation Roles in Electronic Commerce. Workshop 2000: Agent-based simulation. Passau, Germany, May 2-3, (2000) 69-74
18. Labrou, Y.; Finin, T.; and Peng, Y.: The interoper-ability problem: Bringing together mobile agents and agent communication languages. In Ralph Sprague, J., ed., Proceedings of the 32nd Hawaii International Conference on System Sciences. Maui, Hawaii: IEEE Computer Society, (1999)
19. ebXML Project Teams: ebXML Technical Specifications (Architecture, Message Service, Registry Services, Business Process, Requirements) Ver. 1.0.x. [Online]. Available: http://www.ebxml.org (2001)
20. Finin, T., Labrou, Y. and Mayfield, J.: FIPA 97 Specification Foundation for Intelligent Physical Agents. (http://drogo.cselt.stet.it/fipa/spec/fipa97/) (1997)
21. Lee, J. K. and W. K. Lee: An Intelligent Agent Based Competitive Contract Process: UNIK-AGENT. In Proceedings of 13th Hawaii International Conference on System Sciences (1997)

Modelling of Two-Level Negotiation in Hierarchical Hypergame Framework

Toshiyuki Toyota and Kyoichi Kijima

Graduate School of Decision Science and Technology, Tokyo Institute of Technology,
2-12-1 Ookayama, Meguro-ku, Tokyo 152-8552, Japan
{toyota,kijima}@valdes.titech.ac.jp

Abstract. The purpose of this study is to propose two-level hierarchical hypergame model (THHM) and to examine characters in hierarchical ne-gotiation. The THHM captures some essential features of international negotiation between two countries. It usually involves two-level nego-tiation, i.e., international and domestic negotiation. The THHM tries to formulate such a hierarchical negotiation situation by extending the hypergame. The present research is a first contribution to examine two-level negotiation structure in terms of hypergame. We introduce a holis-tic concept of rational behavior as a whole as hierarchical Hyper Nash equilibrium (HHNE). Then, we clarify an existence condition of HHNE.

1 Introduction

The purpose of this study is to propose two-level hierarchical hypergame model (THHM) and to discuss conflict resolution in hierarchical negotiation by using it.

The THHM captures some essential features of international negotiation be-tween two countries. It usually involves two-level negotiations, i.e., international and domestic negotiation. At the international level the diplomat from each country, of course, disputes and negotiates for their diplomatic fruit. At the same time, at the national level the diplomat usually has to deal with domestic con-flicts among the interested parties including the government ministries/agencies and public opinion. That is, the diplomat has to take into account the domestic conflicts when negotiating the counterpart, while the domestic interested par-ties have to understand the international negotiation game. It is, however, quite often that the diplomat misunderstands the domestic parties and vise versa.

The THHM tries to formulate such a hierarchical negotiation situation by ex-tending the hypergame. Since the hypergame is suitable for explicitly discussing misunderstanding between two players, it is appropriate to analyze the situation above.

The present research is an original and the first contribution to examine two-level negotiation structure in terms of hypergame. First, we introduce a holistic concept of rational behavior of the system as hierarchical hyper Nash equilibrium (HHNE). Then, we investigate under what conditions HHNE can exist by assuming a typical structure of THHM.

D.-K. Baik (Ed.): AsiaSim 2004, LNAI 3398, pp. 289–295, 2005.

2 Two-Level Hierarchical Hypergame Model

2.1 Two-Level Negotiation

A main concern of the present research is with clarifying under what conditions a two-level negotiation can be settled. Typical examples of such a negotiation are found in international negotiations between two countries, since an international negotiation often involves domestic negotiation as well. Hence, throughout the arguments we will adopt terminology of international negotiation even to describe a two-level negotiation in general.

At the international level the diplomat from each country, of course, disputes and negotiates for diplomatic fruit. At the same time, at the national level the diplomat usually has to deal with domestic conflicts among the interested parties including the government ministries/agencies and public opinion.

In such a two-level negotiation, the diplomat has to take into account the domestic conflicts when negotiating the counterpart, while the domestic interested parties have to understand the international negotiation game. It is, however, quite often that the diplomat misunderstands the domestic parties and vise versa. Furthermore, there may be another kind of misperception and/or misunderstanding, i.e., that between the domestic parties. For example, in Japan the Ministry of Foreign Affairs sometimes misunderstand the public opinion.

Two-level Hierarchical Hypergame Model (THHM), which we propose in this paper, tries to represent such two kinds of misperception/misunderstanding between the players in an explicit manner by extending hypergame framework. Since the hypergame is a suitable framework for explicitly discussing misunderstanding between two players, it seems quite appropriate to analyze the situation above in terms of it.

2.2 Simple and Symbiotic Hypergame

Before introducing THHM, we need to explain the hypergame framework [4,5]. In the traditional (non-cooperative) game theory, strategy sets and preference ordering of all the players are assumed common knowledge among the players and known to every player. In the hypergame, on the contrary, misperception/misunderstanding of them by the players are explicitly assumed.

We differentiate concept of hypergame into two types in this paper. The simplest one is called simple hypergame and is defined as follows:

Definition 1. Simple hypergame
A simple hypergame of players p and q is a pair of (G_p, G_q).

We have $G_p = (S_p, S_{qp}, \geq_p, \geq_{qp})$ and $G_q = (S_{pq}, S_q, \geq_{pq}, \geq_q)$.

In G_p, S_p denotes a set of strategies of p while S_{qp} denotes a set of strategies which the player p recognizes that q possesses. That is, p perceives that S_{qp} is the set of strategies of q. $\geq_p$ denotes a preference ordering on $S_p \times S_{qp}$ of p while $\geq_{qp}$ is a preference ordering on $S_p \times S_{qp}$ that p assumes that q holds. That is, p perceives that q's preference ordering is $\geq_{qp}$. We define G_q in a similar way.

The definition implies that each player independently perceives the problematic situation without mutual understanding. The rationality of a simple hypergame is defined as follows:

Definition 2. Nash equilibrium of simple hypergame
For a simple hypergame (G_p, G_q) *of players* p *and* q, $(s^{p*}, s^{qp*}) \in S_p \times S_{qp}$ *is a Nash equilibrium of* $G_p = (S_p, S_{qp}, \geq_p, \geq_{qp})$ *if and only if we have*
$(\forall s^p \in S_p)(s^{p*}, s^{qp*}) \geq_p (s^p, s^{qp*})$ *and* $(\forall s^{qp} \in S_{qp})(s^{p*}, s^{qp*}) \geq_{qp} (s^{p*}, s^{qp})$.

This definition shows that if (s^{p*}, s^{qp*}) is a Nash equilibrium of G_p, then p believes that there is no incentive for either of the agents to change their strategy as long as the other does not change its strategy.

While in the simple hypergame, we assume no communication between the players and they make decisions completely independent based on their own subjective $(G_p$ or $G_q)$.

As time passes, the players involved in simple hypergame may begin to understand the other's interpretation of the game. By mutual communication, the players may be able to deeper understand each other and construct a kind of "dictionary" by which each can understand the other's terminology. To deal with such a context, a symbiotic hypergame is defined by introducing functions which represent how each agent interprets the other's game.

Definition 3. 2-person symbiotic hypergame
A 2-person symbiotic hypergame with players p *and* q *is a pair* (G_p, G_q, f, g), *where,* $G_p = (S_p, S_{qp}, \geq_p, \geq_{qp})$, *and* $G_q = (S_{pq}, S_q, \geq_{pq}, \geq_q)$ *are 2-person simple hypergame and we have* $f : S_q \to S_{qp}$ *and* $g : S_p \to S_{pq}$ *hold.*

Function f represents how p perceives a true set of q's strategies while function g represents how q perceives a true set of p's strategies.

Defining 2-person symbiotic hypergame, it is natural to consider overall rationality of (G_p, G_q, f, g). Thus, we define such rationality as follows:

Definition 4. Equilibrium of 2-person symbiotic hypergame
(G_p, G_q, f, g) *is 2-person symbiotic hypergame with players* p *and* q *where* $G_p = (S_p, S_{qp}, \geq_p, \geq_{qp})$, $G_p = (S_{pq}, S_q, \geq_{pq}, \geq_q)$, $f : S_q \to S_{qp}$, *and* $g : S_p \to S_{pq}$. *In this case* $(x^*, y^*) \in S_p \times S_q$ *is a equilibrium if and only if* $(\forall x \in S_p)(x^*, f(y^*)) \geq_p (x, f(y^*))$ *and* $(\forall y \in S_q)(g(x^*), y^*) \geq_q (g(x^*), y)$. $(x^*, y^*) \in S_p \times S_q$ *is called hyper Nash equilibrium.*

When $(x^*, y^*) \in S_p \times S_q$ is hyper Nash equilibrium, p perceives (x^*, y^*) is $(x^*, f(y^*))$, which is a Nash equilibrium in own payoff matrix. Therefore, p has no incentive which p changes a present strategy x^* to the extend not to perceive to change $f(y^*)$. Symmetric arguments hold for q. Hyper Nash equilibrium is a pair of strategies that is interpreted as Nash equilibrium by p and g through f and g, respectively.

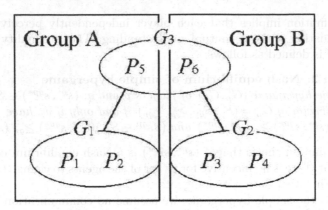

Fig. 1. Two-level Hierarchical Hypergame Model situation.

2.3 Two-Level Hierarchical Hypergame Model

Now, let us define THHM based on the preparation so far. The game situation that THHM tries to formulate consists of three games G_1, G_2, and G_3 as shown in Fig. 1.

Fig. 1 describes an international negotiation between two countries A and B. p_1, p_2 and p_5 are players of A while p_3, p_4 and p_6 are players of B. p_5 and p_6 are a diplomats of A and B, respectively, while p_1, p_2 are domestic interested parties of A, p_3 and p_4 are those of B.

In this paper, we focus on A, namely, relationship between p_5 and p_6 at the international level as well as that among p_1, p_2 and p_5, since a symmetric argument is applicable to B.

Definition 5. THHM
THHM with players p_1, p_2, p_5 and p_6 is $G = (G_1, G_2, G_3, \{k_1, k_2\}, \{j_1, j_2\})$, where $G_1 = (S_1, S_{61}, \geq_1, \geq_{61})$, $G_2 = (S_2, S_{62}, \geq_2, \geq_{62})$, and $G_3 = (S_5, S_6, \geq_5, \geq_6)$. Then, $k_1 : S_1 \to S_5$, $k_2 : S_2 \to S_5$, $j_1 : S_6 \to S_{61}$ and $j_2 : S_6 \to S_{62}$.

k_1 and k_2 represent how p_5 interprets the strategy set S_1 and S_2 of each player of domestic level, respectively. On the contrary, j_1 and j_2 show how p_1 and p_2 perceive the true set of the strategies of p_6, respectively. S_{61} represents a strategy set p_1 believes p_6 should have. $\geq_{61}$ indicates preference ordering on $S_1 \times S_{61}$. The similar arguments are applicable to p_2.

In some cases, we may assume that all players in the same group correctly perceive the strategy set of the counterpart p_6 at international level. It implies that the three players of A recognize the strategies of the opposite side in the same way and correctly. We can express the situation formally as a strong THHM.

Definition 6. Strong THHM
THHM $G = (G_1, G_2, G_3, \{k_1, k_2\}, \{j_1, j_2\})$ is called strong if we have $S_6 = S_{61}$, $S_6 = S_{62}$, and $j_1 = j_2 = 1_{S_6}$, that is, $G = (G_1, G_2, G_3, \{k_1, k_2\}, \{1_{S_6}, 1_{S_6}\})$.

To avoid unnecessary complexity, we will focus only on a strong THHM in the following arguments. The strong THHM assumes that all the players in the

Table 1. G_3: Japan vs. U.S.A.

Japan U.S.A.	YES	NO
YES	(3,4)	(1,1)
NO	(2,3)	(4,2)

Table 2. G_1: The Defense Agency vs. U.S.A.

DA U.S.A.	YES	NO
UF	(2,6)	(1,1)
HS	(5,5)	(3,3)
ND	(4,2)	(6,4)

same group correctly perceive the strategy set of the counterpart p_6 at the international level. It is particularly suitable for describing international negotiations, say, ratifying a treaty. In such negotiations the players at the international level have two strategies, that is, to ratify treaty or not, and the players at the domestic level can be supposed to exactly perceive those strategies.

In order to illustrate the concept of a strong THHM, we adopt an example. Tables 1, 2 and 3 show component games of international negotiation over allegation of Iraq's weapon of mass destruction. The historic background is as follows: The United States took the lead and delivered a military power attack on Iraq on March 20, 2003. Before the attack, the United States negotiated with Japan to agree the attack and to support.

Table 1 illustrates an international negotiation G_3 over a ratification of a treaty between Japan and the United State. The both have "yes" and "no" as their strategies. Tables 2 and 3 indicate domestic negotiation G_1 and G_2 in Japan, respectively. p_1 and p_2 are the Defense Agency and the Ministry of Foreign Affairs in Japan, respectively. Let us suppose S_1={UF, HS, ND} and S_2={F, NF}. UF means to use of the military force. HS stands for humanitarian support by the military, while ND denotes no dispatch. On the other hand, F and NF mean to follow U.S.A. and not to follow U.S.A., respectively. We assume that the mutual perception between international level and domestic level are k_1(UF) $= k_1$(HS) $=$ Yes and k_1(ND) $=$ NO while k_2(F) $=$ Yes and k_2(NF) $=$ NO. Then, k_1 and k_2 are surjective, respectively.

Now, we introduce a holistic concept of rational behavior to a strong THHM as Hierarchical hyper Nash equilibrium.

Definition 7. Hierarchical hyper Nash equilibrium
Let $G = (G_1, G_2, G_3, \{k_1, k_2\}, \{j_1, j_2\})$ be a strong THHM. $(\alpha, \beta, \gamma, \delta) \in S_1 \times S_2 \times S_5 \times S_6$ is called hierarchical hyper Nash equilibrium (HHNE) of G if the following conditions are satisfied.

(1) $(\gamma, \delta) \in S_5 \times S_6$ is Nash equilibrium in G_3.
(2) α is best reply strategy for $j_1(\delta)$ in G_1.
(3) β is best reply strategy for $j_2(\delta)$ in G_2.
(4) $k_1(\alpha) = \gamma$, and $k_2(\beta) = \gamma$.

Table 3. G_2: The Ministry of Foreign Affairs vs. U.S.A.

M of FA U.S.A.	YES	NO
F	(3,4)	(4,1)
NF	(2,3)	(1,2)

Let $(\alpha, \beta, \gamma, \delta)$ be HHNE. **(1)** implies (γ, δ) is Nash equilibrium at the international level. **(2)** indicates that p_1 perceives that α is best reply strategy for $j_1(\delta)$ in p_1's payoff matrix. **(3)** shows that symmetric arguments of **(2)** are applicable to p_2. **(4)** requires a specific way of p_5's interpretation of strategies of the domestic parties p_1 and p_2.

In this way, we can see that HHNE is a natural extension of Nash equilibrium defined by the perception function k_1, k_2, j_1, and j_2.

Let us return to the previous example to illustrate concept of HHNE. From Table 1, the Nash equilibrium in G_3 (γ, δ)=(YES, YES). For δ=YES, the best strategy in G_1 and G_2 is α= HS and β= F, respectively. By the previous assumption of the mutual perception, we have k_1(HS) = YES and k_2(F) = YES. Therefore, we can see the HHNE of this example is (HS, F, YES, YES).

3   Existence Condition of HHNE

As pointed out above, HHNE is a sound concept for characterizing equilibrium of a strong THHM as a whole. Our concern of this section is with finding out conditions under which HHNE exists when a strong THHM is given.

To derive such a condition, we need hierarchical consistency between the international level and the domestic level:

Definition 8. k_1 and k_2 are satisfied hierarchical consistency if and only if $\forall s_6 \in S_6, \forall s_i, \forall s_i' \in S_i, (s_i, s_6) \succeq_i (s_i', s_6) \Rightarrow (k_i(s_i), s_6) \succeq_5 (k_i(s_i'), s_6)$, where $i = 1, 2$.

The definition implies that p_5 correctly perceives the domestic players' preference.

Let us explain the condition by using Tables 1 and 2. First, we have YES $\in S_6$, HS and ND $\in S_1$. Then, it implies (HS, YES) $\succeq_1$ (ND, YES), while $(k_1(\text{HS}), \text{YES}) \succeq_5 (k_1(\text{ND}), \text{YES})$. Therefore, in this case k_1 satisfies the hierarchical consistency. Next, we have YES $\in S_6$, UF and ND $\in S_1$. Then, it is followed that (ND, YES) $\succeq_1$ (UF, YES), while $(k_1(\text{UF}), \text{YES}) \succeq_5 (k_1(\text{ND}), \text{YES})$. It implies that k_1 does not satisfy the hierarchical consistency. Similarly, we can see that k_2 and k_2 satisfy the hierarchical consistency.

Then, we have a main result:

Proposition 1. Let $G = (G_1, G_2, G_3, \{k_1, k_2\}, \{1_{S_6}, 1_{S_6}\})$ be a strong THHM. Assume there is Nash equilibrium $(\gamma, \delta) \in S_5 \times S_6$ at the international level. If k_1 and k_2 are satisfy hierarchical consistency and surjective, there is HHNE in G.

The condition that k_1 and k_2 are surjective means that p_5 supposes the set of strategies of p_1 and p_2 without redundancy.

Proof. There is $(\gamma, \delta) \in S_5 \times S_6$ is Nash equilibrium in G_3, both k_1 and k_2 are surjective. First, we consider G_1, since the hierarchical consistency is satisfied for $\forall s_{11}, \forall s_{12} \in S_1$,

$$(s_{11}, \delta) \succeq_1 (s_{12}, \delta) \Rightarrow (k_1(s_{11}), \delta) \succeq_5 (k_1(s_{12}), \delta).$$

Moreover, α is assumed to the p_1's best reply strategy for δ,

$$(\alpha, \delta) \succeq_1 (s_{12}, \delta) \Rightarrow (k_1(\alpha), \delta) \succeq_5 (k_1(s_{12}), \delta),$$

$(k_1(\alpha), \delta)$ represents Nash Equilibrium in G_3, therefore $k_1(\alpha) = \gamma$.
Similarly, the same results in G_2. Consequently, HHNE $(\alpha, \beta, \gamma, \delta)$ exists. $\square$

Let us apply Proposition 3.1 to our example. We can check the conditions of Proposition 3.1. Indeed, Nash equilibrium of G_3 is (YES, YES) from Table 1. Because k_1 and k_2 satisfy the hierarchical consistency and are both surjective, we have (HS, F) $\in S_1 \times S_2$. Therefore, Proposition 3.1 implies that there is HHNE in G. Indeed, as shown before, (HS, F, YES,YES) is HHNE.

4   Conclusion

We proposed two-level hierarchical hypergame model (THHM) consisting of international and domestic level. It formulates such a hierarchical negotiation situation by extending the hypergame. Then, we introduced a holistic concept of rational behavior as hierarchical hyper Nash equilibrium (HHNE). Finally, we clarified existence condition of HHNE.

Since the THHM does not take conflicts among players at domestic level into consideration, these treatments are open to future research.

References

1. R., D. Putnum: Diplomacy and domestic politics: the logic of two-level games. International Organization 42, 3, Summer (1988) 427-460.
2. K., Iida: When and How Do Domestic Constraints Matter?: Two-level games with uncertainty. Journal of conflict resolution **37** No. 3 Sep. (1993) 403-426.
3. J., Mo: The Logic of Two-level Games with Endogenous Domestic Coalitions. Journal of conflict resolution **38** No. 3 Sep. (1994) 402-422.
4. M., Wang, K. W., Hipel and N. M., Fraser: Solution Concepts in Hypergames. Applied Mathematics and Computation **34** (1989) 147-171.
5. K., Kijima: Intelligent Poly-agent Learning Model and its application. Information and Systems Engineering **2** (1996) 47-61.
6. P., G. Benntt: Bidders and dispenser: manipulative hypergames in a multinational context. European Journal of Operational Research **4** (1980) 293-306.
7. M. Leclerc, B. Chaib-draa: Hypergame Analysis in E-Commerce: A Preliminary Report. Centre for Interuniversity Research and Analysis on Organizations Working Paper 2002s-66 (2002).

A Personalized Product Recommender for Web Retailers

Yoon Ho Cho[1], Jae Kyeong Kim[2,*], and Do Hyun Ahn[2]

[1] School of e-Business, Kookmin University
861-1 Jungnung, Sungbuk, Seoul 136-702, Korea
Tel: +82-2-910-4950, Fax: +82-2-910-4519
www4u@kookmin.ac.kr
[2] School of Business Administration, Kyung Hee University
#1, Hoeki, Dongdaemoon, Seoul 130-701, Korea
Tel: +82-2-961-9355, Fax: +82-2-967-0788
{jaek,adh}@khu.ac.kr

Abstract. This paper proposes a recommendation methodology to help customers find the products they would like to purchase in a Web retailer. The methodology is based on collaborative filtering, but to overcome the sparsity issue, we employ an implicit ratings approach based on Web usage mining. Furthermore to address the scalability issue, a dimension reduction technique based on product taxonomy together with association rule mining is used. The methodology is experimentally evaluated on real Web retailer data and the results are compared to those of typical collaborative filtering. Experimental results show that our methodology provides higher quality recommendations and better performance, so it could be a promising marketer assistant tool for the Web retailer.

1 Introduction

Due to the rapid growth of e-commerce, customers on the Web retailer are often overwhelmed with choices and flooded with advertisements for products. A promising technology to overcome such an information overload is recommender systems. One of the most successful recommendation techniques is Collaborative Filtering (CF) [8, 10] which identifies customers whose tastes are similar to those of a given customer and it recommends products those customers have liked in the past. Although CF is the most successful recommendation technique, it suffers from two major shortcomings.

First, when there is a shortage of ratings, CF suffers from a sparsity problem. Most similarity measures used in CF work properly only when there exists an acceptable level of ratings across customers in common. An increase in the number of customers and products worsens the sparsity problem, because the likelihood of different customers rating common products decreases. Such sparsity in ratings makes the formation of neighborhoods inaccurate, thereby resulting in poor recommendations. In attempt to address the sparsity problem, researchers have proposed many approaches. One of the most widely used approaches is to use

* Corresponding author.

D.-K. Baik (Ed.): AsiaSim 2004, LNAI 3398, pp. 296–305, 2005.
© Springer-Verlag Berlin Heidelberg 2005

implicit ratings. The implicit ratings approach attempts to increase the number of ratings through observing customers' behaviors without their burden of giving subjective customer ratings or registration-based personal preferences. In particular, the clickstream in the Web retailer provides information essential to understanding the shopping patterns or prepurchase behaviors of customers such as what products they see, what products they add to the shopping cart, and what products they buy. By analyzing such information via Web usage mining, it is possible not only to make a more accurate analysis of the customer's interest or preference across all products (than analyzing the purchase records only), but also to increase the number of ratings (when compared to collecting explicit ratings only). Next, CF suffers from a *scalability* problem. Recommender systems for large Web retailers have to deal with a large number of customers and products. Because these systems usually handle high-dimensional data to form the neighborhood, the nearest neighbor CF-based recommender system is often very time-consuming and scales poorly in practice. A variety of approaches to address the scalability problems can be classified into two main categories: dimensionality reduction techniques and model-based approaches. Singular value decomposition (SVD) is a widely used dimensionality reduction technique. It enables us to project high dimensional (sparse) data into low dimensional (dense) one through concentrating most of information in a few dimensions. In model-based approaches, a model is first built based on the rating matrix and then the model is used in making recommendations. Several data mining techniques such as Bayesian network, clustering and association rule mining have been applied to building the model.

This article proposes a recommendation methodology, called Web usage mining driven Collaborative Filtering recommendation methodology using Association Rules (WebCF-AR), to address the sparsity and the scalability problems of existing CF-based recommender systems. Web usage mining is employed as an implicit ratings approach to address the sparsity problem. Web usage mining analyzes customers' shopping behaviors on the Web and collects their implicit ratings. This increases the number of ratings rather than obtaining subjective ratings, thereby reduces the sparsity. The scalability problem is addressed by using the dimensionality reduction technique together with the model-based approach. Product taxonomy is employed as a dimensionality reduction technique that identifies similar products and groups them together thereby reducing the rating matrix. Association rule mining, as a model-based approach, is applied to identify relationships among products based on such a product grouping and these relations are used to compute the prediction score. Association rules are mined from clickstream data as well as purchase data.

2 Related Work

2.1 Web Usage Mining

Web usage mining is the process of applying data mining techniques to the discovery of behavior patterns based on Web data, for various applications. In the advance of e-commerce, the importance of Web usage mining grows larger than before. The overall process of Web usage mining is generally divided into two main tasks: data preprocessing and pattern discovery. The data preprocessing tasks build a server session file where each session is a sequence of requests of different types made by single user during a single visit to a site. Cooley et al. [4] presented a detailed description of data preprocessing methods for mining Web browsing patterns. The pattern discovery tasks involve the discovery of association rules, sequential patterns, usage clusters, page clusters, user classifications or any other pattern discovery method. Usage pattern extracted from Web data can be applied to a wide range of applications such as Web personalization, system improvement, site modification, business intelligence discovery, usage characterization, etc.

2.2 Association Rule Mining

Given a set of transactions where each transaction is a set of items (itemset), an association rule implies a knowledge or pattern in the form of $X \Rightarrow Y$, where X and Y are itemsets; X and Y are called the body and the head, respectively. The support for the association rule $X \Rightarrow Y$ is the percentage of transactions that contain both itemset X and Y among all transactions. The confidence for the rule $X \Rightarrow Y$ is the percentage of transactions that contain itemset among transaction that contain itemset X. The support represents the usefulness of the discovered rule and the confidence represents certainty of the rule. Association rule mining is the discovery of all association rules that are above a user-specified minimum support and minimum confidence. Apriori algorithm is one of the prevalent techniques used to find association rules [1]. Association rule mining has been widely used in the filed of recommender systems [6, 7, 10]. However, existing works have focused on finding association among co-purchased products in purchase data. In our research, association rule mining is used to discover association rules from Web log data as well as purchase data.

2.3 Product Taxonomy

A product taxonomy is practically represented as a tree that classifies a set of products at a low level into a more general product at a higher level. The leaves of the tree denote the product instances, UPCs (Universal Product Codes) or SKUs (Stock Keeping Units) in retail jargon, and non-leaf nodes denote product classes obtained by combining several nodes at a lower level into one parent node. The root node labeled by All denotes the most general product class. A number called level can be assigned to each node in the product taxonomy. The level of the root node is zero, and the level of any other node is one plus the level of its parent.

The product taxonomy plays an important role in the knowledge discovery process since it represents Web retailer dependent knowledge and may affect the results. In many applications, strong association rules are more likely to exist at high levels of the product taxonomy but may likely repeat common knowledge.

For example, an association rule at a high level "80% of customers who buy clothes also buy footwear" may be given to marketers of the Web retailer. On the other hand, rules at a low level may be more interesting, but are difficult to find. For example, an association rule at a low level "40% of customers who buy shirts also buy shoes" could be mixed with many uninteresting rules. Therefore, it is important to mine association rules at the right level of the product taxonomy [2, 5].

3 Methodology

The entire procedure of WebCF-AR is divided into four phases: grain specification, customer preference analysis, product association analysis, and recommendation generation.

3.1 Phase 1: Grain Specification

In this phase, we identify similar products and group them together using product taxonomy so as to reduce the product space. All the products are grouped by specifying the level of product aggregation on the product taxonomy that is provided by the marketer or domain expert. We refer to such a specification as the *grain*. Formally, a grain G is defined as a set of nodes (product or product class) excluding the root node in the product taxonomy that cross with any possible path from a leaf node (product) to the root node at one and only one node. Therefore, every leaf node has its corresponding grain node (it is called as a *grain product class* in this article).

Consider several examples of specifying the grain as shown in Fig. 1, where grains are denoted by shaded regions. Fig. 1(a) shows a case of specifying the grain at a lower level, such that $G = \{$'Outerwear', 'Pants', 'Shirts', 'Shoes', 'Socks', 'Skincare', 'Perfumes', 'Bags', 'Belts', 'Wallets'$\}$. Fig. 1(b) shows a case of specifying the grain at a higher level, such that $G = \{$'Clothes', 'Footwear', 'Cosmetics', 'Accessories'$\}$. For the grain from Fig. 1(a), the grain node of 'UPC00' is 'Outerwear' and that of 'UPC29' is 'Shirts'. However, in the grain of Fig. 1(b), the grain node of both 'UPC00' and 'UPC29' is 'Clothes'. The grain can be specified across different levels of the taxonomy as shown in Fig. 1(c). Choosing the right grain depends on the product, on its relative importance for making recommendations and on its frequency in the transaction data. For example, frequently purchased products might stay at a low level in the taxonomy while rarely purchased products might be drilled up to a higher level. It is usually desirable to specify the grain as a set of nodes with relatively even data distribution since association rule mining produces the best results when the products occur in roughly the same number of transactions in the data [5].

3.2 Phase 2: Customer Preference Analysis

WebCF-AR applies the results of analyzing preference inclination of each customer to make recommendations. For this purpose, we propose a *customer pref-*

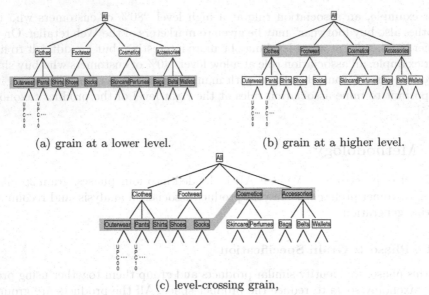

(a) grain at a lower level.   (b) grain at a higher level.

(c) level-crossing grain,

Fig. 1. Examples of specifying the grain.

erence model represented by a matrix. The customer preference model is constructed based the following three general shopping steps in Web retailers.

- *Click-Through:* the click on and the view of the Web page of the product
- *Basket Placement:* the placement of the product in the shopping basket
- *Purchase:* the purchase of the product

In general Web retailers, products are purchased in accordance with the three sequential shopping steps, so we can classify all products into four product groups such as purchased products, products placed in the basket, products clicked through, and the other products. This classification provides an $is-a$ relation between different groups such that purchased products $is-a$ products placed in the basket, and products placed in the basket is-a products clicked through. From this relation, it is reasonable to obtain a preference order between products such that {products never clicked} $\prec$ {products only clicked through } $\prec$ {products only placed in the basket} $\prec$ {purchased products}.

Let p_{ij}^c be the total number of occurrences of click-throughs of a customer i across every products in a grain product class j. Likewise, p_{ij}^b and p_{ij}^p are defined as the total number of occurrences of basket placements and purchases of a customer i for a grain product class j, respectively. p_{ij}^c, p_{ij}^b and p_{ij}^p are calculated from clickstream data as the sum over the given time period, and so reflect individual customer's behaviors in the corresponding shopping process over multiple shopping visits. From the above terminology, we define the customer preference matrix $P = (p_{ij})$, $i = 1, \cdots, m$ (total number of customers), $j = 1, \cdots, n$(total number of grain product classes, i.e., $|G|$), as follows:

$$p_{ij} = \begin{cases} \dfrac{p_{ij}^c}{\sum_{i=1}^m p_{ij}^c} + \dfrac{p_{ij}^b}{\sum_{i=1}^m p_{ij}^b} + \dfrac{p_{ij}^p}{\sum_{i=1}^m p_{ij}^p} & \text{if } p_{ij}^c > 0, \\ 0 & \text{otherwise} \end{cases} \tag{1}$$

p_{ij} implies that the preference of customer i across every products in a grain product j depends on the normalized value of total number of occurrences of click-throughs, basket placement and purchases. It ranges from 0 to 3, where more preferred product results in bigger value. Note that the weights for each shopping step are not the same although they look equal as in Equation (1). From a casual fact that customers who purchased a specific product had already not only clicked several Web pages related to it but placed it in the shopping basket, we can see that Equation (1) reflects the different weights.

3.3   Phase 3: Product Association Analysis

In this phase, we first search for meaningful relationships or associations among grain product classes through mining association rules from the large transactions. In order to capture the customer's shopping inclination more accurately, we look for association rules from three different transaction sets: purchase transaction set, basket placement transaction set and click-through transaction set.

Next, we calculate the extent to which each grain product class is associated with the others using the discovered rules. This work results in building a model called *product association model* represented by a matrix. Given two grain product classes X and Y , let $X \overset{p}{\Rightarrow} Y$, $X \overset{b}{\Rightarrow} Y$, and $X \overset{c}{\Rightarrow} Y$ denote association rules in purchase transaction set, in basket placement transaction set, and in click-through transaction set, respectively. Then, a product association matrix $A = (a_{ij}), i,j = 1, \cdots, n$ (total number of grain product classes) is defined as follows:

$$a_{ij} = \begin{cases} 1 & \text{if } i = j \\ 1 & \text{if } i \overset{p}{\Rightarrow} j \\ .5 & \text{if } i \overset{b}{\Rightarrow} j \\ .25 & \text{if } i \overset{c}{\Rightarrow} j \\ 0 & \text{otherwise} \end{cases} \tag{2}$$

The first condition of Equation (2) captures the association among different products in a product class; a purchase of a product implies he/she has a same preference in other products in the same product class. The multipliers for purchase associations are heuristically set higher than those for basket placement, because we can normally assume that the degree of association in the purchase is more related to the purchasing pattern of customers than those in the basket placement. In the same manner, the multipliers for basket placement associations are set higher than those for click-through.

3.4   Phase 4: Recommendation Generation

In the preceding phases, we have built the customer preference model and the product association model. A personalized recommendation list for a specific customer is produced by scoring the degree of similarity between the associations of each grain product class and his/her preferences and by selecting the best scores. Cosine coefficient [6] is used to measure the similarity. The matching score s_{ij} between a customer i and a grain product class j is computed as followings:

$$s_{ij} = \frac{P_i A^j}{\|P_i\|\|A^j\|} = \frac{\sum_{k=1}^{n} p_{ik} a_{kj}}{\sqrt{\sum_{k=1}^{n} p_{ik}^2} \sqrt{\sum_{k=1}^{n} a_{kj}^2}} \qquad (3)$$

where P_i is the i-th row vector of the $m \times n$ customer preference matrix , and A^j is the j-th column vector of the $n \times n$ product association matrix A. Here, m refers the total number of customers and n refers the total number of grain product classes. The s_{ij} value ranges from 0 to 1, where more similar vector results in bigger value.

Equation (3) implies that all products in a grain product class would have identical matching scores for a given customer when the grain product class is specified at a product class level not a product level. Hence, in that case, we have to choose which products in the grain product class to be recommended to the customer. WebCF-AR selects products with the highest reference frequency. The reference frequency of a product $j(1 \le j \le l), RF_j$, is defined as follows:

$$RF_j = \frac{r_j^c}{\sum_{k=1}^{l} r_j^c} + \frac{r_j^b}{\sum_{k=1}^{l} r_j^b} + \frac{r_j^p}{\sum_{k=1}^{l} r_j^p} \qquad (4)$$

where l is the number of products, r_j^c and r_j^b, and r_j^p are the total number of occurrences of click-throughs, basket placements and purchases of a product j, respectively. This method follows from the hypothesis that the more a product is referred, the higher the possibility of purchase.

4   Experimental Evaluation

4.1   Experiments

For our experiments, we used Web log data and product data from a C Web retailer in Korea that sells a variety of beauty products.

About one hundred Web log files were collected from four IIS Web servers during the period between May 1, 2002 and May 30, 2002. For an application to our experiments, the data preprocessing tasks such as data cleansing, user identification, session identification, path completion, and URL parsing were applied to the Web log files. Finally, we obtained a transaction database in the form of <time, customer-id, product-id, shopping-step> in which the shopping-step is one of the click-through step, the basket-placement step and the purchase step. This database contains 2,249,540 transactions of 66,329 customers. The period between May 1, 2002 and May 24, 2002 is set as training period and the period between May 25, 2002 and May 30, 2002 is set as test period. As the target customers, we selected 116 customers who have purchased at least one product for both periods. Finally, the training set consists of 8960 transaction records created by the target customers for the training period, and the test set consists of 156 purchase records created by them for the test period.

The selected C Web retailer deals in 3216 products. The product taxonomy consists of three levels of hierarchy except the root All. The top level (level 1) contains ten product classes, the next level (level 2) contains 72 product classes, and bottom level (level 3) contains 3216 products.

With the training set and the test set, WebCF-AR works on the training set first, and then it generates a set of recommended products, called recommendation set, for a given customer. To evaluate the quality of the recommendation set, recall and precision have been widely used in the recommender system research. These measures are simple to compute and intuitively appealing, but they are in conflict since increasing the size of the recommendation set leads to an increase in recall but at the same time a decrease in precision [3, 10]. Hence, we used a combination metric that gives equal weight to both recall and precision called $F1$ metric [9]:

$$F1 = \frac{2 \times recall + precision}{recall + precision} \tag{5}$$

To evaluate the scalability issue, we used a performance evaluation metric in addition to the quality evaluation metric. The *response time* and *throughput* were employed to measure the system performance. The response time defines the amount of time required to compute all the recommendations for the training set whereas the throughput denotes the rate at which the recommendations were computed in terms of recommendations per second.

A system to perform our experiments was implemented using Visual Basic 6.0, ADO components and MS-Access. We run our experiments on Windows 2000 based PC with Intel Pentium III processor, having a speed 750 MHz and 1GB of RAM. To benchmark the performance of WebCF-AR against those of other recommender system, we also developed a typical nearest neighbor CF-based recommender system [10] which builds customer profiles using purchase records only, and applies no dimensionality reduction technique to its recommendation procedure. For more details about this benchmark system, please refer Sarwar et al [10]. It was also implemented in Visual Basic 6.0 and was tested in the same experimental environment.

4.2   Results and Discussion

To determine the optimal values of the minimum support and the minimum confidence, we performed an experiment where we varied the two values from 0.1 to 0.9 in an increment of 0.05. From the results, we selected the minimum support value of 0.3 and the minimum confidence value of 0.55 as optimum values for our subsequent experiments.

In order to evaluate the impact of grain specification on the recommendation quality, we performed experiments with three types of grains: grain at level 1 (labeled T1), grain at level 2 (labeled T2) and level-crossing grain (labeled TC) of

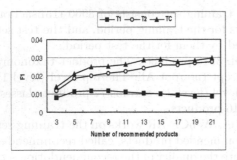 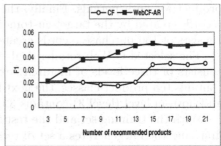

Fig. 2. Impact of grain specification.   **Fig. 3.** Quality comparison of WebCF-AR and benchmark CF algorithm.

the product taxonomy. TC was specified using the grain specification algorithm suggested in section 3.2. For specifying TC, we selected 10 level-2 grains of which the dimensionality reduction threshold varies from 20 to 65 in an increment of 5 and computed F1 for each of them. The average value of F1 was used. From the results in Fig. 2, we can see that TC performs better than the others. These results verify that the grain with an even distribution among products leads to the better quality of recommendations. For the remaining experiments, we used TC as the specified grain.

Finally, we compared WebCF-AR with the benchmark CF algorithm. The best performance of the benchmark CF algorithm is used for this comparison. Our results are shown in Fig. 3. It can be observed from the charts that WebCF-AR works better than the benchmark CF algorithm at all the number of recommended products. With the optimal choice for each of parameters, WebCF-AR performs even better, achieving an average improvement of about 64%.

Table 1 shows the response time and the throughput provided by WebCF-AR and the benchmark CF algorithm. Looking into the results shown in Table 1, we can see that WebCF-AR is about 27 times faster than the benchmark CF algorithm. This point is very important because the number of products and that of customers grow very fast with the widespread use of Web retailers.

Table 1. Performance comparison of WebCF-AR and benchmark CF algorithm.

	Benchmark CF	WebCF-AR
Response time(sec.)	91.53	3.40
Throughput	20.00	519.00

5 Conclusion

The works presented in this article make several contributions to the recommender systems related research. First, we applied the product taxonomy both to reducing the sparsity and to improving the scalability. Secondly, we developed a Web usage mining technique to capture implicit ratings by tracking customers' shopping behaviors on the Web and applied it to reducing the sparsity. Lastly, a Web usage mining technique is used to find rich and interesting association rules from clickstream data and it is applied to improve the scalability.

While our experimental results suggest that WebCF-AR is effective and efficient for product recommendations in the e-business environment, these results are based on data sets limited to the particular Web retailer with a small number of customers and products. Therefore, it is required to evaluate WebCF-AR in more detail using data sets from a variety of Web retailers. It will be an interesting research area to conduct a real marketing campaign to customers using WebCF-AR and then to evaluate its performance.

Acknowledgements

This work was supported by the faculty research program (2004) of Kookmin University in Korea.

References

1. Agrawal, R., Imielinski, T., & Swami, A. (1993). Mining association between sets of items in massive database. Proceedings of ACM SIGMOD International Conference (pp. 207-216).
2. Berry, J.A., & Linoff, G. (1997). Data mining techniques: for marketing, sales, and customer support. New York: Wiley Computing Publishing.
3. Billsus, D., & Pazzani, M.J. (1998). Learning collaborative information filters. Proceedings of the 15th International Conference on Machine Learning (pp. 46-54).
4. Cooley, R., Mobasher, B., & Srivastava, J. (1999). Data preparation for mining world wide Web browsing patterns. Journal of Knowledge and Information Systems, 1, 5-32.
5. Han, J., & Fu, Y. (1999). Mining multiple-level association rules in large databases. IEEE Transactions on Knowledge and Data Engineering, 11, 798-804.
6. Lawrence, R.D. et al. (2001). Personalization of supermarket product recommendations. Data Mining and Knowledge Discovery, 5, 11-32.
7. Lin, W.S., Alvarez, A., & Ruiz, C. (2002). Efficient adaptive-support association rule mining for recommender systems. Data Mining and Knowledge Discovery, 6, 83-105.
8. Resnick, P. et al. (1994). GroupLens: an open architecture for collaborative filtering of netnews. Proceedings of ACM Conference on Computer Supported Cooperative Work (pp. 175-186).
9. Rijsbergen, C.J. (1979). Information retrieval (2nd ed.). London: Butter-worth.
10. Sarwar, B. et al. (2000). Analysis of recommendation algorithms for e-commerce. Proceedings of ACM E-Commerce Conference (pp. 158-167).

Simulating the Effectiveness of Using Association Rules for Recommendation Systems

Jonghoon Chun[1,*], Jae Young Oh[2], Sedong Kwon[2], and Dongkyu Kim[3]

[1] President & CEO, Corelogix Inc, Department of Computer Engineering,
Myongji University, Yongin, Kyunggi-Do 449-728 Korea
Tel:+82-31-330-6441
jchun@mju.ac.kr
[2] Department of Computer Engineering, Myongji University
Yongin, Kyunggi-Do 449-728 Korea
{jyoh,sdkwon}@mju.ac.kr
[3] R&D Director & CTO, Corelogix Inc, Seoul, 151-172 Korea
dkkim@corelogix.co.kr

Abstract. Recommendation systems help overcome information overload by providing personalized suggestions based on a history of users' preference. Association rule-based filtering method is often used for automatic recommendation systems yet it inherently lacks ability to single out a product to recommend for each individual user. In this paper, we propose an association rule ranking algorithm. In the algorithm, we measure how much a user is relevant to every association rule by comparing attributes of a user with the attributes of others who belong to the same association rule. By providing such an algorithm, it is possible to recommend products with associated rankings, which results in better customer satisfaction. We show through simulations, that the accuracy of association rule-based filtering is improved if we appropriately rank association rules for a given user.

1 Introduction

The internet is increasingly used as one of the major channels for sales and marketing. However, it becomes harder and harder for users to find right products out of millions of products that the Internet offers. A typical e-commerce transaction consists of customers gathering information and trying to find items out of 100,000 even millions of goods and services in stock. The main problem the customers face is how to locate an item which meets his individual need. Recommendation systems help overcome this information overload by providing personalized suggestions based on a history of a user's likes and dislikes[9]. Among other technologies, association rule-based filtering method is the most practical choice for automatic recommendation systems. The traditional association rule-based filtering generates association rules by analyzing transactions in database[1,2,3], and one or more products are recommended

---

* Please address all correspondence to: Prof. Jonghoon Chun, Department of Computer Engineering, College of Engineering, Myongji University, Namdong San 38-2, Yongin, Kyunggi-Do, 449-728, Korea; jchun@mju.ac.kr; Tel:+82-31-330-6441.

D.-K. Baik (Ed.): AsiaSim 2004, LNAI 3398, pp. 306–314, 2005.

by selecting useful association rules. In reality, a user may possibly be associated with more than one association rules, but there has been no effort what so ever to try differentiating association rules that a single user might be encountered into.

In this paper, we propose an association rule ranking algorithm. In the algorithm, we measure how much a user is relevant to every association rule by comparing attributes of a user with the group of users who belong to the same association rule, and a product may or may not be recommended according to the resulting rank of the association rule. We believe that the effectiveness of recommendation systems can be improved if we adequately rank the association rules for a given user. In our work, we propose three different calculation methods to measure the relevancy between a user and an association rule. We show through simulations, that the accuracy of association rule-based filtering is in fact better in accuracy than random or simple confidence-based rule selection methods.

Section 2 presents related work. In section 3, we give an algorithm and show how a recommendation system can be designed with the use of association rules along with our ranking algorithm. In section 4, we evaluate our approach with a simulated experiment. The paper ends with a conclusion.

2 Related Work

Content-based filtering method uses ideas based on theories developed in text retrieval and information filtering area[4,6,7]. In order to recommend an item, the system measures the similarity between the contents of item and information request of a user, and the items are recommended in listed format of their relative similarity to information request of a user.

The major drawback of content-based filtering method is that it is not easy to effectively analyze the contents since extracting appropriate features from item's contents is known to be a hard problem. Also since the main basis for recommendation is the comparison between user profile and the item's contents, considerably large amount of time and effort has to put up to maintain user profiles in order to get reliable recommendations.

User-based collaborative filtering is one of the most successful technology for building recommendation systems to date, and is extensively used in many commercial systems[5]. A user-based recommendation system works as follows. For each user, it uses historical information to identify a neighborhood of people that in the past have exhibited similar behavior and then analyze the neighborhood to identify new pieces of information that will be liked by the user.

Recommending products using information on associative knowledge is one of the most successful technologies being used in e-Commerce nowadays[8]. Association knowledge is derived from one or more transactions in which two different products are purchased together. If the frequency of such transactions is more than a certain threshold value, association knowledge is inferred. Let P be a set of products where m is number of products in the set as $P = \{P_1 \ P_2 \ ... , P_m\}$, and let transaction $T \subseteq P$ be a set of products that are purchased together, an association rule which satisfies $X, Y \subseteq P$ and $X \cap Y = 0$ between product set X and Y represents a strong probability for both X and Y existing in the same transaction. The most referred commercial site that uses the association rule based recommendation is the largest on-line book-store Amazon.com. In Amazon.com, 'Customers Who Bought' function provides recommendation information by applying association rules for books that are purchased together. Association rule based filtering has its advantages. Association rules may be reused; in other word, a reasonable association knowledge extracted with one e-Commerce site, may be applied to another similar site with just a little modification.

3 Recommendation System Using Association Rules

A formal representation of an association rule is as follows.

$$\forall_{X,Y,Z} \{P_1(X,Y) \wedge P_2(X,Y) \ldots \wedge P_n(X,Y) => R(X,Z = z_m)\}$$

P : Database predicates

X : a set of users

Y :if V is continuous - valued attribute $(v_i < V < v_j)$

   if V is distinct - valued attribute or items purchased by user $(V = v_k)$.

Z : a set of items $Z = \{z_1, z_2, \ldots z_l, \cdots, z_m\}$

n : number of predicates used in association rules

P1 ... Pn are conditions and R is defined to be operation.

Metrics used for measuring usefulness of association rules are support and confidence and they are defined as follows.

$$\text{Support} = P(A \cap B) = \frac{\text{number of transactions including both A and B}}{\text{number of total transactions}}$$

$$\text{Confidence} = P(B|A) = \frac{\text{number of transactions including both A and B}}{\text{number of transactions including A}}$$

A customer may be associated with many association rules; therefore we need to have a way to prioritize association rules according to their ranks for a given customer. In other words, we need a process to recommend products according to association rule's rank. For example, as shown in figure 1, customer U1 belongs to association rules Rule1, Rule2, and Rule3, thus we need to figure out the most relevant and reasonable association rule to apply. In doing so, we use a ranking algorithm in which a relevancy is computed between a given user and every association rule found.

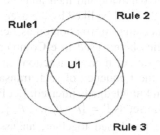

Fig. 1. A customer may be associated with many association rules

We use the matrix W $(m \times n)$ to represent every user's relevancy value for each association rule.

$$W = \begin{bmatrix} W_{11} & W_{12} & \cdots & W_{1(n-1)} & W_{1n} \\ W_{21} & W_{22} & \cdots & W_{2(n-1)} & W_{2n} \\ \vdots & \vdots & W_{ij} & \vdots & \vdots \\ W_{(m-1)1} & W_{(m-1)1} & \cdots & W_{(m-1)(n-1)} & W_{(m-1)n} \\ W_{m1} & W_{m2} & \cdots & W_{m(n-1)} & W_{mn} \end{bmatrix}$$

m: number of customers

n: number of association rules found

wij: Relevancy weight of customer I for association rule j

Consider the following example in figure 2. If we envision an association rule as a cluster which contains a group of customers who all satisfies an association rule, relative relevancies of rule j and k are considered to be different for a given user i.

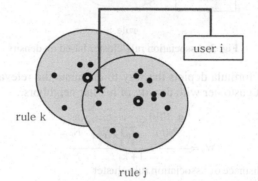
rule k

user i

rule j

Fig. 2. Association rule cluster

It is possible to calculate the relative distance from the center of cluster(association rule) for a given user by considering attributes with range values. The center of cluster may be thought as a representative of customer cluster. In the following, we present three different calculation formulas to determine the relevancy values for a given user.

The following formula 1 is a relevancy calculation method based on relative location of the user within the association rule cluster as shown in figure 2.

$$W_{ij} = \frac{\alpha \left(\dfrac{\vec{u_i} \cdot \vec{r_j}}{\|\vec{u_i}\| \times \|\vec{r_j}\|} \right) \times \dfrac{n_j}{N_j}}{1 + k_{ij}}$$

rule: association rule, if condition then operation   (1)

W_{ij}: relevancy weight(ranking value) of rule $j(r_j)$ for user $i(u_i)$

u_i: user i

r_j: association rule j

k_{ij}: frequency of purchase occurred by previous recommendation

n_j:number of transactions satisfying condition and operation in rule j

N_j:number of transactions satisfying condition in rule j

α:weighting coefficient

Figure 3 represents an alternative way to calculate the relevancy ranking. As shown, if there exist more customers within a certain range, we could consider that the customer is more relavant to that association rule. In figure 3, user i supposed to have higher relevancy value than user j.

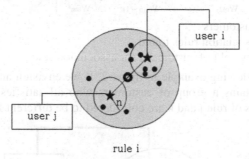

rule i

Fig. 3. Association rule cluster based on density

The following formula depicts the way to calculate the relevancy value considering the location of customer with density of his/her neighbors.

$$W_{ij} = \frac{\left(\dfrac{\alpha \cdot N(l_{ij})}{N(r_i)} \times \dfrac{\vec{u_i} \cdot \vec{r_j}}{\left| \vec{u_i} \right| \times \left| \vec{r_j} \right|} \times \dfrac{n_j}{N_j} \right)}{1 + k_{ij}}$$

l = normalized distnace of association rule cluster

$$l = Dis(r_i) \times \frac{\beta}{100} \tag{2}$$

$N(l_{ij})$: number of customers within distnace of l from user j in rule i

$N(r_i)$: number of customers in rule i

α : weighting coefficient

β : weighting coefficient for l

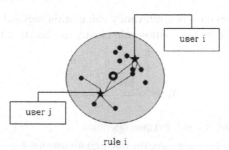

rule i

Fig. 4. Association rule cluster based on average distance

Figure 4 shows another alternative way to calculate the relevancy ranking. As shown, if the average distance from a given customer of n customers is shorter than others, we could consider that the customer is more relevant to that association rule in comparison to other customers. In figure 4, user i supposed to have higher relevancy value than user j because average distance of nearest 4 customers to user i is shorter than that of nearest 4 customers to user j.

The following formula calculates the relevancy value considering the location of customer with average distance of his/her neighbors.

$$W_{ij} = \frac{\alpha \left(\dfrac{m}{\sum\limits_{x=1}^{m} Dis(u_i, u_x)} \times \dfrac{\vec{u_i} \cdot \vec{r_j}}{\|\vec{u_i}\| \times \|\vec{r_j}\|} \right) \times \dfrac{n_j}{N_j}}{1 + k_{ij}}$$

m : normalized numbe r of customers   (3)

$$m = N(ri) \times \frac{\beta}{100}$$

$N(r_i)$: number of customers in rule i

$Dis(u_i, u_x)$: distance between u_i and u_x

α : weighting coefficient

β : weighting coefficient for calculation of m

The following is the recommendation algorithm.

Input

WList : relevancy value matrix between customer and association rule

WList = (W11, W12, ⋯, W1m,

        W21, W22, ⋯, W2m,

        ⋮

        Wn1, Wn2, ⋯, Wnm)

RList : List of products recommended

Default : List of default product recommendation

When customer (U) x login

If !(WList $_x$.exist)

        WList $_x$ = Calculate(FindRule(x))

End If

WList $_x$ = Desc_Sort(W $_{x1}$, W $_{x2}$, W $_{x3}$, ⋯ , W $_{xm}$)

For p = 0 to Count(WLis   t $_x$) - 1

If RList contains WList[p].product

        continue

    Else

        RList.add(WList $_x$[p].product)

```
                    If  Count(RList)  ==  N
                            break;
                  End  If
          End  If
    End For


    If Count(RLis  t) < N
          For  q = 0 to  N - Count(RList) - 1
                    RList.add(  Default[q]  )
          End  For
    End  If
    Recommend    (RList)
```

Algorithm 1. Recommendation

4 Simulation

In this section, we show the effectiveness of the relevancy ranking algorithm that we proposed in the last section. We use the set of transaction data used in [9] for our simulation. The data set contains 3,465 transactions of 73 distinct customers. Over all data have 220 different attributes, among them, there are 43 attributes with range values which are used as the fundamental basis for our distance/density measure of the customer. Product set includes 1008 different products which are classified into three different levels of classifications, namely high, medium, and low. In the high level, there are 4 classes, in the medium level, there are 40 classes, and in the lowest level there are 253 classes. We assume that product data are purchased and recommended using medium level classification. We use open data mining library Weka to extract association rules. We use 10-fold cross-validation method to verify the effectiveness of our recommendation, in which the 90% of simulation data are used as training set and the rest 10% of data are assumed to be future purchases which are used for testing the effectiveness.

Extracted association rules are used to recommend only top N products ranked, first by random selection, secondly by using only confidence values, and thirdly by using our relevancy values computed. We use the following formula to measure the effectiveness of recommendation, in doing so we compared the list of products recommended with the list of products that has been actually purchased.

$$Accuracy = \frac{number\ of\ products\ actually\ purchased(n)}{number\ of\ products\ recommend(N)}$$

As you can see in figure 5, our method outperforms the other two cases. Especially in comparison to the case when confidence-based recommendation was performed, about 5 to 10% of accuracy improvement is accomplished throughout.

Figure 6 shows the performance of our three different relevancy calculation methods, and differences among them are negligible. However, in comparison to confidence- based recommendation, the difference of accuracy gets larger as the number of recommended products decreases; which is good as far as practicability of our approach is concerned considering most commercial web sites only recommending small number of products in general.

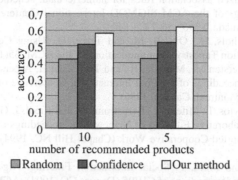

Fig. 5. Recommendation based on random selection, confidence vs. our method

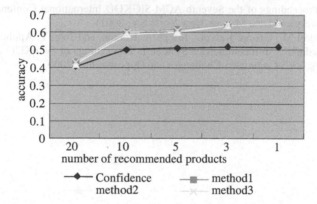

Fig. 6. Comparison among three different relevancy calculation methods

5   Conclusion

In this paper, we have proposed a new personalized recommendation method based on the use of association knowledge. In our method, we have proposed an association rule ranking algorithm to be utilized for personalized ranking of association rules for a given user, and through simulations, we have successfully shown the effectiveness of our method. In order for our approach to be applicable to real-life situations, sound and complete comparison to contents-based filtering, user-based collaborative filtering, and other similar methods must be carried out in depth.

References

1. Agrawal, R., Imielinski, T., and Swami A. Mining association rules between sets of items in large databases. In Proceedings of the ACM SIGMOD Conference on Management of Data(Washington D.C., 1993), 207-216

2. Agrawal, R., and Srikant R. Fast Algorithm for Mining Association Rules. In Proceedings of the 20th VLDB Conference (Santiago, Chile, 1994)
3. Fukuda, T., Morimoto Y., Morishita, S., and Tokuyama T. Data mining using two-dimensional optimized association rules for numeric data: scheme, algorithms, visualization. In Proceedings of the ACM SIGMOD International Conference on Management of Data (Montreal, Canada, 1996), 13-23.
4. Goldberg, D., Nichols, D., Oki, B. M. and Terry, D., Using Collaborative Filtering to Weave an Information Tapestry, Communications of the ACM, 35(12): 61-70, 1992
5. Holsheimer, M., Kersten M., Mannila, H., and Toivonent H. A perspective on database and data mining. In Proceedings of the First International Conference on Knowledge Discovery and Data Mining (Montreal, Canada, 1996).
6. Resnick P., Neophytos, I., Miteth, S., Bergstrom, P., and Riedl, J. GroupLens: An Open Architecture for Collaborative Filtering of Netnews. In Proceedings of CSCW94: Conference on Computer Supported Cooperative Work (Chapel Hill NC, 1994), Addison-Wesley, 175-186.
7. Shardanand, U., and Maes, P. Social Information Filtering: Algorithms for Automating 'Word of Mouth'. In Proceedings of CHI95 (Denver CO, 1004), ACM Press, 210-217
8. Zheng, Z., Kohavi, R. and Mason, L. Real World Performance of Association Rule Algorithms. In Proceedings of the Seventh ACM-SIGKDD International Conference on Knowledge Discovery and Data Mining (New York, NY, 2001)
9. Sarwar, Badrul M., George Karypis, Joseph A. Kontan, and John T. Application of dimensionality Reduction in recommender System. In Proceedings of WebKDD 2000 Workshop Web Mining for E-Commerce-Challenges and Opportunities. 2000

Intuitive Control of Dynamic Simulation
Using Improved Implicit Constraint Enforcement

Min Hong[1], Samuel Welch[2], and Min-Hyung Choi[3]

[1] Bioinformatics, University of Colorado Health Sciences Center,
4200 E. 9th Avenue Campus Box C-245, Denver, CO 80262, USA
Min.Hong@UCHSC.edu
[2] Department of Mechanical Engineering, University of Colorado at Denver,
Campus Box 112, PO Box 173364, Denver, CO 80217, USA
sam@carbon.cudenver.edu
[3] Department of Computer Science and Engineering, University of Colorado at Denver,
Campus Box 109, PO Box 173364, Denver, CO 80217, USA
minchoi@acm.org

Abstract. Geometric constraints are imperative components of many dynamic
simulation systems to effectively control the behavior of simulated objects. In
this paper we present an improved first-order implicit constraint enforcement
scheme to achieve improved accuracy without significant computational burden. Our improved implicit constraint enforcement technique is seamlessly integrated into a dynamic simulation system to achieve desirable motions during
the simulation using constraint forces and doesn't require parameter tweaking
for numerical stabilization. Our experimental results show improved accuracy
in maintaining constraints in comparison with our previous first-order implicit
constraint method and the notable explicit Baumgarte method. The improved
accuracy in constraint enforcement contributes to the effective and intuitive
motion control in dynamic simulations. To demonstrate the wide applicability,
the proposed constraint scheme is successfully applied to the prevention of excessive elongation of cloth springs, the realistic motion of cloth under arduous
collision conditions, and the modeling of a joint for a rigid body robot arm.

1 Introduction

One of the primary goals of a dynamic simulation is to create physically realistic
motions and the intuitive control of a dynamic simulation plays a key role in accom-
plishing the intended realism. Maintaining geometric constraints with utmost accuracy can be an effective tool to control the behavior of simulated objects in many
applications including physically based simulation, character animation, and robotics.
Recently we have established the first-order implicit constraint enforcement to effectively enforce geometric constraints [2]. This paper presents the improved first-order
implicit constraint enforcement that extends our previous work to efficiently and
robustly administrate the motion of objects. The enhanced accuracy in constraint
enforcement enables us to improve the motion control in many applications.

Hard springs and penalty forces [13, 14] are often used as an alternative to geometric constraints but they may create a stiff numerical system and consequently the
simulator may have to use a small integration time step to avoid numerical instability.
Furthermore, the accuracy of the intended behavior control is not warranted since the

D.-K. Baik (Ed.): AsiaSim 2004, LNAI 3398, pp. 315–323, 2005.
© Springer-Verlag Berlin Heidelberg 2005

choice of coefficients can greatly alter the associated object behavior. Constrained dynamics using Lagrange Multipliers are well studied and widely used to enforce geometric constraints. Baumgarte constraint stabilization technique has been successfully used for many applications [3, 4, 7, 8, 9, 15] to reduce the constraint errors and stabilize the constrained dynamic system. Barzel and Barr [8] introduced dynamic constraints to control behaviors of rigid bodies using geometric constraints. Platt and Barr [9] presented reaction constraints and augmented Lagrangian constraints to flexible models. Witkin and Welch [7] applied the explicit Baumgarte method into non-rigid objects to connect them and control the trajectory of motion. Baraff [4] used linear time Lagrange multiplier method to solve constraint system. Ascher and Chin [11] explicated the inherent drawbacks of Baumgarte stabilization method and introduced constraint stabilization techniques. Cline and Pai [6] introduced post-stabilization approach for rigid body simulation which uses stabilization step to compensate the error of integration for each time step. However, it requires additional computational load to reinstate the accuracy and it may cause the loss of natural physical motions of object under complicated situations since the constraint drift reduction is performed independently from the conforming dynamic motions.

The rest of this paper is organized as follows: the brief overview of constraint enforcement is elaborated in section 2. Section 3 elucidates a description of improved first-order implicit constraint enforcement scheme. In section 4 we provide the explanation for our simulated examples and experimental results of our method.

2 Constraint Enforcement

Two main concerns of constraint enforcement to create tractable and plausible motions are physical realism and the controllability of simulation. Our primary contribution of this paper is the enhanced accuracy in constraint enforcement using the improved first-order implicit constraint management scheme which can be applied to any dynamic simulations – character animation, cloth simulation, rigid body simulation, and so on. Without proper control terms or constraints, the dynamic simulations may generate abnormal or undesirable motions. Under this circumstance, it is considered that an accurate and robust constraint management plays an important role in controlling the motion of dynamic simulations. This paper describes a method to achieve exact desired behaviors without any discordant motions of object by employing accurate and effective geometric constraints.

To achieve certain behaviors in a physically-based simulation, the desirable geometric restrictions should be intuitively reflected into the dynamic system. Most constraint-based dynamic systems [1, 4, 6, 7, 8] convert geometric restrictions into constraint forces to maintain the restrictions and they are blended with other external forces. Any relations or restrictions which can be represented as algebraic equations can be applied as constraints in the dynamic simulation system. For instance, animators can enforce various types of constraints to intuitively control desirable initial environments or modify the motions of objects during simulation: distance constraints (node-node or node-arbitrary point), angle constraints, parallel constraints, and inequality constraints [1].

The explicit Baumgarte method [3, 4, 7, 8] uses a classic second order system, similar to a spring and damper, to enforce and stabilize the constraints. This method helps in enforcing constraints within the framework of Lagrange multipliers method, but it requires ad-hoc constant values to effectively stabilize the system. Finding proper problem specific coefficients is not always easy. In order to overcome this problem, we proposed the implicit constraint enforcement approach [2]. It provides stable and effective constraint management with same asymptotic computational cost of explicit Baumgarte method. Moreover, the explicit Baumgarte method is conditionally stable while the implicit method is always stable independent from the step size. This paper is extension of our previous work to improve the accuracy of constraint enforcement. The computational solution is detailed in the implicit constraint using improved first-order scheme section.

3 Implicit Constraint Using Improved First-Order Scheme

One of the most popular methods for integrating a geometric constraint into a dynamic system is to use Lagrange multipliers and constraint forces. The constraint-based formulation using Lagrange multipliers results in a mixed system of ordinary differential equations and algebraic expressions. We write this system of equations using 3n generalized coordinates, q, where n is the number of discrete masses, and the generalized coordinates are simply the Cartesian coordinates of the discrete masses.

$$q = [x_1 y_1 z_1 x_2 y_2 z_2 \cdots x_n y_n z_n]^T$$

Let $\Phi(q(t),t)$ be the constraint vector made up of m components each representing an algebraic constraint. The constraint vector is represented mathematically as

$$\Phi(q(t),t) = [\Phi^1(q(t),t) \Phi^2(q(t),t) \cdots \Phi^m(q(t),t)]$$

where the Φ^i are the individual scalar algebraic constraint equations. We write the system of equations

$$M\ddot{q} + \Phi_q^T \lambda = F^A \tag{1}$$

$$\Phi(q(t),t) = 0$$

where F^A are applied, gravitational and spring forces acting on the discrete masses, M is a 3nx3n diagonal matrix containing discrete nodal masses, λ is a mx1 vector containing the Lagrange multipliers and Φ_q is the mx3n Jacobian matrix, and subscript q indicates partial differentiation with respect to q.

We previously proposed an implicit first-order constraint enforcement scheme to effectively control cloth behavior [2]. That implicit method was first order and was implemented the following way. The equation of motion (equation (1)) along with the kinematics relationship between q and $\dot{q}$ are discretized as

$$\dot{q}(t + \Delta t) = \dot{q}(t) - \Delta t M^{-1} \Phi_q^T(q(t),t)\lambda + \Delta t M^{-1} F^A(q(t),t) \tag{2}$$

$$q(t + \Delta t) = q(t) + \Delta t \dot{q}(t + \Delta t) \tag{3}$$

The constraint equations written at new time thus they are treated implicitly

$$\Phi(q(t + \Delta t), t + \Delta t) = 0 \tag{4}$$

Equation (4) is now approximated using a truncated, first-order Taylor series

$$\Phi(q(t),t)+\Phi_q(q(t),t)(q(t+\Delta t)-q(t))+\Phi_t(q(t),t)\Delta t=0$$

Note that the subscripts q and t indicate partial differentiation with respect to q and t, respectively. Eliminating $q(t+\Delta t)$ results in the following linear system with λ the remaining unknown.

$$\Phi_q(q(t),t)M^{-1}\Phi_q^T(q(t),t)\lambda=\frac{1}{\Delta t^2}\Phi(q(t),t)+\frac{1}{\Delta t}\Phi_t(q(t),t)$$

$$+\Phi_q(q(t),t)\left(\frac{1}{\Delta t}\dot{q}(t)+M^{-1}F^A(q(t),t)\right)$$

Note that the coefficient matrix for this implicit method is the same as the coefficient matrix for the Baumgarte method. This system is solved for the Lagrange multipliers then equations (2) and (3) are used to update the generalized coordinates and velocities.

In what follows we describe a scheme that while formally first-order, contains a correction that improves the accuracy of the scheme without significant computational cost. This scheme may be described as using a second order mid-point rule for the momentum equation while solving the constraint equations implicitly using a first order linearization. The first step calculates all variables including the Lagrange multiplier at the half-time step.

$$\dot{q}(t+\Delta t/2)=\dot{q}(t)-\frac{\Delta t}{2}\left(M^{-1}\Phi_q^T(q(t),t)\lambda-M^{-1}F^A(q(t),t)\right) \tag{5}$$

$$q(t+\Delta t/2)=q(t)+\frac{\Delta t}{2}\dot{q}(t+\Delta t/2) \tag{6}$$

$$\Phi(q(t),t)+\Phi_q(q(t),t)(q(t+\Delta t/2)-q(t))+\Phi_t(q(t),t)\frac{\Delta t}{2}=0 \tag{7}$$

Again note that the subscript t indicates partial differentiation with respect to t. $q(t+\Delta t/2)$ and $\dot{q}(t+\Delta t/2)$ are eliminated from equations (5) through (7) resulting in the linear system that must be solved to obtain the half-step Lagrange multipliers:

$$\Phi_q(q(t),t)M^{-1}\Phi_q^T(q(t),t)\lambda=\frac{4}{\Delta t^2}\Phi(q(t),t)$$

$$+\Phi_q(q(t),t)\left(\frac{2}{\Delta t}\dot{q}(t)+M^{-1}F^A(q(t),t)\right)$$

Equations (5) and (6) are then used to obtain $q(t+\Delta t/2)$ and $\dot{q}(t+\Delta t/2)$. The second step uses the half-step Lagrange multiplier and evaluates the Jacobian using $q(t+\Delta t/2)$ to form the mid-point step:

$$\dot{q}(t+\Delta t)=\dot{q}(t)-\Delta tM^{-1}\Phi_q^T(q(t+\Delta t/2),t+\Delta t/2)\lambda+\Delta tM^{-1}F^A(q(t),t)$$

$$q(t+\Delta t)=q(t)+\Delta t\dot{q}(t+\Delta t/2)$$

This scheme is formally first order as the linearized constraint equation (equation (7)) is solved only to first order. Despite this theoretical first order accuracy the scheme exhibits improved accuracy while requiring solution of the same linear system as the first order scheme described above.

To integrate non-constrained portion of system, the second-order explicit Adams-Bashforth method [10] is used to predict the next time status of the system. The explicit Adams-Bashforth method for solving differential equations is second order and provides better stability than first-order explicit Euler method. The second-order Adams-Bashforth method has the general form:

$$\eta_{i+1} = \eta_i + \frac{1}{2}h(3f(x_i,\eta_i) - f(x_{i-1},\eta_{i-1}))$$

This equation is integrated into ODE (Ordinary Differential Equation) system as a force computation to calculate next status of velocity and then we can estimate the next status of position.

$$\dot{q}(t + \Delta t) = \dot{q}(t) + \Delta t M^{-1}(3F^A(q(t),t) - F^A(q(t - \Delta t),t))$$

$$q(t + \Delta t) = q(t) + \Delta t \dot{q}(t + \Delta t)$$

4 Applications and Experimental Results

One good example of the constraint enforcement for controlling motion is "super-elastic" effect [5] in cloth simulation. Most clothes have their own properties for elongation. Stiff springs can be a candidate solution for this problem, but there still exists numerical instability for linear or non-linear springs when we apply a large integration time step. Provot [5] applied the distance reduction using dynamic inverse constraints for over-elongated springs. However, this approach ignores physical consequences of moving nodes instantly and it may cause loss of energy or original dynamic behavior of motions. In addition, the ordering of node displacement is critical when the internal meshing structures are complex and collision is involved. To prevent excessive elongation of cloth, our approach replaces the springs with implicit constraints when the elongations of springs are over a certain threshold. In our method, adaptive constraint activation and deactivation techniques are applied to reduce computational burden of solving a larger linear system. For instance, only when springs are over elongated, these springs are replaced with implicit constraints. When the constraint force is working to resist the shrinkage, our system deactivates these constraints into springs.

We applied our new technique to a cloth simulation to prevent over-elongation of cloth springs. A piece of cloth is initially in a horizontal position and falling down due to gravity. A relatively heavy ball is moving toward the middle of the falling cloth. Figure 1 illustrates the different motion of cloth without (a) and with (b) constraint enforcement. In figure 1 (a), some springs are excessively stretched and they cause unstable motion of cloth simulation. However, figure 1 (b) shows plausible and stable motion of cloth using implicit constraint enforcement. Although a moving ball hits the middle portion of falling cloth which abruptly creates severe collision forces to specific parts of cloth, an animation [12] using implicit constraint enforcement

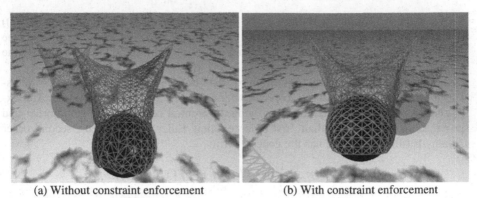

| (a) Without constraint enforcement | (b) With constraint enforcement |

Fig. 1. Mesh view of simulated motion of a cloth patch on the collision between a moving ball and the freely falling cloth under gravity. The cloth is hung by two fixed point-node constraints and the implicit constraint enforcement is used to prevent the excessive elongation of cloth springs.

shows realistic behavior of the cloth and an animation in figure 1 (a) using only springs includes unstable and odd motions of cloth.

To test the accuracy of the constraint enforcement and measure the total amount of constraint drift during simulation, we have tested our improved first-order implicit constraint enforcement system with a robotic link connected with revolute joints. Figure 2 shows 5-link robot arm falling down under gravity. Initially the robot links are in horizontal position and one end of it is fixed. We recorded the constraint error over each time step which is obtained by accumulated absolute value of difference between each original link distance and current link distance.

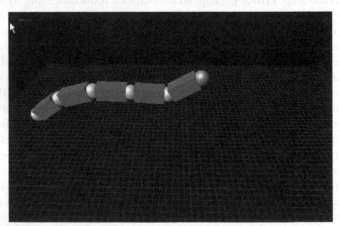

Fig. 2. A simulation of a freely falling 5-link robot arm under gravity. Each Link is constrained using improved implicit constraint enforcement and a red sphere illustrates fixed point-node constraint to hang the robot links.

Figure 3 shows an accumulated experimental constraint error data from the simulation of a falling 5-link robot arm shown in figure 2. We compared implicit constraint enforcement using first-order scheme and the improved first-order scheme and tested these systems using two different categories: integration time step size (0.01 and 0.1 seconds) and mass (1 and 50 unit). Although the accumulated constraint error grows for bigger masses and bigger integration time step sizes, the implicit constraint enforcement using the improved first-order scheme bestows approximately two times higher accuracy. In figure 4, we measure the accumulated constraint error under same condition of figure 2 using explicit Baumgarte method. This graph shows that the implicit constraint enforcement reduces constraint errors effecting comparison with the overall errors of Baumgarte stabilization. The choice of alpha and beta feedback terms has a dominant effect on how well this Baumgarte constraint enforcement performs.

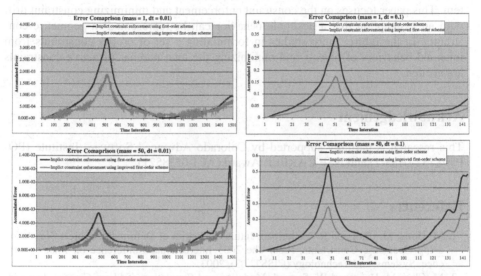

Fig. 3. Accumulated error of implicit constraint enforcement using improved first-order scheme and first order scheme. Different masses and integration time steps are applied to these systems for comparison.

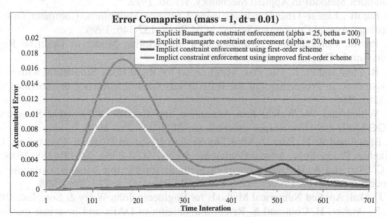

Fig. 4. Accumulated constraint error comparison for explicit and implicit constraint enforcement.

5 Conclusion

This paper describes a new improved first-order implicit constraint enforcement scheme that enhances the accuracy in constraint enforcement. It doesn't require any ad-hoc constants to stabilize the system. While the additional computation cost is relatively minimal, the accuracy of the constraint enforcement is significantly enhanced. This technique can be used to control any dynamic simulation environment using various types of geometric constraints such as angle or distance restrictions. This paper shows how well the control of the over elongation problem of cloth is handled using the implicit constraint enforcement. Adaptive constraint activation and deactivation approach also reduces the computational expense of the numerical system. This paper focuses on the constraint enforcement and minimizing constraint drift with an assumption that a desirable behavior can be represented in geometric constraints. In the future, more studies in formulating a set of proper constraints from intended behaviors of objects and minimizing the total number of active constraints are needed.

Acknowledgement

This research is partially supported by Colorado Advanced Software Institute (PO-P308432-0183-7) and NSF CAREER Award (ACI-0238521).

References

1. M. Hong, M. Choi and R. Yelluripati. Intuitive Control of Deformable Object Simulation using Geometric Constraints, Proc. The 2003 International Conference on Imaging Science, Systems, and Technology (CISST' 03), 2003.
2. M. Choi, M. Hong and W. Samuel. Modeling and Simulation of Sharp Creases. Proceedings of the SIGGRAPH 2004 Sketches, 2004.
3. J. Baumgart. Stabilization of Constraints and Integrals of Motion in Dynamical Systems. Computer Methods in Applied Mechanics. 1:1 36. 1972.
4. D. Baraff . Linear-Time Dynamics using Lagrange Multipliers, Computer Graphics Proceedings. Annual Conference Series. ACM Press, 137-146, 1996.
5. X. Provot. Deformation Constraints in a Mass-Spring Model to Describe Rigid Cloth Behavior. In Graphics Interface. 147-154. 1995.
6. M. B. Cline and D. K. Pai. Post-Stabilization for Rigid Body Simulation with Contact and Constraints. In Proceedings of the IEEE International Conference on Robotics and Automation. 2003.
7. A. Witkin and W. Welch. Fast Animation and Control of Nonrigid Structures. ACM SIGGRAPH 1990 Conference Proceedings. 24. 4. 243-252. 1990.
8. R. Barzel and A. H. Barr. A Modeling System Based on Dynamic Constraints. Computer Graphics Proceedings. Annual Conference Series. ACM Press. 22. 179-188. 1988.
9. J. C. Platt and A. H. Barr. Constraint Methods for Flexible Models. Computer Graphics Proceedings. Annual Conference Series. ACM Press. 22. 4. 279-288. 1988.
10. T. J. Akai. Applied Numerical Methods for Engineers. John Wiley & Sons Inc. 1993.
11. U. R. Ascher, H. Chin and S. Reich. Stabilzation of DAEs and invariant manifolds. Numerische Methematik, 67(2), 131-149, 1994.

12. Computer Graphics Lab. Department of Computer Science and Engineering. University of Colorado at Denver. http://graphics.cudenver.edu

13. A. Witkin, K. Fleischer and A. Barr. Energy Constraints on Parameterized Models. ACM SIGGRAPH 1987 Conference Proceedings. 225-232. 1987.

14. D. Terzopoulos, J. Platt, A. Barr and K. Fleischer. Elastically Deformable Models. ACM SIGGRAPH 1987 Conference Proceedings. 205-214. 1987.

15. D. Metaxas and D. Terzopoulos. Dynamic Deformation of Solid Primitives with Constraints. Computer Graphics. 309-312. 1992.

Air Consumption
of Pneumatic Servo Table System

Takashi Miyajima[1], Kenji Kawashima[2], Toshinori Fujita[3],
Kazutoshi Sakaki[4], and Toshiharu Kagawa[2]

[1] Tokyo Institute of Technology, 4259 Nagatsuta-cho,
Midori-ku, Yokohama 226-8503, Japan
[2] Precision & Intelligence Laboratory,
Tokyo Institute of Technology, Japan
[3] Tokyo Denki University, 2-2, Kanda-Nishiki-cho,
Chiyoda-ku, Tokyo, 101-8457, Japan
[4] Sumitomo Heavy Industries, Ltd. 9-11, Kita-Shinagawa 5-chome,
Shinagawa-ku, Tokyo, 141-8486, Japan

Abstract. We have developed a pneumatic servo table system, which
is a precise positioning system using air power. In this paper, the energy
consumption of the pneumatic servo table system is investigated. To clarify the energy consumption, the air consumption must be clarified. We
analyzed the air consumption of the system and became clear theoret-
ically that the air consumption could be separated to the consumption
to move the table and that to accelerate. The certainty of the analysis
is confirmed with the experimental result using quick response laminar
flow meter. Furthermore the best reference trajectory is derived on the
point of energy saving.

1   Introduction

From a viewpoint of energy saving, to know energy consumption of the system and to control it is very important. On pneumatic instruments, the energy
consumption is dominated by the air consumption. Therefore, to clarify air consumption of the system is critical problem.

Although the analysis of air consumption is important, it is insufficient up
to now. One reason of it is the difficulty of the sensing of the air consumption.
The air flow to drive system is varied dynamically and flow condition becomes
unsteady state, but the measurement of unsteady state flow is very difficult. Furthermore, the nonlinearity of the pneumatic system makes theoretical analysis
difficult [1].

We clarified the air consumption of the pneumatic servo table system [2].
The pneumatic servo table system is one of a precise positioning system using
air power. This system has many great advantages as compared with the electro-
motion system, such as high mass power ratio, low heat generation and low
magnetic field generation.

In this paper we constructed linearized model of the system and analyzed
the air consumption. As a result of the system analysis, the air consumption of

D.-K. Baik (Ed.): AsiaSim 2004, LNAI 3398, pp. 324–333, 2005.

the system is dominated by the moving distance and the maximum acceleration. Then, we reconstructed simulation model considered non-linearity of the system to more acculate analysis.

Furthermore, we measured the air consumption of the system by experiment to confirm that it could be separated to the consumption to move the table and that to accelerate.

Nomenclature

a	acceleration	$[\text{m/s}^2]$
A_1	constant of system closed loop transfer function	$[\text{s}^{-1}]$
A_2	constant of system closed loop transfer function	$[\text{s}^{-2}]$
A_3	constant of system closed loop transfer function	$[\text{s}^{-3}]$
A_t	pressurised area	$[\text{m}^2]$
b	critical pressure ratio	[-]
d	diameter of laminar flow meter	$[\text{m}]$
F_t	force of servo table	$[\text{N}]$
G	mass flow rate	$[\text{kg/s}]$
j	jerk	$[\text{m/s}^3]$
K_n	gain for flow rate	$[\text{m/(s·V)}]$
l	length of laminar flow meter	$[\text{m}]$
M_t	servo table slider mass	$[\text{kg}]$
M_{air}	mass of air consumption	$[\text{kg}]$
P	pressure	$[\text{Pa}]$

R	gas constant	$[\text{J/(kg·K)}]$
s	Laplace operator	$[\text{s}^{-1}]$
t	time	$[\text{s}]$
v	velocity	$[\text{m/s}]$
V	volume of pressure room	$[\text{m}^3]$
x	servo table slider displacement	$[\text{m}]$
δ	parameter of non-choke flow	[-]
Δp	differential pressure of laminar flow meter	$[\text{Pa}]$
ζ	damping ratio	[-]
θ	temperature	$[\text{K}]$
θ_a	room temperature	$[\text{K}]$
μ	viscosity	$[\text{Pa·s}]$
ρ	density of air	$[\text{kg/m}^3]$
ω_n	natural frequency	$[\text{rad/s}]$

suffix

0	equilibrium point
R	right pressure room
L	left pressure room

2   Pneumatic Servo Table System

2.1   Constructs of Pneumatic Servo Table System

Pneumatic Servo Table System. Figure 1 shows the schematic diagram of the pneumatic servo table system. This system is constructed from a pneumatic actuator and two servo valves. A quick response laminar flow meter is attached on the upper stream of the servo valve to measure the air consumption of the system. This system is controlled by a computer which mounted real-time operating system.

Pneumatic Actuator. The pneumatic actuator is shown in Fig. 2. The pneumatic actuator is constructed from a slider and a guide. The guide of the actuator is fixed to the wall and the slider moves along with the guide. The slider position of the pneumatic servo table system is controlled precisely using air power. Internal of the slider is separated to two pressure rooms by a wall. The slider is driven by the differential pressure between two rooms. The friction force of

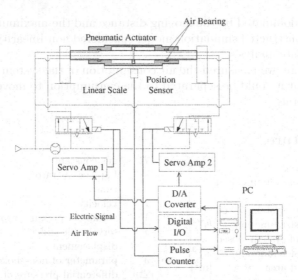

Fig. 1. The Schematic Diagram of the Pneumatic Servo Table System.

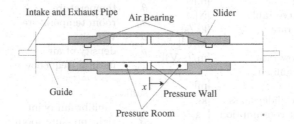

Fig. 2. Schematic Diagram of the Pneumatic Actuator.

the actuator is one of the most critical problems [3]. Therefore, the friction force of the actuator is extremely reduced by an air bearing which attached on the sliding surface.

Quick Response Laminar Flow Meter. To measure the air consumption of the servo table system, we mounted a quick response laminar flow meter [4] on the upper stream of servo valves.

Figure 3 shows the sectional view of the laminar flow meter. The laminar flow meter is constructed from a flow channel filled with narrow tubes and a quick response differential pressure sensor. The gas flow passing through narrow tubes becomes laminar flow because of the viscosity. The relation between the mass flow rate and the differential pressure is given by,

$$G = \frac{\pi d^4}{128 \mu l} \Delta p \tag{1}$$

This flow meter has the dynamic characteristics of 30[Hz]. The experimental result of the frequency response of 30[Hz] is shown in Fig. 4.

Fig. 3. Sectional View of Laminar Flow Meter.

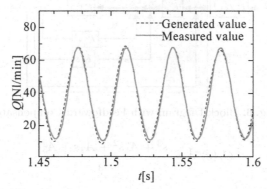

Fig. 4. Dynamic Characteristics of Laminar Flow Meter (f=30Hz).

2.2   Model of System

The friction force on the sliding surface is extremely low because of the air bearing, so the friction force is negligible. The air state condition is the isothermal and the dynamic characteristics of the servo valve is negligible for simplification of the model. Then, the system is linearized on the neighbourhood of equilibrium point as follows:

$$P_n(s) = \frac{K_n \omega_n^2}{s(s^2 + \omega_n^2)} \tag{2}$$

2.3   Control Theorem

PDD² Controller. We applied PDD² control theorem to this system. Figure 5 shows the block diagram of the system.

Then the closed loop transfer function $G_n(s)$ is described as follows:

$$G_c(s) = \frac{A_3}{s^3 + A_1 s^2 + A_2 s + A_3} \tag{3}$$

Feedfoward Compensation. To improve the system dynamics, the inverse model of closed transfer function eq. (4) is attached as a feedfoward compensator. [5] The block diagram of the feedfoward compensator is shown in Fig. 6.

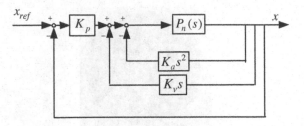

Fig. 5. Block Diagram.

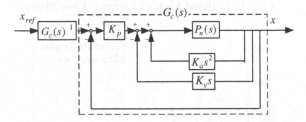

Fig. 6. Block Diagram with Feedfoward Compensator.

$$G_c(s)^{-1} = \frac{s^3 + A_1 s^2 + A_2 s + A_3}{A_3} \tag{4}$$

Reference Trajectory. To realize the feedfoward compensation, a 3-times differentiable reference trajectory is required. Therefore, we designed a reference trajectory of a third order curve.

3 Simulation Model

3.1 Linear Model

On the sec. 2.2, we showed the linearized model of the pneumatic servo table system. The system sensing element is a position sensor. So, we must estimate the velocity and the acceleration to realize PDD^2 control. These variables are estimated using Kalman filter.

The simulation model, which based on the linearized model, is constructed on the simulation software (MATLAB simulink).

3.2 Air Consumption

The equation of motion of the pneumatic servo table system is given by,

$$F_t = M_t \ddot{x} \tag{5}$$

Here, F_t is a force acting on the pressure wall of the pneumatic actuator. F_t is given by,

$$F_t = (P_R - P_L)A_t \tag{6}$$

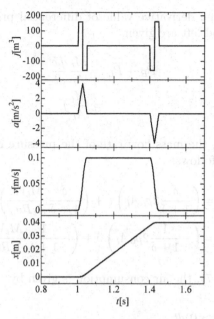

Fig. 7. Reference Trajectory.

From eq. (5) and (6) the acceleration of the slider is obtained as,

$$\ddot{x} = \frac{A_t}{M_t}(P_R - P_L) \tag{7}$$

Under the isothermal condition, the time derivation of the ideal gas equation of both pressure rooms are given by,

$$P_{0R}A_t\dot{x} + V_{0R}\dot{P}_R = R\theta_{0R}G_R \tag{8}$$
$$-P_{0L}A_t\dot{x} + V_{0L}\dot{P}_L = R\theta_{0L}G_L \tag{9}$$

On the neighbourhood of equilibrium point, the pressure and the volume is approximated as follows:

$$P_{0R} = P_{0L} = P_0 \tag{10}$$
$$V_{0R} = V_{0L} = V_0 \tag{11}$$
$$\theta_{0R} = \theta_{0L} = \theta_a \tag{12}$$

From eq. (8) - (12),

$$2P_0A_t\dot{x} + V_0(\dot{P}_R - \dot{P}_L) = R\theta_a(G_R - G_L) \tag{13}$$

On the equilibrium point, the flow charged to the pressure room on the right side becomes choke condition, and the flow discharged from that of the left side becomes non-choke. So, the mass flow rate is given by,

$$G_R = -\delta G_L \tag{14}$$

From eq. (7), the time derivative value of differential pressure of the pressure room on the right and left are given,

$$\dot{P}_R - \dot{P}_L = \frac{M_t}{A_t}\frac{d\ddot{x}}{dt} \tag{15}$$

$$= \frac{M_t}{A_t}j$$

From eq. (13) - (15), the mass flow rate of the pressure room on the right and left are represent as follows:

$$G_R = \left(\frac{\delta}{1+\delta}2\rho_0 A_t\right)v + \left(\frac{\delta}{1+\delta}\frac{M_t V_0}{A_t R\theta_a}\right)j \tag{16}$$

$$G_L = \left(-\frac{1}{1+\delta}2\rho_0 A_t\right)v + \left(-\frac{1}{1+\delta}\frac{M_t V_0}{A_t R\theta_a}\right)j \tag{17}$$

From eq. (16) and (17), the air consumption is given by,

$$M_{airR} = \int G_R(> 0)dt \tag{18}$$

$$= \left(\frac{\delta}{1+\delta}2\rho_0 A_t\right)\int v(> 0)dt + \left(\frac{\delta}{1+\delta}\frac{M_t V_0}{A_t R\theta_a}\right)\int j(> 0)dt$$

$$M_{airL} = \int G_L(> 0)dt \tag{19}$$

$$= \left(-\frac{1}{1+\delta}2\rho_0 A_t\right)\int v(> 0)dt + \left(-\frac{1}{1+\delta}\frac{M_t V_0}{A_t R\theta_a}\right)\int j(> 0)dt$$

It is clear from eq. (18) and (19) that the first term represents the air consumption to expand the volume of the room and the second term represents the air consumption to accelerate it. [6]

The total air consumption M_{air} is the sum of the M_{airR} and M_{airL},

$$M_{air} = M_{airR} + M_{airL} \tag{20}$$

3.3   Consideration of System Nonlinearity

We derived and constructed the linearized model of the system, but the real system has various nonlinearity.

We reconstructed the simulation model for more accurate simulation. The reconstructed model considered the temperature change, the dynamic characteristics of the servo valves, the flow choking condition and the effect of the non-equilibrium point.

The air consumption of the system is also calculated based on the nonlinear system.

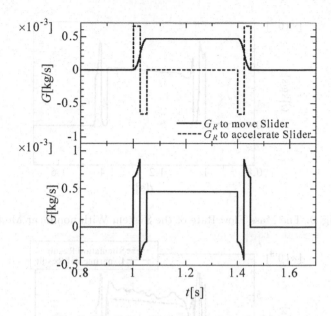

Fig. 8. The Mass Flow Rate of the System.

4   Simulation Result

4.1   Result of Linear Model

The simulation result based on the linear model is shown in Fig. 8.

From Fig. 8 , the mass flow rate of system is represented by the velocity of slider and the jerk. The discharging mass flow rate is the sum of to move slider and to brake slider, then the optimal trajectory can derive in the viewpoint of energy saving.

4.2   Result of Nonlinear Model

The simulation result based on the nonlinear model is shown in Fig. 9.

From Fig. 9, it is clear that the curve of the mass flow rate becomes distorted. On the linear model, the mass flow rate to move slider by uniform velocity becomes constant. But on the nonlinear model, the mass flow rate becomes slightly increase because of the volume change of the pressure room.

5   Experimental Result

We measured the air consumption of the system. The experimental result and the simulation result is shown in Fig. 10. It is confirmed that the simulation result is showed good agreement with the experimental result.

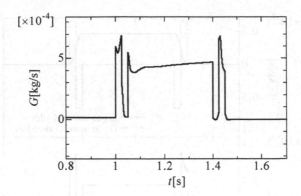

Fig. 9. The Mass Flow Rate of the System With nonlinear Model.

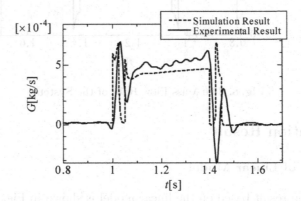

Fig. 10. Experimental Result of System Air Consumption.

6   Conclusion

In this paper, We constructed the simulation model of the pneumatic servo table system to inquire the air consumption of the system.

The air consumption of the system can divide into the air consumption to expand the volume of the pressure room and the air consumption to accelerate the slider.

By the mass flow measurement experiment using the quick response laminar flow sensor, the simulation result is according to the experimental result.

References

1. Uebing M., Vaughan N.D., and Surgenor B.W. On linear dynamic modeling of a pneumatic servo system. The 5th Scandinavian International Symposium on Fluid Power 2,1997.
2. Kagawa T., Tokashiki L.R., and Fujita T. Accurate positioning of a pneumatic servo system with air bearings. *Power Transmission and Motion Control*, pages 257–268, 2000.

3. Brun X., Sesmat S., Scavarda S., and Thomasset D. Simulation and experimental study of the partial equilibrium of an electropneumatic positioning system, cause of the sticking and restarting phenomenon. Proceedings of The 4th JHPS International Symposium on Fluid Power, 1999.
4. Funaki T., Kawashima K., and Kagawa T. Dynamic characteristic analysis of laminar flow meter. SICE Annual Conference in Fukui, 2003.
5. Endo S., Kobayashi H., Kempf C., Kobayashi S., Tomizuka M., and Hori Y. Robust digital tracking controller design for high-speed positioning systems. *Control Engineering Practice*, 4(4):527–535, 1996.
6. Fujita T., Sakaki K., Kawashima K., Kikuchi T., Kuroyama K., and Kagawa T. Air consumption in pneumatic servo system. The 7th Triennial International Symposium on Fluid Control, Measurement and Visualization, 2003.

Empirical Analysis for Individual Behavior of Impatient Customers in a Call Center

Yun Bae Kim[1], Chang Hun Lee[1], Jae Bum Kim[1],
Gyutai Kim[2], and Jong Seon Hong[3]

[1] School of Systems Management Engineering, Sungkyunkwan University, Suwon, Korea
{kimyb,lch714,kjbnhg}@skku.edu
[2] Chosun University, Kwangju, Korea
[3] School of Economics, Sungkyunkwan University, Seoul, Korea

Abstract. Sound scientific principles must be prerequisites for sustaining the levels of service and efficiency of modern call center, and these principles, in turn, better is based on real-world data. For example, in order to determine the least number of agents that could provide a given service level, it is critical to understand customers' impatience while waiting at the phone to be served. In this paper, we use stochastic models to plan call center operations and analyze projected performance focusing on the impatience of customers through simulation experiments.

1 Introduction

Telephone companies, marketing agencies, collection agencies, and others whose primary means of customer contact is via telephone want to make their call centers efficient and productive. The world of call centers is vast: some estimate that 70% of all customer-business interactions occur in call centers; that $700 billions in goods and services were sold through call centers in 1997, and this figure has been expanding 20% annually. There exist an increasing number of multi-media call centers that can provide multi-media service [15]. Call centers are of interest, because of the sheer size of the industry and the operational and mathematical complexity associated with these operations, which makes it difficult for decision makers to understand system dynamics without effective modeling.

Call centers can be beneficially viewed as stochastic systems, within the Operations Research paradigm of queueing models including the telephone, video, Internet, fax and e-mail services. From a view of queueing model, the customers are callers, servers (resources) are telephone agents (operators) or communication equipments, and queues consisted of callers wait for a service provided by system resources. The customers, who are only virtually present, are either being served, or waiting in tele-queue: up to possibly thousands of customers sharing a phantom queue, invisible to each other and the agents serving them. Customers are leaving the tele-queue when the one of two things happen; either being served or abandoning the tele-queue due to impatience that has built up exceeding one's anticipated worth of the service. The M/M/S (Erlang-C) model is a perfect example.

Queues are often an arena where customers establish contact. Especially, "human queues" express preference, complaint, abandonment and even spread around nega-

D.-K. Baik (Ed.): AsiaSim 2004, LNAI 3398, pp. 334–342, 2005.
© Springer-Verlag Berlin Heidelberg 2005

tive impressions. Thus, customers treat the queueing experience as a window to the service-providing party.

In such a queue, reneging is a commonly observed phenomenon from customers leaving a service system before receiving service due to customers' impatience or uncertainty of receiving service. Abandonment is one of the most hotly debated topics in call center management and research. Customer's tolerance for waiting and the probability that this customer hang up and thereby leave the queue must be analyzed in order to effectively model customer abandonment behavior.

Many researchers have examined the challenge of modeling these problems from both an empirical point of view and from an analytic perspective. It seems that Palm [17] was the first to deal with the impatience problem. Barrer [4, 5] analyzed the M/M/1+D system, where the first M denotes the arrival process, the second M, the service process with one server be distributed exponentially and the symbol D stands for the constant tolerance times. Brodi derived and solved for the M/M/1+D system the corresponding integro differential equation [10]. The general GI/GI/1+GI system is treated in [8]. The multi-server M/M/s+D system was analyzed by Gnedenko and Kowalenko [10] by giving an explicit solution of the system of integro differential equations for the work load of the s servers, which yields formulae for performance measures. This method was successfully further applied by Brandt et al. [7] to maximal waiting times as the minimum of a constant and an exponentially distributed time and also [7] for the general M/M/s/+GI system where the arrival intensities may depend on the number of busy servers (cf. also [10]). Later, independently Haugen and Skogan [11] and Baccelli and Hebuterne [2] derived results for the M/M/s/+GI system, too.

Recently, Boots and Tijms [6] proposed an approximate modeling approach for M/G/s+D. Their contribution is to establish heuristic results for the multiserver case with general service time. However, they provided only an alternative formula of the loss probability. Our work is in the same spirit as [6] and investigates the other feasibility of modeling the general multiserver system through simulation experiments.

The M/M/s/n+GI system, i.e., with a finite waiting space and the more general model were not yet treated in the references mentioned above. The reason is the difficulty that the number of customer left behind at a service completion epoch does not provide sufficient information to describe the future behavior systems [14][18]. Thus, the results concerning the reneging probability of the M/M/s/n+M system and the M/G/s/n+M in given in this paper may be new approximate values. Note that our model includes the case of a finite waiting room, whereas this case is excluded in the models mentioned above. In the literature, there are several other known mechanisms where calls leave the system due to impatience: if a call can calculate its prospective waiting time for its service time, then it leaves immediately if this time exceeds its maximal waiting time. This strategy needs more utilization of the waiting places because they will not be occupied by calls that later abandon due to impatience. In this case not all work is useful because a call may leave the system due to impatience during its service. Moreover, this case differs from actual system in that the number of calls in system.

The remainder of this paper is organized as follows. In Section 2, we derive the system equations of the model proposed for M/M/s/n+M focusing on the key output statistics, especially the probability of call abandonment. It is very difficult even for yielding the integro equations that the system is modeling as a closed formula whereas the more general system with the general service time. In Section 3, we try to analyze size the performance of ACD (Automatic Call Distributor) systems based on the model denoted and derived in section 2 through simulations. We compare the results of simulation experiments with the theoretical solutions of system equations. In addition, we discuss how call centers of the general model with general service time make use of simulation models, in order to use for system performance evaluation. Finally, in Section 6, we discuss the results of our research and propose future directions for call center simulation.

2   The Formula for Call Center Systems

2.1   System Equations and Its Solution

In our example, simulated customers arrive at the call center and are served by an agent if one is available. If not, customers join the queue, at which point they are also assigned a "life span" drawn from an exponential distribution.

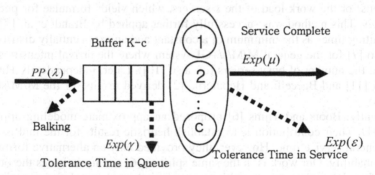

Fig. 1. M/M/c/K + Mq+Ms. This system has impatient calls that they abandon to take service if waiting time or service time is longer than their expectation. Also, this model has the finite buffers and the finite agents. Therefore, a call may be balked when all buffers are occupied. The calls' arrival follows Poisson process.

If a customer's life span expires while one is still waiting in queue, one then abandon the queue. That is, we represent customers' tolerance for waiting in queue as an exponential random variable (as suggested by [9]).

In this model, we need to set the level of tolerance, which we estimate from historical callers' time data in queue. We do not include caller retrial in the example.

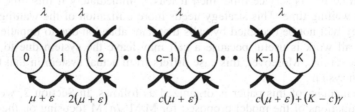

Fig. 2. Markov-Chain of M/M/c/K + Mq+Ms system.

The system equation of the model is drawn from the Markov-Chain of Fig. 2. M/M/c/K queueing system is always stable, and then we shall derive steady-state probability. By FCFS discipline, the first call in the queue will be served at once when more than one server is available.

We define variables as follow. Suppose that following random variables excluded N and p_n don't consider balking calls. Specially, the event of new call's arrival is except from $I_n^{(j)}$ because the reneging probability is not affected.

N : The number of calls in the system.

$p_n = \Pr[N = n]$: Steady-state probability of the number of calls in the system.

R_q : The event that an arriving call will abandon in the queue.

$R_{q,n}$: The event that the call finding n calls at its arrival in the system will abandon in the queue.

R_s : The event that an arriving call will abandon in service.

$R_{s,n}$: The event that the calls finding n calls at its arrival in system will abandon in service.

R_n : The event that the calls finding n calls at its arrival in system will abandon service through the whole process.

R : The event that an arriving call will abandon service through the whole process.

$T_{q,n}$: The waiting time of a call finding n calls in system at its arrival.

T_q : The waiting time of a call.

$I_n^{(1)}$: After a call arrives at n calls in the system, next event is service completion or reneging among busying servers.

$I_n^{(2)}$: After a call arrives at n calls in the system, next event is reneging among the front queueing calls.

$I_n^{(3)}$: After a tagged call arrives at n calls in the system, next event is reneging of the tagged call itself.

p_n is equal with the probability of the number of calls at the time call's entering system by PASTA [13][14]. From the truth, we obtain

$$p_0 = \left[1 + \left(\sum_{i=1}^{c} \frac{1}{i!} \left(\frac{\lambda}{\mu + \varepsilon} \right)^i \right) + \frac{1}{c!} \left(\frac{\lambda}{\mu + \varepsilon} \right)^c \sum_{j=1}^{K-c} \frac{\lambda^j}{\prod_{i=1}^{j} \{ c(\mu + \varepsilon) + i\gamma \}} \right]^{-1}$$

$$p_n = \begin{cases} \dfrac{1}{n!} \left(\dfrac{\lambda}{\mu + \varepsilon} \right)^n p_0 & , \; 1 \leq n \leq c \\[3mm] \dfrac{1}{c!} \left(\dfrac{\lambda}{\mu + \varepsilon} \right)^c p_0 \prod_{i=1}^{n-c} \left(\dfrac{\lambda}{c(\mu + \varepsilon) + i\gamma} \right) & , \; c < n \leq K \end{cases}$$

(2.1)

We induce (2.1) directly from Fig. 2. In (2.1), p_K is the balking probability, the average of the number of calls in the system is $E[N] = \sum_{n=1}^{K} np_n$.

2.2 Empirical Analysis of the Probability of Reneging

Now, let's find the rest performance measures of the system. $\Pr[R_{q,n}]$ is equivalent to

$$\Pr[R_{q,n}] = 0, \quad 0 \leq n < c \tag{2.2}$$

$$\Pr[R_{q,n}] = \sum_{j=1}^{3} \Pr[R_{q,n} \mid I_n^{(j)}] \Pr[I_n^{(j)}]$$

$$= \Pr[R_{q,n-1}] \cdot \frac{c(\mu + \varepsilon) + (n - c)\gamma}{c(\mu + \varepsilon) + (n - c + 1)\gamma} + \frac{\gamma}{c(\mu + \varepsilon) + (n - c + 1)\gamma} \quad , c \leq n < K \tag{2.3}$$

(2.2) is obvious and (2.3) is conditional probability by next event after tagged calls' arrival. Then, $\Pr[R_q]$ is given by

$$\Pr[R_q] = \sum_{n=c}^{K-1} \Pr[R_{q,n}] \cdot \frac{p_n}{(1 - p_K)} \tag{2.4}$$

Also, we are able to calculate $\Pr[R_s]$. First of all, we must find $\Pr[R_{s,n}]$ in order to calculate $\Pr[R_s]$. $\Pr[R_{s,n}]$ is equivalent to

$$\Pr[R_{s,n}] = \frac{\varepsilon}{\mu + \varepsilon}(1 - \Pr[R_{q,n}]) \quad, 0 \leq n < K \tag{2.5}$$

By (2.5), we find that $\Pr[R_s]$ follows as

$$\Pr[R_s] = \sum_{n=0}^{K-1} \Pr[R_{s,n}] \cdot \frac{p_n}{(1 - p_K)} \tag{2.6}$$

Considering (2.4) and (2.6), we find that $\Pr[R_n]$ and $\Pr[R]$ follow as

$$\Pr[R_n] = \begin{cases} \dfrac{\varepsilon}{\mu + \varepsilon} & , 0 \leq n < c \\[2mm] \dfrac{\Pr[R_{n-1}] \cdot \{c(\mu + \varepsilon) + (n - c)\gamma\} + \gamma}{c(\mu + \varepsilon) + (n - c + 1)\gamma} & , c \leq n < K \end{cases} \tag{2.7}$$

$$\Pr[R] = \sum_{n=0}^{K-1} \Pr[R_n] \cdot \frac{p_n}{(1 - p_K)} \tag{2.8}$$

The average of waiting time is able to obtain as the same manner in (2.3) and (2.4). First of all, we must find $E[T_{q,n}]$. $E[T_{q,n}]$ is given by

$$E[T_{q,n}] = \begin{cases} 0 & , 0 \leq n < c \\[2mm] \dfrac{[1 + E[T_{q,n-1}] \cdot \{c(\mu+\varepsilon) + (n-c+1)\gamma\}] \cdot \{c(\mu+\varepsilon) + (n-c)\gamma\} + \gamma}{\{c(\mu+\varepsilon) + (n-c+1)\gamma\}^2} & , c \leq n < K \end{cases} \tag{2.9}$$

By (2.9), we can obtain that

$$E[T_q] = \sum_{n=c}^{K-1} E[T_{q,n}] \cdot \frac{p_n}{(1 - p_K)} \qquad (2.10)$$

Above (2.9) and (2.10) is total calls' average queueing time regardless of the calls' reneging event in queue. Now, let's find the average queueing time of only reneging calls in queue. First, $E[T_{q,n} \mid R_{q,n}]$ is

$$E[T_{q,n} \mid R_{q,n}] = \sum_{j=1}^{3} E[T_{q,n} \mid R_{q,n}, I_n^{(j)}] \cdot \Pr[I_n^{(j)} \mid R_{q,n}]$$

$$= \frac{\Pr[R_{q,n-1}] \cdot \{c(\mu + \varepsilon) + (n-c)\gamma\} + \gamma}{\Pr[R_{q,n}] \cdot \{c(\mu + \varepsilon) + (n-c+1)\gamma\}^2} + \Pr[R_{q,n-1}] \times \qquad (2.11)$$

$$\frac{E[T_{q,n-1} \mid R_{q,n-1}] \cdot \{c(\mu + \varepsilon) + (n-c+1)\gamma\}\{c(\mu + \varepsilon) + (n-c)\gamma\}}{\Pr[R_{q,n}] \cdot \{c(\mu + \varepsilon) + (n-c+1)\gamma\}^2}$$

(2.11) is available in $c \le n < K$. By (2.11), $E[T_q \mid R_q]$ is given by

$$E[T_q \mid R_q] = \sum_{n=c}^{K-1} E[T_q \mid R_q, N = n] \cdot \Pr[N = n \mid R_q]$$

$$= \frac{1}{\Pr[R_q]} \sum_{n=c}^{K-1} E[T_{q,n} \mid R_{q,n}] \Pr[R_{q,n}] \cdot \frac{p_n}{(1 - p_K)} \qquad (2.12)$$

3 Performance Analysis of ACD Systems with Using Simulation

For many services and businesses there is a need to match incoming calls with agents, e.g. in telebanking, teleshopping, information services, mail orders, etc. ACD systems are controlled by software allows managing the incoming calls such that a group of ACD agents can handle a high volume of incoming calls. For a description of ACD systems see [3,7,9,12,15,16]. The ACD feature provides statistical reporting tools in addition to call queueing. ACD historical reports allow to identify times when incoming calls abandon after long waits in the queue because too few agents are allocated, or times where agents are idle. One might wish to use these data to determine an optimal operating strategy with respect to minimal cost (or maximal revenue) and various service quality requirements, e.g. for the abandon probability and for quantiles of waiting times.

There is a tradeoff between various performance measures, e.g. between the number of 'abandoned' and 'blocked' calls, i.e., calls which are rejected by the system when all lines are occupied. Thus, it is a non-trivial problem to determine the optimal operational strategy. For optimizing ACD systems, cf., e.g. [7], it is necessary to have stable and fast algorithms for computing the relevant performance measures.

In this section, we compare theoretical values with simulation results. And, we will test various situations that service time follows general distribution. As an example, we consider the following basic parameter set:

- mean inter-arrival time $1/\lambda$: 0.2 sec.
- mean service time $1/\mu$: 1 sec.
- mean tolerance time in queue $1/\gamma$: 5 sec.
- mean tolerance time in service $1/\varepsilon$: 20 sec.
- number of agents : 5
- the number of buffers(include servers' capacity) : 20

We explain the following performance measures again; $E[N]$ is the average number of calls in the system. $\Pr[R_q]$ is the probability of abandonment in queue. $\Pr[R_s]$ is the reneging probability in service. $\Pr[R]$ is the reneging probability through the whole process. $E[T_q]$ is the average waiting time in queue. $E[T_q \mid R_q]$ is the average tolerance time only for reneging calls in queue. p_K is the balking probability.

In Table 1., it is remarkable that the difference between $E[\Gamma] = 1/0.2$ and $E[T_q \mid R_q]$ is very large. Table 1. shows that the distribution of only reneging calls' queueing time is not equal with that of the whole calls' tolerance time.

Table 1. The comparison of theoretical results and simulation results.

System performance	Theoretical result	Simulation result	Relative error(%)
$E[N]$	6.57116	6.57134	0.0027
$\Pr[R_q]$	0.08978	0.08979	0.0038
$\Pr[R_s]$	0.04334	0.04334	0.0059
$\Pr[R]$	0.13313	0.13313	0.0006
$E[T_q]$	0.44893	0.44897	0.0109
$E[T_q \mid R_q]$	0.56116	0.56129	0.0226
p_K	0.00119	0.00119	0.3256

(Simulation running time is 20 million sec.)

We can find results that visualize the reneging probability in queue and the balking probability according to $\rho(= \lambda/c\mu)$ in Fig. 3. Fig. 3. shows results when service time S is general distribution. In general distribution, service time is fixed as Lognormal (1, 1).

In Fig. 3., we see that M/M/c/K and M/G/c/K are nearly equal in the system performances only if E[S] and Var[S] are the same respectively. Exp(1) and LN(1, 1) are special cases when COV $\left(= \sqrt{Var[S]}/E[S]\right)$ is 1.

Fig. 4. shows simulation results that represent the reneging probability in queue as COV and ρ varies. Service Time S is gamma distribution and others initial conditions are identical. Simulation running time is 1 million seconds respectively.

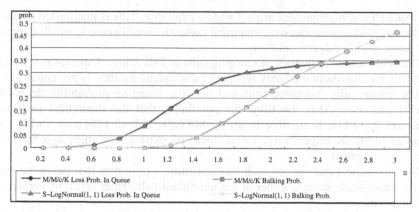

Fig. 3. The comparison of performance measures in S ~ Exp(1) and LN(1,1).

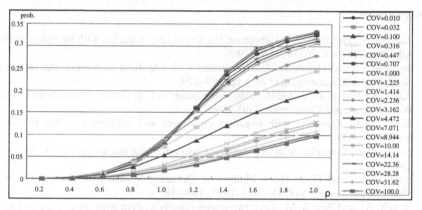

Fig. 4. The comparison of performance.

It shows that the reneging probability in queue is very different as COV of the service time changes. Roughly, the reneging probability in queue decrease as COV increase where COV > 1 and ρ is fixed. Such as this pattern is the same in Lognormal Distribution. But, the values of reneging probability itself differ from Gamma case.

Large COV means large Var[S] where E[S] is fixed. As Var[S] increases, the frequency of extreme values increases. Extremely high service time is hardly realized due to the tolerance time in service. Most of low service times give rise to reduce the average service time. Consequently, large Var[S] means really low E[S] and the low reneging probability in queue. In actual simulation experiments, the potential E[S] of the reneging call in service is lager as Var[S] is lager. So, if we should intend to find the probability in M/G/c/K + Mq+Ms with impatient calls, we will consider the characters of service time distribution besides average and variance. (i.e. skewness).

4 Conclusions

It is surprising, perhaps astonishing when considering the data-intensive hi-tech environment of the modern call centers, that operational data at the appropriate resolution for research and management are not widely available. This is manifested by the lack of documented, comprehensive, empirical research of call centers, which is precisely our prime goal in this present study.

We find various properties of reneging calls in call center. Especially, it is significant that the distribution of only reneging calls' queueing time is not equal with that of the whole calls' tolerance time and that the reneging probability decreases as Var[S] increases. Also, we show the possibility that the theoretical performance measures in M/M/c/K are able to use in M/G/c/K if both E[S] and Var[S] are the same as M/M/c/K respectively. To use these properties, this study is a help to call center management.

We also expect that our research will constitute a prototype that paves the way for future larger-scale studies, either at the individual call-center level, or perhaps even industry-wide.

References

1. Baccelli, F., Boyer, P., and Hebuterne, G., Single server queues with impatient customers, Adv. Appl. Probab. 16 (1984) 887–905.
2. Baccelli, F., and Hebuterne, G., On queues with impatient customers, Performance '81, North-Holland, Amsterdam (1981), pp. 159–179.
3. Bapat, V., Pruitte Jr., and E.B., Using Simulation in Call Center, Proceedings of the 1998 Winter Simulation Conference, 1395-1399.
4. Barrer, D.Y., Queueing with impatient customers and indifferent clerks, Oper. Res. 5 (1957) 644–649.
5. Barrer, D.Y., Queueing with impatient customers and ordered service, Oper. Res. 5 (1957) 650–656.
6. Boots, N.K., and Tijms, H., A Multiserver Queueing System with Impatient Customers, Management Science Vol. 45, No 3.(1999) 444-448.
7. Brandt, A., and Brandt, M., On a two queue priority system with impatience and its application to a call center, Methodology and Computing in Applied Probability 1(1999) 191-210
8. Daley, D. J., General customer impatience in the queue GI/G/1, J. Appl. Probab. 2 (1965) 186–205.
9. Garnett, O., Mandelbaum, A., and Reimann, M. L., Designing a Call Center With Impatient Customers, Manufacturing and Service Operations Management 4 (2002): 208 -227.
10. Gnedenko, B.V., and Kovalenko, I.N., Introduction to Queueing Theory, 2nd ed, Springer-Velrag, New York (1989)
11. Haugen, R.B., and Skogan, E., Queueing systems with stochastic time out, IEEE Trans. Commun. COM-28 (1980) 1984–1989.
12. Hoffman, K. L. and C. M. Harris. 1986. Estimation of a Caller Retrial Rate for a Telephone Information System. European Journal of Operational Research 27:207-214.
13. Kleinrock, L., Queueing Systems, Vol. I: Theory, Wiley, New York (1975).
14. Lee, H.W., Queueing theory, 2nd ed. (in Korean), Sigma Press (1998)
15. Mandelbaum, A., Sakov, A., and Zeltyn, S., Empirical Analysis of a Call Center, Technion, Israel Institute of Technology (2001)
16. Mehrota, V., Fama, J., Call center Simulation Modeling: Methods, Challenges, and Opportunities, Proceeding of the 2003 Winter Simulation Conference, 135-143
17. Palm, C., Methods of judging the annoyance caused by congestion, Tele No. 2 (1953) 1–20.
18. Tijms, H., Stochastic Models: an algorithmic approach, Wiley, New York (1994)

FSMI: MDR-Based Metadata Interoperability Framework for Sharing XML Documents

Hong-Seok Na[1] and Oh-Hoon Choi[2]

[1] Department of Computer and Information Science, Korea Digital University,
139 Chungjeongno 3-ga, Seodaemun-gu, Seoul 120-715, Korea
hsna99@kdu.edu
[2] Research Institute for Computer Science & Engineering Technology, Korea University,
1, 5-ka, Anam-dong, Sungbuk-gu, 136-701, Korea
pens@swsys2.korea.ac.kr

Abstract. XML and metadata are used to represent the syntax, structure and semantic of information resources. But, various metadata sets, developed by independent organizations without any standards or guidelines, make it difficult to share XML encoded information resources. In this paper, we propose a framework named FSMI(Framework for Sharable Metadata Interoperability) to enable metadata level standardization and XML document sharing between applications. In FSMI, users of each system describe the difference between their XML schema and standard metadata registered in metadata registry, then they can interchange their XML documents with other system by resolving the difference. FSMI overcomes the limitations of other approaches with respect to exactness, flexibility and standardization, and provides an environment for business partners of different metadata to share XML encoded information resources.

1 Introduction

Extensible Markup Language(XML)[1] is a simple, very flexible text format derived from SGML(ISO 8879). Originally designed to meet the challenges of large-scale electronic publishing, XML is also playing an increasingly important role in the exchange of a wide variety of data on the Web and elsewhere. With XML, application designers can create sets of data element tags and structures that define and describe the information contained in a document, database, object, catalog or general application, all in the name of facilitating data interchange[2].

Metadata is a data about data. And it is essential to information sharing[3]. It describes attribute, property, value type and context on data[4]. There are many standard sets of metadata – Dublin Core for information resources, MARC for library information, MPEG-7 for multimedia information and so on.

XML schema is a specification defining the structure of tags and plays a role of metadata on XML documents. XML documents, the unit of data exchange, are based on schemas(XML schema definitions). Thus, XML documents based on same schema definition have same structure and tags. And, it is possible to share data between applications if they use standardized schema definition. Such standardized schema definitions include CML DTD[5] used in chemistry domain, Math DTD[6] in mathematics, ONIX[7] in digital library domain, and so on.

D.-K. Baik (Ed.): AsiaSim 2004, LNAI 3398, pp. 343–351, 2005.

However, in many cases, each system(organization or application domain) has developed their own metadata(schema definition) and use XML documents based on the schema. There is no interoperability with other systems without manual converting processes. For example, if we represent "author" of a book named "Database Design", Dublin Core which is a metadata standard for describing general properties of information resources, uses a term "creator", but ONIX uses a term "author".

In this paper, we first categorize heterogeneity between XML documents – syntactic, structural and semantic heterogeneity, and second design a framework(FSMI, Framework for Sharable Metadata Interoperability) to increase the interoperability of XML encoded information resources using different metadata sets.

In this framework, we use ISO 11179 metadata registry[8] as a common metadata model and propose a metadata description language(MSDL, Metadata Semantic Description Language) that represents the exact meaning of XML tags by describing the structural and semantic differences with standard metadata in metadata registries. With MSDL, users can describe the syntactic, structural and semantic difference of their XML schema to other's schema, and based on this description, applications can convert other system's XML document to a document with their own schema.

2   Related Works

We first define information heterogeneity, then introduce some other approaches that enable the interoperability of XML documents having different schema definitions.

2.1   Information Heterogeneity

Table 1. shows the type of heterogeneity between standard metadata based on ISO 11179 MDR and XML schema definitions. Semantic heterogeneity means that the two xml tags has same meaning but different names. Structural heterogeneity means that two sets of xml tag have same meaning but different structures including composition, decomposition and rearrangement. Syntactic heterogeneity means the difference of data values in XML tags. Conflicts in this heterogeneity include measurement unit, data type and data code sets.

Table 1. Three Types of Heterogeneity on XML documents

	Description and Examples
Semantic Heterogeneity	Different terminologies have same semantics in context. ex) title - book name, price – cost
Structural Heterogeneity	Structurally different with same semantics ex1) name - first name, last name ex2) first name, last name – name ex3) last name - first name, last name
Syntactic Heterogeneity	Semantically, structurally same, but have different data type, measurement unit and code sets ex1) {kr, jp, cn} - {korea, japan, china} ex2) mile - kilometer ex3) integer – float

2.2   Other Approaches for Sharing Information Resources

The most straightforward approach for XML document interoperability is 1:1 mapping. In 1:1 mapping, each system tries to map their XML schema to other's schema. Microsoft's Biztalk[9] is the representative software. Users of Biztalk create a mapping table for each others application, and interchange their XML documents. After receiving, it converts the documents according to receiver's understandable format. Maintaining integrated global schema is another method. EAI solutions analyze each application's data and processes, create global mapping tables and rules, and enable each system to exchange XML encoded data.

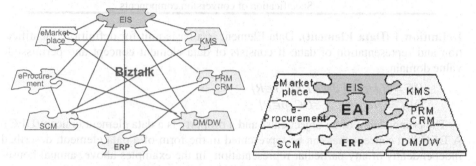

Fig. 1. XML Schema Interoperability in Biztalk and EAI

Also, many organizations develop standard metadata(XML schema) and member of the domain use it as a data interchange format. The representative examples are ebXML[10] and Rosettanet[11] in electric commerce domain, and ONIX in digital contents(e-book) domain. These approach is very simple and powerful when all members use it as their data standard.

These three approaches have advantages and disadvantages in exactness, flexibility, initial construct and maintenance cost. In the last section, we will compare our proposed approach to them.

3   FSMI

We propose a framework(FSMI, Framework for Sharable Metadata Interoperability) to enable matadata level standardization and XML document transformation between applications using different metadata and schemas. The key idea of FSMI is that users of each system describe the difference between their XML schema and standard metadata registered in metadata registry. Then they interchange their XML documents with other system as resolving the difference of other's document using the descriptions. As shown in Table 2, FSMI consists of three interfaces - metadata registration and service, metadata description, and XML document transformation.

3.1   Interface 1: Metadata Registration and Service

We use data element model of ISO 11179 as a standard metadata model for comparison between different XML schemas. Data element is a basic unit of identification, description and value representation unit of a data.

Table 2. FSMI Interfaces – functions and specifications

	Description of functions and specifications
Interface 1	Metadata Registration and Service - Registering standard metadata to 11179-based registry - XML based metadata service specification
Interface 2	Metadata Description - MSDL specification - MSDL document registration and service specification
Interface 3	XML document transformation - Specification of conversion rules - Specification of conversion components

Definition 1 (Data Element). Data Element is a basic unit for definition, identification and representation of data. It consists of data element concept and permissible value domain.

$$DE = \{(dec, r)|\ dec \in DEC,\ r \in R\}$$
$$DEC = \{(oc, p)|\ oc \in OC,\ p \in P\}$$

The combination of an object class and a property is a data element concept(DEC). A DEC is a concept that can be represented in the form of a data element, described independently of any particular representation. In the examples above, annual household income actually names a DEC, which has two possible representations associated with it. Therefore, a data element can also be seen to be composed of two parts: a data element concept and a representation. a data element is composed of three parts as Fig. 2[8].

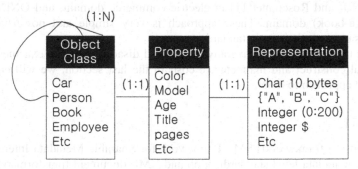

Fig. 2. Structure of Data Element

Object class(OC) is a set of ideas, abstractions, or things in the real world that can be identified with explicit boundaries and meaning and whose properties and behavior follow the same rules.Examples of object classes are cars, persons, books, employees, orders, etc. Property(P) is what humans use to distinguish or describe objects. Examples of properties are color, model, sex, age, income, address, price, etc.

The most important aspect of the representation(R) part of a data element is the value domain. A value domain is a set of permissible (or valid) values for a data element. For example, the data element representing annual household income may have the set of non-negative integers (with unit of dollars) as a set of valid values. Fig.2 shows some examples related on Definition 1.

OC={person, car, employee, book, product, ...}
P={name, age, title, cost, ...}
R={char 30byte,{blue, red, green},{0..100},...}
DEC={personname, carname, booktitle, productcost, ...}
DE={(personname, char 30byte), (carname, char 30byte),
(productcost,dollar),...}

Fig. 3. Examples on Data Element

A metadata registry(MDR) supports registration, authentication and service of such data elements. Each data elements have an unique identifier, all users and programs using the metadata registry identify a data element by the identifier. A data element also contains it definition, name, context, representation format, and so on.

3.2 Interface 2: Metadata Description

The key idea of this paper is that users of each system describe the difference between their XML schema and standard metadata registered in metadata registry. Then they interchange their XML documents with other system as resolving the difference of other's document using the descriptions.

We designed MSDL(Metadata Semantic Description Language) as a tool for describing the difference between local XML schema and standard metadata in MDR. MSDL consists of namespace on the location of MDR and local schema, correspondent information(MAP) and transformation rules for converting local document to/from standard document.

Definition 2 (MSDL). A language for describing the difference between local schema and standard metadata. It uses XML syntax.

$MSDL=(mdrNameSpace,xmlNameSpace,MAP,CODESET)$

Where, *mdrNamespace* and *localNamespace* must appear once in a MSDL document, but *MAP* appears as much as the number of tags to be converted.

Definition 2.1 (MAP). MAP defines semantic, structural and syntactic conflicts between tags in XML documents and metadata in MDR.

$MAP=\{(mdrE,localE,maptype)|\ mdrE \subset DE,\ localE \subset XS,\ maptype \in MapT\}$

Where, mdrE is a subset of standard metadata in MDR, and has one or more elements. localE is a subset of metadata defined in local XML schema, and also has one or more elements.

Definition 2.2 (Mapping Type). MapT defines structural conflicts and syntactic conflicts between metadata with same semantics.

$MapT=\{(sct,vct\ |\ sct \in SCT,\ vct \in VCT)$
$SCT=\{substitution, composition, decomposition, rearrangement\}$
$VCT=\{(CodeSetmdr,\ CodeSetlocal),\ (MUnit,\ cvalue),(TypeCasting,\ cfunction)\}$

Where, $number(CodeSetmdr) = number(CodeSetlocal)$,
Munit = measurement unit(m, km, kg, etc.),
Cvalue = constant for measurement unit chage,
cfunction = name of system support function for type casting

SCT(Structural Change Type) has one of the four elements – substitution, composition, decomposition, rearrangement. Where, substitution is for resolving semantic conflict, and other three elements are for structural conflict. For example, "book title" and "book name" have same meaning but different name, in this case, "substitution" is applied. On the other hand, a "name" tag may correspond to two tags "first name" and "last name", in this case, "composition" or "decomposition" is applied.

VCT(Value Change Type) has one of the three elements – code set conflict, measurement unit conflict and data type conflict. In code set conflict, we consider only the case of two code sets having same number of elements.

A MSDL document must have one or more mapping type to describe the difference between standard metadata and local schema. Fig. 4 show an example of MDSL document mapping local element "Book price" to standard metadata element "Price". In this example, two elements have semantic conflict and syntactic(measurement unit) conflict.

```
...
<map>
    <mdrElement>
        <DataelementName> <element>Price</element> </DataelementName>
        <elementID>DE024041</elementID>
        <dataType>
            <nonCodeSetDataType>string</nonCodeSetDataType>
        </dataType>
        <measurementUnitID>MU000000</measurementUnitID>
    </mdrElement>
    <localElement>
        <elementName>Price</elementName>
        <elementPath>/BooK/Price</elementPath>
        <dataType>
            <nonCodeSetDataType>string</nonCodeSetDataType>
        </dataType>
        <measurementUnitID>MU000001</measurementUnitID>
    </localElement>
    <mappingRule>
        <mappingType>substitution</mappingType>
    </mappingRule>
</map>
```

Fig. 4. An Example of MSDL Document

3.3   Interface 3: XML Document Transformation

MSDL just specifies information about conflicts of local schema and standard metadata. Thus, it is necessary to develop a system which transform imported XML document to self-understandable one using MSDL. FSMI interface 3 defines a system for transforming XML encoded source data to fit other system's XML document using each system's predefined MDSL documents. Fig. 5 shows the transforming process.

XML document converting process consists of two steps. The first step is to generate a standard document with standard data element in MDR from source document,

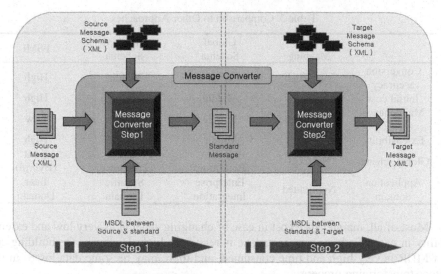

Fig. 5. XML document transform process

local XML schema and MSDL of the schema. Then, in the second step, standard document is converted to target document using target document's XML schema and MSDL.

In the step 1 and step 2, FSMI applies mapping rules in MSDL to resolve structural and syntactical heterogeneity (conflicts). Generally, these two types of heterogeneity co-exist in a document, it is necessary to apply mapping rules with combined manner.

4 Comparison

1:1 mapping approach is relatively exact because experts of each system directly map to other system's schema. But, the mapping cost increases exponentially as the number of related systems. Especially, in case of schema change, the maintenance cost will be high.

Global schema can cover large parts of each system's characteristics of schema through analysis process, and the maintaining cost is very low in case of schema change because systems only have to change related parts of global schema. But the initial building cost of global schema may be high.

If all systems use standardized XML schema, then there are no syntactic, structural and semantic conflicts. It is very ideal status on data sharing and integration. But It is impossible to make a standard system(schema) that support all the case of system, and necessary to consider another processes for data that standard does not cover.

Also, a domain may have more than one metadata standards. For example, in digital content domain have metadata standard including ONIX, MARC, DC, MPEG7 and in electric commerce domain, they have standard including cXML, ebXML, Rosettanet, eCo, and so on.

Table 3 shows the characteristics of each approaches including FSMI. With FSMI, conversion accuracy is relatively high, because all the needed part of schema for sharing, can be covered in MDR using registering process.

Table 3. Comparison to Other Approaches

	1:1 Mapping	Global Schema	Standard Schema	FSMI
Conversion accuracy	Very High	High	Middle	High
Initial cost	Low	Middle	Very high	High
Maintenance cost	Very High	Middle	Low	Low
Extensibility	Low	Middle	Middle	High
Global schema	Not exist	Exist (Bottom-up)	Exist (Top-down)	Exist (Hybrid)
Application domain	Not limited	Enterprise Integration	Specific Domain	Inter Domain

Most of all, maintenance cost in case of changing schema, is very low and extensibility in case of adding new metadata, is relatively high. However, the building cost of MDR is very high and time consumed, and there may be some data losses in two steps transforming process.

5   Conclusion and Future Works

XML is de facto standard for representing information resources in distributed environment. But it is necessary to share metadata of XML document defining syntax, structure and semantic of the document.

In this paper, we propose a framework, FSMI as a tool for describing difference between standard metadata and local XML schemas to support interoperability of XML documents. With MSDL in this framework, users can describe the syntactic, structural and semantic difference of their XML schema to other's schema, and based on this description, can convert other system's XML document to a document with their own schema.

Users who wish to exchange their document, just describe the difference of their XML schema with standard metadata in MDR, and make a MSDL document on the schema, then other system can convert the document to understandable ones.

FSMI overcomes limitations of 1:1 mapping, global schema and standard schema approaches, providing low maintenance cost on schema changes, high extensibility in increase of systems, and domain independent document sharing.

In the future, we have plans to develop the converting system using MSDL, and extend it to support legacy solutions including EAI, B2Bi, Web Service, and so on.

References

1. W3C : Extensible Markup Language. http://www.w3.org (2004)
2. Rik Drummond, Kay Spearman. : XML Set to Change the Face of E-Commerce. Network Computing, Vol.9, N.8. 5 (1998) 140-144
3. Sheth, A. : Changing focus on Interoperability: From System, Syntax, Structure to Semantics. Interoperating Geographic Information Systems. Kluwer Academic Publishers. (1999)
4. Jane Hunter : MetaNet - A Metadata Term Thesaurus to Enable Semantic Interoperability Between Metadata Domains. Journal of Digital Information, Vol. 1, Num. 8. (2001)

5. Chemical Markup Language, Version 1.0, Jan.1997.
6. W3C : Mathematical Markup Language (MathML) 1.0 Specification. (1998)
7. ONline Information eXchange : http://editeur.org/
8. ISO : Information Technology - Metadata Registries. ISO/IEC 11179, Part 1 – 6 (2003)
9. Microsoft : Microsoft Biztalk Server. http://www.microsoft.com/biztalk
10. http://www.ebxml.org
11. http://www.rosettanet.org
12. David Bearman, Eric Miller, Godfey Rust, Jennifer Trant, Stuart Weibel. : A Common Model to Support Interoperable Metadata. D-lib Magazine, Vol. 5, Num. 1. (1999)

Simulation-Based Web Service Composition: Framework and Performance Analysis

Heejung Chang[1], Hyungki Song[1], Wonsang Kim[1], Kangsun Lee[1,*],
Hyunmin Park[1], Sedong Kwon[1], and Jooyoung Park[2]

[1] Department of Computer Engineering, MyongJi University, South Korea
ksl@mju.ac.kr
[2] Department of Computer Science and Engineering, Ewha Womans University, South Korea

Abstract. As Service Oriented Architecture is expected to grow in the future e-business, web services are believed to be the ideal platform to integrate various business artifacts disparate platforms and systems. QoS(Quality of Service) of web services in terms of performance, reliability, and availability becomes the key issue when web services model complex and mission-critical applications. In this paper, we propose a simulation-based framework that enables web services to be composed based on their QoS properties. The proposed framework iteratively analyzes the QoS of the composite web services with multiple perspectives until the QoS requirements are met to various service users and providers.

1 Introduction

As economy business processes transcend departmental as well as organizational boundaries, web services are expected to provide the ideal platform to automate these processes by integrating disparate platforms and systems. [1] A web service is a software application identified by a URI(Uniform Resource Identifier), whose interfaces and binding are capable of being defined, described and discovered by XML artifacts and supports direct interactions with other software applications using XML based messages via internet-based protocols.[2] Figure 1 shows the core technologies of web services. Web service providers use the Web Service Description Language(WSDL) to describe the services they provide and how to invoke them. The service providers then register their services in a public service registry using universal description, discovery, and integration (UDDI). Application programs discover services in the registry and obtain a URL for the WSDL file that describes the services using the XML-based Simple Object Access Protocol (SOAP) in either asynchronous messaging or remote procedure call (RPC) mode. Once deployed, web services are easily aggregated into composite services (or *web process*) specified by service flow languages, such as WSFL(Web service Flow Language), XLANG, BPEL4WS(Business Process Execution Language for Web Services), and DAML-S.[3-6].

Web service composition enables business organizations to effectively collaborate in their business processes. An example is a "Travel Planner" system which aggre-

---

* Corresponding author.

D.-K. Baik (Ed.): AsiaSim 2004, LNAI 3398, pp. 352–360, 2005.

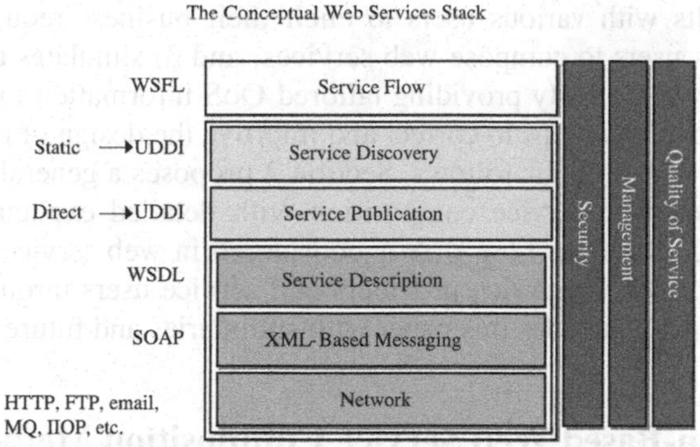

The Conceptual Web Services Stack

WSFL	Service Flow		
Static ⟶ UDDI	Service Discovery		
Direct ⟶ UDDI	Service Publication	Security	Management
WSDL	Service Description		
SOAP	XML-Based Messaging		
HTTP, FTP, email, MQ, IIOP, etc.	Network		

Fig. 1. Enabling Technologies for Web Services

gates multiple component web services for flight booking, travel insurance, accommodation booking, car rental, and itinerary planning executed sequentially or concurrently.[3] In order to meet business goals in competing business markets, the composed web process should satisfy not only functionalities but also QoS(Quality of Service) criteria. For example, in the Travel Planner example, the overall duration of the composite service execution should be minimized to fulfill customer expectations.

QoS of web process becomes the key issue when web services model complex and mission-critical applications [7-8]. In order to properly assess QoS of the composite web services, we need to have 1) QoS analysis methods, and 2) QoS models specific for the composite web services. In general, two areas of research are available for *QoS analysis*: mathematical methods, and simulation. While mathematical methods [8-9] have been used to assess overall QoS properties of web services in a cost effective way, these methods make explicit assumptions on usage patterns, and workload of web services, and therefore, sometimes lack reality, such as, abrupt change of usage patterns, and unexpected network overheads. Another alternative is to utilize simulation techniques. Simulation plays an important role by exploring "what-if" questions during the composition of web services.[10] *QoS model* of web processes is a combination of several qualities or properties of a service, such as availability, security, reliability and performance. [8,10]. Such a model makes possible the description of Web services and web processes according to their timeliness, cost of service, and reliability. As pointed out in Ref [8], QoS of web services have to be evaluated from various perspectives of the providers and users of web services.

- *Service users* who invoke web services to create all or part of their applications
- *Service providers* who create, expose and maintain a registry that makes those services available

For example, in the Travel Planner example, providers of web services (such as the airline-booking web service) are interested in detailed performance on individual web services, while the users of these services (such as travel agent site) are interested in the overall performance of the composite web services. Most of the existing researches lack explicit support to provide various perspectives of QoS for both services providers and service users.

In this paper, we propose a simulation-based methodology to compose web services and guarantee QoS with multiple levels of perspectives. The proposed methodology 1) consults with various users to elicit their business requirements and QoS criteria, 2) helps users to compose web services, and 3) simulates the composed web services to analyze QoS. By providing tailored QoS information to different types of users, our methodology helps to correct and improve the design of the web process.

This paper is organized as follows. Section 2 proposes a general framework to enable QoS-enabled web service composition with detailed explanation on core elements. Section 3 illustrates how QoS is considered in web service composition with multiple perspectives for service providers and service users through Travel Planner example. Section 4 concludes this paper with summaries and future works.

2   Simulation-Based Web Service Composition Methodology

Figure 2 shows the general framework to enable QoS-considered web service composition. Service users browse and search web services to satisfy their business goals and QoS requirements with the help of *Browser* and *Searcher*. Through Web *Process Composer*, service users graphically specify what kinds of web services to have, how these web services interact with each other, and where QoS constraints are imposed. *Web Process Description Handler* translates graphical representation of the web process into web specification languages (ex. WSDL, BPEL4WS, etc.) with QoS extension. *Web Process Interpreter* generates executable codes for the web process, and *Simulation Model Generator* produces the corresponding simulation models to analyze the QoS of the web process. *Execution Handler* executes the generated model and sends results to QoS Monitor. *QoS Monitor* analyzes outputs and evaluates QoS of the web process. Then, the evaluation results are feedback to Web Process Composer. The simulation is repeated until the desired QoS is achieved. Modeling and Simulation can utilize reuse repository; QoS-proven web services are stored in *Web Service Repository*, and be reused later for creating other web services.

During the web service composition, Web Process Composer is responsible to interact with service users and service providers to achieve the desired QoS of the composite web services. As in Refs [8], service users and service provides have different QoS perspectives:

- *Service providers*: In order to offer best-effort QoS policy, service providers may establish bounds on various QoS metrics in their service level agreements (SLAs). Examples of conditions that an SLA may contain include the average and worst-case response time and average throughput for the web services, collectively and individually. Also, information on how throughput and response time varies with load intensity (eg. number of simultaneous users, message sizes, request frequency) is useful for performance-critical applications.

- *Service users*: In order to accomplish QoS requirements for business needs, service users are mainly interested in the average response time and average throughput, collectively. Service users often set a QoS objective function and try to solve the function with some constrains to satisfy. For example, service users may want to maximize the number of simultaneous users as long as 95% of the simultaneous users get responses within 3 seconds. QoS objective functions may be more complicated when conflicting QoS properties are considered at the same time. For example, service users may want to maximize the number of simultaneous users and minimize the developmental cost at the same time as long as availability, reliability, and security levels are managed under the specified thresholds.

Web Process Composer maintains various scenarios to meet the different QoS perspective of web service users. Figure 3 and Figure 4 show part of simulation scenario profiles with inputs, outputs and types of queries. For each scenario, simulations are performed to determine if the desired QoS has been achieved or not. Since web services are inherently nondeterministic due to network delays and usage patterns, simulation is a useful technique to analyze the composite web services more realistically.

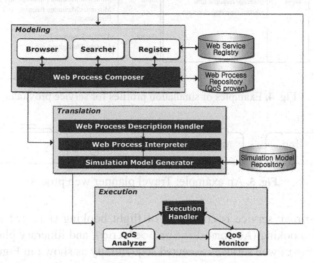

Fig. 2. Simulation-based Web Service Composition: Framework

Scenario 1		Scenario 3	
		Objective	Throughput Percentile
Scenario 2	Overall throughput	Scenario 4	What is the maximum number
Objective	Overall response time	Objective	Response Time Percentile
Type of Query	What is the overall performance of web process as the number of simultaneous users increases?	Type of Query	What is the maximum number of users to satisfy the given response time and percentile?
Input	•Minimum / Maximum number of simultaneous users •Simulation step	Input	•Minimum / Maximum number of simultaneous users •Simulation step •Response time threshold
Output	•Average response time •Worst- case response time	Output	•Optimal number of users •Average response time •Average throughput

Fig. 3. Examples of simulation profiles for service users

3 QoS Analysis: Travel Planner Example

In this section, we illustrate how performance properties of web services are assessed for service users and service providers through Travel Planner example:

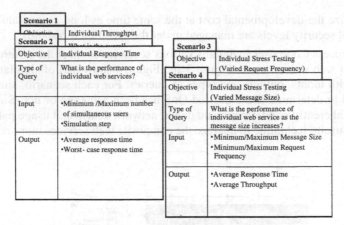

Fig. 4. Examples of simulation profiles for service providers

Fig. 5. An example: Travel planner web process

`TravelPlanner` service is composed of flight booking (`FlightBooking`), accommodation booking (`AccomodationBooking`), and itinerary planning (`ItineraryPlanner`) web services executed sequentially as shown in Figure 5.

Assume that multiple versions of a web service are available for the stacked rectangles in Figure 5: `AccomodationBooking` service and `ItineraryPlanner` service can be performed with two and three different web services, respectively.

Each of the services has different levels of availability and reliability as explained later in Section 3.2. Availability of a service is the probability that the service is accessible as defined in equation (1).

$$\text{Availability} = \texttt{T(S)} / \Delta, \tag{1}$$

where `T(S)` is the total amount of time (in seconds) in which service S is available during the last delta seconds. Delta is a constant set by an administrator of the service community. Reliability is of a service is the probability that a request is correctly responded within the maximum expected time frame as defined in equation (2).

$$\text{Reliability} = \texttt{N(S)} / \texttt{I}, \tag{2}$$

where `N(S)` is the number of times that the service S has been successfully delivered within the maximum expected time frame(which is published in the Web service description), and `I` is the total number of invocations.

Performance is assessed in terms of response time and throughput by a series of simulations. Section 3.1 and Section 3.2 illustrate the simulation phase.

3.1 Performance Analysis for Service Providers

Below is the list of questions specifically useful to service providers.

- What is the response time of the individual web services?
- What is the throughput of the individual web services?
- What is the trace of response time for the individual web services as message size increases?

By using the simulation profiles in Figure 4, response time and throughput are analyzed for the individual web services with varying user load. Figure 6 show that response time increases as the number of simultaneous user's increases. We also experiment how response time and throughput vary as message size increases. As shown in Figure 7, response time drastically increases when message size exceeds 15K. When message size becomes 15K, we further analyze response time with varying request frequency. As in Figure 7, more requests cause late response time. Therefore, it is advised to chunk messages in an affordable size to avoid overheads from the frequent requests.

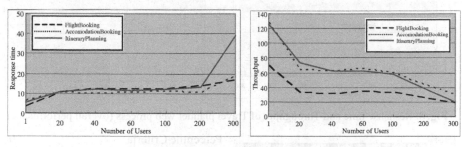

Fig. 6. Response time and Throughput of Travel planner: Individual web service

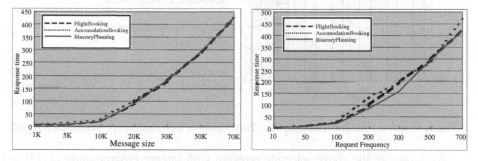

Fig. 7. Trace of response time with varied message size and request frequency

3.2 Performance Analysis for Service Users

Below is the list of questions specifically useful to service users.

- What is the overall response time of the composite web services?
- What is the overall throughput of the composite web services?
- How many simultaneous users can be handled if 90% users can get a response within 15 ms in average?

Figure 8 uses simulation profiles illustrated in Figure 3 and shows how response time and throughput change as the number of simultaneous users increases. Figure 9 answers how many users can be managed if 90% of simultaneous requests are responded within 15 milliseconds. The analysis suggests that we can maximally manage 40 simultaneous users to satisfy percentile and response time criteria.

Moreover, service users may want to select providers on the basis of QoS. In our example, we have 2 candidate providers for `AccomodationBooking` and 3 candidate providers for `ItineraryPlanner`. Assume that services users want to ask the following question:

- In my new web process, which web services should be included to minimize response time as long as availability and reliability requirements are managed under the given thresholds? (Assume that availability >= 0.95, and reliability >= 0.90, number of simultaneous users = 300)

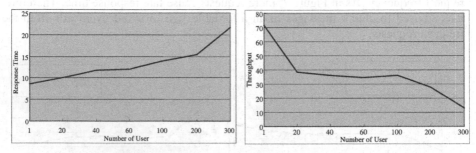

Fig. 8. Response time and Throughput of Travel planner web process (overall performance)

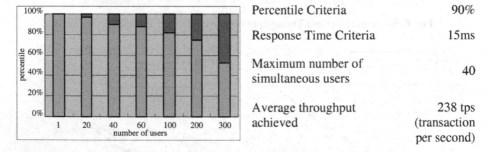

Percentile Criteria	90%
Response Time Criteria	15ms
Maximum number of simultaneous users	40
Average throughput achieved	238 tps (transaction per second)

Fig. 9. Response time and Throughput of Travel planner web process (percentile considered)

We formulate the above selection problem with Integer-Programming [11] based approach. Let a binary integer variable L_i denote the decision to select or not to select the web service, S_i.

$$L_i = \begin{cases} 1 & \text{if web service } S_i \text{ is selected} \\ 0 & \text{otherwise} \end{cases} \tag{3}$$

Then, the objective function for the selection problem is defined as in equation (4)

$$\text{Minimize} \sum T_i L_i \tag{4}$$

Subject to

$$\prod (A_i L_i) \geq A_c \tag{5}$$

$$\prod (R_i L_i) \geq R_c \tag{6}$$

T_i, A_i, and R_i are the response time, availability, and reliability of web service S_i, respectively. A_c, and R_c are the threshold of availability and reliability, respectively. Equation (5) and Equation (6) are the cumulative function of availability, and reliability, respectively, for the composite web services.

Table 1. Availability and Reliability of web services in Travel Planner

Abbrev.	Web Service Name	Availability	Reliability	Response Time
S_1	FlightBooking	0.99	0.98	18.78
S_{21}	AccomodationBooking1	0.98	0.98	16.35
S_{22}	AccomodationBooking2	0.98	0.95	17.22
S_{31}	ItineraryPlanner1	0.98	0.98	38.79
S_{32}	ItineraryPlanner2	0.95	0.95	30.33
S_{33}	ItineraryPlanner3	0.99	0.99	48.10

Table 1 shows availability and reliability of individual web services specified by users. Availability and reliability may take probability distributions instead to enhance randomness. For example, Availability of AccomodationBooking1 can take *weibull* distribution (eg. α =2, β = 1). In Travel Planner example, we only take the average values for simplicity. Response times are assessed from simulations when the number of simultaneous users is 300.

In the Travel Planner example, the optimal web process is determined by solving Equation (4) with the constraints defined in Equation (5) and (6). The optimal QoS is achieved by assembling S_1, S_{21}, S_{31} services by considering simulation results and QoS data given in Table 1.

4 Conclusion

We proposed a general framework to guarantee QoS of web services. The key to our method is to use simulation technology to analyze the QoS of web services; Users specify the service flow of web services with QoS concerns. Then, simulations are carried out according to the interesting scenario profiles. Finally, simulation results are sent back to users in order to find and improve QoS problems. We realized that various types of QoS information would be useful to meet multiple perspectives of users and providers of web services. Several simulations could answer simple QoS concerns, for example, the overall response time and throughput, while more complicated QoS concerns could be defined and solved by IP-based approaches. More detailed analysis on the individual web services has been done for service providers to provide Service Level Agreement.

Web service composition can be done either statically or dynamically. The decision of whether to make a static or dynamic composition depends on the type of process being composed. If the process to be composed is of a fixed nature wherein the business partners/alliances and their service components rarely change, static composition will satisfy the needs. However, if the process has a loosely defined set of functions to perform, or if it has to dynamically adapt to unpredictable changes in the environment, then the web services to be used for the process are decided at run-time [12]. Our methodology is suitable for static composition considering the complexity of simulation and the IP-based selection. However, if there is an efficient heuristic method that solves the selection problem in a tractable time, the web services are selected on the fly. We would like to continue out efforts to compose web services dynamically.

Acknowledgement

This work was supported by grant No. R05-2004-000-11329-0 from Korea Science & Engineering Foundation.

References

1. Preeda Rajasekaran, John A. Miller, Kunal Verma and Amit P. Sheth, *Enhancing Web Services Description and Discovery to Facilitate Composition,* In Proceedings of the First International Workshop on Semantic Web Services and Web Process Composition (SWSWPC'04), San Diego, California , 2004, (to appear)
2. W3C, *Web services Architecture Requirements*, W3C Working Draft, http://www.w3.org/TR/2002/WD-wsa-reqs-20020429
3. Leymann, F, *Web service flow language (WSFL) 1.0,* http://www-4.ibm.com/software/solutions/webservices/pdf/WSFL.pdf, Sep. 2002
4. Thatte.S, *XLANG: Web Services for Business Process Design,* http://www.gotdotnetcom/team/xml_wsspecs/ xlang-c/default.htm, Spec. 2002
5. Curbera, F., Goland Y., Klein. J., Leymann F., Roller D., Thatte S., and Weerawarana S., *Business Process Execution Language for Web services,* http://msdn.microsoft.com/webservices/default.asp?pull='library.enus/dnbiz2k2/html/bpel1 -0.asp, Sep. 2002
6. Ankolekar, A., Burstein, M. Hobbs, J. Lassila, O., Martin, D. McDermott, D., McIlraith S., Narayanan, S. Paolucci, M. Payne, T., and Sycara K, *DAML-S: Web Service Description for the Semantic Web*, In International Semantic Web Conference, Sardinia, Italy, pp. 348-363
7. Liangzhao Zeng, Boualem Benatallah, Marlon Dumas, *Quality Driven Web Services Composition*, In Proceedings of WWW2003, May 20 – 24, 2003, Budapest, Hungary, ISBN 1-58113-680-3/03/0005, pp. 411 - 421
8. Daniel A. Menasce, *QoS Issues in Web Services*, Nov/Dec., 2002, IEEE Internet Computing, pp. 72 – 75
9. Cardoso, J., *Stochastic Workflow Reduction Algorithm,* http://lsdis.cs.uga.edu/proj/meteor/QoS/SWR_Algorithm.htm
10. Gregory Silver, John A. Miller, Jorge Gardoso, Amit P. Sheth, *Web service technologies and their synergy with simulation*, In Proceedings of the 2002 Winter Simulation Conference, 606 – 615
11. D.P. Ravindran., J.J. Solberg., *Operations Research*, Wiley, New York, 1987
12. Senthilanand chandrasekaran, *Composition, performance analysis and simulation of web services*, Master Thesis, 2002, University of Georgia

Research and Development on AST-RTI

Yongbo Xu[1], Mingqing Yu[1], and Xingren Wang[2]

[1] The First Institute of aeronautics of Air Force, Xinyang China
[2] Advanced Simulation Technology Aviation Science and Technology Key laboratory,
Beijing University of Aeronautics & Astronautics, Beijing, China
xu_yong_bo@163.com

Abstract. High level architecture, as a new standard of distributed interactive simulation, facilitated simulation components interoperability and reusability, Run time infrastructure is software implementation of HLA, whose run-time performance is important for simulation applications. We have developed an alternatives version of RTI, named AST-RTI, which complies with HLA Interface Specification version 1.3. All services have been realized with excellent performance. In this paper, the principle of RTI design will be proposed, and the time management, resource storage as well as communication mechanism in AST-RTI will be discussed.

1 Introduction

Distributed Simulation architecture has been changed from DIS (Distributed Interactive Simulation); ALSP (Aggregate Level Simulation Protocol) to HLA (High Level Architecture). As a new advanced simulation standard, HLA is much different from the others. It has standard criterion and provides two desirable properties: reusability and interoperability for applications. RTI (Run-Time Infrastructure) is software that conforms to interface specification of HLA. It divides the simulation operation layer from lower communication layer, and provides runtime support environment for a variety of different simulation scenarios with different time advance types.

There are many versions of RTI, for example , Defense Modeling and Simulation Office (DMSO)RTI1.3; RTI NG ;open source code lightweight RTI of GMU George Masion University – and commercial versions pRTI; MÄK RTI etc. Advanced Simulation Technology Aviation Science and Technology Key laboratory in Beijing University of Aeronautics and Astronautics (BUAA) has developed AST_RTI.The application showed the design and implementation had improved the RTI running performance effectively. In this Paper, design of software architecture; time management and resource storage and communication mechanism will be discussed.

1.1 The Whole Structure of AST_RTI

The Interface Specification (IF)of HLA can be incarnated in these three layers in interoperability architecture: the application layer, the model layer and service layer. In application, we found some problems in DMSO's RTI 1.3v5, such as the slow initiation, error block in some cases. Its performance in real-time system will lower in complex systems. For this reason, we developed AST_RTI, which has distributed

D.-K. Baik (Ed.): AsiaSim 2004, LNAI 3398, pp. 361–366, 2005.
© Springer-Verlag Berlin Heidelberg 2005

real-time scalable network framework of loose and tight coupled topology. It can be transplanted on Windows/UNIX. It is implemented in distributed-concentrative type. The layer design thought is carried out in network layer. The virtual network layer encapsulates the communication in component or library, and then the service layer is separated from network layer.

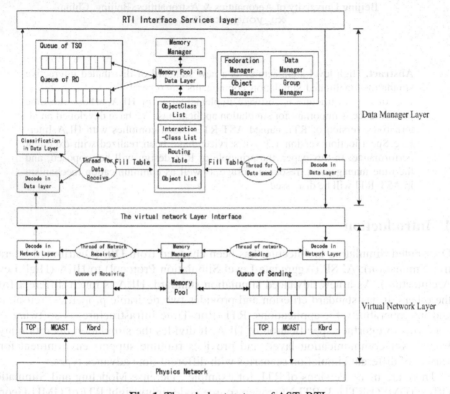

Fig. 1. The whole structure of AST_RTI

AST-RTI has hierarchical structure, and it can be divided to three layers: interface services layer; data management layer and virtual network layer. It is shown in Fig. 1. This thought can reduce the design complexity and be well for maintenance and reusability. Different from DMSO RTI, we define RTIAmbassador as abstract class for interface extension. The functions are also encapsulated to a couple of classes, such as federation management class; declare management class; time management class, etc. the logic management, just as kernel in operating system, control calls what functions in right sequence. It will be coupled if functions be called, with different management class used at same time. In AST_RTI, the loose-coupling components are used for extension and reusability.

1.2 Thread Model

The implementation mechanisms are different in different versions of RTI. According to type of task handling, it can be divided to two types: single-threaded and multi-threaded. Type of single-threaded can handle tasks in a single queue, but may be blocked. Type of multi-threaded can assign time slices for every task, reducing CPU's idle time. It should be pointed out that the efficiency in multi-threaded may be not higher than single thread. Multi-threaded can reduce block chance for distributed simulation applications, which handles multi-tasks. There are three types of thread models: single-threaded; asynchronous IO (AIO) and parallel. The first two are used in DMSO RTI 1.x NG.

Single-threaded process switches threads between RTI internal services and application services, with function tick() being called. AIO process can handle internal services in main thread and handle network tasks in network thread. It will read network data in local queues when application call tick(), and reflect callback functions based on the results. CPU idle waiting time can be reduced in AIO process, even with more time cost in thread switches. In these two mechanisms, application layer should call tick() function to get time slices. Parallel models were used in AST-RTI. There are a couple of threads working in RTI. The complication thread calls for RTI services, and then RTI can do its own work, reflect callbacks, with it internal threads coordinated with complication thread. The function tick() is no more needed. Full multi-threaded was implemented in AST-RTI, with thread pool used. The optimized thread methods can increase the efficiency and lower the cost of threads switches.

2 Time Management

There are two aspects in time management: coordinate and advance. "Coordinate" means time-constrained federates never receive Time-stamp-order (TSO) messages with time stamp earlier than their current time. "Advance" means federates advance their local logic time after they have handled all relevant TSO messages. Time-regulating federates can send TSO messages with timestamp greater than its current time, and time-constrained federates should receive these messages.

Time-constrained federates must query Lower Bound Time Stamp (LBTS), and LBTS will be changed if time-regulating federates advance their logic time. Frequently LBTS query/change will make time management bottleneck in large scale systems. In this paper the method of LBTS query is changed from active query to passive update. Time-constrained federates no more query LBTS to FEDEX because query results may be same. It will greatly lower LBTS communication times when RTI informs time-constrained federates only if LBTS has been changed. For time-constrained federates, LBTS will be changed if and only if the slowest one of all time-regulating federates has advanced its time. So, the frequency of LBTS transfer must be lower or equal to iterative frequency of the slowest federate of all time-regulating federates.

3 Resource Storage

RTI needs to handle both local data and remote data in distributed simulation nodes. The data consistency must be kept in distributed environment. We defined optimized data structure in AST_RTI, including static data and dynamic data. Static data consists of information of object classes; interaction classes; attributes and parameters from system fed file. Dynamic data has information of data of system in runtime, such as information of federation; federates; object instances and value of attributes and parameters. Handles of static data are unique in system and can not be changed in runtime, while handles of dynamic data may be changed in runtime.

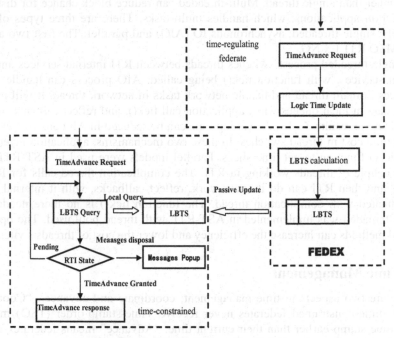

Fig. 2. The implementation flow

Data index operation from name to handle, or from handle to name, is required in declare management and object management. Ancillary services have been defined in HLA for these functions. Time cost will be high if RTI reads the fed file many times. So, static data structure is defined in AST_RTI for efficient data storage and rapid data access, as shown in fig.3. The mapping structure of static data is maintained in memory, including handles and names of them for index. Federates will read the fed file only once when application is initiated, so that many times of hard disk access will be avoided. Static data structure and dynamic data structure work together to achieve efficient data access.

4 Communication Mechanism

RTI components consist of the Central RTI Component (CRC) and the Local RTI Component (LRC). CRC has two parts: RTIEX and FEDEX. RTIEX is the server that manages all federations, and FEDEX is the server that manages all federates in the federation. LRC is embedded into federates, providing services interface for applications.

There are two kinds of data transfer flows in RTI: control stream and data stream. As shown in fig.4.a, FEDEX will be bottleneck of system if it handles all these data. The data strategy that distinguishes control stream and data stream has been implemented in AST-RTI, shown in fig.4.b. FEDEX can only handle management of communication channels and federation. In runtime, a great deal of data transfer occurred among different LRC, not between LRC and FEDEX.

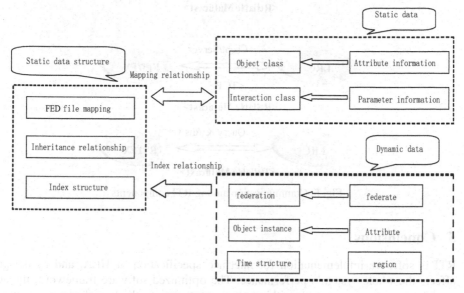

Fig. 3. The resource storage

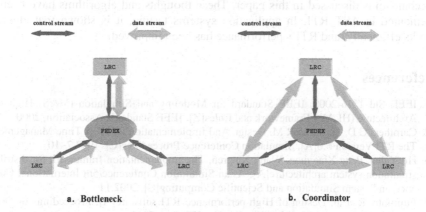

a. Bottleneck   b. Coordinator

Fig. 4. The role of FEDEX

The communication scheme among RTI components in AST-RTI is shown in fig.5. Data using data transportation schemes "reliable" is transferred in multicast channels. Data of "best effort " is transferred in reliable multicast channels. TCP connections are used between LRC and FEDEX for federation management. The global management information is transferred in reliable multicast channels between LRC and RTIEX. It should be pointed out that multicast services based on TCP/IP protocol uses UDP protocol, and then they are not reliable. The lost-resend mechanism and stream control has been used in AST-RTI for reliable multicast.

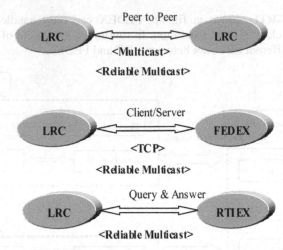

Fig. 5. Communication among RTI components

5 Conclusions

RTI is software implementation for interface specification of HLA, and its design affects the whole system's performance. The optimized software framework; thread model; time management algorithm; resource storage methods and communication mechanism is discussed in this paper. These thoughts and algorithms have been implemented in AST_RTI. In application systems testing, it is shown that algorithm works efficiently, and RTI's performance has been improved.

References

1. IEEE Std 1516-2000, IEEE Standard for Modeling and Simulation (M&S) High Level Architecture (HLA) – Framework and Rules[Z]. IEEE Standards Association, 2000
2. Carothers C D, Fujimoto R M. Design And Implementation Of HLA Time Management In The RTI Version F.0[A]. Simulation Conference Proceedings[C]. 1997.7~10
3. Hu Yahai, Peng Xiaoyuan, Wang Xingren, The implementation framework for distributed simulation system architecture[A], Asian Simulation Conference/5th International Conference on System Simulation and Scientific Computing[C], 2002.11
4. Fujimoto, R.et al, Design of High-performance RTI software[A], Proceedings of Distributed Simulations and Real-time Applications[C], Aug.2000

HLA Simulation Using Data Compensation Model

Tainchi Lu, Chinghao Hsu, and Mingchi Fu

Dept. of Computer Science and Information Engineering, National Chiayi University
No. 300, University Rd., Chiayi, Taiwan 600, R.O.C.
tclu@mail.ncyu.edu.tw

Abstract. In this paper we aim to incorporate wireless and mobile network with IEEE standard 1516 high level architecture (HLA). The proposed HLA federation not only provides seamless data communication but also supports the higher host interoperability for large-scale simulations. As the contemporary wireless LAN is concerned, it inherently sustains data communication suspension and intermittent message loss as long as a phenomenon known as handoff occurs. Accordingly, we devise data compensation models (DCM) to cope with data loss problem. In particular, the DCM neither aggravates any extra computing loads in federates nor exhausts memory storage in the period of mobile client's short-term disconnection. As a consequence, the simulation results indicate that the proposed architecture of HLA federation supports the applicable mobility management, seamless data communication, and low system resource consumption for wireless and mobile distributed applications.

1 Introduction

Nowadays, demands for Internet technology have been rapidly growing to cooperate with distributed computation, wireless communication, and mobility management [1]. The modern networking usage for most people is not restricted with cable connections, desktop computers, and in fixed places. The motivation for us to integrate wired/wireless LAN with IEEE 1516 High Level Architecture (HLA) is to support mobile users to join the HLA distributed federation [2] in any time and any where by any device with seamless data transmission. High Level Architecture [9], it is an integrated and general-purpose distributed architecture, which is developed by the U.S. DOD and proved by the IEEE to provide a common interoperable and reusable architecture for a large-scale distributed modeling and simulation. Therefore, we perform the mobility management for mobile users by means of using three significant components corresponding to the HLA distributed federation, namely IP-based mobility management agent (IP-MMA), federate agent (FA), and data compensation model (DCM). No matter where the mobile users move to, they can still gain lower network latency through the dispatching and forwarding operations of the IP-MMA, and get higher data communication efficiency by the FA's DCM.

However, an intermittent disconnection of the mobile client is possibly occurred as long as mobile clients roam to the other place and have handoff procedures between different network areas [3-8]. Note that when the client is under disconnection, the other clients are still exchanging update messages with each other via the FAs and HLA RTI. Consequently, this mobile client is forced to sustain state inconsistency,

D.-K. Baik (Ed.): AsiaSim 2004, LNAI 3398, pp. 367–375, 2005.

message loss, and so forth. For most sophisticated systems, to cache update messages in the buffer of a server is a straightforward approach to overcoming the data loss problem. Nevertheless, the limited memory capacity of the server suffers from an overflow problem for lack of garbage collection. Due to the above reason, we propose two kinds of DCMs, one is conservative backup-starting-point (BSP) scheme and the other is optimistic time-stamp-ordered (TSO) scheme, to continuously reposit the necessary update messages in FA's shared buffer and periodically remove the stale data as well as performing the garbage collection. When a FA occasionally receives a disconnection exception from a mobile client, it makes use of the DCM to backup the messages until the client promptly resumes its network connection.

Moreover, we take advantage of HLA time management service to guarantee data-centric consistency and absolute event order by means of generating time-stamp-ordered (TSO) messages. The HLA time management concerns with the mechanisms for controlling the time advancement of each federate, such as time-steeped and event-driven mechanisms. As the message latency is concerned, a message by specifying a time stamp could be queued in a federate's TSO priority buffer to wait for processing, because the federate would be too busy to handle with the numerous messages in real time. As a consequence, it would have an influence on the data compensation rate of the DCM TSO scheme.

2   Related Work

In the past decade, a variety of checkpoint-recovery and message logging methods has been proposed for distributed systems, such as sender-based logging, receiver-based logging, and distributed snapshot [10]. These methods provide an efficient management of the recovery information and bring the system from its present erroneous state back to a previously correct state. However, these schemes cannot be directly implemented without any sophisticated alteration for the wireless and mobile distributed environments. Note that one of the low bandwidth of wireless networks, poor signal quality of trouble spots, and limited battery capacity of mobile devices could lead to a tough problem for system developers. Consequently, the above issues make it hard to use the schemes to embrace a large number of control signals and a vast amount of information in an interactive message. In 2003, Taesoon proposed an efficient recovery scheme [11] based on message logging and independent checkpoint-recovery to provide fault-tolerance for the mobile computing systems. Based on Taesoon's scheme, while a mobile host moves within a certain range, recovery information of the mobile host is not moved accordingly. If the mobile host moves out of the range, it transfers the recovery information into the nearby support station. Hence, the scheme controls the transfer cost as well as the recovery cost.

In comparison with [11], we also make good use of the checkpoint-recovery and message logging methods to recover from a fault for both fixed and mobile hosts. In addition to employing the HLA distributed simulation framework, the proposed distributed federation comprises a located-based mobility management, agent-based service, and data compensation model.

3 System Architecture

In this section, we propose the architecture of an HLA distributed federation and further describe three crucial components of the architecture: federate agent (FA), IP-based mobility management agent (IP-MMA), and data compensation model (DCM).

3.1 Federation and Federate

With respect to the official specification of HLA, each participating simulation is referred to as a joining federate f. Likewise, the collection of federates which are interconnecting through the run-time infrastructure (RTI) is referred to as an HLA federation F. In other words, the federation F is a set up federates and its definition is shown in (1).

$$F = \{f_{ij}\}, \quad where \quad i, j \in Z^+ \tag{1}$$

We designate several individual sub-networks within a local area network (LAN) as a well-defined network area A_i and further build up specific HLA federates, which are referred to as federate agents (FAs) f_{ij} to provide message exchanging and data buffering services. The FAs are widely spread across the whole specified areas to aggregate an HLA federation F. The relationship among an area A_i, a federate agent f_{ij}, and a federation F is expressed in (2) and (3).

$$A_i = \bigcup_{j=1}^{J_i} f_{ij}, \quad where \quad 1 \le i \le I \quad and \quad I, J_i \in Z^+ \tag{2}$$

$$F = \bigcup_{i=1}^{I} A_i, \quad where \quad I \in Z^+ \tag{3}$$

Furthermore, the process interconnections among FAs are simply achieved by the HLA run-time infrastructure, which provides the six major services for simulators to coordinate the operation and exchange update messages. Within an area A_i, an IP-MMA is established to manage the FAs' working and handle with the connecting routes of client hosts. As long as all FAs initially declare to join in the federation, they first need to have registration at their specified IP-MMA in the identical area. The IP-MMA keeps the registration information of each FA in its FA lookup table in order to perform the underlying dispatching operation regarding to the client hosts.

We illustrate the proposed HLA distributed federation in Fig. 1. Fig. 1 shows one HLA federation and three individual areas, which each area comprises one IP-MMA and several FAs. In practice for our experimental environment, we adopt 1000Mbps Gigabit Ethernet to communicate three areas with each other and set up the IEEE 802.11b wireless networks in the range of each area. For instance, in the area 1, the mobile client 1 attempts to connect to the IP-MMA 1-1 for being dispatched to a FA by means of associating with a WLAN access point and following the IEEE 802.11b protocol. The relevant TCP/IP network settings, such as IP address, DNS, and gate-way, are dynamically specified by the DHCP server which is set up in the area 1. The IP-MMA 1-1 first makes use of the forwarding client mechanism to identify whether the client 1's IP address is fell into the network domain of the area 1 or not. If so, the

IP-MMA 1-1 refers to its local FA lookup table to select an eligible FA with low loading value and dispatches the FA to the client 1 for further connection. After the client 1 successfully connected to the FA 1-1, it can propagate its update messages to the whole HLA federation by means of the FA 1-1's message relaying services. Note that the FA 1-1 sends these messages with time stamps to the federation via HLA RTI time management service. The RTI can guarantee that no matter where the other FAs locate inside and outside the area 1 can receive the broadcasting update messages from the FA 1-1 in order and with consistency. Accordingly, the fixed and mobile client hosts with reliable network connection are capable of receiving the client 1's update messages via the fore mentioned message propagation model. Besides the forwarding client mechanism, we adopt the data compensation model (DCM) to overcome the data loss problem, which is taken place in the period of the mobile client's intermittent disconnection phase.

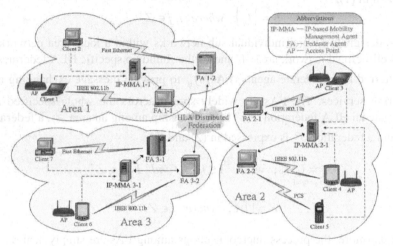

Fig. 1. A demonstrated HLA distributed federation

3.2 Dispatching Operation

In this sub-section, we take a look at the underlying dispatching operation regarding to an IP-based mobility management agent (IP-MMA). The IP-MMA is set up within an independent network area to not only dispatch client hosts to connect with HLA federate agents but also provide substantial mobility services to the moving clients. Fig. 2 shows the connection diagram corresponding to the IP-MMA and a mobile client in a single area. In the following, the six steps are required to accomplish the underlying dispatching operation but except the process of the forwarding client mechanism.

Step 1: An HLA federation is declared to create and start a well-defined distributed simulation, all the active federate agents associated with the federation need to make a registration at their specified IP-MMA.

Step 2: The IP-MMA automatically configures each FA's current connecting circumstance to make sure that every FA is alive with reliable communication.

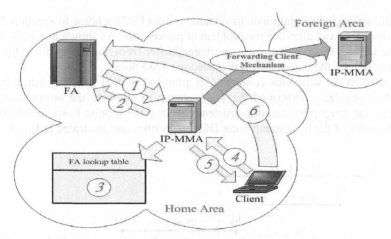

Fig. 2. A client connects with the appointed FA by means of the underlying dispatching operation

Step 3: The IP-MMA records the current registration information of the FAs in its local FA lookup table for the sake of executing the client forwarding and dispatching operations.

Step 4: The client can arbitrarily connect to any IP-MMAs that are explicitly belonging to the HLA federation. By means of performing the mechanism of forwarding client, the IP-MMA can forward a mobile client to an appropriate area, which is closer to the current geographical location of the client.

Step 5: To consider the case that the IP-MMA and a mobile client collocated in the identical area. The client connects to the specified IP-MMA for the sake of waiting for IP-MMA's dispatching operation. Consequently, the IP-MMA checks the client's present care-of-address against its FA lookup table to find out an appropriate FA. Furthermore, the IP-MMA writes the client's current connecting information to a message log in order to persistently keep track of its up-to-date connecting status.

Step 6: Thereafter, the client is dispatched by the IP-MMA to connect to the specified FA, and then it successfully joins in the HLA federation for participation.

3.3 Data Compensation Model

We propose a data compensation model (DCM) to deal with packet loss and data inconsistency problems when mobile hosts are in the process of handoff. For example, a DHCP server dynamically assigns the relevant TCP/IP network settings (e.g. IP address, DNS, and gateway) to a mobile host, which is moving from the preceding cell with outdated network settings, an intermittent disconnection is probably occurred at once. In other words, the intermittent disconnection is taken place because a handoff procedure turns up through the cells or an IP address of the mobile host is substituted to another care-of-address. In the meanwhile, all interactive messages sent from federate agents are entirely lost during client's network disconnection phase. As a result, we devise an optimistic time stamp order (TSO) scheme to speedup the buff-ering operation and alleviate the problem of packet loss. As shown in Fig. 3, the TSO scheme keeps backup messages in the shared buffer from the TSO point to the current time and deletes the data which is preceded the TSO point.

As long as the wallclock is allowed to grant to the next one, the shared buffer of the FA is regularly performed the TSO scheme to alleviate the buffer overflowing problem and keep information consistency among the mobile hosts. In addition, the pseudo codes of the time stamp order DCM algorithm are illustrated in Fig. 4.

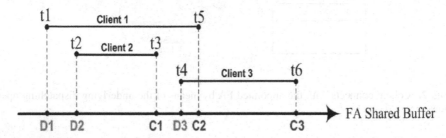

Fig. 3. The DCM_TSO employs time-stamp-ordered messages to flag the states

4 Simulation Results

In this experiment, we first assume that there are not any communication interrup-tions or network disconnections occurred in the mobile nodes. In addition, we take the buffer size as a major factor to evaluate the time stamp order (TSO) scheme. With respect to the optimistic TSO scheme, it relies on the value of a maximum delay fac-tor ε to periodically remove the unnecessary interactive messages from the shared buffer. The value of ε has been estimated to be eight in our experimental environ-ment. Moreover, we further take into account the double and half values of the esti-mated ε to compare with each other.

In contrast with the preceding experiment, we suppose that a mobile node could move to any network cell with handoff process within an area range. Moreover, the buffer size is also regarded as a crucial factor to compare the conservative BSP scheme with the optimistic TSO scheme and single buffer method. We arrange a client connection schedule in advance to take a look at the variances in the shared buffers, which are maintained by different buffering approaches. The schedule is given in Table 1. There are total ten mobile clients simulated in the environment. For instance, the client 1 disconnected from a FA at wallclock time 1 and resumed its network connection at wallclock time 9.

The simulation result is shown in Fig. 5. In Fig. 5, we can see that both of the sin-gle buffer method and TSO scheme reposit backup messages in their shared buffers when an intermittent disconnection occurs to a mobile client. Moreover, concerning to the conventional single buffer method, it creates an individual buffer for each dis-connected mobile client; hence, it relatively requires more buffer capacities to support more disconnected mobile clients. In contrast, the TSO scheme only creates a shared buffer for all disconnected mobile clients and makes good use of HLA time

---

DCM_TSO ( )

D01 Create $B_{i,j}$

D02 $D_m \leftarrow 0$

D03 $C_n \leftarrow 0$

D04 $isDisconnected \leftarrow$ FALSE

D05 **while** System is Alive

D06   **do switch** messages received from the RTI

D07   **case** m_{iact}

D08   **if** $isDisconnected =$ TRUE

D09   **then** $B_{i,j} \leftarrow m_{iact}$

D10   **case** m_d

D11   **then** DISCONNECTION (m_d)

D12   **case** m_{rc}

D13   **then** RECONNECTION (m_{rc})

DISCONNECTION (m_d)

D01 $m \leftarrow m + 1$

D02 Associate D_m with the UID field of the m_d

D03 Mark the D_m in $B_{i,j}$

D04 $isDisconnected \leftarrow$ TRUE

RECONNECTION (m_{rc})

R01 $n \leftarrow n + 1$

R02 Associate C_n with the UID field of the m_{rc}

R03 Mark the C_n in $B_{i,j}$

R04 **for** search mark in $B_{i,j}$

R05   **do if** client's UID of D_m is the same as client's UID of C_n

R06   **then** reconnection client get m_{iact} of $B_{i,j}$ between D_m and C_n

R07   Delete mark D_m and C_n in $B_{i,j}$

R08 **if** any mark is no exist in $B_{i,j}$

R09 **then** $isDisconnected \leftarrow$ FALSE

R10 **else then** $LBIS \leftarrow$ minimum index value of D_m

R11 Keep the data from $LBIS$ in $B_{i,j}$ and delete the preceding data

---

Fig. 4. DCM_TSO algorithm

management service to facilitate the operation of data reposition to the FAs. Likewise, the BSP scheme also maintains a shared buffer for all disconnected mobile clients and it periodically perform the operation of garbage collection by relying on the maximum delay factor ε.

Table 1. The pre-defined client connection schedule

Client Number	Disconnection Time		Client Number	Disconnection Time	
	From	To		From	To
Client 1	1	8	Client 6	25	33
Client 2	3	7	Client 7	26	31
Client 3	6	12	Client 8	32	39
Client 4	19	28	Client 9	50	55
Client 5	21	27	Client 10	52	58

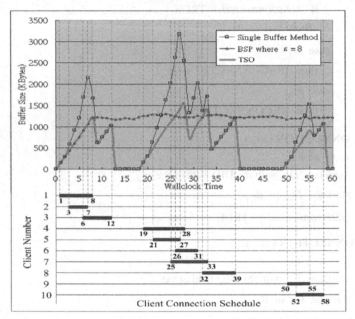

Fig. 5. The buffer variance of the shared buffer associated with the pre-defined client connection schedule

5   Conclusions

In this paper, we have proposed the wireless and mobile distributed federation, which combines wired/wireless local area network with IEEE 1516 high level architecture to provide client hosts with high-efficient message exchanging services and ubiquitous network access. Particularly, the joining participants can optionally hold personal communication devices, PDA, notebook computers, or desktop PCs to directly access the network by depending on their current physical network states. With respect to mobile hosts, when they move from their home network area to another foreign area, a long-term disconnection is unavoidably occurred to disrupt the consecutive data communication. Comprehensibly, even though the mobile host possibly resumes its network connection later, it cannot help but have a connection with the preceding network area due to its outdated connection information is still retained. In order to support location-based service and improve data-exchanging efficiency for a mobile host located in a new foreign area, we have designed the forwarding client mechanism to facilitate mobility management to the HLA federation. The simulation result has revealed that the forwarding client mechanism explicitly works out when mobile clients roam between the two heterogeneous network areas. Moreover, the data compensation model (DCM) has been presented to solve an intermittent data inconsistency problem when mobile clients have handoff across the interior cells within an area. Finally, we have shown that the DCM provides the conservative data compensation service for clients and spends less memory space than the traditional shared buffer method.

Acknowledgement

This project is supported by National Science Council, Taiwan, R.O.C., under contract NSC93-2213-E-415-006.

References

1. J.S. Pascoe, V.S. Sunderam, U. Varshney, and R.J. Loader: Middleware Enhancements for Metropolitan Area Wireless Internet Access. Future Generation Computer Systems, Vol. 18, No. 5. (2002) 721-735.
2. Tainchi Lu, Chungnan Lee, Ming-tang Lin, and Wenyang Hsia: Supporting Large-Scale Distributed Simulation Using HLA. ACM Transaction on Modeling and Computer Simulation, Vol.10, No. 3. (2000) 268-294.
3. Lawrence Y. Deng, Timothy K. Shin, Ten-Sheng Huang, Yi-Chun Liao, Ying-Hong Wang, and Hui-Huang Hsu: A Distributed Mobile Agent Framework for Maintaining Persistent Distance Education. Journal of Information Science and Engineering, Vol. 18, No. 4. (2002) 489-506.
4. A. Kahol, S. Khurana, S. Gupta, and P. Srimani: An Efficient Cache Maintenance Scheme for Mobile Environment. In Proceedings of International Conference on Distributed Computing Systems (2000) 530-537.
5. X.Z. Gao, S.J. Ovaska, and A. V. Vasilakos: Temporal Difference Method-based Multi-Step Ahead Prediction of Long Term Deep Fading in Mobile Networks. Computer Communications, Vol. 25, No. 16. (2002) 1477-1486.
6. Zhimei Jiang, Kin K. Leung, Byoung-Jo J. Kim, and Paul Henry: Seamless Mobility Management Based On Proxy Servers. In Proceedings of IEEE Wireless Communication and Networking Conference, Vol. 2. (2002) 563-568.
7. Jon Chiung-Shien Wu, Chieh-Wen Cheng, Nen-Fu Huang, and Gin-Kou Ma: Intelligent Handoff for Mobile Wireless Internet. ACM Mobile Networks and Applications, Vol. 6. (2001) 67-69.
8. I-Wei Wu, Wen-Shiung Chen, Ho-En Liao, and Fongray Frank Young: A Seamless Handoff Approach of Mobile IP Protocol for Mobile Wireless data Networks. IEEE Transactions on Consumer Electronics, Vol. 48, No. 2. (2002) 335-344.
9. U.S. Department of Defense Modeling and Simulation Office (DMSO), High Level Architecture. Available through the Internet: http://www.dmso.mil/.
10. Andrew S. Tanenbaum and Maarten van Steen: Distributed System: Principles and Paradigms. Prentice-Hall, NJ. (2002) 361-412.
11. Taesoon Park, Namyoon Woo, and Heon Y. Yeom: An Efficient Recovery Scheme for Fault-Tolerant Mobile Computing Systems. Future Generation Computer Systems, Vol. 19, No. 1. (2003) 37-53.

Modeling and Simulation of Red Light Running Violation at Urban Intersections

Jinmei Wang[1,2], Zhaoan Wang[1], and Jianguo Yang[1]

[1] School of Electrical Engineering, Xi'an Jiaotong University, Xi'an, Shanxi, 710049
wjm@mailst.xjtu.edu.cn
{zawang,yjg}@mail.xjtu.edu.cn
[2] School of Physics & Electrical Information Engineering,
Ningxia University, Yinchuan, Ningxia, 750021

Abstract. More occupant injuries occurred in red light running conflicts, compared to other urban conflict types. And red light running behavior of vehicles happens at urban intersections is caused by high approach speeds coupled with aggressive driving. It includes inadvertently red light running that is easy to arouse the sympathizing of traffic managers other than advertently behavior of driver that should be punished. The reason and characteristic of inadvertently red light running is related with width of urban intersection, idea speed, vehicle dynamics nature, yellow interval of signal and dilemma zone. Further more, to enhance inspection and punishment upon red light running violations and to setup rational signal phases and intervals obviously can decrease red light running behaviors. And that safety and reasonable signal phase and intervals should be designed at intersections, also enforcement tolerance for photo red light programs should be considered would reduce dilemma zone. A model of dilemma zone upon inadvertently red light running has been presented, also an example at partial intersections in Xi'an. According to the above ideas, models suitable for red light running at urban intersections are established. The model has been calibrated and tested with data from local area of Xi'an, west in China.

1 Introduction

Excessive approach speed at signalized intersections often causes drivers to enter the intersection during the yellow light phase and sometimes even during the red light. The latter action is considered as a red light violation. Generally, the red light running crashes were characterized by a higher percentage of total injuries than the urban signalized intersection crashes (54 percent vs. 44 percent, respectively) [1]. According to the traffic laws, law enforcement officials hardly issue a penalty to a driver for not stopping during a yellow change interval. But drivers entering intersections when the signal turns red (i.e. a red light running violation) have to undertake punishment. But in practice, these violations are often affected by the duration of the yellow and all-red intervals of the traffic signal [2], i.e. inadvertently red light running behavior of drivers is yet in existence.

Drivers approaching a signalized intersection are required to make a decision whether to stop or to continue through the traffic signal. This decision is often related to many factors, some of which include:

D.-K. Baik (Ed.): AsiaSim 2004, LNAI 3398, pp. 376–385, 2005.

◆ Vehicle approach speed;
◆ Color of the traffic signal when noticed by the driver;
◆ Location of the vehicle with respect to the traffic signal when the yellow light was observed;
◆ Weather condition (rain, snow, sleet); pavement condition (wet, icy, snowy) and type of vehicle;
◆ Nature driver behaviors (aggressive, non-aggressive).

Gazis, Herman and Maradudin observed the behavior of drivers confronted by a yellow signal light [3]. They analyzed the problem facing drivers when confronted with an improperly timed yellow interval and concluded that some drivers could be too close to the intersection to stop safely or comfortably and yet too far from the intersection to pass safely through. This is known as the dilemma zone.

Determining a proper clearance interval is critical to mitigating traffic crash problem at a signalized intersection, and as such it must be carefully used in conjunction with site-specific data. When the signal changes from green to yellow at an intersection, a decision is required [4].

Retting and Greene studied the influence of traffic signal timing on red light running at signalized intersections [5]. The purpose of the study was to examine the effects of signal timing design on red light compliance as a result of an increase in change intervals to values recommended by the Institute of Transportation Engineers (ITE). The ITE procedure computes yellow interval timing as "a function of approach speed and grade, along with assumed values for perception-reaction time, deceleration rate, and acceleration due to gravity." Further, the procedure defines the length of the all-red interval as a function of approach speed and width of the intersection roadway that must be clear. The study conclude that increasing the length of the yellow signal toward the ITE recommendation significantly decreased the chance of red light running and that the length of the all-red interval did not seem to affect red light running.

In allusion to above reasons, a comparative study is formulated in the research. The red light running violations at different intersections in Xi'an were shot, counted, calculated and analyzed. And the timing of traffic signal period at urban intersection is reasonable, especially whether the all-red interval is designed, and whether there is red light running automated camera that significantly influences the red light running violations is affirmed.

Otherwise, in the model we developed below, we do not argue that red light violation cameras and other enforcement mechanisms do not impact voluntary and deliberate red light violations. We augment a simple model that describes the relationship between the width of the intersection; the ideal velocity, and the length of the yellow phase; the length of the vehicle and the driving characteristics (acceleration, deceleration) and dilemma zone. The model takes velocity and driving characteristics as given and then derives an optimal yellow phase and defines the dilemma zone under perfect information. Based on the model, the computer simulation has been done. And the model has been calibrated and tested with data from local area of Xi'an, west in China.

2 Comparative Study of Red Light Running at Urban Intersections

2.1 Research Method

At present, four main reasons about red light running violations have been indicated in existing research: the first is short of effective supervising and punishing to violations; the second is that the safety decision-making traffic codes of drivers are incomplete; the third is that the clearance interval at intersection is not quit reasonable; the forth is that some drivers always are too aggressive.

In addition, red light running violations largely occur during off-peak hours when traffic flow is lower than its capacity, approach speeds are high, and arrival of traffic at intersection is random. So data collection was performed during off-peak hours under good weather condition, and red light running violations were continuously shot by the camcorder at fields.

A comparative study is adopted in the research, i.e. we would compare red light running rates with difference traffic control method at urban intersections, and then we analyze them through statistical mathematics method. Hence, all the features of intersection must be as similar as possible, such as similar road condition and traffic flow etc.

2.2 Data Gain and Taking Video Process

All three selected intersections are located in the southeast side of Xi'an. These three intersections have comparably geometric features, traffic capacity, and traffic volume. But their traffic control ways are different. The first intersection is designed with yellow interval and red interval, neither all-red clearance interval nor red light running automatic camera. The second intersection is designed with all-red clearance interval, but the red light running automated camera is not installed. The third intersection is designed with both all-red clearance interval and red light running automated camera. Due to former red light running violation data has not been collected, the differences between testing field and controllable field couldn't be compared. We make use of a comparable research technology, i.e. the difference of red light running violation will be compared under the condition whether all-red light interval is designed, and the condition whether red light running violation automated camera is installed.

Traffic volumes for three intersections during off-peak hours are show as tabel.1.

Table 1. Average traffic volume for intersections during off-peak hours

No.	Intersection name	Traffic volume during off-peak hours					
		9:30 – 10:30	10:30 – 11:30	15:00 – 16:00	16:00 – 17:00	17:00 – 18:00	Average
1	Xingqin road – Xianning road	4287	4381	4630	4178	4467	4387
2	Youyi road – Taiyiroad	4183	4415	4524	4131	4642	4379
3	Xianning road – Taiyiroad	4301	4237	4589	4246	4637	4402

A digital camcorder SONY TRV18E (30 frame/s) was installed on the piles stood on the intersections of Xingqin road – Xianning road and Youyi road – Taiyi road. The traffic condition from east to west and south to north were shot. Since traffic administer department had installed red light running violation automated camera at intersection of Xianning road – Taiyi road, video information of that site were offered by traffic administer department directly. Two photograph groups with two persons each did the shooting work for five weekdays from Monday to Friday at every intersection toward east-west and south-north direction. Therefore video information about traffic condition and signal light phase during off-peak hours such as 9:30 – 10:30, 10:30 – 11:30, 15:00 – 16:00, 16:00 – 17:00, 17:00 – 18:00, five hours every day, had been gained. Red light running violations were continuously shot by the camcorder at fields. A total of 25 hours of field data hours was collected at each of three sites during off-peak periods on typical weekdays.

2.3 Data Compare and Analyze

A total of 25 hours red light running violations data was collected at each intersection. Average red light running violations a hour at respective intersections with only yellow light interval and all-red clearance interval and automated camera are shown as tabel.2.

Table 2. Average red light running violation for intersections

No.	Intersection name	Average red light running violation off-peak hours					Average
		9:30 – 10:30	10:30 – 11:30	15:00 – 16:00	16:00 – 17:00	17:00 – 18:00	
1	Xingqin road – Xianning road	21	26	32	25	19	24.6
2	Youyi road – Taiyiroad	17	21	26	15	27	21.3
3	Xianning road – Taiyiroad	8	7	12	5	11	8.7

Statistical analysis was performed to evaluate the difference in the red light running violations at three treatment intersections. The chi-square test of significance was used as the analysis tool. At an alpha-level of 0.05 and four degrees of freedom, $\chi^2_{critical} = 5.991$, $\chi^2_{21} = 0.238$, $\chi^2_{31} = 7.716$, $\chi^2_{32} = 6.208$. χ^2_{21}, χ^2_{31}, χ^2_{32} is respectively denoted chi-square volume of the second intersection, the third intersection toward the first intersection and that of the third intersection toward the second one. It is obviously that $\chi^2_{21} < \chi^2_{critical}$, $\chi^2_{31} > \chi^2_{critical}$, $\chi^2_{32} > \chi^2_{critical}$. Therefore it is not remarkable that only all-red clearance interval phase is designed to decrease red light running violations. And it is remarkable that both all-red clearance interval and red light running automated camera is designed could decrease violations effectively. Because enforcement of using red light violation camera system can provide spot images of violation behavior as powerful proofs to punish committed drivers, such enforcement may remarkably modify behaviors of drivers. Especially it can restrict those aggressive drives effectively.

3 Model Research and Analysis of Inadvertently Red Light Violations

For advertently red light violations, the use of automated red light cameras can provide judgement and enforcement to punish the drivers who is in violation. However, these initiatives have failed to reduce the number of red light violations to zero. The drivers who always abide by traffic rules shall find that sometimes they get into dilemma zone. They have to move ahead unceasingly with a risk of red light violation, since an emergency brake would result in further venture. The characteristic of this red light violation is inadvertently, so we call it "inadvertently red light running phenomenon".

Many researchers have noted that drivers cannot predict the onset of a yellow signal [5], and traffic signals that provide insufficient yellow intervals cause some drivers to run red lights inadvertently [6]. So we recommend the consideration of a minimum enforcement tolerance for photo red light programs [7]. Presently, the enforcement tolerance has not been adopted in Xi'an, China.

3.1 Dilemma Zone With Regard to Red Light Running

The Fig.1 above illustrates what happens when an automobile approaching an intersection sees the yellow light. Drivers who are in the "Can't Go" zone as the light turns yellow know they are too far back and won't be able to reach the intersection before the light turns red – they must stop. Drivers who are in the "Can't Stop" zone know they're too close to the intersection to stop safely – they must proceed. But when the yellow time is inadequate, there is place in between both zones where the driver can neither proceed safely, nor stop safely. Engineers call this the "Dilemma Zone" [8].

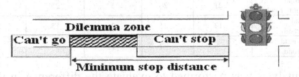

Fig. 1. Sketch Map of Dilemma zone

3.2 Mathematics Model of Calculating Yellow Time

First, we calculate the minimum distance (x) from the stop bar required for a vehicle to be able to stop:

$$x = v_0 \bullet t + \frac{v_0^2}{2a} \tag{1}$$

Where:

v_0 : initial velocity of approaching vehicle, in meter per second

t: reaction time, typically assumed as 1.0 sec
a: deceleration rate, typically assumed as 3m/sec^2
x: minimum stopping (reacting + braking) distance, in meter

For a vehicle to travel to the stop bar that is just closer than x meter away from the stop bar when the yellow is displayed, the time it will take will be the minimum duration of yellow. An equation for the distance that must be covered in terms of velocity and required yellow time is:

$$x = v_0 \bullet Y \tag{2}$$

Since both equation (1) and (2) are based on the total distance to the stop bar, x, we can set them equal to each other and solve for the yellow time, and insert an inequality sign to denote that this time is a minimum:

$$v_0 \bullet t + \frac{v_0^2}{2a} = v_0 \bullet Y \tag{3}$$

$$Y \geq t + \frac{v_0}{2a} \tag{4}$$

Based on an Institute of Transportation Engineers (ITE) recommended practice, incorporates the effect of approach grades into the calculation for yellow or change interval timing [5]:

$$Y \geq t + \frac{v}{2a + 2Gg} \tag{5}$$

Where: G is acceleration due to gravity (9.8m/sec²); g is approach grade (fraction).

Actually, the minimum stopping distance is thus a physical property that depends on the vehicle (responsiveness and possible deceleration rate), the driver (chosen velocity, reaction time and preferred deceleration rate), and the roadway (affects possible deceleration rate). The minimum stopping distance thus depends on the laws of physics and the capabilities and tolerances of the driver, and has absolutely nothing to do with the yellow time.

3.3 Generation and Analysis of Dilemma Zone Based on Yellow Interval

Current practice is to use a perception-reaction time of one second when calculating yellow signal timing. It is recommended that perception-reaction time be no less than 2.5 seconds to accommodate older drivers. 85th percentile perception-reaction time values ranging is from 1 second to 1.8 seconds. Approximately 90 percent of all drivers decelerate at rates greater than 3.4 m/s². Such decelerations are within the driver's capability to stay within his or her lane and maintain steering control during the braking maneuver on wet surfaces. As follows, we include a few examples to help illustrate the generation theory of option and dilemma zone:

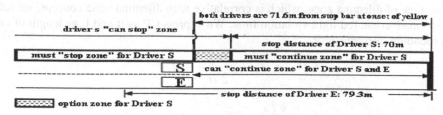

Fig. 2. Option zone illustration

Consider two drivers traveling at the speed limit of 64 km/hour towards a signalized intersection. Shown in fig.2, Driver S can meet the standard assumptions, i.e. reaction time of one second and a deceleration rate of 3.4 m/s², therefore, he can stop in a total (reaction + braking) distance of 70m. Driver E takes reaction time of 1.5

seconds and deceleration rate of 3.4 m/s². Therefore, he has to stop in a total distance of 79.3m far from the stop bar.

Both Driver S and Driver E are 71.6m from the intersection when the signal turns yellow. The yellow interval is 4.0 seconds. Driver S would come to rest 1.6m behind the stop bar after red. However, since Driver E is closer to the stop bar than that driver's stopping distance, he would be unable or unwilling to stop.

A yellow interval of 4.0 seconds for 64km/hour implicitly assumes that every driver who is 70m or closer to the stop bar at the onset of yellow will correctly choose to continue, while every driver who is further away than 71.6m can stop and will correctly choose to stop.

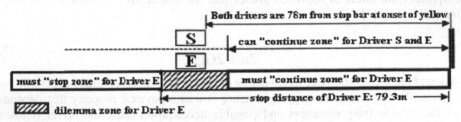

Fig. 3. Dilemma zone illustration

Shown in Fig.3, both Driver S and Driver E are 78m from the intersection when the signal turns yellow. As before, the yellow interval is 4.0 seconds. Since Driver S is further away from the stop bar than that driver's stopping distance, he should choose to stop. He would come to rest 8m behind the stop bar after red. Whereas Driver E is closer to the stop bar than that driver's stopping distance, he would be unable or unwilling to stop. Driver E will commit a red light running violation 0.39 seconds into the red interval. For Driver E with an increase in the reaction time of 0.5 seconds, yet a dilemma zone appears.

4   Further Discuss About Dilemma Zone and Inadvertently Red Violation

The time of dilemma zone, which is correlative with dilemma zone concept, we refer to as inadvertent red light violation time. We express CP as it and L as length of corresponding dilemma zone.

$$CP = t_i + \frac{v_0}{2a_i} - Y \tag{6}$$

$$l = \frac{v_0^2}{2a} - (Y - t_i)v_0 \tag{7}$$

The subscript i denotes that the reaction time and deceleration rate in the equation are for a particular driver rather than standard values.

Calculated by above formulation, the length of dilemma zone and inadvertent red light violation time caused by various conditions is shown in Tabel.3.

Shown in Tab.3, For 32km/hour speed limit and the traffic control signal with 2.0 seconds yellow interval that is widely existed in urban intersections in Xi'an, reaction time of one seconds and deceleration of $3.0m/s^2$ would cause 4.2m dilemma zone and 0.48s inadvertent red light running time. When the reaction time is added 0.5 seconds, 8.7m dilemma zone and 0.98s inadvertent red light running time would be caused.

Table 3. Length of dilemma zone and inadvertent red light violation time

Speed limit (km/hour)	Yellow time (s)	Reaction time (s)	Deceleration $(m/s^{2)}$	Dilemma zone (m)	Inadvertent red light violation time(s)
64	4.0	1.5	3.4	2.0	0.11
64	4.0	1.5	3.0	8.2	0.46
64	4.0	1.0	2.8	3.1	0.18
32	2.0	1.5	3.0	8.7	0.98
32	2.0	1.0	3.0	4.2	0.48
16	2.0	1.5	3.0	1.1	0.24

5 Microcosmic Simulation of Vehicle Behavior Cross Intersection

5.1 System Analysis and Design

We make microcosmic simulation to vehicle that is crossing intersection. The simulation system is made use of OOP (Object Oriented Programming) method and OOP language (Visual C++). And intersections of Youyi Road and Taiyi Road, Xianning Road and Taiyi Road in Xi'an are the study objects. During each simulation interval of real time, the time scanning technology is adopted. Vehicles generation at intersection is random. They head for any other intersection that is not jumping-off point. If vehicles generated wouldn't get across the closest intersection, then generation should stop once saturation state appears. Vehicles generation should start again when non-saturation state appears. There are three main function modules in whole simulation system. They are signal control module, vehicles generation module and vehicle movement control module. These three modules are independent severally. Each module is fitted respective timer.

The variational state along with times of each group of signal at each intersection is offered in signal control module. Vehicles that are ready to get across intersection during off-peak hour are generated randomly by vehicle generation module. The movement of vehicles can be interrupted through timer in vehicle movement control module. Also the speed in running can be changed by parameters adjusting.

5.2 Model Parameters and Simulation

We define the mathematics model of vehicle behavior getting across intersection during yellow interval by equation (1), (2), (3), (4), (5), (6) and (7) presented in section two. The model describes the relationship between parameters of width of the intersection; ideal velocity, and length of the yellow phase, length of the vehicle and the driving characteristics (acceleration, deceleration) and dilemma zone at the same time. Velocity and driving characteristics are taken as given in the model. Standard value and initial value may be input or adjusted by man-machine interface.

The actual values at urban intersection in Xi'an is regarded as standard values, i.e. speed of 32 km/hour and yellow time of 2.0 seconds, reaction time of one second and 1.5 seconds and standard deceleration of 3.0m/s^2.

Fig.4 is the sketch map of decision-making process of vehicle movement control module in getting across intersection in simulation system. In the figure, A denotes stopping before the stop bar; B denotes putting in with violation and getting across intersection; C denotes putting in with violation and stopping inside intersection; D denotes getting across intersection safety.

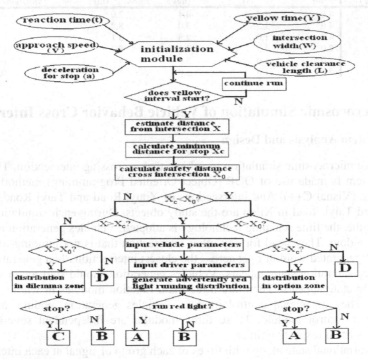

Fig. 4. Decision-making model of vehicle behavior at intersection

5.3   Simulation Results Analysis

We have simulated the vehicle behaviors at intersections of Youyi Road and Taiyi Road, Xianning Road and Taiyi Road in Xi'an. The count results of red light violations simulated are shown in Tabel.4. It is obviously that almost 80 percent of red light entries occur within the first 1.5 seconds of the red light indication. It quite corresponds with the case on fifth row in Table.3.

Table 4. Distributing of red light violation times

Youyi Road – Taiyi Road		Xianning Road – Taiyi Road	
Hourly average violations	Hourly average violations after 1.5seconds of red	Hourly average violations	Hourly average violations after 1.5seconds of red
24.6	5.2	8.7	2.2
79% entered on 1.5 seconds of red		74.7% entered on 1.5 seconds of red	

6 Conclusions

Although the public policy initiatives such as automated red light violation camera have reduced advertently red light violation behavior, they can't reduce inadvertently violation behavior. The fact that shorter yellow time is designed in some cities conceal motivation to increase incomes for local traffic manage departments.

Dilemma zone and possibility that vehicles run red light inadvertently have been discussed in the paper. The red light running violations at different intersections in Xi'an were shot, counted, calculated and analyzed. Based upon these, a microcosmic model and simulation was made for vehicle behavior getting across intersections. The simulation results correspond with video practices highly. The main conclusions are: to enhance inspection and punishment upon red light running violations can restrict aggressive driving behavior effectively; to setup rational yellow intervals can decrease dilemma zone, consequently reduce inadvertently red light running phenomena. Furthermore, some drivers cannot predict the onset of a yellow signal, and insufficient yellow intervals would cause some drivers to run red lights inadvertently. So we recommend the consideration of a minimum enforcement tolerance for photo red light programs. In this way, the benefits of public, drivers and government manage department can be balanced so that more safe, convenient and effective traffic policies can be made.

References

1. Yusuf M. Mohamedshah, Li WanChen, Association of Selected Intersection Factors With Red Light Running Crashes[J], Institute of Transportation Engineers, 2000(8):47-59
2. Tapan K. data, Kerrie Schattler, and Sue Datta, Red Light Violation and Crashes at Urban Intersection[J], Transportation Research Record 1734, 2000:52-58.
3. Gazis,D., R. herma, and A. Maradudin. The Problem of the Amber Signal Light in Traffic Traffic Engineering Journal, Research Laboratories, General Motors Corporation, Warren, Michigan 1995
4. Tarawneh, Tarek M., Singh, Virendra A., McCoy, Patrick T., Investigation of effectiveness of media advertising and police enforcement in reducing red-light violations, Transportation Research Record 1693, 1999, p 37-45
5. Retting, Richard A and Michael A Greene, Influence of Traffic Signal Timing on Red-Light Running and Potential Vehicle Conflicts at Urban Intersections [J]. Transportation Research Record 1595, 1997.pp: 1-7
6. Retting, Richard A; Williams, Allan F. and Michael A. Greene, Red-Light Running and Sensible Countermeasures: Summary of Research Findings [J]. Transportation Research Record 1640, 1998.pp: 23-26
7. Smith D M, McFadden J, Passetti K A. A Review of Automated Enforcement of Red Light Running Technology and Programs [J], Transportation Research Record, n 1734, 2000, p: 29-37
8. Chiu liu, Herman, Robert, Gazis, Denos C. A Review of The Yellow Interval Dilemma [J]. Transportation Research Part A: Policy and Practice, Volume: 30, Issue: 5, September, 1996, pp: 333-348.

Improved Processor Synchronization
for Multi-processor Traffic Simulator

Shunichiro Nakamura[1], Toshimune Kawanishi[1], Sunao Tanimoto[1],
Yohtaro Miyanishi[2], and Seiichi Saito[3]

[1] Department of electrical & electronics engineering, Nippon Institute of Technology
4-1 Gakuendai, Miyashiromachi, Saitama, 345-8501, Japan
{nakamu,tanimoto}@nit.ac.jp
[2] The Department of Media Architecture, Future University Hakodate
116-2 Kamedanakanomachi, Hakodate, 041-8655, Japan
miyanisi@fun.ac.jp
[3] Information Technology R & D Center, Mitsubishi Electric Corp.
5-1-1 Ofuna, Kamakura, 247-8501, Japan
ssaito@isl.melco.co.jp

Abstract. We are developing a traffic simulator which executes real-time simulation, so as to make ad hoc control of simulation possible. In these circumstances it is very preferable that entire range of the simulated area should be always displayed simultaneously. For this purpose we are attaching multi-
processing feature to the simulator to widen displaying area. We presented it in
Asian Simulation Conference 2002. But after that presentation some problems
were detected. Those were caused in the synchronization method between processors. To fix those problems we made an improvement to the synchronization
method. It was implemented successfully. The evaluation result shows that even
for the performance aspect the improved synchronization method gets higher
level. We describe the new synchronization method and its evaluation result in
this paper.

1 Introduction

For the past several decades, traffic congestion in cities and suburbs has continued to
be a problem. To achieve smooth traffic flow, many approaches using personal computers have been proposed and many simulators are available [1]. The present authors
have been developing a traffic simulator that incorporates unique functions. In fiscal
2000, our group developed a standalone traffic simulator [2], and in fiscal 2002, announced a multi-processor model of the simulator [3]. Since then, a complete synchronization function has been added to the simulator in order to synchronize processors in the multi-processor system. This improved synchronization feature is reported
here.

2 Proposed Method to Widen Screen Display

Traffic simulators generally output information as an animation or as statistical information (cumulative volume of traffic, average distance from the car ahead, etc.).
Animations make the information easier to understand and interpret by users. Fur-

D.-K. Baik (Ed.): AsiaSim 2004, LNAI 3398, pp. 386–391, 2005.

thermore, the traffic simulators themselves may be real-time or non-real-time processes. Despite the disadvantage of requiring much greater computing power, the former is generally considered superior because the user can observe the effect of arbitrary changes instantly. For the road simulation shown in Fig. 1, real-time simulation is useful for evaluating how a traffic light timing change at the Kamiuma intersection will affect traffic conditions at the Ohara intersection (both positive and negative).

Current computers are very efficient, and even a microscopic simulator can provide simulation over a comparatively wide simulation area [4]. However, while the calculation is possible, real-time animation display may be not. The minimum size of recognition on an ordinary display has been empirically found to be about 1000 m × 700 m for a simulation area, with 2 × 4 pixels for the display of one car. To allow for a wider range, the simulation area is often much larger than the display, making it necessary for users to zoom in on a target point. However, this method is not suitable for observing how a change applied to one point will affect others, as in the example mentioned above. To extend the simulation area, our group proposed the use of multiple personal computers for parallel simulation [3]. This allows for the regular observation of traffic conditions without complicated zooming. Figure 2 shows a sample installation of the system. An advantage of this system is that a flexible area such as that shown in Fig. 1 can be set as the simulation area, instead of being limited to a square or rectangular area, which also means that the number of personal computers required for simulation can be reduced. Each personal computer is connected via a 100 M/bps Ethernet over UDP/IP with speed priority.

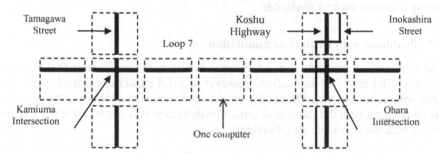

Fig. 1. Wide-area simulation

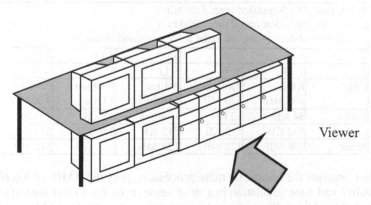

Fig. 2. Example of computer installation

3 Synchronization

3.1 Original Synchronization Method

Figure 3 shows the synchronization method used in the simulator at the stage of nnouncement in 2002 [3].

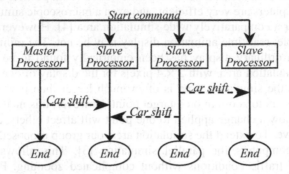

Fig. 3. At-start synchronization method (Conventional)

Synchronization is established by a batch command issue at the start, but later synchronization between processors is not available. This method was originally adopted because while processor clock errors occurred over longer periods of time, the error during a simulation was negligible.

3.2 Problems with Real-Time Simulation

The synchronization method in 3.1 above is based on the assumption that the system realizes strict real-time simulation. However, careful checking clarified that no real-time simulation is strict. Depending on the loads on each processor, the processing speed deviates with the passage of time. The deviation was evaluated using five processors with the following specifications:

Table 1. Time for each processor

1. No. cars: 34 Simulated time: 5:00 min				
2. No. cars: 500 Simulated time: 5:00 min				
Actual time of each processor under the above conditions				
1	5:13	5:01	5:00	5:28	5:28
2	7:24	5:30	5:12	5:30	5:30
CPU	K6-2 400 MHz	Pentium 3 550 MHz	Athlon 650 MHz	Pentium 3 1 GHz	Pentium 4 2.26 GHz
Prim.	64 KB	32 KB	128 KB	32 KB	20 KB
Sec.	512 KB	512 KB	512 KB	256 KB	512 KB
Mem.	256 MB	128 MB	128 MB	256 MB	512 MB

Table 1 presents the results. Of these processors, only the AMD Athlon (650 MHz) could realize real-time simulation in a strict sense, even for a short time of only 5 min simulating 34 vehicles. The other processors were already delayed by this time. Increasing the number of cars to 500 resulted in even greater delays. To our interest, a high-performance processor like an Intel Pentium 4 (2.26 GHz) does not provide smaller delays. If the number of vehicles and other loads differ between processors, processing times will differ even more.

3.3 Improved Synchronization Method

Considering the situation mentioned above, a complete synchronization method was devised, as outlined in Fig. 4.

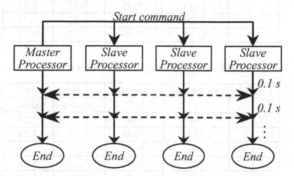

Fig. 4. Complete synchronization method

In this method, the master processor sends a simulation start command to each slave processor at 0.1 s intervals, corresponding to the minimum time unit of the simulator. After 0.1 s of simulation, each slave processor returns a report to the master processor. On receiving reports from all slave processors, the master processor issues a 0.1 s processing instruction. This scheme clearly imposes a communication overhead, and it is anticipated that this overhead will reduce the processing speed.

4 Evaluation

Two processors were evaluated, configured as master and slave, and the synchronization methods were evaluated for different numbers of cars (loads). The evaluation results for the conventional synchronization method are shown in Table 2, indicating the actual time taken for a 5 min simulation.

The difference between the speed of processors by the original synchronization method is 28 s for only 64 cars. This is a significant problem. Furthermore, as the number of cars increases, the difference from the simulated time increases for both the master and slave processors. In contrast, the complete synchronization method (Fig. 3) ensures that the simulation is competed simultaneously by the master and slave processors. Even when the number of cars is increased, the difference from the simulated time does not increase. This is contrary to our expectation that the proposed synchronization method would incur significant processing overhead compared to the original synchronization method. The reason for this better-than-expected performance is currently unclear, but will be further studied in the future.

Table 2. Original synchronization method

Test No.	No. of cars	Simulated time	Actual time	
			Master	Slave
1	64	5:00	5:28	5:00
2	330	5:00	5:28	5:10
3	600	5:00	5:34	5:27
4	900	5:00	5:55	5:45
5	1170	5:00	6:08	5:59
6	1740	5:00	6:28	6:32
7	2880	5:00	7:27	7:35
8	3420	5:00	7:32	8:31
9	4290	5:00	8:12	9:13

Table 3. Proposed synchronization method

Test No.	No. of cars	Simulated time	Actual time	
			Master	Slave
1	64	5:00	5:28	5:28
2	330	5:00	5:28	5:28
3	600	5:00	5:28	5:28
4	900	5:00	5:28	5:28
5	1170	5:00	5:28	5:28
6	1740	5:00	5:28	5:28
7	2880	5:00	5:28	5:28
8	3420	5:00	5:28	5:28
9	4290	5:00	5:28	5:28

The proposed complete synchronization method allows for completely reproducible simulation that is unaffected by the processing speed of each processor. Even under the same conditions, the original synchronization method could not ensure totally reproducible simulations because priority was not given to reproducibility. Now, with the reproducibility offered by the proposed method, a fast forward function can be realized in a multi-processor environment, a feature that could not be implemented using the original synchronization method due to wildly fluctuating processing speed during fast forward. The complete synchronization method ensures correct synchronization even during fast forward processing. The fast forward function will be implemented in the near future.

5 Conclusion

The synchronization method originally incorporated into the authors' simulation system at the time of the announcement in fiscal 2002 [3] has been upgraded to a complete synchronization method that forces synchronization in each time-step of the simulation. This method has improved reproducibility and made a fast forward function feasible. This study also verified that simulation using the complete synchronization method, while expected to incur significant overhead, is not slowed at all compared to the original system.

References

1. Hotta M. , Yokota T. , et al. : " Congestion Analysis and Evaluation by Traffic Simulation ", ITS YOKOHAMA '95 Proceedings, pp2062-2067 (1995)
2. Igarashi T. , Nakamura S. , et al. : " Development of traffic simulator NITTS " , 20th Simulation Technology Conference of Japan Society for Simulation Technology, pp295-298 (2001)
3. Nakamura S. , Igarashi T. , et al. : " Development of A Traffic Simulator Which Uses Multi-Processing Method ", The 5th International Conference on System Simulation and Scientific Computing, pp305-308 (2002)
4. Masao Kuwahara : Traffic simulation by wide area network, Automobile technology, Vol.52, No.1, pp28-34 (1998)

Evaluation of Advanced Signal Control Systems with Sensing Technology by Simulation

Mohd Nazmi[1], Nadiah Hanim[1], and Sadao Takaba[2]

[1] Graduate Student, School of System Electronic, Graduate School of Engineering,
[2] Professor, Department of Information Networks (Department of Computer Science),
Tokyo University of Technology,
1404-1 Katakura, Hachioji, 192-0982 Tokyo, Japan
Tel/Fax: +(81)-426-37-2493
takaba@cc.teu.ac.jp

Abstract. This paper describes a new data collection system for determining individual vehicle routes from images detector sequence. These sensing technologies are installed at the traffic signal to gain a microscopic flow from upstream intersection. From the investigation, the techniques for analyzing the images can be applied for calculation of the traffic information. The image sensors are set for microscopic detection via the direction of vehicles, and for every lane at specified perception range. Two cases study has been carried out to analyze the method. We try to solve the problem of signal control system for the FAST system and complex intersection. Using image sensor for (i) control traffic flow in FAST system, (ii) manage the control system in complicated intersection, is highly proposed. Arena simulations are used to analysis the investigation data and monitoring the traffic signal control system performance.

1   Introduction

Introduce the real time signal control at improve traffic congestion, distributing traffic and setup the network intersection. This research is the continuation of the last paper presented at "SeoulSim 2001". Using the video image processing is a technology that analyzes video images through specialized systems for advanced traffic management systems, and traffic data collection applications. Various operational tests have verified imaging sensors as practical alternatives to conventional sensors for signalized intersection control. Our primary purpose is to manage the congestion, improving safety and provide an equal access on every signalized intersection. First step of study is to setup a networking intersection for real-time control system. Therefore, a new style of traffic data collection, online communication for twin or network intersection is introduced. This project is focusing on enhancing deployable image sensing technology to provide more traffic information and improve the system accuracy. The system enhancements which have been implemented will make extensive use of ITS field to ensure deployment success.

D.-K. Baik (Ed.): AsiaSim 2004, LNAI 3398, pp. 392–402, 2005.

2   Outline of Traffic Control System

2.1   Basic Systems Concept

In traffic control, the split, cycle length, and offset timing parameters are continuously and incrementally adjusted by the control system. There are no fixed coordination timing plans. Traffic control requires a sensor that monitors traffic in real time. As traffic changes, the splits, offsets, and cycle lengths are adjusted by several seconds. In network intersection, the neighboring intersections conditions are very important to know. Our systems lead the communication between the intersections to exchange the information data. From the gain data, we could predict the condition of each intersection. Therefore, we could prepare the signal parameter and timing in advance. The goal is always to have the traffic signals operating in a manner that optimizes traffic flow.

2.2   System Model

The algorithm and the model of the traffic signal control have been designed. This model generate the real-time traffic control system such as the calculation, predicting time simulation and memorizing the new data. Our targets are to make an installation and operation at every intersection signal controller system by using this system and the performance will be as same as the traffic main center system. We are trying to reduce the facilities of the traffic main center so that the signal controller can manage the system by our installation system capability.The model structure (signal controller) is based at three main systems. They are (1) Memory and database. (2) Predictor controller system. (3) System controller / analysis.

Each model has their algorithm and the way to run the system by using the real-time data collection that is collects from inside the intersection by vehicles detector (image sensor). The layout of our traffic signal control system structure is referred as shown in the fig. 1.

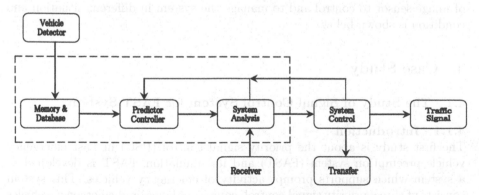

Fig. 1. Traffic Signal Control Model.

3   Image Sensing System

3.1   Detection Concept

Cameras Installation at the intersection and obtaining specified data of vehicles by analyzing these images are widely used. This type of detectors is installed at the intersection. Two categories of them are considered. First, they are installed to observe vehicles at the inflow area of the intersection. Vehicles flow into the intersection from an incoming direction and to different outgoing directions is measured as shown in fig. 2. Second, they are installed to observe vehicles at the outflow area of the intersection. Vehicles coming from different directions and flowing out to the same direction are measured as shown in fig. 3.

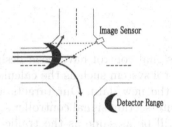
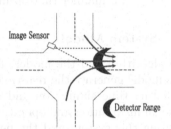

Fig. 2. Flow-in Measurement.   **Fig. 3.** Flow-out Measurement.

From these measurements, number of vehicles which is going to left turn, right turn and straight through can be counted. With their summation, total number of vehicles, total traffic volume on each incoming and outgoing links can be calculated. Vehicles stopping at the intersections due to blocking of outgoing links are also can be measured. The above concept has completely explained and presented in the SeoulSim 2001 conference. As for the applied use of the image sensor related to our control system, two cases study are proposed. The usage of image sensor to control and to manage the system in different situation and condition is shown below.

4   Case Study

4.1   The Study of Signal Control System for FAST System

4.1.1   Introduction
The first study is about the priority signal control (PSC) in Fast emergency vehicle preemption system (FAST) and its simulation. FAST is developed as a system which supports prompt activity of emergency vehicles. This system consists of three; priority signal control, course guidance, and emergency vehicles approach information warning. It solves many problems at the time of the urgent accident occur.

4.1.2   The FAST Merit

The benefits of the FAST system are (i) aim to shot the arriving time to the spot, (ii) improvement in the rate of an arrest and the rate of life save, and (iii) prevent any collision at the intersection.

4.1.3   Information Gathering in FAST Sytem

The optical beacon is used as communication interface in FAST system. The information is transmitted to a traffic control center to give a priority to pass the intersection by extended the green time. By using the image sensor, the real-time system could detect directly the emergency vehicle and change the signal parameter itself without using the main center.

4.1.4   Calculation

1. Light Traffic

   First condition for light traffic before performing the PSC is the signal in green. Then the PSC will perform a process to green light extension. In this case, we can say that vehicle will stop when the signal is in red, and its become red at t=0. A form of signal time is from green to red.

   (a) When the signal with PSC, the traffic will flow smooth and the vehicle can move with the initial density ρ_1 and speed μ_1 (fig. 4).

   (b) When the signal without PSC, the vehicle will stop at red time, will start flow with $\rho_1{:}\mu_1$ after the time of $\frac{t_s}{2}$ (fig. 5).

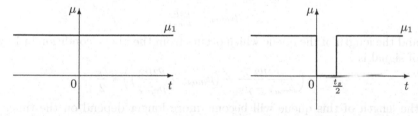

Fig. 4. Signal with PSC in heavy traffic.   **Fig. 5.** Signal with PSC in heavy traffic.

Second condition for light traffic by performing the PSC; the signal will process to short the red time. The time which becomes red, t=0 and the time which becomes green, $t=\tau_1$. A form of signal time is from red to green.

(a) When the signal with PSC, there will be a queue of vehicle at the signal. When this queue is clear, vehicle will move at $\rho_1{:}\mu_1$ condition. At this time, the length of vehicle queue is $\left(\frac{\rho_1\mu_{max}}{\rho_{max}}\right)\tau_1$. A time for clearance of the queue is t_1. This length is divided by μ_{max}-$\left(\frac{\rho_1\mu_{max}}{\rho_{max}}\right)$ which is the speed of length extinction (fig. 6)

$$T = \frac{\frac{\rho_1\mu_{max}\tau_1}{\rho_{max}}}{\mu_{max} - \frac{\rho_1\mu_{max}}{\rho_{max}}} = \tau_1 + \frac{\tau_1}{\frac{\rho_{max}}{\rho_1} - 1} = \frac{\tau_1}{1 - \frac{\rho_1}{\rho_{max}}} \qquad (1)$$

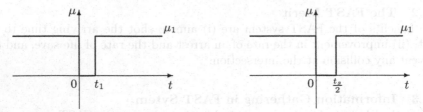

Fig. 6. Signal with PSC in heavy traffic.   **Fig. 7.** Signal with PSC in heavy traffic.

And the time for vehicle which start to move, t_1 is

$$t_1 = \tau_1 + \frac{\tau_1}{1 - \frac{\rho_1}{\rho_{max}}} \tag{2}$$

(b) When the signal without PSC, the vehicle will stop at red time, and begin to move at $\rho_1:\mu_1$ when green time is $\frac{t_s}{2}$ (fig. 7).

2. Heavy Traffic

In heavy traffic, each cycle of the signal will form a queue of vehicle and it is increasing every cycle. The length of the queue of the vehicle can be found as follows. When the signal time is red, the vehicle length speed will increase at

$$\frac{\rho_2\mu_2}{\rho_{max} - \mu_2} \text{ and}$$

when the signal time become green, the vehicle length speed will decrease at

$$\mu_{max} - \frac{\rho_2\mu_2}{\rho_{max}}$$

and the length of the queue which occurs from the above condition in 1 cycle of signal is

$$\left(\frac{\rho_2\mu_2}{\rho_{max} - \mu_2} - \left(\mu_{max} - \frac{\rho_2\mu_2}{\rho_{max}} \right) \right) * \frac{t_s}{2} \tag{3}$$

the length of this queue will become more longer depend on the time. We assume this length as X.

First condition for heavy traffic before performing the PSC is the signal is in green. Then the PSC will perform a process to green light extension. The signal time which become to red is t=0. A form of signal time is from green to red.

(a) When the signal with PSC, even the green time is extended, the queue of vehicle are already occur. Vehicle will move at $\rho_2:\mu_2$ after the clearance of vehicle queue. The time until the vehicle start to move is t_2. The queue length at that time is X_a (fig. 8). Where,

$$t_2 = \frac{X_a}{\mu_{max} - \frac{\rho_2\mu_2}{\rho_{max}}} \tag{4}$$

(b) When the signal without PSC, the vehicle will stop at red time, will start flow with $\rho_2 : \mu_2$ after the time of $\frac{t_s}{2}$ (fig. 9)

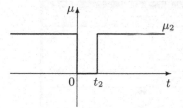

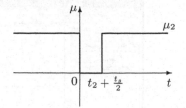

Fig. 8. Signal with PSC in heavy traffic.   **Fig. 9.** Signal with PSC in heavy traffic.

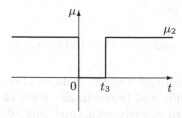

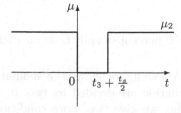

Fig. 10. Signal with PSC in heavy traffic.   **Fig. 11.** Signal with PSC in heavy traffic.

Second condition for heavy traffic is by performing the PSC; the signal will process to short the red time. The time which becomes red, t=0 and the time which becomes green, t=τ_2. A form of signal time is from red to green.

(a) When the signal with PSC, there are already vehicle queue occur. Vehicle will move at ρ_2:μ_2 after the clearance of vehicle queue. The time until the vehicle start to move is t_3. The queue length when the signal change to red is X_b (fig. 10), and the length until that time is $X_b + \left(\frac{\rho_2 \mu_2}{\rho_{max} - \rho_2} \right) \tau_2$ and

$$t_3 = \frac{X_b + \left(\frac{\rho_2 \mu_2}{\rho_{max} - \rho_2} \right) \tau_2}{\left(\mu_{max} - \frac{\rho_2 \mu_2}{\rho_{max}} \right)} \tag{5}$$

(b) When the signal without PSC, the vehicle will stop at red time, will start flow with ρ_2:μ_2 after the time of $\frac{t_s}{2}$ (fig. 11).

4.1.5 Simulation
The simulation was performed by nine condition of heavy traffic by using Arena Simulation software. Below, the nine condition for heavy traffic.

1. Single sided and one lane passing
2. Without any trespassing
3. Never runs through the opposite lane
4. Ignore any vehicle acceleration
5. The initial traffic density at the time of light traffic is ρ_1
6. The initial traffic density at the time of heavy traffic is ρ_2
7. $\rho_1 < \rho_2$
8. The time of 1 cycle of the signal is t_s ($t_s = 120$ sec)

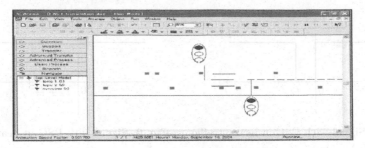

Fig. 12. Simulation Model.

9. 50 percent of split time for each red time and green time (60 sec each)

Above figure shows the simulation sample image and the main condition of simulation are divided by two. It is light traffic and heavy traffic. Each of the traffic, we give two more condition such with priority signalized control and without priority signalized control. Finally every control we set the signal time from green to red (G ⇒ R) and red to green (R ⇒ G). With all these condition, we run the every simulation for 10 times. The results of simulation are shown at table 1 below.

4.1.6 Simulation Result

The simulation have been done by using the above all nine conditions. The results are shown in the following tables (table 1). From the result below, we identify that the effect of light traffic of the PSC is larger than that of heavy traffic. Some the cases where the drivers could not pass the intersection in the cycle 4 and 6 even the signal are generated to undersize and expand the green time. When the priority signal is controlled in light traffic ($\rho_0 \leq \frac{\rho_{max}}{2}$), the queue can be occasionally done according to the situation of the signal.

Table 1. Simulation Result.

Frequency	Time Until Vehicle Passes a Signal (sec)										
	1X	2X	3X	4X	5X	6X	7X	8X	9X	10X	Mean
Light Traffic (G-R) with control	28	29	28	27	30	29	30	28	31	29	28.9
Light Traffic (G-R) without control	65	64	63	63	64	66	65	64	63	64	64.1
Light Traffic (R-G) with control	34	33	37	34	34	37	33	34	33	37	34.6
Light Traffic (R-G) without control	65	65	64	63	66	64	63	65	64	63	64.2
Heavy Traffic (G-R) with control	64	27	31	28	64	30	29	62	63	30	42.8
Heavy Traffic (G-R) without control	69	70	71	68	68	69	71	70	70	68	69.4
Heavy Traffic (R-G) with control	40	67	38	64	41	38	65	63	40	39	49.5
Heavy Traffic (R-G) without control	71	72	70	74	74	73	72	71	72	72	72.1

4.2 The Study of Signal Control System for Complicated Intersection

4.2.1 Introduction

The second study is concerning the simulation for relief of traffic congestion according to signal control in the complicated intersection. Intersections are areas where most conflicts between various roadway users occur. Intersections with numerous streets entering from different direction can create misunderstanding for users. Such intersections should be avoided and designed instead as simple right angle intersections whenever possible. For an already existing complicated intersection, the best performances of signal control are needed for maintaining the traffic to prevent congestion. Our researches carry out on a complicated intersection with many problems during every morning and evening rush hours. For one year we experience, integrate and factually determine the exact cause of these traffic problems. Arena Simulation is used for this purpose to verify the traffic signal improvement.

4.2.2 Investigation and Results

The research are give attention to a problems occur at complicated intersection where two main road adjoin to each other. The signal priority is given to the large main road. As a result, a heavy congested will occur every rush time especially in the morning at a small main road. Investigations are done to find the problem and we are focus at the intersections signal control system. Fig, 13 is the map of the problem intersection.

Investigations are done at the model intersection. Investigations are done for one hour every morning and evening rush time, and also afternoon time. The numbers of vehicle at I, II, III, V lanes are taken and the signal parameters are recorded.

1. By referring to the map, Main road A is large road. Signal priority is given for main road A at lane IV. Therefore, the waiting times for vehicles to flow

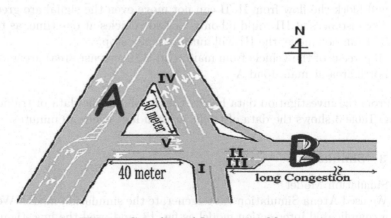

Fig. 13. Complicated Intersection Map.

Table 2. Vehicle every 1 hour.

	7:30 8:30	8:30 9:30	12:00 13:00	17:00 18:00
I	369	230	232	315
II	339	403	428	383
III	130	137	127	76
V	402	215	177	158
VI(I+II)	708	633	660	698
II+III	469	540	555	459

Table 3. Vehicle every 30 minutes.

	7:30 8:00	8:00 8:30	8:30 9:00	9:00 9:30	12:00 12:30	12:30 13:00	17:00 17:30	17:30 18:00
I	191	178	150	80	110	122	145	170
II	172	167	209	194	215	213	195	188
III	65	65	66	71	59	68	40	36
V	205	197	110	105	96	81	97	61
VI(I+II)	363	345	359	274	325	335	340	358
II+III	237	232	275	265	274	281	235	224

Table 4. Signal Parameter for each Intersection.

SIGNAL I

SIGNAL II & III

SIGNAL IV

Offset Time = 35s

Signal Num.	G	Y	R	⇒
Signal I	30	3	87	–
Signal II & III	60	3	42	15
Signal IV	40	3	77	–

out from IV are long as results. A traffic congestion always happen at main road B in the morning and evening rush time.

2. Vehicles from main road B which flow to the north of main road A area will face two signals. There are gaps for offset time between two intersections, as a result vehicles can not pass the intersection smoothly.
3. The flow of the vehicles of V obstructs the flow of the vehicles of II.
4. The distance between IV and intersection are short (about 50 meters). When vehicles from I flow in to the IV, the road is filled with vehicles and the queue will block the flow from II. II can not move even the signal are green time.
5. The distances at III could fill only for two vehicles at one time, as the result if II can not move, the III will automatically stop.
6. 80 percent of the vehicle from main road B (from east area) are going to the north area at main road A

From the investigation data below, table 2 shows the data of traffic every 1 hour. Table 3 shows the data of traffic at each lane every 30 minutes.

4.2.3 Simulation

1. Simulation Model
   We used Arena Simulation 7.0 to generate the simulation model. We created a complicated intersection model as fig. 13, and used the investigation data

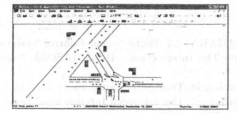

Fig. 14. Simulation Model.

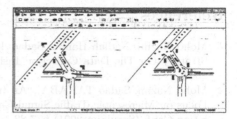

Fig. 15. Comparison Model.

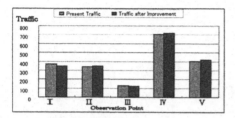

Fig. 16. Comparison of Traffic Flow.

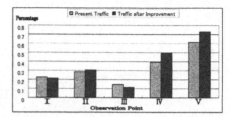

Fig. 17. Comparison of Smooth Flow.

as the simulation input. The signal parameters are taken from the exact data (table 4). Below, fig. 14 shows the simulation model with the present traffic conditions. Fig. 15 shows the comparisons of present traffic condition and traffic condition after the improvement of signal parameter.

2. Simulation Result

By comparing the present traffic condition and traffic condition after the improvement of signal control, we can say that (i) vehicle can flow more than present traffic condition (fig. 16). (ii) The vehicle flows also more smooth that the present traffic condition (fig. 17).

5 Conclusions

The two cases studies are very important to the research. In order create the best performance of adaptive signal network controller by image sensor as data collection tool. We are try to find any problem which will occur at the intersection and their signal control system.

The FAST system study is mainly to find any problem which will happen to the signal control system where there are interruptions from the user side. With the intention of solving the FAST difficulties, image sensors are required to install in the study model. Instead of detecting the emergency vehicles the sensor also can be used for monitoring the traffic condition.

We also have learned the change of the system when there are interruption form internal side systems at the complicated intersection study. We can say that, the image sensor is more effective to apply in the complicated intersection to control the traffic flow and to manage the system change.

References

1. Mohd Nazmi, Nadiah Hanim, Sadao TAKABA, "A Traffic Signal Control System Strategy On The Data Collection Inside The Intersection," IEEJ-JIASC03, Aug. 2003.
2. Mohd Nazmi, Sadao TAKABA, "An Intelligent Transport Management System. – Focus on Microscopic Traffic Signal Control –," The 1st Seoul International Simulation Conf. (SeoulSim 2001) p77-82, Seoul Nov. 2001.
3. Sadao TAKABA, "A Method of Special Flow Measuring Using Moving Pictures and Its Application," JSST, Jun 1985

Proposal and Evaluation of a System Which Aims Reduction in Unexpected Collision at Non-signal Intersection

Hiromi Kawatsu, Sadao Takaba, and Kuniaki Souma

Dept. of Information Networks,
Tokyo University of Technology,
1404-1, Katakura, Hachioji, Tokyo, 192-0982, Japan
Tel +81-426-37-2111(ext.2526) / Fax +81-426-37-2493
{kawatsu,takaba}@cc.teu.ac.jp

Abstract. Focusing attention to the non-signal intersection in a residential area, a new traffic control system after securing a pedestrians' safety is proposed. Reduction in the unexpected collision accident at the intersection where pedestrians are the main accident case is expectable with use of this system. In the model intersection, smoothness of the vehicles in the proposed system was evaluated by simulation, and the effectiveness of the system was shown.

1 Introduction

In the 20th century, the traffic management is very important from the viewpoint of safety and smoothness in the road traffic which became base of the society. The traffic control system which used the computer was born in 1960's. It has been applied to the arterial roads where a large amount of traffic is treated and the improvement of safety and smoothness is advanced. However, the development of the traffic control system for the non-signal intersection in the residential area is on the way in the near future. Half of the unexpected collision has been occurred at an urban non-signal intersection [1], as the safety measures there are not enough. With this intention, the system which gives priority to the improvement of safety in the non-signal intersection is proposed. We aims to show the effectiveness by the simulation.

2 Outline of System Which Is Proposed

Figure 1 shows an intersection rivet which is a device to notify for the drivers the existence of the frontage intersection, when the view is not clear such as during night time or in rain. It reminds the driver to decrease the speed of vehicle when approaching the intersection. This leads to the decrease in the accident [2]. It is necessary to confirm safety with stopping temporarily or going slowly before entering the intersection, because the driver cannot notice other vehicles and

D.-K. Baik (Ed.): AsiaSim 2004, LNAI 3398, pp. 403–409, 2005.

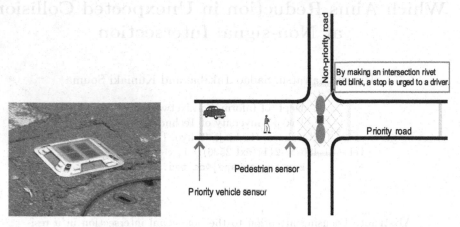

Fig. 1. An intersection rivet.   **Fig. 2.** Image of the system.

pedestrians presence until he comes near to the intersection even in the place where the intersection rivet has been set up. However, when there is little traffic, the caution by the drivers becomes insufficient, and a lot of unexpected collisions by non safety confirmation occurred in the intersection. Moreover, there is danger which causes the accident because of defective view with a high wall even if safety is once confirmed.

A safety system is proposed by using the intersection rivet more effectively which is the traffic equipment already used at present. This system is arranged the pedestrian sensors and the priority vehicle sensors besides the intersection rivet as shown in Fig. 2. The stop is urged to the driver by making the intersection rivet red blink when there is a vehicle or a pedestrian which approaches the intersection from the other sides. The stop is urged (1) To the driver in the priority vehicle and the non-priority vehicle when there is a pedestrian within the range of perception of the pedestrian sensor (2) To the driver in the non-priority vehicle when there is a priority vehicle within the range of perception of the priority vehicle sensor. The driver can notice the presence of the pedestrians and the other vehicles before he comes near to the intersection, and the complication[1] generated by defective view and non safety confirmation can consequently be avoided. It will lead to the reduction in the unexpected collision. Moreover, it can be also expected that the number of accidents which cause by the pedestrian who abruptly rushes into the intersection could be decreased. It has a big meaning to reduce two main accident types; one is unexpected collision which is the major accident in the non-signal intersection, and the other is abruptly rushed accident where the pedestrian becomes the first person of the accident in most cases [1].

---

[1] Complication is defined as the state with the possibility to collide with the vehicle and the pedestrian from the other side when the vehicle went into the intersection without stopping.

3   Evaluation of the System

The improvement of the safety at the intersection can be expected by introducing the proposed system. However, it is important to verify how smoothness of the vehicle is kept in realizing the system. Real traffic was observed in the model intersection, and the values were used as input parameters for the simulation. Differences in the time required to pass the intersection were compared with several conditions in the simulation, and the smoothness of the vehicle in the system was evaluated.

3.1   Observation in Model Intersection [3]

Pictures were taken with a video camera to understand the traffic situation of the model intersection, then the number of vehicles and of pedestrians and waiting time were obtained.

 − Observation Place
   Non-signal intersection in Nishi-Tokyo City Yanagisawa 3.
   There is a high wall at the private house in a corner of this intersection, and the view is bad. Therefore, the driver cannot notice the presence of other vehicles and pedestrians before reach the intersection.
 − Observation Date
   January 15(Wed), 16(Thu) and 17(Fri), 2003.

Numbers of vehicles and pedestrians are shown in Table 1 and waiting time is shown in Table 2. In the model intersection, there is more traffic during 15:00 - 16:00 than that during 8:00 - 9:00 which is the rush time to work and school. During the observation, it is found that non-priority vehicles with going slowly or stopping temporarily were only about 60% of all. As a matter of fact, there were a lot of offenders in drivers who run this non-priority road.

Table 1. Number of vehicles and of pedestrians.

	8:00 - 9:00	15:00 - 16:00
Priority vehicle	71	157
Non-priority vehicle	35	70
Pedestrian	36	60

Table 2. Waiting time at the intersection (sec).

	8:00 - 9:00		15:00 - 16:00	
	Ave	Max	Ave	Max
Priority vehicle	0.3	18	1.7	9
Non-priority vehicle	5.7	23	2.6	16
Pedestrian	0.3	7	0.8	7

3.2   Simulation

In this system the stop is urged by making the intersection rivet red blink; (1) To the driver in the priority vehicle and the non-priority vehicle when there is a pedestrian within the range of perception of the pedestrian sensor (2) To the driver in the non-priority vehicle when there is a priority vehicle within the range of perception of the priority vehicle sensor. The intersection rivet is always blinked to yellow to clarify the position of the intersection in the other case.

In the simulation, it was assumed that traffic of the priority vehicle, non-priority vehicle, and pedestrian was shown in **3.1** and Table 1. The range of perception of the sensor and the condition of the movement of the vehicle were changed in three patterns.

The simulation condition in pattern I, pattern II, and pattern III is shown in Tables 3 and 4. The pedestrian sensing position was set at entrance of the intersection in pattern I. In patterns II and III, the pedestrian sensing was begun 3.5m upstream from entrance of the intersection in order to prioritize the pedestrian. It is safe to pass without going slowly at the time of a yellow blinking, when the system is working ideally and all drivers of non-priority vehicles follow the instruction. In pattern III, moreover, non-priority vehicles as well as the priority vehicles are permitted to go at the time of yellow blinking.

Table 3. Parameters and their values.

	Pattern I	Patterns II and III
Width of road	5.0m for each road	
Pedestrian sensing position	Entrance of the intersection	3.5m upstream from entrance of the intersection
Priority vehicle sensing position	20m upstream from entrance of the intersection	
Speed of pedestrian	1.4m/s	
Speed of vehicle	Usually:30.0km/h=8.3m/s, Going slowly:15km/h=4.2m/s	

3.3   Simulation Result

The result of the simulation conducted under the conditions of **3.2** is shown in Fig. 3 and Table 5. Figure 3(a) shows the ratio of Non-stop, Slow-down and Stop for all vehicles and pedestrians. Figure 3(b) shows the time extra taken to pass the intersection by going slowly and stopping temporarily. The time required in order that vehicles may run through from the position 20m before the intersection to the end of the intersection is 3.01 seconds at the speed of 30 km/h. On the other hand, it takes 5.36 seconds when going slowly. When the driver goes slowly and stops temporarily, time required to run through the intersection turns into which added waiting time to it. The time extra taken by going slowly with/without stopping temporarily are given by subtracting 3.01 seconds from time when the vehicle runs through. Since the non-priority vehicle

Table 4. Movement of priority vehicles, non-priority vehicles, and pedestrians.

	Patterns I and II	Pattern III
Priority vehicles	While the red blinking, go slowly on entering the intersection and wait. Go without slow down in case of yellow blinking.	
Non-Priority vehicles	Go slowly on entering the intersection at any time and wait while the red blinking.	While the red blinking, go slowly on entering the intersection and wait. Go without slow down in case of yellow blinking.
Pedestrians	Pass in the road shoulder zone outside the intersection, and pass freely inside the intersection (Including diagonal). Stop and cross after safety confirmation when there are approaching vehicles.	

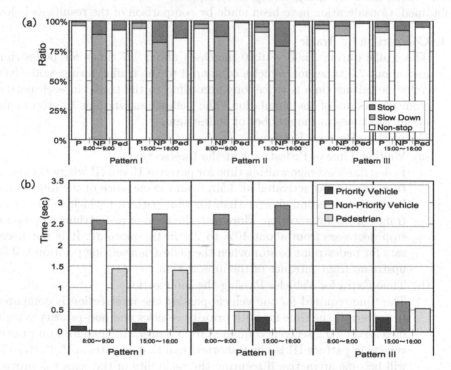

Fig. 3. Simulation result, (a) The ratio of Non-stop, Slow-down and Stop, and (b) Time extra taken to pass the intersection and waiting time at the intersection.

has to go slowly when entering the intersection at any time in patterns I and II, the white part of Fig. 3(b) indicates the time extra taken by going slowly. For the pedestrian, the time stands for the average of waiting time of the person who had waited at the intersection. Table 5 shows each of the maximum waiting time of the priority vehicle, non-priority vehicle and pedestrian.

Table 5. Maximum waiting time at the intersection (sec).

	Pattern I		Pattern II		Pattern III	
	8:00 - 9:00	15:00 - 16:00	8:00 - 9:00	15:00 - 16:00	8:00 - 9:00	15:00 - 16:00
Priority vehicle	3.06	6.70	8.80	12.6	8.08	13.3
Non-Priority vehicle	8.50	11.2	12.9	20.3	11.1	14.9
Pedestrian	3.35	3.73	1.12	1.53	1.18	1.73

4   Comparison and Consideration

In the simulation, waiting time at the intersection for pedestrians and passing time extra taken by going slowly and stopping temporarily for vehicles were obtained. Consideration have been made by comparison of the results as below.

1. Comparison by Rraffic
   The traffic during 15:00 - 16:00 increases about 1.7 times for pedestrians and about 2.0 times for vehicles compared to the traffic during 8:00 - 9:00. Vehicles waiting time is twice long according to the traffic in each pattern from the results of the simulation. The pedestrian who has to stop is also about two times in the number of pedestrians.
2. Comparison Between Patterns
   (a) Waiting Time of Pedestrians at the Intersection
       Pedestrian's average waiting time for patterns II and III where the pedestrian sensor was activated at 3.5m from the entrance of the intersection is about one second shorter than for that pattern I which senses at entrance of the intersection. Moreover, the ratio of pedestrians who had to stop decreases from about 10% to 2% in the morning. It is not necessary for pedestrians to stop when the pedestrian sensing position is 3.5m upstream from entrance of the intersection.
   (b) Time Extra for Vehicles Passing the Intersection
       The time required for the vehicle passing the intersection is compared. In pattern II the time for both priority vehicles and non-priority vehicles increased by 0.2 seconds compared to in pattern I. Time for non-priority vehicle in pattern III is much shorter than that in pattern II. Pattern III will become attractive if securing the reliability of the sensor is proven.
3. Comparison in Maximum Waiting Time at the Intersection
   The maximum waiting time at the intersection is compared. Table 5 shows that they are about six seconds longer in pattern II and III than in pattern I for the priority vehicle. We found that maximum waiting time for non-priority vehicle in pattern III increased by about three seconds compared to that in pattern I. The maximum waiting time for pedestrians is shortened about two seconds. There is no problem for a small increase of the waiting time of the vehicle, when giving a priority to the pedestrian in a non-signal intersection.

It is obvious that the safety is improved in this system, because the drivers can notice other vehicles and pedestrians presence beforehand when entering the intersection. Since the result of simulation indicates that the system can keep smoothness for the vehicles, the validity of the proposed system is confirmed.

5 Conclusion

The traffic control system which gives priority to the improvement of safety in the non-signal intersection in a residential area is proposed. In the model intersection, smoothness of the vehicles in the proposed system was evaluated by simulation, and the effectiveness of the system was shown. In the future, effectiveness of the system in case of the increased traffic both in vehicles and in pedestrians will be investigated.

References

1. Traffic Engineering Handbook 2001, Corporation Traffic Engineering Society(2001)
2. The Kanagawa Prefecture Police, http://www.police.pref.kanagawa.jp/index.htm (2004)
3. Kuniaki Souma, Safety improvement in non-signal intersection, 2002 Bachelor Thesis at Tokyo University of Technology(2003) (in Japanese)

Dynamic Channel Reservation Scheme
for Mobility Prediction Handover
in Future Mobile Communication Networks

Hoon-ki Kim[1], Soo-hyun Park[2], and Jae-il Jung[3]

[1] Dept. of Software Engineering, Dongyang Technical College,
62-160, Kochuk-dong, Kuro-ku, Seoul, 152-714, Korea
kimhk@dongyang.ac.kr
[2] School of Business IT, Kookmin University
861-1 Jeungreung Dong Sungbuk Gu, Seoul, 136-702 Korea
shpark21@kookmin.ac.kr
[3] Division of Electrical and Computer Engineering, Hanyang University,
17, Haengdang-dong, Sungdong-ku, Seoul, 133-791, Korea
jijung@hanyang.ac.kr

Abstract. This paper proposes efficient channel assignment scheme for mobility prediction handover. In future mobile communication networks, the radius of the cell becomes smaller so as to utilize the limited channel efficiently. Because of the smaller radius of the cell, it is difficult to maintain Quality of Service (QoS) on account of occurring the call drop phenomenon and the insufficient channel during the terminal movement. In order to prevent these phenomena, the mobility prediction handover schemes are proposed. However, the problem is the bad effectiveness of the channel because existing channel assignment schemes ignore the mobility prediction handover. The proposed scheme maintains the dropping probability of handover calls, reduce the blocking probability of new calls, and increase utilization of channel.

1 Introduction

Currently, mobile operators are extending 2.5G networks with the goal of moving toward more advanced technologies made possible by the so called third generation mobile communication systems (3G) where massive investments are being made for new spectrum licenses and network deployment [1]. The new paradigm behind 3G is the widespread provision of multimedia flavor to the "anytime and anywhere" concept. But the 3G UMTS (Universal Mobile Telecommunication Systems), as currently defined, will not be able to support broadband services since the maximum transmission rate is limited to 2 Mb/s. Moreover, 2Mb/s are only achievable at a very short range and under restricted mobility conditions [2].

To support such broadband services, it is important to effectively utilize the scarce wireless resources by employing microcells or picocells. However, this accompanies the increase of handover frequency and complicated mobility patterns among users. Moreover, the fluctuation of the handover calls and new calls varies widely depending on both time and location [3].

Handover is a process whereby a mobile station communicating with one wireless base station is transferred to another base station during a call (session). If handover occurs frequently and QoS is not guaranteed, then the quality of the call can be de-

D.-K. Baik (Ed.): AsiaSim 2004, LNAI 3398, pp. 410–418, 2005.

graded, or the call drop can occur. Because of that reason, researches on channel assignment schemes have been made to reduce dropping probability of handover calls, and on handover algorithm that, in order to ensure QoS for handover, reserve channels where the movement is predicted [4-6]. Those channel assignment schemes, however, are not considered for mobility predicted handover. Because the predicted handover call is pre-assigned with a channel in mobility predicted handover, the dropping probability of handover call decreases far below the required level, while the blocking probability of new call increases, decreasing overall channel utilization.

In this paper, efficient channel assignment scheme for mobility prediction handover is proposed. The proposed scheme adjusts the guard channel size in accordance with the handover prediction ratio, maintains the dropping probability of handover calls, decreases the blocking probability of new calls, and increases the channel utilization. The rest of the paper is organized as follows. Section 2 describes the existing channel assignment schemes, and Section 3 presents our channel assignment scheme for mobility prediction handover. Section 4 evaluates the performance, and finally, section 5 gives conclusions.

2 Channel Assignment Scheme

QoS provisioning of services and efficient utilization of limited bandwidth are important issues in mobile communication networks. This section describes the fully shared scheme, guard channel scheme, and dynamic channel reservation scheme.

2.1 Fully Shared Scheme (FSS)

FSS (Fully Shared Scheme) can be represented as an example of nonprioritized scheme. In FSS, the BS handles the call requests without any discrimination between handover and new calls. All available channels in the BS are shared by handover and new calls. Thus, it is able to minimize rejection of call requests and has the advantage of efficient utilization of wireless channels. However, it is difficult to guarantee the required dropping probability of handover calls [7].

2.2 Guard Channel Scheme (GCS)

GCS can be represented as an example of prioritized schemes. GCS gives higher priority to handover calls than new calls. In GCS, guard channels are exclusively reserved for handover calls, and the remaining channels, called normal channels, can be shared equally between handover and new calls.

Thus, whenever the channel occupancy exceeds a certain threshold, GCS rejects new calls until it goes below the threshold. In contrast, handover calls are accepted until the channel occupancy goes over the total number of channels in a cell. It offers a generic means to decrease the dropping probability of handover calls but causes reduction of total carried traffics. The reason why total carried traffic is reduced is that fewer channels except the guard channels are granted to new calls. The demerits become more serious when handover requests are rare. It may bring about inefficient spectrum utilization and increased blocking probability of new calls in the end because only a few handover calls are able to use the reserved channels exclusively. The use of guard channels requires careful determination of the optimum number of these channels, knowledge of the traffic pattern of the area, and estimation of the channel occupancy time distributions [7-9].

2.3 Dynamic Channel Reservation Scheme (DCRS)

DCRS can be represented as an example of prioritized schemes. In DCRS, both handover and new calls share equally the normal channels, which are radio channels below the threshold. The guard channels, the remaining channels above the threshold, are reserved preferentially for handover calls in order to provide their required QoS. Those channels, however, can also be allocated as much as the request probability for new calls instead of immediately blocking, unlike GCS. Thus handover calls can use both normal and guard channels with probability one if these channels are available. New calls use normal channels with probability one, but guard channels can be used for new calls according to the request probability (RP). It contributes to reducing the blocking probability of new calls and improving the total carried traffic [10-11].

The RP reflects the possibility that the BS permits new calls to allocate the wireless channel among the guard channels. It is dynamically determined by the probability generator in which the RP is computed considering the mobility of calls, total number of channels in a cell, threshold between normal channels and guard channel, and current number of used channels. Among these factors, the mobility of calls is most important. The mobility of calls in a cell is defined as the ratio of the handover call arrival rate to the new call arrival rate [10].

The formula to calculate RP for new calls in DCRS is as shown in Eq. 1 [7].

$$RP = MAX \left\{ 0, \alpha \left[\frac{C-i}{C-T} \right] + (1-\alpha) \cos \frac{2\pi (i-T)}{4(C-T)} \right\}^{1/2} \tag{1}$$

where, C is the number of total channel in the cell, T is the threshold of normal channel and guard channel, i is the number of channels being used, and α indicates the call traffic pattern calculated in the following formula.

$$\alpha = \frac{\text{number of new calls generated in the current BS}}{\text{number of handover calls generated in the current BS}}$$

3 Mobility Prediction Dynamic Channel Reservation Scheme (MPDCRS)

In DCRS scheme, the mobility predicted handover is not considered. If we add the feature of mobility prediction handover to DCRS (called predicted DCRS scheme), channel reservation for the predicted handover can be made on a normal channel. Therefore, the number of dropped handover calls is fewer and the dropping probability of handover calls is lower than the expected level. On the other hand, because many normal channels are reserved for predicted handover calls, the blocking probability of new calls gets higher than a required level, and therefore, utilization of the total number of channel gets worse. It is because there is no adjustment of the guard channel size.

In this paper, we propose MPDCRS as an efficient channel assignment scheme for mobility prediction handover. The MPDCRS adjusts the size of guard channel in accordance with prediction ratio of handover calls as well as considers the feature of the mobility prediction handover. We also make channel reservation for the predicted handover on a normal channel.

Adjusting the guard channel size, the proposed MPDCRS scheme maintains the dropping probability of handover calls in a required level, decreases blocking probability of new call and increases overall channel utilization. Fig. 1 shows channel allocation of MPDCRS. In MPDCRS, the handover call and the new call share normal channels. In this scheme, handover calls are always assigned if the cell has free channels irrespective of normal channel or guard channel. On the other hand, new calls are assigned with the channel under the same conditions as handover calls in normal channel. In guard channels, which are above the threshold, new calls are not directly rejected but are assigned with guard channels depending on the request probability (RP) of the new call calculated by the probability generator. RP for new calls is the parameter that determines assignment of guard channels to the new call. In guard channels, the new calls are allowed depending on the probability value calculated by RP.

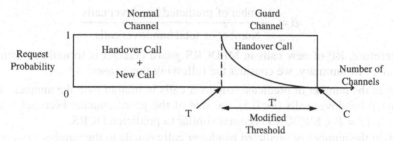

Fig. 1. Channel allocation of MPDCRS

Because of reservation on normal channel for predicted handover calls, MPDCRS suggests that the number of the guard channels should be decreased, and therefore, overall channel utilization is increased. In MPDCRS, the adjustment of guard channel size is conducted by determining the modified threshold (T'). T' divides channels into normal channel and guard channel and is calculated by the handover prediction ratio. T' can move between T and C. T is the threshold in DCRS. We assume that all base stations (BSs) are provided with the overall handover prediction ratio through an entire network, and the BS adjusts the size of the guard channel with the handover prediction ratio.

Eq. 2 is used to calculate RP of new calls in MPDCRS.

$$RP = MAX\left\{0, \alpha\left[\frac{C-i}{C-T'}\right] + (1-\alpha)\cos\frac{2\pi(i-T')^{1/2}}{4(C-T')}\right\} \tag{2}$$

Eq. 3 is the formula to calculate the modified threshold in MPDCRS.

$$T' = T + \beta \cdot (C - T) \tag{3}$$

where, $\beta(0 \le \beta \le 1)$ is the handover prediction ratio calculated in the following formula.

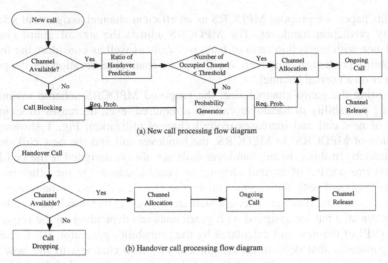

(a) New call processing flow diagram

(b) Handover call processing flow diagram

Fig. 2. Call flowchart in MPDCRS

$$\beta = \frac{\text{Number of predicted handover calls}}{\text{Number of total handover calls}}$$

Therefore, RP of new calls in MPDCRS guard channel is formulated using Eq. 2 and Eq. 3. In summary, we consider the following three cases:

- when the number of predicted handover calls is smaller than the number of unpredicted handover calls ($\beta < 0.5$), the size of the guard channel becomes smaller. If $\beta = 0$ ($T = T'$), MPDCRS becomes similar to predicted DCRS,
- when the number of predicted handover calls equals to the number of unpredicted handover calls ($\beta = 0.5$), T' is in the middle of T and C,
- when the number of predicted handover calls is bigger than the number of unpredicted handover calls ($\beta > 0.5$), the guard channel becomes much smaller (T' moves to C). If $\beta = 1$, MPDCRS becomes the FSS.

Fig. 2 is the call flow chart for new calls and handover calls in MPDCRS. In the case of new call, if the guard channel is used, the scheme estimates the guard channel threshold with the handover prediction ratio, and then, determines whether to block the new call depending on the probability value of RP. In the case of handover call, however, in order to ensure QoS to the user, the available channel is assigned if there is free channel.

4   Performance Evaluation

We compare performance among MPDCRS, unpredicted DCRS and predicted DCRS in terms of dropping probability of handover calls, blocking probability of new calls, and channel utilization. We assume that the average number of mobile users is much bigger than the number of channels in the BS so that the net call arrivals to the BS are approximated as a Poisson process [8],[10]. Also, we assume that the total number of channel in a BS is 60 and all the wireless channels have the same fixed capacity [7].

The input parameters in our simulation are:

1. λ: call arrival rate. Call arrives according to a Poisson process of rate λ [4],[5],
2. μ: mean call completion rate. Call holding time is assumed to be exponentially distributed with a mean of $1/\mu$ [5],
3. η: portable mobility. User residual time is assumed to be exponentially distributed with a mean of $1/\eta$ [5].

Several performance metrics are considered:

1. P_d: dropping probability for a handover call,
2. P_b: blocking probability for a new call attempt,
3. U_c: channel utilization of BS.

Figures 3, 4, and 5 illustrate the effect on MPDCRS. The mean call holding time $1/\mu$ is 6 min, and the mean user residual time $1/\eta$ is 3 min. We assume that handover calls are 50% of total calls, and predicted handover calls are 25%, 50% and 75% of handover calls.

Fig. 3 shows the dropping probability of handover call. The figure shows that the dropping probability of the mobility predicted DCRS handover is much lower than that of the unpredicted DCRS handover. In the case of MPDCRS, the dropping probability is between that of the unpredicted DCRS and that of the predicted DCRS. The effect of predicted DCRS and MPDCRS becomes greater as the handover prediction ratio gets higher.

When handover prediction ratio is 25%, P_d of predicted DCRS is decreased by 44 ~ 64% compared to unpredicted DCRS in low load. In MPDCRS, P_d is decreased by 18 ~ 22% compared to unpredicted DCRS. P_d of predicted DCRS is decreased by 22 ~ 24% compared to unpredicted DCRS in high load. P_d of MPDCRS is decreased by 14 ~ 16% compared to unpredicted DCRS in high load.

When handover prediction ratio is 50%, P_d of predicted DCRS is decreased by 41 ~ 64% compared to unpredicted DCRS in low load. In MPDCRS, P_d is decreased by 30 ~ 40% compared to unpredicted DCRS in low load. P_d of predicted DCRS is decreased by 26 ~ 32% compared to unpredicted DCRS in high load. P_d of MPDCRS is decreased by 4 ~ 12% compared to unpredicted DCRS in high load.

The effect of predicted DCRS and MPDCRS becomes greater as the handover prediction ratio gets higher. It is due to the channel reservation for the predicted handover on a normal channel.

Fig. 4 shows the blocking probability of the new call. The figure shows that the blocking probability of the mobility predicted DCRS handover is a little lower than that of the unpredicted DCRS handover. In the case of MPDCRS, the blocking probability of the new call becomes much lower. The effect of MPDCRS becomes better as the handover prediction ratio gets higher.

When handover prediction ratio is 25%, P_b of new call in predicted DCRS is increased by 4 ~ 5% compared to unpredicted DCRS handover in low load. In MPDCRS, P_b is decreased by 6 ~ 11% compared to unpredicted DCRS in low load. P_b of predicted DCRS is increased by 3 ~ 6% compared to unpredicted DCRS in high load. P_b of MPDCRS is decreased by 5 ~ 9% compared to unpredicted DCRS in high load.

When handover prediction ratio is 50%, P_b of new call in predicted DCRS is increased by 8 ~ 10% compared to unpredicted DCRS handover in low load. In

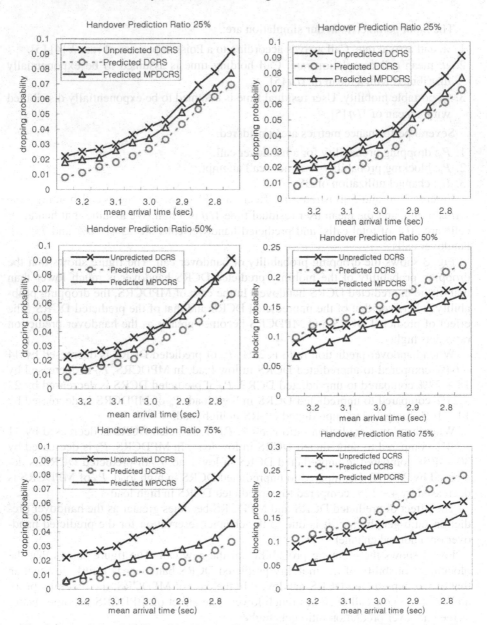

Fig. 3. Dropping probability of handover call   **Fig. 4.** Blocking probability of new call

MPDCRS, P_b is decreased by 20 ~ 26% compared to unpredicted DCRS in low load. P_b of predicted DCRS is increased by 13 ~ 16% compared to unpredicted DCRS in high load. P_b of MPDCRS is decreased by 8 ~ 12% compared to unpredicted DCRS in high load.

The effect of MPDCRS becomes better as the handover prediction ratio gets higher. It is due to the adjustment (reduction) of guard channel size.

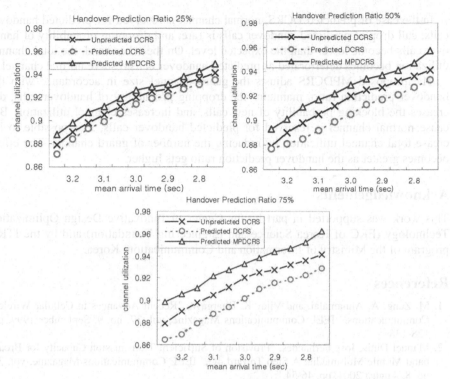

Fig. 5. Channel utilization

Fig. 5 shows the channel utilization. The figure shows that the channel utilization of the mobility predicted DCRS handover is lower than that of the unpredicted DCRS handover. In MPDCRS, the channel utilization becomes higher than predicted DCRS. The effect of MPDCRS becomes better as the handover prediction ratio gets higher.

When handover prediction ratio is 25%, U_c of predicted DCRS is decreased by 1% compared to unpredicted DCRS handover in low load. In MPDCRS, U_c is increased by 1% compared to the unpredicted DCRS in low load. U_c of predicted DCRS is decreased for 1% compared to unpredicted DCRS in high load. U_c of MPDCRS is increased by 1% compared to unpredicted DCRS in high load.

When handover prediction ratio is 50%, U_c of predicted DCRS is decreased by 0 ~ 1% compared to unpredicted DCRS handover in low load. In MPDCRS, U_c is increased by 1 ~ 2% compared to unpredicted DCRS in low load. U_c of predicted DCRS is decreased by 1% compared to unpredicted DCRS in high load. U_c of MPDCRS is increased by 1 ~ 2% compared to unpredicted DCRS in high load.

The effect of MPDCRS becomes better as the handover prediction ratio gets higher. It is due to the adjustment (reduction) of guard channel size and the reduction in blocking probability of the new call.

5 Conclusion

In this paper, we proposed MPDCRS as an efficient channel assignment scheme for mobility prediction handover.

In the case of predicted DCRS, normal channel is reserved for predicted handover calls, call drop of predicted handover calls is rare, and dropping probability of handover calls become lower than the expected level. On the other hand, the total channel utilization becomes deteriorated as predicted handover calls reserve normal channel.

The proposed MPDCRS adjusts the guard channel size in accordance with the handover prediction ratio, maintains the dropping probability of handover calls, decreases the blocking probability of new calls, and increases channel utilization. Because normal channel is reserved for predicted handover calls, it is possible to increase total channel utilization by reducing the number of guard channel. The effect becomes greater as the handover prediction ratio gets higher.

Acknowledgements

This work was supported in part by the Center of Innovative Design Optimization Technology (ERC of Korea Science and Engineering Foundation) and by the ITRC program of the Ministry of Information and Communication, Korea.

References

1. M. Zeng, A. Annamalai, and Vijay K. Bhargava, "Recent Advances in Cellular Wireless Communications," IEEE Communications Magazine, vol. 37, no. 9, September 1999, pp. 128-138
2. Manuel Dinis, Jose Fernandes, "Provision of Sufficient Transmission Capacity for Broadband Mobile Multimedia: A Step Toward 4G," IEEE Communications Magazine, vol. 39, no. 8, August 2001, pp. 46-54
3. In-Soo Yoon, Byeong Gi Lee, "DDR: A Distributed Dynamic Reservation Scheme That Supports Mobility in Wireless Multimedia Communications," IEEE JSAC, pp. 2243-2253, November, 200
4. William Su and Mario Gerla, "Bandwidth allocation strategies for wireless ATM networks using predictive reservation," GLOBECOM 1998. The Bridge to Global Integration. IEEE, Vol 4, pp 2245 -2250, 1998.
5. Ben Liang, Zygmunt J. Haas, "Predictive Distance-Based Mobility Management for PCS Networks," IEEE Journal on Selected Areas in Communication, pp880-892, June 1995.
6. Ming-Hsing Chiu, Mostafa A. Bassiouni, "Predictive Schemes for Handoff Prioritization in Cellular Networks Based on Mobile Positioning," IEEE Journal on Selected Areas in Communications, pp510-522, March 2000.
7. Daehyoung Hong, Stephen S. Rappaport, "Traffic Model and Performance Analysis for Cellular Mobile Radio Telephone Systems with Prioritized and Non-Prioritized Handoff Procedures," IEEE Trans. Vehic. Tech., Vol. VT-35, no. 3, pp 77-92 Aug. 1986.
8. S. Tekinay and B. Jabbari, "A Measurement-Based Prioritization Scheme for Handovers in Mobile Cellular Networks," IEEE JSAC, pp. 1343-50, Vol. 10, no. 8. October, 1992.
9. Yi-Bing Lin, Seshadri Mohan, Anthony Noerpel, "PCS Channel Assignment Strategies for Hand-off and Initial Access," IEEE Personal Communications, pp. 47-56, Third Quarter, 1994.
10. Young Chon Kim, Dong Eun Lee, Bong Ju Lee, Young Sun Kim, "Dynamic Channel Reservation Based on Mobility in Wireless ATM Networks," IEEE Communication Magazine, pp. 47-51, November, 1999.
11. K.C. Chua, B.Bensaou, W.Zhuang, S.Y.Choo "Dynamic Channel Reservation(DCR) Scheme for Handoffs proiritization in Mobile Micro/Picocellular Networks,"ICUPC'98, IEEE 1998 International conference on Universal Personal Communications Record, Vol 1, pp 383-387, 1998

A Novel Mobility Prediction Algorithm
Based on User Movement History
in Wireless Networks

Sedong Kwon, Hyunmin Park*, and Kangsun Lee

Department of Computer Engineering, MyongJi University,
San 38-2 Namdong, YongIn, Kyunggido, Korea 449-728
Tel: +82-31-330-6442, Fax: +82-31-330-6432
{sdkwon,hpark,kslee}@mju.ac.kr

Abstract. For an efficient resource reservation, mobility prediction has been reported as an effective means to decrease call dropping probability and to shorten handoff latency in wireless cellular networks. Several early-proposed handoff algorithms making use of a user's movement history on a cell-by-cell basis work on the assumption that the user's movement is restricted to the indoor locations such as an office or a building. These algorithms do not present a good performance in a micro-cell structure or a metropolis with complicated structure of roads. To overcome this drawback, we propose a new mobility prediction algorithm, which stores and uses the history of the user's positions within the current cell to predict the next cell.

1 Introduction

One of the most important features in a wireless cellular communication network is the user mobility [1]. Since a mobile user is free to move anywhere, the scheme relying on local information (e.g., information where the user currently resides) cannot guarantee Quality of Service (QoS) requirements of a connection throughout its lifetime. The seamless mobility requires an efficient resource reservation and a support for the handoff from one cell to another. When a handoff occurs to a neighboring cell where there is not enough bandwidth, the handoff call is dropped. It is possible to reserve resources at all neighboring cells for handoff users. However, this scheme leads to an overall wastage of bandwidth, and makes new calls suffer from a severe blocking problem even at light loads.

In order to optimize the efficiency of a resource reservation, an accurate prediction algorithm for a user movement is required [2]. Several mobility prediction algorithms [3–9] have been proposed to decrease the call dropping probability and to shorten the handoff latency. It is possible to trace and predict the user's movement from its present location and velocity, via different coverage zones within a cell [3], or the GPS (Global Positioning System) [4]. Another alternative is to employ the user's mobility history and stochastic models for movement

---

* Author to whom correspondence should be addressed.

D.-K. Baik (Ed.): AsiaSim 2004, LNAI 3398, pp. 419–428, 2005.

prediction. Besides the user's mobility history, other information like connection patterns and stay intervals can also be utilized for analysis [7].

Some of the basic prediction algorithms [5] based on mobility patterns of the users are introduced as follows. The Location Criterion algorithm [5] predicts the next cell, which has the highest probability of the departure history in the current cell. The Direction Criterion algorithm [5] takes into account the user's direction of travel, which is identified by the current and previous cell locations, and uses the departure history of this direction to improve the accuracy. The cell with the highest departure rate is predicted as the next. The Segment Criterion algorithm [5] further extends the Direction Criterion algorithm. All the past cell-by-cell movement patterns are used to predict the next cell. The Direction Criterion algorithm shows the best performance among all these basic prediction algorithms [5]. The Sectorized Mobility Prediction algorithm [9] utilizes a cell-sector numbering scheme and a sectorized mobility history base (SMHB) to predict the movements of users. The SMHB algorithm presents a high prediction accuracy for both regular and random movements of users. However, our simulation results show that the prediction accuracy of the SMHB algorithm decreases in the realistic environment with a complex road structure in the handoff area of the cell.

This paper proposes a new prediction algorithm, which is called the Zoned Mobility History Base (ZMHB) algorithm. It employs a cell structure that has six sectors and each sector is further divided into three zones by handoff probability, and exploits cell-zone numbering and intra-cell-basis movement history, which stores a record of the user's movement in the current cell. This paper compares the ZMHB algorithm with some other existing prediction algorithms. Our simulation results show that the ZMHB algorithm provides the better prediction accuracy. The remainder of this paper is organized as follows. In Section 2, ZMHB algorithm and its cell structure is introduced. Section 3 describes the simulation environment for comparing the ZMHB and the other prediction algorithms, and the detailed performance results are presented. Finally, conclusion and future works are given in Section 4.

2   Cell Structure and Zoned Mobility History Base (ZMHB) Algorithm

Several early proposed mobility prediction algorithms [5–9] make use of a user's movement history on a cell-by-cell basis. They work on the assumption that the user's movements are restricted to the indoor locations such as an office or a building, and provide a poor performance when no past individual cell-by-cell basis history is available (e.g., a call first occurs in the current cell). So, we propose a new prediction algorithm, which stores and uses the history of a user's positions within the current cell, not the cell-by-cell history.

In this paper, a hexagonal cell structure is employed as shown in Fig. 1. A cell is divided into three regions by handoff probability [9]. The most inner part of the cell where the probability of handoff is negligible is called the No-Handoff

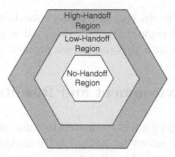

Fig. 1. A Cell Structure.

region. The middle part of the cell where the probability of handoff is low is defined as the Low-Handoff region. Finally, the most outer part of the cell has the high probability of handoff and is called the High-Handoff region. A user can receive the beacon signal above the threshold value for the handoff from the neighboring cells. There are two types of the handoff, a hard handoff and a soft handoff, whose characteristics are related to resource reservation and release. They may be characterized as "break before make" and "make before break", respectively [1]. The soft handoff is assumed throughout this paper.

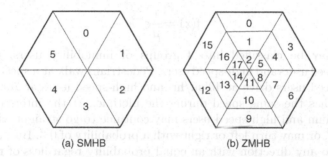

Fig. 2. A Cell Structure in SMHB and ZMHB Algorithm.

Fig. 2(a) shows a cell structure of the Sectorized Mobility History Base (SMHB) algorithm [9], which stores the positions of the user on a sector-by-sector basis instead of a cell-by-cell basis to predict the next cell. The previous research experiments on the SMHB algorithm have not allowed a move into another High-Handoff region of an adjacent sector in the same cell after the prediction, and so have not explored the effect caused by a complicated geographical feature of the High-Handoff region in a metropolis. This paper takes into account such complex situations. In Fig. 2(b), a cell structure for our proposed prediction algorithm, the Zoned Mobility History Base (ZMHB) algorithm, is demonstrated. In ZMHB, each sector in SMHB is further divided into three zones by handoff probability as in Fig. 1. The ZMHB algorithm keeps and utilizes the zone-by-zone movement history patterns of the user in a cell. The numbers in Fig. 2(b) depict region numbers. If [(a region number) modulo 3] is equal to zero, this region is a High-Handoff region. If a new user call is generated in this region or a mobile user call moves into this region, we predict the next cell.

3   Simulation Environment and Results

The proposed ZMHB algorithm is compared to some other prediction algorithms in an environment with seven hexagonal cells making use of Visual C++ as the simulation tool. The whole cellular system composed of seven hexagonal cells(Fig. 4 & 7), and the diameter of each cell may be varied from 200m up through 1km. As in [10], the border cells are connected so that the whole cellular system forms a similar structure to a sphere.

Calls are generated by a Poisson process with a rate λ in each cell, and can appear anywhere with an equal probability. The Poisson distribution has the probability mass function,

$$P[X=x] = e^{-x}\frac{\lambda^x}{x!} \tag{1}$$

where x is the number of call occurrences and λ is the average number of calls. The lifetime of each call is exponentially distributed with a mean, μ (60-300 seconds). The exponential distribution with a mean μ has the probability density function

$$f(x) = \frac{1}{\mu}e^{-\frac{x}{\mu}} \tag{2}$$

Calls may be classified into 4 groups of immobility users, pedestrians, medium-speed users and high-speed users. Pedestrians walk at a speed of 4km/hr, medium-speed users drive at 40km/hr, and high-speed users at 70km/hr. Each traffic call has the same speed during the lifetime. At the intersection of two roads, medium and high speed users may continue to go straight with a probability of 0.7, or may turn left or right with a probability of 0.3. But, a pedestrian can move to any direction with an equal probability regardless of roads or offroads. Table 1 shows the call generation rate of each user group on and off a road.

The statistics of about 1,000 calls per each cell are collected for a simulation experiment. For each data point, ten experiments are performed. For simplicity, the factors employed in the experiments are collectively provided as follows: F(Factor) = [Average number of call occurrences in the unit time, Lifetime(sec), Total number of call occurrences in a cell (approximate numbers), Cell Diameter(m)].

Table 1. Traffic call generation rates on and off a road.

	Immobility Users	Pedestrian	Medium-Speed Users	High-Speed Users
Road	10%	10%	40%	40%
Offroad	40%	40%	15%	5%

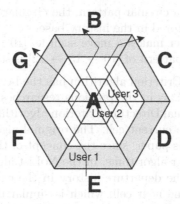

Fig. 3. Example of Movement Paths.

In Fig. 3, the gray area depicts the High-Handoff region. When a new user appears in this area or a mobile user enters this region, the prediction algorithm is executed to predict the next cell. However, if the user moves into another High-Handoff region in the same cell after the prediction for the user's movement, another prediction is not performed again. If the user makes a handoff to a next neighboring cell, this movement pattern is stored in the history base. However, if the user moves back to the inner part of the present cell, this movement pattern is neglected. In order to predict the next cell, SMHB and ZMHB algorithms make use of the previously recorded mobility patterns of the users. Both algorithms store the user's movement pattern in their history bases as follows. The region numbers in Fig. 2 are employed.

Stored pattern for User 1:

- SMHB algorithm: cell E → cell A[3 → 4 → 5] → cell G
- ZMHB algorithm: cell E → cell A[10 → 13 → 14 → 17 → 16 → 15] → cell G

In ZMHB algorithm, though User 1 passes through the 9th region of the cell A from the cell E, region number 9 is not included in the movement pattern because of the soft handoff.

Stored pattern for User 2:

- SMHB algorithm: cell A[2 → 1 → 0] → cell B
- ZMHB algorithm: cell A[7 → 8 → 5 → 2 → 1 → 0] → cell B

Stored pattern for User 3:

- SMHB algorithm: cell A[2 → 1 → 0] → cell C
- ZMHB algorithm: cell A[7 → 4 → 1 →0] → cell C

The followings are assumed to make use of the history base.

1) Search time of the history base + The application time of the prediction ≤ The time in the High-Handoff region of the present cell.
   The users have a time enough for a prediction and a resource reservation before a handoff.

2) If the user moves in a circular pattern, the circular part is removed from the movement pattern stored in the history base.

For example, if a user makes a move such as $[10 \rightarrow 13 \rightarrow 10 \rightarrow 13 \rightarrow 12]$, only the movement pattern of $[10 \rightarrow 13 \rightarrow 12]$ is stored in the history base.

Since the Direction Criterion algorithm has the best performance among the basic prediction algorithms in the previous researches [5], it is adopted for the comparison. In the original Direction Criterion algorithm, the direction is defined as the previous and the current cell. The original algorithm will be called the Weight algorithm in this paper, since the concept of the direction criterion will be employed in the other algorithms. In the NoList algorithm, the cell with the highest probability of the departure history in the current handoff region of a sector is predicted as the next cell, which is similar to the Location Criterion algorithm. If the direction criterion concept is incorporated in this algorithm, the direction can be identified as the previous cell location and the current handoff region of the present cell. The Direct algorithm predicts the only neighboring cell adjacent to the present sector as the next cell. The SMHB algorithm stores the position of the user on a sector-by-sector basis in the history base and predicts the most likely subsequent cell. The ZMHB algorithm keeps the movement patterns of the user on a zone-by-zone basis in the history base and predicts the most likely subsequent cell. In the SMHB and the ZMHB algorithm, if a new call occurs in the handoff area or the current user's movement pattern is not found in the history base, the NoList algorithm is applied for the prediction. The accuracies are measured as follows:

$$\text{Accuracy for handoffs} = \frac{\text{No. of successful handoffs}}{\text{No. of handoffs}} \tag{3}$$

$$\text{Accuracy for reservation} = \frac{\text{No. of successful handoffs}}{\text{No. of predictions}} \tag{4}$$

In equation (3), the prediction accuracy for handoffs is defined by the ratio between the number of successfully predicted handoffs and the total number of actual handoff calls. The accuracy for handoffs provides the information about the dropped handoff calls caused by the prediction misses. Equation (4) measures the prediction accuracy for reservation as the ratio between the number of successfully predicted handoffs and the total number of the predicted calls. The predicted calls include not only the actual handoffs but also no-handoff calls, which are the cases that the user moves back to the inner part of the current cell or the call lifetime ends in the handoff region after the predictions. The accuracy for reservation gives the information about the wasted resources since the resource is reserved in the predicted cells after the predictions regardless of the handoffs.

Previous Cell Location Not in the History Base

In this section, it is assumed that the information of the previous cell location is not used to predict the next cell. Fig. 4 shows the road structure of the simulation environment 1, and the black and the white areas represent the roads and the off-roads, respectively.

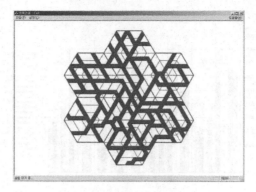

Fig. 4. Simulation Environment 1.

$F = [0.1, 120(s), 1000, 200(m)]$

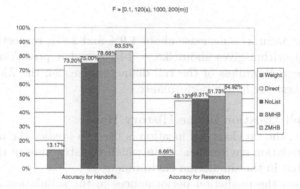

Fig. 5. Comparison of the prediction accuracies in simulation environment 1.

Fig. 5 compares the accuracies of the prediction in the simulation environment 1. The Weight algorithm predicts the next cell based on the highest departure rate in the present cell, irrespective of signal strengths in the handoff region. Since there exist six neighboring cells and the departures to all neighboring cells are equally distributed, the accuracy is about 1/6. Since the number of predictions includes the number of no-handoff calls, the accuracies for reservation in all algorithms are less than 60% and the ZMHB algorithm shows the best performance. The accuracy of the ZMHB algorithm for handoffs is about 5.3 times as good as that of the Weight algorithm, and the ZMHB algorithm gives about 6.2 ∼ 14.1% higher prediction accuracy, compared to the others.

Fig. 6 demonstrates the prediction accuracies in a simulation environment 1 when the cell diameter increases by 200m. The Weight algorithm is excluded in this simulation result because its prediction accuracy is negligible compared to the others. With the assumption of the same average call lifetime, the probability that the user call is ended in the handoff region increases, as the cell diameter increases. So the prediction accuracies for the resource reservation in all algorithms decrease about 5.4 ∼ 23.1% with the cell diameter increase. The prediction accuracies of the Direct and the NoList algorithm for handoffs slightly decrease below 1% as the cell diameter increases. For the SMHB and the ZMHB

F = [0.1, 120(s), 1000, 200(m) - 1000(m)]

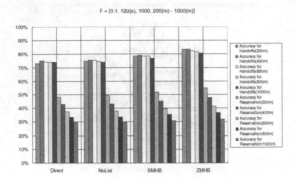

Fig. 6. The prediction accuracies vs. the cell diameter (200m increment) in simulation environment 1.

algorithm, the accuracies decrease about 1.9% and 2.8%, respectively. However, the ZMHB algorithm gives about 3.8 ~ 14.1% higher prediction performance than the others, regardless of the cell diameter size. So, the ZMHB algorithm provides the best prediction performance.

Previous Cell Location in the History Base
In this section, the previous cell location is used to predict the next cell. Fig. 7 shows the simulation environment 2, in which the structure of the roads is different from that in the simulation environment 1.

Fig. 8 shows the prediction performances in the simulation environment 2. As shown in Fig. 8, incorporating the direction concept using the information of the previous cell location for the NoList algorithm improves the prediction accuracy, which is similar to the previous researches on the Location Criterion algorithm and Direction Criterion algorithm [5]. The NoList(P) algorithm gives about 3.9 ~ 5.7% higher prediction accuracy than the NoList. However, for the SMHB and the ZMHB algorithm, which use the intra-cell movement history, the information of the previous cell history does not improve the prediction performance. Whether the previous cell history is included in the history base or not, the ZMHB algorithm demonstrates about 8.9 ~ 14.7% better prediction performance than the others.

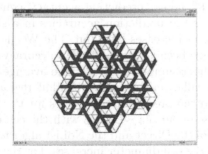

Fig. 7. Simulation Environment 2.

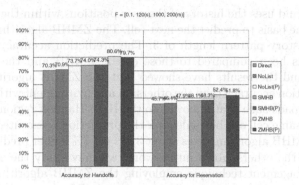

Fig. 8. The accuracies of the algorithms using the previous cell history in simulation environment 2.

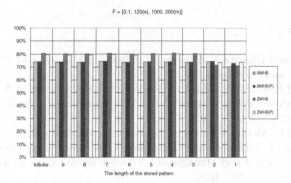

Fig. 9. The accuracies of the SMHB and the ZMHB algorithm for handoffs vs. the length of the stored pattern in simulation environment 2.

Fig. 9 compares the SMHB and the ZMHB algorithm as the length of the stored history pattern varies. For example, if the length of the stored pattern is restricted to 3 and a user moves in such a pattern as [present cell[10 → 13 → 14 → 17 → 16 → 15] → next cell], only the pattern of [present cell[17 → 16 → 15] → next cell] is stored in the history base. As shown in Fig. 9, if the length of stored history pattern in the SMHB is below 2, the prediction accuracy decreases about 5.3%. In the ZMHB, if the length is below 3, the accuracy decreases about 12.4%. However, if the length is equal to or above 3, the ZMHB algorithm shows a similar accuracy to the case of the indefinite length. In the worst case, if the length of the stored pattern is restricted to 1, the ZMHB algorithm shows the same prediction performance as the NoList algorithm.

4   Conclusions and Future Works

Mobility prediction is an effective means to provide Quality of Service guarantees and to decrease a call dropping probability in wireless cellular environments. We have proposed a new mobility prediction algorithm, called the ZMHB algorithm, which stores and uses the history of the user positions within the current cell on a zone-by-zone basis to predict the next cell. The ZMHB algorithm requires only the stored history pattern length of 3. The prediction accuracy of the ZMHB algorithm has been compared to those of the SMHB and the other existing algorithms, and the results have showed that the ZMHB algorithm provides a better prediction accuracy than the others. In an environment with a complicated structure of roads such that a move into the High-Handoff region of an adjacent sector in the same cell can be made after the prediction is executed in the present sector, the ZMHB algorithm gives about 3.8 ~ 14.7% higher prediction accuracy, compared to the others. For a future work, we may study the efficiency of the resource management technique employing the ZMHB algorithm. Since a cell may be partitioned into even smaller regions, it will be interesting to investigate its effect on the prediction accuracy.

References

1. Q. A. Zeng and Dharma P. Agrawal. *Handbook of Wireless and Mobile Computing.* John Wiley and Sons Inc, 2002.
2. S. Choi and K. G. Shin. Adaptive Bandwidth Reservation and Admission Control in QoS-Sensitive Cellular Networks. *IEEE Transactions on Parallel and Distributed Systems*, 13(9):882–897, 2002.
3. H. Kim and C. Moon. A Rerouting Strategy for Handoff on ATM-based Transport Network. *IEEE 47th Vehicular Technology Conference*, pages 285–289, 1997.
4. S. Bush. A Control and Management Network for Wireless ATM Systems. *IEEE ICC'96*, pages 459–463, 1996.
5. R. De Silva J. Chan and A. Senevirance. A QoS Adaptive Mobility Prediction Scheme for Wireless Networks. *IEEE GLOBECOM'98*, pages 1414–1419, 1998.
6. V. Bharghavan and J. Mysore. Profile Based Next-cell Prediction in Indoor Wireless LAN. *IEEE SICON'97*, 1997.
7. J. Chan et. al. A Hybrid Handoff Scheme with Prediction Enhancement for Wireless ATM Network. *APCC'97*, pages 494–498, 1997.
8. P. Bahl T. Liu and I. Chlamtac. Mobility Modeling, Location Tracking and Trajectory Prediction in Wireless ATM Networks. *IEEE JAC*, 16(6):922–936, 1998.
9. A. Jennings R. Chellappa and N. Shenoy. The Sectorized Mobility Prediction Algorithm for Wireless Networks. *ICT April 2003*, pages 86–92, 2003.
10. M. Naghshineh and M. Schwartz. Distributed Call Admission Control in Mobile/Wireless Networks. *IEEE J. Selected Areas in Comm*, 14(4):711–717, 1996.

Performance Evaluation of Threshold Scheme for Mobility Management in IP Based Networks

Yen-Wen Lin, Hsin-Jung Chang, and Ta-He Huang

Department of Computer Science and Information Engineering,
National Chiayi University, Chiayi, Taiwan, R.O.C.
ywlin@mail.ncyu.edu.tw

Abstract. *Mobile IP* is the most widely known mobility management scheme in wireless networks. However, some inherent drawbacks of this protocol are partially solved. Although *Route Optimization* tries to solve these problems, it cannot manage mobility binding efficiently. In this paper, three mobility binding management schemes, including *Always Push Scheme*, *On Demand Scheme*, and *Threshold Scheme*, are proposed, evaluated and compared. The simulation results show that the mobility binding updating strategy significantly impacts the overall performance of mobile systems. *Threshold Scheme* proposed in this paper outperforms both the *Mobile IP* and *Route Optimization* for mobility management in IP based networks.

1 Introduction

It becomes increasingly important to support IP mobility as the surprisingly explosive growth of Internet and the impressive advances of the mobile communication technology. The Mobile IP (MIP) protocol [1] was originally proposed to provide mobility to the existing services networking with IP technology of Internet. In MIP, all packets destined to the mobile node (MN) are firstly routed through the home agent (HA) of this MN and, consequently, results in a sub-optimal routing. It is the notorious *triangle routing problem*. The Route Optimization Protocol in Mobile IP (MIPRO) protocol [2] proposed by the IETF aims to solve the triangle routing problem. In MIPRO, if the correspondent node (CN) has an entry in the binding cache of the care-of-address (CoA) of the MN, packets sent from the CN can be directly tunneled to the MN without explicit intervention of the HA. In Mobile IPv6 [3] has similar binding cache design as in the MIPRO, however, binding update messages are sent from the MN to the CN rather than from the HA to the CN.

Though MIPRO solves the triangle routing problem of the original MIP protocol, it has its own problems. In MIPRO, significant control messages such as *Binding Update Message* and *Binding Warning Message* have been generated to achieve the route optimization, which cause extra signaling, and processing load. Related researches [4], [5], [6] including *regional registration* [4], *local anchoring* [5], and *hierarchical management* [6], have been proposed to remedy the drawbacks of MIPRO.

A great deal of effort has been made on managing the mobility in IP based networks. What seems to be lacking, however, is the efficiency of maintaining the mobility binding. Our proposal is motivated by the following observations. In MIPRO, route optimization is fulfilled by keeping a copy of mobility binding in the local cache at each CN ever communicating with the MN. Therefore, the CN can find the location

D.-K. Baik (Ed.): AsiaSim 2004, LNAI 3398, pp. 429–438, 2005.
© Springer-Verlag Berlin Heidelberg 2005

of the MN by checking the binding information recorded in the local cache and sends datagram accordingly. If, fortunately, the binding in the cache is correct, the CN directly sends datagram via an optimal route without intervention of the HA. However, in case the binding is stale, the datagram will be sent to previous location of the MN and will be sent back to the HA. Eventually, the HA will forward the datagram to the MN. Meanwhile, the HA will send *Binding Update Message* to the CN to refresh its binding information for future use. As a result, if the binding information in the cache is not updated in time, the datagram sent from the CN will be delivered via a sub-optimal route which may be even longer than the triangle routing in original MIP. That causes costly data routing cost and extra binding update message cost. In MIPRO, to keep the binding information in the local cache consistent with the copy of HA, additional signaling load resulting from flooding control messages including the *Binding Update Message* and *Binding Warning Message* is inevitable. The binding update strategy in MIPRO should be more deliberate to further improve the overall performance by reducing unnecessary signaling overhead. In this paper, mobility management schemes are proposed to improve the problems of MIPRO protocol. The performance of these schemes are carefully evaluated and compared.

The rest of this paper is organized as follows: In Section 2, we describe the proposed schemes. Simulation results are presented and analyzed in Section 3.

2 Mobility Management Schemes

Before going further, a few terms are defined. The FA_i is the serving FA of i-th subnet. Specially, the FA_{CN} is the serving FA of the CN. The FA_{prev} is the serving FA before the MN moves. The FA_{curr} is the serving FA the MN currently locates at. Besides, we use the term "cache hit" here as "Correct location information of the MN can be found in the mobility binding cache.", and the term "cache miss" here as "Cannot find the correct location information of the MN in the mobility binding cache."

In MIPRO, the binding cache kept at each CN cannot be shared with others intent to communicate with the same MN. In our design, the binding cache is maintained at the FA_{CN}. The method makes the binding cache can be shared with other CNs currently located at the coverage of the same FA. Both the space/processing cost for binding cache and the associated signaling cost can be remarkably reduced. Also, in MIPRO, the CN can send datagram to the MN via the optimal route if cache hit. However, unexpected extra traffic overheads are generated if cache miss happens. That is, the datagram may be delivered via an even longer path than that of the notorious triangle routing in original MIP scheme. We can presume that the cache hit ratio of the mobility binding cache consequentially impacts the performance of the system. We develop a series of experiments and get the conclusion that the frequency of updating the binding cache seriously dominates the system performance. The simulation results suggest that higher updating frequency makes the binding cache more precise, and accordingly, generates higher cache hit ratio and reduced data routing cost. That is, the higher the updating frequency the less the data routing cost. In the following, we develop three binding updating schemes including *Always Push Scheme*, *On Demand Scheme*, and *Threshold Scheme*, to find out the best suitable mobility binding updating strategy with the most profitable affiliated costs.

2.1 Always Push Scheme

In view of that a mobility binding cache maintaining scheme with higher cache hit rate reduces related data routing time. For obtaining high cache hit rate, the first binding cache updating scheme we propose is called *Always Push Scheme*. In this scheme, each time the MN changes its location, the HA receives a *Registration Request Message* from the MN, meanwhile, the HA sends *Binding Update Message* to the FA_{CN}. That is, the FA_{CN} almost all the time keeps updated binding in the local cache. It implies lower cache miss rate and higher binding updating cost.

2.2 On Demand Scheme

As suggested in *Always Push Scheme*, the HA may issue too many *Binding Update Message* than actually needed and consequently generates unnecessary traffic cost. To save such superfluous spend, the HA should issue *Binding Update Message* only when it's necessary. To fulfill this concept, the second binding updating scheme we have proposed is called *On Demand Scheme*. In this scheme, the HA does not actively update the binding cache of each FA_{CN} each time the MN registers its new location. The HA will send the FA_{CN} *Binding Update Message* only when the cache miss happens. That is, the CN cannot find "correct" location information in the local cache at the FA_{CN} when the CN sends the datagram to the MN. This scheme, as compared to *Always Push Scheme*, yields lower binding updating cost and higher cache miss rate.

2.3 Threshold Scheme

In the spectrum of binding updating frequency, *Always Push Scheme* is at one end while *On Demand Scheme* is at the other end. Between these two extreme ends, there are many binding updating strategies with different updating frequencies. The third binding cache updating scheme we have proposed is called *Threshold Scheme*. In this scheme, the binding cache will be updated only when it is rewarding. This scheme can dynamically decide whether or not to update the mobility binding cache. We use the term LCMR, as expressed in Equation (1), to quantify the usage of a binding cache of a FA_{CN} and the movement behavior of a MN. The denotation $LCMR_{i,j}$ represents the ratio of the $Call_{i,j}$ and the $Move_i$, where $Call_{i,j}$ is the number of the user i called by users located at the area j during the observed time; and $Move_i$ represents the number of movement of the user i during the observed time.

$$LCMR_{i,j} = \frac{Call_{i,j}}{Move_i} \tag{1}$$

The higher $LCMR_{i,j}$ is the more rewarding to update the binding cache of the serving FA of area j. Moreover, since the *Binding Updating Message* is sent from the HA to the FA_{CN}, we also consider the distance between the HA and the FA_{CN} as a part of updating message cost. The longer the distance is the costlier to update the binding cache. Take above two factors into consideration, we coin a new parameter $Eager_{FA}$ to judge whether it is rewarding to update the FA's binding cache. As suggested in its name, the $Eager_{FA}$ implies how eager a FA needs the binding cache to be updated.

$$Eager_{FA} = \frac{LCMR_{MN,FA}}{d_{FA,HA}} \tag{2}$$

In the Equation (2), the parameter $LCMR_{MN, FA}$ represents, in the observed duration, the ratio of the number of calls issued from the CNs at the coverage of FA to the MN to the number of the movement of the MN. Meanwhile, the parameter $d_{FA, HA}$ represents the distance between the FA and the HA. The parameter $Eager_{FA}$ of a certain FA is maintained at the HA and is accordingly updated either when the MN moves or when a CN sends datagram to the MN. Also, we devise a new parameter, called the system threshold ($Threshold_{sys}$) as the evaluating threshold of whether or not to update a binding cache. For now, the value of $Threshold_{sys}$ is determined by experiment that will be introduced later. When the $Eager_{FA} > Threshold_{sys}$, each time the MN moves, the HA immediately sends the *Binding Update Message* to the FA_{CN} (as in *Always Push Scheme*). On the contrary, when the $Eager_{FA} < Threshold_{sys}$, the HA sends *Binding Update Message* to the FA_{CN} only when the binding cache miss happens (as in *On Demand Scheme*). Involved steps are described as follows.

Step 1: The MN receives *Agent Advertisement Message* from the FA_{curr} when it moves in the coverage of the FA_{curr}; the MN gets the new CoA.

Step 2: The MN sends *Registration Request* with the *Previous Foreign Agent Notification Extension* to the FA_{curr}; the FA_{curr} thence knows related information about the FA_{prev}.

Step 3: FA_{curr} delivers *Registration Request* to the HA. FA_{curr} also sends *Binding Update Message* to the FA_{prev} to update new location binding of the MN for future possible datagram forwarding.

Step 4: The HA updates the MN's binding after receiving *Registration Request* from the FA_{curr} and sends *Registration Reply* back. Also, the HA sends *Binding Update Message* to the FA_{CN} to update binding in its local cache when $Eager_{FA} > Threshold_{sys}$.

Step 5: The FA_{curr} delivers *Registration Reply* to the MN.

2.4   Routing Datagram to the MN

To send datagram to the MN, the CN needs to find out current location of the MN. If correct location information can be found in the local cache of the FA_{CN}, the CN can directly send datagram to current location of the MN via optimal route. Otherwise, if the cached binding is stale, the datagram will be sent to previous location. The FA_{prev} will detect this condition. Either the FA_{prev} or the HA is responsible of forwarding the datagram to the MN. In case that the FA_{prev} knows current location of the MN (assisted by the *Previous Foreign Agent Notification Extension* [2]), the FA_{prev} will forward received datagram to the FA_{curr}. Otherwise, the datagram will be sent to the HA and be forwarded to current location of the MN later.

3   Simulations and Results

The aim of the following simulations is to evaluate the performance of various mobility binding management schemes in IP based networks. Three binding cache updating schemes proposed in this paper, including *Always Push Scheme*, *On Demand Scheme*, and *Threshold Scheme*, are evaluated and compared. Also, the simulations compare the proposed *Threshold Scheme* with well-known Mobile IP (MIP) and the Route Optimization (MIPRO).

Table 1. Simulation parameter settings

Parameter	Setting
Simulation Time	10000.0 time units
Number of MNs	1-10
Number of CNs	1-100
Number of HAs	1-10
Number of FAs	1-10
Distance between FA_{CN} and MN's HA	3-15
Mean Send Duration Time of CN	0.1-10 time units (exponential distribution)
Mean Stay Time of MN	1.0 time units (exponential distribution)
Cache Size	1-9 entries

The Network Model: Grid configuration is taken for the logical wireless network. Each cell represents a subnet which is managed by one and only one routing agent (either a HA or a FA). The distance between nodes residing at two different cells p and q with coordinates (x_p, y_p) and (x_q, y_q) is $|x_p - x_q| + |y_p - y_q|$.

The Mobility Model: The initial location of a MN in the grid is randomly selected. The MN stays in the same cell for the duration *StayTime* as an *Exponential Distribution* with mean given by *MeanStayTime*. On expiry of the *StayTime*, the MN randomly moves to one of the four neighboring cells. That is, the probability of the four neighboring cells being chosen by the MN to move is equal.

The Source Model: The location of a CN is also generated at random. The behavior of a CN communicating with the MN is approximated by a *Poisson Process*. The *SendDurationTime* between two successive datagram sent from a CN to the MN is given by an *Exponential Distribution* with mean of *MeanSendDurationTime*.

Table 1 summarizes the parameter settings in the simulations. The performance metrics for comparing various schemes include the data routing cost, the update message cost, and the total cost. The data routing cost is defined as the cost of a CN finding out current location of the MN and delivering datagram to the MN, which is measured by computing the hop count along the path from the CN to current location of the MN. The update message cost is defined as the cost of maintaining the mobility binding, which is measured by computing the hop count for sending the binding update messages. The total cost represents the overall performance, which is defined as the sum of the data routing cost and the update message cost.

3.1 Evaluating *Always Push Scheme*

The aim of this experiment is to observe the performance of *Always Push Scheme* over various LCMR and different $d_{HA, FA}$. In Fig. 1 (a), the data routing cost increases with increased LCMR. It is because that larger LCMR implies more datagram sent from the CNs of the same FA to the MN. It yields higher data routing cost. However, the relationship between the data routing cost and the $d_{HA, FA}$ is not obvious. It is because that the data routing cost (from the CN to the MN) is not directly relative to the $d_{HA, FA}$ (from the HA to the FA_{CN}). Also, in Fig. 1 (b), the update message cost keeps constant when LCMR increases. It is because that, in *Always Push Scheme,* each time the MN moves, the HA will immediately send *Binding Update Message* to the serving FA_{CN} to update its cache. Therefore, the update message cost is not relative to the

LCMR. In Fig. 1 (b), the update message cost of smaller $d_{HA, FA}$ is less than that of larger one ($3 < 6 < 9 < 12 < 15$). It is because that, the *Binding Update Message* is sent from the HA to the FA. Then, the update message cost is obviously relative to the $d_{HA, FA}$. Furthermore, Fig. 1 (c) shows the result of summing up the effects of Fig. 1 (a) and Fig. 1 (b).

Summary: In *Always Push Scheme*, larger LCMR generates higher both data routing cost and total cost. However, update message cost is not obviously dependant on the LCMR. Also, larger $d_{HA, FA}$ yields higher update message cost and total cost. But, the data routing cost is not obviously dependent on the $d_{HA, FA}$.

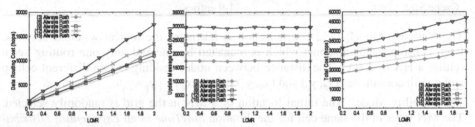

Fig. 1. The Performance of *Always Push Scheme* (a) Data Routing Cost (b) Update Message Cost (c) Total Cost

3.2 Evaluating *On Demand Scheme*

The aim of this experiment is to observe the performance of *On Demand Scheme* over various LCMR and different $d_{HA, FA}$. In Fig. 2 (a), the data routing cost increases with increased LCMR. It is because that larger LCMR implies more datagram sent from the CNs of the same FA to the MN. It yields higher data routing cost. Besides, the relationship between the data routing cost and the $d_{HA, FA}$ is not obvious. It is because that the data routing cost (from the CN to the MN) is not directly relative to the $d_{HA, FA}$ (from the HA to the FA_{CN}). Also, in Fig. 2 (b), the update message cost increases when LCMR increases. It is because that, in *On Demand Scheme*, the HA will send *Binding Update Message* to the serving FA to update its cache only when cache miss happens (cannot find correct binding in the cache). Larger LCMR implies more datagram sent from the CN to the MN. Consequently, it causes higher probability of cache miss and associated update message cost. Besides, in Fig. 2 (b), the update message cost of smaller $d_{HA, FA}$ is less than that of larger one ($3 < 6 < 9 < 12 < 15$). It is because that the *Binding Update Message* is sent from the HA to the FA that is obviously relative to the $d_{HA, FA}$. Moreover, because that the total cost is the sum of the data routing cost and the update message cost. Fig. 2 (c) shows the result of summing up the effects of Fig. 2 (a) and Fig. 2 (b).

Summary: In *On Demand Scheme*, larger LCMR generates higher data routing cost, update message cost, and total cost. Also, larger $d_{HA, FA}$ yields higher update message cost and total cost. But, the data routing cost is not obviously dependent on the $d_{HA, FA}$.

3.3 Determining the *Threshold_{sys}*

The aim of this experiment is to find out the threshold filtering out unrewarding cache updates, used in *Threshold Scheme*. Remember that, in *Threshold Scheme*, the HA

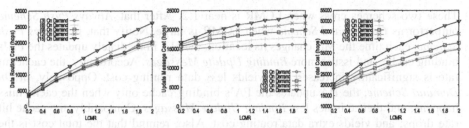

Fig. 2. The Performance of *On Demand Scheme* (a) Data Routing Cost (b) Update Message Cost (c) Total Cost

sends *Binding Update Message* to the FA_{CN} to update binding in its local cache only when the $Eager_{FA} > Threshold_{sys}$. The value of $Threshold_{sys}$ is thus very important and makes a great impact on the system performance. We thence determine the $Threshold_{sys}$ by experiment. In Fig. 3, the total cost decreases with increased $Threshold_{sys}$ when the $Threshold_{sys}$ is less than 0.12, while the total cost increases with increased $Threshold_{sys}$ when the $Threshold_{sys}$ is greater than 0.12. It is because that when the $Threshold_{sys}$ is small the system will issue too many *Binding Update Messages* and generate higher update message cost. On the contrary, when the $Threshold_{sys}$ is large the system will issue too few *Binding Update Messages*, though generate less update message cost, which generates higher data routing cost due to increased cache miss rate.

Summary: In virtue of the total cost of the system is smallest when the $Threshold_{sys}$ is 0.12, we take 0.12 as the system pre-decided $Threshold_{sys}$ value in *Threshold Scheme* and is used in the following simulations.

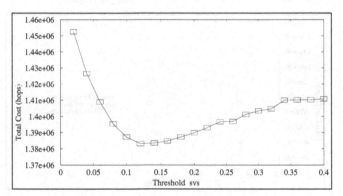

Fig. 3. Determining the $Threshold_{sys}$

3.4 Comparing *Always Push Scheme, On Demand Scheme*, and *Threshold Scheme*

The aim of this experiment is to evaluate the performance of *Always Push Scheme, On Demand Scheme*, and *Threshold Scheme*. In Fig. 4, when the LCMR is small, *On Demand Scheme* generates less total cost than that of *Always Push Scheme*. While when LCMR increases, *On Demand Scheme* generates increasingly heavier total cost.

These two schemes cross when LCMR is near 1.2. After that, *Always Push Scheme* outperforms *On Demand Scheme* when LCMR is large. Notify that, in *Always Push Scheme*, each time the MN changes its location, the HA immediately updates the FA's binding cache; and issues more *Binding Update Messages*. Accordingly, the cache hit rate is significantly increases; and yields less data routing cost. Oppositely, in *On Demand Scheme*, the HA updates the FA's binding cache only when the cache miss happens; and thence issues less *Binding Update Messages*. Compatibly, the cache hit rate drops; and yields extra data routing cost. Also, remind that the total cost is the sum of the data routing cost and the update message cost. When LCMR is small, the update message cost dominates the total cost. Thus, *On Demand Scheme* outperforms *Always Push Scheme*. On the contrary, when the LCMR is large, the data routing cost leads the total cost. Consequently, *Always Push Scheme* towers over *On Demand Scheme*.

Besides, as shown in Fig. 4, without respect to the LCMR, *Threshold Scheme* yields less total cost than both *Always Push Scheme* and *On Demand Scheme*. It is interesting to find that, when the LCMR is small, the performance of *Threshold Scheme* resemble that of *On Demand Scheme*, while, when the LCMR is large, the performance of *Threshold Scheme* approximates that of *Always Push Scheme*. Refresh that, in *Threshold Scheme*, the HA updates the FA's binding cache only the expression $Eager_{FA} > Threshold_{sys}$ is sustained; that matches the strategy of *Always Push Scheme*. Otherwise, the HA will update the FA's binding cache when cache miss happens, that matches the strategy of *On Demand Scheme*. As a result, *Threshold Scheme* keeps the total cost low no matter the LCMR is.

Summary: *Threshold Scheme* outperforms *Always Push Scheme* and *On Demand Scheme* without respect to the value of LCMR.

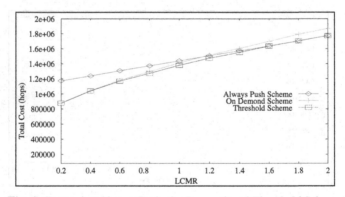

Fig. 4. Comparing *Always Push, On Demand,* and *Threshold Scheme.*

3.5   Comparing the *Threshold Scheme* with MIP and MIPRO

The aim of this experiment is to comparing the performance of *Threshold Scheme* with well known MIP Scheme and MIPRO Scheme. When $d_{FA, HA}$ is small ($d_{FA, HA} = 3$), the update message cost becomes trivial. In Fig. 5 (a), the *Threshold Scheme* generates higher cache hit ratio than that of the MIPRO. It is because that, in MIPRO the HA updates the CN's binding cache only when the cache miss happens, which causes higher cache hit ratio. Oppositely, in *Threshold Scheme*, the HA updates the binding cache when it's rewarding, which generates higher cache hit ratio. In Fig. 5 (b), due to triangle routing, the MIP generates highest data routing cost. Since both MIPRO and *Threshold Scheme* keeps a copy of binding in the local cache, the CN can sends datagram directly to the MN without intervention of the HA, they generate less data routing cost than that of MIP. Since, in MIPRO, it issues *Binding Update Message* only when cache miss happens. Thus, its cache miss rate is higher than that of *Threshold Scheme*. Thence, MIPRO needs extra data routing cost resulting from sub-optimal routing (forwarded by the HA). In this case, *Threshold Scheme* performs best. In Fig. 5 (c), since there is no mobility binding cache updating cost in MIP, it generates least update message cost. Also, *Threshold Scheme* performs better than MIPRO. It is because that the rewarding evaluation mechanism designed in *Threshold Scheme*, it issues less *Binding Update Message* than MIPRO, thus, less update message cost. In Fig. 5 (d), *Threshold Scheme* generates least total cost. It means that *Threshold Scheme* outperforms the other two schemes.

Summary: Since *Threshold Scheme* has highest cache hit ratio, it outperforms both MIP Scheme and MIPRO Scheme.

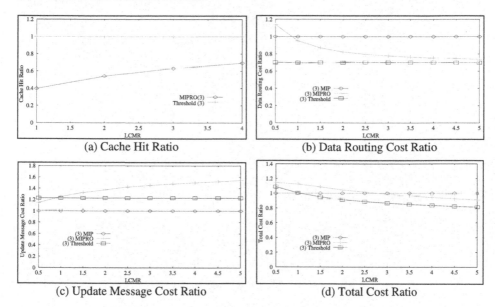

(a) Cache Hit Ratio   (b) Data Routing Cost Ratio

(c) Update Message Cost Ratio   (d) Total Cost Ratio

Fig. 5. Comparing *Threshold Scheme*, MIP, and MIPRO.

4 Conclusions

In this paper, three mobility binding management schemes are proposed, evaluated and compared. *Threshold Scheme* proposed in this paper can dynamically decide whether or not to update the binding cache according to the rewarding evaluation mechanism. The simulation results also suggest that *Threshold Scheme* outperforms well known *MIP* and *MIPRO* for managing the mobility binding in IP based networks.

Acknowledgement

This work was supported in part by the R.O.C. Taiwan National Science Council under grant no. NSC-92-2213-E-415-003

References

1. Perkins, C.,: IP Mobility Support for IPv4. RFC 3344. (2002)
2. Perkins, C., Johnson, D.: Route Optimization in Mobile IP. IETF. (2000)
3. Johnson, D., Perkins, C., Arkko, J.: Mobility Support in IPv6. IETF. (2002)
4. Gustafsson, E., Jonsson, A., Perkins, C.: Mobile IP Regional Registration. IETF. (2000)
5. Ho, J., Akyildiz, I.: Local Anchor Scheme for Reducing Signaling Costs in Personal Communications Networks. IEEE/ACM Trans. Networking, vol. 4, no. 5, (1996) 709-725
6. Soliman, H., Castelluccia, C., El-Malki, K., Bellier, L.: Hierarchical MIPv6 Mobility Management. IETF. (2002)

Effect Analysis of the Soft Handoff Scheme
on the CDMA System Performance

Go-Whan Jin[1] and Soon-Ho Lee[2]

[1] School of Technical Management Information System,
Woosong University, Taejon, Korea
gwjin@woosong.ac.kr
[2] Graduate School of Management,
Korea Advanced Institute of Science and Technology, Seoul, Korea
iskra@kgsm.kaist.ac.kr

Abstract. In a general CDMA cellular system environment, extending the relative portion of soft handoff region to the whole cell area is bound to enlarge the cell coverage area, which in turn increases the new call arrival rate of the cell. Performance analysis on the CDMA system, however, was conducted in the several papers with the new call arrival rate in a cell given fixed, based on the incorrect setting for a general environment that the relative portion of the soft handoff region can be increased while keeping the new call arrival rate of the cell constant. We exactly analyze the effect of the soft handoff scheme on the CDMA system performance, rending the opposite result to the previous studies. To highlight the accuracy of our analysis, the well-known performance measures are also obtained with the traffic model and Markovian queuing model.

1 Introduction

Recently, code-division multiple-access (CDMA) has become a most promising technology for the current cellular networks due to its various advantages. One of the most important merits of the CDMA cellular system is its soft handoff capability for higher communication quality. Once in the soft handoff region, a mobile is usually linked via multiple signal paths to both the current base station and the target base stations, guaranteeing a smooth transition without the chance of the ongoing call being dropped, as in the hard handoff scheme. Also, by using the soft handoff scheme and a proper power control strategy as proposed in IS-95, the required transmitted power as well as interference can be reduced to improve the communication quality and system's capacity [8, 11].

A host of studies on the soft handoff have been reported in the literatures. Some studies investigated the effect of the size of the soft handoff region on the CDMA system capacity [1, 2, 3]. Also, for the hard handoff mechanism, there are several studies on the load analysis and the system performance [4, 6, 10]. Drawing our attention among the soft handoff studies is the one on the performance analysis of a CDMA cellular system, which is known difficult due to the soft handoff process, during which a mobile is connected to two or more base stations.

Recently, the system performance of the soft handoff scheme of the CDMA cellular system given with a fixed number of available channels is analyzed in the several

D.-K. Baik (Ed.): AsiaSim 2004, LNAI 3398, pp. 439–449, 2005.
© Springer-Verlag Berlin Heidelberg 2005

papers. Worth noting is the performance study by Su *et al.* on the soft handoff in the CDMA cellular system [7]. They used the Markov chain model to analyze the soft handoff process. With the new call arrival rate in a cell given fixed, their analysis is centered on investigating the performance trend with respect to the relative portion of the soft handoff region to the whole cell area. In particular, they assumed that the soft handoff region can be extended within a cell coverage area which is given fixed and presented that the larger the area of the soft handoff region, the lower the blocking and handoff refused probabilities.

But in the general CDMA system environment, this relative portion cannot be enlarged while keeping the call arrival rate fixed, the assertion of which is elaborated on. Note that expanding the soft handoff region is bound to enlarge the cell coverage area. Therefore, the new call arrivals are affected by the size of soft handoff region in the general CDMA system environment. In this paper, the modification on the new call arrival rate, on the contrary to the previous study, leads to the conclusion that measures such as blocking and handoff refused probability become poorer as the size of the soft handoff region gets larger.

The rest of this paper is organized as follows: Section 2 explains our system model, providing with necessary explanations. A system analysis is described in Section 3. Section 4 investigates numerical results under various system parameters. Finally in the last section, concluding remarks and suggestions for future work are given.

2   System Model

We assume that mobiles are located evenly over the entire service area, which consists of an infinite number of identical cells. Also, it is considered that one mobile unit carries one call only, i.e., there is no bulk handoff arrival. It is assumed that every handoff request is perfectly detected in our model and the channel assignment is instantly accomplished if the channel is available. In the soft handoff scheme, although there is no available channel, a handoff request must not be denied immediately, but be put into a queue list.

Whereas a cell is matched with a single hexagonal boundary in the case of hard handoff, the soft handoff region of a cell is automatically defined as the overlapped area, i.e., as the area of a cell overlaid with its neighboring cells. The inner boundary of the soft handoff region is determined by the neighboring cell boundaries, while the cell boundary itself becomes its outer boundary. This observation makes it self-evident that the soft handoff region in the general CDMA system environment cannot be made larger without enlarging the cell coverage area.

Both new call arrivals and handoff arrivals are assumed to be Poisson-distributed with mean Λ_n and Λ_h respectively. Let α be the ratio of the soft handoff region to the whole cell area. As illustrated in Figure 1, extending the soft handoff region is bound to enlarge the cell coverage. Therefore, α cannot be increased without enlarging the cell coverage area, i.e., without increasing the new call arrival rate Λ_n. In this paper, we consider an accurate system model, in which the size of cell coverage area changes according to α and the cell residual time is computed based on the geometrical model.

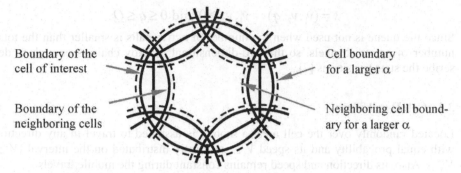

Fig. 1. Cell boundaries

Each cell will reserve C_h channels out of total C available channels exclusively for handoff calls, because disconnecting an on-going conversation is more detrimental than blocking a new call. Let Q and T_c be the allowable maximum queue length and the call duration time respectively. Assume that T_c is exponentially distributed with mean μ_c^{-1}.

We may divide a cell into three regions: one normal region and two handoff regions. The normal region is surrounded by the handoff region. The soft handoff region is bisected into two sub-regions according to the hexagon boundary in the hard handoff scheme. Under the hard handoff scheme, if employed, the new call arrival rate to a cell of hexagonal type, denoted by Λ_{hard}, would be given fixed from its fixed cell size. Assumed that the location of a new call is uniformly distributed all over a cell, the new call arrival rates in three different regions are as follows:

$$\Lambda_n = \Lambda_{hard} / (1 - \alpha/2),$$
$$\Lambda_{n1} = (\alpha/2) \cdot \Lambda_n,$$
$$\Lambda_{n1'} = (\alpha/2) \cdot \Lambda_n,$$ (1)
$$\Lambda_{n2} = (1 - \alpha) \cdot \Lambda_n.$$

Λ_{n1} and $\Lambda_{n1'}$ denote the new call arrival rates in the inner handoff region and in the outer handoff region respectively. Λ_{n2} represents the new call arrival rate in the normal region.

We assume that when a new call arrives in the handoff region, it can ask for channel assignment from both the cell of interest and the neighboring cell. Even when the new call is blocked from the cell of interest, it may successfully get a channel assignment from the neighboring cell and become a handoff arrival for the cell of interest.

3 System Analysis

From the description of the system model, we can use the birth-death process in order to analyze the system performance as in [5]. Let v_1 and v_2 be the number of active calls in the handoff region and the number of active calls in the normal region respectively. Also, q denotes the number of mobile units in queue. Using these notations, the state in the process is defined as follows:

$$s = (v_1, v_2, q) \quad v_1, v_2 \geq 0 \text{ and } 0 \leq q \leq Q.$$

Since the queue is not used when the number of active calls is smaller than the total number of active channels, so the two-dimensional Markov chain is enough to describe the state transitions [7].

3.1 Cell Dwell Time

Located randomly over the cell area, a mobile is assumed to travel in any direction with equal probability and its speed V is uniformly distributed on the interval [V_{min}, V_{max}]. Also, its direction and speed remains constant during the mobile travels.

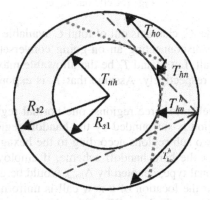

Fig. 2. The dwell times of the normal and the handoff regions

Let T_{nh} be the dwell time in the normal region. T_{hn} and T_{ho} denote the dwell times in the handoff region when a call moves into the normal region and moves out the cell respectively. Also, T_{hn}^h and T_{ho}^h represent the dwell times in the case of the handoffed call from outside the cell. To facilitate the analysis, the area between the interest cell and neighboring cell boundaries is equally sized with a doughnut by two co-centric circles with radii R_{s2} and R_{s1} such that the size of the region between two circular boundaries is equal to the corresponding one of Figure 1, as shown in Figure 2. We derive the PDFs of the dwell times in the normal and handoff regions as illustrated in Figure 2.

By extending the results by [3], we obtain the probability density functions for T_{nh} and T_{hn}, T_{hn}^h, T_{ho} and T_{ho}^h, the description of which are given for typical model due to lack of space. We can derive the probability density function of T_{nh} as follows:

$$f_{T_{nh}}(t) = \begin{cases} \int_{V_{min}}^{V_{max}} v \cdot \sqrt{(2R_{s1})^2 - (tv)^2} \Big/ (\pi R_{s1}^2) \cdot 1/(V_{min} - V_{max}) dv, \\ \qquad\qquad \text{for} \quad 0 \leq t \leq (2R_{s1})/V_{max} \\ \int_{V_{min}}^{(2R_{s1})} v \cdot \sqrt{(2R_{s1})^2 - (tv)^2} \Big/ (\pi R_{s1}^2) \cdot 1/(V_{min} - V_{max}) dv, \\ \qquad\qquad \text{for} \quad (2R_{s1})/V_{max} \leq t \leq (2R_{s1})/V_{max} \end{cases} \tag{2}$$

The probability density functions of T_{hn} and T_{ho} are given by

$$f_{T_{hn}}(t) = \begin{cases} \int_{V_{min}}^{V_{max}} v \cdot 2R_{s1}/\left(\pi\left(R_{s2}^2 - R_{s1}^2\right)\right) \cdot 1/(V_{min} - V_{max})\,dv, \\ \qquad \text{for } 0 \leq t \leq (R_{s2} - R_{s1})/V_{max} \\[4pt] \int_{V_{min}}^{V_{max}} v \cdot 2R_{s1}/\left(\pi\left(R_{s2}^2 - R_{s1}^2\right)\right) \cdot 1/(V_{min} - V_{max})\,dv + \int_{(R_{s2}-R_{s1})/t}^{V_{max}} f_1(t,v)\,dv, \\ \qquad \text{for } (R_{s2} - R_{s1})/V_{max} \leq t \leq \sqrt{R_{s2}^2 - R_{s1}^2}\big/V_{max} \\[4pt] \int_{V_{min}}^{\sqrt{R_{s2}^2-R_{s1}^2}/t} v \cdot 2R_{s1}/\left(\pi\left(R_{s2}^2 - R_{s1}^2\right)\right) \cdot 1/(V_{min} - V_{max})\,dv + \int_{(R_{s2}-R_{s1})/t}^{\sqrt{R_{s2}^2-R_{s1}^2}/t} f_1(t,v)\,dv, \\ \qquad \text{for } \sqrt{R_{s2}^2 - R_{s1}^2}\big/V_{max} \leq t \leq (R_{s2} - R_{s1})/V_{min} \\[4pt] \int_{V_{min}}^{\sqrt{R_{s2}^2-R_{s1}^2}/t} v \cdot 2R_{s1}/\left(\pi\left(R_{s2}^2 - R_{s1}^2\right)\right) \cdot 1/(V_{min} - V_{max})\,dv + \int_{V_{min}}^{\sqrt{R_{s2}^2-R_{s1}^2}/t} f_1(t,v)\,dv, \\ \qquad \text{for } (R_{s2} - R_{s1})/V_{min} \leq t \leq \sqrt{R_{s2}^2 - R_{s1}^2}\big/V_{min} \end{cases} \tag{3}$$

where

$$f_1(t,v) = v \cdot \frac{\sqrt{-\left(R_{s2}^2 - R_{s1}^2\right)^2 + 2\left(R_{s2}^2 + R_{s1}^2\right)(tv)^2 - (tv)^4}}{\pi\left(R_{s2}^2 - R_{s1}^2\right)tv} \cdot \frac{1}{(V_{max} - V_{min})}$$

$$f_{T_{ho}}(t) = \begin{cases} \int_{V_{min}}^{V_{max}} f_2(t,v)\,dv, \quad \text{for } 0 \leq t \leq (R_{s2} - R_{s1})/V_{max} \\[4pt] \int_{V_{min}}^{V_{max}} f_2(t,v)\,dv + \int_{(R_{s2}-R_{s1})/t}^{V_{max}} f_1(t,v)\,dv \\ \qquad \text{for } (R_{s2} - R_{s1})/V_{max} \leq t \leq \sqrt{R_{s2}^2 - R_{s1}^2}\big/V_{max} \\[4pt] \int_{V_{min}}^{V_{max}} f_2(t,v)\,dv + \int_{(R_{s2}-R_{s1})/t}^{\sqrt{R_{s2}^2-R_{s1}^2}/t} f_1(t,v)\,dv + \int_{\sqrt{R_{s2}^2-R_{s1}^2}/t}^{V_{max}} f_3(t,v)\,dv \\ \qquad \text{for } \sqrt{R_{s2}^2 - R_{s1}^2}\big/V_{max} \leq t \leq 2\sqrt{R_{s2}^2 - R_{s1}^2}\big/V_{max} \\[4pt] \int_{V_{min}}^{2\sqrt{R_{s2}^2-R_{s1}^2}/t} f_2(t,v)\,dv + \int_{(R_{s2}-R_{s1})/t}^{\sqrt{R_{s2}^2-R_{s1}^2}/t} f_1(t,v)\,dv + \int_{\sqrt{R_{s2}^2-R_{s1}^2}/t}^{2\sqrt{R_{s2}^2-R_{s1}^2}/t} f_3(t,v)\,dv \\ \qquad \text{for } 2\sqrt{R_{s2}^2 - R_{s1}^2}\big/V_{max} \leq t \leq (R_{s2} - R_{s1})/V_{min} \\[4pt] \int_{V_{min}}^{2\sqrt{R_{s2}^2-R_{s1}^2}/t} f_2(t,v)\,dv + \int_{V_{min}}^{\sqrt{R_{s2}^2-R_{s1}^2}/t} f_1(t,v)\,dv + \int_{\sqrt{R_{s2}^2-R_{s1}^2}/t}^{2\sqrt{R_{s2}^2-R_{s1}^2}/t} f_3(t,v)\,dv \\ \qquad \text{for } (R_{s2} - R_{s1})/V_{min} \leq t \leq \sqrt{R_{s2}^2 - R_{s1}^2}\big/V_{min} \\[4pt] \int_{V_{min}}^{2\sqrt{R_{s2}^2-R_{s1}^2}/t} f_2(t,v)\,dv + \int_{\sqrt{R_{s2}^2-R_{s1}^2}/t}^{2\sqrt{R_{s2}^2-R_{s1}^2}/t} f_3(t,v)\,dv \\ \qquad \text{for } \sqrt{R_{s2}^2 - R_{s1}^2}\big/V_{min} \leq t \leq 2\sqrt{R_{s2}^2 - R_{s1}^2}\big/V_{min} \end{cases} \tag{4}$$

where

$$f_2(t,v) = v\,\frac{\sqrt{\left(2R_{s2}^2\right) - (tv)^2}}{\pi\left(R_{s2}^2 - R_{s1}^2\right)}\,\frac{1}{(V_{max} - V_{min})}$$

$$f_3(t,v) = v\,\frac{2R_{s1}}{\pi\left(R_{s2}^2 - R_{s1}^2\right)(V_{max} - V_{min})}$$

The probability density functions of T_{hn}^h and T_{ho}^h are also given by

$$
f_{T_{hn}^h}(t) = \begin{cases}
\int_{R_{s2}-R_{s1}/t}^{V_{max}} \dfrac{v \cdot \left(\left(R_{s2}^2 - R_{s1}^2\right)^2 - (tv)^4\right) \cdot v/\left(\left(V_{max}^2 - V_{min}^2\right)/2\right)}{\left(2R_{s2}(tv)^2 \sqrt{2R_{s2}(tv)^2 - \left(-R_{s1}^2 + R_{s2}^2 + (tv)^2\right)^2}\right)} dv, \\
\qquad \text{for } (R_{s2}-R_{s1})/V_{max} \le t \le \sqrt{R_{s2}^2 - R_{s1}^2}/V_{max} \\[2ex]
\int_{R_{s2}-R_{s1}/t}^{\sqrt{R_{s2}^2-R_{s1}^2}/t} \dfrac{v \cdot \left(\left(R_{s2}^2 - R_{s1}^2\right)^2 - (tv)^4\right) \cdot v/\left(\left(V_{max}^2 - V_{min}^2\right)/2\right)}{\left(2R_{s2}(tv)^2 \sqrt{2R_{s2}(tv)^2 - \left(-R_{s1}^2 + R_{s2}^2 + (tv)^2\right)^2}\right)} dv, \\
\qquad \text{for } \sqrt{R_{s2}^2 - R_{s1}^2}/V_{max} \le t \le (R_{s2}-R_{s1})/V_{min} \\[2ex]
\int_{V_{min}}^{\sqrt{R_{s2}^2-R_{s1}^2}/t} \dfrac{v \cdot \left(\left(R_{s2}^2 - R_{s1}^2\right)^2 - (tv)^4\right) \cdot v/\left(\left(V_{max}^2 - V_{min}^2\right)/2\right)}{\left(2R_{s2}(tv)^2 \sqrt{2R_{s2}(tv)^2 - \left(-R_{s1}^2 + R_{s2}^2 + (tv)^2\right)^2}\right)} dv, \\
\qquad \text{for } (R_{s2}-R_{s1})/V_{min} \le t \le \sqrt{R_{s2}^2 - R_{s1}^2}/V_{min}
\end{cases}
\tag{5}
$$

$$
f_{T_{ho}^h}(t) = \int_{V_{min}}^{V_{max}} v \cdot 2R_{s2}/\sqrt{(2R_{s2}/tv)^2 - 1} \cdot v/\left((V_{max}^2 - V_{min}^2)/2\right)dv,
$$
$$
\text{for } 0 \le t \le 2\sqrt{R_{s2}^2 - R_{s1}^2}/V_{max}
$$
$$
+ \int_{V_{min}}^{2\sqrt{R_{s2}^2-R_{s1}^2}/t} v \cdot 2R_{s2}/\sqrt{(2R_{s2}/tv)^2 - 1} \cdot v/\left((V_{max}^2 - V_{min}^2)/2\right)dv,
$$
$$
\text{for } 2\sqrt{R_{s2}^2 - R_{s1}^2}/V_{max} \le t \le 2\sqrt{R_{s2}^2 - R_{s1}^2}/V_{min}
\tag{6}
$$

3.2 Steady-State Probabilities

In this model, a state transition can occur in five conditions as follows: new call arrival, call completion, handoff arrival, handoff departure, and region transitions (from the normal region to the handoff region and vice versa). An example with $C = 5$, $C_h = 2$, and $Q = 3$ is shown in Figure 3.

Let $p(i, j, q)$ be the steady-state probability of the state $s = (i, j, q)$. The sum of the steady-state probabilities is given by

$$
\sum_{i=0}^{C} \sum_{j=0}^{C-i} \sum_{q=0}^{Q} p(i,j,q) = 1.
\tag{7}
$$

Let μ_D^{-1} and μ_d^{-1} be the mean dwell times in the normal region and in the handoff region respectively. The rate of handoff departure μ_{do} and the rate of region transition μ_{dk} is also obtained from the weighted sum of $E[T_{ho}]$ and $E[T_{ho}^h]$, $E[T_{hn}]$ and $E[T_{hn}^h]$, respectively. From the chosen linearly independent flow equilibrium equations and (7), we can solve this two dimensional birth-death process. Finally, we can get the steady-state probabilities as follows:

$$
\Lambda_h = \sum_{i=0}^{C} \sum_{j=0}^{C-i} (j \cdot \mu_D) \cdot p(i,j,0)
$$
$$
+ \sum_{i=0}^{C} \sum_{q=1}^{Q} (j \cdot \mu_D) \cdot p(i,j,q)\Big|_{j=C-i} \cdot
\tag{8}
$$
$$
+ \Lambda_{n1} \cdot (1 - P_B) + \Lambda_{n1'} \cdot (1 - P_B)
$$

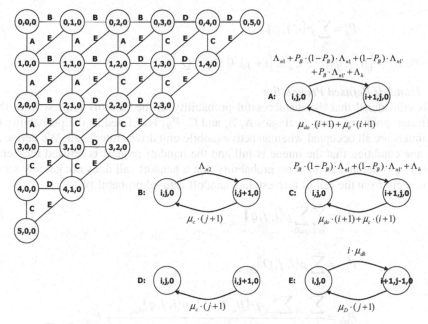

Fig. 3. The birth-death processes of a cell with $C = 5$, $C_h = 2$, and $Q = 3$

In equation (8), the last term represents the rate of new calls which can only get a channel assignment from the neighboring cell and become a handoff arrival for the cell of interest. P_B is the blocking probability from the cell's point of view, which will be described in next subsection.

3.3 Performance Measurement

The definitions of performance measurement are the same as those of [7] and some equations are modified slightly.

A   Carried Traffic and Carried Handoff Traffic

Both the carried traffic and the carried handoff traffic per cell are defined as the average number of occupied channels in the cell of interests and the average number of occupied channels in the handoff region of the cell of interests respectively.

$$A_c = \sum_{i=0}^{C} \sum_{j=0}^{C-i} (i+j) \cdot p(i,j,0) + \sum_{i=0}^{C} \sum_{q=1}^{Q} C \cdot p(i,j,q)\Big|_{j=C-i} \tag{9}$$

$$A_{ch} = \sum_{i=0}^{C} \sum_{j=0}^{C-i} i \cdot p(i,j,0) + \sum_{i=0}^{C} \sum_{q=1}^{Q} i \cdot p(i,j,q)\Big|_{j=C-i} \tag{10}$$

where Q is the queue length.

B   Blocking Probability

A new call will be blocking by a cell when the total number of used channels is larger than or equal to $C - C_h$, so the blocking probability from the cell's point of view is defined as follows:

$$P_B = \sum_{B_0} p(i, j, q) \tag{11}$$

where $B_0 = \{(i, j, q) | C - C_h \le (i + j), 0 \le q \le Q\}$.

C Handoff Refused Probability

It is considered that the unsuccessful probability of the handoff request from three different points of view as classes A, B, and C. P_{HA} is defined as the probability that channels are all occupied when an active mobile unit drives into a cell. We define P_{HC} for the condition that the queue is full and the handoff request is refused forever. At last, P_{HB} is defined as the total probability that a handoff call does not get any service ultimately from the cell of interest, i.e., handoff refused probability.

$$P_{HA} = \sum_{i=0}^{C} \sum_{q=0}^{Q} p(i, j, q)\Big|_{j=C-i} \tag{12}$$

$$P_{HC} = \sum_{j=0}^{C} p(i, j, Q)\Big|_{i=C-j} \tag{13}$$

$$P_{HB} = \frac{\sum_{j=0}^{C} \sum_{q=1}^{Q} q \cdot (\mu_c + \mu_d) \cdot p(i, j, q)\Big|_{i=C-j}}{\Lambda_h \cdot (1 - P_{HC})} + P_{HC} \tag{14}$$

D Channel Efficiency

We assume that there are at most two different sources in diversity reception in this analysis. Consider three different conditions in which an active call in the handoff region will carry two channels as follows:

- F_{2n}: the generated rate of new calls with two channels,
- F_{2h}: the arrival rate of handoff calls with two channels,
- F_{2t}: the transition rate of active calls from the normal region which will hold two channels.

$$\begin{aligned} F_{2n} &= (\Lambda_{n1} + \Lambda_{n1'}) \cdot (1 - P_B)^2, \\ F_{2h} &= \Lambda_h \cdot (1 - P_{HB}), \\ F_{2t} &= \Gamma \cdot (1 - P_{HB}). \end{aligned} \tag{15}$$

where

$$\Gamma = \sum_{i=0}^{C} \sum_{j=0}^{C-i} (j \cdot \mu_D) \cdot p(i, j, 0) + \sum_{i=0}^{C} \sum_{q=1}^{Q} (j \cdot \mu_D) \cdot p(i, j, q)\Big|_{j=C-i}$$

The total channel efficiency in a cell can be defined as

$$E = 1 - \frac{1}{2} \cdot \left(\frac{\kappa_2}{F_1} \right) \tag{16}$$

where

$$F_1 = (\Lambda_{n1} + \Lambda_{n1'}) \cdot (1 - P_B) \cdot P_B + \Gamma \cdot P_{HB},$$

$$\kappa_2 = A_{ch} \cdot \frac{F_{2n} + F_{2h} + F_{2t}}{F_{2n} + F_{2h} + F_{2t} + F_1}.$$

4 Numerical Result

Consider a CDMA cellular system consisting homogeneous cells with radius R_{s2} = 1 *mile* (calculating R_{s1}), where the portion of the soft handoff region to the entire cell area α is set at 0.1. As α increases, we can obtain R_{s2} and R_{s1}, shown in Figure 4. The system parameters used in the numerical examples are C = 12, C_h = 2, μ_c = 0.01, Q = 4, V_{min} = 1 *mile*, and V_{max} = 80 *miles*. Table 1-5 summarize the computation results obtained in relations between performance measurements and new call arrival rates for the different ratio α.

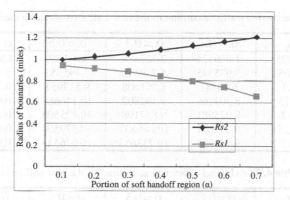

Fig. 4. Cell boundaries of soft handoff according to α

Note that the higher the α, the larger the new call arrival rate of a cell, it is obvious to observe that a higher value of α renders not only a larger blocking probability but also a larger handoff refused probability, listed in Tables. It is unfortunate to note that our results are opposite to those in [7, 8, 11]. As soft handoff region become larger, however, the blocking probabilities in this note and [7] employing the limited physical resource, i.e., predefined number of channels, should increase because a call use two or more channels in the soft handoff region, while the outage probabilities in [8] and [11] become smaller since the channel capacity is increased by the channel diversity reducing interference level in the handoff region. Since an active call in the handoff region may be connected via more than a single channel, a higher value of α will yield the negative effect of lowering channel efficiency. As α increases, however, both carried traffic in a cell and carried traffic in the handoff region become larger by increasing of the cell residual time and the call arrival rate.

Table 1. Blocking probability according to the new call arrival rate

Λ_n	$\alpha = 0.1$	$\alpha = 0.3$	$\alpha = 0.5$	$\alpha = 0.7$
0.025	0.00002	0.00013	0.00116	0.01455
0.05	0.00526	0.05141	0.20128	0.48145
0.1	0.11296	0.32078	0.55502	0.73675
0.15	0.29390	0.53559	0.71619	0.83206
0.2	0.43916	0.65914	0.79528	0.87762
0.25	0.54268	0.73473	0.84106	0.90404
0.3	0.61726	0.78466	0.87060	0.92121

Table 2. The carried traffic according to the new call arrival rate in a cell

Λ_n	$\alpha = 0.1$	$\alpha = 0.3$	$\alpha = 0.5$	$\alpha = 0.7$
0.025	1.840424	2.322486	3.093469	4.512081
0.05	3.935345	5.749546	7.495552	9.18302
0.1	6.84881	8.391608	9.512951	10.317204
0.15	8.304766	9.392802	10.146688	10.681445
0.2	8.975981	9.837108	10.425508	10.846596
0.25	9.332607	10.080837	10.579504	10.939969
0.3	9.548787	10.233653	10.676543	10.999775

Table 3. The carried traffic according to the new call arrival rate in the handoff region

Λ_n	$\alpha = 0.1$	$\alpha = 0.3$	$\alpha = 0.5$	$\alpha = 0.7$
0.025	1.840424	2.322486	3.093469	4.512081
0.05	3.935345	5.749546	7.495552	9.18302
0.1	6.84881	8.391608	9.512951	10.317204
0.15	8.304766	9.392802	10.146688	10.681445
0.2	8.975981	9.837108	10.425508	10.846596
0.25	9.332607	10.080837	10.579504	10.939969
0.3	9.548787	10.233653	10.676543	10.999775

Table 4. The channel efficiency according to the new call arrival rate

Λ_n	$\alpha = 0.1$	$\alpha = 0.3$	$\alpha = 0.5$	$\alpha = 0.7$
0.025	0.98972	0.95054	0.90308	0.85870
0.05	0.98875	0.95211	0.91619	0.88689
0.1	0.98907	0.95077	0.91531	0.89091
0.15	0.98888	0.95072	0.91884	0.89726
0.2	0.98863	0.95111	0.92159	0.90114
0.25	0.98839	0.95159	0.92361	0.90372
0.3	0.98818	0.95204	0.92512	0.90557

Table 5. The handoff refused probability for the class B according to the new call arrival rate

Λ_n	$\alpha = 0.1$	$\alpha = 0.3$	$\alpha = 0.5$	$\alpha = 0.7$
0.025	0.000000	0.000000	0.000002	0.000085
0.05	0.000001	0.000180	0.003197	0.023770
0.1	0.000069	0.002911	0.018714	0.054074
0.15	0.000312	0.007862	0.032381	0.071786
0.2	0.000647	0.012504	0.041547	0.082077
0.25	0.000999	0.016364	0.047911	0.088766
0.3	0.001343	0.019534	0.052557	0.093458

5 Conclusion

This study has investigated the effect of soft handoff scheme on the CDMA cellular system performance. In the general CDMA cellular system, extending the relative portion of soft handoff region to whole cell area is tightly bound to enlarge the cell coverage area, i.e., increases the new call arrival rate of a cell. Considered in this way, we have modified the system performance model. To highlight the accuracy of our analysis, the well-known performance measures are also obtained with the traffic model and Markovian queuing model. The modification on the new call arrival rate leads to the conclusion that, opposite to that made by the previous study, measures such as blocking and handoff refused probability become poorer as the size of the soft handoff region gets larger.

References

1. Y. W. Chang and E. Geraniotis: Accurate Computations of Cell Coverage Areas for CDMA Hard and Soft Handoffs. Proceedings of IEEE Vehicular Technology Conference (1996) 411-415
2. T. Chebaro and P. Godlewski: Average External Interference in Cellular Radio CDMA Systems. IEEE Transactions on Communications, Vol. 44, (1996) 23-25
3. M. Chopra, K. Rohani, and J. Reed: Analysis of CDMA Range Extension Due to Soft Handoff. Proceedings of IEEE Vehicular Technology Conference (1993) 917-921
4. R. A. Guerin: Channel Occupancy Time Distribution in a Cellular Radio System. IEEE Transactions on Vehicular Technology, Vol. 35 (1987) 89-99
5. D. Hong and S. S. Rappaport: Traffic Model and Performance Analysis for Cellular Mobile Radio Telephone Systems with Prioritized and Nonprioritized Hand-off Procedures. IEEE Transactions on Vehicular Technology, Vol. 35, (1986) 77-92.
6. D. L. Pallant and P. G. Taylor: Channel Occupancy Time Distribution in a Cellular Radio System. Operations Research, Vol. 43, No. 1 (1995) 33-42
7. Szu-Lin Su, Jen-Yeu Chen, and Jane-Hwa: Performance Analysis of Soft Handoff in CDMA Cellular Networks. IEEE Journal on Selected Areas in Communications, Vol. 14, No. 9 (1996) 1762-1769
8. S. C. Swales et al.: Handoff Requirements for a Third Generation DS-CDMA Air Interface. IEE Colloquium, Mobility Support Personal Communications (1993)
9. Dong-Wan Tcha, Suk-yon Kang, and Go-Whan Jin: Load Analysis of the Soft Handoff Scheme in a CDMA Cellular System. IEEE Journal on Selected Areas in Communications, Vol. 19, No. 6 (2001) 1147-1152
10. H. Xie and D. J. Goodman: Mobility Models and Biased Sampling Problem. Proceedings of IEEE International Conference of Universal Personal Communications, (1993) 803-807
11. A. J. Viterbi, A. M. Viterbi, K. S. Gilhousen, and E. Zehavi: Soft Handoff Extends CDMA Cell Coverage and Increases Reverse Link Capacity. IEEE Journal on Selected Areas in Communications, Vol. 12, No. 8 (1994) 1281-1288

Derivation of Flight Characteristics Data
of Small Airplanes Using Design Software
and Their Validation by Subjective Tests

Sugjoon Yoon[1], Ji-Young Kong[1], Kang-Su Kim[1],
Suk-Kyung Lee[1], and Moon-Sang Kim[2]

[1] Department of Aerospace Engineering, Sejong University 98 Gunja-Dong,
Gwangjin-Gu, Seoul, 143-747 Republic of Korea
[2] School of Aerospace and Mechanical Engineering, Hankuk Aviation University200-1,
Whajon-dong, Koyang-city, Kyungki-do, 412-791 Republic of Korea

Abstract. It is very difficult to acquire high-fidelity flight test data for small airplanes such as typical unmanned aerial vehicles and RC airplanes because MEMS-type small sensors used in the tests do not present reliable data in general. Besides, it is not practical to conduct expensive flight tests for low-priced small airplanes in order to simulate their flight characteristics. A practical approach to obtain acceptable flight data, including stability and control derivatives and data of weights and balances, is proposed in this study. Aircraft design software such as Darcorp's AAA is used to generate aerodynamic data for small airplanes, and moments of inertia are calculated from CATIA, structural design software. These flight data from simulation software are evaluated subjectively and tailored using simulation flight by experienced pilots, based on the certified procedure in FAA AC 120-40B, which are prepared for manned airplane simulators. Use of design S/W for generation of parameter values representing flight characteristics turns out valid. In this study a practical procedural standard is established for procuring reliable data replicating flight characteristics of an airplane.

1 Introduction

In general, parameter values representing flight characteristics of an airplane are derived from either flight tests or dedicated design software such as DATCOM [1]. However, it is practically very difficult to obtain reliable data from flight tests for small airplanes such as RC (Remote Control) airplanes and UAV's (Unmanned Aerial Vehicles), which have very limited payload capacities. High-fidelity sensors used in typical flight tests of manned airplanes are relatively big in volume and weight for small airplanes, which causes change of original flight characteristics and results in the measurement of data with significant errors. MEMS sensors may be considered to be alternatives to conventional ones. But their fidelity and reliability are much lower than those of conventional sensors, and their test results are not accurate enough to be used in typical flight simulation.

The purpose of this study is to establish a practical procedural standard for procuring reliable data replicating flight characteristics of an airplane, which can be used in

D.-K. Baik (Ed.): AsiaSim 2004, LNAI 3398, pp. 450–457, 2005.
© Springer-Verlag Berlin Heidelberg 2005

flight simulation and design of a flight control system of a small airplane. In this study Darcorp's AAA (Advanced Aircraft Analysis) software [2], which has been widely used in the conceptual design of an airplane, is adopted for derivation of aerodynamic data and structural design software, Dassault's CATIA [3], for computation of moments of inertia. Then the design data obtained from AAA and CATIA are implemented in a proven flight simulation code and subjectively validated based on the test procedure regulated in FAA's AC120-40B [4]. RC airplanes such as Extra 300s and 40% scale Night Intruder UAV of Korea Aerospace Industry are used in this paper as test beds for validation of the proposed procedure. The standardized procedure has been applied to derivation of flight characteristics data of several other airplanes, and turns out to be satisfactory.

2 Derivation of Flight Characteristics Data

2.1 Derivation of Aerodynamic Data

In this study seven RC airplane models, including high-wing Trainer 40, UT-1, low-wing Extra 300s (Fig. 1), and 40% scale Night Intruder (Fig. 2) are selected and their aerodynamic data are derived from Darcorp's AAA design software. Among them only Extra 300s and 40% scale Night Intruder are used as test beds in this paper. Aerodynamic data generally vary depending on the position in the flight envelop. However, a flight envelop of an RC airplane is usually very small, and its flight characteristics is about the same in the whole flight envelop. Thus steady state level flight is assumed as a single reference flight condition in the derivation of aerodynamic parameters of an RC airplane. Geometric dimensions are obtained by either measuring real RC airplanes or reading manufacturer's design drawings. Total mass and the center of gravity of an airplane are computed by measuring mass and position of each component such as fuselage, main wing, tail wing, fuel, landing gear, servo, and so on. Aerodynamics depends on the external shape of an airplane. Especially, airfoils of main and tail wings are critical in computation of   aerodynamics. These features are used as important inputs to AAA. In computing thrust forces of engines manufacturer's data are essential.

Fig. 1. Extra 300s   Fig. 2. 40% scale Night Intruder

While a typical AAA design process is illustrated in Fig. 3, the customized design process applied to this study is described in Table 1. Fig. 4 shows the result of the 9th process in the table, and Table 2 summarizes major flight characteristics data obtained from AAA for Extra 300s and 40% scale Night Intruder.

Table 1. Customized design process for derivation of stability and control derivatives

1. input present velocity, altitude, weight
2. input engine type, configuration of main and tail wings, number of landing gears, etc
3. measure and input sizes of fuselage, main wing, tail wing, etc
4. input derivative values related to airfoils of main and tail wings
5. measure and input weights of fuselage, main and tail wings, fuel, landing gear, etc
6. set and input required flight performance values such as stall velocity, landing distance, maximum cruise speed, etc
7. input propulsion performance values such as maximum thrust, fuel consumption rate, etc
8. set other required values for stability and control derivatives
9. AAA returns stability and control derivatives for a designed airplane

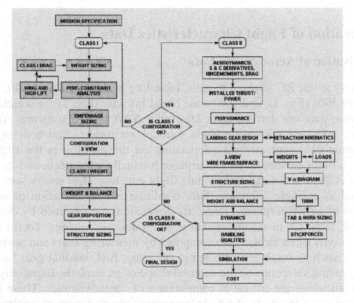

Fig. 3. Typical AAA design process

Input Parameters

Altitude	500	m	λ_w	0.53		λ_h	0.52		f_{gap_h}	1.00	
U_1	80.00	$\frac{km}{hr}$	$\Lambda_{c/4_w}$	4.0	deg	$\Lambda_{c/4_h}$	3.0	deg	$D_{f_{max}}$	0.14	m
$c_{l_{\alpha_w}(M=0)}$	6.0161	rad^{-1}	X_{apex_w}	0.26	m	X_{apex_h}	0.63	m	$\Delta \bar{X}_{ac_f}$	0.0000	
f_{gap_w}	1.00		$Z_{c/4_w}$	0.26	m	$Z_{c/4_h}$	0.12	m	$\Delta \bar{X}_{ac_{nacelle}}$	0.0000	
S_w	0.39	m^2	S_h	0.09	m^2	η_h	1.000		$\Delta \bar{X}_{ac_s}$	0.0000	
AR_w	5.73		AR_h	3.69		$c_{l_{\alpha_h}(M=0)}$	6.0161	rad^{-1}	X_{cg}	0.33	m

Output Parameters

M_1	0.066		$C_{L_{\alpha_w}}$	4.3388	rad^{-1}	$\bar{X}_{ac_{wf}}$	0.2421		$\bar{X}_{ac}$	0.3051	
$\bar{q}_1$	288.21	$\frac{N}{m^2}$	$C_{L_{\alpha_{wf}}}$	4.3394	rad^{-1}	X_{ac_h}	0.69	m	S.M.	19.02	%
$\bar{X}_{cg}$	0.1069		X_{ac_w}	0.37	m	$\bar{X}_{ac_h}$	1.4400		C_{L_α}	4.5802	rad^{-1}
$\bar{c}_w$	0.27	m	$\bar{X}_{ac_w}$	0.2421		$C_{L_{\alpha_h}}$	3.6579	rad^{-1}	C_{m_α}	-0.9076	rad^{-1}
n_{mgc_w}	0.041	m	$X_{ac_{wf}}$	0.37	m	$\frac{d\varepsilon}{d\alpha}$	0.7273				

Fig. 4. AAA GUI showing stability and control derivatives of Extra300s

Table 2. Flight characteristics data for RC airplanes (NI: Night Intruder)

Aero Coeff.	Extra 300s	40%scale NI	Aero Coeff.	Extra 300s	40%scale NI
CL0	0.4119	0.6043	CMAdot	-2.2830	-3.9936
CLA	4.5802	5.0169	CMQ	-4.0303	-12.3658
CLQ	5.7712	5.5208	CLB	-0.0027	-0.0769
CD0	0.089	0.0434	CLP	-0.6167	-0.6375
CDA	0.9061	0.2117	CLR	0.3744	0.3692
CYB	-0.4035	-0.5463	CLDR	-0.0102	-0.0243
CYP	-0.0337	-0.0435	CNB	0.123	0.1134
CYR	0.3154	0.2839	CNP	-0.0037	-0.0100
CM0	-0.0553	0.0000	CNR	-0.1469	-0.1326
CMA	-0.9076	-1.0026	CLDE	0.5216	0.1922

2.2 Computation of Weight and Balances

Principal moments of inertia of an airplane are computed by two different methods: one is to conduct experiments with a real RC airplane, and the other one is to use Dassault's CATIA design software. However, product of inertia can be obtained only by CATIA design. Two different data sets of moments of inertia are examined by implementing them in the in-house flight simulation software, which has been developed and examined for simulation of several military UAV systems. Experienced pilots fly the simulation models with different data sets, and validate the models subjectively based on the test procedures regulated in FAA 120 40B.

2.2.1 Experimental Method
Moments of inertia of an RC airplane can be obtained by measuring weight, cable length, distance between cables, and period of oscillation in the experiment illustrated in Fig. 6. The mathematical relation [5] between these parameters is as follows:

$$I = W_M T^2 \frac{D^2}{16\pi^2 L} \tag{1}$$

where,

I = moment of inertia (kg/m^2)
W_M = weight of an airplane (Kgm/sec^2)
T = period of oscillation (sec)
D = distance between cables (m)
L = cable length (m)

The average period of oscillation is obtained by measuring five consecutive periods after the coupling effect between principal axes diminishes significantly. Fig. 5 shows three experimental settings for three principal axes, while Table 3 contains experimental values for Extra 300s.

Table 3. Experimental values of 3 principal moments of inertia for Extra 300s

I_{xx}	I_{yy}	I_{zz}
0.147 kg/m^2	0.204 kg/m^2	0.044 kg/m^2

2.2.2 Computational Method Using CATIA
CATIA, 3D structural design software from Dassault, returns every moment of inertia and center of gravity once 3D configuration design is completed and material density

is input. The design process of Extra 300s is illustrated in Fig. 6, and resulting moments of inertia are listed in Table 4. Moments of inertia for 40% scale Night Intruder is also included in the table. Use of CATIA requires more efforts and time in the learning and design process than the experimental method, even though it can avoid the structural damage in the cable connection part of an airplane, which may not be avoidable for preparation of the experiment.

Table 4. Computational values of moments of inertia for Extra 300s and 40% scale NI (unit: kg/m^2)

	Ixx	Iyy	Izz	Ixz
Extra 300s	0.175	0.179	0.342	-0.0008248
40% scale NI	10.832	22.080	24.220	-4.902

Fig. 5. Experimental settings for measuring 3 principal moments of inertia (I_x, I_y, I_z)

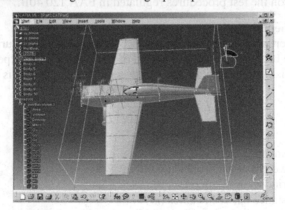

Fig. 6. CATIA design process of Extra 300s

The values of moments of inertia in Table 3 and 4 are noticeably different. Two different sets of parameter values are implemented in proven flight simulation software and subjectively validated based on the test procedures regulated in FAA's AC120-40B. Pilots turn out to prefer data values derived from CATIA because the flight characteristics with the computed data set resembles the real one better than the flight model with moments of inertia from experiments. In order to obtain a more accurate data set, the RC plane should be heavily damaged for cable connection. Because structural damage of an expensive airplane was not allowed in the experiment, the cable hook could not be located at the right position. This constraint causes couplings among oscillations with respect to three principal axes, and results in measurement errors in moments of inertia.

3   Test and Evaluation of Design Parameters

3.1   Flight Simulation S/W (RC Virtual Flight)

The main purpose of derivation of valid aerodynamic data is to simulate the flight characteristics of an airplane without flight test data. On the contrary, the data set mostly from computer simulation software is validated subjectively by flying a simulated airplane in this study. Therefore, reliable flight simulation software is required for the purpose. UAV-HILS laboratory at Sejong University has developed comprehensive flight simulation software named "RC Virtual Flight" [6], [7] since 1998. RC Virtual Flight has been upgraded and applied to various research projects sponsored by Korean government and industry. Table 5 summarizes its features, while Fig. 7 captures some of its GUI windows.

Fig. 7. Snapshots of RC virtual flight simulation

Table 5. Features of RC virtual flight

Features	Remarks
fixed-wing and rotary-wing models	Pioneer, F16, ChangGong-91, Extra300s, UH60, S61, AH1, Yamaha Rmax, etc. full flight envelope 6 DOF nonlinear math models
3D terrain libraries	domestic airports such as Kimpo, Incheon, Kimhae, Jeju, etc. domestic RC runways such as Amsadong, Yoido, Oedo, etc. military UAV runways
atmospheric environment	turbulence, side wind, gust, wind shear, etc.
sound effects	engine, propeller noise Doppler effects 5.1 channel
lesson plan	RC flight training lessons edited by experience RC Flight Instructors military UAV training lessons
Maintenance and upgrade	off-the-shelf hardware products object oriented program based on C++, Matlab, and Simulink open architecture
remote controller	USB interface RC trainer jack interface compatible with either simulated or real hardware

3.2   Subjective Tests in FAA AC120-40B

Validation procedure of the simulation data follows the subjective test rules in FAA AC 120-40B, which is prepared for manned airplane simulators. The test procedure and its applicability to RC airplanes are summarized in Table 6. Flight profiles of RC and manned airplanes are about the same. Thus typical flight profiles, such as taxing, taking-off, cruising, loitering, and landing, and their relevant tests have to be applicable to RC airplanes, too. Test items for the flight profiles and systems, which are neither comprised nor critical in RC flight, must be discarded. Besides, test procedures with respect to a visual system and a motion platform of a manned flight simulator do not have to be applied to the flight data validation process of an RC airplane.

Table 6. Subjective test items in FAA 120-40B and their applicability to RC airplanes

Subjective Tests		Applicability
Check Items	Details	
pre-flight	check system equipments such as switches, indicators, etc.	O
pre-takeoff	engine start	O
	taxi	O
takeoff	normal takeoff	O
	abnormal/emergency Takeoff	O
In-flight	climb	O
	cruise	O
	descent	O
approaches	non-precision	X
	precision	X
landing	normal landing	O
	abnormal/emergency landing	O
post landing	landing roll and taxi	O
flight phase	airplane and power plant systems operation	O
	flight management and guidance system	X
	airborne procedures	O
	engine shutdown and parking	O
visual system		X
motion system		X
special effect		X

3.3 Subjective Test and Evaluation by Experienced RC Pilots

Following the procedure regulated in FAA AC 120-40B, the design parameter values obtained from AAA and CATIA are subjectively evaluated by experienced RC pilots. The design data such as moments of inertia, stability derivatives, etc are implemented in RC Virtual Flight and tuned based on the indications of experienced pilots. Table 7 shows a part of pilots' indications and corrected parameters with respect to Extra 300s.

Table 7. Pilots' indications and corrected parameters with respect to Extra 300s

Pilots' Indications	Corrected Parameters
Rudder effectiveness on roll is too large.	CLB (C_{L_β})
Rudder effectiveness on yaw is too small.	CNDR ($C_{N_{\delta_r}}$)
Airspeed is too sensitive to thrust.	CDu (C_{D_u})
Pitch angle becomes too large during the trimmed level flight as the airspeed increases.	CMu (C_{M_u})

4 Conclusions

Darcorp's AAA is used for aerodynamic design and Dassault's CATIA for derivation of weights and balances, while a subjective test procedure is extracted from FAA AC 120 40B. Seven RC airplanes, comprising Extra 300s and Night Intruder 40% scale, are selected as test beds. Use of design S/W for generation of parameter values representing flight characteristics turns out valid. In this study a practical procedural standard is established for procuring reliable data replicating flight characteristics of an airplane, which can be used in flight simulation and design of flight control systems of small airplanes.

Acknowledgement

This research(paper) was performed for the Smart UAV Development Program, one of the 21st Century Frontier R&D Programs funded by the Ministry of Science and Technology of Korea.

References

1. USAF Stability and Control DATCOM, Flight Control Division, Air Force Flight Dynamics Laboratory, Write-Patterson Air Force Base, Fairborn, OH.
2. Jan Roskam,: Airplane Design, Roskam Aviation and Engineering Corporation (1986)
3. CATIA Reference Manual.
4. FAA AC120-40B.: Airplane Simulator Qualification (1991)
5. R.C. Nelson.: Flight Stability and Automatic Control, McGraw-Hill (1998)
6. Yoon S.: Development of a UAV Training Simulator Based on COTS Products and Object-Oriented Programming Languages, Proceedings of AIAA Modeling & Simulation Conference, Aug. 3-5 (2001)
7. Yoon S. and Nam K.: Development of a HLA-based Simulation Game for RC Airplanes and UAVs, Proceedings of AIAA Modeling and Simulation Conference, Aug. (2004)

KOMPSAT-2 S/W Simulator

Sanguk Lee, SungKi Cho, and Jae Hoon Kim

Communications Satellite Development Group, ETRI
Gajung-dong 161, Yusong-gu, Daejeon 305-350, Korea

Abstract. In this paper, we present design and implementation features and validation of KOMPSAT-2 simulator (SIM). The SIM is implemented on PC server to minimize costs and troubles on embedding onboard flight software into simulator. Object oriented design methodology is employed to maximized S/W reusability, and XML is used for S/C characteristics data, TC, TM and Simulation data instead of expensive commercial DB. Consequently, we can reduce costs for the system development, efforts on embedding flight software. A simulation for science fine mode operation is presented to demonstrate the overall validation of SIM. SIM has been verified officially by various tests such as SIM subsystem test, MCE system integration test, interface test, site installation test, and acceptance test. They were performed successfully.

1 Introduction

Electronics and Telecommunications Research Institute (ETRI) developed Mission Control Element (MCE) for KOrean Multi-Purpose SATellite-1 (KOMPSAT-1) which lifetime was three years was launched in December 1999. It has been being used for operation of KOMPSAT-1 normally so far. Also, we have developed MCE for KOMPSAT-2, which is equipped Multi-spectral Camera (1m panchromatic and 4m multi-band) and delivered the MCE to Korean Aerospace Research Institute (KARI). The Korean Ground Station of KOMPSAT-2 consists of MCE and IRPE (Image Reception and Processing Element). The MCE provides the mission analysis and planning and satellite control capabilities that allow the operators to carry out the KOMPSAT-2 mission. Space-ground communications are based on the CCSDS standard format. Figure 1 shows the simplified functional schematics of the system. KOMPSAT-2 MCE is consisted of four subsystems such as TTC (Tracking, Telemetry, and Command Subsystem), SOS (Satellite Operation Subsystem) [1], MAPS (Mission Analysis and Planning Subsystem) [2], and SIM (Satellite Simulator Subsystem) [3]. For KOMPSAT-2, TTC provides the S-band uplink and downlink communications interface with satellite, the CCSDS command and telemetry processing, and tracking capabilities. SOS provides spacecraft command generation and execution, satellite state of health monitoring, and housekeeping telemetry data processing. MAPS provides mission planning, incorporates user requests, defines configurations, and prepares mission schedules. MAPS, also, provides KOMPSAT-2 operation support functions such as orbit determination and prediction and antenna position data generation for tracking.

1.1 SIM Functional Description

SIM is one of the KOMPSAT-2 MCE subsystems and Fig. 2 shows the functional architecture of SIM. SIM is a software system of simulating the dynamic behavior of

D.-K. Baik (Ed.): AsiaSim 2004, LNAI 3398, pp. 458–466, 2005.

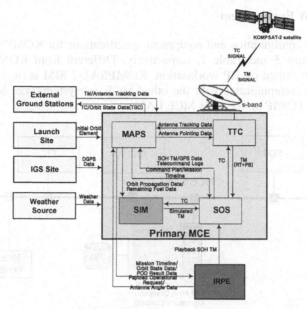

Fig. 1. Schematics of the KOMPSAT-2 MCE

KOMPSAT-2 by use of mathematical models. SIM is utilized for command verification, operator training, satellite control procedure validation, and anomaly analysis.

SIM models each of satellite subsystems as accurate as possible with the constraint of real-time operation condition, and display status of satellite including orbit and attitude in alphanumeric and graphic method.

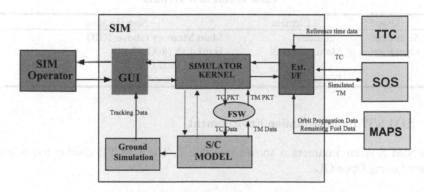

Fig. 2. SIM Functional Architecture

SIM operates in real-time, and receives telecommands, distributes them to corresponding subsystems, and sends the simulated results to SOS in telemetry format. SIM is capable of operating in on-line as well as stand-alone mode. SIM also operates in real-time and pre-defined variable speed simulation for analysis purpose. In addition, SIM supports simulation for the spacecraft status by various events and initialization data.

1.2 SIM H/W Environment

The hardware configuration and equipment specifications for KOMPSAT-2 SIM are shown in Figure 3 and Table 1, respectively. Different from KOMPSAT-1 SIM, which was developed on HP workstation, KOMPSAT-2 SIM is developed on a PC server, which communicates with the other MCE subsystems, i.e. SOS, TTC, and MAPS, using TCP/IP protocol via MCE LAN.

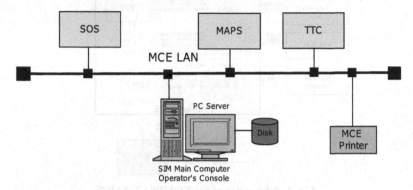

Fig. 3. H/W Configuration and KOMPSAT-2 SIM

A PC server as platform including in Window 2000 as an operating system was used as H/W platform of KOMPSAT-2 SIM. The PC server also contains VR graphic display of the KOMPSAT-2 attitude and orbit motion.

Table 1. SIM H/W Element

Usage	Element	Specification
Simulator Operation	Simulator Main computer	- Main Memory (above 1GB)
		- Hard disk (80G *2)
		- 2.6GHz Intel CPU
	Display device	- LCD monitor (21")

1.3 SIM S/W Implementation Environment

The SIM S/W environment is shown in Table 2. The SIM VR display too is implemented using Open GL.

Table 2. SIM S/W Environment

Element	Specification
Operating System	- MS Windows 2000
Programming Language	- C++ : GUI, Models
	- C : FSW embedded
Data Management	- XML
Library	- Open GL : VR Display

2 SIM Design

Design of SIM is carried out using OOA (Object-Oriented Analysis) / OOD (Object-Oriented Design) methodology and is described as the use case model, domain model, user interface design, logical view, implementation view, process view, and deployment view[4][5]. The UML (Unified Modeling Language) notation is common language to specify, construct, visualize, and document for designing object-oriented software system by using Rational Rose™. The standard OOA/OOD procedure has been applied to design KOMPSAT-2 SIM to maximize reusability, extensibility, and reliability. Object-Oriented Process is proceeded by Use Case Modeling, Domain Modeling, Logical View Design, Implementation View Design, Process View Design, and Deployment View Design.

2.1 Use Case Modeling

Use Case Modeling is expressed as Use Case Diagram, which describes system requirement in the viewpoint of user's points of view. Use Case Diagram is to describe external view of the system. Then, each Use-Case description, a basic flow, and an alternative flow are presented. Also pre-condition/post conditions of each Use Case are described. Following is an example of Use Case Modeling of "Login" of KOMPSAT-2 SIM. The last item entitled as "Source" is SIM subsystem specification ID to trace specification for design phase.

- Use-Case Specifications
     Login
- Description
     Only the authorized operator allows using SIM.
- Flow of Events
     Basic Flow
          This Use-Case is used as an operator starts SIM.
          1. SIM requires operator's ID and Password input
          2. An operator inputs ID and Password.
          3. SIM verifies ID and Password and displays a main window for the authorized operator.
     Alternative Flow: Invalid ID/Password
          If an operator is not authorized, SIM displays error massage and backs to show Login Window.
- Pre-Conditions
     None
- Post-Conditions
     None
- Source
     SIM32400A : Secure Operation

2.2 Domain Model

Domain Model describes how the Use Case is realized in Use Case Model. Domain Modeling can be expressed by Class Diagram, which describes how these classes interact with each other. Figure 4 shows class diagram of Login.

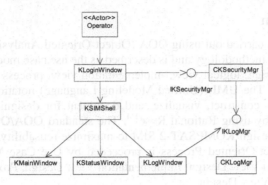

Fig. 4. Class Diagram of Login

2.3 Logical View

Logical view of SIM decomposes into conceptual packages or logical components and describes connections between them. Figure 5 shows Logical View of SIM subsystem which is consisted of Kernel, UI, OBC (OnBoard Computer), RDU (Remote Drive Unit), ECU (Electric Control Unit), Models, TCProcess, SDTProcess, SecurityMgr, TMMgr, InitDataMgr, TCMgr, SDTMgr, EXTINTFACE, Tread, and Socket packages.

For the KOMPSAT-2 SIM, the onboard flight software is embedded into SIM as an independent process that has thread for interface with scheduler, and 1553B bus is emulated by COM (Common Object Model). For example, we go through into more detailed design with CES (Conical Earth Sensor) as a sensor model. CES model belong to Sensor subpackage of Models Package, which is one of SIM subsystem packages. Figure 5 shows SIM top-level.

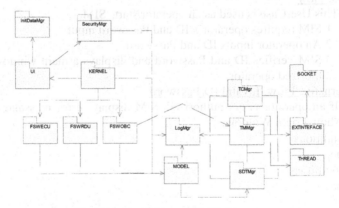

Fig. 5. Logical View of SIM Subsystem

2.4 Implementation View Design

The Architecture of SIM from Implementation view of the system is described here. Implementation view describes the actual software module, their relations, and contents along with consideration of the requirements as shown in Figure 6.

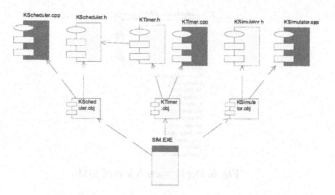

Fig. 6. Implementation View of Control Package

2.5   Process View Design

SIM has the only one EXE file and number of DLL (Dynamic Link Library)s, which are running like independent processes as shown in Figure 7. Hereafter, EXE and DLLs are named after process. Process view describes execution structure of the SIM system along with consideration of the requirements related to performance, reliability, expandability, system management, and synchronization.

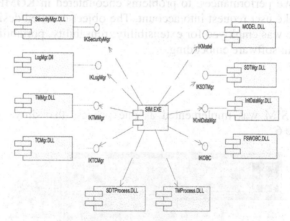

Fig. 7. Process View of SIM

2.6   Deployment View Design

In deployment view, SIM architecture is described in the physical point of view. Figure 8 shows nodes, links between nodes, and their platform.

3   SIM Implementation and Characteristics

3.1   KOMPSAT-2 Flight Software Embedding

The PC server was used for avoiding byte&bit ordering problem due to infrastructure difference. Object-Oriented Programming technique was used to minimize flight

Fig. 8. Deployment View of SIM

software revision. For example, reserved I/O functions like inp (), inpw (), outp (), and outpw () will be functions named after the same function name using function overloading technique[6]. Some system calls on VRTX operating system, also, will be replaced with functions produced by function overloading. Also, the same technique will be used for scheduler onboard and process manager. Data exchange between DLLs will be implemented with Component Object Model (COM). Key Parameter Data (KPD) in EEPROM was emulated with the file that contains KPD data to support KPD upload. For KOMPSAT-2 SIM, some new features and changes was made to improve performances, to problems encountered in KOMPSAT-1 development, and to take user request into account. The object-oriented design and programming technique was employed for extensibility, reusability, portability, and less code changes in flight software embedding.

3.2 SIM GUI

KOMPSAT-2 SIM was implemented as one process that drove several DLLs as shown in figure 6.

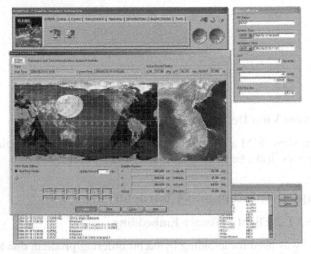

Fig. 9. Main Window Configuration of KOMPSAT-2 SIM

Those DLLs are implementation of elements packages for SIM in implementation and process views. Figure 9 is the main window configuration of the implemented KOMPSAT-2 SIM GUI. For user's comprehension of satellite operation, KOMPSAT-2 SIM provides ground track view, celestial track view, and VR display.

3.3 VR Display for Comprehension

Specially, VR display can display satellite orbit and attitude motion dynamically not only using dynamic simulation data but also using real-time TM data from satellite and playback TM data.

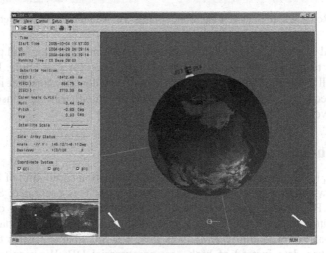

Fig. 10. 3D display of spacecraft motion

To do so, a preprocessor to convert from spacecraft TM into data required for 3D Display is introduced. The preprocessor can be divided into two. The one is real-time preprocessor and the other is playback processor. MCE of KOMPSAT receives real-time TM and playback TM during pass time. Real-time TM is distributed into SOS GUI and other network computers to display TM and subsystem engineers monitor state of health of spacecraft. High-speed playback data is stored into playback data file for later analysis purpose. Figure 10 shows VR display window for KOMPSAT-2 SIM.

4 Kompsat-2 SIM Validation

The overall functions, which are defined on KOMPSAT-2 SIM specifications, have been validated via SIM subsystem acceptance test and MCE system acceptance test. Functional verification of H/W unit model was performed via independent unit test.

During the both tests, overall function has been verified by simulation of several operation modes such as solar array deployment and its test, sun point mode, Earth search mode, attitude hold mode, normal operation mode, and so on. In this paper, simulation results of science fine mode operation are presented for KOMPSAT-2 SIM S/W validation as an example. In KOMPSAT-2 normal operation, science fine mode is used for high resolution Multi-Spectral Camera (MSC) mission. In this mode, Star

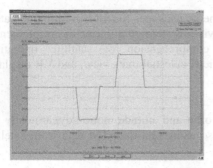

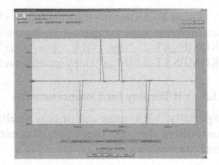

Fig. 11. Roll Angle in Science Fine Mode   **Fig. 12.** Attitude Errors in Science Fine Mode

trackers and reaction wheels are main attitude sensor and actuator, and pointing accuracy is less than 0.015 for roll and pitch and 0.025 deg for yaw angle. Figure 11 and 12 show roll angle and attitude errors in –30 and +30 degree maneuver in science fine mode, respectively. Requirements for the mode are satisfied in the senses of pointing and maneuver.

5   Conclusions

The design, implementation characteristics and validation simulation of KOMPSAT-2 simulator, which is a subsystem of KOMPSAT-2 MCE is presented. SIM is implemented on PC server to minimize costs and troubles on embedding onboard flight software into simulator. OOA/OOD methodology is employed to maximized S/W reusability and expendability. XML was used for S/C characteristics data, TC, TM and simulation data instead of high cost commercial DB. We significantly reduce costs and efforts for the system development. All requirements defined in KOMPSAT-2 MCE and SIM requirement document were verified through official tests such as subsystem test, system integration test. The overall functions were validated simulation results of science fine mode operation. KOMPSAT-2 MCE system was delivered to KARI and installation and acceptance test were performed successfully.

References

1. Mo, Hee-Sook, Lee, Ho-Jin, and Lee, Seong-Pal: Development and Testing of Satellite Operation System for Korea Multipurpose Satellite- I. ETRI Journal 1 (2000), 1-11.
2. Won, Chang-Hee, Lee, Jeong-Sook, Lee, Byoung-Sun, and Eun, J._W.: Mission Analysis and Planning System for Korea Multipurpose Satellite-I. Etri Journal, 3 (1999) 29-40.
3. Choi, W. S., Lee, Sanguk, Eun, J. W., Choi, H, and Chae D. S.: Design Implementation and Validation of the KOMPSAT Spacecraft Simulator. KSAS International Journal 2 (2000) 50-67.
4. Lee, Sanguk, Cho, Sungki, Kim, J. H., and Lee, Seong-Pal : Object-Oriented Design of KOMPSAT-2 simulator including onboard Flight Software. 14th International Conference in System Science. Wroclaw, Poland. (2001)
5. Odell, J. J. and Martin, J.: Object Oriented Method: A Foundation. Prentice-Hall, (1995)
6. Lee, Sanguk, Choi, W. S., and Chae, D. S.: Implementation of KOMPSAT Simulator Interfaces Between Flight Software and S/C Subsystem Model. KSAS J. 3 (1999) 125-131.

The Robust Flight Control of an UAV Using MIMO QFT: GA-Based Automatic Loop-Shaping Method

Min-Soo Kim[1] and Chan-Soo Chung[2]

[1] Sejong-Lockheed Martin Aerospace Research Center,
Sejong University, 98 Kunja, Kwangjin, Seoul, Korea 143-747
mskim@sejong.ac.kr
[2] Electrical Engineering Department,
Soongsil University, 1-1 Sangdo, Dongjak, Seoul, Korea 156-743
chung@ssu.ac.kr

Abstract. This paper presents a design method of the automatic loop-shaping which couples up manual loop-shaping method to genetic algorithm (GA) in multi-input multi-output (MIMO) quantitative feedback theory (QFT). The QFT is one of the robust control design methods to achieve a given set of performance and stability criteria over a specific range of plant parameter uncertainty. The design procedure of the QFT is developed in a constructive manner and involves several essential steps. In QFT controller design process, the most important step is the loop-shaping in frequency domain by manipulating the poles and zeros of the nominal loop transfer function. However, the loop-shaping is currently performed in computer aided design environments manually, and moreover, it is usually a trial and error procedure. To solve this problem, automatic loop-shaping method based on GAs is developed. Robust control techniques including the QFT are particularly useful for the design Unmanned Aerial Vehicle (UAV) because dynamics of UAV vary substantially throughout the flight envelope. The QFT loop-shaping is successfully applied to the enhancement of the robust stability and disturbance rejections performance of the UAV based on specific margin specifications.

1 Introduction

Quantitative Feedback Theory (QFT) has been successfully applied to many engineering systems since it was developed by Horowitz [1-3]. The basic idea in QFT is to convert design specifications of a closed-loop system and plant uncertainties into robust stability and performance bounds on the open-loop transmission of the nominal system, and then design a controller by using the gain-phase loop-shaping technique. The QFT method is a robust control design technique that uses feedback of measurable plant outputs to generate an acceptable response considering disturbance and uncertainty. This method uses quantitative information such as the plant's variability, the robust performance requirements, control performance specifications, the expected disturbance amplitude, and attenuation requirements. The controller is determined through a loop-shaping process based on a Nichols Chart. This Nichols Chart contains some information of the phase and gain margins, stability bounds, performance bounds, and disturbance-rejection bounds. Lastly, the pre-filter is de-

D.-K. Baik (Ed.): AsiaSim 2004, LNAI 3398, pp. 467–477, 2005.
© Springer-Verlag Berlin Heidelberg 2005

signed in order to meet the tracking performance requirements using a Bode diagram where the closed-loop frequency response is shaped. The QFT technique has been applied successfully into many classes of linear problems including single-input single-output (SISO), multi-input multi-output (MIMO) plants, and various classes of nonlinear and time-varying problems. In the last ten years, the application of QFT to flight-control systems design has been studied extensively [4-7].

The Genetic Algorithm (GA) is a stochastic global search method that mimics the metaphor of natural biological evolution [8],[9]. GA operates on a population of potential solutions by applying the principle of survival of the fittest to produce better and better approximations to a solution. At each generation, a new set of approximations is created by the process of selecting individuals according to their level of fitness in the problem domain and breeding them together using operators borrowed from natural genetics. This process leads to the evolution of populations of individuals that are better suited to their environment than the individuals that created from just as in the natural adaptation.

The QFT design procedure often falls in a trial and error process although the simulation package is usually used. To solve this manual loop-shaping design problem, computer based design techniques have been recently investigated and developed.

1. The use of Bode Integrals within an iterative approach to loop-shaping by Horowitz and Gera [10].
2. Thompson and Nwokah [11] proposed an analytical approach to loop-shaping, if the templates may be approximated by boxes or an initial QFT controller already exists.
3. A linear programming approach to automatic loop-shaping was proposed by Bryant and Halikias [12].
4. An automatic technique was proposed by Chait [13], which overcomes the non-convexity of the bounds on the open-loop transmission, while the design is based on the closed-loop bounds.

The major disadvantage of four approaches is the inability in solving a complicated nonlinear optimization problem. Further, these conventional methods often based on the unpractical assumptions and often lead to very conservative designs.

In this paper, we present further advancements and improvements in automatic loop-shaping method for the given specifications using GA as below.

1. Ability to deal with the various design specification: The proposed algorithm can be applied to more various design specifications which include robust stability specifications, tracking specifications, margin specifications, minimum high frequency gain, etc. because the objective function to minimize in GAs is based on the only QFT bounds.
2. Meaningful controller elements: The controller consists of poles, zeros, damping factors, natural frequencies and so on. So we can construct a complex or simple controller based on the handling   the addition/elimination elements.
3. Fine tuning: The proposed method provides fine tuning based on controller's elements.
4. Flexible fitness function: The proposed algorithm provides a compromise among contradicted specifications because distance measures in fitness function are used.

2 Automatic Loop-Shaping Method in MIMO QFT

Given a nominal plant and QFT bounds, the loop-shaping problem is that synthesize a controller which achieves closed-loop stability, satisfies the QFT bounds and has minimum high-frequency gain. The usual approach to this problem involves manual loop shaping method in the frequency domain by manipulating the poles and zeros of the nominal loop transfer function.

The manual loop-shaping procedure may be replaced by the computer based optimization procedures. However, such a nonlinear, non-convex multi-objective optimization problem can be hardly solved using conventional techniques. One approach to solve this problem is GA which was originally motivated by the mechanisms of natural selection and evolutionary genetics.

2.1 Definition of Automatic Loop-Shaping

Given the nominal plant $P_0(s)$ and the finite QFT bounds, design $G(s)$ such that
1. The system achieves internal stability.
2. The system satisfies its bounds.
3. The system has a minimum bandwidth.
4. The system has a minimum high frequency gain.

The controller should be synthesized automatically such that all QFT bounds and the stability of the closed-loop nominal system are satisfied and the given performance index is minimized.

In this paper we provide an automatic loop-shaping algorithm that couples up advantages of a classical manual loop-shaping method to those of GAs. The manual loop-shaping method uses constructive design processes that add gain, pole/zeros, lead/lags, and complex pole/zeros to initial nominal loop to derive a nominal loop shape in an iterative fashion. This method is automated in a QFT Toolbox [7] which simplified the iteration process and allowed for higher order approximations. But there was no guarantee of convergence and rational function approximation was needed to obtain an analytical expression for the loop. So we utilize the advantage of GAs.

From the manual loop-shaping method, the characteristic and/or advantage of the proposed method are follows:
1. Can apply to SISO as well as MIMO systems.
2. Use a gain, simple pole/zeros, and complex pole/zeros as optimization variables.
3. Can diminish the range of gain values based on computer aided design method.
4. Can place optimization variables (pole/zeros) in a specific region.
5. Provide fine tuning.

From GA, the characteristic and/or advantage of the proposed method are follows:
1. Need not predetermine the order of the nominator/denominator.
2. Can use constructive design method using termination condition.
3. Coding with a floating-point representation, multi-population.

2.2 Design Procedure of Automatic Loop-Shaping

The following algorithm shows procedure of the proposed method of automatic loop-shaping.

```
Decide variables to optimize;
Initialize controller type G₀(s);
Add frequency array;
Find optimal solution of loop-shaping in GA routine;
(If termination condition is reached
 Stop program;
 End if)
For i=1 to M,
    Add variables and decide controller type Gᵢ(s);
    Find optimal solution in GA routine with Gᵢ(s);
    (If termination condition is reached
            Stop program
    End if)
Next i;
```

This method is based on the constructive manners and each operation is iteratively performed until a termination criterion is satisfied or the iteration number reaches a positive integer M.

There is an important factor, termination condition, in the proposed algorithm.

$$\left(\sum_{i=1}^{m} J_{stability_i} = 0\right) \& \left(\sum_{j=1}^{n} J_{tracking\_bound_j} \le \varepsilon_T\right) \& \left(J_{min\_BW} \le \varepsilon_{BW}\right) \& \left(\Delta J_{high\_freq} \le \varepsilon_{high\_freq}\right) \quad (1)$$

where ε_T and ε_{BW} are margin factors, and $\Delta J_{high\_freq} = \left| J_{high\_freq_t} - J_{high\_freq_{t-1}} \right|$.

When a certain specification contradicts other specifications, this margin factor can provide a compromise points. For example, when tracking specifications and minimum bandwidth are contradict each other, a compromise point is given as a $J_{tracking\_bound} + J_{min\_BW} \le \dfrac{\varepsilon_T + \varepsilon_{BW}}{2}$. Suppose the initial controller type is defined as a

$G_0(s) = k$, then the successive controller types are $G_1(s) = k\dfrac{1}{\left(1 + s/p\right)}$,

$G_2(s) = k\dfrac{\left(1 + s/z\right)}{\left(1 + s/p\right)}$, $G_3(s) = k\dfrac{\left(1 + s/z\right)}{\left(1 + s/p_1\right)\left(1 + s/p_2\right)}$ or $G_3(s) = k\dfrac{\left(1 + s/z\right)}{\left(1 + \left(2\zeta/\omega_n\right)s + \left(s^2/\omega_n^2\right)\right)}$, and

so on. The controller type of the jth design procedure is as follows, which the variables to be optimized are selected by $\{k, z_1, z_2, \cdots, z_l, p_1, p_2, \cdots, p_m, \zeta_1, \zeta_2, \cdots, \zeta_n, \omega_{n_1}, \omega_{n_2}, \cdots, \omega_{n_n}, \cdots\}$.

$$G_j(s) = k\dfrac{\left(1 + s/z_1\right)\left(1 + s/z_2\right)\cdots\left(1 + s/z_l\right)}{\left(1 + s/p_1\right)\cdots\left(1 + s/p_m\right)\left(1 + \left(2\zeta_1/\omega_{n_1}\right)s + \left(s^2/\omega_{n_1}^2\right)\right)\cdots\left(1 + \left(2\zeta_n/\omega_{n_n}\right)s + \left(s^2/\omega_{n_n}^2\right)\right)} \quad (2)$$

The choice of the range of the parameters is important in the automatic loop-shaping, which determines the size of the searching space.

2.2.1 Range of Gain k

If there is no prior information about the range of the gain then the minimum is chosen as close to zero, and the maximum of all parameters can be chosen as a reasonably large number. But, we can know rough values of the controller's gain based on computer aided design environments, for example, the QFT Toolbox for MATLAB [7].

2.2.2 Searching Space of Optimized Variables

All poles and zeros are negative values with $[-10^{-3} \ -10^{3}]$ because of the stable minimum phase controller requirement. And, the damping factor ζ_i is chosen from 0.1 to 10 since the large ζ_i doesn't give influence on the change of $L(s)$ and the natural frequency ω_{n_i} has bounds $[10^{-4} \ 10^{4}]$.

2.2.3 Fitness Function

The fitness in GA for the QFT design should reflect the stability and performance requirements, and the performance index, given by

$$J = \gamma_S \sum_{i=1}^{m} J_{stability\,i} + \gamma_T \sum_{j=1}^{n} J_{tracking\_bound\,j} + \cdots + \gamma_{BW} J_{\min\_BW} + J_{high\_freq} \qquad (3)$$

where γ_S , γ_T and γ_{BW} are a weighting factor. Here m and n denote the number of the sampled frequency of stability bounds (or margin bounds) and tracking bounds respectively. $J_{stability\,i}$, $J_{tracking\_bound\,j}$, $J_{\min\_BW}$, and J_{high_freq} represent performance indexes of the stability specification, tracking bounds, a minimum bandwidth, and a maximum high frequency gain, respectively.

The first, the stability of the closed-loop nominal system is simply tested by checking the existence inside of a given stability bound at each sampled frequency. The function is

$$J_{stability\,i} = \begin{cases} 1 & \text{If inside the bound at } \omega_i \\ 0 & \text{Otherwise} \end{cases} \qquad (4)$$

At equation (3), γ_s makes a heavy penalty with a large value when the nominal loop invades the stability bounds.

The second, tracking bounds index and minimum bandwidth index are defined by

$$J_{bound\_spec\,i} = \begin{cases} 0 & \text{If the QFT bound at } \omega_i \text{ is satisfied} \\ d_{\min} & \text{Otherwise} \end{cases} \qquad (5)$$

where $d_{\min}$ denote the shortest distance to the tracking bounds at each frequency.

The third, the minimum bandwidth index is simply tested by calculating the distance of the two points between the left-lower frequency point on stability bounds (or margin bounds) and the nearest point on the nominal loop.

The last, considering the effects of high frequency sensor noise and the high frequency dynamics, the cost function to be minimized is the high frequency gain of the open-loop transmission $L(s)$ where the high frequency gain of the controller given by

$$\lim_{s \to \infty} s^q L(s) \tag{6}$$

where the relative order of the transfer function $L(s)$ is q.

2.3 Controller Design Procedures of MIMO QFT with Margin Specifications

Let $L(s)$ denote a transfer function of SISO system. The margin of $L(s)$ is $m(\omega)$ if $\left|1+L(j\omega)\right|^{-1} = m(\omega)$. A phase margin means that at the frequency ω such that $\left|L(j\omega)\right| = 1$ and a gain margin means that at the frequency ω for which $\arg L(j\omega) = -180°$.

Let P is a 2×2 LTI plant which belongs to a given set {P}. The design objective is how to compute bounds on the controller $\mathbf{G} = diag(g_1, g_2)$ such that the given margin specifications will be satisfied for a given set of plants {P}, the following margin specifications are satisfied $\left|1+L_k(j\omega)\right|^{-1} \le m_k(\omega)$, where $k = 1, 2$ and $\omega \ge \omega_{mb}$. ω_{mb} is a transition frequency between low and high frequencies. Generally, $\omega_{mb} = 0$ is a satisfactory choice.

The design problem of MIMO system summarizes two sequential procedures [5].

At the first step, g_1 is designed to satisfy bounds calculated to meet various other specifications along with the following inequalities.

$$\left|1+\frac{g_1}{\pi_{11}}\right|^{-1} \le m_1(\omega) \quad , \quad \left|1+g_1 p_{11}\right|^{-1} \le m_1(\omega) \tag{7}$$

for all $\mathbf{P} \in \{P\}$, $\omega \ge \omega_{mb}$, and the matrix transfer function $P = [p_{ij}]$ and its inverse is $P^{-1} = [\pi_{ij}]$.

At the second step, g_2 is designed to satisfy bounds calculated to meet the given specifications along with the following inequalities.

$$\left|1+L_2\right|^{-1} = \left|1+\frac{g_2}{\pi_{22}-\dfrac{\pi_{12}\pi_{21}}{\pi_{11}+g_1}}\right|^{-1} \le m_2(\omega) \tag{8}$$

$$\left|1+L_1\right|^{-1} = \left|\frac{1+g_1 p_{11}+g_2(p_{22}+g_1 \det P)}{1+g_2 p_{22}}\right|^{-1} \le m_1(\omega)$$

for all $\mathbf{P} \in \{P\}$ and $\omega \ge \omega_{mb}$.

3 Simulations

3.1 Problem Definitions

The plant is the lateral-directional dynamic characteristics of an UAV [5]. The plant model is described by its state space equations with state vector consisting of the perturbed side velocity v, roll rate p, yaw rate r, roll angle Φ, heading angle Ψ, side force Y and side-slip angle β, and with output vector $y = \begin{bmatrix} v & p & r & \Phi & \Psi & Y & \beta \end{bmatrix}^T$ as follows:

$$\begin{cases} \dot{x} = Ax + Bu \\ y = Cx + Du \end{cases} \tag{9}$$

The input vector is made up of the aileron angle δ_a, rudder angle δ_r, and the gust velocity V_{gust} which is an external input signal to the system. The values of the A, B, C and D matrices are

$$A = \begin{bmatrix} -0.1463 & 5.4663 & -266.7630 & 9.8079 & 0 \\ 0.1302 & -2.5800 & 0 & 0 & 0 \\ 0.0900 & 0 & -0.2946 & 0 & 0 \\ 0 & 1.0000 & 0.0205 & 0 & 0 \\ 0 & 0 & 1.0002 & 0 & 0 \end{bmatrix} \quad B = \begin{bmatrix} 0 & 0.1263 & -0.1463 \\ -0.9020 & 0 & 0.1302 \\ 0 & -0.2933 & 0.0900 \\ 0 & 0 & 0 \\ 0 & 0 & 0 \end{bmatrix}$$

$$C = \begin{bmatrix} 1.0000 & 0 & 0 & 0 & 0 \\ 0 & 57.2958 & 0 & 0 & 0 \\ 0 & 0 & 57.2958 & 0 & 0 \\ 0 & 0 & 0 & 57.2958 & 0 \\ 0 & 0 & 0 & 0 & 57.2958 \\ -0.0149 & 0 & 0.0174 & 0 & 0 \\ 0.2146 & 0 & 0 & 0 & 0 \end{bmatrix} \quad D = \begin{bmatrix} 0 & 0 & 0 \\ 0 & 0 & 0 \\ 0 & 0 & 0 \\ 0 & 0 & 0 \\ 0 & 0 & 0 \\ 0 & 0.0129 & -0.0149 \\ 0 & 0 & 0 \end{bmatrix} \tag{10}$$

The plant uncertainty exists in the A, B matrices. The uncertainty of A and B are

1. The variation of a_{11}, a_{22} and a_{33} resides between 50% and 200%.
2. The variation of a_{21} and a_{31} takes values from 75% to 150%.
3. The variation of b_{21} and b_{32} takes values from 80% to 120%.
4. The parameters are correlated such as $b_{13} = a_{11}$, $b_{23} = a_{21}$, and $b_{33} = a_{31}$.

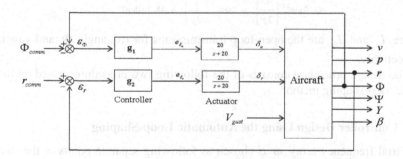

Fig. 1. Block diagram of lateral/directional control systems

The plant to be controlled is 2×2 plant from the inputs $\begin{bmatrix} \delta_a & \delta_r \end{bmatrix}^T$ to the outputs $\begin{bmatrix} \Phi & r \end{bmatrix}^T$. The actuator dynamics are modeled by a first order low pass filter with a simple pole at $s = -20$ for both channels. This transfer function of the actuator is given by $G_{act}(s) = \dfrac{20}{s + 20}$.

The closed-loop specifications are the following:

1. Margin specifications have at least 10[dB] gain margin and at least 45° phase margin. This margin specification is represented by equation (11).

$$M = 20\log\left\{\frac{\gamma}{\gamma - 1}\right\} \; [\text{dB}] \quad , \quad \phi = 2\sin^{-1}\left\{\frac{1}{2\gamma}\right\} \; [\text{deg}] \tag{11}$$

2. As a bandwidth specifications, the crossover frequency of the open-loop for Φ command is at least at $\omega = 10$ [rad/s] and the first cross-over frequency of the open-loop for r command is at about $\omega = 2.5$ [rad/s].

3. For a step input disturbance rejection is given like

$$\left|\frac{\Phi}{V_{gust}}\right| \leq 0.15 \; \left.[\text{deg}]\middle/[m/s]\right. \;\; , \;\; \left|\frac{r}{V_{gust}}\right| \leq 0.1 \; \left.[\text{deg}/s]\middle/[m/s]\right. \; \text{for } t > 1 \text{ [sec]}$$

$$\left|\frac{\beta}{V_{gust}}\right| \leq 0.1 \; \left.[\text{deg}]\middle/[m/s]\right. \; \text{for } t > 0.5 \text{ [sec]} \tag{12}$$

4. Noise rejection specification is given

$$\left|\frac{\Phi}{\Phi_{noise}}\right| \leq 0.1 \;, \;\; \left|\frac{\Phi}{r_{noise}}\right| \leq 0.03s \;, \;\; \left|\frac{r}{\Phi_{noise}}\right| \leq 0.02s^{-1} \;, \;\; \left|\frac{r}{r_{noise}}\right| \leq 0.2 \tag{13}$$

When the noise specifications will contradict the other specifications, the lowest specified bandwidth will be chosen as the design goal. And the margin specifications is replaced by

$$\left|\frac{1}{1 + L_i(j\omega)}\right|^{-1} \leq 3 \;\; [\text{dB}]$$

$$M = 20\log\left\{\frac{\gamma}{\gamma - 1}\right\} = 20\log\left\{\frac{1.41}{1.41 - 1}\right\} = 10.73[\text{dB}] \tag{14}$$

$$\phi = 2\sin^{-1}\left\{\frac{1}{2\gamma}\right\} = 2\sin^{-1}\left\{\frac{1}{2\gamma}\right\} = 48.7 \text{ [deg]}$$

where L_1 and L_2 are the open-loop transmissions for roll angle Φ and yaw rate r, respectively.

The design procedure consists of the following two procedures based on the automatic loop-shaping method.

3.2 Controller Design Using the Automatic Loop-Shaping

The trial frequency array ω is chosen as following separate points in the frequency spectrum. And the fitness function in GA for the margin specifications is given by.

$$\omega = [1, 2, 3, 4, 4.1, 4.2, 4.3, 4.4, 4.5, 5, 6, 8, 10, 15, 20, 30, 40, 50, 60] \text{ [rad/s]} \tag{15}$$

$$J = \gamma_S \sum_{i=1}^{l} J_{\text{margin}_i} + \gamma_{BW} J_{\text{min\_BW}} + J_{\text{high\_freq}} \tag{16}$$

where, as a weighting factor, $\gamma_S = 10^6$ and $\gamma_{BW} = 10^4$ are used.

We try to find a lower order controller for the loop transmission L_1 and specify a second order controller which is defined as equation (17) based on the proposed automatic loop-shaping algorithm.

$$g(s) = k \frac{\left(1 + \frac{s}{z_1}\right)\left(1 + \frac{s}{z_2}\right)}{\left(1 + \frac{2\zeta s}{\omega_n} + \frac{s^2}{\omega_n^2}\right)} \tag{17}$$

The designed controller g_1 is

$$g_1(s) = -1.657 \frac{\left(1 + \frac{s}{8.001}\right)\left(1 + \frac{s}{22.758}\right)}{\left(1 + 2 \times \frac{0.500s}{60.191} + \frac{s^2}{60.191^2}\right)}$$

$$= -\frac{32.974s^2 + 1014.261s + 6004.290}{s^2 + 60.193s + 3622.972} \tag{18}$$

Fig. 2 shows the results of the loop-shaping for L_1 to calculate the controller g_1 with automatic loop-shaping using GA. The loop transmission L_1 for the nominal case is calculated from the A, B, C, and D matrices and is satisfied with the margin specifications. The result of the manual loop-shaping shows that its 0[dB] frequency is approximately 10 [rad/s].

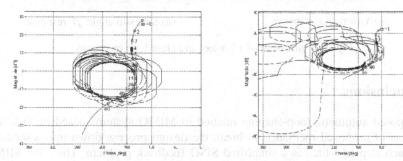

Fig. 2. Nominal $L_1(j\omega)$ and bounds in automatic loop-shaping   **Fig. 3.** Nominal $L_2(j\omega)$ and bounds in automatic loop-shaping

Next, the second procedure is that g_2 is designed to satisfy inequality (8) where $m_i(\omega) = 3$[dB]. The calculated bounds and the nominal second loop L_2 are shown in Fig. 3. The designed controller g_2 is

$$g_2(s) = -0.873 \frac{\left(1 + \frac{s}{2.754}\right)\left(1 + \frac{s}{23.083}\right)}{\left(1 + 2 \times \frac{0.814s}{120.601} + \frac{s^2}{120.601^2}\right)}$$

$$= -\frac{199.749s^2 + 5161.035s + 12700.019}{s^2 + 196.309s + 14544.600} \tag{19}$$

3.3 Time Domain Simulation

The optimization procedures for loop roll angle Φ and loop yaw rate r based on GA allow obtaining controllers satisfying given specifications respectively. A gust input of 1[m/s] is shown in Fig 4-(a). And its results are shown in Fig. 4 from all 128 plant parameter variations. Each response is within the time response specifications.

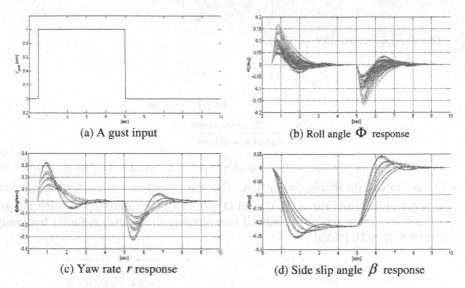

(a) A gust input   (b) Roll angle Φ response

(c) Yaw rate r response   (d) Side slip angle β response

Fig. 4. A gust input of 1 [m/sec] and close loop responses

4   Conclusions

We proposed automatic loop-shaping method in MIMO systems based on GA. The basic idea of proposed method is to break the design process down into a series of steps, each step of which is a simplified SISO feedback problem. Thus, the MIMO feedback problem is transformed into the loop-shaping problems of a nominal transfer function under constraints, named bounds. Each one of the closed loop specifications is transformed into the form of bounds whose intersections are the constraints on the nominal loop transmission to be shaped at each of the sequential design steps. A solution to the original problem is simply a combination of the solutions of all the steps.

This automatic loop-shaping method tested a robust MIMO flight control systems of an UAV and we could verify the effectiveness and performance of the proposed algorithm.

Acknowledgement

This research was supported by the Korea Research Foundation under grant No. 2002-005-D20002.

References

1. Horowitz, I. M., Sidi, M.: Synthesis of Feedback Systems with Large Plant Ignorance for Prescribed Time Domain Tolerance. Int. J. Control 16 (2), (1972) 287-309
2. Horowitz, I. M.: Synthesis of Feedback Systems with Nonlinear Time Uncertain Plants to Satisfy Quantitative Performance Specifications. IEEE Trans. on Automatic Control 64, (1976) 123-130
3. D'Azzo, J. J. and Houpis, C. H.: Linear Control System Analysis and Design: Conventional and Modern. 3rd Edition, McGraw-Hill, New York (1988) 686-742
4. Houpis, C. H and Rasmussen, S. J.: Quantitative Feedback Theory: Fundamentals and Applications, Marcel Dekker Inc. New York, (1999)
5. Yaniv, O.: Quantitative Feedback Design of Linear and Nonlinear Control Systems, Kluwer Academic Publishers Boston (1999)
6. Grimble, M. J.: Industrial Control System Design, John Wiley & Sons, Ltd. New York (2001) 379-418
7. Borghesani, C., Chait, Y., and Yaniv, O.: Quantitative Feedback Theory Toolbox: For Use with Matlab, Math-Works (1994)
8. Goldberg, D. E.: Genetic Algorithms in search, Optimization and machine Learning, Addison Wesley Publishing Company (1989)
9. Wright, A. H.: Genetic Algorithms for Real Parameter Optimization, In Foundations of Genetic Algorithms, J. E. Rawlins (eds.), Morgan Kaufmann (1991) 205-218
10. Horowitz, I. and Gera A.: Optimization of the loop transfer function, Int. J. of Control, Vol. 31 (1980) 389-398
11. Thompson D. F. and Nwokah O. D. I.: Analytic loop shaping methods in quantitative feedback theory, J. of Dynamic Systems, Measurement, and Control, Vol. 116 (1994) 169-177
12. Bryant G. F. and Halikias G. D.: Optimal loop-shaping for systems with large parameter uncertainty via linear programming, Int. J. of control, Vol. 62, no. 3 (1995) 553-568
13. Chait Y.: QFT loop shaping and minimization of the high-frequency gain via convex optimization, Proceedings of the Symposium on Quantitative Feedback Theory and other Frequency Domain methods and Applications (1997) 13-28
14. Kim, M. S.: Automatic Loop-Shaping for QFT Controller Synthesis of MIMO Systems, Ph.D Thesis, Dept. of Electrical Engineering, Soongsil University, Seoul, Korea (2003)

Design and Implementation
of an SSL Component Based on CBD*

Eun-Ae Cho[1], Young-Gab Kim[1], Chang-Joo Moon[2], and Doo-Kwon Baik[1]

[1] Software System Lab. Dept. of Computer Science & Engineering,
Korea University 1, 5-ga, Anam-dong, Seongbuk-gu, Seoul, 136-701, Korea
{eacho,ygkim,baik}@software.korea.ac.kr
[2] Center for Information Security Technologies (CIST),
Korea University 1, 5-ga, Anam-dong, Seongbuk-gu, Seoul, 136-701, Korea
mcjmhj@korea.ac.kr

Abstract. SSL is one of the most popular protocols used on the Internet for secure communications. However SSL protocol has several problems. First, SSL
protocol brings considerable burden to the CPU utilization so that performance
and speed of the security service is lowered during encryption transaction. Second, SSL protocol can be vulnerable for cryptanalysis due to the fixed algorithm being used. Third, it causes a problem of mutual interaction with other
protocols because of the encryption export restriction policy of the U.S. Fourth,
it is difficult for developers to learn and use cryptography API for SSL. To
solve these problems, in this paper, we propose an SSL component based on
CBD. The execution of the SSL component is supported by Confidentiality and
Integrity component. It can encrypt data selectively and use various mechanisms such as SEED and HAS-160. Also, it can complement the SSL protocol's
problems and, at the same time, take advantage of component. Finally, in the
performance analysis, we present a better result than the SSL protocol as the
data size is increased.

1 Introduction

In recent years, SSL (secure socket layer) protocol has been mainly used as a security
protocol for secure communications over the Internet with OpenSSL by Eric A.
Young, JSSE (java secure socket extension) by Sun Microsystems, etc. While SSL is
the most common and widely used protocol between web browsers and web servers[1], it has several problems: First, SSL protocol brings considerable burden to the
CPU utilization so that performance of the security service in encryption transaction
is lowered because it encrypts all data which is transferred between server and client[2][3]. Second, SSL protocol can be vulnerable for cryptanalysis due to the fixed
algorithm being used. So, developer cannot use the other mechanisms such as SEED
and HAS-160. Third, it causes a problem of mutual interaction with other protocols
due to the encryption export restriction policy of the U.S[4]. Finally, it is difficult for
developers to learn and use cryptography API (application program interface) for
SSL. Hence, we need a new method which is different from the existing one to use
SSL protocol more efficiently in design and implementation of applications.

In this paper, we propose an SSL component based on CBD (component based development) in order to solve the problems mentioned above. The component is im-

---

* This work was supported by the Ministry of Information & Communications, Korea, under
the Information Technology Research Center (ITRC) Support Program.

D.-K. Baik (Ed.): AsiaSim 2004, LNAI 3398, pp. 478–486, 2005.

plemented on the application level where it can encrypt and decrypt data. Users can choose various algorithms of encryption. The SSL component provides convenience to the developers who are not accustomed to security concepts. It can also easily provide SSL services when interlocked with other business components because of its component based implementation. The SSL component is reusable and increases the productivity and it decreases the cost[5]. In addition, as component developers need platform-independent methods to support their developments regardless of platform's cryptography APIs, it can support the component platform independently irrespective of the kind of subordinate encryption APIs. As a result, the component makes it easy to interact and interlink between protocols.

In this paper, we propose the requirements for SSL component and design it based on CBD. We have implemented internally SSL handshake protocol and SSL record protocol in order to perform the same functions of the existing SSL implementations. Further, we designed the main security component – Integrity and Confidentiality service component – that supports the part of SSL component. Here, we add standard algorithms used in Korea for cryptography such as SEED[6] and HAS-160[7] and developers using the component can select the algorithm type and encoding/decoding type. In other words, we provide variety cryptography mechanisms of SSL and implementations to encrypt/decrypt data selectively, thus rendering data processing to be more efficient. We used the Rational Rose 2000 as a design tool for the component and sequence diagram[8], and implemented the SSL, Confidentiality and Integrity component with EJB (enterprise java beans)[9]. Lastly, we tested the implementations of each component with the appropriate scenario for SSL component and J2EE (java 2 platform, enterprise edition)[10] server according to our proposed design. We also tested the efficiency against the standard SSL protocol.

The remainder of this paper is organized as follows. Chapter 2 explains the analysis of requirements for SSL component and presents the proposed design of SSL component, which solves the aforementioned problems. Chapter 3 presents the implementation of SSL and the main security component according to the design of Chapter 2, and then provides the test through the proper scenario. Further, performance evaluation results compared with SSL protocol are presented in Chapter 3. Finally, Chapter 4 concludes the paper with a brief discussion on the future work.

2 Design of SSL Component

In this chapter, we propose the requirements for the SSL component derived from the problems of the SSL protocol, and design the SSL component and the main security component (Confidentiality and Integrity service component). Then, we define the main methods of the SSL component, and describe the whole movement and flow of the messages between Confidentiality component, Integrity component, and SSL component using a sequence diagram.

2.1 Requirements for SSL Component

We classified requirements for SSL component into two parts: first are requirements for solving the SSL protocol problems, and second are requirements for the implementation of SSL component based on CBD.

1) Requirements for Solving the SSL Protocol Problems

Our purpose is to provide both convenience and compatibility and at the same time, the performance should be improved without affecting the security issue.

– By encrypting data selectively, the amount of data processed by CPU should be reduced without affecting the security problem.
– The SSL component should be standardized in order not to depend on a specific mechanism, but it should be able to provide proper security mechanism according to the encryption level.
– The security requirements such as confidentiality and integrity for SSL component platform should be supported apart from the subordinate cryptography API in a system.
– Programming which is related to security should be required at a minimum in the code of application component, which exist or will be developed.

2) Requirements for Implementation of SSL Component Based on CBD

To implement the SSL component, which supports confidentiality and integrity in J2EE platform environment, the following requirements are needed:

– It should be ensured that the reusability of SSL component using standardized security component interface(IConfidentiality, IIntegrity) in business components.
– An application component developer should be able to detect the subordinate security component within the component code and set the desired security component by the algorithm name.
– When message protocol (e.g. handshake, record protocol, etc.) is implemented, it has to support the order of the messages which is defined in SSL standard.
– In this paper, SSL component focused on implementation of confidentiality and integrity function by using anonymous Diffie-Hellman algorithm for key exchange in the server's handshake protocol.
– It should create the MAC and perform encryption/decryption using the confidentiality/integrity component implemented with standardized security component interface in the record protocol.
– It should maintain and/or manage the session state parameter and the connection state parameter between client and server environment, where it has an end-to-end connection in SSL component. Thus, SSL component should be implemented with EJB code in the form of the session bean that its inner states consist of a session state and a connection state.
– It has to support Korea standard cryptograph algorithm such as SEED and HAS-160.

2.2 Design of SSL Component

To satisfy the function mentioned in the requirements above, we designed and implemented the component based on EJB. The whole SSL component based on CBD is composed of main security components such as confidentiality and integrity component as shown in Fig. 1.

Table 1 shows the major interface of SSL component based on the requirements. According to SSL component interface, we designed EJB Component, which is called SSLComponent to perform the same function as SSL protocol.

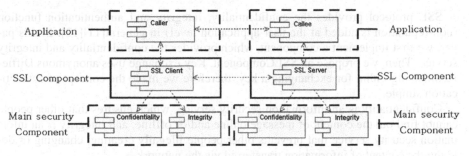

Fig. 1. Component Diagram of SSL Component

Table 1. Description of the main method

Component Name	Main Method	
	Name	**Description**
SSL Component	startHandshake()	To exchange the session key and algorithm, server and client start handshake protocol.
	clientHello()	To start handshake protocol, client calls server.
	Request-KeyExchange()	To exchange the key, client transfers its public key to server.
	getPubKey()	To create the public key for the other part.
	encryptSSLMessage()	After handshake, both server and client encrypt the real data to SSL message using the set key and algorithm.
	decryptSSLMessage()	After handshake, both server and client decrypt SSL message to the real data using the set key and algorithm.
	finishHandshake()	Both server and client send the current cipher spec, and set up the security connection.
	decisionAlgorithm()	When helloMessage is sent, server decides the algorithm of confidentiality and integrity component for current session.
	createMasterSecret()	Both server and client create the master key to share the algorithm and key for SSL.
	preMaster()	Both server and client create the pre-master key using their own private key and the other part's public key which is came from requestKeyExchange().
	setAlgorithm()	To set a proper algorithm.
Integrity Component and Confidentiality Component	getAvailable-Algorithms()	To get all available algorithm presently.
	getAlgorithm()	To get the algorithm that is currently set.
	setNameType()	To set the type of the algorithm name either OID(object identifiers) or Name.
	getNameType()	To get the type of the current algorithm name.
	initialize()	To receive the key value and initiation value by the parameter and then performs initialization.
	update()	To receive the message made for byte array and then encrypts/decrypts the data as much as inner buffer allows.
	Finalize()	To pass the result for all message encryption including the remainder of message in fixed buffer.
	getLength()	To get the length of the encrypted message.
	setEncodingType()	To set the encoding type either Raw or Base64.
	getEncodingType()	To get the current encoding type.

SSL protocol provides the confidentiality, integrity and authentication function (non-reputation is added at the user application level) in general[11][12]. In this paper, we first implement a component, which provides the confidentiality and integrity service. Then, we propose an SSLComponent. Key exchange uses anonymous Diffie-Hellman algorithm for exchanging a key, therefore we made the concept of authentication simple.

Confidentiality is an information security service, which ensures that other people cannot find out the content of message online and/or offline, and integrity is an information security service that prevents other people illegally creating, changing or deleting the content of information transferred via the network.

Fig. 2 shows the actions between SSLComponent and confidentiality component in the sequence diagram. It shows diagram that a message is encrypted or decrypted for enabling confidentiality. In Fig. 2, on the left is the encryption process and on the right is the decryption process for message. The kind and order of the message received and sent is the same in the confidentiality component. First, SSLComponent creates the home interface of confidentiality component, and checks available algorithms, and then exchange current algorithm and finally selects the algorithm to be used. Type of algorithm is selected can be chosen either as by the algorithm name or algorithm OID chosen by the developer. After the algorithm selection, we can also choose the encoding type either as Raw or Base64. After these processes, SSLComponent sends the encrypted message through the process such as *initialize*, *update* and *finalize* process.

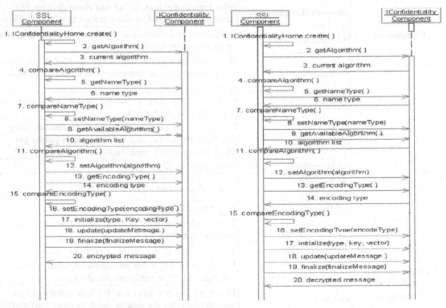

Fig. 2. Sequence diagram of confidentiality component

Fig. 3 presents the message process between SSLComponent and Integrity component with the sequence diagram. It acts on the same message and order as confidentiality component.

We designed the EJB component called SSLComponent that behaves according to the SSL protocol and according to the definition of SSLComponent interface. We implemented the SSL record protocol and SSL handshake protocol as a part of the SSLComponent, which provides the confidentiality and integrity service.

Fig. 4 shows the whole message flow according to the role of each SSLComponent by the sequence diagram. We can classify the roles into SSLServer and SSLClient.

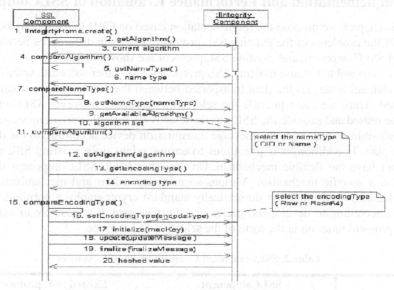

Fig. 3. Sequence diagram of integrity component

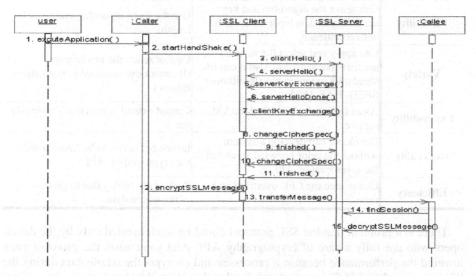

Fig. 4. Sequence Diagram of SSLComponent

Through the steps 2 to 11, SSLClient and SSLServer initialize the logical connection and execute the handshake protocol. The steps 5 to 7 show the asymmetric key exchange between SSLClient and the SSLServer. We omitted the key exchange method because anonymous Diffie-Hellman was used. The steps 9 to 11 are the process to complete the security connection. The steps 12 to 15 are SSL record protocol.

3   Implementation and Performance Evaluation of SSLComponent

In this chapter, we propose the implementation based on CBD to overcome the limitation of the problems of the existing SSL protocol. The main differences between proposed SSLComponent and existing SSL protocol are shown in Table 2.

It is impossible to reuse existing SSL protocol within other software, since once the SSL channel is set, all the data transported between the server and the client, is encrypted. Thus, we can't provide the selective encryption. However SSLComponent can be reused and provide the SSL service selectively for only certain messages. Thus it is possible to customize the message transmission device according to the development plan. In addition, it is precarious to ensure safety since existing SSL protocol doesn't have the flexible mechanism. On the other hand, SSLComponent does not rely on a specific mechanism. Various security algorithms and mechanisms can be selected and used including domestically standard cryptography algorithms used in Korea according to developers' preference. Furthermore, it is possible to extend the SSL protocol function in the form of the selective SSL service.

Table 2. SSLComponent Compared with SSL protocol

	SSLComponent	Existed SSL protocol
Reusability	Can be reused.	Cannot be reused.
Flexibility	Can select the algorithm and key according to developer's intention with confidence.	Use the fixed algorithm and fixed key length.
Variety	Can apply and select the various security mechanisms.(Also it can use standard Korean crypto algorithms-SEED, HAS-160.)	Cannot select the mechanism. (It cannot use standard Korean algorithms.)
Extensibility	As an option, it can extend the SSL service.	Cannot extend connection previously set.
Generality	Developer can develop a system without the prior knowledge related to the cryptography API.	Developer needs to be familiar with the cryptography API.
Efficiency	Can reduce the CPU overhead and time.	Reduce efficiency due to processing of the whole data.

There is a problem that the SSL protocol could be implemented only by the developers who are fully aware of cryptography API. And sometimes the protocol even lowered the performance because it processes and encrypts the whole data during the SSL process. But SSLComponent can be developed by the developers who are not even aware of API because the component uses a common and standard form. In addition, using the proposed component, CPU utilization overhead can be reduced as well as the time overhead, because the developer can decide the API and the algorithm selectively.

SSLComponent were built on the test server. Test server has the following configuration: Xeon 2.2GHz Processor, 512KB L2 ECC Cache, 512MB ECC SDRAM Memory, 36.4GB Pluggable Ultra320 SCSI 10k Universal Disk and Compaq 7760 10/100/1000 TX PCI NIC LAN Card.

We have chosen a scenario of online shopping (buying goods) to compare the proposed SSLComponent efficiency with existing SSL protocol. The criterion for comparison is the process time, which is needed for safe transmission of the data. For performance evaluation, we tested both SSLComponent and existing SSL protocol on the same server hardware.

Although same server hardware was used, computing environments were different for the two protocols being compared. In other words, SSL protocol is mainly used as https in web server like Apache, while SSLComponent is executed in an application server like J2EE based on Java. The Java environment may have the disadvantage in the time factor due to virtual machine it is running on. Thus we had to consider those differences for the clear and correct performance evaluation between both. We have tested SSL protocol in Apache server environment and SSLComponent in J2EE server environment for performance measurement. Fig. 5 shows the comparison result between the pure Apache Tomcat 5.0 and the pure J2SDKEE1.3.1 server. For equivalent application of execution environment, we take their difference into consideration and perform our tests.

The difference value is calculated by subtracting Apache web server process value from J2SDKEE1.3.1 server. The data is represented by function (1)

$$y=159.62\ln(x)-766.3 \tag{1}$$

y is the value of subtracting Apache web server from J2SDKEE1.3.1 server and x is the value of data size.

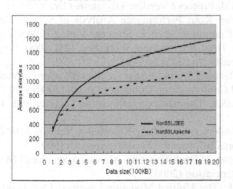

Fig. 5. J2EE Compared with Apache

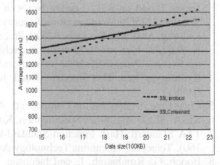

Fig. 6. SSLComponent Compared with Https

Fig. 6 shows the comparison between the execution result of SSLComponent in J2EE server and the execution result of https protocol using SSL protocol in Apache server. As shown in the graph, server's process time is proportioned to the data volume. When data volume is small, data processing time is a little bit longer than component performance time, but the more data volume is increased, the more the time gap is narrowed. When data volume is very big, the processing time of SSLComponent is shorter than https. On the other hand, the processing time for a small data volume is not significantly improved. However, SSLComponent does not have any worse performance efficiency than the SSL protocol, and the processing time for large data is better than SSL protocol.

4 Conclusion and Future Works

In this paper, we proposed and implemented the model to provide the SSL service through the system based on CBD. SSLComponent model based on CBD extends the existing SSL protocol through the use of partial message encryption and Korean domestic standard cryptography algorithms which wasn't provided in the existing SSL protocol. We also overcame the limitation of SSL protocol(e.g. it cannot be reused, has atomic property, cannot use Korean domestic standard algorithm, cannot extend, and etc.) through the software development which has the component concept. The SSLComponent can be reused, encrypted selectively, applied for Korean domestic standard algorithm and it can be extended.

As showed in the chapter 3, when the size of data becomes small we can reduce the need for CPU resources by using the remote connection of a component. On the other hand, when the data volume size is increased, proposed SSLComponent becomes more efficient, since it can encrypt the selected data based on CBD. In this paper, we used anonymous Diffie-Hellman algorithm, which is quite simple. As for future work, we need to propose and design the complex authentication part more clearly. Further, we need to present and implement the standard for the component that would provide other security services such as non- reputation and availability.

References

1. A. Freier, P Karlton, and P. Kocher: The SSL Protocol Version 3.0, Internet Draft (1996)
2. Xiaodong Lin, Johnny W. Wong, Weidong Kou: Performance Analysis of Secure Web Server Based on SSL. Lecture Notes in Computer Science, Springer-Verlag Heidelberg, Volume 1975/2000, Information Security: Third International Workshop, ISW 2000, Wollongong, Australia, December 2000. Proceedings (2003) 249-261
3. K. Kant, R. Iyer and P. Mohapatra: Architectural Impact of Secure Socket Layer on Internet Servers. Proc. IEEE 2000 International Conference on Computer Design (2000) 7-14
4. Kyoung-gu, Lee: TLS Standard Trend. KISA, The news of Information Security, vol. 19 (1999)
5. Chris Frye: Understanding Components. Andersen Consulting Knowledge Xchange (1998)
6. KISA: SEED Algorithm Specification. Korea Information Security Agency (1999)
7. TTA Standard: Hash Function Standard-Part 2: Hash Function Algorithm Standard(HAS-160). Telecommunications Technology Association (2000)
8. Booch, G., Rumbaugh, J., and Jacobson, I.: The Unified Modeling Language User Guide. Addison Wesley Longman (1999)
9. Enterprise Java Beans Specification Version 2.0 Final Release. Sun Microsystems Inc (2001)
10. Sun, Java 2 Platform Enterprise Edition Specification, Version 1.4, Sun Microsystems Inc (2004)
11. William Stallings: Cryptography and Network Security. Principles and Practice, 3rd edn, Prentice Hall (2002)
12. R. W. Badlwin et C. V. Chang: Locking the e-safe. IEEE Spectrum (1997)

A Scalable, Ordered Scenario-Based
Network Security Simulator

Joo Beom Yun[1], Eung Ki Park[1], Eul Gyu Im[1], and Hoh Peter In[2]

[1] National Security Research Institute,
62-1 Hwa-am-dong, Yu-seong-gu,
Daejeon, 305-718, Republic of Korea
{netair,ekpark,imeg}@etri.re.kr
[2] Dept. of Computer Science and Engineering,
Korea University, Seoul, 136-701,
Republic of Korea
hoh_in@korea.ac.kr

Abstract. A network security simulator becomes more useful for the study on the cyber incidents and their defense mechanisms, as cyber terrors have been increasingly popular. Until now, network security simulations aim at damage estimation of incidents in small-size networks or performance analysis of information protection systems. However, a simulator is needed to handle large-size networks since attacks in these days are based on large-size global networks such as the Internet. The existing simulators have limitations for simulating large-scale Internet attacks. In this paper, a scalable, ordered scenario-based network security simulator is proposed. Our proposed simulator is implemented by expanding the SSFNet program to client-server architectures. A network security scenario is applied to test effectiveness of our proposed simulator.

1 Introduction

Cyber attacks are increasingly taking on a grander scale with the advance of attack techniques and attack tools, such as Nessus Scanner [1], Internet Security Systems(ISS) Internet Scanner [2], COPS security checker [3], N-Stealth [4], and so on. Attacks may be launched from a wide base against a large number of targets with the intention of acquiring resources as well as disrupting services [5] of a target host. Research on cyber attacks and their effects is fundamental to defend computer systems against cyber attacks in real world [6–8].

Network security simulation is widely used to understand cyber attacks and defense mechanisms, their impact analysis, and traffic patterns because it is difficult to study them by implementing such dangerous and large-scale attacks in the real environments. Network security simulation is begun from IAS (Internet Attack Simulator) proposed by Mostow, et al. [9]. Several scenarios are simulated using IAS. The simulations on password sniffing attacks and effects of firewall systems are reported in [10]. Smith and Bhattac Harya [11] used simulation techniques to analyze the overall performance of a small-size network by

D.-K. Baik (Ed.): AsiaSim 2004, LNAI 3398, pp. 487–494, 2005.

changing the position of a firewall system. Simulation is also used to estimate traffics by changing the network topology [12, 13]. However, these simulators have limitations to represent the real-world network environments and configuration, and realistic security scenarios. For example, the existing simulators are not scalable enough to represent realistic attack and defense mechanisms in a large-scale network.

This paper proposes a scalable, ordered scenario-based network security simulator to overcome the limitations of existing network security simulators. Our proposed network security simulator was developed based on the following requirements. First, a real-world network security problems needs to be modeled and simulated in a scalable way. For example, the simulator needs to simulate cyber attack and defense mechanisms in a very large Internet-based, global network environment. Second, network security simulation components can be reusable. Once network security simulation components are defined, they can be reusable for other scenarios with minimum reconfiguration or parameter modification. Third, the simulator can simulate various cyber intrusion and defense scenarios. Thus, the simulator needs to be general enough to apply for various scenarios. Fourth, the simulator needs to have easy-to-use features. For example, it must provide a graphical user interface to edit cyber intrusion scenarios and to observe results clearly.

This paper is presented as follows: The context of the work is presented in Section 2. The overview of our proposed network security simulator is described in Section 3. A case study with Slammer worm is discussed in Section 4 to test effectiveness of the proposed tool, followed by conclusions in Section 5.

2   The Context: An Object-Oriented Network Security Simulator

An object-oriented network security simulation model was proposed by Donald Welch et al. [10]. The following object classes were presented with their attributes and behaviors:

Node Address: *Node Address* is an object which is used to identify a network node such as a workstation, a server, and a router. It has attributes such as processor speed, memory capacity, and provided services. Its behaviors represent actions performed in the node.

Connection: *Connection* represents a physical link among the nodes. Examples of its attributes are a connection speed, connection details, and reliability of the connection.

Interaction: *Interaction* shows an exchange of information among nodes. "Interaction" in [10] represents only an important event instead of showing packet-level information.

Infotron: *Infotron* is the smallest piece of interesting information. This can be a whole set of databases or a particular database

An example of the object-oriented network security simulation model with a scenario of password sniffing is described in Figure 1. The object-oriented simulation model has several benefits such as easy-to-model, inheritance, information hiding, and reusability. For example, the object classes used to design the scenario can be inherited to driven classes or reused for other scenarios.

However, it is difficult to model a sequence of events with this approach because cyber attack scenarios are represented into actors and actions as shown in the left box in Figure 1. In addition, the "interaction" in the right box is not intuitively understood and the picture does not represent its behaviors in detail. For example, "Sniff" behavior in the picture does not have detailed information. It is needed to instruct a program to execute sniffing, sniffing network packet, and acquiring credit card information.

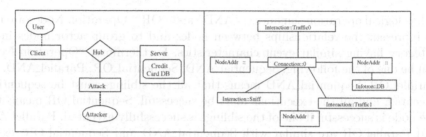

Fig. 1. Donald Welch, et al.'s simulation.

Thus, we propose a new network security simulation model and a tool to overcome these limitations. The new model and the tool are described in the next section.

3   Our Proposed Work: A Scalable, Ordered Scenario-Based Network Security Simulator

Our proposed network security simulations are based on a modified version of "Attack Tree" [14] to describe scalable, ordered scenarios. "Attack Tree" consists of a root node and subnodes. The root node represents the final objective of a certain attack, and subnodes are preceding nodes to reach the root node. Attack Tree is helpful to understand the hierarchical structure of subnodes. However, it is difficult to understand the sequence of each subnode. It is needed to present the sequence of subnodes when a cyber attack scenario is simulated. Our proposed network security simulator adapts the attack tree by adding numbers which represents execution orders. An example of the modified attack tree is presented in Figure 2.

Since a sequence number is assigned to each node, it is easy to understand the order of simulation execution and to map each step into a position of the attack tree. Two-type nodes are defined in the modified attack tree: "Actor node" and "Operation node". "Actor Node" is a terminal node of the tree and it executes events. A node name is described as an event name. "Operation Node" has one or more subnodes, and determines the order of subnode execution.

```
I Sniffing Scenario (Scenario)
  1.1 Sniffing Start(SequentialAND)
      1.1.1 Connecting Attacker PC (Actor)
      1.1.2 Start Sniffing Program (Actor)
  1.2 Sniffing Network Packet (SequentialAND)
      1.2.1 Sniff all packets passing through a Hub (Actor)
  1.3 Acquiring a credit card information (SequentialAND)
      1.3.1 Listen User sfConnection (Actor)
      1.3.2 Sniff a credit card information (Actor)
```

Fig. 2. An Example of Our Proposed Simulator: Credit Card Information sniffing scenario.

It has logical operation semantics: "AND" and "OR". Operation Nodes are used to represent the relationships between nodes and to group actor nodes into a category having similar event characteristics. Furthermore, "Operation Node" can be one of the following: Sequential_AND, Sequential_OR, Parallel_AND, and Parallel_OR. Sequential_AND means that all the siblings must be sequentially executed, and all the executions must be successful. Sequential_OR means that the node is successful if one of the siblings is successfully executed. Parallel_AND and Parallel_OR are similar with Sequential_AND and Sequential_OR, except that the subnodes are executed in parallel.

The overall architecture of our proposed network security simulator is shown in Figure 3. It has client-server architecture to improve simulation performance and to execute processes in parallel or distributed fashions. The simulation client is a Graphic User Interface (GUI) program that shows scenarios and their results. The simulation server has a simulation engine, DML [15] Generator, and Daemon.

The snapshot of our proposed network security simulator is shown in Figure 4. Network topologies and network configurations are edited and displayed using GUI. A simulation scenario of sniffing credit card information (upper right) is executed on the network. The "red line" represents the important credit information communicated between nodes. Our proposed simulator can show the

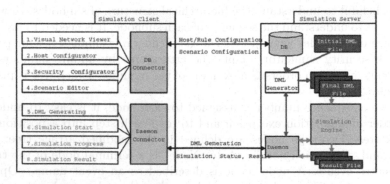

Fig. 3. The Architecture of Our Proposed Network Security Simulator.

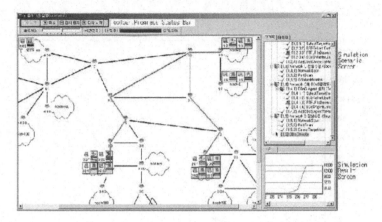

Fig. 4. Simulation Execution Screen.

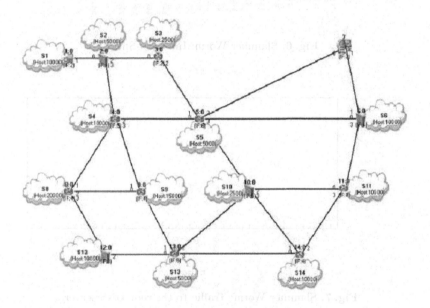

Fig. 5. Network Environment.

simulation process graphically and can express more than 100,000 nodes. The simulation results are shown in a graph (below right). The objects shown in Figure 4 were implemented by expanding the classes of the SSFNet simulator.

4   A Case Study: Slammer Worm

In this section, simulation results of the Slammer worm propagation is reported to show effectiveness of our proposed simulator. Network environments used in the simulation are shown in Figure 5. A network model of our simulation

Infection Speed

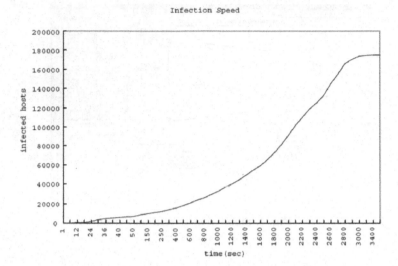

Fig. 6. Slammer Worm: Infection Speed.

Traffic

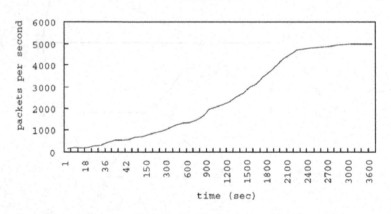

Fig. 7. Slammer Worm: Traffic to the root DNS server.

is a huge network consisting of 14 sub-networks and 175,000 nodes. All the hosts have a MS-SQL server running, so they are vulnerable to the Slammer worm. Simulation is started with 100 infected hosts. Simulation time is 1 hour. Slammer worm randomly queries a "name resolution" to a root DNS server in our simulation. This causes the DNS server down. We have observed the number of infected hosts and packets per second to the root DNS server.

The Slammer worm is bandwidth-limited. Its propagation speed is roughly linear with the bandwidth directly available to the already infected hosts. When a high number of hosts are infected, there can also be additional limitations because of ISP and backbone bandwidth limits [16]. We assume a vulnerable population of 175,000 hosts. Figure 6 shows how many hosts have been infected by the slammer worm. The initially infected population was 100 hosts distributed over the different speeds according to sub-network size. The simulation deviates slightly from the observed propagation speed of the Slammer in [17], because of simulation parameters. The initial doubling time is about 7 seconds and the doubling time after a few minutes is lengthened. The simulation then reaches an infection level of 90% after about 2600 seconds. Figure 7 shows traffics to the DNS server. It is counted on the root DNS server in the sub-network 7. We limited input packets of the root DNS server to 5,000. Traffic plots are similar to infection speed plots. However, traffic plots have reached quickly to a saturation point rather than infection speed plots. This means that the root DNS server is down before the full infection. This phenomenon is similar to what happened in Korea on January 25 2003.

To demonstrate the feasibilies of our proposed simulator, we have shown simulation results. The infection speed and traffic is similar to observed data in [17]. Our simulator has high scalability because it can model more than 100,000 nodes. Also it can represent ordered simulation scenarios like Figure 2.

5   Conclusions

A scalable, ordered scenario-based network security simulator is proposed. The simulator is implemented based on SSF(Scalable Simulation Framework) and SSFNet [18]. The benefits of the simulator are: 1) it enables users to simulate scenarios that require sequences of activities; 2) scalable simulations are possible because it can handle up to 100,000 nodes; 3) it has easy-to-use GUI to model the latest cyber terror scenarios such as DDoS and worm propagation; and 4) it allows users to reuse previously developed object classes to create new scenarios. We have shown the simulation results of the slammer worm. Also, the simulation results were compared with those of observed data [17].

References

1. Deraison, R.: Nessus Scanner, http://www.nessus.org/.
2. Internet Security Systems Internet Scanner. http://www.iss.net/.
3. Farmer, D., Spafford, E.: The cops security checker system. In: Proceedings of the Summer Usenix Conference. (1990)
4. N-Stealth: Vulnerability-Assessment Product.
   http://www.nstalker.com/nstealth/.
5. Householder, A., Houle, K., Dougherty, C.: Computer attack trends challenge inernet security. In: Proceedings of the IEEE Symposium on Security and Privacy. (2002)
6. McDermott, J.P.: Attack net penetration testing. In: Proceedings of the 2000 workshop on New security paradigms, ACM Press (2000) 15–21
7. Steffan, J., Schumacher, M.: Collaborative attack modeling. In: Proceedings of the 17th symposium on Proceedings of the 2002 ACM symposium on applied computing, ACM Press (2002) 253–259

8. Lee, C.W., Im, E.G., Chang, B.H., Kim, D.K.: Hierarchical state transition graph for internet attack scenarios. In: Proceedings of the International Conference on Information Networking 2003. (2003)
9. John R. Mostow and John D. Roberts and John Bott: Integration of an Internet Attack Simulator in an HLA Environment. In: Proceedings of the 2001 IEEE Workshop on Information Assurance and Security, West Point, NY (2001)
10. Donald Welch, Greg Conti, Jack Marin: A framework for an information warfare simulation. In: Proceedings of the 2001 IEEE Workshop on Information Assurance and Security, West Point, NY (2001)
11. Smith, R., Harya, B.: Firewall placement in a large network topology. In: Proceedings of the 6th IEEE Workshop on Future Trends of Distributed Computing Systems. (1997)
12. Breslau, L., Estrin, D., Fall, K., Floyd, S., Heidemann, J., Helmy, A., Huang, P., McCanne, S., Varadhan, K., Xu, Y., Yu, H.: Advances in network simulation. IEEE Computer **33** (2000) 59–67 Expanded version available as USC TR 99-702b at http://www.isi.edu/~johnh/PAPERS/Bajaj99a.html.
13. Technology, O.: Opnet modeler (Mar 2001)
14. Schneier, B.: Attack tree secrets and lies. In: John Wiley and Sons. (2000) 318–333
15. SSF Research Network http://www.ssfnet.org/SSFdocs/dmlReference.html: Domain Modeling Language Reference Manual.
16. Arno Wagner, Thomas Dubendorfer, B.P.R.H.: Experiences with worm propagation simulations. ACM (2003) 33–41
17. Moore, D., Paxson, V., Savage, S., Shannon, C., Staniford, S., Weaver, N.: Inside the slammer worm. IEEE Security & Privacy **1** (2003) 33–39
18. SSF Research Network http://www.ssfnet.org/homePage.html: Scalable Simulation Framework.

A User Authentication Model
for the OSGi Service Platform

Chang-Joo Moon[1] and Dae-Ha Park[2,*]

[1] Center for Information Security Technologies (CIST), Korea University,
Anam-dong, Sungbuk-gu, Seoul, Korea
mcjmhj@korea.ac.kr
[2] Korea Digital University, 18F, 139, Chungjeongno 3-ga, Seodaemun-gu, Seoul, Korea
summer69@kdu.edu

Abstract. In this paper, we propose a new user authentication model for the
OSGi service platform environment. The authentication model is extended
from Kerberos. Since a user can use an authentication ticket repeatedly, the
concept of SSO (Single-Sign on) is supported. With a login service for all bundle services in a gateway, the convenience that service providers do not need to
develop proprietary authentication modules can be provided. In addition, using
symmetric-key cryptosystem helps gateways to use their limited resources efficiently during user authentication tasks.

1 Introduction

The OSGi (*Open Services Gateway Initiative*) is an organizational group with the
objective of developing open specifications to deliver multiple services over wide
area network (WAN) to local area networks (LANs) and devices, and accelerating the
demand for products and services based on those specifications in the embedded
markets such as home appliances. The OSGi services platform specification (*OSGi
specification*, briefly) [11] describes a service framework and defines reference architectures and a set of standard service interfaces to support for developing OSGi-
compliant service gateways (*gateway*, briefly). The gateway operator (*operator*,
briefly) plays important roles in remote management of gateways. The service users
can use various services installed into the gateway by connecting with wired or wireless client devices.

Development of a user authentication model for the OSGi environment is essential
to provide authentication for service users to access gateways and service bundles to
be installed in gateways. To develop appropriate user authentication model, we
should consider some constraints that are different from general open distributed
component service environments (e.g., CORBA, EJB, and DCOM). The constraints
include mobility of service bundles, efficiency of gateway recourses, diversity of
multiple services, and usability of users and operators.

Current OSGi specification defines 'User Admin Service' as a standard service for
performing user authentication. To use a service in a gateway, a user should request
authentication to the service installed in framework. Then, using an authentication

---

* Dae-Ha Park is the corresponding author. This work was supported by the Ministry of Information & Communications, Korea, under the Information Technology Research Center
(ITRC) Support Program

D.-K. Baik (Ed.): AsiaSim 2004, LNAI 3398, pp. 495–504, 2005.

server, the service performs authentication based on authentication information received form the user. The authentication information may be from various ways such as identifier and password, OTP (one-time password), biometrics or public-key certificates. However, whenever a user wants to access other services, he or she should repeatedly request authentication for each services. Therefore, the main disadvantage of current OSGi user authentication is that users should always involve in generating new authentication information whenever they wish to use several services in a gateway or a specific service in several gateways.

We insist that the authentication model for users who wish to use gateway services with remote connection from external network should satisfy the following three requirements.

* **Requirement 1:** The convenience for service users should be considered. A user may use a specific service located in multiple gateways as well as multiple services in a gateway. Therefore, we need SSO (single-sign on) functionality that supports to use multiple services through only one authentication with single authentication information (e.g., password).
* **Requirement 2:** The mobility of components (i.e., service bundles) dynamically installed in a gateway should be considered. Therefore, there need a common user authentication mechanism for all services in a gateway rather than an authentication mechanism for each bundle.
* **Requirement 3:** The constraints for limited gateway resources (e.g., slow CPUs, insufficient memory or network latency) should be considered. Therefore, it is better to use symmetric-key cryptosystem than to use public-key cryptosystem if possible. In addition, we need a way to reduce user authentication information that is stored and managed in gateway memory as small as possible.

In open distributed service environments, there exist various user authentication protocols for supporting to connect services provided by servers. According to [8], we can classify the protocols into four types; using symmetric-key cryptosystem without TTP (e.g., ISO/IEC one-pass symmetric-key unilateral authentication protocol [7]), using symmetric-key cryptosystem with TTP (e.g., Neumann-Stubblebine protocol [4] and Kerberos protocol [1]), using public-key certificate in PKI environment (e.g., X.509 certification protocol [2]), and using both symmetric-key and public-key cryptosystems (e.g., EKE protocol [14]).

Among these protocols, we apply Kerberos to the user authentication model with some modification suitable to the OSGi environment. Since the protocol supports repeated authentication through reuse of authentication ticket, it is adequate to implement SSO functionality satisfying the first requirement. On the other hand, since with a login service in a gateway each service of bundles does not need to implement detailed login tasks, we can satisfy the component mobility in the second requirement. Since we use symmetric-key cryptosystem to generate authentication tickets and to provide message confidentiality, the effect of system resource constraints can be reduced comparing to use public-key cryptosystem and can satisfy the third requirement.

The rest of this paper consists as follows. In section 2, we describe the user authentication model with some entities and software components for the OSGi environment. Section 3 shows the prototype implementation and the result of performance test. Finally, section 4 mentions conclusion remarks.

2 User Authentication Model for the OSGi Environment

2.1 OSGi Entities for User Authentication

Figure 1 shows entities of the user authentication model for the OSGi service platform environment.

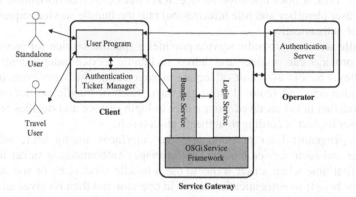

Fig. 1. Entities of user authentication model

The OSGi entities participating in the user authentication model are as follows.

* **Operator:** In the OSGi environment, an operator usually manages multiple gateways. The operator can register identifiers for users of each gateway and can manage a user authentication server.
* **Gateway:** Each gateway installs all the bundles that provide services to its users. A gateway has a login service that provides user authentication for its bundle services.
* **Client:** A client, which is a terminal connected to gateways and operators, provides interfaces to a user and manages authentication tickets including user authentication information.

The users are classified into standalone users and travel users according to the number of gateways which they can access. The standalone users can connect only one gateway and use several services whereas travel users can connect several gateways and use certain services. For example, a standalone user connect to one gateway through network, and then turns on and off the electric lamp using the bundle service for electric lamp control. A travel user goes round each gateway inspecting an electric meter using the bundle services for checking electricity usage.

2.2 Software Components

We invent software components that play as actors for modeling entities. These are an authentication server, a login service, bundle services, and user programs.

* **Authentication server:** This plays the same roles as an authentication server in Kerberos authentication service. It performs authentication task according to a user authentication request and issues an authentication ticket as the result. The authentication ticket plays similar roles to TGT (ticket granting ticket) in Kerbe-

ros, but contains user role information, which is not in TGT. The role information in authentication ticket represents the scope of user permissions [12] that is useful to decide whether the user can access services or system resources in a gateway.

♦ **Login service:** This is similar to TGS (ticket granting server) in Kerberos in that it confirms the authentication ticket issued by authentication server. However, unlike TGS, it does not issue service ticket but deliver authorization information (i.e., user identifier and role information) into the bundle service requested as a result of authentication.

♦ **Bundle service:** A bundle service provides a specific service for users. Each service provider can produce and deliver its bundle service independently. Operator handles a bundle as a unit of service management, which performs dynamic installation into and removal from gateways. Each of bundle services delivers authentication ticket received from a user to login service and decides authorization for user request according to authentication result.

♦ **User program:** User program provides interfaces among user, authentication server and each service bundles, and manages authentication ticket for the user. The first time when a user wants to use a bundle service, he or she authenticates him or herself to authentication server in operator and then receives an authentication ticket as a result of successful authentication. The user stores and maintains the ticket. The user delivers the ticket to the bundle service that passes it to login service to decide whether he or she can use the service. When the user wishes to use other bundle services in the same gateway, he or she can reuse the ticket without repeated issuing of new authentication tickets from authentication server.

2.3 User Authentication Model

Figure 2 shows the entire process of the proposed authentication model that contains the formalized messages of user authentication model.

<Notation>
ID_U	Identifier of a user U
ID_{BS}	Identifier of a bundle service BS
ID_{SG}	Identifier of a service gateway SG
$K_{a,b}$	A shared secret key for A and B
$\{M\}K_{a,b}$	Encrypted message M with Ka,b
N_A	Random nonce generated by A
$TS_\#$	Timestamp
$Ticket_{auth}$	Authentication ticket
$Auth$	Authorization request
Subject	IDs and role information for authorized users
RS_U	A set of roles for the authenticated user U
$\|$	Concatenation

<Protocol phases>
① AS_REQ : $ID_U \| ID_{SG} \| N_U$
② AS_REP : $\{K_{u,sg} \| ID_{SG} \| N_U \| TS_1\}K_{u,op} \| Ticket_{auth}$
③ BS_REQ : $ID_U \| ID_{BS} \| Ticket_{auth} \| Auth$
④ LS_REP : $\{ID_U \| TS_2\}K_{u,sg}$, Subject
⑤ BS_REP : $\{ID_U \| TS_2\}K_{u,sg}$

<Authentication ticket and authorization request>

$$Ticket_{auth} = \{K_{u,sg} \parallel ID_U \parallel RS_U \parallel TS_1\}K_{op,sg}$$

$$Auth = \{ID_U \parallel ID_{BS} \parallel TS_2\}K_{u,sg}$$

<Protocol scenario>

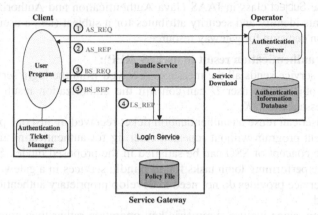

Fig. 2. User authentication model

The user authentication model consists of five phases. The details of each phase are as follows.

① Requesting user authentication (AS_REQ):

Before using a specific bundle service, a user should authenticate him or herself to user authentication server located in operator. The user U sends his or her identifier ID_U, the gateway's identifier ID_{SG}, and fresh random nonce N_U to the authentication server AS. We assume that user identifier and password are already registered in operator and can be accessible from authentication server.

② Issuing authentication ticket (AS_REP):

An authentication server in OP prepares to send a session key $K_{u,sg}$ and an authentication ticket $Ticket_{auth}$ to U. The session key $K_{u,sg}$ is encrypted with a secret key $K_{u,op}$ derived from user password registered within the operator. The authentication ticket $Ticket_{auth}$ is encrypted with another secret key $K_{op,sg}$ shared between a login service in SG and the authentication server in OP. $K_{op,sg}$ can be shared by using key sharing mechanism during the gateway's bootstrapping.

③ Requesting service authorization (BS_REQ):

U sends $Ticket_{auth}$ to a specific bundle services BS that it wishes to use. Being encrypted the authentication ticket can assure confidentiality and integrity on the way of network transmission. In addition, the user sends encrypted authorization request $Auth$ with a session key $K_{u,sg}$. The received bundle service relays $Auth$ to the gateway's login service. Since both $Ticket_{auth}$ and $Auth$ are encrypted with $K_{op,sg}$ and $K_{u,sg}$ respectively, untrusted bundle services cannot modify them.

④ Sending authorization result to bundle service (LS_REP):

The login service decrypts $Ticket_{auth}$ with $K_{op,sg}$ and verifies whether the requesting user is authentic by decrypting $Auth$ with $K_{u,sg}$ contained in $Ticket_{auth}$. If authentication succeeds, the login service performs authorization by checking whether a set of roles RS_U contained in $Ticket_{auth}$ is implied in a set of roles assigned to BS according to the gateway's security policy. It encrypts the result of authorization with $K_{u,sg}$ and sends it to the bundle service by combining with a Subject object that contains the privileges for using required system resources. The Subject object, which is an instance of the Subject class in JAAS (Java Authentication and Authorization Service) [3], represents identity and security attributes for a subject (i.e., a user) and supports authorization decision for gateway resources.

⑤ **Sending authentication result to user** (S_REP):

The bundle service sends all the messages received from the login service except for the Subject object. The user U can confirm the authorization result by decrypting received messages with $K_{u,sg}$.

Since a user can reuse an authentication ticket received by just one password entering into client program without repeated entering for authenticating multiple bundle services, the concept of SSO can be satisfied in the proposed model. Since there is a login service performing login tasks for all bundle services in a gateway, the convenience that service provides do not need to develop proprietary authentication modules are provided.

In addition, since we use symmetric-key operation rather than complex and slow public-key operation to encrypt protocol messages, it helps gateways to use faster and smaller system resources for user authentication tasks.

2.4   Symmetric-Keys in the Model

The proposed authentication model requires three symmetric-keys to be shared among OSGi entities. Table 1 describes these keys by comparing with others.

Table 1. Comparisons of symmetric-keys in the model

	$K_{u,op}$	$K_{u,sg}$	$K_{sg,op}$
Entities	User and operator	User and gateway	Gateway and operator
Purpose	Encryption of $K_{u,sg}$	Encryption of user authentication information	Encryption of authentication ticket
Creation	User	Authentication server	Key generator in operator
Term of validity	Until user changes	Life cycle of authentication ticket	Until authentication server changes

- $K_{u,op}$ is a user password which is registered to authentication server when the user is enrolled firstly. No additional sharing mechanisms are required because the key is not transferred through network.
- $K_{u,sg}$ is encrypted with $K_{u,op}$ as a part of the authentication information and passed to a user whenever authentication of the user succeed. The key is passed to the login service with being protected by the authentication ticket.
- $K_{sg,op}$ are two types. The one is used for a standalone user's authentication ticket and the other is used for travel user's authentication ticket. The operator creates both types of $K_{sg,op}$ and then stores them into manifest files of each bundles just before they installed in gateways.

The bundle manifest file containing $K_{sg,op}$ may not be secure when the login service bundle is downloaded in the gateway. Separately user authentication model, much of the work is performed at the bootstrapping process of the gateway. Therefore we suppose that safe channel is created between the gateway and the operator by certain security mechanisms.

3 Implementation and Evaluation

To implement prototype for the user authentication mechanism, we use Windows 2000 servers as operating systems for both operator and service provider. For the environment that supports Java language and its libraries, we choose JDK 1.3. We use Oracle 8i for bundle repository that contains information about user authentication. As a gateway framework, we use Sun Microsystems JES (Java Embedded Server) [16] that is compatible to current OSGi specification. As a hardware platform for gateway, we use Ericsson E-Box [6] that provides relatively poor system resources (e.g., 200 MHz CPU, 64 MB RAM, 10/100M Ethernet Adapter) than other workstations. As a Java-based cryptographic library we use J/LOCK [15] developed in Security Technologies Inc. Figure 3 shows implementation environment for the prototype.

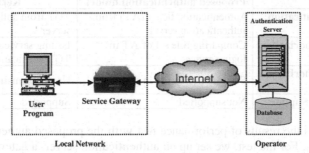

Fig. 3. The prototype implementation environment

Figure 4 is the class diagram representing connections between a bundle service LightService and the login service LoginService. User login tasks are performed in getSubject() method of LiginServiceImpl class that implements LoginService interface. LSActivator class is used to register LoginService into framework. The getSubject() method in LoginServiceImpl class checks the validation of authentication tickets. Then, it returns a Subject object right after getting authorization information about the user's privileges to use the service through Policy class [10] in Java 2 security architecture.

Table 2 presents the comparisons of proposed user authentication model with Kerberos authentication mechanism. Our authentication model performs service authorization for a user by comparing role set of authentication ticket, while Kerberos issues service tickets for each service to use. In our model, each service gateway manages the authorization information in the form of security policy file, while TGS (ticket-granting server) manages the authorization information in Kerberos. The way of our model to manage authorization enables login service supported. In this paper, however, we do not consider the support for inter-realm (i.e., the support for authentication ticket issued by other operators).

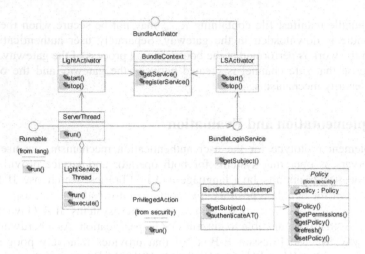

Fig. 4. Class diagram for login service

Table 2. Comparisons of proposed authentication model with Kerberos

	Proposed authentication model	Kerberos
User Authentication Information	Authentication ticket (AT) from authentication server	TGT from authentication server
Service Authorization Mechanism	Comparing role set of AT in login service	Issuing service ticket with TGT verification in TGS
Service Authorization Information	Security policy file in gateway	User information in TGS
Inter-realm Support	Not supported	Supported

Figure 5 is the results of performance test with the proposed authentication model and Kerberos. For the test, we set up an authentication server, a gateway, a user program on each node and connect them into a local network. In case of Kerberos, both the authentication server and the ticket server are placed on the operator's side.

The x-axis is the number of service requests and the y-axis is the required time for performing the request. In case of Kerberos, a user connects to the TGS which is placed the operator's side to acquire the service ticket whenever the user requests authorization for services. On the other hand, the proposed authentication model treats user authorization requests by comparing role set in authentication ticket without connecting the operator every times unlike Kerberos does. Therefore the required time for authorization is decreased like as the experiment result of Figure 5.

The proposed authentication model removes risk factors [13] that could happen in authentication process. The proposed model can be safer from main-in-the-middle attacks such as replay attack on the authentication ticket because of timestamp and encrypted authentication information. We use SEED[9], Korean domestic symmetric-key algorithm [10], for encryption of the authentication ticket and messages to make the proposed model safer from differential cryptanalysis, linear cryptanalysis, and other key-related attacks. However, our model is still vulnerable to the password guessing attack and the login spoofing attack since it uses a plaintext user password to create the authentication ticket.

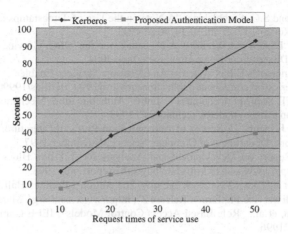

Fig. 5. Results of performance test

The symmetric-key is used to encrypt the authentication ticket and the messages in proposed model because the OSGi gateway has limited system resources unlike traditional distributed services environment. If the gateway has enough system resources in the future, it would be better to change the proposed model to the PKI-based authentication model.

4   Conclusion

In this paper, we proposed a new user authentication model for supporting SSO functionalities that are useful for safe and easy user authentication in the OSGi service platform environment.

Since a user can reuse an authentication ticket received by just one password entering into client program without repeated entering for authenticating multiple bundle services, the concept of SSO can be satisfied. Since there is a login service performing login tasks for all bundle services in a gateway, the convenience that service providers do not need to develop proprietary authentication modules is provided.

In addition, since we use symmetric-key operation rather than complex and slow public-key operation to encrypt protocol messages, it helps gateways to use faster and smaller system resources for user authentication tasks.

More formalized specification and verification of the user authentication model are presented in [5] as a part of entire security architecture for the OSGi service platform environment.

References

1. B.C. Neuman, T. Ts'o, "Kerberos: An Authentication Service for Computer Network", IEEE Computer Magazine 32(9), pp.33-38, September 1994.
2. CCITT Recommendation, "X.509 - The Directory Authentication Framework", December 1988.
3. C. Lai, L. Gong, "User Authentication and Authorization in the Java Platform", Computer Security Applications Conference, December 1999.

4. C. Neuman and S.G. Stubblebine, "A Note on the Use of Timestamps as Nonces", Operating System Review, 27(2), pp.10-14, April 1993.
5. D.H. Park, "ESAOP: A Security Architecture for the OSGi Service Platform Environment", PhD Dissertation, Korea University, June 2004.
6. Ericsson, "Ericsson's E-box system – An Electronic Services Enabler", http://www.ericsson.com/about/publications/review/1999_01/files/1999015.pdf
7. ISO/IEC, "IT Security techniques - Entity Authentication Mechanisms Part 2: Entity authentication using symmetric techniques", 1993.
8. J. Clark and J. Jacob, "A Survey of Authentication Protocol Literature: Version 1.0", University of York, Department of Computer Science, November 1997.
9. KISA, "A Report on Development and Analysis for 128-bits Block Cipher Algorithm (SEED)", http://www.kisa.or.kr/technology/sub1/report.pdf, 1998.
10. M. Pistoia, et al., "Java 2 Network Security, 2nd edition", Prentice Hall, 1999.
11. OSGi, "OSGi Service Platform – Release 3", http://www.osgi.org/, March 2003.
12. R. S. Sandhu, et al., "Role-Based Access Control Models", IEEE Computer 29(2), pp.38-47, February 1996.
13. S.M. Bellovin and M. Merritt, "Limitations of the Kerberos Authentication System", Proc. of the Winter 1991 USENIX Conference, January 1991.
14. S.M. Bellovin and M. Merritt, "Encrypted Key Exchange: Password-Based Protocols Secure Against Dictionary Attacks", Proc. of IEEE Symposium on Research in Security and Privacy, pp.72-84, May 1992.
15. STI – Security Technologies Inc., "Java Cryptography Library – J/LOCK", http://www.stitec.com/product/ejlock.html.
16. Sun Microsystems, "Java Embedded Server", http://wwws.sun.com/software/embeddedserver/.

A Security Risk Analysis Model for Information Systems

Hoh Peter In[1,*], Young-Gab Kim[1], Taek Lee[1], Chang-Joo Moon[2],
Yoonjung Jung[3], and Injung Kim[3]

[1] Department of Computer Science and Engineering, Korea University
1, 5-ga, Anam-dong, SungBuk-gu, 136-701, Seoul, Korea
{hoi_in,always,comteak,mcjmhj}@korea.ac.kr
[2] Center for the Information Security Technology, Korea University
1, 5-ga, Anam-dong, SungBuk-gu, 136-701, Seoul, Korea
mcjmhj@korea.ac.kr
[3] National Security Research Institute
62-1 HwaAm-dong, YuSeong-gu, 305-718, Daejeon, Korea
{yjjung,cipher}@etri.re.kr

Abstract. Information security is a crucial technique for an organization to survive in these days. However, there is no integrated model to assess the security risk quantitatively and optimize its resources to protect organization information and assets effectively. In this paper, an integrated, quantitative risk analysis model is proposed including asset, threat and vulnerability evaluations by adapting software risk management techniques. It is expected to analyze security risk effectively and optimize resources to mitigate the risk.

1 Introduction

As information communications (e.g., emails, chatting, messengers) or transactions (e.g., e-commerce, m-commerce) through the Internet are exponentially increasing, cyber security incidents (from a virus through email messages to cyber terrors to critical business information systems) are sharply increasing. A trend in cyber security vulnerabilities and virus attacks is shown in Figure 1.

In addition, according to the data of Common Vulnerabilities and Exposures (CVE) operated by MITRE, the number of incidents related with security vulnerabilities was 2,573 in April, 2, 2003. This is enormously increased in comparison with the 1,510 incidents in May, 7, 2001.

It is urgently needed to analyze the risk of the security vulnerabilities and prevent them effectively. The advantages of the security risk analysis are as follows:

- Enabling to develop secure information management
- Monitoring critical assets of organization and protecting them effectively
- Supporting effective decision-making for information security policies
- Establishing practical security policies for organizations
- Providing valuable analysis data for future estimation

However, it is quite challenging to develop a general, integrated security risk analysis model.

---

* Corresponding author.

D.-K. Baik (Ed.): AsiaSim 2004, LNAI 3398, pp. 505–513, 2005.
© Springer-Verlag Berlin Heidelberg 2005

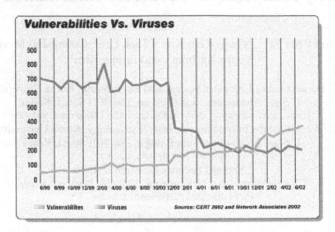

Fig. 1. Increasing Trend in Security Vulnerabilities & Viruses [CERT]

In this paper, a security risk analysis model is proposed by adapting a software risk management model used for the NASA mission-critical systems. Since security is one of software quality attributes, security risk needs to be investigated in the context of software risk. Security risk can also be defined by multiplication of loss (or damage) and the probability that the attacks occur by reinterpreting the concepts of loss and probability in the definition of software risk. However, the definition of loss and measurement of probability need to be adapted in the context of cyber security. In the risk analysis model, the loss (or damage) of assets in an organization due to the cyber security incidents is measured based on assets, threats, and vulnerability as shown in Figure 2. The detailed definition and model will be explained in Section 2. The related work is presented in Section 3, and the conclusion in Section 4.

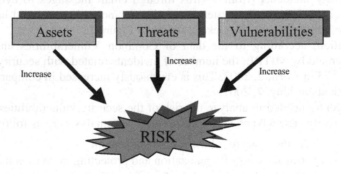

Fig. 2. Security Risk and Related Elements

2   Security Risk Analysis Model

The proposed security analysis model is shown in Figure 3. The four steps are proposed as a security risk analysis process. STEP 1 identifies the assets, threats and vulnerabilities of the target organization (Step 1.2 ~ 1.4) and evaluates them (Step 1.5). The outputs of Step 1 (i.e., assets, vulnerabilities, and threats) are used to analyze security risk in STEP 2. In STEP 3, suitable risk mitigation methods are applied to decrease the potential threats for the assets. STEP 4 summarizes the initial risks (i.e., before the risk mitigation methods are applied), the types and cost of the risk mitigation methods, the residual security risks (i.e., after the mitigation methods are applied), and their Return-On-Investment (ROI). These helps the decision-makers optimize resources for minimizing security risks. The detailed steps are described in the following subsections.

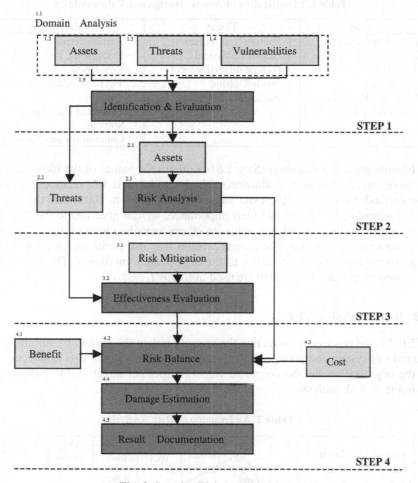

Fig. 3. Security Risk Analysis Model

2.1 Identification and Evaluation of Assets, Threats and Vulnerabilities (STEP 1)

Domain analysis in Step 1.1 is a domain-specific process for customizing the security risk analysis to improve the model accuracy. The types of assets, vulnerabilities, threats, and their probabilities are analyzed per domain. These of military or financial institutes are different from those of academic institutes.

Step 1.2 to Step 1.4 analyzes core assets of an organization (Step 1.2), potential threats (Step 1.3) and vulnerabilities (Step 1.4). The analysis includes what kinds of threats and vulnerabilities exist for a specific asset and the probabilities of threats that will occur. Table 1 shows a general classification of assets, threats, and vulnerabilities based on IT infrastructure.

Table 1. Classification of Assets, Threats and Vulnerabilities

Asset	Threat	Vulnerability
1. Information/Data 2. Documents 3. Hardware 4. Software 5. Human Resource 6. Circumstances	1. Human/Non-human 2. Network/Physical 3.Technical/Environment 4. Inside/Outside 5. Accidental/Deliberate	1. Administering Documents, Personnel, Regulation 2. Physical Circumstances or Facilities 3. Technical Hardware, Software, Communication/Network

Identification & Evaluation (Step 1.5) evaluates the values of the identified assets. Relative and absolute asset evaluation methods can be used. In a relative asset evaluation method, for example, the asset value ranges from 1 to 100. After the assets are sorted according to the order of their importance, we can give 100 to the first important asset in the list, and relative values to others according to the relative importance by comparing the first one. The sorting process may use a criterion or a set of criteria (e.g., monetary value or criticality to the organization missions). The results of the evaluated asset-value in this step are used in STEP 2.

2.2 Risk Analysis (STEP 2)

STEP 2 is the process of evaluating the security risk of the system by summing up all the risks of system components with considering the existing threats in the core assets of the organization and the degree of vulnerabilities per threat. Table 2 illustrates an example of Risk analysis.

Table 2. An Example of Risk Analysis

Proba-bility	Threats Model		Assets Model / Vulnerability Model	AM_1 (PC)	AM_2 (Security Policy)	AM_3 (System S/W)	...
0.5	TM_1 (DoS)	→	VM_1 (unprotected major communication facilities)	0.9	0.6	0.8*	
0.7	TM_2 (Virus)	→	VM_2 (unfit network management)	0.6	0.5	0.9	
0.4	TM_3 (Cracking)	→	VM_3 (unprotected storage devices)	0.3	0.4	0.6	
	...		...				

As shown above, the security risk can be calculated by the relationship among the Asset Model (AM), Vulnerability Model (VM) and Threats Model (TM). TM and VM identify probability of potential threats and effect of vulnerability. We can calculate the security risk against each asset using the following equation:

$$RISK = Loss * Probability$$

The risk is defined as multiplication of loss and probability. In this point, the "Loss" means the decline of the asset value when an asset is exposed to some vulnerabilities. "Probability" means the probability of threat-occurrence from the corresponding vulnerabilities.

Let us take an example. There is TM_1 (e.g., DoS attack) as a threat against asset AM_3 (e.g., System S/W) shown in Table 2. The threat-occurrence probability is 0.5. When TM_1 occurs, VM_1 is an vulnerability to which asset AM_3 is exposed (in other words, TM_1 may be occurred due to VM_1). Therefore, 0.8 means the decline of the asset-value (AM_3) in this example. In this way, we can calculate the sum of risk related to the system S/W asset (column AM_3). The following result means that.

$$Total\ Risk\ of\ AM_3 = 100 * (0.8 * 0.5 + 0.9 * 0.7 + 0.6 * 0.4) / 3$$
$$= 100 * 1.27 / 3$$
$$= 42.3$$

After all, when asset AM_3 is exposed to threat TM_1, the asset-value is decreased from 100 to 57.7 by the decline-rate 42.3 due to the corresponding vulnerability.

Note that one threat can be related to several vulnerabilities. Therefore, an efficient mapping algorithm is needed between threat and vulnerability in this step. The asset-value should also be considered as the value of money against the asset for estimating the amount of total damage in the organization.

2.3 Risk Mitigation and Effectiveness Evaluation (STEP 3)

STEP 3 is a process that shows the list of current security countermeasure in the organization, and selects suitable mitigation methods against the threats, then shows the effectiveness of the mitigation methods. Therefore, the manager who has charge of risk management can decide risk mitigation methods appropriate for the organization. The combination of risk mitigation methods affects on both the probability of threat-occurrence and the vulnerability-rate of assets, finally reduces the whole risk of assets in the organization. Table 3 shows an example that presents the effectiveness of risk mitigation methods.

Table 3. The effectiveness of Risk Mitigation Methods

Mitigation Method \ Vulnerability Model	Vaccine	Smart Card	Firewall	...
VM_1 (unprotected major communication facilities)	0.2	0.6	0.1*	
VM_2 (unfit network management)	0.6	0.5	0.5	
VM_3 (unprotected storage devices)	0.3	0.2	0.1	
...				

As shown table 3, we can reduce the vulnerability-rate of VM_1 (e.g., unprotected communication facilities) to 0.1* with the risk mitigation method, firewall. Generally applying a risk mitigation method to some vulnerabilities can reduce the rate of not only one vulnerability but also several related vulnerabilities simultaneously. Furthermore we can get the rate of risk reduction effectively with considering which vulnerabilities can be affected by selecting some risk mitigation methods (step 3.2). As a result, in the above example, the whole amount of risk reduction after applying the risk mitigation method firewall is as follows: (under the assumption the asset value is 100)

$$RISK = 100 * (0.1 * 0.5 + 0.5 * 0.7 + 0.1 * 0.4) / 3$$
$$= 100 * 0.44 / 3$$
$$= 14.7$$

The general classification of risk mitigation methods, which can be applied to IT infrastructure, is as followings:

- Access Control: Recognition of a organism, Physical Media(Smart Card), Authentication System based on Password, etc.
- Password Control: Public Key Infrastructure(PKI), etc.
- Internet Security Control: Invasion Detection System(IDS), Firewall, Vaccine, etc.
- Application Security Control: database security, system file security, etc.
- Physical/Environmental Security Control: facilities security, institution security, etc.

The manager of the organization can select and combine the risk mitigation methods under consideration of what asset is important relatively and how mush budget for risk management the organization has (Step 3.1).

2.4 Damage Estimation and Reporting Result (Step 4)

Risk Balance (Step 4.2) concludes the final estimation against the risk calculated Step 2 and 3. In this step, The questions are investigated for the risk management such as *What kind of threats can be reduced? What are residual risks if the risk mitigations are applied? What is the ROI of each risk mitigation?*

The ROI is calculated by the ratio of the benefit and the cost. The benefit is calculated from the following equation: (initial risk) – (residual risk after the risk mitigation method is applied). An ROI example is as follows:

[In STEP 2] Initial risk of AM3 (System S/W) before installing firewall

$$Total\ Risk\ of\ AM3 = 100 * (0.8 * 0.5 + 0.9 * 0.7 + 0.6 * 0.4) / 3$$
$$= 100 * 1.27 / 3$$
$$= 42.3$$

[In STEP 3] Residual Risk of AM3 (System S/W) after installing firewall

$$RISK = 100 * (0.1 * 0.5 + 0.5 * 0.7 + 0.1 * 0.4) / 3$$
$$= 100 * 0.44 / 3$$
$$= 14.7$$

$$Benefit = (Initial\ risk) - (Residual\ risk)$$
$$= 42.3 - 14.7$$
$$= 27.6$$

[In STEP 4] ROI (Return On Investment)

$$= 27.6/4 * 100 = 690$$

(assuming that the cost of firewall is 4 units)

Damage Estimation (Step 4.4) calculates the total damage against the core asset in the organization based on the each risk calculated in step 2. The damage against the asset is a sum of the recovery cost, which is used to recover the asset before the threat is occurred. The recovery cost can be calculated as follows:

Total Cost = Acquisition Cost + Operation Cost + Business Opportunity Cost

The Acquisition Cost is an early cost of buying or installing an asset. The Operational Cost is a cost against human resource, who is able to recover the damaged asset. Finally the Business Opportunity Cost is the business loss by losing opportunity to earn monetary gains.

Result Documentation (Step 4.5) is a process that prints the result such as the risk per each asset, the list of mitigation method, the Return on Investment, the damage and so on. Through this step, the organization can decide their risk management policy

2.5 Implementation of the Security Risk Analysis Model

Based on the security risk analysis model, a security risk analysis program is developed as shown in Figure 4.

3 Related Work

As information security management becomes more important recently, advanced information security organizations, such as BSI in Britain [7] or NIST in USA [2, 3, 4, 5, 11], have been developing guidelines or methodologies for the national defense by themselves, and also 'Guidelines of the Management of IT Security' as a standard in ISO/IEC [9].

In the past, many researches have been done for security risk assessment [1, 6, 8, 10]. The representatives of security risk assessment are as follows: *Information Security Assessment Methodology (IAM), Vulnerability Assessment Framework (VAF), and Operationally Critical Threat, Asset, and Vulnerability Evaluation (OCTAVE).*

IAM is a security assessment method to analyze potential vulnerabilities of an organization, and use for an educational program of the US Department of Defense with the experiences of National Security Agency (NSA) for 15 years.

VAF is a methodology for vulnerability assessment which was developed by KPMG Peat Marwick LLP with the commission of Critical Infrastructure Assurance Office in 1998. With the minimal essential infrastructure of the relevant organization, it analyzes the vulnerability data aggregated from selected assets, and then calculates the vulnerability grade as a result of qualitative assessment.

OCTAVE is a security assessment method which was developed by Software Engineering Institute of Carnegie Mellon University in USA. The security assessment consists of three steps: making file of threat scenarios based on assets, recognizing the vulnerabilities about major facilities, and assessing the risk and developing security strategies.

They are partially supporting the risk analysis process with a focus of vulnerability analysis and assessment. However, there is no integrated risk analysis based on software risk theory and the quantitative approach, which are used in this paper.

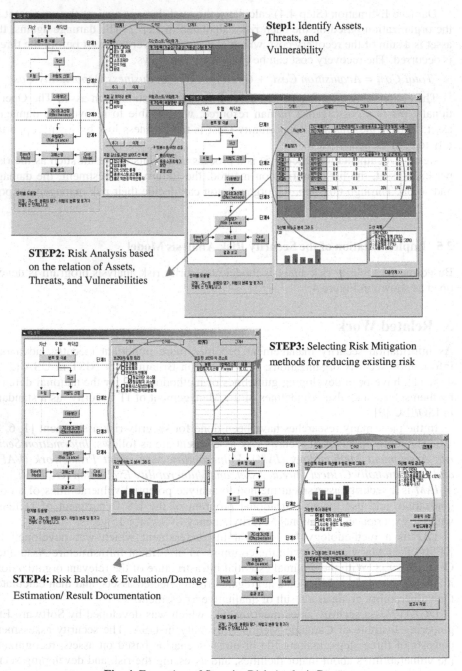

Step1: Identify Assets, Threats, and Vulnerability

STEP2: Risk Analysis based on the relation of Assets, Threats, and Vulnerabilities

STEP3: Selecting Risk Mitigation methods for reducing existing risk

STEP4: Risk Balance & Evaluation/Damage Estimation/ Result Documentation

Fig. 4. Execution of Security Risk Analysis Program

4 Conclusion

Information security is a crucial technique for an organization to survive in these days. However, there is no integrated model to assess the security risk quantitatively and optimize its resources to protect organization information and assets effectively. In this paper, an integrated, quantitative risk analysis model based on definition of software risk is proposed including asset evaluation, threat evaluation, and vulnerability evaluation. Domain analysis model, one of components in the proposed model, makes it possible to analyze security incidents and estimate the damage cost per domain.

Our contribution is to apply the concept and techniques of software risk management for security risk analysis. As a result, the security risk can be evaluated quantitatively based on loss of the assets and its probability. In addition, we consider risk mitigation techniques to reduce the relevant risks and understand residual risks

In future, we will focus on the following: development of detailed guidelines of each model with examples, development of an automatic analysis system or tool for risk analysis and risk mitigation with visualization feature, optimization of proposed security policies in the proposed model, and case studies to improve the proposed model.

References

1. GAO, "Information Security Risk Assessment - Practices of Leading Organizations," Case Study 3, GAO/AIMD-00-33, 1999. 11.
2. NIST, "Guide for Selecting Automated Risk Analysis Tools", NIST-SP-500-174, Oct. 1989.
3. FIPS-191, "Specifications for Guideline for The Analysis Local Area Network Security," NIST, Nov. 1994.
4. NIST, "Risk Management Guide for Information Technology Systems," NIST-SP-800-30, 2001.10.
5. FIPS-65, "Guidelines for Automatic Data Processing Risk Analysis, NIST, 1975
6. GAO, Information Security Risk Assessment - Practices of Leading Organizations, Exposure Draft, U.S. General Accounting Office, August 1999.
7. BSI, BS7799 - Code of Practice for Information Security Management, British Standards Institute, 1999.
8. CRAMM, "A Practitioner's View of CRAMM," http://www.gammassl.co.uk/.
9. ISO/IEC JTC 1/SC27, Information technology - Security technique - Guidelines for the management of IT security (GMITS) - Part 3: Techniques for the management of IT security, ISO/IEC JTC1/SC27 N1845, 1997. 12. 1.
10. R.S. Arnold, S. A. Bohner, "Impact Analysis – Towards a Framework for Comparison," Proceedings of Conference on Software Maintenance, IEEE CS Press, Los Alamitos, CA, pp. 292-301 1993.
11. G. Stoneburner, A. Goguen, A. Feringa, "Risk Management Guide for Information Technology Systems," Special Publication 800-30, National Institute of Standards and Technology, October 2001.

Full Fabrication Simulation of 300mm Wafer Focused on AMHS (Automated Material Handling Systems)

Youngshin Han, Dongsik Park, Sangwon Chae, and Chilgee Lee

School of Information and Communication Engineering, Sungkyunkwan University,
300 Chunchun-dong, Jangan-gu, Suwon, Kyunggi-do 440-746, S. Korea
yshan@ece.skku.ac.kr

Abstract. Semiconductor fabrication lines are organized using bays, equipment, and stockers. Most 300mm wafer lines use AMHS (Automated Material Handling System) for inter-bay and intra-bay lot transportation. In particular, the inter-bay AMHS moves lots between stockers, whereas intra-bay AMHS moves lots between stockers and tools, or between tools within the same bay. The key concern for manufacturer is how to raise productivity and to reduce turn-around time. This paper presents a simulation to reduce turn around time and raise productivity by reducing the delivery time (which affects AMHS).

1 Introduction

The major Semiconductor manufacturers are changing production environments from 200mm wafer to 300mm wafer. 300mm lots with 25 wafers is too heavy for an operator to carry. The semiconductor companies that have processes of research / analysis for 300mm FAB are adopting automation. From this process, FAB lines with high powered automation systems encountered some difficulties related to automation components. These semiconductor companies found the need for reduction of expenses and took a policy to actualize the reduction of expenses with FAB automation. However, adopting the process of 300mm FAB does not mean they can reduce expenses. One interesting occurrence with adopting the process of cost-down automation is that most semiconductor companies are moving toward using 300mm automation programs. To succeed with a 300mm program we are not concerned with the individual automation components, but instead take all individual automation components and use them as one whole system. Many wafer FABs have been laid out as bays, and each bay is supplied with stockers [1]. In a large wafer FAB, inter-bay and intra-bay automated material handling systems (AMHS) have been widely used to transfer lots. In particular, the inter-bay AMHS moves lots between stockers, whereas intra-bay AMHS moves lots between stockers and tools, or between tools within the same bay. Most companies are interested in reducing the average cycle time in order to increase productivity and to improve on-time delivery. In material handling problems, average cycle time is affected by the time that lots wait in queue for transport and the actual transit time. By definition, the lot-delivery time is the period from when a lot sends a request for transport until the lot is transferred to its final destination [2]. This paper shows a simulation to reduce turn-around time and raise productivity by reducing the delivery time (which affects AMHS).

D.-K. Baik (Ed.): AsiaSim 2004, LNAI 3398, pp. 514–520, 2005.

2 Computer Modeling and Simulation

2.1 Necessity of Simulation Model

The FAB line consists of 500-700 processes and the equipment internal to the line contains over 600 complex process systems.

2.2 Benefits of Simulation

Generally, we gain some advantages through the simulation of FAB lines. Using the simulation, the system can be investigated appropriately even though there are little changes which are not normally easy to see. Specific benefits of FAB line simulation are listed below:

- Can be used to system efficiency.
- Gives the ability to determine if the system is working the way it should be.
- Reduces the time of system construction.
- Issues of important systems can be considered in an early stage.
- Can result in cost reduction of the system construction.

2.3 Simulation Procedure

For this paper, the simulation procedure is going below:

Data Collection
To construct a computer model of a system, it is necessary to understand the system properly, and the data for the objects and parameters that compose the computer model are also needed. All the data used in this paper was taken from a real FAB line that is currently running in a domestic facility. This real data was preprocessed prior to use for our purpose. The actual data needed to make a reasonable computer model are stated below.

Implementation Sequence of Model
In this paper, the target system of the model is a real FAB line of a domestic semiconductor manufacturer and currently running. After receiving the real preprocessed data provided by the manufacturer, the construction for the computer model should be made through a sequence like the following:

Initial modeling of FAB line
The purpose of initial modeling of FAB line is to verify the constructed computer model in relation to the simulation purpose.

Detail modeling of FAB line
After verifying the initial modeling of the FAB line, a detailed model for the simulation should be constructed to find problems with the model and perform experiments for making several processing schemes and changing the layout to reduce defects.

Execution of the Model
To execute the computer model in simulation, predefined scheduling should be made according to the initial data and the information related to FAB-In. In this paper, the initial data was acquired from the results of 30 days of input data with the constructed computer model in our simulation. With the initial data, the warm-up period could be reduced.

Collection of Simulation Results
The desired simulation results such as FAB-In lots per day, cycle time, delivery time, and vehicle utilization should be considered and determined to construct the computer model prior to execution. After building up the computer model according to these factors, the resulting data can be acquired after completion of the simulation.

Analysis of the Resulting Data
The resulting data such as delivery time, cycle time, and vehicle utilization can be used to analyze the performance of a currently running system and design effective vehicle scheduling algorithms.

3   Assumptions

The assumptions used in the model are as follows:

- Basic unit of the model is equipment.
- Basic unit of a moving entity is a lot.
- 2 device types are used.
- Capacity of a stocker is infinite.
- Up/Down is applied to equipment. (MTTR (Mean Time To Repair) / MTBF (Mean Time Between Failure)).
- There are buffers in equipment.
- Buffer capacity of equipment without batch jobs is number of ports.
- Buffer capacity of wet equipments with batch size of 1 is 1 otherwise it is 6.
- For the rest of equipment with buffer capacity, the buffer capacity is twice the batch size.
- Interval of entering lots is constant.
- Insert 4 lots per batch.

4   Rules

The rules used in the model are as follows:

- Use OHT for transferring between bays within a cell.
- Apply reservation rules to prevent deadlock in equipment ports.
- If the in-port of equipment is full, the lot will wait in a stocker.
- Vehicles move using the principle of shortest path movement.
- Equipment will be applied if there are the same types of equipments in the same bay.
- Optimize using list (Eq_List) of available equipment.
- Processes never stop while in service.

5   Input Parameters

- Vehicle (OHT/OHS) number
- Vehicle (OHT/OHS) speed
- Stocker robot cycle time
- Operator moving time

Fig.1. shows 300mm AMHS proto-type layout. 300mm AMHS is consists of the following: Cell, Stocker, Bay, OHT, OHS.

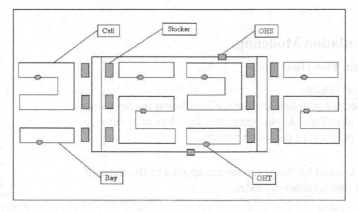

Fig. 1. 300mm AMHS Proto-type Layout

6 Operation Logic for Vehicles

Since there are many unpredictable factors, it is difficult to find the optimal transfer time. Therefore, we concentrate on developing dispatching algorithms for each problem to minimize delivery time.

The notation used in the proposed algorithms are listed below:

- $d_{x,y}^{jth}$: Distance between source stocker x to destination equipment y within a cell which is similar to transport of lots by j-th vehicle (j is the number of vehicles within a cell).
- d_x^y : Shortest distance from source (or destination) stocker x to equipment y.
- UV^{jth} : Utilization of returned lot by j-th vehicle at determination point.

6.1 Vehicle Selection

The vehicles(both idle and moving within a cell) that are returning lots are randomly selected. The random selection algorithm for selecting a vehicle is the built-in algorithm of AutoMod^{TM} 9.1 in the vehicle procedure. Even if the vehicle was selected by the vehicle selection algorithm, the vehicle within a cell with the lowest utilization will be selected. If $UV^{jth} \geq UV^{ith}$ is true, the i-th vehicle is selected, otherwise, the j-th vehicle is selected.

6.2 Selecting the Shortest Path

If the vehicle was selected using the random selection algorithm, the vehicle with the shortest distance from source stocker to destination stocker will be selected. So, if $d_{x,y}^{jth} \geq d_{x,y}^{ith}$, the i-th vehicle of the same cell will be selected, otherwise, the j-th vehicle will be selected. If $d_{x,y}^{jth} \geq d_{x,y}^{ith}$ and $UV^{jth} \geq UV^{ith}$, the i-th vehicle will be selected and if

$d_{x,y}^{jth} \le d_{x,y}^{ith}$ and $UV^{jth} \le UV^{ith}$, the j-th vehicle will be selected. Also, if $d_{x,y}^{jth} \ge d_{x,y}^{ith}$ and $UV^{jth} \le UV^{ith}$, the j-th vehicle will be selected, otherwise, the i-th vehicle will be selected.

7 Simulation Modeling

7.1 Main Flowchart

Generator Vehicle
The generator supplies lots for the simulation model using constant distribution, but, generates four lots at a time to ease the progress of the batch process.

FAB-in
The lots created by the generator are applied to the input of the FAB line simulation model.

Bay Processing
Bay processing involves transporting lots from cells to equipment.

FAB-out
If the current step becomes last the step, it is applied to the output of the FAB line simulation model.

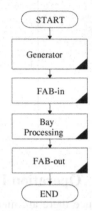

Fig. 2. Main Flowchart

7.2 Simulation Model

Our simulation models are implemented with discrete-event simulation. All the experiments are executed on a Pentium 4 personal computer with Microsoft Windows 2000. Fig.3. shows a representative layout of a 300-mm wafer FAB that contains double loop inter-bay and intra-bay systems, bays, equipment and stockers. Most 300mm wafer lines use AMHS for inter-bay and intra-bay lot transportation. In particular, the inter-bay AMHS moves lots between stockers, whereas intra-bay AMHS moves lots between stockers and tools, or between tools within the same bay. In general, the tools in a wafer FAB can be categorized into cells: diffusion, photo, etching, thin-film, and inspection. The layout is categorized into cells, in which the etching cell occupies a bay, the thin-film cell occupies several bays, and the photo cell several occupies bays.

8 Experimental Scenario

8.1 FAB-in Lots per Day

After testing the simulation model, FAB-in lots per day results for each of two devices are as follows:

Table 1. Fab-in lots per day result of each of two devices

	Input (lot / day)	Mean Value (wafer / month)
A-Type	322	15K
B-Type	114	18K
SUM	436	33K

8.2 Turn Around Time

Fig.4. shows turn-around time. From the result of the simulation model for two types, the model is approaching steady-state. The lot of A-Type converges at around 40 days and the lot of B-Type converges around 35 days. Even though the throughput is increased, utilization of vehicles is high and the average delivery time is shortened.

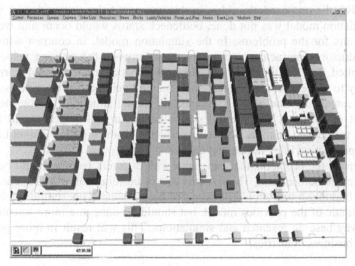

Fig. 3. Representative 3D Graphic Layout of a 300mm Wafer FAB

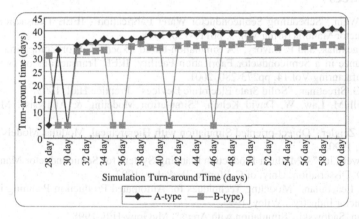

Fig. 4. Turn-around time of two days

9 Conclusion and Future Work

Most companies are interested in reducing the average cycle time in order to increase productivity and improve on-time delivery. In material handling problems, average cycle time is affected by the time that lots wait in queue for transport plus the transit time. By definition, the lot-delivery time is the period from when a lots sends a request for transport until the lot is transferred to its final destination. AMHS has not been applied fully in any location so many problems that will occur with AMHS have to be predicted and addressed by running simulations. We can see that the best algorithm is that which reflects the important factor: decreasing delivery time, which increases productivity. The utilization of vehicles within a cell is equally distributed. This will contribute to designing a more efficient system. We've performed many simulations, and by trial and error, we've balanced the load situation. This gave better results with the proposed algorithm in this paper. If balancing with trial and error in the simulation model was not done, bottleneck status would occur and we would not have results for the problems. In the simulation model, in contrast with the actual semiconductor manufacturer, communication between the moving vehicles has not been applied. The simulation tool used was AutoMod, and in this software tool, there is no way to model the communication of moving vehicles for discrete events. Therefore we cannot say that we've simulated the actual situation. If AutoMod provided this kind of function, the results of simulations in this paper would have better matched the actual situation. The semiconductor production line cannot be compared to other production lines in size and cost.

Therefore, the problems that can occur with different factors should be monitored and prevented before hand. The size and cost of building a new line is enormous. The plans of building a line should be prepared before hand. This should not be predicted using statistic of the previous model, but should be done in a more scientific and objective way base on research and simulations. This will result in more profitable results for business industries.

References

1. L.M.Wein, "Scheduling Semiconductor Wafer Fabrication", IEEE Transactions on Semiconductor Manufacturing Vol. 3, pp115-130, 1988.
2. G. Mackulak, and P. Savory, "A Simulation-based Experiment for Comparing AMHS performance in a Semiconductor Fabrication Facility. IEEE Transactions on Semiconductor Manufacturing Vol.14, pp273-280, 2001.
3. Ben G. Streetman, "Solid State Electornic Devices", Prentice Hall, 1995.
4. Averill M. Law, W. David Kelton, "Simulation Modeling & Analysis", McGraw-Hill, 1991.
5. B. P. Zeigler, "Object-oriented Simulation with Hierarchical, Modular Models", Academic Press, 1990.
6. Chihwei Liu, "A Modular Production Planning System for Semiconductor Manufacturing", Ph. D. Dissertation, Univ. of California, Berkeley, 1992.
7. R. C. Leachman, "Modeling Techniques for Automated Production Planning in the Semiconductor Industry", Wiley, 1993.
8. Kelton, Sadowski, "Simulation with Arena", McGraw-Hill, 1998.
9. Robert F. Pierrer, "Semiconductor Device Fundamentals", Addison Wesly.
10. "Dynamic Factory Modeling Person Guided Vehicle (PGV) Report", www.sematech.org, 2000.
11. "Evaluation of Valves Used for Delivering Semiconductor Process Chemicals.", www.sematech.org, 1994.
12. "Metrics for 300 mm Automated Material Handling Systems (AMHS) and Production Equipment Interfaces_ Revision 1.0", www.sematech.org,1998.

Modeling and Testing of Faults in TCAMs

Kwang-Jin Lee[1], Cheol Kim[1], Suki Kim[1], Uk-Rae Cho[2], and Hyun-Guen Byun[2]

[1] Department of Electronics Engineering, Korea University,
Anam-dong 5 -1, Sungbuk-Ku, Seoul 136 - 701, Korea
[2] Samsung Electronics, San #16 Banwal-Ri,Taean-Eup,
Hwasung-City, Gyeonggi-Do 449-711, Korea

Abstract. This paper proposes novel fault models for ternary content address-able memories (TCAMs) regarding both physical defects and functional fault models. Novel test algorithms which guarantee high fault coverage and small test length for detecting the faults in high density TCAM cell array is also pro-posed. The proposed T_{S-T} algorithm is suitable for detecting physical faults with compact test patterns and provides 92% fault coverage with small test length. Also, the proposed T_{M-T} algorithm makes it possible to detect various bridging faults among adjacent cells of high density CAMs. The test access time is (10 * l * 3 + 3) * write time and (12 * l * 3) * compare time in a n word by l bit TCAM.

1 Introduction

As lookup memories are required to have fast search time for fast address classifica-tion or packet filtering, TCAMs are being spotlighted because they are suitable for search algorithms such as LPM(Longest Prefix Match). In particular, high perform-ance network routers necessitate TCAMs with higher density for the desired fast lookup performance in large routing tables. Thus, testing these large TCAMs is be-coming an important issue. But, test methods and algorithms for CAM are focused on only BCAMs (Binary Content Addressable Memories) up to now[1],[2],[4].

In this paper, we propose novel TCAM fault modeling suitable for high density TCAMs with various bridge faults among adjacent TCAM cells and efficient TCAM test algorithm for those faults.

Section II of this paper introduces an TCAM circuit and its operations. Section III describes the functional fault modeling based on physical defects. A new algorithm for an efficient TCAM test is described in section IV. An verification of the proposed algorithm is mentioned in section V. Finally, conclusion is presented in section VI.

2 TCAM Overview

Fig.1 shows a general TCAM cell where storage section stores '0', '1', 'x' values. And comparison section compares storage value with search data, which is composed of transistors T7, T8, T9, and T10 [3].

The mapped relation between stored data and logic value according to data cell and mask cell is shown in Table.1.

D.-K. Baik (Ed.): AsiaSim 2004, LNAI 3398, pp. 521–528, 2005.
© Springer-Verlag Berlin Heidelberg 2005

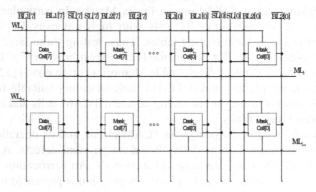

Fig. 1. TCAM cell scheme

Table 1. Stored data at a TCAM cell

Data cell		Mask cell		Stored data	Logic value
BL	/BL	BL	/BL		
Low	High	Low	High	00	X (don't care)
Low	High	High	Low	01	0
High	Low	Low	High	10	1

Fig. 2 shows an example of TCAM cell array structure where search data lines (SL, $\overline{SL}$) and read/write data lines (BL, $\overline{BL}$) are placed per a TCAM cell. The match line (ML) is placed per a word to notify comparison results.

Fig. 2. An example of TCAM cell array

3   Proposed TCAM Fault Model

3.1 Fault Models with only Single TCAM Cell

Several functional fault models which can occur at a storage section in single TCAM cell are similar to conventional SRAM cell case.

The functional faults at a storage section are as follows:

Cell Stuck at Fault (SAF) : A TCAM cell has stuck at fault if the logic value of the cell is not changed by any operation at the cell or by influences from others

Cell Stuck Open Fault (SOF): A TCAM cell has stuck open fault if it is not possible to access the cell by any operation in a cell and this fault is one type of Address decoder faults

Transition Fault (TF): A TCAM cell has transition fault if it fails to transitions $0 \rightarrow 1$ or $1 \rightarrow 0$

Coupling Fault (CF): TCAM cell has Coupling Fault if a write operation of a cell influences the value of another cell. coupling faults can be classified as follows: inversion fault, idempotent fault, bridging fault, state fault. we consider only a bridging fault which means that two cells are shorted together

Data retention fault (DRF): TCAM cell has Data Retention Fault if the cell fails to retain its logical value after some units of time

And functional faults related to the comparison section are as follows:

Conditional Match Fault (CMF): Under certain condition according to cell states and search data, the match function can be performed correctly

Conditional Mismatch Fault (CMMF): Under certain conditions according to cell states and search data, the mismatch function can be performed correctly

Unconditional Match Fault (UCMF): The comparison operation will be always a match regardless of values of search data

Unconditional Mismatch Fault (UCMMF): The comparison operation will be always a mismatch regardless of values of search data[1], [2].

Table 2 shows mapping table between physical defects and functional faults.

Table 2. Functional level fault model example

Functional fault	Physical defect
SAOF	WL,DS,MS SAOF / T6 stuck on fault
$0 \rightarrow 1$ TF	BL SAOF / T1 stuck on fault
CMF	SL SAOF / SL($\overline{SL}$)-WL bridging fault
UCMF	SL-$\overline{SL}$ / BL($\overline{BL}$)-ML bridging fault

3.2 Fault Models Among Adjacent TCAM Cells

Fault models among adjacent TCAM cells are necessary to detect bridging faults at functional tests. These fault models include coupling faults which can occur by bridging faults in single TCAM cell. If a bridging fault occurs between Jcell and (J+1)cell, effects of the fault can be expressed with related cells as shown in both Table 3 and Table 4.

Table 3. Fault models of row coupled cells

Bridging	Functional fault	
$(J_{cell} - J+1_{cell})$	J_{cell}	$J+1_{cell}$
BL–WL	CMF/ $1 \rightarrow 0$ TF	SOF
BL($\overline{BL}$)–ML	UCMF	UCMF
SL($\overline{SL}$)–WL	CMF/ SOF	CMF/ SOF
WL–ML	UCMMF	UCMMF

Table 4. Fault models of column coupled cells

Bridging	Functional fault	
$(J_{cell} - J+1_{cell})$	J_{cell}	$J+1_{cell}$
BL1–BL2	CMF/ $0 \rightarrow 1$ TF	CMF/ $0 \rightarrow 1$ TF
$\overline{BL1}$–BL2	$1 \rightarrow 0$ TF	$1 \rightarrow 0$ TF
BL–SL($\overline{SL}$)	CMF/ $0 \rightarrow 1$ TF	CMF/ $0 \rightarrow 1$ TF
$\overline{BL}$ – SL($\overline{SL}$)	$1 \rightarrow 0$ TF	$1 \rightarrow 0$ TF

4 Test Algorithm

4.1 Notation for Test Algorithm

Notations for functional test algorithm of both single TCAM cell and bridging faults among adjacent TCAM cells are expressed as follows

W_1 : write '1' to a cell

W_0 : write '0' to a cell

W_X : keep previous state of a cell

$W_{a,b}$: write 'a' to a data cell and 'b' to a mask cell

C_1 : compare '1'

C_0 : compare '0'

$F^{\Updownarrow}$: word parallel access.

$F^{\Leftrightarrow}$: bit parallel access.

$F|_0^l$: move a base cell moved from 0 bit to l bit sequentially

4.2 Proposed Algorithm for Single TCAM Cell

The sequential analysis about each of the faulty single TCAM cell has led to a lot of tests pattern for each fault. However, the proposed T_{S-T} algorithm can detect all testable faults independently except WL-SA1F.

$$
\begin{aligned}
T_{S-T} \\
\{ W_{0,0} \ W_{0,X} \ C_0 \ C_1 \ W_{0,1} \ W_{0,X} \ C_1 \ C_0 \ W_{0,0} \ C_0 \ C_1 \\
W_{0,0} \ W_{X,0} \ C_0 \ C_1 \ W_{1,0} \ W_{X,0} \ C_1 \ C_0 \ W_{0,0} \ C_0 \ C_1 \}
\end{aligned}
\tag{1}
$$

If physical defects occur in single TCAM cell, $W_{0,X}$ (or $W_{X,0}$) and comparison operations can detect faults.

4.3 Proposed Algorithm for High Density TCAMs

For all single TCAM cells in high density cell array and various bridging faults among adjacent TCAM cells, the T_{S-T} test algorithm should be extended because the algorithm is single cell based test algorithm which has limits to detect bridging faults among adjacent cells in same word on word based TCAM cell array. Therefore we propose new test algorithm, T_{M-T} algorithm which can cover both single TCAM cell functional faults and various bridging faults among adjacent TCAM cells. T_{M-T} algorithm is based on both T_{M-col} test algorithm for a n word by 1 bit and T_{M-row} test algorithm for a 1 word by l bit.

Fig.3 shows the $T_{M\text{-col}}$ test algorithm which can cover not only single TCAM cell test but also bridging faults detect among row coupled cells with 95% fault coverage. We can use word parallel access for reducing test access time when test patterns are injected: In the best case, $(10 * l) *$ write time and $(12 * l) *$ compare time. But we have to trade off reducing access time with power consumption.

$$T_{M\text{-col}}$$

$$\{W_{0,0}{}^{\updownarrow} \ W_{0,X}{}^{\updownarrow} \ C_0{}^{\updownarrow} C_1{}^{\updownarrow} \ W_{0,1}{}^{\updownarrow} \ W_{0,X}{}^{\updownarrow} \ C_1{}^{\updownarrow} \ C_0{}^{\updownarrow} \ W_{0,0}{}^{\updownarrow} C_0{}^{\updownarrow} \ C_1{}^{\updownarrow} \tag{2}$$

$$W_{0,0}{}^{\updownarrow} \ W_{X,0}{}^{\updownarrow} \ C_0{}^{\updownarrow} \ C_1{}^{\updownarrow} \ W_{1,0}{}^{\updownarrow} \ W_{X,0}{}^{\updownarrow} \ C_1{}^{\updownarrow} \ C_0{}^{\updownarrow} \ W_{0,0}{}^{\updownarrow} \ C_0{}^{\updownarrow} \ C_1{}^{\updownarrow}\}$$

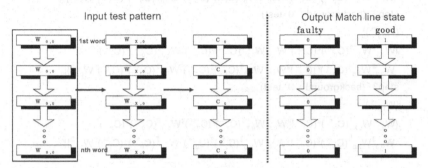

Fig. 3. $T_{M\text{-col}}$ algorithm for n word * 1 bit

Fig. 4 shows $T_{M\text{-row}}$ test algorithm which covers bridging fault between selected base cell and adjacent cell. After $T_{S\text{-T}}$ test pattern is applied on a base cell, the base cell is shifted to next bit cell and the operation is repeated. This algorithm has 93% fault coverage in row coupled cells test and $(10 * l *) *$ write time and $(12 * l *) *$ compare time.

$$T_{M\text{-row}}$$

$$\{W_{0,0} \ W_{0,X} \ C_0{}^{\leftrightarrow}C_1{}^{\leftrightarrow} \ W_{0,1} \ W_{0,X} \ C_1{}^{\leftrightarrow} C_0{}^{\leftrightarrow} \ W_{0,0} \ C_0{}^{\leftrightarrow} C_1{}^{\leftrightarrow} \tag{3}$$

$$W_{0,0} \ W_{X,0} \ C_0{}^{\leftrightarrow} C_1{}^{\leftrightarrow} \ W_{1,0} \ W_{X,0} \ C_1{}^{\leftrightarrow} C_0{}^{\leftrightarrow} \ W_{0,0} \ C_0{}^{\leftrightarrow} C_1{}^{\leftrightarrow}\}$$

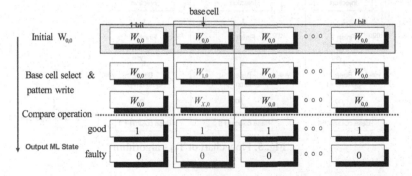

Fig. 4. $T_{M\text{-row}}$ algorithm for 1 word * l bit

The proposed $T_{M\text{-T}}$ test algorithm uses $T_{M\text{-COL}}$ test algorithm in case of write operation and uses both $T_{M\text{-COL}}$ test algorithm and $T_{M\text{-ROW}}$ test algorithm in case of comparison operation. It is performed with a base column shifted from 0 column to (l-1) column repeatedly as shown in flowchart of T_{M-T} algorithm of Fig.5. Through the procedures, proposed algorithm (T_{M-T}) can detect both single cell faults and bridging faults among adjacent TCAM cells with 92% fault coverage.

T_{M-T} algorithm

step1 (background : 'X' test)

$(W_{0,0}{}^{\leftrightarrow})^{\$}$

$\{W_{0,0}{}^{\$}W_{0,x}{}^{\$}(C_{0}{}^{\leftrightarrow})^{\$}(C_{1}{}^{\leftrightarrow})^{\$}W_{0,1}{}^{\$}W_{0,x}{}^{\$}(C_{1}{}^{\leftrightarrow})^{\$}(C_{0}{}^{\leftrightarrow})^{\$}W_{0,0}{}^{\$}(C_{0}{}^{\leftrightarrow})^{\$}(C_{1}{}^{\leftrightarrow})^{\$}$

$W_{0,0}{}^{\$}W_{x,0}{}^{\$}(C_{0}{}^{\leftrightarrow})^{\$}(C_{1}{}^{\leftrightarrow})^{\$}W_{1,0}{}^{\$}W_{x,0}{}^{\$}(C_{1}{}^{\leftrightarrow})^{\$}(C_{0}{}^{\leftrightarrow})^{\$}W_{0,0}{}^{\$}(C_{0}{}^{\leftrightarrow})^{\$}(C_{1}{}^{\leftrightarrow})^{\$}W_{0,0}{}^{\$}\}_{0}^{l-1}$

step2 (background : '0' test)

$(W_{0,1}{}^{\leftrightarrow})^{\$}$ 　　　　　　　　　　　　　　　　　　　　　　　(4)

$\{W_{0,0}{}^{\$}W_{0,x}{}^{\$}(C_{0}{}^{\leftrightarrow})^{\$}(C_{1}{}^{\leftrightarrow})^{\$}W_{0,1}{}^{\$}W_{0,x}{}^{\$}(C_{1}{}^{\leftrightarrow})^{\$}(C_{0}{}^{\leftrightarrow})^{\$}W_{0,0}{}^{\$}(C_{0}{}^{\leftrightarrow})^{\$}(C_{1}{}^{\leftrightarrow})^{\$}$

$W_{0,0}{}^{\$}W_{x,0}{}^{\$}(C_{0}{}^{\leftrightarrow})^{\$}(C_{1}{}^{\leftrightarrow})^{\$}W_{1,0}{}^{\$}W_{x,0}{}^{\$}(C_{1}{}^{\leftrightarrow})^{\$}(C_{0}{}^{\leftrightarrow})^{\$}W_{0,0}{}^{\$}(C_{0}{}^{\leftrightarrow})^{\$}(C_{1}{}^{\leftrightarrow})^{\$}W_{0,1}{}^{\$}\}_{0}^{l-1}$

step3 (background : '1' test)

$(W_{1,0}{}^{\leftrightarrow})^{\$}$

$\{W_{0,0}{}^{\$}W_{0,x}{}^{\$}(C_{0}{}^{\leftrightarrow})^{\$}(C_{1}{}^{\leftrightarrow})^{\$}W_{0,1}{}^{\$}W_{0,x}{}^{\$}(C_{1}{}^{\leftrightarrow})^{\$}(C_{0}{}^{\leftrightarrow})^{\$}W_{0,0}{}^{\$}(C_{0}{}^{\leftrightarrow})^{\$}(C_{1}{}^{\leftrightarrow})^{\$}$

$W_{0,0}{}^{\$}W_{x,0}{}^{\$}(C_{0}{}^{\leftrightarrow})^{\$}(C_{1}{}^{\leftrightarrow})^{\$}W_{1,0}{}^{\$}W_{x,0}{}^{\$}(C_{1}{}^{\leftrightarrow})^{\$}(C_{0}{}^{\leftrightarrow})^{\$}W_{0,0}{}^{\$}(C_{0}{}^{\leftrightarrow})^{\$}(C_{1}{}^{\leftrightarrow})^{\$}W_{1,0}{}^{\$}\}_{0}^{l-1}$

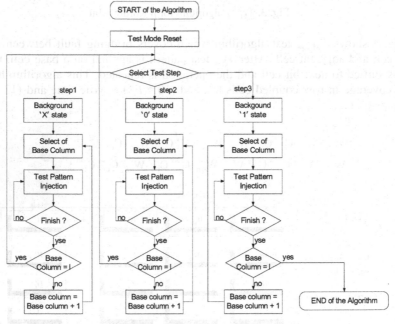

Fig. 5. Flow chart of T_{M-T} algorithm

5 Verification

The proposed algorithm of Fig. 5 can be verified by H/W implementation as shown in Fig. 6. This paper uses the behavior modeling of TCAM cell array which is implemented by C code for verification of proposed algorithm. As a result, the proposed algorithm guarantees high fault coverage as shown in Table.6, whereas most of conventional test algorithms can detect only a few bridging faults among adjacent cells for functional test[1]. It has average 87 % fault coverage in case of single cell and 94% fault coverage in case of adjacent cells. Test length in a n word by l bit TCAM can be reduced to $(30 * l + 3)$ write operation and $(36 * l)$ compare operation since it uses cell access method with word-parallel and bit-serial manner.

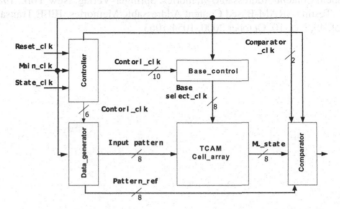

Fig. 6. TCAM block modeling for verification

Table 5. Comparison of fault coverage

Physical fault		Proposed algorithm	Conventional algorithm
Single cell fault case	Line / node Stuck at fault	96 %	94 %
	Transistor stuck on/ open fault	88 %	89 %
	Bridging fault in single TCAM	79 %	75 %
Multi-cell fault case	Bridging fault among adjacent cells	94%	Not testable

6 Conclusion

This paper proposes novel fault model for TCAMs regarding both physical defects and functional fault models. And novel test algorithm which has low test access time and high fault coverage for high density TCAMs is proposed. T_{S-T} algorithm is suitable for detecting physical fault with compact test pattern and provides 87% fault coverage with low test access time. Also T_{M-T} algorithm allows us to detect various bridging faults among adjacent TCAM cells with 94% fault coverage. As a result, proposed algorithm provides 92% fault coverage. Test length can be reduced to $(30 * l + 3)$ write operation and $(36 * l)$ compare operation since it uses cell access method with word-parallel and bit-serial manner.

References

1. P. R. Sidorowicz.: approach to modeling and testing memories and its application to CAM Formal Framework for Modeling and Testing Memories. IEEE VLSI Test Symposium, April 1998, 411-416
2. Kun-Jin Lin and Cheng-Wen Wu.: Testing Content-Addressable Memories Using Functional Fault Models and March-Like Algorithms" IEEE Transactions on computer aided design of integrated circuit and systems, vol 19, NO. 5 May 2000, 577-588
3. T. Kohonen. Content-Addressable Memories. Springer-Verlag, New York, 1980
4. J. Zhao. :Testing SRAM-Based Content Addressable Memories. IEEE Transaction on computer, vol. 49, NO. 10, October 2000, 1054-1063

Transient Test of a NSSS Thermal-Hydraulic Module for the Nuclear Power Plant Simulator Using a Best-Estimate Code, RETRAN

Myeong-Soo Lee[1], In-Yong Seo[1], Yo-Han Kim[1],
Yong-Kwan Lee[1], and Jae-Seung Suh[2]

[1] Korea Electric Power Research Institute, 103-16 Yuseong-Gu Daejeon, Korea
{fiatlux,iyseo,johnkim,leeyk}@kepri.re.kr
http://www.kepri.re.kr/index.html
[2] Future and Challenge Tech., San 56-1, Sillim-dong, Gwanak-gu, Seoul, Korea
jssuh@kepri.re.kr

Abstract. Korea Electric Power Research Institute (KEPRI) developed real-time simulation Thermal hydraulic model (ARTS) which is based on the best-estimate code, RETRAN. It was adapted Kori unit 1 nuclear power plant analyzer (NPA). This paper describes the result of the some of steady and transient tests of the Kori unit 1 NPA. The simulated results are compared and reviewed with the plant data and calculated reference data.

1 Introduction

KEPRI (Korea Electric Power Research Institute) developed a nuclear steam supply system thermal-hydraulic module for the Westinghouse type nuclear power plant simulator[1] using a practical application of a experience of ARTS code development[2-4]. Using this ARTS module, transient test of Kori unit 1 NPA is accurately described.

The ARTS code was developed based on RETRAN [5], which is a best estimate code developed by EPRI (Electric Power Research Institute) for various transient tests of NPP (Nuclear Power Plants). Robustness and the real time calculation capability have been improved by simplifications, removing of discontinuities of the physical correlations of the RETRAN code and some other modifications. And its scope for the simulation has been extended by supplementation of new calculation modules such as a dedicated pressurizer relief tank model and a backup model. The supplement is developed so that users cannot recognize the model change from the main ARTS module.

Regarding the transient analysis of ARTS module, only few results have been published up to now [3]. The aim of this paper is, therefore, to present a RETRAN based NSSS T/H module ARTS for transient operation of the Kori unit 1 NPA, and the subsequent simulation of the NPP operator training.

2 Transient Tests of ARTS

The various test simulations and analysis of these results have been performed to confirm the real-time calculation capability, robustness and fidelity of the ARTS

D.-K. Baik (Ed.): AsiaSim 2004, LNAI 3398, pp. 529–535, 2005.
© Springer-Verlag Berlin Heidelberg 2005

code. While the results have to be compared with experiment results and/or the independent calculation results of a best-estimate code, the qualitative analyses have been performed for the transients selected from the items out of the acceptance test procedure (ATP). The tests have been performed in two different test environments. First, the non-integrated standalone test (NIST) is done to evaluate the self-integrity of the ARTS code. Since ARTS takes the boundary conditions required to calculate the NSSS T/H phenomena through a global vector in simulator environment, the simplified models to provide these boundary conditions have been developed for NIST. After validating the self-integrity of ARTS, the integrated test with other simulator software has also been performed in simulator environment.

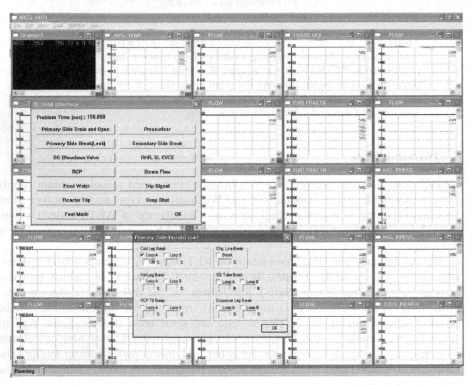

Fig. 1. NIST Environment for ARTS

2.1 Development of the NIST Environment

The ARTS code was developed in Windows operating system as that of simulator. The simulation scope of ARTS is the reactor coolant, pressurizer, PRT and steam generators. To perform the NIST environment, we developed the simple models for main steam line system, safety injection, feedwater system, residual heat removal system, chemical and volume control system, reactor protection and control system, etc. And a GUI (graphic user interface) to simulate operator actions and malfunctions and produce the graphical output is also developed using QuickWin program to perform the NIST effectively. Figure 1 shows the developed NIST environment.

Table 1. Comparison of Initial Steady State Variables with Full Power Normal Operation Data

Major T/H variables	Design values	Calculation results	Difference (%)
Core Power (MWt)	1723.50	1723.40	0.01
Pressurizer Pressure (kg/cm^2)	158.19	158.05	0.09
Pressurizer Level(%)	55.00	54.81	0.35
RCS Flow (kg/sec)	8557.60	8562.56	-0.06
Hot Leg Temperature (K)	592.55	592.10	0.14
Cold Leg Temperature (K)	556.05	556.24	-0.07
Average Temperature (K)	574.25	574.17	0.03
Steam Line Pressure (kg/cm^2)	59.12	59.14	-0.03
Steam Generator Level (%)	56.20	56.86	-1.17
Feedwater Flow (kg/sec)	947.50	948.37	-0.09
Steam Flow (kg/sec)	947.50	948.01	-0.05

2.2 NIST Test

All transients for NIST started from a full power condition. The full power condition for NIST is obtained after 3600 sec of a null transient. Table 1 shows the comparison between the calculated results and the design values for nominal operation condition. All major system variables are bounded within an acceptable range. The followings are the test items for normal operation condition.

- Stability tests for the various operation modes, e.g., 100%, 75%, 50% power conditions
- Heat balance tests for the various operation modes
- Operation tests from 100% to 75% power, 75% to 50%, 50% to 2%, etc.

The overall performance of ARTS in these tests was satisfactory especially in terms of robustness and fidelity.

2.3 Transient Test

Transient test has been performed to evaluate the robustness and fidelity of the ARTS module. The limiting cases of each transient type in ATP are selected for the test items for NIST.

- Manual reactor trip
- All feedwater pump trip
- All main steam isolation valve trip
- All reactor coolant pump (RCP) trip
- Single RCP trip
- Turbine trip
- Road rejection
- Offsite power loss
- Steam line break
- Pressurizer PORV open

– Road rejection
– Road increase
– Feedwater line break
– Steam dump fail open
– Steam generator safety valve fail open, etc.

The test results were satisfactory and the robustness and fidelity of ARTS were validated. For large-break LOCA, the ARTS module was switched over to the backup module with little discontinuities.

The ARTS was also tested in the simulator environment for evaluate the real-time simulation capability in standalone mode. The simulation time and the calculation results were almost identical with those in Windows environment.

2.4 Site Acceptance Test

After successful completion of the NIST, we integrated the ARTS module with other simulator modules developed using GSE tool. The SAT was performed with a complete set of the full-scope simulator, which includes all hardware and software. The SAT consists of over several hundred-test items and it was done according to the ATP. The test was completed with a few minor discrepancies.

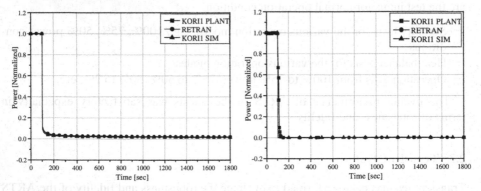

Fig. 2. Reactor Power (Turbine Trip)   **Fig. 3.** Turbine Power (Turbine Trip)

In order to check the validity of ARTS module, several transient cases were considered: (a) a transient by a manual turbine trip (b) a transient by a pressurizer safety valve (SV) open (see Fig. 8 and Fig. 9) (c) a transient by loss of coolant accident (LOCA) (see Fig. 10 and Fig. 11) (d) a transient by a main steam line break (MSLB).

In case of turbine trip, we compared the results of ARTS with plant data and calculated RETRAN values for the turbine trip accident (see Fig. 2 through Fig. 7). During the entire simulation time (1800 sec), the turbine trip is inserted at 100sec. The reactor trip is immediately occurred due to turbine trip. The trend of ARTS is similar to the trend of plant. A different of charging control makes the different pressurizer pressure trend.

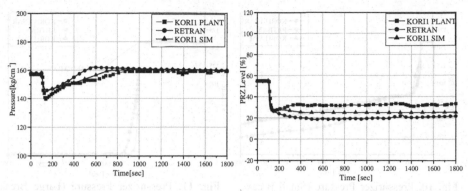

Fig. 4. Pressurizer Pressure (Turbine Trip)   **Fig. 5.** Pressurizer Water Level (Turbine Trip)

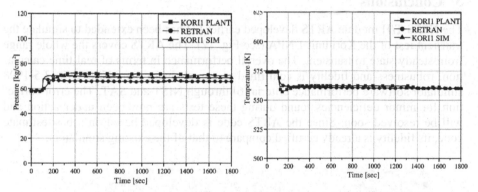

Fig. 6. Steam Generator Pressure (Turbine Trip)   **Fig. 7.** RCS Average Temperature (Turbine Trip)

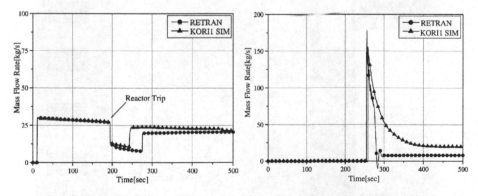

Fig. 8. Steam Inlet Flow of PRT   **Fig. 9.** Steam Outlet Flow of PRT

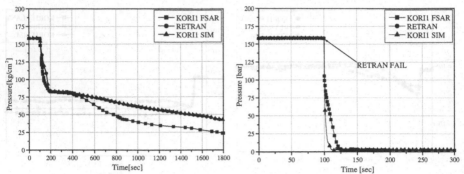

Fig. 10. Pressurizer Pressure (Small Break LOCA)

Fig. 11. Pressurizer Pressure (Large Break LOCA)

3 Conclusions

The NSSS T/H module ARTS developed earlier [3] has been extended to simulate the transient test of the Kori unit 1 NPA (see Fig. 12). The ARTS covers the whole range from steady state to transient. The overall performance in terms of real-time calculation, robustness and fidelity is confirmed to comply with the ANSI/ANS -3.5-1998 simulator software performance criteria [6] through NIST and SAT. The error correction for minor deficiencies found during these tests are now been under action and will be resolved soon. Since the ARTS code is developed based on a best-estimate code, its fidelity is already certified compare to that of the existing simulators.

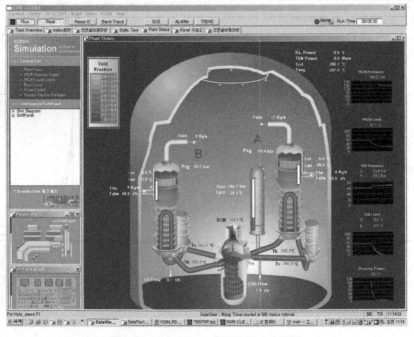

Fig. 12. Main View of Kori Unit 1 NPA

References

1. Suh, J.S., et al.: The Development of Virtual Simulator for Kori #1 Power Plant, KEPRI (2002)
2. Lee, M.S., et al.: Upgrade of the KNPEC-2 Simulator, 01-KEPRI-251, KEPRI (2001)
3. Kim, K.D., et al.: Development of An NSSS Thermal-Hydraulic Program for the KNPEC-2 Simulator Using a Best-Estimate Code: Part I. Code Development, Proc. KNS Spring Meeting (2001)
4. Suh, J.S., Development of a Nuclear Steam Supply System Thermal -Hydraulic Module for the Nuclear Power Plant Simulator Using a Best-Estimate Code, RETRAN, Ph.D. Thesis, Department of Nuclear Engineering, Hanyang University (2004)
5. M. P. Paulsen et al., RETRAN 3D code manual, EPRI NP-7450, Electric Power Research Institute (1998)
6. Nuclear Power Plant Simulators for Use in Operator Training and Examination, ANSI/ANS-3.5-1998, American Nuclear Society (1998)

Simulating Groundwater Transport Process
Using a Vertical Heterogeneity Model: A Case Study

Samuel S. Lee[1] and Hoh Peter In[2,*]

[1] Civil Engineering, FERC San Francisco Regional Office, USA
Samuel.lee@ferc.gov
[2] Department of Computer Science & Engineering, Korea University, Seoul, Korea
hoh_in@korea.ac.kr

Abstract. It is important to simulate a groundwater transport process, e.g., pollutant migration, through the vadose zone and subsequent mixing within the saturated zone to assess potential impacts of contaminants in the subsurface in preliminary stages. It is challenging to simulate heterogeneous soil characteristics and non-uniform initial contaminant concentration. This paper proposes a vertically heterogeneous model of combining heterogeneous vadose zone transport model and saturated zone mixing model using one-dimensional finite difference method to simulate the transport processes of liquid-phase advection, liquid- and vapor-phase dispersion, sorption, and decay of the contaminant. The proposed model was verified by comparing to an analytical solution and laboratory soil column experiments.

1 Introduction

One-dimensional finite difference scheme was employed for solving the vadose zone transport equation, and a mass-balance principle was used for the mixing calculation in the saturated aquifer underneath the soil columns. Although several computer codes (VLEACH, VLEACHSM 1.0a, EPACML, etc.) are available which can incorporate the heterogeneity of the soil properties (e.g., volumetric water content and corresponding infiltration rate), often many sites do not have the same degree of sophistication in the field-measured data. Even when there is a site with a reasonable amount of field data, a screening level of estimation is often necessary before conducting a comprehensive simulation. Moreover, often times when the pollutant migrates vertically to the ground water, a quick estimation of this mixing is necessary.

This paper proposes a vertically heterogeneous simulation model to meet these needs by combining a one dimensional heterogeneous vadose zone transport model and saturated zone mixing model.

In 1993, Ravi and Johnson [1] developed one dimensional transport program called VLEACH, which handles only vertical migration of pollutant in a homogeneous soil column. Later, Lee [2] developed VLEACHSM by adding the liquid-phase dispersion, decay terms in the vadose zone, and the saturation zone mixing into VLEACH. Our proposed model is improved further by implementing the heterogeneous soil property and the Graphic User Interface (GUI). GUI relieves a user from tedious and error-prone processes of input data preparation and output visualization. In addition,

---

* Corresponding author.

D.-K. Baik (Ed.): AsiaSim 2004, LNAI 3398, pp. 536–544, 2005.
© Springer-Verlag Berlin Heidelberg 2005

this model allows the specification of two different types (Dirichlet's and Cauchy's) of boundary conditions at the top of soil column. Using a simple mass-balance technique, the saturated zone module estimates the concentration of contaminants by mixing of leachate from the vadose zone with groundwater. This module uses the effluent concentration at the bottom of the soil column, which is estimated from the vadose zone module. A complete mixing of the leachate with the groundwater is assumed.

2 Background

To understand vertically heterogeneous simulation model, the model components are formalized and described. The model of Vadose Zone Transport is presented in Section 2.1, and Saturated Zone Mixing in Section 2.2.

2.1 Vadose Zone Transport

Considering the three equilibrium phases of pollutants in an unsaturated soil column, its one-dimensional governing transport equation is expressed as follows:

$$\theta_w \frac{\partial C_w}{\partial t} + \theta_a \frac{\partial C_a}{\partial t} + \rho_b \frac{\partial C_s}{\partial t} \frac{\partial}{\partial z}\left(\theta_w D_w \frac{\partial C_w}{\partial z}\right) + \frac{\partial}{\partial z}\left(\theta_a D_a \frac{\partial C_a}{\partial z}\right) - \frac{\partial}{\partial z}(q_w C_w) - \mu_w \theta_w C_w - \mu_a \theta_a C_a - \mu_s \theta_b C_s \quad (1)$$

where, C_w = concentration of a contaminant in liquid (water) phase [mg/L],

$\quad C_a$ = concentration of a contaminant in vapor (air) phase [mg/L],

$\quad C_s$ = concentration of a contaminant in solid phase [mg/Kg],

$\quad \theta_w$ = volumetric water content (volume of water / total volume) [m³/m³],

$\quad \theta_a$ = air-filled porosity (volume of air / total volume) [m³/m³].

$\quad q_w$ = water flow velocity (recharge rate) [m/yr],

$\quad D_w$ = dispersion coefficient for the liquid phase contaminant

$\qquad$ in the pore water [m²/yr],

$\quad D_a$ = gaseous phase diffusion coefficient in the pore air [m²/yr],

$\quad \mu_w$ = first order decay rate of a contaminant in water phase [l/yr],

$\quad \mu_a$ = first order decay rate of a contaminant in gaseous phase [l/yr],

$\quad \mu_s$ = first order decay rate of a contaminant in solid phase [l/yr],

$\quad \rho_b$ = bulk density of the soil [gr/cm³], and

$\quad z$ = vertical coordinate with positive being downward, t = time [yr].

Note that the total porosity (n) equals the sum of the water filled porosity and the air filled porosity. The air flow velocity (q_a) is assumed to be zero in this study. For simplicity, it is assumed that $\mu_w = \mu_a = \mu_s = \mu$.

Instantaneous equilibrium (partitioning) of the contaminant among the phases is assumed according to the following linear relationships:

$$C_s = K_d\, C_w \qquad (2)$$
$$C_a = HC_w \qquad (3)$$

where K_d [ml/g] is the distribution coefficient between the solid phase and liquid phase, and H [dimensionless] is the partition coefficient between the air phase and water phase. Using an empirical relationship, K_d can be expressed as $K_d = K_{oc} \cdot f_{oc}$, where K_{oc} [ml/g] is the organic carbon-water partition coefficient and f_{oc} [g/g] is the fraction organic carbon of the soil.

The dimensionless form of the Henry's partition coefficient, H, can be determined from the more common form having the units of atmospheres-cubic meters per mole (atm-m³/mol) using the following equation

$$H = K_H/(RT) \tag{4}$$

where K_H [atm-m³/mol] is the dimensional form of Henry's Law constant, R is the universal gas constant ($R = 8.2 \times 10^{-5}$ atm-m³/mol·K), and T is the absolute temperature ($^{\circ}K = 273.16 + {}^{\circ}C$).

The dispersion coefficient in the unsaturated zone is regarded as a linear function of the pore water velocity as:

$$D_w = \alpha_L \frac{q_w}{\theta_w} \tag{5}$$

where α_L is the longitudinal dispersivity [feet] of the vadose zone.

The gas phase diffusion coefficient (D_a) in the porous medium is calculated by modifying the free air diffusion coefficient using the Millington model [3]:

$$D_a = D_{air} \frac{(n - \theta_w)^{7/3}}{n^2} \tag{6}$$

where D_{air} is the diffusion coefficient of the contaminant in the free air, and n is the total porosity of the soil.

By substituting equation (2) and (3) into (1), and neglecting the air flow velocity ($q_a = 0$), the governing transport equation can be simplified as:

$$\theta \frac{\partial C_w}{\partial t} = D \frac{\partial^2 C_w}{\partial t^2} + \left(\frac{\partial D}{\partial z} - q_w \right) \frac{\partial C_w}{\partial z} - \left(\frac{\partial q_w}{\partial z} + \mu \theta \right) C_w \tag{7}$$

where, $\theta = \theta_w + \theta_a H + \rho_b K_d$ and $D \equiv \theta_w D_w + \theta_d D_a H$

To solve the above equation (7), an initial conditions and additional equations are necessary. For the detailed, see [12].

2.2 Saturated Zone Mixing

After estimating the liquid phase solute concentration (C_w) at the bottom of the soil column, the mixed concentration in the aquifer can be calculated using a mass-balance technique as below (USPEA, 1989 [4] and Summers et al., 1980 [5]):

$$C_{mix} = \frac{\left(C_{aq} q_{aq} A_{aq} + C_w q_w A_{soil} \right)}{\left(q_{aq} A_{aq} + q_w A_{soil} \right)} \tag{8}$$

where C_{aq} is the concentration of horizontal groundwater influx, q_{aq} is the Darcy velocity in the aquifer, A_{aq} is the cross-sectional aquifer area perpendicular to the groundwater flow direction, and A_{soil} is the cross-sectional area perpendicular to the vertical infiltration in the soil column. The aquifer area (A_{aq}) is determined by multiplying the horizontal width of the soil column with the vertical solute penetration depth.

3 A Vertical Heterogeneity Model: Our Proposed Approach

Based on the equations shown in the background, new models such Vadose Zone Leaching (Section 3.1) and Saturated Zone Mixing (Section 3.2) are presented with Numerical Stability (Section 3.3).

3.1 Vadose Zone Leaching

The governing solute transport equation (7) is solved using the finite difference method. Differential equations dealing with liquid contaminant concentration C_w as a function of time and depth are converted into the finite difference equations dealing with the corresponding variable C_i^k centered on time between two time steps:

$$C_w \rightarrow \frac{\left(C_i^{k+1} + C_i^k\right)}{2}, \qquad \frac{\partial C_w}{\partial t} \rightarrow \frac{\left(C_i^{k+1} - C_i^k\right)}{\Delta t}$$

where Δt is the time increment, the subscript i refers to the discretized soil column cell and the superscript k refers to the time level. The subscript w is dropped for simplicity. Converting the other terms into finite difference form, the governing equation can be written as:

$$\left(-M_i + M_i' - N_i\right)C_{i-1}^{k+1} + \left(1 + 2M_i + N_i' + L_i\right)C_i^{k+1} + \left(-M_i - M_i' + N_i\right)C_{i+1}^{k+1}$$
$$= \left(M_i - M_i' + N_i\right)C_{i-1}^k + \left(1 - 2M_i - N_i' - L_i\right)C_i^k + \left(M_i + M_i' - N_i\right)C_{i+1}^k$$

(9)

where the dimensionless constants M_i, M_i', N_i, N_i', and L_i are:

$$M_i \equiv \frac{\Delta t}{2\left(\Delta z\right)^2}\frac{1}{\theta_i} D_i , \quad M_i' \equiv \frac{\Delta t}{2(\Delta z)^2}\frac{1}{\theta_i}\frac{D_{i+1} - D_{i-1}}{4}$$

$$N_i \equiv \frac{\Delta t}{4\Delta z}\frac{1}{\theta_i} q_i , \; N_i' \equiv \frac{\Delta t}{4\Delta z}\frac{1}{\theta_i}(q_{i+1} - q_{i-1}), \; L_i \equiv \frac{\Delta t}{2}\mu$$

Similarly, the finite difference form of the initial condition for the liquid phase solute concentration is

$$C_i^1 = \frac{\left(C_s^1\right)_i}{K_d} \; (K_d > 0, \; 2 \le i \le n-1), \text{ or } C_i^1 = \left(C_w^1\right)_i \; (K_d = 0, \; 2 \le i \le n-1)$$

The finite difference forms of the top boundary conditions for the soil column are:

First Type Top Boundary Condition,

$$C_i^k = \frac{C_s^k(z=0)}{K_d}\exp\{-\gamma(k-1)\Delta t\} \; (k=1, 2, \ldots)$$
$$C_s^k \neq 0, \quad \text{when } t \le t_0$$
$$C_s^k = 0, \quad \text{when } t > t_0$$

(10)

Third Type Top Boundary Condition,

$$C_1^{k+1} - \frac{\Psi'}{\Phi'}C_2^{k+1} = -\frac{\Omega'}{\Phi'}C_1^k + \frac{\Psi'}{\Phi'}C_2^k + \frac{q_w C_0}{\Phi'}\exp(-\gamma t)$$

(11)

Where $\Phi' = \frac{D(2M + L + 1)}{4(\Delta z)(M + N)} + \frac{q_w}{2}$, $\Psi' = \frac{DM}{4(\Delta z)(M + N)}$, $\Omega' = \frac{D(2M + L - 1)}{4(\Delta z)(M + N)} + \frac{q_w}{2}$.

M, N, and L were defined in equation (9). The second type bottom boundary condition is used in this model as follows:

$$\frac{C_n^{k+1} - C_{n-1}^{k+1}}{\Delta z} = 0 \tag{12}$$

The above finite difference form of simultaneous equations are programmed in C++ to solve for the value of C_i^{k+1} by the Thomas algorithm [7].

3.2 Saturated Zone Mixing

Based on the mass balance principle of equation (8), the mixed concentration is estimated as:

$$C_{mx1} = \frac{q_{w1}L_1 C_{w1} + q_{aq}\widetilde{H}_{d1}C_{aq}}{q_{mx1}\widetilde{H}_{d1}}, \quad C_{mx2} = \frac{q_{w2}L_2 C_{w2} + q_{aq}\widetilde{H}_{d1}C_{aq} + q_{mx1}\widetilde{H}_{d1}C_{mx1}}{q_{mx1}\left(\widetilde{H}_{d1} + \widetilde{H}_{d2}\right)}$$

where $C_{mx(i)}$ is the "Mixed Concentration in Groundwater."

3.3 Numerical Stability

Often, the efficiency of a numerical technique is limited due to the instability, oscillation, and mass-balance problems. Several methods have been proposed to determine the stability criteria of finite difference calculation (e.g., Fourier expansion method, matrix method, and other [8]). The Fourier expansion method, developed by von Neumann, relies on a Fourier decomposition of the numerical solution in space neglecting boundary conditions. It provides necessary conditions for stability of constant coefficient problems regardless of the type of boundary condition [9]. The matrix method, however, takes eigenvectors of the space-discretization operator, including the boundary conditions, as a basis for the representation of the spatial behavior of the [9, 10]. Based on the von Neumann method, Crank-Nicolson scheme of finite difference equation can be derived as:

$$\Delta z < \frac{2D}{q_w} \tag{13}$$

$$\Delta t < \frac{2\theta(\Delta z)^2}{\sqrt{(2D)^2 - \left(q_w\Delta z\right)^2}} \tag{14}$$

According to the stability criteria expressed in equations (13) and (14), it is clear that the combined dispersion coefficient (D) must be greater than zero. In natural soil conditions, it is rare to have a value of D as zero or close to zero. Specifically, if there is downward infiltration in the vadose zone, hydrodynamic dispersion of contaminant is inevitable. In addition, the air diffusion coefficients of selected organic compounds are must greater than zero.

Note that the numerical criteria presented in equations (13) and (14) are general guidelines and may not work for all situations. To ensure a stable result, the output should be checked thoroughly. If the results show any oscillation or negative concentration, a smaller time step and/or cell size than the values from the above equations should be tried. After having "stable" results, additional runs are recommended to check convergence of the simulation results. In this case, even a smaller time step and/or cell size can be used to check whether the solution converges or not. Calculating more precise diffusion and dispersion coefficients such as D shown in equation (13) given a specific situation is another research topic (i.e., out of scope in this paper).

4 Simulation: A Case Study

A hypothetical example with two soil columns is simulated to demonstrate the impact of soil heterogeneity.

4.1 Simulation Setting

The example depicts two soil columns arranged perpendicular to the ground-water flow direction (see Figure 1). The vadose zone soil is divided into four soil layers whose total porosity decreases (from 0.44 to 0.38) and water-filled porosity increases (from 0.26 to 0.32) along with the depth shown in Figure 1. The bulk density is adjusted according to the total porosity change (contribution from the water content change is disregarded). The soil column 1 has 1st type top boundary condition. The soil column 2 has 3rd type top boundary condition. This assumed set of parameters are derived based on a filed geologic situation where the total porosity of soil decreases along with the depth due to gravitational pressure while the soil becomes wetter (water-filled porosity increases) along with the depth because water sinks down to the lower layers. Recharge rate q is kept constant (0.3048 $m^3/yr/m^2$) in order to keep water-filled porosity of each layer constant. Organic content f_{oc} is also kept constant (0.005 g/g).

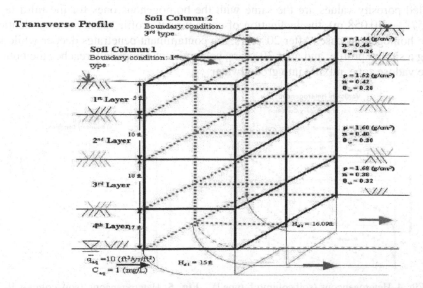

Fig. 1. Profile Example Problem. Soil column 1 is the 1st type of boundary condition. Soil column 2 is the 3rd type boundary condition.)

4.2 Simulation Results and Their Validation

To demonstrate the soil heterogeneity effect in the column the homogeneous soil columns were simulated (Figures 2 and 3). Note that, for homogeneous soil, uniform values soil property were used (bulk density 1.6, total porosity n = 0.4, and water filled porosity = 0.3).

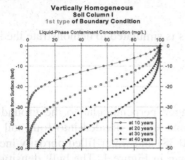 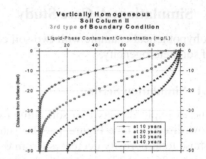

Fig. 2. Homogeneous (Soil Column I;Type I)   **Fig. 3.** Homogeneous (Soil Column II;Type III)

In both Soil Column 1 and Soil Column 2, the effects of putting four heterogeneous layers, instead of one vertically homogenous layer, are obvious as shown in see Figures 4 and 5. The liquid-phase contaminant went deeper at the heterogeneous soil after the same period of time (71 mg/L at –4.572 m in Soil Column 1 at 10 years compared with 28 mg/L for the homogenous case). That is because the total porosity values for the first (0 ~ -1.524 m) and second (-1.524 ~ -4.572 m) layers are larger than the homogeneous value of 0.4. In addition, since the total porosity and water-filled porosity values are the same with the homogenous ones for the third layer (-4.572 ~ -10.058 m), the inclination of contaminant profile mostly seems identical to the homogenous case. After 20 years, the contaminant penetrates deeper while keeping a similar profile, and after 30 years, contaminant completely reaches the bottom of the vadose zone, mixing into groundwater.

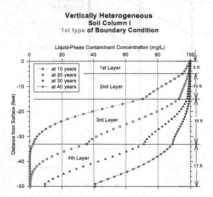 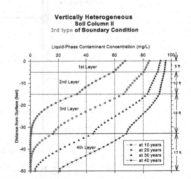

Fig. 4. Heterogeneous (soil column I, type I)   **Fig. 5.** Heterogeneous (soil column II, type III)

The simulation results of mixed concentration at the ground water were shown in Figure 6. A homogeneous soil column was simulated and the results were compared with an analytical transport solution [11] (see Figure 7). The close matched of the results indicates the simulation works correctly in homogeneous case. For the heterogeneous case, the simulation results showed reasonable match with column experiment data [12], which is available through the Internet (http://www.vadose.net)

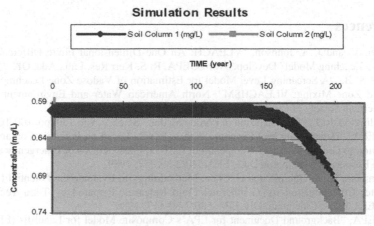

Fig. 6. Simulation results (mixed ground water)

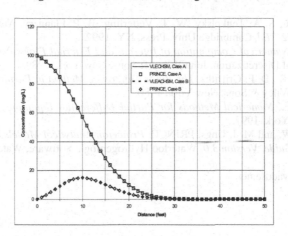

Fig. 7. Verification, compared from [7]

5 Conclusion

An integrated model was developed. The proposed model can handle vertical heterogeneity of soil. The model can deal with vertically heterogeneous soil columns which require more parameters to describe the soil properties of each layer.

The simulation model was verified by matching between the simulation results and an analytical solution (for homogeneous column) and between the simulation results and the soil column experiment (for heterogeneous soil).

Acknowledgments

The research leading to this paper has been funded by the United State Environmental Protection Agency. The author acknowledges Sam Lee in Dynamec Co., Dan Urish in University of Rhode Island for providing VLEACHSM 1.0a program and all laboratory equipments.

References

1. Ravi, V. and J. A. Johnson, "VLEACH: An One-Dimensional Finite Difference Vadose Zone Leaching Model" Developed for USEPA, R. S. Kerr Res. Lab., Ada, OK. 1993.
2. Lee, S. B., "A Screening Level Model for Estimation of Vadose Zone Leaching and Saturated Zone Mixing: VLEACHSM", North American Water and Environment Congress, American Society of Civil Engineers. 1996.
3. Millington, R. J., "Gas Diffusion in Porous Media," Science, vol. 130, pp. 100-102. 1959.
4. USEPA, "Determining Soil Response Action Levels Based on Potential Contaminant Migration to Ground Water: A Compendium for Examples", Office of Emergency and Remedial Response, Washington D.C., EPA/540/2-89/057, 1989.
5. Summers, K., S. Gherini, and C. Chen, Methodology to Evaluate the Potential for Ground Water Contamination for Geothermal Fluid Releases, Prepared by Tetra Tech Inc. for USEPA/IERL, Cincinnati, OH, EPA-600/7-80-117, 1980.
6. USEPA, "Background Document for EPA's Composite Model for Landfills (EPACML)", Prepared by Woodward-Clyde Consultants for USEPA, Office of Solid Waste, Wash. D.C., 1990.
7. Press, W. H., S. A. Teukolsky, W. T. Vetterling, B. P. Flannery, *Numerical Recipies in FORTRAN, 2nd Ed.* Cambridge Univ. Press, NY. 1992.
8. Hirsch, C., *Numerical Computation of Internal and External Flows*, Vol. 1, Fundamentals of Numerical Discretization. John Wiley & Sons, New York. 1989.
9. Mitchel, A. R., D. F. Griffiths, *The Finite Difference Method in Partial Differential Equations*, John Wiley & Sons, New York. 1980.
10. Ames, W. F., *Numerical Methods for Partial Differential Equations*, 2nd Ed., Academic Press, New York. 1997.
11. Cleary, R. W. and M. J. Ungs, PRINCE: *Princeton Analytical Models of Flow and Transport, User Guide, Version 3.0*, Waterloo Hydrogeologic Software, Waterloo, Ontario, Canada. 1994.
12. http://www.vadose.net

Vendor Managed Inventory
and Its Effect in the Supply Chain

Sungwon Jung[1], Tai-Woo Chang[2], Eoksu Sim[1], and Jinwoo Park[1]

[1] Department of Industrial Engineering, Seoul National University,
Seoul, 151-744, S. Korea
{jsw25,ses}@ultra.snu.ac.kr, autofact@snu.ac.kr
[2] Postal Technology Research Center, ETRI,
POB 106, Yuseong-Gu, Daejeon, 305-600, S. Korea
keenbee@etri.re.kr

Abstract. Many companies have tried to get the competitive advantage
through the effective supply chain management. To manage the supply
chain effectively, many strategy based on the shared information among
the entities in supply chain has been developed. Although the supply
chain members agree with the idea that the policy based on the shared
information will be beneficial to improve the effective supply chain, they
are reluctant to share their information owing to some barriers such as
the expensive technology investment, personnel training and lack of mutual trust. The benefit from the sharing information may vary according
to the supply chain environment. According to an environment of supply
chain, the profit gained through the policy based on the shared information may be too low to collect the invested amounts or high enough to
give proper compensation. To gain this insight, we analyze the impact of
the Vendor Managed Inventory, which is one of the popular supply chain
strategy based on the share information, in various supply chain environment. The simulation experiment shows in which environment their
benefit from the policy based on the shared information will be increased
(or decreased).

1   Introduction

A supply chain is a system consisting of material suppliers, production facilities, distribution centers, and customers who are all linked together via the
downstream feed-forward flow of materials(deliveries) and the upstream feedback
flow of information. Nowadays, the field of supply chain management has gone
through drastic changes. The remarkable development in information technology makes it possible for members of the supply chain to share their information
and cooperate to improve the efficiency of their supply chain. Vendor Managed
Inventory (VMI) is one of the supply chain strategies to get the competitive
advantage through the effective supply chain. In VMI, the vendor or supplier is
given the responsibility of managing the customer's stock based on the shared
information between them. In practical, many supply chain members are reluctant to adopt the VMI, although they agree with the idea that the integrated

D.-K. Baik (Ed.): AsiaSim 2004, LNAI 3398, pp. 545–552, 2005.

decision will be beneficial to improve their supply chain performance, owing to some barriers such as the expensive technology investment, personnel training and lack of mutual trust. [6] The best way to encourage for them to share their information and reflect it to make a better decision is to show how much it will be beneficial to their interest. According to an environment of supply chain, the profit gained through VMI may be too low to collect the invested amounts or high enough to get a proper compensation. In this study, we analysis the performance of adopting VMI in various supply chain environment. The purpose of this study is to find the insight about in which environment this profit will be increased or decreased.

Recently, many researches have focused on the impacts of information sharing in supply chain. Chen [2] studied the benefits of information sharing in a multi-echelon serial inventory system by computing the difference between the costs of using echelon reorder points and installation reorder points. He observed that information sharing reduced costs by as much as 9%, but average only 1.75% Some of his observations were that benefits decreased with increase in the number of stages, and were lowest at moderate values of penalty cost. Cachon and Fisher [1] studied the benefits of information flow in one warehouse multi-retailer systems. They assumed infinite capacity but supplies to the retailers were limited by the on-hand inventory at the supplier. They show how the supplier can use such information to better allocate stock to retailers and observed that the benefits of information sharing under these settings were quite small, average about 2.2%. Gavirneni [3] considered a simple supply chain containing one capacitated supplier and a retailer facing i.i.d. demands for a single product which is similar to our model. Although the model they assumed was similar to ours, they did not consider the VMI environment. He considered two modes of operation at the retailer's. One is using the traditional policy and the other is the policy based on the shared information. He performed the detailed computational study. The computational results showed that the policy based on the shared information reduced the cost 10.4% on the average and this reduction in costs is higher at higher capacities, higher supplier penalty cots, lower retailer penalty costs, moderate values of set-up costs, and at lower end-customer demand variance. In our study, we compare the traditional supply chain model in which vendor considers only his production plan and the VMI supply chain model where vendor given the responsibility of managing the customer's stock using simulation experiment. We will show that using VIM strategy in fact result in a reduction in the total supply chain cost, and analyze how the reduction in cost is affected by various supply chain parameters such as capacity, demand variance, and cost parameters.

The rest of the paper is organized as follows: In the next section, we describe the model and its assumptions. Then, we explain our experiment design in section 3 and show the result of the experiment in section 4. Finally, we close by summarizing the results and suggestions for future research.

2   Model Description

We consider a simple supply chain containing one capacitated manufacturer (vendor) and three distribution centers (customer). We assume the demand at distribution centers are forecasted at the first day of the planning period and this forecasted demand information is used for the inventory control at distribution centers and the production planning at manufacturer. The difference between the traditional and VMI supply chain is in the responsibility of the inventory control at distribution centers. In VMI scenario, distribution centers offer the manufacturer the information of the forecasted demand at each site and give the right of controlling their inventory. In returns, we assume distribution centers could be guaranteed the minimal service level for their customers and save the cost of handling the inventory. To grasp the benefit of such system, we develop the simulation model shown in Fig. 1 and compare the simulation results of those two scenarios with adjusting the various supply chain environment parameters. In the following sections, we will show the detailed model.

2.1   Scenario 1: Traditional Supply Chain Model

In the traditional supply chain, distribution centers and manufacturer make inventory decisions and production plan separately. We assume manufacturer request order information from the distribution centers to make a production plan for next planning horizon. To give an order information, distribution centers forecast the demand from their customers and make their own inventory decision following the (r,Q) policy. That is, when the inventory at each distribution center falls below the r, he orders the order quantity Q. We set the reorder point r about the two times the standard deviation (i.e., a 97.5% targeted service level) of the demand during the lead time. The order quantity Q is determined through EOQ process as $Q = \sqrt{2\tilde{D}k/h}$ where $\tilde{D}$ is the average demand and k, h is the fixed ordering cost and inventory holding cost respectively. Once the period-by-period needs of each of the distribution centers are determined, the total demand for each period at the manufacturing facility can be calculated by offsetting each distribution center's requirement by the lead time, and the aggregating over all the distribution centers. Manufacturer also makes a production plan to satisfy the order requirement from the distribution centner at the first day of the planning period. We assume the order requirements which can not be satisfied with the manufacturer's capacity are sent to the distribution centers using an expediting (e.g., overtime) strategy where the cost of expediting incurs. Once the plans are computed for the entire planning horizon, daily simulation begins until the end of the current planning period. Once the next planning period is reached, the planning cycle is performed again.

2.2   Scenario 2: VMI Supply Chain Model

In the scenario with VMI supply chain model, each distribution center forecasts his demand at the first day of the planning period. The forecasted daily demand

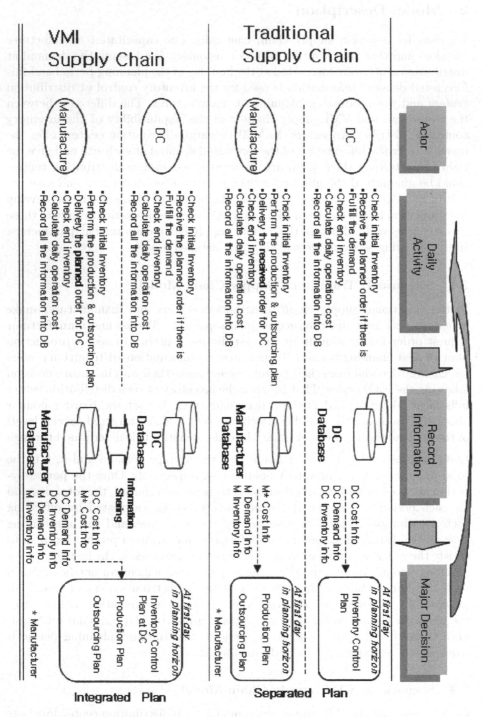

Fig. 1. The simulation model for traditional and VMI supply chain.

and inventory status information at each distribution center is sent to the manufacturer. The manufacturer makes a consolidated plan including the inventory control at distribution centers and production at manufacturer based on the shared information between them. The other assumptions are assumed similar such as in the traditional model. After finishing the computation for the entire planning horizon, daily simulation begins until the end of the current planning period. Once the next planning period is reached, the planning cycle is performed again. These procedure repeated until the end period of the simulation.

3   Simulation Experiment Design

We consider three environment factors in this simulation experiment. The first factor is the capacity tightness faced by the manufacturer. Capacity tightness refers to how tight the manufacturer's daily production capacity is relative to the sum of the total distributions daily demands. The other factors are cost paraments, which include the ratio of the fixed ordering cost to inventory holding cost at distribution centers and the expedition cost at manufacturer.

The experimental set-up for our study is as follows. We assume the demand at distribution centers during the time period occurs following the poisson distribution with average 20 (d_{avg}). The standard deviation of the end demand (d_{std}) is set to take values 4 which means the variance to mean ratio is 20%. If the value from demand distributions is below zero, we assume the demand is zero. The holding cost at distribution centers (h_{DC}) is 1 while the fixed ordering cost (k_{DC}) is allowed to take values 40, 160 and 360 which leads to the economic order quantity at distribution centers as 40, 80 and 120 respectively. The holding cost (h_M) and setup cost (k_M) are 2 and 500 respectively. The expedition cost at manufacturer (e_M) is allowed to take values 50, 100 and 150. We assume the capacity at manufacturer $(capa_M)$ to take values 235, 270 and 305 which leads to the capacity ratio to total average demand at manufacturer as 1.3, 1.5, and 1.7. The planning horizon (p_h) and planning frequency $(p_f)$is set as 20 days (one month) and 10 days respectively. The planning horizon means the number of periods for which plans are made and the planning frequency means the time interval between which plans are made. As the performance measure, we select the cost reduction percentage in VMI model compared with the one in traditional model. The detailed experiment parameters setting is shown in table 1.

For our experiment, we develop the simulation model using C++ language. ILOG CPLEX engine is used to obtain a solution of the production plan at manufacturer side in those two scenarios. Many global companies in practice

Table 1. Parameter setting for experiment.

p_h	p_f	d_{avg}	d_{std}	h_{DC}	k_{DC}	$capa_M$	h_M	k_M	e_M
					40	235			50
20	10	20	4	1	160	270	2	500	100
					360	305			150

adopt the ILOG CPLEX engine to solve their production planning problem. ILOG CPLEX component libraries make it possible to embed ILOG CPLEX engine seamlessly in our model. The production plan of the manufacturer in the two scenario is formulated as a mixed integer problem known as a NP-hard problem. It is well known that the time to get the optimal solution of the NP-hard problem is required exponentially as the problem size grows. [7] So, if the percent of the solution quality compared with the lower bound of the optimal solution is above 95% or the time to find solution takes over 20 minutes, we stop finding the optimal solution and select the best solution among the obtained ones. At each case, the simulation is run for three years. After initial warm-up period of six month, statistics are collected for those years and the average cost reduction percentage is calculated.

4 Computational Results

In our simulation experiments, all of each of the factor combinations are simulated. The results shows the cost reduction in VMI is 5.8% on the average. Analyzing the result, we can find out that the average cost of distribution centers in VMI model is 5% 10% higher than then one in traditional model while the cost reduction at manufacturer in VMI model is 30% 40%. It means the main cost reduction in VMI model comes from the manufacturer's. The observations from our experiment are detailed below.

4.1 The Effect of the Capacity

Fig. 2 contains the average cost reduction percentage as a function of the capacity at manufacturer. It shows the cost reduction is high at higher values of manufacturer capacity. It can be explained that the high flexibility of capacity at manufacturer make it possible to react the situation more effectively. If the capacity is too tight, it means that there are not have many production planning alternatives. Thus when the manufacturer has higher capacity, he is able to make more effective integrated production plan which leads to the significant cost reduction.

Similar result has been reported by Zhao et al. [8] although he does not consider the VMI environment. - A higher capacity level leads to greater benefit of sharing information. From our experiment result, we can also say that high level capacity at manufacturer is required to increase the benefit of VMI system.

4.2 The Effect of the Cost Parameters

Fig. 3 contains the average cost reduction percentage as a function of the setup cost at distribution center and expedition cost at manufacturer. The result shows that the high setup cost at distribution center and the low expedition cost at manufacturer reduce the benefits of VMI systems. As mentioned above, the benefits of VMI system comes from the cost reduction at manufacturer. The

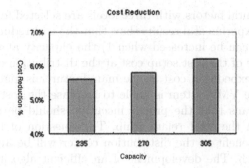

Fig. 2. Average cost reduction as a function of the capacity.

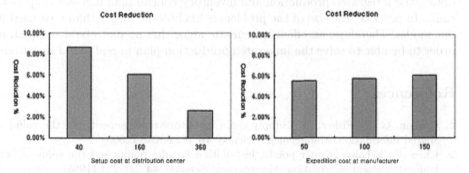

Fig. 3. Average cost reduction as a function of cost parameters.

cost occurred at distribution center in VMI model increases because distribution center in traditional model follows optimal policy. High setup cost at distribution center means the importance of the inventory control decision at distribution center which reduces the impact of the cost reduction at manufacturer. We can explain the effect of the low expedition cost at manufacturer in similar way. Low expedition cost means the decrease of the benefit through efficient production plan at manufacturer which leads to the decrease of the impact of the cost reduction. Similar result also can be found from previous research. [5] found that cost saving through information sharing is higher when a supplier's holding cost is smaller. However, he does not consider the VMI environment.

5   Conclusions

The primary purpose of this study is to gain insight into which supply chain environments allow the benefit obtained from the VMI system to be increased or decreased. In this study, we consider both the traditional model in which the distribution centers and the manufacturer make separate inventory decisions and production plans, and the VMI model in which the manufacturer makes a consolidated plan, including inventory control at the distribution centers and production at the manufacturer, based on the information shared between them.

Three environmental factors with three levels are selected for the simulation experiments. The experimental results show that the cost reduction obtained using the VMI system can be increased when 1) the capacity at the manufacturer is high, 2) the value of the cost setup cost at the distribution center is low, and 3) the value of the expedition cost at the manufacturer is high. We also conclude that adopting the VMI system is liable to increase the cost at the distribution centers. This means that the proper incentive should be given to the distribution centers in the VMI relationship. The question of how to guarantee a certain level of benefit to the distribution center will be an important subject for future research. The development of an efficient algorithm will also be an important topic. When the size of the problem increases, the time required to obtain the integrated production and inventory control plan increases exponentially. In practice, the size of the problem is likely to be bigger than that used in our model. Therefore, an efficient heuristic algorithm needs to be developed, in order to be able to solve the integrated production plan in real world situations.

References

1. Cachon, G. P., Fisher, M.: Supply chain inventory management and the value of shared information. Management Science. **46** 1032-1048 (2000)
2. Chen, F.: Echelon reorder points, installation reorder points and the value of centralized demand information. Management Science. **44** 221-234 (1998)
3. Gavirneni, S.: Information flows in capacitated supply chains with fixed ordering costs. Management Science. **48** 644-651 (2002)
4. Lee, H. L., So, K. C. and Tang, C. S.: The value of information sharing in a two-levle supply chain. Management Science. **46** 626-643 (2000)
5. Moinzadeh, K.: A multi-echelon inventory system with infromation exchange. Management Science. **48** 414-426 (2002)
6. National Research Council: *Surviving Supply Chain Integration: Strategies for Small Manufacturers* (Washington, DC: National Academy Press 2000)
7. Papadimitriou, C. and Steiglitz, K.: *Combinatorial optimization: algorithms and complexity* (Prentice-Hall, Inc. 2000)
8. ZHAO, X., XIE, J. and LAU, R. S. M.: Improving the supply chain performance use of forecasting models versus early order commitments. International Journal of Production Research. **39** 3923-3939 (2001)

Dynamic Performance Simulation of a Tracked Vehicle with ADAMS Tracked Vehicle Toolkit Software

H.B. Yin and Peng-Li Shao

China North Vehicle Research Institute, P.O. Box 969-58, Beijing 100072, China
yinhb201@163.com, shao9@sina.com

Abstract. This paper introduces usage of ADAMS in modeling and simulation of tracked vehicle performance. And how to make full use of the ADAMS Toolkit (ATV) in modeling and simulation is analyzed, further applications of the ATV software in simulating dynamic performance of a tracked vehicle are made. The following three typical examples are simulated: (1) making the vehicle climb a ramp with slope of 32 degrees, (2) making the vehicle get over a 2.8m ditch, (3) making the vehicle cross a 0.8m high obstacle wall. Through analysis of simulation results of the prototype and later physical performance testing, the conclusion can be made that the simulation results are consistent well with the physical testing data. In the process of simulation, a DLL file is built. And the problem of "sliding down" the ramp with slope of 32-degree while it is climbing up is encountered, through making various efforts, this "sliding down" problem is finally solved successfully....

1 Introduction

The virtual prototyping software enables engineers to build a computer model with moving parts, and then simulate its motion behaviors and optimize the design before building the first physical prototype. Thousands of engineers from nearly all-manufacturing industry today rely on Adams serial programs to help them achieve complete virtual prototyping solutions to their challenging engineering problems. ADAMS' Tracked Vehicle Toolkit (ATV) is an add-on module of Adams serial products. It can be used to assist civilian and military tracked vehicle designs. Users with the competitive software can greatly reduce their reliance on costly physical prototypes, dramatically shorten their product development cycles, and significantly improve the quality of their new product designs. In this paper, three typical examples are simulated: the vehicle climbing a ramp with a slope of 32 degrees, the vehicle getting over a 2.8m ditch, the vehicle crossing a 0.8m high obstacle wall. Through the simulations, good results are obtained.

2 Features of ATV

The ATV is capable of dealing with multiple track systems in one model with 3D dynamic visual behaviors and contact calculations including sponson (track-shelf) impacts. It has hard and soft soil interaction models with track systems. Many interfaces are supported in ADAMS ATV and users can build track systems with ATV and then use other ADAMS tools to extend their models. All ADAMS products can be

D.-K. Baik (Ed.): AsiaSim 2004, LNAI 3398, pp. 553–558, 2005.
© Springer-Verlag Berlin Heidelberg 2005

used in conjunction with the ATV toolkit. It is easy to run ATV together with other ADAMS modules such as ADAMS/Hydraulics and ADAMS/Controls. And other engineering simulation tools such as MATLAB, MATRIX or EASY5 can be used through interface of ADAMS/Controls in co-simulation. If more accurate simulation is needed, flexible parts of larger bodies can be coupled to the ATV model through the ADAMS/Flex interface.

There is a shared standard database system for ATV. It is used for repeatedly re-trieving parametric element model and modifying related data in modeling. After finishing the modeling, the complete model can be saved again to a new database for further simulation. Still, it has more road data files available in the shared database.

3 Application of ATV to Modeling

The vehicle hull and track system are two relatively independent large parts in model-ing. The hull refers to a global coordinate system. The track system's physical pa-rameters refer to the hull's reference coordinate. After the hull has been built, the track system can then be built. The track system is the parent entity that includes sus-pensions, wheels, ground model, belt and tracks. All ground contact parameters are defined in the track system.

It is convenient to retrieve and modify model elements of the track system with ATV, and then users can assemble the tracks and force subroutines, which can auto-matically calculate all the forces between wheels, tracks and the ground. In this proc-ess, users can follow a routine provided in the ATV toolkit to get correct simulation models. Some models of ATV database and debugging problems are introduced in the following.

3.1 Model of the Hull

The basic entity in ATV to build a track model is the hull. In ATV it is defined as the part to which the wheels in the track system are connected. The hull acts as the parent structure for the track system, i.e. when the track system is created, the hull is associ-ated with it and all the wheels in the track system will automatically be connected to the hull. After that, the track system has been built; the hull can also be exchanged with a modal flexible body imported through the ADAMS/Flex module. The hull can have interaction with the ground and the tracks.

3.2 Model of the Track System

The track system is just like a structure support for all sub-elements in it. All wheels, suspensions, tracks and the ground model belong to it. The track system position and orientation are determined relative to the hull reference system. The track system must specify what type of ground should be used.

3.3 Model of Road Files

A series of coordinate values of many points, which define the road profile, are used to build the road files. All these points are sewed together to build the road profile according to the principle that three points form a triangular patch. The road files can also be generated through other ways. The three road files, which depend on the needs to predict the general towing performance of the tracked vehicle ahead of the detailed design process, are built in this paper.

3.4 Model of the Drive Torque on Sprockets

There are two drive modes. One is the motion drive and the other is the torque drive. The spline function is used to fit the curve relationship between the torque on sprockets and the vehicle velocity in order to reflect the real towing ability of the vehicle physical prototype. The variations of velocity and resistance in reality are taken into account for the towing torque applied on sprockets.

4 Debugging

4.1 Problem of Memory Modes

There are five modes to manage memory in solving the model. However, a memory mode problem occurs when it solves a large model which includes two or more tracks in one model, and the problem is solved at last after a DLL file is generated and some memory setting changes are made.

4.2 Problem in Climbing up 32 Degrees' Ramp

In the simulation process of climbing up 32 degrees' ramp with the hard road, the vehicle first climbs along the ramp, and when the whole vehicle hull is over the ramp, it begins to slowly "slide down" the ramp. Can the "sliding down" really reflect that the vehicle has not the ability to climb up the ramp? The cause for this pheonominon is found after many similar simulation cases are calculated. The problem exists in the parameter setting of the road type. Some parameters have to be set again according to the Bekker theory[5] which governs the dynamic interaction between the ground and the track system. At last, the possibilities that cause the "sliding down" phenomenon, which arises from the ground coherent force, are excluded.

5 Three Working Conditions Simulated

Considering the complexity of the terrain and ground, the vehicle will encounter not only natural ditches but also artificial obstacles such as perpendicular walls. The road files of three types are built in order to predict the performance of the tracked vehicle's maneuverability. The following three working conditions are designing targets for general performances of the vehicle[2]. The ground files consist of the three files: the file for a ramp with a slope of 32 degrees, the file for a 2.8 m wide ditch, and the file for a 0.8 m high obstacle wall.

5.1 Climbing Ramp with a Slope of 32 Degrees

With the towing torque of the first gear and the initial velocity of 0.5 m/s, the vehicle's towing performance is simulated when climbing up the ramp with a slope of 32 degrees. Many parameters such as the torque on sprockets, the track forces and the forces on road wheels etc. are obtained. From table 1, Fig.1 and Fig.2, we can conclude that the towing torque of the first gear is correctly applied on the sprockets and the results obtained from the simulation are accurate and reliable.

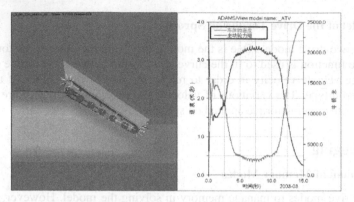

Fig. 1. Torque vs. velocity when climbing

Table 1. The simulated vs. the design sprocket torque [2]

Design	V	0.234	0.4727	0.7324	1.005	1.2907	1.618
Values	T	21960	20000	18200	16700	14990	13065
Simulated	V	0.2458	0.4946	0.7611	1.004	1.043	1.6258
Values	T	20110	19800	18030	16706	16469	13019

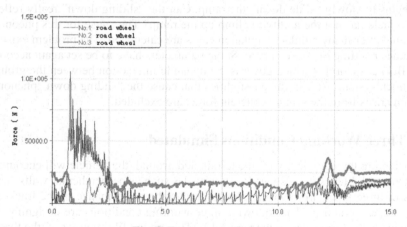

Fig. 2. Forces on road wheels

5.2 Getting over a 2.8m Wide Ditch

The third towing torque of the third gear is applied on the sprockets. The initial velocity is set to 1 m/s in order to alleviate the impacts. The analytical simulation results are obtained as shown in Fig. 3 and Fig. 4.

5.3   Crossing a 0.8m High Obstacle Wall

Considering the dynamic impacts and towing ability, the drive torque and initial velocity are the same as those of climbing the ramp with 32 degrees. The vehicle's maneuverability can be examined through the simulation. The simulated results are shown in Fig. 5 and Fig. 6.

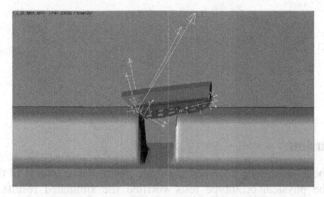

Fig. 3. Dynamic demo. of getting over a ditch

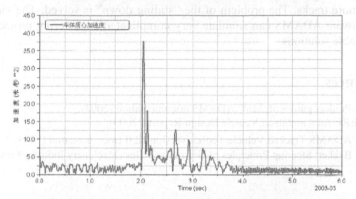

Fig. 4. Acceleration of the vehicle's mass center

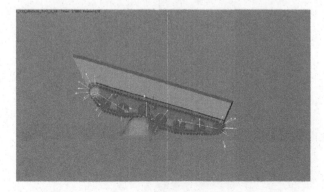

Fig. 5. Dynamic demo of crossing an obstacle

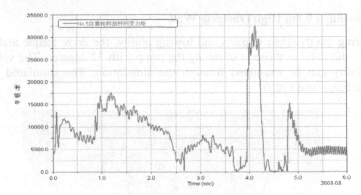

Fig. 6. Torque on the torsion bar

6 Conclusion

From the above simulation analyses, the three performance targets of the vehicle are reached. The physical prototype tests verified the simulated results. A DLL file (memory management) is built, and can be used for large-scale models, which include two or more tracks. The problem of the "sliding down" is solved. The virtual prototype software ADAMS can simulate very complicated engineering projects and provide reliable solutions.

References

1. MSC: Product guides of MSC.ADAMS documentation (2003)
2. NOVERI: An Amphibian Vehicle Design Report (2003).
3. MSC: ADAMS Tracked Vehicle Toolkit version 12.0 Documentation (2003).
4. M. G. Bekker: Steering Principles of Ground vehicles (Machinery Industry Press, 1994)

Evaluation of Operation Policies
for Automatic Delivery Sequence Sorting Systems

Joon-Mook Lim[1] and Tai-Woo Chang[2]

[1] Department of Industrial and Management Engineering,
Hanbat National University, Daejeon, 305-719, S. Korea
jmlim@hanbat.ac.kr
[2] Postal Technology Research Center, ETRI, Daejeon, 305-350, S. Korea
keenbee@etri.re.kr

Abstract. The development of automatic delivery sequence sorting system entails enormous cost and time, yet it is still an essential element that determines
the future of the postal service. In this paper, we select executive alternatives
that consider restrictive conditions both in terms of physical environments as
well as technical factors such as the location where the mail is sorted, the location where the addresses are read, and the types of video coding for unread addresses. Then, we evaluate the alternatives by performing cost and performance
analyses on processing time, machine utilization and throughput. Lastly, we
perform an analysis on the overall business to reflect on the qualitative factors.

1 Introduction

Nowadays customer demands for more rapid and accurate postal delivery in Korea
have increased. In order to meet such demand of the customers, the Korea Post is
making tremendous efforts to improve postal logistics. In connection therewith, it
recently established automatic facilities and adopted information systems. Although
the Post has set up 22 mail centers to improve the productivity of the mail sorting
process, the delivery sequence sorting processes are still done manually. For this
reason, the Post is considering introducing automatic delivery sequence sorting system.

The development of this automatic delivery sequence sorting systems (ADSS) will
entail enormous cost and time, but yet, we cannot overlook the fact that it is still an
essential element that determines the future of the postal service [1]. The number of
such costly facilities should be determined based on the forecasted mail volume and
the efficiency of the operation policies should be evaluated through simulation and
systematic methods before prior to deploying them in practice [2].

In this study, we propose several alternatives for the location of the sorting systems. These systems were evaluated by conducting analyses in terms of cost and
performance. In addition, we performed analysis on the overall economy to reflect on
the qualitative factors such as economic efficiency, expansibility and maintainability.

This paper is organized as follows. In Section 2, we examine the mail flow in Korea and propose alternatives for deployment of the delivery sequence sorters. Section 3 discusses cost analysis and in Section 4, we describe models for simulation and
discuss the evaluation carried out on performance of the alternatives. Following the

D.-K. Baik (Ed.): AsiaSim 2004, LNAI 3398, pp. 559–567, 2005.
© Springer-Verlag Berlin Heidelberg 2005

evaluation of the alternatives qualitatively with AHP (Analytic Hierarchy Process) in Section 5, conclusions and suggestions on further studies will be presented in Section 6.

2 Alternatives for Operation of ADSS

The general flow of mail processing in Korea is illustrated in figure 1. Based this figure, an alternative would be to locate ADSS at the mail centers or delivery post offices. If the ADSS is installed at delivery post offices, it could replace the present manual sorting process partly. On the other hand, it is necessary to extend the facilities at mail centers for installing ADSS.

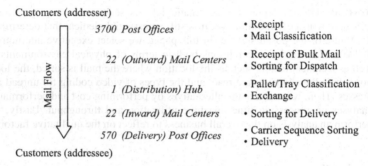

Fig. 1. Mail Processing Flowchart

We compared a distribution plan of ADSS over delivery post offices (PO) with a centralization plan of ADSS in mail centers (MC) as shown in table 1.

Table 1. Pros and cons of each deployment plan

	Distribution plan	Centralization plan
pros	1. preserve the present mail flow 2. information-independent PO 3. minimize business failure in initial deployment of ADSS	1. change the existing processes, especially the time schedules 2. maximize the utilization of ADSS 3. easy to maintain a carrier sequence DB
cons	1. require more ADSS 2. consume more time to recognize address and to sort mail owing to duplicated OCR 3. require operators in all PO where ADSS is installed	1. increase the amount of work in MC 2. difficult to secure the additional building site to install ADSS in MC 3. difficult to manage a large-sized carrier sequence DB 4. difficult to deploy ADSS by stages

2.1 Distinction Criteria

In this study, we choose three distinction criteria concerning physical and technical environments for selecting executive alternatives. The first criterion is to decide where to locate ADSS, as mentioned above.

The second criterion is to decide where to read the mail addresses. The sequence sorting process consists of two functions. One is recognizing address and printing a barcode for the delivery point according to the result of optical character recognition (OCR). The other is sorting the mail according to those barcodes. Therefore, it is important to decide whether the OCR should be done at mail centers or in delivery post offices. Since all the sorters at mail centers can read only postcodes now, it will be necessary to remodel the sorting machines in order to ensure that they can read multi-line addresses.

The last criterion is to decide what type of video coding for address will be used. Video coding means that an operator looks over a mail image, of which a machine fails in OCR and types the address string into a video terminal. Video coding is consequently a form of back-up task, incurring unnecessary additional labor cost. In this study, therefore, we considered three possible forms of video coding system (VCS) as follows.

– Networked VCS: video coding is done in a center which networks post offices or mail centers and collects the images from all the facilities
– Standalone VCS: video coding is done in each site where OCR is done
– None: all letters of which a machine fails in OCR are sorted by hand

2.2 Selection of Executive Alternatives

There are 12(=2x2x3) possible alternatives according to the criteria presented in above section. However, if the impracticable circumstances whereby sequence sorting precedes the address recognition are excluded, this would bring down to 9 executive alternatives as depicted in table 2.

Table 2. Practical alternatives for operation policies

Reading location		Sorting location		Video coding style			Alternatives	
MC	PO	MC	PO	Network	Standalone	None		
o		o		o			A1	Hybrid
o		o			o		A2	
o		o				o	A3	
	o	o		o			A4	Distributed
	o	o			o		A5	
	o	o				o	A6	
o			o	o			A7	Centralized
o			o		o		A8	
o			o			o	A9	

The executive alternatives can be classified into three categories. The cases that addresses are recognized in mail centers and letters are sorted by barcodes correspond to the Hybrid alternatives. Whether all functions of delivery sequence sorting are processed at mail centers or in post offices determines the category of alternatives, Distributed or Centralized.

3 Cost Analysis

In this study, we evaluated the alternatives through two types of quantitative analyses: Analysis on the cost incurred which are composed of fixed cost such as introduction fee of machines and variable cost such as labor cost, and the other on the performance.

In order to do a cost comparison, it was essential that the criteria that were applied to all the alternatives were the same. However, it is almost impossible to evaluate them on the assumption that the Post installs ADSS in all mail centers or delivery post offices under the same conditions. Therefore, we made the following assumptions when evaluating the alternatives with the nature and the characteristics of them as it is.

- The subject of our analysis is a network that consists of one mail center and ten dependent delivery post offices.
- The total volume of machinable letters is 500,000 a day. The spread of mail volume in 10 post offices (PO) follows two patterns. One is 3 PO with 40,000 pieces, 4 PO with 50,000 pieces, and 3 PO with 60,000 pieces. And the other is 2 PO with 40,000 pieces, 6 PO with 50,000 pieces, and 2 PO with 60,000 pieces.
- The available time to process mail with sequence sorters is eight hours a day.

Table 3 shows the total cost and ranking of the alternatives by patterns. Although there is a difference between the two patterns, the alternatives in a category centralized showed to incur lower costs. On the other hand, the distributed alternatives incur greater costs than the hybrid ones, that is, A6 costs less than A3 does. The total cost incurred by A5 is higher than the costs incurred for other alternatives.

Table 3. Comparison of the alternatives by patterns

Alternatives		Pattern 1		Pattern 2	
		Total Cost	Ranking	Total Cost	Ranking
Hybrid	A1	23,840,533	6	24,629,016	6
	A2	23,840,533	6	24,629,016	6
	A3	23,391,998	5	22,603,515	5
Distributed	A4	26,068,443	8	26,068,443	8
	A5	30,314,116	9	30,314,116	9
	A6	21,261,918	4	20,292,818	3
Centralized	A7	20,022,329	1	20,022,329	1
	A8	20,022,329	1	20,022,329	1
	A9	20,933,393	3	20,964,445	4

4 Performance Evaluation

We simulated the alternatives in order to carry out a quantitative analysis on the alternatives. In this simulation, models for each alternative were built, after which we developed models using a package for simulation, AweSim! [4] and a programming tool, Visual Basic.

4.1 Model and Plan

We grouped the 9 alternatives into three categories, hybrid, distributed and centralized. Although each group could be subdivided according to the types of video coding, we made these changes be reflected in each simulation experiment.

Figure 2 shows a simulation model for a category, distributed (A4, A5 and A6). If letters arrive at a delivery post office as in the present process, ADSS sorts mail according to the carrier sequence information that it acquires after recognizing the addresses from mail images.

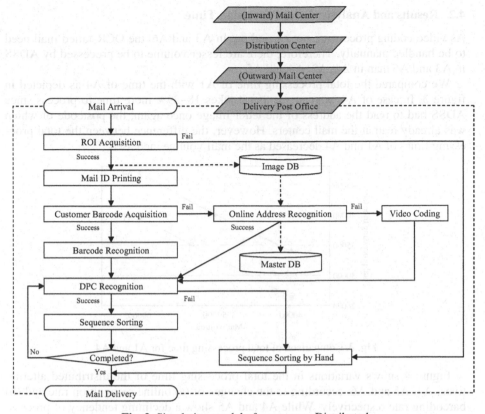

Fig. 2. Simulation model of a category, Distributed

The purpose of carrying out simulations in this study is to evaluate the alternatives based on the performances that could be achieved according to the various factors and to analyze the sets of factors that would guarantee optimal performance. Therefore, we defined the decision criteria and conducted experiments according to the defined sets of factors. We set some basic parameters: barcoding rate=0.4, online recognition rate=0.5, number of video coding terminals=4 and number of letter pieces to process=40,000. Then, we reviewed the performance variances according to the changes in the parameters.

The decision criteria included total processing time for sorting daily mail volume and machine utilization. In addition, we examined the adequate throughput of the machine for each alternative.

Among the alternatives, A7, A8 and A9 were excluded because the result of the simulation for these alternatives could be easily estimated based on the simulation results for A1, A2 and A3. Moreover, since the process of A1 is in accord with that for A2, with the exception of the type of video coding used, we only conducted experiments for A2 only.

4.2   Results and Analysis: Total Processing Time

As video coding process are not carried out in A3 and A6, the OCR-failed mail need to be handled manually. Therefore, there are lesser volume to be processed by ADSS in A3 and A6 than in A1, A2, A4 and A5.

We compared the total processing time of A1 with the time of A4 as depicted in figure 3. In case of A4, we can see that it takes 18~29% more time to process since ADSS had to read the address of the letter image once again, the postcode of which was already read at the mail centers. However, the difference between the total processing times of A1 and A4 decreased as the mail volume increased.

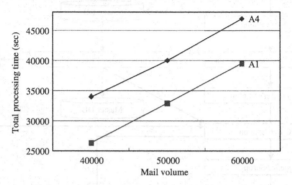

Fig. 3. Comparison of total processing time for A1 and A4

Figure 4 shows variations in the total processing time of the distributed alternatives, A4, A5 and A6, according to the changes of (a) online recognition rate and (b) barcoding rate respectively. While A4 and A5 show a declining tendency of processing time as the rates increases in both cases, A6 does not. This could be due to the fact that the mail volume to be processed by ADSS increases as the rates increase since A6 does not have a video coding process.

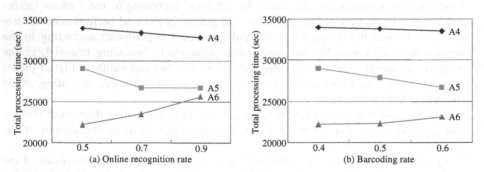

Fig. 4. Comparison of total processing time (a) by online recognition rate and (b) by barcoding rate for A4, A5 and A6

4.3   Results and Analysis: Machine Utilization

Figure 5 shows the machine utilization by mail volume for each alternative. From this figure, we can see clear differences between the alternatives.

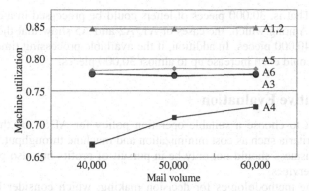

Fig. 5. Comparison of machine utilization by mail volume

Such a result that the machine utilization of A1 and A2 is high could be explained by the fact that ADSS in a delivery post office only sorts mails and its unit processing time decreases when address recognition is performed at the mail centers. The other result whereby the machine utilization of A4 is quite low could be due to the networked VCS. In this case, the utilization becomes low since ADSS has idle time as it must wait until it receives the address data from the VCS.

4.4   Results and Analysis: Throughput

The simulation in this study is aimed at estimating appropriate daily mail volume by alternatives. After fixing the values of parameters as follows: barcoding rate=0.4, online recognition rate=0.5 and number of video coding terminals=4, we carried out simulations on the throughput for given mail volumes. We analyzed the throughput with the average value of total processing time through 5 repeated experiments.

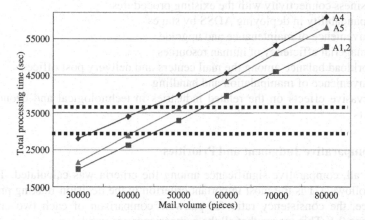

Fig. 6. Comparison of total processing time of A1, A2, A4 and A5 by mail volume

The thick dotted lines in figure 6 denote the available processing time per day for ADSS in two cases, i.e., 8 and 10 hours. According to figure 6, the points (alternatives) under the lines mean that it is possible to process the given mail volume with one ADSS. That is, 30,000 pieces of letters could be processed in a day for A1, A2, A4 and A5. Among which, the cases of A1, A2 and A5 show that the volume grows up to about 40,000 pieces. In addition, if the available processing time is extended to 10 hours, it could even increase up to almost 50,000 pieces.

5 Qualitative Evaluation

It is difficult to choose a suitable operation policy for ADSS on the basis of only qualitative criteria such as cost minimization and machine throughput. This is because the postal business should not only be in pursuit of profits but also provide satisfactory public services.

One of the methodologies for decision making, which consider both qualitative and quantitative criteria, is the analytic hierarchy process (AHP). The core concept of AHP is that it considers various decision criteria with weighting factors and allows various subjects of decision making to participate in the process.

AHP proceeds on decomposition, comparative judgment and comparative priorities [5]. As stated in section 2.2, the alternatives are classified into three categories, Hybrid, Distributed and Centralized, and they become the subjects of this evaluation. A similar analysis was done by Lee *et al*. However, they considered only operations at the mail center [6].

5.1 Decomposition into Hierarchy

The decision criteria, which could be applied in evaluating the alternatives for efficient operation of ADSS, compose of the second layer of the process, while the alternatives are laid in the lowest layer.

We considered the following eight factors as decision criteria.

C1. Technological consistency of the legacy postal automation
C2. Business connectivity with the existing procedures
C3. Expandability in deploying ADSS by stages
C4. Convenience of maintenance and upgrade
C5. Managerial efficiency of human resources
C6. Workload balance among the mail centers and delivery post offices
C7. Convenience of manipulation and handling
C8. Pervasive effects on the related industries in technological and economical aspects

5.2 Comparative Judgment and Priorities

First of all, comparative significance among the criteria was calculated. The results are as follows. C1 is the most important criterion in the decision-making process. For reference, the consistency ratio for pairwise comparison of each two criteria was lesser than 0.1. This means that all the comparisons are valid.

(0.3281, 0.1701, 0.2082, 0.0944, 0.0403, 0.0297, 0.0767, 0.0525)

Comparative significance among the alternatives by each criterion was also computed as shown in table 4. As stated in the comparisons above, the consistency ratios for pairwise comparison of alternatives were smaller than 0.1 for each criterion.

Table 4. Comparative significance among the alternatives by criterion

	C1	C2	C3	C4	C5	C6	C7	C8
Hybrid	0.2157	0.2014	0.3338	0.0833	0.1062	0.1744	0.3338	0.1989
Distributed	0.7231	0.7071	0.5247	0.1932	0.2605	0.3261	0.5907	0.5460
Centralized	0.0612	0.0915	0.1416	0.7235	0.6333	0.4995	0.0755	0.2550

Comparative priorities could be obtained by product of a matrix as shown in table 4 by an inverse matrix of the vector shown above. The significance of the each alternative produced that Hybrid=0.2279, Distributed=0.5791 and Centralized=0.1930. Based on these results, we concluded that the distribution policy would be a preferred alternative.

6 Conclusion

The primary purpose of this study is to propose the most optimum operation policy for ADSS by evaluating the executive alternatives.

In selecting the alternatives, we considered nine candidates excluding impracticable ones and classified them into three groups: Hybrid, Distributed and Centralized. We conducted both quantitative and qualitative analyses on said alternatives. The quantitative analysis was executed by comparing cost and performance of each alternative through simulations. Since the project of ADSS is intended for public service, we considered the qualitative factors and conducted a comparative evaluation in company with the experts including government officials like postmen. We used AHP as the analysis tool.

Combining these various evaluations together, the distribution policy where ADSS is installed and operated at the delivery post office has been chosen for a phase-in introduction of the systems in Korea.

References

1. Postal Automation Team: Development of Automatic Delivery Sequence Sorting Systems, ETRI (2001)
2. Postal Automation Team: A Survey for the Efficient Policy of Automatic Delivery Sequence Sorting, ETRI (2001)
3. Postal Services Policy and Planning Bureau: Japan's Postal Service 2001, Ministry of Public Management, Home Affairs, Posts and Telecommunications (2001)
4. Pritsker, A. B. et al.: Simulation with Visual SLAM and AweSim, John Wiley & Sons, New York (1997)
5. Saaty, T. L.: Fundamentals of Decision Making and Priority Theory with the Analytic Hierarchy Processing, University of Pittsburgh, Pittsburgh (1994)
6. Lee, H., Nam, Y., Kim, T.: A Study on the Optimum Operation Policy of Mail Center, Postal Information Review, Vol. 40 (2000) 73–92

Meeting Due Dates
by Reserving Partial Capacity in MTO Firms

Seung J. Noh[1], Suk-Chul Rim[2], and Heesang Lee[3]

[1] School of Business, Myongji University, 120-728 Seoul, Korea
sjnoh@mju.ac.kr
[2] School of Industrial and Information Engineering,
Ajou University, 442-749 Suwon, Korea
scrim@ajou.ac.kr
[3] School of Systems Management Engineering,
Sungkyunkwan University, 440-746 Suwon, Korea
leehee@skku.edu

Abstract. In many make-to-order production systems customers ask due date confirmed and met. Unexpected urgent orders from valuable customers usually require short lead times, which may cause existing orders in the production schedule to be delayed so that their confirmed due dates cannot be met. This imposes significant uncertainty on the production and delivery schedule in a supply chain. In this paper, we propose a new concept of capacity management, called the reserved capacity scheme, as a viable tool for accommodating urgent orders while meeting the confirmed due dates of all the scheduled orders. Simulation results show that the reserved capacity scheme appears to outperform simple FCFS policy in terms of the total profit attained.

1 Introduction

In the make-to-order (MTO) based manufacturing environment, customers usually place orders with fixed due dates. Due dates tend to be getting shorter these days mainly due to the upstream uncertainties in volatile market demand. It is also observed that prospective customers appreciate short lead times[3]. MTO firms capable of prompt and fast delivery tend to have a powerful marketing advantage over competing firms[2]. Thus MTO firms that can promise and deliver short lead times are likely to win more orders, and may earn higher profits than their competitors[10]. It is also pointed out that if an MTO firm promises its customer an overly-optimistic delivery date it risks penalties for tardy deliveries and diminished prospects for future business[3]. Thus, due date promising and on-time delivery are becoming one of the most important success factors in the market.

Promising and meeting due dates, however, is a difficult task and requires efficient capacity and lead time management[11]. Available-to-promise (ATP) logic has been widely used to confirm delivery dates by examining all finished and in-process goods[9]. In the MTO based manufacturing environment, however, suppliers usually have few standard products and volatile difficult-to-predict demands on a variety of end items. This makes it very hard to confirm due dates for incoming work orders solely by available inventory along with ATP logic. Especially when orders have

D.-K. Baik (Ed.): AsiaSim 2004, LNAI 3398, pp. 568–577, 2005.
© Springer-Verlag Berlin Heidelberg 2005

short lead times and production facilities are highly congested, decisions on whether or not to accept an order becomes very important in the context of due date promising for existing and incoming orders.

Researches on the due dates can be classified into two categories: order acceptance[7, 8, 12] and shop floor scheduling[1, 2, 4, 5, 6]. One important issue that has not received enough attention in the literature is the fact that suppliers usually have their own group of important customers who are significantly distinguished from other customers in terms of the sales volume and/or the strategic value, and whose requested due date must be met. In practices, those important customers often place orders with very short lead times even when the production facilities are highly congested. Overtime can be considered as an alternative to handle such urgent orders. However, such a remedy, unlike in the case of assembly industry, is not applicable to process industry.

In today's business environment, due to the increasing use of business solutions such as ERP or SCM, confirming due date becomes more and more important. Especially in the electronic commerce environment where sales, quotation, bidding, and contract are all processed in real time, confirming the due date of an order upon its arrival will become mandatory. Therefore, suppliers are asked to establish a well-defined order acceptance and scheduling policy that enables them to accommodate valuable urgent orders even in a congested period without causing any delays in producing scheduled orders. To the best of our knowledge, little work has focused on this issue. This motivated our study.

We propose an order acceptance and scheduling mechanism, called the reserved capacity scheme, which can help MTO firms effectively cope with urgent orders while meeting the due dates of all orders already in the schedule.

2 Capacity Reservation Scheme

As we propose a new concept, the production system we consider has been kept as simple as possible. While in the future we hope to extend our analysis to more complex systems, we start with a simple case of single production line that runs continuously for 24 hours a day; and orders arrive at any time in a day. An example is multi-item continuous production process of a typical petrochemical plant.

We consider situations where demands exceed short-run production capacity from time to time so that incoming orders are often conflicted with existing work orders in the production schedule. For instance, suppose that an order due on day D arrives when the production line is fully scheduled up to day D. Then, either the incoming order has to be refused with significant loss of potential revenue, or some of the orders already in the schedule have to be delayed with possible tardiness penalty. To resolve such a dilemma, we present in this paper the *reserved capacity scheme*, which guarantees the due dates of all accepted orders; while allowing certain level of urgent orders from important customers to be accepted in the schedule.

For the convenience of our discussion, but without loss of generality, we call an *urgent order* as the one whose due date is within 3 days from its arrival. All orders whose due dates are longer than 3 days are called the *regular orders*.

Daily production capacity is divided into two separate time segments (see Fig. 1.). The term *reserved capacity* is defined as a certain portion of the daily capacity reserved only for urgent orders. The remaining portion of the daily capacity is called *regular capacity*. Note that reserved and regular capacity do not designate specific time segments of a day. They simply indicate portions (e.g., 30%) of the daily capacity. Regular orders are assigned only to the regular capacity. Fig. 1 shows an example of a six-days schedule in which urgent (shaded block) and regular (white block) orders are assigned to the reserved (thick-lined time segment) and the regular (thin-lined time segment) capacity, respectively. Note that the regular orders that appear in days 1 to 3 are the ones that were assigned to the regular capacity at least three days ago.

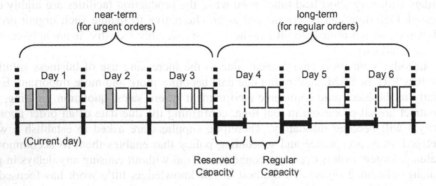

Fig. 1. An example of a six-day schedule using the reserved capacity

We say that the due date of an order is met if its production is completed by the end (24 o'clock) of the due date. An order is accepted if its due date can be met. Otherwise the order is refused. Day 1 is the current day. The reserved capacity is maintained on a daily basis. We assume the followings: 1) Each customer order comes with a fixed due date which is a hard constraint; if we cannot meet the due date we refuse the order. 2) An urgent order is due on day 1 (current day), day 2, or day 3; and regular orders are due on days 4, 5, or 6. No orders with due dates after day 6 are considered. 3) Urgent orders have twice as large marginal profit as regular orders; and setup time between two consecutive jobs are ignored (or can considered as part of the processing time). 4) Any order completed earlier than its due date is immediately shipped out to the customer without causing any inventory charge. 5) Production lot size equals to order size. 6) Lot splitting or pre-emption of an order is not allowed.

Let α denote the portion of the predetermined reserved capacity in a day (e.g. 30%); and let s_{kj} and r_{kj} denote the size (in units of time) of the j-th urgent and regular order scheduled on day k, respectively. Then, S_k and R_k, the remaining reserved and regular capacity of day k, respectively, are defined as in equations (1) and (2). We also define the remaining capacity of day k, denoted as DR_k, as in (3); and the cumulative remaining capacity from day D down to day 1, denoted as CR_D, as in (4).

$$S_k = \alpha \times 24 \text{ hours} - \sum_j s_{kj}, \quad \text{for } k = 1, 2, 3, \tag{1}$$

$$R_k = (1-\alpha) \times 24 \text{ hours} - \sum_j r_{kj}, \quad \text{for } k = 4, 5, 6, \tag{2}$$

$$DR_k = S_k + R_k, \text{ for } k = 1, ..., D, \tag{3}$$

$$CR_D = \sum_{k=1}^{D} (S_k + R_k). \tag{4}$$

Recall that we assume regular orders are due on days 4 to 6, and urgent orders on days 1 to 3. Once an order arrives, a response must be given to the customer as to whether the due date can be met or not. Consider an arriving order due on day D ($1 \le D \le 6$) with a processing time of t^*. A pseudo-code of the order acceptance rule is given in the following.

```
For regular orders (4 ≤ D ≤ 6)

Begin
   for k = D down to 4
      if(Rₖ ≥ t*) then
            Place the order at the end of day k.
            Return(update_1).
         end_if
   Reject the arriving order.
   end
end

For urgent orders (1 ≤ D ≤ 3)

begin
   for k = D down to 1
       if(DRₖ ≥ t*) then
            Place the order at the end of day k.
            Return(update_1).
         end_if
   end
   if(CR_D ≥ t*) then
            Place the order at the end of day D.
            Return(update_2).
   else Reject the arriving order.
   end_if
end

function(update_1)

   Slide all the scheduled orders in day k to the
   left by t*.
   Update Sₖ, Rₖ, DRₖ, and CR_D , Stop.
end_function

function(update_2)

   Sequentially slide scheduled orders in day D down
   to day 1 to the left by t* until no sliding is
   needed.
   Update Sₖ, Rₖ, DRₖ, and CR_D (k = 1 to D), Stop.
end_function
```

After finishing the last job scheduled on a certain day, we may find some of the daily capacity remaining idle. Then, we naturally look to the jobs scheduled on the next day and try to slide them to 'today'. This is because expensive facilities should not remain idle until the next day. Therefore, a mechanism is needed to advance (slide backwards to earlier times) jobs scheduled on the next day. Akkan[1], in a different context of job scheduling, used the term compaction for the sliding mechanism in which a set of jobs scheduled on the following day is advanced to 'now'. We will use the term 'advancing' with a slightly different definition. The term advancing is defined as sliding the first job scheduled on the next day to 'now' while the jobs thereafter remain unchanged. The reason we define the term advancing in this fashion is that we prefer to leave earlier part in time unscheduled for possible arrival of urgent orders with very short lead times. If no gap exists between the last job on 'today' and the first job on the next day, then no sliding

In a particular day when all the scheduled urgent and regular orders are completed, the first urgent order of size t^* scheduled in the following day (day 2) is advanced and processed. (If no urgent order is scheduled on the following day, then the first regular order in the schedule is advanced.) At that moment, DR_2 is increased by t^*, which will contribute to accepting new urgent orders in day 2. Although such advancing repeats, the remaining capacity in days 2 and thereafter will not indefinitely grow because new arriving orders will occupy the remaining capacity.

Fig. 2 (a) shows the schedule at the moment the last job scheduled on the current day (day 1) is completed at time t_0. (Recall that the regular order already processed on day 1 is the one accepted and scheduled at least three days ago.) Then, the first job (job A) scheduled on day 2 is advanced to time t_0 as shown in Fig. 2 (b). At that moment the remaining capacity of day 1 (day 2) is decreased (increased) by the amount of the processing time of job A. Note that the second job (job B) scheduled on day 2 and thereafter are not advanced since we want to maintain rooms for possible arrival of an urgent order (job C) due on day 1 during the processing of job A, as shown in Fig. 2 (b).

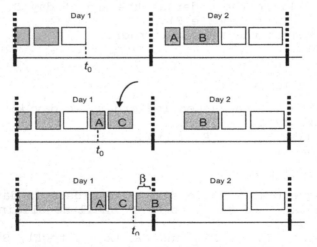

Fig. 2. An illustration of the advancing mechanism

An advanced job may be placed over the two consecutive days as shown in Fig. 2 (c). In such a case, the remaining capacity of day 2 is increased by some amount as denoted by β. Note that although the reserved capacity and regular capacity are separately operated for days 4 to 6 as shown in Fig. 1, and consequently some segments are unscheduled, the production continues without idle time as jobs are advanced one by one as shown in Fig. 2 (c). Production will pause only if there is no job waiting in the rest of the scheduling horizon.

In the job advancing mechanism illustrated so far, we only consider the policy where jobs are advanced one by one in the order they were scheduled. Alternatively, one may want to measure the incremental effect of more sophisticated advancing policies: what if the job with the shortest processing time is advanced first? It is expected that such an SPT-based rule will maintain more rooms in day 1, thereby facilitating the acceptance of arriving urgent orders due on day 1 at a cost of losing some arriving urgent orders due on day 2. Such a rule will be beneficial for the case where lead times of arriving urgent orders are extremely short.

One clear distinction of our work from other previous ones on job scheduling resides in the fact that we intentionally reserve a portion of the daily capacity unscheduled so that urgent orders can be accepted and inserted even in 'today'. This mechanism better fits the internet business environment where urgent orders arrive at any time in a day, asking immediate confirmation of shipping date.

3 Simulation Modeling

In the previous section we introduced the concept of reserving some of the daily capacity for urgent orders. In this section, we will examine how such a rather complicated scheme will ever contribute to improving the system performance by using computer simulation. As this paper is rather conceptual in nature, we will simply compare the performance of the reserved capacity scheme with the simple First-Come First-Served (FCFS) policy.

As a baseline, FCFS policy does not reserve any capacity for urgent orders, nor distinguish urgent orders from regular ones. All the orders are simply assigned to the schedule in the order of their arrival. When an order arrives with a due date D, the order is put to the first available time slot in the production scheduled by searching the schedule backwards from D down to day 1. This logic is simple and clear to both the customers and the supplier. It encourages the customers to ask longer due date since urgent orders may not be accepted. On the other hand, some important customers who frequently place urgent orders may be dissatisfied.

We use Arena 3.5 for the discrete event simulation of the proposed system. We examine three cases where the average percentage of urgent orders (PUO) is 20%, 25%, and 30% of the total number of arriving orders, respectively. For each case, the portion of the reserved capacity (PRC) varies from 20% to 50% of the daily capacity with an increment of 5%.

We use the following parameter values. Orders arrive according to a Poisson distribution with a mean interarrival time of three hours. The processing times of orders are uniformly distributed between four and six hours. The due dates of arriving urgent orders are distributed by 10%, 30%, and 60% on day 1, 2, and 3, respectively.

The due dates of arriving regular orders are evenly distributed over days 4 to 6. Simulation is conducted for 10,000 hours with initial 168 hours truncated.

The outputs of the simulation contain the net profit and the acceptance rate of urgent and regular orders under the FCFS policy and the reserved capacity scheme. The net profit is defined as the total profit attained from the accepted orders less the opportunity loss from the rejected orders. The profit attained from an order is usually dependent upon type and size of the order. In the experiment, however, we use an averaged size-independent fixed profit since the processing time is assumed to vary only from 4 to 6 hours. We let p_u (p_r) denote the profit obtained when an urgent (regular) order is processed. Similarly, we let q_u (q_r) denote the loss when an urgent (regular) order is lost due to lack of available capacity. It is known that urgent orders usually yield more profit than regular orders (Li and Lee, 1994). For simplicity, we assume that $p_r = q_r = 1$, and consider two cases where $p_u = 2$ and $q_u = 2$; and $p_u = 2$ and $q_u = 3$. The reason we consider the last case is that rejecting an urgent order may diminish prospects for future business, and in turn incur potential opportunity loss higher than the lost sales.

4   Simulation Results

Fig. 3 shows the net profit attained by using the reserved capacity scheme. The percentage of the reserved capacity (PRC) varies from 0% through 50% with $p_u = 2$, $q_u = 2$. Each line in the figure corresponds to the level of the net profit attained under PUO=20%, 25%, and 30%, respectively. The case where PRC=0% corresponds to the FCFS policy. The cases of PRCs smaller than 20% are not considered in the simulation because those PRCs are too small to accept an urgent order of a mean processing time of five hours.

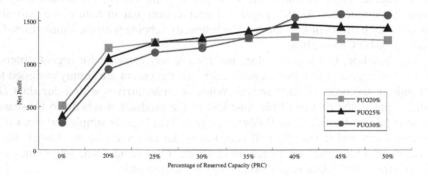

Fig. 3. Comparison of net profits for pu = 2 and qu = 2 under three levels of PUO

In Fig. 3, the net profit increases as PRC and PUO increase. That is, the more the urgent orders arrive, the larger the capacity should be reserved for the urgent orders. It is observed, however, that if excessive portion of the daily capacity is reserved for urgent orders, then the net profit slightly decreases since more of regular orders are rejected. A desirable level of the reserved capacity can be observed for each curve. For instance, for the cases where PUO=20% and 25%, reserving 40% of the daily capacity yields the highest net profit. For the case where PUO=30%, reserving 45%

yields the maximum net profit, which is almost five times higher than under the FCFS policy. Recall that PRC=0% implies the FCFS policy.

If we conduct the computer simulation for more accurate values of PRC and PUO, for instance, with an increment of every 1% instead of every 5% in PRC, then we can get close to the optimal PRC level. If we do this for various profit structures, then we may be able to express the optimal PRC level as a function of PUO and profit parameters, although the optimality will still depend on the distributions of the order arrival and processing times. We leave this as a future research topic.

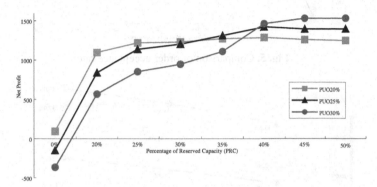

Fig. 4. Comparison of net profits for pu = 2 and qu = 3 under three levels of PUO

Fig. 4 shows the results of the case where $p_u = 2$ and $q_u = 3$. Compared with Fig. 3, the left half part of the net profit line is shifted downward. This implies that if the loss caused by rejecting urgent orders is more serious, FCFS policy may result in negative net profit. By comparing Fig. 3 and 4, we note that the reserved capacity scheme is tolerant of the changes in the opportunity loss values. That is, the two cases result in similar net profits for relatively high PRC levels. It should be pointed out that the profit/loss parameters be determined by evaluating not only the profit but also potential marketing and management strategies.

Fig. 5 shows the acceptance rate of urgent and regular orders for PUO=20%, 25%, and 30%. The level of PRC also varies from 0% to 50%. In the figure it is observed that the acceptance rate of urgent orders increases as the PRC level increases. This result agrees with the intuition that the larger the reserved capacity is, the more urgent orders will be accepted. It is also noted that an excessive reserved capacity (e.g., PRCs higher than 25% with PUO=20%) contributes little to accommodate more urgent orders. The same observation is made for the cases where PUO=25% and 30%.

The results of the acceptance rate can be used for estimating how much of the daily capacity to be reserved for various PUO levels. Even though our results are specific and parameter dependent, we can make the following observations. For instance, when 20% (25%, 30%) of the arriving orders are urgent ones, reserving 25% (35%, 40%) of the daily capacity will suffice for accommodating over 90% of the urgent orders. More specifically, by interpolating the acceptance rate of urgent orders, we can estimate a desirable level of reserved capacity as shown in Fig. 6. For example, if one wishes to strategically accommodate 95% of the urgent orders (i.e., 95% target service level) under PUO = 30%, then the amount of reserved capacity should be set to at least 43% of the daily capacity.

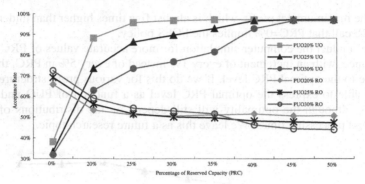

Fig. 5. Comparison of order acceptance rate

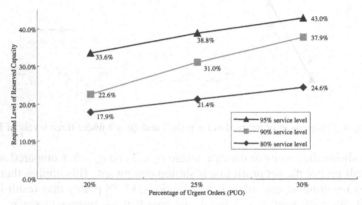

Fig. 6. Required reserved capacity for achieving target service level of urgent orders under three levels of PUO

5 Conclusions

In this paper we introduced a concept of capacity management to cope with volatile demands with short lead times for MTO firms; and investigated its impact on the system performance by using computer simulation. We define the reserved capacity as a portion of daily production capacity reserved only for urgent orders. By conceptually partitioning the daily production capacity into the reserved capacity and the regular capacity, both urgent and regular orders can be confirmed of their due dates at the time they arrive. Simulation results show that the reserved capacity scheme appears to outperform simple FCFS policy in terms of the acceptance rate of urgent orders and the total profit attained. Moreover, the reserved capacity scheme can contribute to stabilizing the production and delivery schedules in a supply chain by confirming and meeting due dates of all work orders.

We propose an order acceptance and scheduling mechanism, called the reserved capacity scheme, which can help MTO firms effectively cope with urgent orders while meeting the due dates of all orders already in the schedule.

Since this study addresses a new concept of order acceptance and operational schedule, lots of issues are to be investigated. Possible extensions may include the case where large-sized orders are to be split into two or more smaller lots, or the case where multiple lines of possibly unequal capacities are available. Another potentially fruitful area of subsequent research is an extension with determining the optimal level of reserved capacity for various levels of parameters: proportion of urgent orders, average profit ratio of urgent orders to regular orders, and probability distribution of due dates of urgent and regular orders.

Order confirmation by the use of the reserved capacity scheme can contribute to traditional business settings where customers and suppliers negotiate the due date. More appropriate business setting may be the one on the internet, where many anonymous customers place orders to many potential suppliers in an extremely competing market environment. As e-commerce grows rapidly, due date confirmation is becoming a key competitive factor for MTO firms. In order for the suppliers and the customers in a supply chain to synchronize their procurement, production, and delivery, due date confirmation will be critical for both parties. Suppliers should be able to not only meet all the confirmed due dates, but also cope with uncertain demands of short lead times. In this regard, the reserved capacity scheme will help MTO firms in various ways. The operation of the reserved capacity scheme we propose is generally applicable to any order based manufacturing industry. Detailed implementation should be in accordance with the characteristics of specific supply chain.

References

1. Akkan, C.: Finite-capacity scheduling-based planning for revenue-based capacity management. European Journal of Operational Research 100, (1997)170-179.
2. Duenyas, I.: Single Facility Due Date Setting with Multiple Customer Classes. Management Science 41, (1995) 608-619.
3. Easton, F., Moodie, D.R.: Pricing and lead time decisions for make-to-order firms with contingent orders. European Journal of Operational Research 116, (1995) 305-318.
4. Grey, M., Tarjan, R., wilfong, G.: One Processor scheduling with Symmetric Earliness and Tardiness Penalties. Math. Operations Research 113, (1998) 330-348.
5. Hendry, L.C., Kingsman, B.G., Cheung, P.: The effect of workload control(WLC) on performance in make-to-order companies. Journal of Operations Management 16, (2001) 63-75.
6. Katayama, H.: On a two-stage hierarchical production planning system for process industries. International Journal of Production Economics 44, (1996) 63-72.
7. Kate, Hans A.: Towards a better understanding of order acceptance. International Journal of Production Economics 37, (1994) 139-152.
8. Kingsman, B., Hendry, L., Mercer, A., de Souza, A.: Responding to customer enquiries in make-to-order companies: Problems and solutions. International Journal of Production Economics 46-47, (1996) 219-231.
9. Kise, H., Ibaraki, T., Mine, H.: A Solvable Case of the One-Machine Scheduling Problem with Ready and Due Times. Operations Research 26, (1998) 121-126.
10. Li, L., Lee, Y.S.: Pricing and Delivery-time Performance in a Competitive Environment. Management Science 40, (1994) 633-646.
11. Ozdamar, L., Yazgac, T.: Capacity driven due date settings in make-to-order production systems. International Journal of Production Economics 49, (1997) 29-44.
12. Wester, F.A.W., Wijngaard, J., Zijm, W.H.M.: Order acceptance strategies in a production-to-order environment with setup times and due-dates. International Journal of Production research 30, (1992) 1313-1326.

Design a PC-CAVE Environment
Using the Parallel Rendering Approach

Jiung-yao Huang[1], Han-Chung Lu[2], Huei Lin[3], and Kuan-Wen Hsu[3]

[1] Department of Communication Engineering, National Chung Cheng University
Chaiyi 62107, Taiwan
comjyh@ccu.edu.tw

[2] Department of Computer Science and Information Engineering,
Fortune Institute of Technology
Haohsiung 842, Taiwan
lu0908@center.fjtc.edu.tw

[3] Department of Computer Science and Information Engineering, Tamkang University
Tamsui 251, Taiwan
amar@ms15.hinet.net

Abstract. In this paper, we present a frame synchronization mechanism for a parallel rendering system to create an immersive virtual environment like CAVE. Our goal is to create a synchronized multi-display parallel rendering system built by the PC cluster. We implemented the proposed frame synchronization algorithm in both hardware and software approaches to investigate whether a special hardware is needed for the PC-CAVE architecture. Our experiments show that the hardware approach can achieve more accurate frame synchronization effect but the software approach provides better flexibility in building such a system.

1 Introduction

The CAVE™ is a high-resolution and wider-view virtual environment for multiple participants [1]. It was designed by the Electronic Visualization Laboratory at the University of Illinois at Chicago in 1992 and first presented in SIGGRAPH93. The CAVE system used a surrounding multiple-screen projection technique to provide a large-scale display for the user. The large physical size of each display screen (such as a room-sized wall) of the CAVE system allowed the user to interact with the rendered objects at their natural sizes. This function can be critical for perception and evaluation of 3D models [2].

In order to achieve high performance, the prototype of CAVE was built with four SGI high-end workstations to render images, respectively, on the left, front, right, and floor display walls to create a surrounded image [1]. In addition, they used the fiber optic reflective memory, which was a cache shared device, to synchronize frames update among these display walls. Our research goal is to constructed on the PC cluster with off-the-shelf components. The rationale of designing a PC-CAVE system is to employ the parallel rendering technique with the image on each display wall rendered by a PC. The key issue of this approach is for the frame synchronization mechanism to tightly synchronize the frame rates among the parallel rendering PCs.

D.-K. Baik (Ed.): AsiaSim 2004, LNAI 3398, pp. 578–588, 2005.
© Springer-Verlag Berlin Heidelberg 2005

Molnar et al. [3] defined the parallel rendering mechanism as a sorting problem. They first pointed out that the parallel rendering system required a feed forward pipeline mechanism which was composed of two main tasks: the geometry processing and the rasterization. Hence, the parallelism of the rendering was achieved by redistributing, or called sorting, the data among rendering processors. They classified parallel rendering mechanisms into sort-first, sort-middle, and sort-last methods according to the location of the data redistribution.

The sort-first method means redistributing the raw primitives whose screen-space coordinate is unknown early in the rendering pipeline and during the geometry processing. By a pre-transformation stage, the raw primitives are redistributed to the appropriate rendering processors. Hence, the sort-first method only requires low communication bandwidth, especially when the frame-to-frame coherence[4] is employed. The Scalable Display Wall at Princeton University [2] is an example of the sort-first technique.

In the sort-middle approach, screen-space primitives are redistributed between the geometry processing and the rasterization. The sort-middle method requires each geometry processor to be able to directly communicate with all of the rasterization processors and this limits its scalability [5].

Finally, the sort-last method defers the sorting until the end of the rendering pipeline. That is, the sorting is executed after primitives have been rasterized into pixels, samples, or pixel fragments. A rendering processor transfers these pixels to the compositing processors which resolve the visibility of pixels from each rendering processor. Blanke et al. [6] used the sort-last method to design their parallel rendering architecture. They used a special hardware named the Metabuffer to composite the final image for display.

After having studied these three parallel rendering approaches, the sort-first method was chosen as the groundwork of our research. The reasons are as follows. First of all, the sort-first method requires lower data transferring bandwidth than the others, especially when the frame-to-frame coherence is manageable. Secondly, the sort-first method preserves the entire rendering pipeline so that we can easily optimize the system performance by using the off-the-shelf 3D graphic acceleration boards. In the following sections, the parallel infrastructure of the PC-CAVE system is given first. The frame synchronization mechanism then follows. At last, the experiments of both hardware and software implementation of the proposed synchronization mechanism are given to study the efficiency of the proposed synchronization mechanism.

2   The Sort-First Parallel Rendering Architecture

For the sort-first method, two issues that may handicap the system must be carefully avoided. One is the load-imbalance when the objects are clumped together in a region. The other is the overhead of pre-transformation of the objects before they are sorted. We employ a master/slave architecture to design our sort-first parallel rendering system. Our goal is to research an inexpensive CAVE system on a PC cluster with off-the-shelf 3D graphic acceleration boards. Since the 3D graphic acceleration board already has the geometric acceleration function on it, we can take advantage of this feature to design the sort-first parallel rendering system. For example, Fig. 1 illustrates the rendering pipeline of the graphics system [7]. The rendering pipeline can be further classified into the Geometry subsystem and Rasterization subsystem. The off-the-shelf 3D graphic acceleration board often hardwires the right part of the geometry subsystem to speed up the computation of each frame. Hence, we can bisect the rendering pipeline of the Geometry subsystem into two parts. One is executed in the master host and the other is executed in the rendering slave.

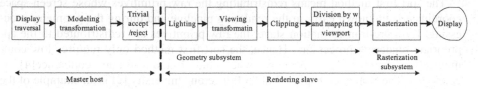

Fig. 1. The rendering pipeline

That is, as illustrated in Fig. 1, after the master host updates the scene tree by the "Display traversal" module, the master host performs the view-dependent culling for each rendering slave. In other words, the master host "sorts" the visible 3D primitives of the scene for each rendering slave based upon their respective view volumes. The master host then sends the sorted 3D primitives to each render slave for it to compute and rasterize its frame image. Since all the rendering slaves are equipped with 3D graphic acceleration boards, the slaves can fully utilize the hardwired accelerating function to speed up their frame displays.

The distribution of the sorted 3D primitives among the rendering slaves is another issue that has to be carefully studied when a CAVE system is designed by a PC cluster. In order to prevent the master host from overwhelm by the distribution of the sorted 3D primitives to the rendering slaves, similar to [2] and [8], we replicated the entire scene database to all of the rendering slaves. In this way, instead of transmitting the geometric information of a 3D primitive, we only need to distribute the unique identification numbers of the sorted 3D primitives. We call this sequence of primitives' identification number as the *V-list (view volume list)*. Depending upon the cases, each element on the V-list contains different types of values. Table 1 summarizes the contents of the V-list in various cases. Notice that we assume that the rendering slave already has the geometric information of each 3D primitive in a separate file in its local storage device. Furthermore, we also assume that the animation sequence of the 3D primitive is saved in the local storage device of the rendering slave.

Table 1. The contents of the V-list

Case	Transmitted Data
State change of a primitive	Object ID, Attribute type list, Value list
Add a new primitive	Object ID, Attribute type, Parent ID, File name of this primitive
Delete a primitive	Object ID
Animated object	Object ID , Time slice

3 The Rationale of Frame Synchronization

After the parallel rendering architecture has been decided, the next issue is the mechanism to synchronize the frame rates among the paralleled rendering slaves to avoid the "tearing" effects across screen boundaries of the resulted surrounded display [9].

3.1 The Concept

Our study on parallel rendering is based on the *sequence of parallel computations*[10], which allows asynchronous execution among computational modules, called phases, and coordinate messages interchange among them, called the phase interface. In this way, the phases of the problem may be solved in parallel whereas these phases are scheduled in a sequential manner. Hence, our parallel rendering research can be illustrated as in Fig. 2.

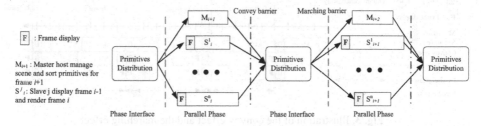

Fig. 2. A sequence of parallel rendering phases

That is, our parallel rendering system is composed of the following sequences of computation:

- Parallel phase: All nodes, including the master host and the rendering slaves, process their tasks for a frame simultaneously.
- Phase interface: The master host distributes the sorted 3D primitives among the rendering slaves for the next frame.

3.2 Parallel Phase

For each frame, all nodes of the parallel rendering system process their tasks asynchronously in the parallel phase. The master node computes or sorts the V-lists for all rendering slaves according to their view volumes. At the same time, the slaves render their frames according to their V-lists that were received from the master host at the pervious parallel phase. It means that, when the slaves are rendering the n_{th} frame in the parallel phase, the master is dealing with the sort processing for the $(n+1)_{th}$ frame. In this way, the master will not lie idle and so the system efficiency is improved.

To synchronize the frame images rendered by different computers, the slave who has finished its rendering task should not rasterize that frame immediately otherwise the tearing effect will occur. Meanwhile, the master should not continue its sorting process after it has finished processing the current frame otherwise a stalling delay will be accumulated. That is, they all have to wait for the slowest one. This action is called *Convoy effect*. We name it this because all the other nodes must convoy the slowest one and cannot proceed arbitrarily. When the slowest one completes, it means the convoy effect also finishes. The system is then switched to the phase interface. So, we can say that the end of the convoy effect is the end of the parallel phase.

3.3 Phase Interface

The end of the parallel phase implies that all of the slaves have completely rendered their current frame and are ready to rasterize the images. It also means that the master host has already finished sorting the scene objects of the next frame for each slave. Hence, it is time for the slaves to receive their V-lists of the next frame and the master host to accept the user's input for the following frame.

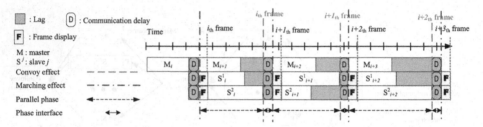

Fig. 3. Illustration of the convoy effect and the marching effect

After the phase interface is completed, the system will return to the parallel phase by notifying the slaves to rasterize the pervious frame images and to render their next frames. This action is called *Marching effect*. Hence, the end of the marching effect is the end of the phase interface. This also implies that the synchronization of this frame is completed successfully. Fig. 3 shows how the convoy effect and the marching effect work together.

3.4 The Controller

In order to achieve a tight coordination among the hosts, a special process, called the *controller*, is proposed to serialize and manipulate the convey effect and the marching effect. When the controller receives the acknowledgment signals from all the rendering slaves, it learns that all the slaves have completed their current parallel phases. It means that the controller is notified that the frame image has already been rasterized and ready to be displayed. The controller then informs this event to the master. Upon receiving this event, the master begins to forward immediately the V-lists of the next frame to each respective slave. The master then responds the controller with an advance signal for it to trigger the rendering slaves to display their respective images and the frame synchronization for the current frame is completed. Hence, the frames of the parallel rendering system are fully synchronized by looping the Parallel phase and Phase interface among the hosts.

4 The Frame Synchronization Algorithm

To verify the proposed frame synchronization mechanism in a parallel rendering environment, a synchronization layer is designed and implemented to handle the convoy effect and the marching effect on each host. That is, this synchronization layer is executed on each host as the middle layer to coordinate the parallel phase and the phase interface among them. Fig. 4 depicts the result of adding the synchronization layer to the system architecture.

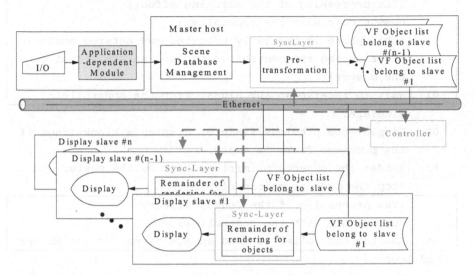

Fig. 4. The synchronized master-slave sort-first parallel rendering system architecture

Hence, combined with the synchronization layer and the controller, the algorithm for the master host can be illustrated as follows.

```
Loop:
     {The processing of the Marching effect}
     communicate with the controller:
     {This phase is initially executed and entered again
     upon receiving an ACK signal from the controller}
01   Ask the controller to send an ADV signal to all
     rendering slaves;
     Simulate:
02   Receive and process the user's input;
03   Manage the scene data;
04   Pre-transform the scene data;
05   Sort the visible objects/primitives according to the
     view volume of each rendering slave(V-lists are
     generated for each rendering slave);
```

```
      {The processing of the Convoy effect}
      communicate with the controller:
  06  Wait for the ACK signal from the controller
End-loop
```

Similarly, the algorithm for the slave can be illustrated as follows.

```
Loop:
      {The processing of the Marching effect}
      communicate with the controller:
      {This phase is initially executed and entered again
      after asking the controller to send an ACK signal to
      the master}
  01  Wait to receive an ADV signal from the controller;
      Simulate:
  02  Display the previous frame or display a blank image if
      the previous frame is empty;
  03  Render the objects/primitives sent by the master and
      stop the process before display;
      {The processing of the Convoy effect}
      communicate with the controller:
  04  Ask the controller to send an ACK signal to the master
End-loop
```

Finally, the algorithm for the controller is as follows.

```
Loop:
      {The marching effect processing center}
      communicate with the master:
  01  Send an ADV signal to all slaves upon receiving the ADV
      request from the master, otherwise wait for the
      master's command;
      {The convoy effect processing center}
      communicate with the slaves:
  02  Send an ACK signal to the master if all slaves have re
      sponded an ACK signal, otherwise wait for the
      remainder(s) slaves which are still busy;
End-loop
```

From the algorithm of the controller, we can see that the controller only needs a counter to collect the ACK signals from all the rendering slaves to know when the convoy effect is finished. If all rendering slaves successfully return an acknowledgement signal, it will then inform the master host about this event.

Similarly, the controller will trigger the Marching effect on the rendering slaves upon receiving the request from the master. All the slaves then display their images before continuing to render their next frames. Obviously, the controller acts only as a messenger and does not carry out any computation. Hence, it will not become the bottleneck of the system performance.

5 Experimental Results

Two questions were raised when the synchronization mechanism was implemented. One was the accuracy of the proposed mechanism. The other was whether the controller should be implemented in hardware or in software. To answer these questions, two experiments were designed and conducted. The first experiment was to compare the frame images of the same system configuration with and without the proposed frame synchronization mechanism. The second experiment was to hardwire the controller first. The efficiency of this hardware version of controller[11] was then compared with the software one that was implemented on the local Ethernet network. The parallel rendering system for all of these experiments were constructed by one master and three slaves to render, respectively, the front, right, and left screen displays. All of them were running on Windows 2000 Professional system.

5.1 The Accuracy

In order to increase the frame delay among the rendering slaves, all of the slaves were purposely equipped with different levels of computation and rendering capabilities. For example, Slave 1 had the CPU of Intel Celeron 1GHz and running NVIDIA Ge-Force4 MX440 64MB graphic acceleration board. Slave 2 was equipped with Intel Pentium IV 2.4GHz CPU and NVIDIA GeForce3 Ti 500 64MB graphic acceleration board where as Slave 3 had Intel Pentium IV 1.7GHz CPU and NVIDIA GeForce3 Ti 500 64MB graphic acceleration board. The master host had the same capability of computation as Slave 2. Fig. 5 is the snapshot of the test scene that contained 14187 polygons and 17325 vertices for this experiment.

Fig. 5. Test scene #1 – The Helicopter

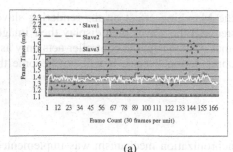
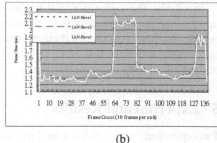

(a)   (b)

Fig. 6. The comparisons of with and without the frame synchronization mechanism. (a) Without the frame synchronization mechanism; (b) With the frame synchronization mechanism

Fig. 6(a) shows the result of the frame rates of the system that does not have the frame synchronization mechanism. The resulted chart shows that the frame rates of rendering slaves were significantly divergence. Even when Slave 2 and Slave 3 were finished, Slave 3 was still struggling to render the remaining scene data.

For this experiment, the algorithm of the controller was software implemented in the master host. That is, the controlled was programmed in the master host to transmit the ACK and ADV signals over the local Ethernet network. As shown in Fig. 6(b), the resulted frame rates on these three rendering slaves were very identical and the image display was fairly smooth. However, the system performance dropped due to the convoy effect that stalled all nodes to wait for the slowest one, i.e. Slave 1 in this case. Hence, it was obvious that the proposed mechanism successfully synchronized the frame display of the parallel rendering system.

5.2   Software vs. Hardware

All nodes in this experiment were equipped with the same peripherals. Different from the previous experiment, the controller was implemented both in hardware and software. The hardwired controller is called the Synchronization Box [11]. The Synchronization Box is a special hardware that uses the parallel port on each host to transmit the ADV and ACK signals. The software implementation of the controller is the same as that of the previous experiment.

Fig. 7 shows the snapshot of the test scene which was a building composed of 2079 polygons and 3877 vertices. The scenario of this experiment was to quickly rotate the view cameras of the rendering slaves in the corners of the building.

Fig. 7. Test scene #2 – The virtual building

The purpose of rotating the view cameras was to detect the tearing effect which could be easily observed on the edges or corners of two walls. The results of the experiment are charted in Fig. 8.

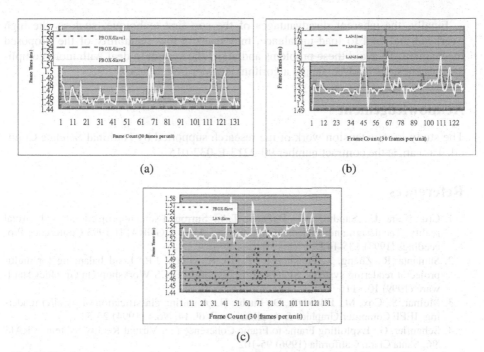

Fig. 8. The hardware vs. software approach. (a) The hardware approach – synchronization by the Synchronization Box [11]; (b) The software approach – send synchronization signals over the local Ethernet network; (c) Average performance comparison

Comparing Fig. 8(a) with Fig. 8(b), we notice that the synchronized frame rate and system performance are both significant. The hardware approach only induced less than 0.5 frame of out-of-sync error whereas the software approach suffered from around 1.5 frames. Furthermore, Fig. 8(c) shows that the system performance that used the Synchronization Box was about 3 frames higher than the software approach. This difference may be induced by the protocol of the local Ethernet network which is much more complicated than the parallel port.

6 Conclusion and Future Works

This paper presents a frame synchronization mechanism for the parallel rendering system on a PC cluster that employed the sort-first technique to distribute the scene data. The proposed mechanism is based on the sequence of the parallel computation method. The proposed mechanism successfully avoids the tearing effect occurring from either the discrepancy in the computation capacity or the different complexity of the scene data on the rendering slaves.

According to the result of our experiments, we find out that using the special synchronization hardware or device can react to the synchronization signals in no time which achieves the best outcome. However, this proprietary hardware has the shortcoming of less flexibility.

Finally, the inherent disadvantages of the sort-first technique, such as the high overhead and tend to load-imbalance, may cause a low scalability of the proposed mechanism. To avoid these problems and allow the system to deal with more complicated scenes, the load-balancing algorithm is under investigation.

Acknowledgement

The study is an extension work of the research supported by National Science Council, Taiwan, at the contract number 90-2213-E-032-015.

References

1. Cruz-Neira, C., Sandin, D.J., DeFanti, T.A.: Surround-screen projection-based virtual reality: The design and implementation of the CAVE. SIGGRAPH 1993 Conference Proceedings (1993) 135-142
2. Samanta, R., Zheng, J., Funkhouser, T., Li, K., Singh, J.P.: Load balancing for multi-projector rendering systems. SIGGRAP/EUROGRAPHICS Workshop On Graphics Hardware (1999) 107-116
3. Molnar, S., Cox, M., Ellsworth, D., Fuchs, H.: A sorting classification of parallel rendering. IEEE Computer Graphics & Applications, Vol. 14, No.4 (1994) 23-32
4. Schaufler, G.: Exploiting Frame to Frame Coherence in a Virtual Reality System. VRAIS '96, Santa Clara, California (1996) 95-102
5. Mueller, C.: The sort-first rendering architecture for high-performance graphics. ACM SIGGRAPH Special Issue on 1995 Symposium on Interactive 3-D Graphics (1995) 75-82
6. Blanke, W., Bajaj, C., Fussell, D., Zhang, X.: The metabuffer: A scalable multi-resolution multidisplay 3D graphics system using commodity rendering engines. Technical Report Tr2000-16, University of Texas at Austin (2000) Available at http://www.cs.virginia.edu/~gfx/Courses/2002/BigData/papers/ [Accessed 2004/9/1]
7. Molnar, S., Fuchs, F.: 18.3 Standard Graphics Pipeline. Computer Graphics : Principles and Practice in C, 2nd edn. Addison-Wesley Pub Co. (1997) 866-873
8. Samanta, R., Funkhouser, T., Li, K., Singh, J.P.: Hybrid sort-first and sort-last parallel rendering with a cluster of PCs. SIGGRAP/EUROGRAPHICS Workshop on Graphics Hardware (2000) 97-108
9. Using Wildcat's Genlock and Multiview Option in Visual Computing Applications. Available at http://www.3dlabs.com/product/technology/Multiview.htm [Accessed 2004/9/1]
10. Chandy, K., Misra, J.: Asynchronous Distributed Simulation via a Sequence of Parallel Computations. Communications of the ACM, Vol. 24, No.11 (1981) 198-206
11. Huang, J.Y., Bai, H.H.: The synchronization algorithm for constructing CAVE system on the PC cluster. The Seventh International Conference on Distributed Multimedia Systems (DMS'2001), Tamsui, Taiwan, (2001) 135-141

VENUS: The Online Game Simulator Using Massively Virtual Clients

YungWoo Jung[1], Bum-Hyun Lim[1], Kwang-Hyun Sim[1], HunJoo Lee[1],
IlKyu Park[1], JaeYong Chung[1], and Jihong Lee[2]

[1] Electronics and Telecommunications Research Institute,
Networked Virtual Environment Research Team, Digital Content Research Division,
161 Gajeong-dong, Yuseong-gu, Daejon, 305-350, Korea
{tristan,ibh63427,shimkh,hjoo,xiao,jaydream}@etri.re.kr
[2] Chungnam National University,
Department. of Mechatronics Engineering., Chungnam National University,
220 Kung-dong, Yuseong-gu, Daejon, 305-764, Korea
jihong@cnu.ac.kr

Abstract. In this paper, we present an efficient method for simulating
massively virtual clients in an online game environment. Massively multiplayer online games and other multi-user based networked applications
are becoming more attractive to the gamer players. Such kind of technology has long been researched in the area called Networked Virtual
Environments. In the game development process, a set of beta tests is
used to ensure the stability of online game servers. A set of testing processes consumes a lot of development resources such as cost, time, and
etc. The purpose of the *VENUS* (Virtual Environment Network User
Simulator) system is to provide an automated beta test environment to
the game developers to efficiently test the online games to reduce development resources.

1 Introduction

Recently, the online game market is growing quickly and becoming a big business [1]. The successful story of Ultima Online [2], Dark Age of Camelot [3]
and EverQuest [4] shows the possibility of the online game market. But this
market value raises the keen competition between online game companies. As
online game market changes, the stability of the game server has become a very
important requirement for playing a game. If some user has been lost his/her
avatar's experience, weapon or item frequently due to its unstable game server
during the game, it will become a major flaw for turning game players' face
away regardless of high quality of graphic or sound of the game. In general, the
game developers use a set of beta tests to archive the stability of online game
server. The general process of a beta test takes several steps as follow. First, the
game company gets a list of candidate testers to perform the beta test throughout the advertisement. Second, the selected beta testers (i.e. game players) play
the game and send the feedback information to the game developers. Third,

D.-K. Baik (Ed.): AsiaSim 2004, LNAI 3398, pp. 589–596, 2005.
© Springer-Verlag Berlin Heidelberg 2005

the game developers analyze the error reports from the feedback and debug the game application. However, this ordinary testing process consumes a lot of development resources such as cost, time, and etc. In many cases, the beta testers' error reports give a vague clue of an error to the developers and make it so hard for the developers to repeat the error environment in exactly the same condition that the beta tester had. This kind of testing strategy is hard to apply to the console based online game compared with PC gaming area. Unlike PC games, the console based online games cannot use the same testing strategy. Further, the console based games are distributed on read-only media (i.e. CD, DVD, Rom-Pack, etc). If the critical error or bug is found in the distributed games, then they have to be recalled. The server testing technology using a virtual client has been studied very actively on file server [5], web server [6] and etc. Online game server testing introduces the new research issues into this area. [7–9]. The purpose of the *VENUS* (Virtual Environment Network User Simulator) system is to provide a beta test environment to the game developers to efficiently test the online games to save development resources. The advantage of the VENUS system is generality and scalability. In this paper, the VENUS system will be illustrated by a three categories. First, several structural characteristics of our system will be explained. We will explain the components in the VENUS system, the generalized communication protocol and a dual layered management structure of the VENUS system for supporting massive multi-player simulation. Second, the application area of the VENUS system will be explained. For this, we will show the two experimental examples of the VENUS system. Third, we will conclude our paper with future work.

2   Architecture of the VENUS System

2.1   Components of the VENUS System

Fig. 1 illustrates the five elements of the VENUS system. A *Virtual Client* (VC) is an activity entity of a single online game user. The game servers cannot distinguish a VC from a human game player. A VC gives a set of stresses to a game server or a collection of game servers during the online game process. To support making a VC by user of the VENUS system, the VENUS system provides VC Engine. VC Engine defines a communication protocol between a VC agent and a VC. VC Engine supports two types of VC. The one is a process based VC and the other is a thread based VC. The difference of VC types is illustrated in Table 1. VC Agent is a bridge between a *Central Engineering Station* (CES) and a VC. A VC agent receives a control command from a CES and sends control commands to a VC. At the same time, VC Agent supports intelligent control of VCs. There exists one VC agent per VC host. The maximum number of VCs on a VC host is dynamically controlled by the VC agent based on a performance index of the VC host. CES is the main user interface program. The simulator user can control the VENUS system using CES. It also supports a script interface to provide more programmatic and complex simulation. In addition, CES can be able to save the monitoring data of game server to a database. The database

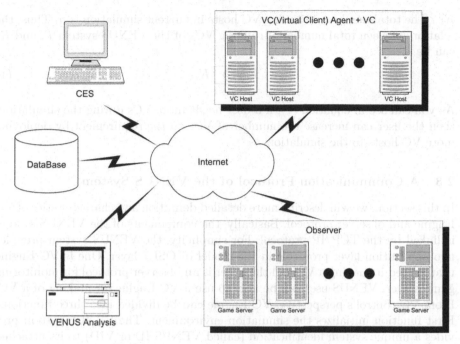

Fig. 1. Block diagram of the VENUS system.

Table 1. The difference of a VC making style.

VC Type	Process based VC	Thread based VC
Number of VC in a VC host	Low	High
Convenience of VC implementation	High	Low
Response Time of VC	Slow	Fast
Form of VC Engine	Socket Protocol	DLL Library

saves a set of monitored data during the simulation and provides monitoring data to the analysis tool in the VENUS system. The monitored data includes not only a performance index on game server but also internal game data on game server. To support flexible data monitoring, a VENUS user can be able to use user defined data in the Observer protocol. The VENUS Analysis is a kind of database searching tool. A user can analyze the monitored data from the database with this tool.

2.2   A Dual Layered Management Structure of the VENUS System

Because a VC host has a limit in number of running VCs, the VENUS system provides a dual layered management structure for supporting massively multi-player simulation. A dual layered management structure is composed of CES and VC Agent. We consider a VC host i with a limit in number of running VCs

K_i . The total number of running VC hosts in current simulation is n. Thus, the relation between total number of running VCs of the VENUS system T, and K_i can be

$$T = \sum_{i=1}^{n} K_i \tag{1}$$

As you can see in equation (1), if a user needs more VCs during the simulation, then the user can increase the number of VCs to the requirement by deploying more VC hosts to the simulation.

2.3   A Communication Protocol of the VENUS System

In this section, we will describe more detailed definition and characteristics of VC Engine and observer protocol. Basically, the components of the VENUS system is linked by the TCP/IP protocol. For simplicity, the VENUS system provides two application layer protocols in the model of OSI 7 layer. One is VC Engine that is used in making a VC and the other is an observer protocol for monitoring game server. VENUS user can be able to use a VC Engine for making of a VC. From VC control's perspective, VC Engine can be divided into three functions. First function initializes the simulation environment. The VENUS system provides a unique system identification (called VENUS ID or VID) to its attached VCs and controls them to login/logout of the game. Second function is related to VC operations in the online game process. A VC can be instructed to move in the game world, changed the characteristics of the avatar and made to perform various actions of avatar. Third function is related to *Quality of Service* (QoS) for VC. After sending any service request from VC to the online game server, the service response time can be measured by this function. To monitor a game server, a VENUS user can be able to use the observer protocol. The major functions of the observer protocol are reporting the performance index (CPU, memory, network bandwidth usage) of game server and transferring the user defined data for checking the run time error in the game server. Because reporting the performance index of a game server is a default action of the observer protocol, the user can get the performance index automatically. However, the method of transferring the user defined data is somewhat different. A user must set the user defined data type and the length of user defined data to CES using an observer protocol during the initialization of the VENUS system. Based on this information, CES makes database table dynamically and saves user defined data to the database. Fig. 2 illustrates the process of setting the user defined data and its example.

3   Experimental Result of the VENUS System

3.1   Application of the VENUS System on Simple Online Game

For usability test of the VENUS system, we apply the VENUS system to a game server. As can be seen from the Fig. 3, the testing environment is composed of

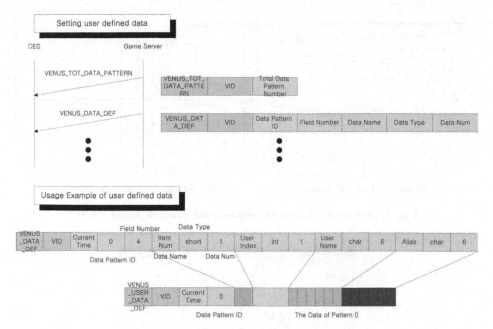

Fig. 2. Block diagram of the VENUS system.

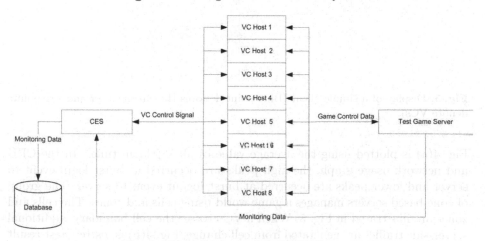

Fig. 3. Block diagram of the VENUS system.

the CES, a database, eight VC hosts and a test game server. Our process based VC can be instantiated 150 times per VC host, the total number of VCs during this simulation is 1200. A test game server group is composed of one login server and four zone servers. The purpose of this test is testing the stability of the login server, the stability of avatar's moving algorithm in zone servers and the efficiency of AOI [10] (Area of Interest) algorithm to reduce message between game players. We put a login server to the stress test that is composed of eight hundreds VCs sending a login requests at once. The response time graph in

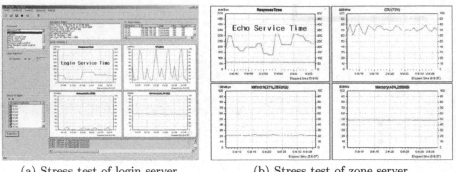

(a) Stress test of login server.   (b) Stress test of zone server.

Fig. 4. Stress test of a simple online game example using VENUS.

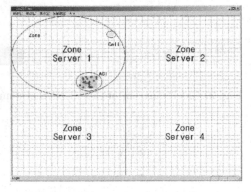

Fig. 5. Display of a simple client (Blue point denotes the current user and red points denote VCs).

Fig. 4(a) is plotted using the average value of all VC login times. In the CPU and network usage graph, the high peaks are occurred at burst login event to server and lower peaks are occurred at burst logout event to server. The group of zone based servers manages a game world using cells and zones. The cells and zones are illustrated in Fig. 5. When a VC crosses the cell boundary, additional server-side traffics are generated from cell change. Fig. 4(b) is a stress test result of this phenomenon in the zone based server. Because the initial positions of VCs are not evenly distributed and the cell events are randomly changed, the usage of CPU and network fluctuates near the average.

3.2 Application of the VENUS System on Commercial MMORPG

Herrcot [11] is an MMORPG game in its open beta test period. We put this game to the simulation at the third closed beta test. The login procedure of Herrcot is through login, world and zone server. The response time in Fig. 6(a) illustrates the processing time of each server at the login time. Fig. 6(b) is a scene from

(a) Stress test result from login server.

(b) Game screenshot during VC login to login server.

(c) All VC is controlled to be seated.

(d) VC is controlled to move to specified position.

Fig. 6. Stress test of commercial MMORPG server using VENUS.

Herrcot when VCs logged in to game. Fig. 6(c) is a scene from Herrcot when VCs are received the seat command. Fig. 6(d) is a screen shot captured when the VCs move to a specified position.

4   Conclusions

In this paper, we proposed an online game simulator to efficiently test online game environment. The purpose of the VENUS system is to support a beta test environment for the game developers to efficiently test the online games to reduce development resources. With VC Engine in the VENUS system, the user can easily simulate any kind of online games for beta test. If enough computer systems are used with VC Engine, the user can make a large number of VCs to simulate the online game environment. Further, it is very helpful for the game developers because it stores an internal game data from the game server to the database. The VENUS system is applied to two different online games to test its usefulness. These tests show the efficiency of the VENUS system and possible points of improvements. A couple of issues should be elaborated on further and have been under our investigation. Throughout the various kinds of simulations, we will provide powerful and efficient functionalities in the VENUS system. Our future works are to improve our VENUS system and make it as a standard model for online game testing.

References

1. DFC Intelligence: The Online Game Market 2003, DFC Intelligence, (June 2003).
2. Electronic Arts: http://www.ea.com.
3. MYTHIC Entertainment: http://www.darkageofcamelot.com.
4. Sony Online Entertainment: http://everquest.station.sony.com.
5. David Arneson, Thomas Ruwart: A Test Bed for a High-Performance File Server., Putting all that Data to Work, Proceedings, Twelfth IEEE Symposium on Mass Storage Systems, Apr (1993) 26–29.
6. Elbaum, S., Karre, S., Rothermel, G.: Improving web application testing with user session data, Proceedings. 25th International Conference on Software Engineering, **3-10**, May (2003) 49–59.
7. Larry Mellon: Automated Testing of Massively Multi-Player Systems: Lessons Learned from The Sims Online, GDC 2003, Spring.
8. Adobbati, R., A.N., Scholer, A., Tejada, S., Kaminka, G.A., Schaffer, S., Sollitto, C.: Gamebots: A 3D Virtual World Test-Bed for Multi-Agent Research, Proceedings of the Second International Workshop on Infrastructure for Agents, MAS, and Scalable MAS, Montreal, Canada, 2001.
9. Wu-chang Feng, Francis Chang, Wu-chi Feng, Jonathan Walpole: Provisioning Online Games: A Traffic Analysis of a Busy Counter-Strike Server, Proceedings of the second ACM SIGCOMM Workshop on Internet measurment workshop, Nov, 2002.
10. K.H.Shim, etc.: Design and Analysis of State Update Rate Control Schemes for Performance Improvement in Networked Virtual Environments, Proc. of the SCI 2002, August, 2002.
11. NaonTech Co., Ltd: http://www.herrcot.co.kr.

Error Metric for Perceptual Features Preservation in Polygonal Surface Simplification*

Myeong-Cheol Ko[1], Jeong-Oog Lee[1], Hyun-Kyu Kang[1], and Beobyi Lee[2]

[1] Dept. of Computer Science, School of Natural Science, Konkuk University,
322 Danwol-dong, Chungju-si, Chungcheongbuk-do 380-701, Korea
{cheol,ljo,hkkang}@kku.ac.kr
[2] Dept. of Anatomy, School of Medicine, Konkuk University
322 Danwol-dong, Chungju-si, Chungcheongbuk-do 380-701, Korea
beobyi.lee@kku.ac.kr

Abstract. Polygonal surface models are generally composed of a large amount of polygonal patches. Highly complex models can provide a convincing level of realism but rather cause problems in real-time applications, such as virtual reality or 3D simulation system focusing on real-time interactivity. Therefore, it is useful to have various simplified versions of the model according to the performance of the system. In this paper, we present a surface simplification algorithm, which can excellently preserve the characteristic features of the original model, even after drastic simplification process. In addition, the proposed algorithm is efficient in memory usage and useful in real-time rendering applications requiring progressive and incremental rendering of surface data.

1 Introduction

Owing to the remarkable development of computer graphics technologies, the 3D graphical models are now generally used in many computer graphics applications. Highly complex models can provide a convincing level of realism but the full complexity is not always required and desirable in some applications, such as simulation and virtual reality systems focusing on real-time interactivity. In such applications, it is acceptable to decrease the fidelity or realism of the model to increase runtime efficiency.

The simplification algorithm starts with an original model and iteratively removes elements from the model until the desired level of approximation is achieved. To decide the order of elements for removal during the simplification process, most of the existing algorithms use the error metric based on distance optimization. That is, the order is determined so that the distance between the meshes before and after removal operation is minimized. However, it is difficult to describe the local geometric characteristics of surface using the metric and finally the detailed local shapes of the original model are not retained. This is due to the fact that although the distance-based metric can minimize the numerical errors introduced during the simplification process, this does not guarantee the degree of loss for geometric information caused by simplification [4].

---

* This work was supported by the faculty research fund of Konkuk University in 2004.

D.-K. Baik (Ed.): AsiaSim 2004, LNAI 3398, pp. 597–606, 2005.
© Springer-Verlag Berlin Heidelberg 2005

In this paper, we present an error metric, which can efficiently describe and reflect the geometric features and changes before and after simplification, and then excellently preserves the original surface features. In addition, we implement a surface simplification algorithm utilizing the error metric.

2 Related Works

Most of the existing simplification algorithms locally perform a sequence of simple topological operations in each simplification step to remove and update certain geometric elements. Previous works can be classified depending on the type of the topological operations. Possible topological operations are vertex decimation, edge decimation, triangle decimation and vertex clustering [3]. Out of these approaches, the following three classes, as illustrated in the Fig. 1, are broadly relevant to our work. We explain the characteristics of the algorithms in each class by focusing on the error metric employed in each algorithm.

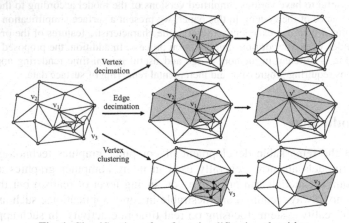

Fig. 1. Classification of surface simplification algorithm.

Vertex Decimation. This method removes a vertex with the decimation cost lower than the predefined threshold in each simplification step and then re-triangulates the hole. In [5], the distance between a given vertex and the average plane composed of its surrounding vertices is used for error metric. The Hausdorff distance between the simplified and the original mesh is used as error metric in [8]. And in [7], the curvature of local surface around a given vertex is used as error metric. In [17], for the planes of triangles adjacent to a given vertex in current mesh, the maximum distance from the corresponding planes of the intermediate mesh, and its accumulation are used for error metric. This approach needs robust re-triangulation method, as in [15], to fill the resulting hole caused by vertex removal, and mainly focuses on connectivity of a surface rather than geometric characteristics. So, the quality of the approximation largely depends on the re-triangulation operation.

Edge Decimation. This method is based on the iterative edge contraction, which replaces an edge with a vertex. Neighboring triangles, which contain both vertices of the edge, are removed from the mesh and remaining triangles are adjusted so that they are appropriately linked to the new vertex. In [9] and [12], energy function using the sum of the squared distance between the sample vertices from the original mesh and the simplified mesh is used for error metric. The quadric error metric is used in [14]. This metric is derived from the quadratic equation representing the measure of sum of squared distance between vertex and associated planes. And in [10], the degree of change of geometric properties, such as area and volume, between successive meshes is used for error metric. In this approach, each algorithm has to determine the positioning policy for a new vertex to replace the edge after collapse operation. The new vertex may or may not exist in the original mesh. The error metric used in each algorithm provides the clues for positioning the generated vertex. We will treat more carefully about the vertex placement policy in the following chapter.

Vertex Clustering. The basic idea of this method is that when we project a mesh to a plane, those vertices concentrated in sufficiently small area can be grouped and then represented by a single representative vertex. In [22], a mesh is divided into uniform 3D grid, namely cluster, and each vertex in the same cluster merged into a representative vertex. Each vertex has a weight, which is calculated by the size of neighboring triangles and the maximum dihedral angle between the adjacent edges. The representative vertex is determined according to the size of the weight value. In [23], they used the Octree data structure as the space partitioning method. Each vertex is graded as the degree of importance calculated by the curvature and length of adjacent edges. In the simplification process, every vertex in a cell is merged into a single most-important vertex. When we use the space partitioning method based on uniform sized grid, the approximation quality can become various according to the location of the representative vertex in each cell. In [19], they improved the partitioning problem by using floating-cell approach, that is, every representative vertex is adjusted to locate in the center of cell. Even though this approach can guarantee reliable and high-speed simplification, the quality of simplified mesh is generally quite low, since it is over-dependent on the criteria determining the cluster size and the distance between clusters.

As we see so far, there is no unique algorithm, which is applicable to every case. Therefore, it is desirable to choose appropriate simplification approach according to the characteristics of surface models or of application areas. After an analysis of existing studies, we decided to use the half-edge contraction scheme, which is one of the edge decimation-based approaches, as the topological operation. This scheme is well fit for our purpose focusing on the real-time rendering applications. The concrete meaning of the half-edge contraction will be discussed in the following chapter.

3 Simplification Framework

One of the goals of this work is to generate various simplified models that are applicable to real-time rendering applications[2]. In this chapter, we describe the structural characteristics of the proposed algorithm for satisfying the goal.

3.1 Notation

Before describing the proposed algorithm, we will briefly introduce some notations. We assume that a polygonal surface model is simply a set of triangular polygons in three-dimensional *Euclidean* space R^3. An arbitrary polygonal surface model, or tri-angular mesh, $M=\{V,T\}$ is a set of vertex set $V=\{v_1,v_2,v_3,...,v_m\}$ and triangle set $T=\{t_1,t_2,t_3,...t_n\}$. M_i is a mesh in the ith stage of iterative simplification process and M_{i+1} is the successive mesh of M_i. An edge $\bar{e}$ is denoted by a set of vertices $\{u,v\}$, where $u,v \in V$. Every edge has a direction. A directional edge $\vec{e}$ is denoted by a set of ordered vertex pairs $\vec{e}(u,v)$. This means that the vertex u will be merged into v after collapse operation. $P(u)$ is a set of planes of the triangles that meet at vertex u and $P(\bar{e})$ is, generally two, a set of planes adjacent to the edge $\bar{e}(u,v)$. Then, $P_i(\bar{e})$ and $P_{i+1}(\bar{e})$ are a set of $\{P(u)-P(\bar{e})\}$ in mesh M_i and M_{i+1}, respectively.

3.2 Half-Edge Contraction

As previously noted, each edge collapse-based simplification algorithm has to present a positioning policy of new created vertex in the simplification process. The positioning policy must be chosen carefully, since it determines the quality of the approximation. Generally, there are three cases of placement policies of new generated vertex v' after the edge $\bar{e}(u,v)$ is collapsed, such as midpoint, optimal point and endpoint approaches.

The midpoint approach creates a new vertex at the midpoint of two vertices of an edge. This is intuitive and unbiased to the positions of the two vertices. However, the drawback is that the volume of the original object becomes smaller as simplification steps proceed. The optimal point scheme generates a new vertex at the optimal position on the contour curve connecting two vertices of an edge. This scheme is known as the method that can create a high-quality approximation. However, finding an optimal position costs a great deal of time and requires extra memory space to store the new vertex. The endpoint approach places a new vertex by merging one of the two endpoints of the edge to the other. Since no new vertex is created, the endpoint approach gives no additional memory burden and rapid calculation is possible. More-over, in most cases, the original shape is well preserved [11], although the method does not use the optimal placement policy.

Our algorithm is based on the half-edge contraction manner, which has the same meaning with the endpoint approach except some implementation details. In terms of implementation, we allow every edge to have directionality. When the directional edge $\vec{e}(u,v)$ is contracted, $P(\bar{e})$ and the start vertex u are removed from the current mesh M_i and $P_i(\vec{e})$ are adjusted to $P_{i+1}(\vec{e})$ in M_{i+1}. With the use of directional edge, we can effectively estimate the degree of geometric deviation after the contraction by temporarily reconstructing the local planes in next step.

4 Error Metric for Feature Preservation

Every surface model has its own shape and features, which determine the cognitive characteristics of the model. The simplification process must be conducted in such a way that the shape and characteristics are preserved [1]. Then, we need to consider the meaning of geometric features and the methodology for preserving such features.

4.1 Geometric Features

In the height fields, or terrain, it is comparatively easy to specify and distinguish the surface features. Points on the surface, such as peaks, ridges, pits and channels, are generally classified as important points. In the case of arbitrary mesh, however, the criteria for defining characteristic feature are not clear, as in the case of terrain. Therefore, we analogize the distinct features by looking for such features as peak points, sharp edges and coplanar faces. The degree of pointedness and sharpness of each primitive generally can be computed by measuring and comparing the local curvature around a vertex, or the normals of adjacent faces. This means that whether a certain primitive should be preserved or not is determined by the importance. In most cases, this is a reasonable approach. However, one thing we must consider, when determining the geometric features, is the relative importance by the influence of each primitive on the local area.

4.2 Error Metric

Error metric is a measure representing the degree of deviation, or error, of the approximation from the original model. The proposed simplification algorithm uses a greedy approach based on iterative edge contraction. That is, because the geometric changes in the greedy simplification always happens in the adjacent area of local element, we can estimate the degree of deviation after collapse operation, by describing the characteristics of local surface appropriately.

In order to estimate the deviation error during the simplification, we define the error metric based on two geometric criteria, as shown in the following equation (1).

$$E(M_i, M_{i+1}) = R(\Delta\mu(P_i, P_{i+1}), \mu(P_i))$$ (1)

Where, $P_{i/i+1}$ denotes a set of local planes in i^{th} and $i+1^{th}$ stages, respectively. μ is a function to compute some geometric quantity, and thus $\Delta\mu(P_i, P_{i+1})$ returns the degree of geometric variation between two geometric states, P_i and P_{i+1}, and $\mu(P_i)$ returns the geometric characteristic of current state P_i. According to our experiments, we observe that these are the minimum set of criteria that are needed to detect the degree of geometric deviation and they work complementary to each other. The proposed algorithm assigns the contraction cost of an edge calculated from equation (1) to corresponding edge. We discuss the details of the equation continuously.

Geometric Variations. As the factors for calculating geometric variations $\Delta\mu$ after edge contraction, we consider the squared distance and the change of surface orientation between the local planes of two successive meshes, respectively. The variation of surface orientation is independent from that of scalar, and thus it is possible to control overall decimation cost, for example, by assigning higher cost to the element when its variation for orientation is high even though it has high reduction probability due to little change in scalar, and vice versa.

The distance $d(t, v)$ between an arbitrary vertex v and plane t can be calculated using equation (2). In the equation, $\bar{v}$ is a vector between v and an arbitrary vertex on the plane t, n_t is a normal vector of t, and $\cdot$ is an inner product of two vectors.

$$d(t,v) = \frac{|\vec{v} \cdot n_t|}{\| n_t \|} \qquad (2)$$

Then, when the edge $\vec{e}(u,v)$ is contracted, the distance between M_i and M_{i+1} can be calculated by the sum of squared distance between the plane set $P_i(\vec{e})$ in current mesh M_i and the vertex v of the edge in descendent mesh M_{i+1} using equation (3).

$$D(M_i, M_{i+1}) = D(P_i, P_{i+1}) = D(P_i(\vec{e}), v) = \sum_{t_i \in P_i(\vec{e})} \left(d(t_i, v) \right)^2 \qquad (3)$$

It must be noted that $D(P_i(\vec{e}), v)$ is the sum of distance-to-plane measurement. This is similar to [14] but the difference is, in this case, only $P(u)$, which is the super-set of $P_i(\vec{e})$ and not $P(u) \cup P(v)$ is considered.

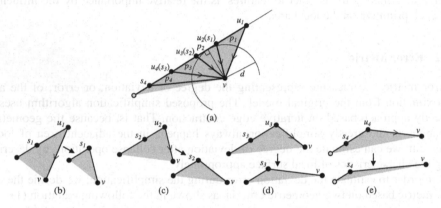

Fig. 2. Cases when the decimation costs based on $\Delta\mu$ distance metric are the same.

Describing the geometric variation of local surface using just the distance measure is not sufficient. The decimation costs of every edge $\vec{e}(u_1,v) \sim \vec{e}(u_4,v)$, in Fig. 2 (a), based on the distance metric are the same because the distances between the vertex v and the planes p_1 to p_4 are all equally d. However, the degree of geometric deviation after simplification is different, as shown in (b) to (e). Considering the orientation change of local surface, as shown in equation (4), can solve this problem. Remaining planes after contraction of edge $\vec{e}(u,v)$ in M_{i+1} are exclusion of $P(\bar{u})$ from $P(u)$, which is $P_{i+1}(\vec{e})$. So, we only consider the variation from its previous planes, $P_i(\vec{e})$ in M_i, when calculating the change of local surface orientation.

$$\Delta N(M_i, M_{i+1}) = \Delta N(P_i(\vec{e}), P_{i+1}(\vec{e})) = \sum_{t \in P_i(\vec{e}),\, t' \in P_{i+1}(\vec{e})} \left(|1 - n_t \cdot n_{t'}| / 2 \right) \qquad (4)$$

Geometric Characteristics. As previously mentioned, since the geometric features of mesh are constructed with small sized geometric elements, which are concentrated in small areas, they have a small quantity of D. The component of orientation variation ΔN may be used for supplementing the error metric. However, there still remained problems in this approach. Fig. 3 represents the cases when the decimation costs based on $\Delta\mu$ component are the same, but the geometric shapes are different, as in (a) to (d). This means that $\Delta\mu$ alone it is possible to compute the same decimation cost for those surface areas having different geometric characteristics. Therefore, to distinguish those areas, we introduce additional components μ and its sub factors, such as local curvature and edge's length.

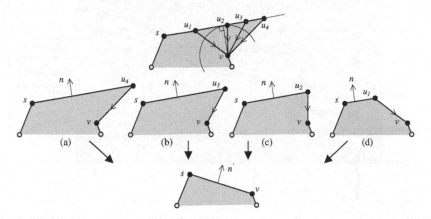

Fig. 3. Cases when the decimation costs based on of error metric are the same.

The local curvature κ is calculated by the sum of inner product between the plane sets $P(u)$ and $P_i(\vec{e})$, which are adjacent to the start vertex u of edge $\vec{e}(u,v)$ and to the edge itself, as shown in equation (5). Equation (5) improves the curvature element of the edge cost formula based on max operator in [13].

$$\kappa(\,P(u),\,P_i(\vec{e})\,) = \sum_{t\in P(u)}\,\underset{t'\in P_i(\vec{e})}{Min}\,(\,|\,1-n_t\cdot n_{t'}\,|/2\,) \tag{5}$$

Generally, short edges are relatively less important, since they make a low impact on local surface. Consequently, they should be assigned with low decimation cost. This means that the length of the edge needs to be considered as the additional component of error metric. We complete the final form of our error metric by multiplying the edge's *Euclidian* length to the summation of equations (3), (4) and (5).

5 Experimental Results

For the measurement of numerical accuracy of the simplified model we use a public tool, namely *Metro*[16]. Fig. 4 represents the comparison graph of each simplified result from seven simplification algorithms including our method for the *Fandisk* model. In the figure, *Ko* means the proposed algorithm. The *Fandisk* model is characterized by the uniform size of triangular patches, and apparent distinction of the differences between low and high curvature regions. At the simplification ratio of 99%, the best results are given by the *Mesh optimization*, and *QSlim* and *Ko* follow. The *Mesh optimization*, however, shows low accuracy when the simplification ratio is less than 90%. In addition, this method keeps a large amount of points sampled from original surface to calculate the degree of deviation from the original model. Consequently, it has high cost in memory usage and time for calculating the optimal location of the new vertex [18]. Therefore, we can assume that the results from *QSlim*, *Ko* and *JADE* are all fairly acceptable.

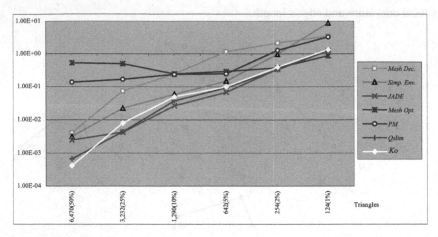

Fig. 4. Average errors for *Fandisk* model (*Mesh Dec.*[5], *Simp. Env.*[6], *JADE*[17], *Mesh Opt.*[20], *PM*[12], *Qslim*[14]).

Another two surface models used in accuracy comparison are *Venus* and *Cow*. These models are characterized by various sizes of triangular patches and sophisticated surface curvature. In this case, we compared the three algorithms, *QSlim*, *JADE* and *Ko*, which rank high in the previous comparison.

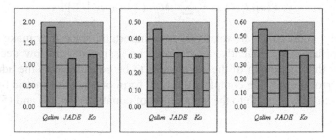

Fig. 5. Maximal, mean and mean square errors for *Venus* model.

Fig. 5 shows the numerical results of the three methods by using *Venus* model, at the simplification ratio of 99.7%. The proposed algorithm obtains the minimal mean and mean square errors. The best result in terms of maximal error is given by *JADE*, which is based on the global error management scheme. Similar to the previous *Mesh optimization*, however, this method costs high in memory usage, since it keeps the history of the removed vertices for the global error management. Fig. 6 represents the visual results of the simplified models for *Cow* and *Venus* obtained from the three algorithms, respectively. In both cases, it is clearly observed that *Ko* retains the best high curvature regions of the original model at the same simplification ratio.

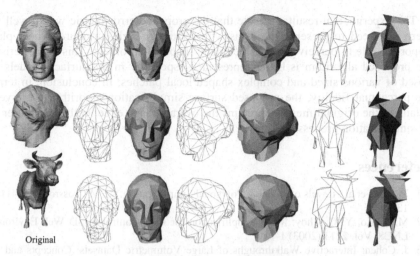

Original

Fig. 6. Visual results for *Venus*(500t) and *Cow*(150t) models (Top: *QSlim*, middle: *JADE*, bottom: *Ko*).

As shown above, the proposed algorithm is slightly better or worse than the existing works, in terms of the overall numerical results. However, in terms of the visual aspect, it shows the excellent results. This indicates that the numerical result does not guarantee the accuracy or quality of the simplified model, since the most important measure of fidelity is not geometric but perceptual[21].

Fig. 7 represents the distribution of the approximation errors for the *Venus* model. Lighter color indicates more errors occurred in the area and darker color indicates less error. The high curvature regions, such as brow, eyes, nose, lips and ears provide perceptual clues to understanding the object. In the result of the *QSlim*, large approximation errors are distributed broadly, since it uses the optimal vertex positioning policy. *JADE* generates approximation errors especially in relatively high curvature regions. These high curvature regions, on the other hand, are excellently preserved in the result of the proposed algorithm.

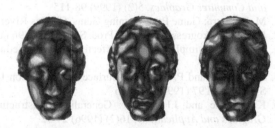

Fig. 7. Distribution of approximation errors(*QSlim*, *JADE*, *Ko*).

6   Conclusions

There is no unique simplification method to generate the best results for every surface model and purpose. This means that it is desirable to select an appropriate method according to the various usages in the applications. The goal of this paper is to propose an error metric to retain the characteristic features of the original model even after drastic simplification process, in terms of perception.

The experimental results indicate that the proposed error metric works well when there is a need to preserve the areas with relatively high curvature, which play an important role to perceive the characteristic shape of an object. This characteristic of the proposed algorithm is better represented especially in the surface models composed of various sized and complex shaped local patches. In conclusion, in terms of the perceptual fidelity, the proposed surface simplification algorithm preserves the details of the original mesh and retains the overall shape excellently, even after drastic simplification process.

References

1. M. Krus et al.: Levels of Detail & Polygonal Simplification. ACM Crossroads, 3(4) (1997) 13-19
2. M. C. Ko, Y. C. Choy: Mesh Simplification for QoS Control in 3D Web Environment. LNCS, Vol. 2713 (2003) 1-11
3. J. Cohen: Interactive Walkthroughs of Large Volumetric Datasets: Concepts and Algorithms for Polygonal Simplification. Proc. SIGGRAPH 2000 Course Notes (2000)
4. J. Cohen: Advanced Issues in Level of Detail: Measuring Simplification Error. Proc. SIGGRAPH 2001 Course Notes (2001)
5. W. J. Schroeder et al.: Decimation of triangle meshes. Proc. SIGGRAPH'92, 26(2) (1992) 65-70
6. J. Cohen et al.: Simplification Envelopes. Proc. SIGGRAPH'96 (1996) 119-128
7. G. Turk: Re-tiling Polygonal Surfaces. Proc. SIGGRAPH'92 (1992) 55-64
8. R. Klein et al.: Mesh reduction with error control. Proc. Visualization'96 (1996) 311-318
9. H. Hoppe: New quadric metric for simplifying meshes with appearance attributes. Proc. IEEE Visualization'99 (1999) 59-66
10. P. Lindstrom et al.: Evaluation of Memoryless Simplification. IEEE Trans. *Visualization and Computer Graphics*, 5(2) (1999) 98-115
11. M. DeLoura: Game Programming Gems. Charles River Media (2001)
12. H. Hoppe: Progressive meshes. Proc. SIGGRAPH'96 (1996) 99-108
13. S. Melax: A simple, Fast, and Effective Polygon Reduction Algorithm. Game Developer, (1998) 44-49
14. M. Garland and P. Heckbert: Surface Simplification Using Quadric Error Metrics. Proc. SIGGRAPH'97 (1997) 209-216
15. K.J. Renze and J.H. Oliver: Generalized Unstructured Decimation. IEEE *Computer Graphics and Applications*, 16(2) (1996) 24-32
16. P. Cignoni et al.: Metro: measuring error on simplified surfaces. Computer Graphics Forum, 17(2) (1998) 167-174
17. A. Ciampalini et al.: Multiresolution Decimation based on Global Error. *The Visual Computer*, Springer International, 13(5) (1997) 228-246
18. P. Cignoni et al.: A Comparison of Mesh Simplification Algorithms. *Computers & Graphics*, Pergamon Press, 22(1) (1998) 37-54
19. K.L. Low and T.S. Tan: Model Simplification using Vertex Clustering. Symposium on Interactive 3D Graphics (1997) 75-81
20. H. Hoppe et al.: Mesh optimization. Proc. SIGGRAPH'93 (1993) 19–26
21. D. P. Luebke: A Developer's Survey of Polygonal Simplification Algorithms. IEEE *Computer Graphics &Applications*, 21(3) (2001) 24-35
22. J. Rossignac and P. Borrel: Multi-resolution 3D approximations for rendering complex scenes. Modeling in Computer Graphics, Springer-Verlag (1993) 455-465
23. D. Luebke and E. Carl: View-Dependent Simplification of Arbitrary Polygonal Environments. Proc. SIGGRAPH'97 (1997) 199-208

Modeling and Simulation of Digital Scene Image Synthesis Using Image Intensified CCD Under Different Weathers in Scene Matching Simulation System

Liang Xiao[1], Huizhong Wu[1], Shuchun Tang[2], and Yang Liu[2]

[1] Department of computer science and technology,
Nanjing university of science and technology, Nanjing, 210094, P.R. China
xtxiaoliang@163.com, wuhuizh@mail.njust.edu.cn
[2] Forth Institute of the Second Artillerist of the Chinese People's Liberation Army,
Beijing, 100085, P.R. China

Abstract. The synthesis of digital real-time image in scene matching has great significant and application in military. We first propose a practical scene matching simulation system. Then most factors effecting on real time image synthesis including the change of missile's poses, hypsography, meteorology and the characteristics of image intensified CDD is investigated. The key mathematical models including eradiate aberrance and camera's aberrance of digital scene image synthesis are discussed. An overall designing framework for simulation system is proposed. Furthermore, a powerful simulated real time image synthesis system named by SMRT-SS oriented to Image Intensified CCD camera is developed. Finally some simulation results are given.

1 Introduction

The scene matching algorithm is the vita technique in missile guidance system. The research on adaptive, robust and real time scene matching algorithm is an effective approach to improve the capacity of accurate missile guidance system[1]. However, the acquisition of the real-time images under different weathers is very difficult task or at great expense. Hence the simulation of digital real-time image in scene matching has great significant and application in military.

Currently, traditional approaches for the simulation of digital real-time image in scene matching can be classified into two classes[2][3].

(1) The approaches based on hardware in the loop simulation(HILS). In HILS, the reference image obtained by missile's CCD camera is projected on the big screen projection, and then the projected image is regarded as the real-time image. The advantage of this method is that it is close to the real environment of missile's flight and the process of CCD's shot. However, it is difficult to construct the weather's effect. Another disadvantage is that it needs a lot of hardware.

(2) The second approach is the pure computer simulation (PCS). The basic method is using the satellite images or aerial images to create the standard reference image and then apply geometry aberrance, eradiate aberrance and camera's aberrance to

D.-K. Baik (Ed.): AsiaSim 2004, LNAI 3398, pp. 607–616, 2005.

simulate the real time image. The approach has two advantages compared with the HILS. Firstly, it is very convenient and flexible for simulating the real time images. Secondly, it can fully escape from the support of hardware. However, the validity of this approach depends on the validity of the mathematical models of solar and atmospheric models and sensor models.

The research described in this paper falls into the second class. We will establish two key mathematical models including eradiate aberrance and camera's aberrance of digital scene image synthesis and proposed the simulation framework.

2 Simulation Framework Design for Scene Matching Simulation System

Taking into account the conditions of our laboratory and combining the acquisition of data resource and the simulation control strategies, the simulation framework for scene matching is established described as Figure 1. (i) Select some suitable matching areas such as mountain region, mound, farmland and residential district, etc, then create some reference images for scene matching.(ii) According to the meteorological condition of the simulated time and the missile's currently geographic position, some distortion is applied to the reference image in order to simulate the possible difference between reference image and real time image under different weathers.(iii) According to the real size of the real time image, take a truncation in the distorted reference image to create a sub-image as the simulated real time image. At the same time, the coordinate of sub-image is recorded. (iv) The scene matching algorithm is applied between reference image and the simulated real time image to obtain the coordinate of the best matching point. (v) Change the location of the sub-image and repeat step 3 and step 4, take a statistical analysis of the probability of scene matching.

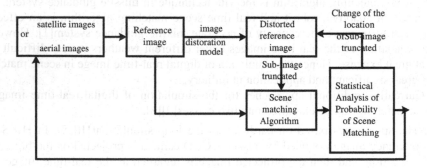

Fig. 1. The simulation framework for scene matching

It is clear that in the above simulation framework, the key technique is the modeling and simulation of digital scene image synthesis, i.e., how to make distortion on the reference image according to the meteorological condition of the simulated time and the missile's currently geographic position. In the following, we will discuss the key techniques and developed the simulation system named by SMRT-SS: Simulated Real Time Image Synthesis System.

3 Solar and Atmospheric Modeling

In this section, the modeling of solar illumination and the atmospheric effects present in optical aerial imaging system is discussed. And then we review the software of atmospheric computing kernel.

The atmospheric effect on spectral radiance consists of two main mechanisms, scattering and absorption. Scattering is mainly due to the presence of particles in the atmosphere, while absorption comes about due to the energy transfer from the optical radiation to molecular motion of atmospheric gases. Both of these effects are wavelength dependent. The exoatmospheric spectral irradiance, $E_{\lambda,Exo}$, is attenuated and scattered by the atmosphere before reaching the surface as the direct spectral irradiance $E_{\lambda,Direct}$. Some of this scattered radiation also reaches the surface as $E_{\lambda,Diffuse}$, the diffuse spectral irradiance (or skylight irradiance). The reflected spectral radiance $L_{\lambda,Surface}$ passes through the atmosphere and is attenuated by the spectral transmittance $T_{\lambda,Atm}$ of the atmosphere. Also, some of the solar irradiance that is scattered by the atmosphere finds its way into the sensor field of view as $L_{\lambda,Path}$, the path spectral radiance. This path radiance also includes that which may have been reflected off of the nearby surface (adjacency effect) before being scattered into the sensor field of view, as well as the background radiation of the atmosphere.

These factors contribute to the spectral radiance of the scene, as received by the sensor, in a manner described by equation [4][5]

$$L_{\lambda,Sensor} = \frac{1}{\pi}\left\{ \cos(\theta_{solar})E_{\lambda,Direct} + E_{\lambda,Diffuse}\right\}\cdot R_{\lambda}\cdot T_{\lambda,Atm} + L_{\lambda,Path} \tag{1}$$

Here, R_{λ} is the spectral reflectance of the surface. In the most general sense it is the Spectra Bidirectional Reflectance Factor (SBRF) that gives the reflectance for all angles of incidence and viewing. The other factors also depend upon the angles of illumination and viewing as well as the quality of the atmosphere. In our simulation system, the angle of solar θ_{solar} is replaced by the incident angle θ_{inci}, which can be determinate by the angles of solar and the slope angle of scene. Hence the DEM (Data Evaluation Map) is required.

In our simulation system, the solar and atmospheric model is implemented with the use of the computer code LOWTRAN. The program LOWTRAN has evolved over the years from simply an atmospheric transmittance code to one that is now capable of computing direct solar irradiance and multiply scattered atmospheric radiance. LOWTRAN uses radiative transfer theory to compute the transmittance and radiance in each of 32 layers of the atmosphere. Well documented data tables embedded within the program give accurate spectral transmittance and radiance values at minimum wave number intervals of 20 cm-1. LOWTRAN is a moderate resolution atmospheric propagation model in program. LOWTRAN is currently supported by the Air force Geophysics Laboratory(AFGL)[6][7]. In order to calculate the atmospheric spectral irradiance, such as direct spectral irradiance, the diffuse spectral irradiance, spectral transmittance and the path spectral radiance, four calls to LOWTRAN are set up within the input file. The first call calculates the direct solar spectral irradiance at the surface. The second calculates the transmittance of the path from the surface to the sensor. The third and fourth calls calculate the path radiance seen by the sensor for surface albedoes of 0 and 1. In our simulation system, the computing kernel of LOWTRAN-7 is embedded into our software as a dynamic link library (DLL) and the input file of LOWTRAN-7 can be generated according the task of the simulation in real time.

4   Synthetic Simulation and Modeling of Image Intensified CCDs(IICCD)

Characterization of the image Intensified CCDs (IICCD) is important for the synthetic purpose. Some of the parameters are used to create a "new" low-light-level sensor in the synthetic image generation environment. Some work has already been published on modeling and performance characteristics of IICCD. The intensified sensor that was modeled here is of the Generation III intensifier with a GaAs photocathode[8][9]. Key effects to be considered in the evaluation of an image intensified CCD can include, but are not limited to, light quantum parameters such as luminance gain, automatic brightness control specification, photocathode sensitivity (luminous sensitivity and radiant sensitivities), signal-to-noise(S/N), and modulation transfer function (MTF). Geometric parameters include resolution, magnification, nonlinearities, and blooming effects. In this research we concerned ourselves with parameters such as gain, signal to noise ration, shot noise, pre-amplifier noise, blooming, and MTF. All the technique details can be found in the references [10-12].

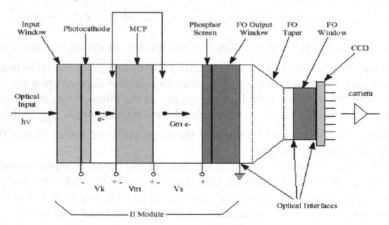

Fig. 2. Schematic design of a fiber-optically (FO) coupled IICCD assembly

4.1   The Structure of IICCD [10]

Figure 2 shows a detailed schematic design of a fiber-optically (FO) coupled IICCD assembly. Here we see an input window, a photocathode, a micro-channel plate, a phosphor screen and an output window. We also see a fiber optic taper, instead of a simple unity magnification FO window. The photocathode on the vacuum side of the input window converts the input optical image into an electronic image at the vacuum surface of the photocathode in the II. The MCP is used to amplify the electron image pixel-by-pixel. The amplified electron image at the output surface of the MCP is reconverted to a visible image using the phosphor screen on the vacuum side of the output. In our simulation system, all the mechanism of functional device in IICCD will be described by mathematical models and simulated by pure computer software.

4.2 Component MTFs of the IICCD

Based on the results of literature [12], in IICCD system, there is an image chain associated with MTFs. The modulation transfer function, MTF, is essentially a spatial frequency response of an optical imaging system and can be determinated in many ways. From a visual standpoint, high values of MTF correspond to good visibility, and low values to poor visibility. In this research we only apply the most important component MTFs with the assumption of circular symmetry. When individual components are taken into account, a first order system MTF can be given by[11-12]

$$MTF_{sys} = MTF_{opt} \cdot MTF_{II} \cdot MTF_{fo} \cdot MTF_{ccd} \qquad (2)$$

Furthermore, the MTF of the intensifier system can be further broken down into sub MTFs which include the MTF of the II-to-CCD coupling, first-optical-to-fiber-optic interfaces, fiber-optic-to CCD interface, etc. Use of the MTF in the characterization of an imaging system requires first a Fourier transform of the object into its spectrum then multiplication of the object spectrum by the appropriate MTF followed by an inverse Fourier transform to obtain the modified profile of the image.

4.2.1 Optics MTF – MTF_{opt}

At the front end of most intensifiers are optics used to collect light while bring the object to focus at the image plane. Furthermore, the entrance pupil is usually that of a circular pupil of diameter a and the incoherent optical transfer function is proportional to the autocorrelation of the pupil function. Let $\rho = \sqrt{u^2 + v^2}$ denote the spatial frequency variable and $v_c = \dfrac{a}{\lambda d}$ is the cutoff frequency transmitted by this system (here d is the distance between pupil plane and the image plane, λ is the cental wavelength), then the optics MTF can be modeled as

$$MTF_{opt}(\rho) = \frac{2}{\pi - 2}\{\cos^{-1}[\frac{\rho}{v_c}] - \sin[\cos^{-1}(\frac{\rho}{v_c})]\} \qquad (3)$$

4.2.2 II Tube MTF – MTF_{II}

In general, the MTF here is representative of the II tube itself and the MTFs associated with intensifier tubes can usually be described by a mathematical function with only two parameters. The form of this function is as follows:

$$MTF_{II} = e^{-(\xi/\xi_c)^n} \qquad (4)$$

where ξ_e is the frequency constant and n is the MTF index. The shape of the MTF curve is described by the MTF index, and large value of n are associated with a rapid decrease in MTF at the frequency constant ξ_e.

4.2.3 Fiber Optic MTF – MTF_{fo}

As we know, in many of today's IICCDs, the phosphor screen is deposited on a fiber optic bundle which in effect re-samples the images. The bundle is then tapered down to accommodate a CCD. Since the fiber optics are circular in nature we can use the cylinder function to describe the transmittance. The Hankel transform of this yields the sombrero function. Furthermore, the sombrero function can be rewritten in terms of a first-order Bessel function of the first kind,

$$MTF_{fo} = \left| \frac{2 J_1 (2 \pi d \rho)}{2 \pi d \rho} \right| \tag{5}$$

where d is the fiber optic coupler's core-to-core pitch and $\rho = \sqrt{u^2 + v^2}$ is the spatial frequency in lp/mm.

4.2.4 CCD MTF – MTF_{ccd}

Since a single pixel in the focal plane of CCD is of size b and d, and is modeled in the space domain as a 2D rectangular function, then the MTF of the CCD can be modeled as

$$MTF_{ccd} = \frac{\sin(\pi \xi b)}{\pi \xi b} \cdot \frac{\sin(\pi \eta d)}{\pi \eta d} \tag{6}$$

4.3 The Noise Simulation Model of IICCD System

In general, the dominant noise sources in IICCD systems are shot noise (photon or quantum noise), and readout noise [12]. In our simulation system, the shot noise is viewed as the Poisson noise, while readout noise is treated as "additive" Gaussian noise with mean, μ, and standard deviation, σ. The mean is computed based on the mean value of the image. The standard deviation is computed from a desired signal-to-noise (SNR) ratio. To simulate the effects of such noise, the noise function in MATCOM tools are used. MATCOM tools can be easily compiled with Visual C++ and have powerful mathematical software packet.

4.4 Implementing of the Sensor Model

The application of the sensor model is done as a post-processing operation to the radiance imagery. The model itself was written in Visual C++ and is designed to read in a double precision radiance field image and output a copy of the input with some of the artifacts described earlier. The overall procedure is seen below

$$I_{Final}(x,y) = \left[FFT^{-1} \left\{ \left\{ FFT \left(L(x,y) + \sigma_p \right) \right\}_{(u,v)} \cdot MTF_{sys} \right\}_{(x,y)} + \sigma_N \right] \cdot G + B \tag{7}$$

The first step is implementing the shot noise σ_p on the image $L(x,y)$. We then compute the magnitude of the fourier transform of this result so that we can multiply it by the MTF of the image intensifier, fiber optic bundle, CCD. Finally, additional pre-amplifier noise is added.

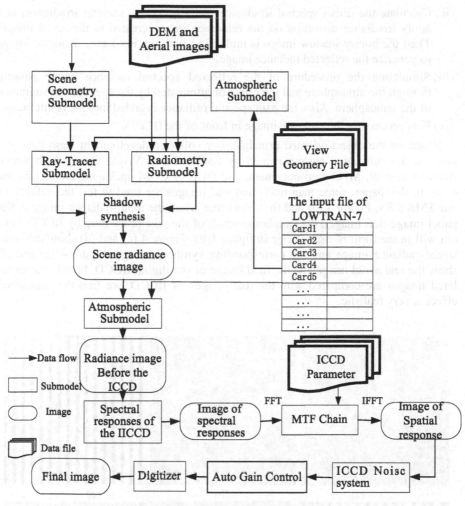

Fig. 3. The diagram of the SMRT-SS

5   SMRT-SS: Simulated Real Time Image Synthesis System

The overall approach to synthetic modeling of a low-light-level imaging sensor is illustrated in Figure 3. The SMRT-SS input parameters include the Data Evaluation Map (DEM) and the corresponding aerial images, the missile's pose parameter file, ICCD parameter file and the weather's file (the input file of LOWTRAN-7), etc. The system of SMRT-SS consists of several key sub-models including scene geometry submodel, ray-tracer submodel, radiometry submodel, atmospheric submodel and IICCD submodel,etc. The simulation procedure can be mainly listed below:

(i) Based on the method of ray-tracer technique, we can calculate the binary image of the shadow caused by the DEM data.

(ii) Calculate the direct spectral irradiance and the diffuse spectral irradiance, and apply irradiance distortion on the reference image created in the aerial image. Then the binary shadow image is multiplied with the irradiance distorted image to generate the reflected radiance image.

(iii) Simulating the procedure of the reflected spectral radiance image passing through the atmosphere and the effect of attenuated by the spectral transmittance of the atmosphere. Also, the path spectral radiance is added into the result image.

(iv) Post processing the radiance image in front of the IICCD.

Based on the object-oriented principle, our software development team have developed the simulation system named by SMRT-SS in Visual C++ 6.0, in which ActiveX control, MFC muti-document, and DLL or COM packaging techniques are used. In this paper, some man-made and real images are used to test the validity of our SMRT-SS. Figure 4 (a) and (b) shows that when the input radiance image is flat panel image (left image), the simulation result of the post-processing by IICCD camera will in the form of right image in figure 4(b). Figure 4 (c) and (d) show the sinusoidal radiance image and the corresponding synthesis image. Figure 4 (e) and (f) show the real aerial image and the final image of our digital IICCD. Finally the simulated images are compared with the real images of IICCD, we find the simulation effect is very realistic.

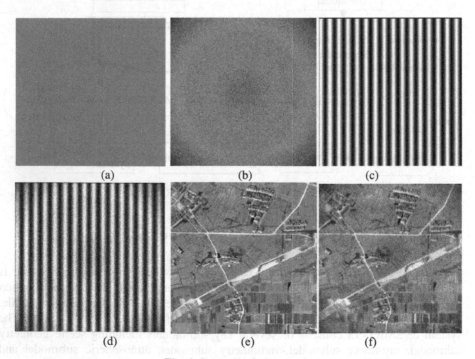

(a)   (b)   (c)

(d)   (e)   (f)

Fig. 4. Some simulation results

6 Conclusions

We outlined the simulation framework for real time image synthesis in the simulation system of scene matching. Most factors effecting on aviation imaging of natural environment and cultural characteristics are analyzed. Mathematical models of digital scene image synthesis are established according to essential principals of atmosphere transporting theory. We propose a designing framework for simulation system. Aviation imaging mechanism oriented to Image Intensified CCD camera is studied and an aviation imaging simulation system named by SMRT-SS is developed using the computing kernel of LOWTRAN-7. Using digital elevation map(DEM), geometry model with material and weather information, the system can simulate the imaging process dynamically and a lot of digital scene images can be generated according to different weathers.

Acknowledgements

This work was partially funded by the China National Defence Research Foundation, NUST initiate foundation 2004 and the National Special Foundation of Doctoral Programs Grant No. 200288024 (in Chinese).

References

1. Min-Shou Tang: The application of correlation matching technique in image guidance. NAIC-ID(RS)T:0382-96.
2. Wang Yanli, Li Tiejun, Chen Zhe: Study on global simulation technology of scene matching navigation system. Journal of system simulation, 2004,vol 16, No.1,pp 108-112. (in chinese)
3. Yang xiao-gang, Miao Dong, Cao fei: A practical scene Matching simulation method. Journal of system simulation, 2004,vol 16, No.3,pp 363-366.(in chinese)
4. Schott, J.R., Remote Sensing: The Image Chain Approach. Oxford University Press, NewYork, (1997).
5. John P. Kerekes, David A. Landgrebe: Modeling, simulation, and analysis of optical remote sensing systems. TR-EE 89-49, school of Electrical Engineering, Purdue University West Lafayette, Indiana 47907, USA.
6. Kneizys, F.X., et al.: Atmospheric Transmittance/Radiance: Computer Code LOWTRAN 6. AFGL-TR-83-0187, ERP No. 846, Air Force Geophysics Laboratory, Hanscom Air Force Base, MA, August 1983
7. Kneizys, F.X., E.P. Shettle, L.W. Abreu, J.H. Chetwynd, G.P. Anderson, W.O. Gallery, J.E.A. Selby, and S.A. Clough: Users' Guide to LOWTRAN 7. AFGL-TR-88-0177, ERP No. 1010, Air Force Geophysics Laboratory, Hanscom Air Force Base, MA, August 1988.
8. Frenkel,A., Sartot, M.A., Wlodawski, M.S.: Photon-noise-limited operation of intensified CCD cameras. Applied Optics, vol. 36, no. 22, p. 5288-97, (1997).
9. Ientilucci, E.J., Brown, S.D., Schott, J.R., Raqueno, R.V.: Multispectral simulation environment for modeling low-light-level sensor systems. Proc. SPIE, 1998, 3434: 10-19.
10. Ruffner J.W, Piccione D, Woodward K. Development of a night-driving simulator concept for night vision image-intensification device training. Proc. SPIE, 1997 3088:90-197.

11. Moran W.W, Ulich B.L, Elkins W.P, Strittmatter R.L, DeWeert M.J.: Intensified CCD (ICCD) dynamic range and noise performance. Proc. SPIE, 1997, 3173: 430-457.
12. Ientilucci Emmett J: Synthetic Simulation and Modeling of Image Intensified CCDs. MS thesis, Center for Imaging Science in the College of Science, Rochester Institute of Technology,USA,2000,3.

Performance Comparison
of Various Web Cluster Architectures*

Eunmi Choi[1], Yoojin Lim[2], and Dugki Min[3]

[1] School of Business IT, Kookmin University,
Chongnung-dong, Songbuk-gu, Seoul, 136-702, Korea
emchoi@kookmin.ac.kr
[2] School of Computer Science and Electrical Engineering, Handong Global University,
Heunghae-eub, Puk-ku, Pohang, Kyungbuk, 791-940, Korea
[3] School of Computer Science and Engineering, Konkuk University,
Hwayang-dong, Kwangjin-gu, Seoul, 133-701, Korea
dkmin@konkuk.ac.kr

Abstract. This paper investigates system attributes and the corresponding performances of three Web cluster architectures, which have different
traffic managers: the Alteon Web switch, the ALBM Traffic Manager, and
a Linux Virtual Server. The Alteon Web switch is a hardware solution
and the others provide software solutions to traffic managers that control incoming traffic in web clusters. In order to compare performances
of these Web cluster architectures, we perform a set of experiments with
various load scheduling algorithms by using a number of classified workloads. The experimental results compare the peak performances of the
Web clusters and show the effects of workload types and load scheduling
algorithms on performance trends.

1 Introduction

Web clusters have been used as a scalable server platform for web service applications [1–3]. A web cluster is composed of a number of server nodes interconnected by one or more traffic managers. A new server node could be added
to or removed from the cluster on demand. Incoming Internet traffic to a web
cluster is distributed to the server nodes according to the workload distribution
algorithm of the traffic manager. This paper investigates system attributes of
three web cluster architectures which have different traffic managers: the Alteon
Web switch, the ALBM Traffic Manager, and a Linux Virtual Server. The Alteon Web switch [4] is a hardware L4 switch developed by Nortel Network corp.
The Alteon switch distributes the incoming traffic according to its server load
balancing algorithms on top of the WebOS. Since the Alteon Web switch is popular as a HW L4 switch in a commercial market, we use it as a representative.
Compared to the HW L4, a software L4 switch is economical and cost-effective.

---

* This work was supported by the Korea Science and Engineering Foundation
(KOSEF) under Grant No. R04-2003-000-10213-0. This work was also supported
by research program 2004 of Kookmin University in Korea.

D.-K. Baik (Ed.): AsiaSim 2004, LNAI 3398, pp. 617–624, 2005.
© Springer-Verlag Berlin Heidelberg 2005

The ALBM Traffic Manager is a software L4 switch that has been developed as a part of the Adaptive Load Balancing and Management (ALBM) cluster [5]. The ALBM cluster is a reliable, scalable, and manageable web cluster and has capabilities of adaptive management and adaptive load balancing. The Linux Virtual Server(LVS) is also a software L4 switch that has been developed by LVS project [6–8]. On top of a Linux OS, the LVS provides a number of workload scheduling algorithms.

In this paper, we compare the performances of the web clusters constructed on top of the above three traffic managers. Among them, the Alteon Web switch is the most expensive and hardware-based solution. Does the Alteon Web switch achieve the best performance among them for any types of workload? Is there any performance gain to use an appropriate load scheduling algorithm depending on characteristics of workloads? Does the adaptive load balancing mechanism, provided in the ALBM cluster, have the effect on the performance of the system? In order to answer these questions, we perform a set of experiments with various classes of workloads; fine-grain workload, e-commerce workload, and dynamic mixed workload. They are classified according to the required resource amount and the resource holding time, i.e., service duration.

This paper is organized as follows. Section 2 introduces three web cluster architectures that are considered in our experiments. Section 3 shows performance results of those web clusters for various workloads. We summarize in Section 4.

2   Web Cluster Architecture

For our experiments, we construct three web cluster testbeds that have different architectures. One is based on a H/W L4 switch, the others are based on S/W network appliances. In this section, we introduce our experimental testbeds with their architectures.

2.1   Alteon Web Cluster

The Alteon Web cluster testbed has an architecture as shown in Figure 1(a). The core component of this architecture is an Alteon Web switch. Alteon Web switches [4] are hardware switches developed by Nortel Network corp, supporting Layer 2/3 and Layer 4-7 switching. With Alteon Web OS software, Alteon Web switches provide server load-balancing, application redirection, and advanced filtering, aggregating 10/100 Mbps Ethernet ports. As in Figure 1(a), the Alteon Web cluster testbed has a star architecture based on an Alteon Web switch, i.e., all server nodes are hooked directly to an Alteon Web switch that is connected to the Internet. The Alteon Web switch receives all incoming requests and distributes them to server nodes balancing the workload. As for server load balancing, the Alteon Web switches provide several algorithms, such as, Round-Robin, Least Connection, Weighted Round-Robin, and Weighted Least Connection. The NAT(Network Address Translation) mechanism is used for routing. The H/W switch-based clusters are easy to deploy and maintain.

2.2 LVS Web Cluster

The LVS(Linux Virtual Server) [6] is an open project to develop software server load balancer in the Linux kernel level. The LVS provides three routing mechanisms, such as, NAT (Network Address Translation), IP Tunneling, and Direct Routing. As for server load balancing algorithms, it provides Round-Robin (RR), Weighted RR (WRR), Least Connection (LC), and Weighted LC (WLC) algorithms. The commercial products based on the LVS are Turbo Cluster Server [9] and RadHat HA Server [10].

As shown in Figure 1(b), the software load balancer where the LVS is running is attached as a network appliance to a L3 switch. All server nodes in the cluster are also interconnected to the same L3 switch. The LVS load balancer has the virtual IP address of the cluster and all incoming requests are delivered to it at first. The LVS load balancer forwards incoming requests to an appropriate server according to its scheduling and routing mechanism, balancing loads among servers.

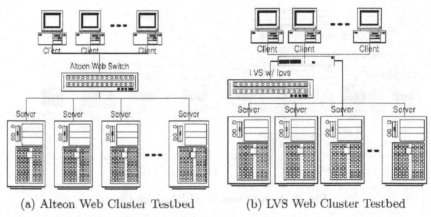

(a) Alteon Web Cluster Testbed   (b) LVS Web Cluster Testbed

Fig. 1. Sub-components of Event Notification Service.

2.3 ALBM Web Cluster

The ALBM web cluster has a similar architecture to that of a LVS cluster as shown in Figure 2; it has a software load balancer, called "Traffic Manager(TM)." As a front-end component having the virtual IP address of the cluster, the TM interfaces the rest of cluster nodes, making the distributed nature of the architecture transparent to Internet clients. All inbound packets to the system are received and forwarded to application servers through the TM. It provides network-level traffic distribution service by keeping on balancing the servers' loads. On each of server nodes, a middleware service, called 'Node Agent' (NA), is deployed and running. The NA is in charge of management operations of the ALBM cluster, such as cluster formation, membership management, and adaptive management and load balancing.

The management station, called 'M-Station', is a management center of the entire ALBM cluster system. All administrators' commands typed through management console are received and delivered to the M-Station. Communicating with NAs, the M-Station performs cluster management operations, such as cluster creation/removal and node join/leave. The M-Station also collects the current configurations and states of the entire system, and provides the information to other authenticated components in the system. Using the server state information provided by NAs, it performs proactive management actions according to the predefined policies. Besides, the M-Station checks the state of the system in service-level, and carries on some actions when the monitored service-level quality value is significantly far behind with the service-level QoS objectives. All the components in the ALBM system are interconnected with public and private networks. The public network is used to provide public services to Internet clients, while the private network is used for secure and fast internal communications among components on management purpose.

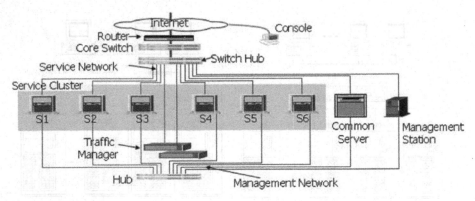

Fig. 2. Architecture of ALBM Web Cluster.

3   Experimental Results

This section shows experimental results of three Web cluster systems with the Alteon Web switch, ALBM TM, and LVS. The workloads are generated by a number of tens of client machines are interconnected with the cluster of server nodes in the same network segment. We use the Web Bench Tool [11] to generate workload stress to the system. The system features used in the performance experiments are as follows. The Alteon Web switch uses Network Address Translation (NAT) mechanism by assign 3 ports to clients among 8 ports of 100MB. The LVS uses ipvs 0.8.0 and Direct Routing(DR) mechanism. Each server node has PIII-900MHz dual CPU and 512 MB memory. Each client has PIV 1.4GHz CPU and 256MB memory. The network bandwidth is 100MB. The employed load scheduling algorithm is the Least Connection, since the Least Connection algorithm generally shows better performances than the Round-Robin algorithm [5].

In our experiments, we use a number of workload sets: fine-grain workload, e-commerce workload, and dynamic mixed workload. The fine-grain workload consists of simple requests of html documents, requesting a static web page in a small size. The e-commerce workload is used by a general workload generated by Web-tree request in Web-bench tool. The dynamic mixed workload contains workload of fine-grain requests and workload of various memory requests. Since most of coarse grain web requests require some portion of resource, such as memory, in a server node, a large number of requests may reserve a large amount of resources, yielding overloading state of the server. Based on these kinds of workloads, we can test various cases of performance on the three Web cluster architectures.

3.1   Fine-Grain Workload

By generating a huge number of fine-grain requests from clients, we can compare the peak performance of the three kinds of switches. Figure 3 shows the result. The x-axis shows the number of client machines generating requests and the y-axis shows the number of requests serviced per second, i.e. throughput. The more clients are participated, the more stress is given. A workload generated from clients is a request to receive a 0k html document. Thus, this type of workload does not take time for the service. The Alteon Web switch achieved the two to three times better performance (18000 requests/sec) than others due to its hardware capability.

Figure 4 shows the performance speedup by increasing the number of server nodes in the Web cluster system. Each point of the figure is obtained from the peak performance achieved under a given number of server nodes with fine-grain requests. The Alteon Web switch produces a steady performance speedup, while the ALBM TM and LVS show the bottleneck of the switches by having a saturated amount of services although more server nodes are available.

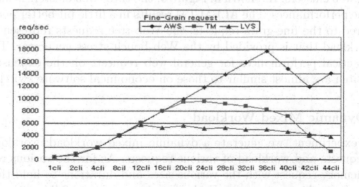

Fig. 3. Performance Comparison for Fine-Grain Requests.

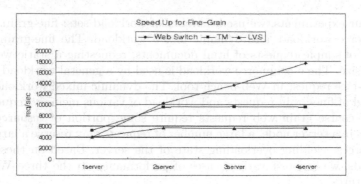

Fig. 4. Performance Speedup for Fine-Grain Workload.

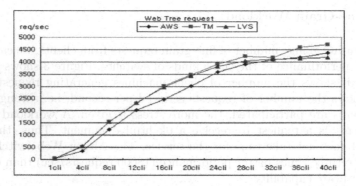

Fig. 5. Performance Comparison for Web-Tree Requests.

3.2 e-Commerce Workload

The Web-bench provides a general workload pattern, called Web-tree, which is mixed with various Html documents, CGI, and images files. By using these Web-tree requests, we make a general E-Commerce workload to compare the three kinds of web clusters. As shown in Figure 5, the three web clusters achieve almost the same performances. The ALBM TM results in a little bit better performance. Compared to the fine-grain requests, the Web-tree requests represent a general web workload that is provided by the Web-bench stress generator. This means that practical performances for general web requests on the expensive Alteon Web switch are almost similar to those on economical software switches.

3.3 Dynamic Mixed Workload

In this experiment, we generate a dynamic mixed workload: workload of fine-grain requests and workload of various requests. As for the various request, we use a memory request with the various sizes and durations to hold the memory resource from the server node. Each of various requests follows a uniform distribution between 5 and 15MB for its memory size and a uniform distribution between 0 and 20 seconds for its duration time. We mix 60% of fine-grain requests

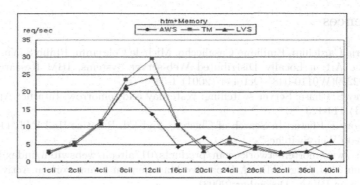

Fig. 6. Performance Comparison for Dynamic Mixed Requests.

and 40% of various memory requests. We may expect the better performance from the Alteon Web switch because it has the high capability on the fine-grained requests. However, the ALBM TM results in the better performance than others as in Figure 6. The Alteon Web switch is in the worst case. It is because coarse-grain workload requires server resources by reserving some amount of resources and performing operation in a server node. Discreet workload scheduling is more effective than fast simple traffic assignment. It is necessary to consider servers' states when scheduling workload, as in the ALBM TM. This becomes more crucial since recent Web requests tend to be coarse-grain workload due to Web services and service-oriented operations.

4   Conclusion

We perform a set of experiments to compare the performances of three web cluster architectures constructed by three types of traffic managers: the Alteon Web switch, the ALBM Traffic Manager, and the LVS. Among them, the Alteon Web switch is a hardware solution, and the others are software network appliances. According to our performance results, the Alteon web cluster shows the best performance for fine-grain workload, but similar or worse performances for the other workloads that have various resource requirements in service size and time. As for a Web-tree workload that is provided by Web-bench tool as a general E-commerce workload, the ALBM web cluster shows a better performance than the others, since the Web-tree workload includes mixed requests with various resource requirements. For the dynamic mixed workload, the ALBM web cluster achieves the best performance than Alteon and LVS web clusters. Even the economical SW L4 switch could achieve the better performance in coarse-grain workload than an expensive HW L4 switch. Considering servers states when scheduling workloads in the L4 switch level yields the more effective performance.

References

1. Valeria Cardellini, Emiliano Casaliccho, Michele Colajanni, Philip S. Yu: The State of the Art in Locally Distributed Web-server Systems. IBM Research Report, RC22209(W0110-048) October (2001) 1-54
2. Jeffray S. Chase: Server switching: yesterday and tomorrow. Internet Applications (2001) 114-123
3. Gregory F Pfister: In Search of Clusters, 2nd Ed. Prentice Hall PTR (1998)
4. "Alteon Web Switches",
   http://www.nortelnetworks.com/products/01/alteon/webswitch/index.html
5. Eunmi Choi, Dugki Min: A Proactive Management Framework in Active Clusters, LNCS on IWAN, December (2003)
6. Wensong Zhang: Linux Virtual Server for Scalable Network Services. Linux Symposium 2000, July (2000)
7. Wensong Zhang, Shiyao Jin, Quanyuan Wu: Scaling Internet Service by LinuxDirector. High Performance Computing in the Asia-Pacific Region, 2000. Proceedings. The Fourth International Conference/Exhibition, Volume: 1, (2000) 176 -183
8. "LVS documents", http://www.linuxvirtualserver.org/Documents.html
9. TurboLinux, Turbo Linux Cluster Server 6 user guide, http://www.turbolinux.com
10. RedHatLinux, Piranha white paper,
    http://www.redhat.com/support/wpapers/piranha/index.html
11. Web Bench Tool, http://www.etestinglabs.com

Simulating Session Distribution Algorithms for Switch with Multiple Outputs Having Finite Size Queues

Jaehong Shim[1], Gihyun Jung[2], and Kyunghee Choi[3]

[1] Department of Internet Software Engineering, Chosun University,
Gwangju, 501-759, South Korea
jhshim@chosun.ac.kr

[2] School of Electrics Engineering, Ajou University,
Suwon, 442-749, South Korea
khchung@ajou.ac.kr

[3] Graduate School of Information and Communication, Ajou University,
Suwon, 442-749, South Korea
khchoi@ajou.ac.kr

Abstract. We propose and simulate session distribution algorithms for switch with multiple output links. The proposed algorithms try to allocate a new session to an output link in a fashion that output link utilization can be maximized but packet delay difference between sessions in a service class be minimized. The simulation proves that SCDF (Shortest Class Delay First) shows the best performance in terms of maximizing link utilization and balancing packet delay difference between sessions.

1 Introduction

Even though physical network bandwidth has increased rapidly, many network users still complain about slow service. It is mainly due to the fact that the performance increase in network equipments located at the edge of and on network cannot catch up the bandwidth increase of network. Firewall and e-mail gateway are included in the typical equipments that require network bandwidth sacrifice. One popular and intuitive way to resolve the bottleneck introduced by network equipments and to fully utilize network bandwidth is to hire multiple low performance equipments served in multiple links or to hire one very expensive high performance equipment.

A simple example of network configuration with multiple links is illustrated in Fig. 1. The bandwidth of each link is usually same but could be different. The firewall hired in the configuration is assumed to have not enough performance to filter all packets coming through the input link (This is true in many real networks). The L4 switch (switch 1 in Fig. 1) classifies packets coming through the input link based on their sessions and service classes, and then distributes the classified packets into N firewalls. The packets filtered by firewalls are fed into the L4 switch (switch 2) at which the output link is connected.

The L4 switch (switch 1) placed between a high speed network and multiple lower speed networks needs a distribution mechanism to fairly distribute packets from the input link connected to the high speed network to the multiple shared output links

D.-K. Baik (Ed.): AsiaSim 2004, LNAI 3398, pp. 625–634, 2005.
© Springer-Verlag Berlin Heidelberg 2005

connected to the lower speed networks. A strong constraint the distribution mechanism must satisfy is that all packets belonging to the same session should be transferred through the *same* shared output link. When the mechanism receives a packet from the input link, it determines one out of multiple shared output links if the packet is the first one of a new session. Otherwise, the packet should be fed to the shared output link through which the session the packet belongs to has transferred.

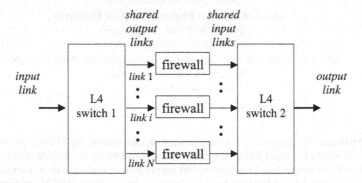

Fig. 1. A typical network configuration utilizing multiple links for packet processing

There are a couple of important issues in designing the distribution mechanism. The first issue is how to maximally utilize a given physical bandwidth of multiple shared output links. The second issue is how to fairly balance packet delay difference between sessions that belong to a service class but are served through different output links. (By a service class, we mean a group of sessions that have common characteristics like service types, protocol or traffic pattern.) Balancing packet delay difference is very closely linked to QoS along with minimizing packet loss [1, 2].

Many previous works [3-5] proposed algorithms to deal with the issues to occur when multiple sessions share a *single* output link. They seldom dealt with the issues for distributing multiple sessions into multiple output links. And they neither consider the constraint that packets that belong to the same session must be served through the same output link. One of main reasons that a clear idea has not been proposed for improving the performance of switch with multiple output links is that it is very hard to drive a mathematical model for the switch. As an alternative, we propose several heuristic session distribution algorithms in this paper. For comparing their performance, the switch is modeled and simulated with the proposed algorithms.

2   Session Distribution Algorithms

Assume that we have a switch that handles M service classes with an input and N shared output links. That is, the switch processes packets belonging to M different service classes coming from a large bandwidth input network (for example, *input link* in Fig. 1) and then forwards the packets to one of N links, where the forwarded link is decided by a session distribution algorithm the switch hires.

The bandwidth (or link rate) of link i among N output links is denoted as B_i where $1 \le i \le N$ and their sum as $B\ (= \sum_{i=1}^{N} B_i)$. B_i may or may not be same as B_j where $i \ne j$ and $1 \le i, j \le N$. For simplicity we use normalized link rate $r_i\ (= B_i / B)$ and their sum r $(= \sum_{i=1}^{N} r_i = 1)$. From now on, normalized link rate and normalized bandwidth are used throughout this paper. The reserved normalized bandwidth of service class C_k is R_k where $1 \le k \le M$. Two different sessions in a service class may be transmitted through two different output links. Then the sum of all reserved bandwidths, $R\ (= \sum_{k=1}^{M} R_k)$ for all M service classes must be less than or equal to the total bandwidth that the switch allows. That is, relationship $R \le r$ must be hold for graceful service. Another variable, the *shared ratio* of C_k is defined as $\phi_k = \dfrac{R_k}{R}$.

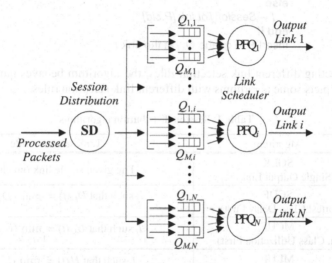

Fig. 2. Scheduling model for switch with multiple output links

A scheduling model for switch with multiple output links is illustrated in Fig. 2. The model is divided into two parts: *Session Distribution module* (SD) and *Packet Fair Queuing algorithm* (PFQ). A SD cooperates with PFQ running in each output link. The SD distributes sessions into multiple output links at the moment a session initiation packet like *SYN* packet is encountered. All the packets of a session have to be transferred through the same link determined by the SD. Then PFQ, one of GPS [6] related scheduling algorithms [4, 7-9] designed for *single* output link, schedules the packets assigned to each link. PFQ fairly allocate the link bandwidth to sub service classes distributed to the link and guarantees the reserved bandwidth of sub service classes. It maintains finite-size packet queues of which number is equal to the number of service classes that the switch serves. That is, if the number of service classes that a switch has to serve is M, M queues are maintained in each link. The set of sessions of sub service class $C_{k,i}$, which belong to service class C_k and are allocated to link i (l_i), is queued in $Q_{k,i}$, the kth queue of l_i, in the order the packets arrive. A bandwidth, $\phi_k \cdot r_i$, is reserved for $C_{k,i}$. Reserved bandwidth of C_k means the bandwidth required for C_k to get the minimum quality of service.

The proposed session distribution algorithms are mainly for the SD and aim to solve two issues: maximizing link utilization and minimizing packet delay difference (or balancing packet delay) between the sessions that belong to a service class but are served through different output links. The algorithms look like;

> Session_Distribution_Algorithm (packet P) :
>
>   Let P be a packet to be passed to one of output links
>   Let $P.sid$ be a session ID of P
>   Assume $P.sid$ belongs to C_k
>   **if** (P is the first packet of $P.sid$) **then**
>   find a link f by the *link selection rule*
> $SessionToLink[P.sid] \leftarrow f$
>   **else**
> $f \leftarrow SessionToLink\ [P.sid]$
>   **end if**
>   Insert P to queue $Q_{k,f}$ at the link f

By selecting different link selection rules, the algorithm behaves quite differently. Table 1 depicts some algorithms with different link selection rules.

Table 1. Session distribution algorithms

Algorithm	Link selection rule
SGLK (Single Output Link)	The given single link (no choice)
SCDF (Shortest Class Delay First)	l_f such that $D_{k,f}(t) = \min\limits_{1 \le i \le N} (D_{k,i}(t))$
MCUF (Min. Class Utilization First)	l_f such that $U_{k,f}(t) = \min\limits_{1 \le i \le N} (U_{k,i}(t))$
MLUF (Min. Link Utilization First)	l_f such that $U_f(t) = \min\limits_{1 \le i \le N} (U_i(t))$
SLDF (Shortest Link Delay First)	l_f such that $D_f(t) = \min\limits_{1 \le i \le N} (D_i(t))$

SGLK represents the case that the switch has only one output link (that is, $N = 1$) and all service classes are served through the link. It is not necessary to consider distributing sessions to multiple output links. Thus the overhead for session distribution is eliminated and its performance is the maximum to which other algorithms can reach.

When a new session that belongs to service class C_k arrives at time t, SCDF distributes the session to the link with the shortest average estimated delay for C_k at the time. $D_{k,i}(t)$, *average estimated delay* of $C_{k,i}$, is defined as $D_{k,i}(t) = \dfrac{L_{k,i}(t)}{\phi_k r_i}$. $L_{k,i}(t)$ is the average queue length of $C_{k,i}$ at time t and calculated by a recursive equation $L_{k,i}(t+1) = \alpha\, T_{k,i}(t) + (1-\alpha)\, L_{k,i}(t)$ where $T_{k,i}(t)$ is the current queue length of $C_{k,i}$, and $0 < \alpha < 1$ [10].

MCUF distributes the session to the link with the smallest traffic load for C_k at the time where the traffic load means the bandwidth requirement. $U_{k,i}(t)$ is defined as the traffic load of $C_{k,i}$ *at time t*.

MLUF distributes the session to the link with the minimum link load at the time. The link load $U_i(t)$ of l_i at time t is defined as $U_i(t) = \sum_{k=1}^{M} U_{k,i}(t)$. And *SLDF* distributes the session to the link with the shortest link delay *at the time. The average estimated link delay, $D_i(t)$, of l_i at time t is defined as* $\sum_{k=1}^{M} D_{k,i}(t)$.

3 Simulation

To observe link utilization, packet delay difference and the impact of queue length to link utilization, we performed an empirical study with a hypothetical switch. Since the algorithms do not show any significant performance differences when all output links are underutilized (the later empirical study reveals it.), we focus on the cases that more than one output links are pushed to be over-utilized. When one or more output links are monopolized by one or more sessions and fully utilized, the monopolized links may hurt the utilizations of other non-monopolized links. In addition, the monopolization makes the packet delay of sessions served in the monopolized links quite different from the delay of sessions served in other non-monopolized links.

The hypothetical switch utilized in the empirical study is equipped with a single input link and three output links l_1, l_2 and l_3. l_1 roles as a link monopolized by an abnormal session. And l_2 and l_3 represent other non-monopolized links. The sum of queue sizes in each output link is proportion to its normalized bandwidth r_i and the length of queue allocated to a sub service class in each link is again proportional to the share ratio of the class. The normalized bandwidths of output links r_1, r_2, r_3 are 1/3 each. Two service classes C_1 and C_2 are defined and their share ratios, ϕ_1, ϕ_2, are 0.6 and 0.4, respectively. We set $\alpha = 0.1$ for estimating average delay since other empirical study showed that most proposed algorithms have the best performances at the value.

Fig. 3-1 illustrates *the normalized input traffic (defined as input traffic / physical bandwidth) of* C_1 and C_2, and their sum. The average of input traffic is shown in Fig. 3-2. 6,000 sessions are generated for C_1 and C_2 and their connection request intervals and life-times are exponentially distributed [2, 5, 9]. The packet arrival rate of each session follows a Poisson distribution with mean equal to the average bandwidth requirement divided by the number of sessions being served in the class at a moment. To eliminate the impact of packet processing overhead, we assume all packets have the same type and length. A session, S_{abnorm}, that belongs to C_2 is served at l_1 through-

out the simulation. It requires a bandwidth of 0.133. That is equal to the reserved bandwidth for $C_{2,1}$, (=0.4/3). S_{abnorm} mimics an abnormal service like an extremely large secure file transfer or a DDoS attack.

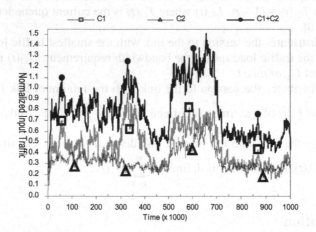

Fig. 3-1. Generated normalized input traffic amount for C_1 and C_2

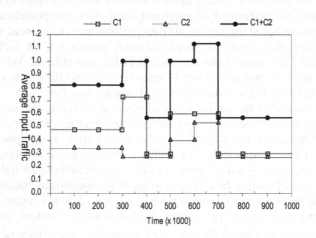

Fig. 3-2. Average normalized input traffic amount for C_1 and C_2

To see how the algorithms work in various input loads (bandwidth requirements), the empirical study is done with three input load patterns, which are 1) the bandwidth requirements of both of C_1 and C_2 are smaller than their share ratios as in interval [0,300], [400,500] and [700,1000], 2) the total required bandwidth is equal to the physical bandwidth but the bandwidth requirement for one service class (C_1) is greater than its share ratio but the bandwidth requirement for other service class (C_2) is smaller than its share ratio as in [300,400], 3) the overall bandwidth requirement is equal to or greater than the given physical bandwidth but the bandwidth required for one class (C_1) is nearly equal to its share ratio and that for other class (C_2) is equal to or greater than its share ratio as in [500,700].

WF^{2}Q+ [9] runs in each output link as the PFQ. When WF^{2}Q+ selects the next packet for service at time t, it chooses, among the set of packets that have started receiving service in the corresponding GPS [6] system at time t, the first packet that would complete service in the corresponding GPS system. It is known to have the tightest delay bounds and low time-complexity among link schedulers [5].

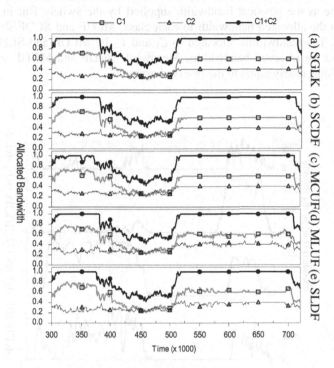

Fig. 4. Allocated bandwidths

Fig. 4 shows the bandwidths allocated to C_1 and C_2 when SGLK, SCDF, MCUF, MLUF and SLDF are hired as the SD module. The sum of packet queue lengths in the switch is assumed 2620 packets large (approximately 20 ms delay is expected for a 1 Kbyte packet coming from a Giga-bit speed network to pass through the switch). When a required bandwidth is below the given physical bandwidth (during intervals [0, 300], [400, 500], and [700, 1000]), it seemed no problem to service all sessions except some cases that the allocated bandwidths nearly reach to the full physical bandwidth. Thus we focus on the output in [300,700]. In interval [300,400], the sums of bandwidths allocated to two links by SCDF, MLUF and SLDF are nearly same as that by SGLK. But the overall bandwidth utilized by MCUF is smaller. Since S_{abnorm} seizes 0.133 bandwidth of l_1, making $U_{2,1}(t)$ (the traffic load of C_2 in l_1) 0.133, which is the half of full required bandwidth of C_2 (0.267 in Fig. 3-2), MCUF tries to evenly allocate new sessions of C_2 to l_2 and l_3 (except l_1) and makes $U_{2,2}(t)$ and $U_{2,3}(t)$ approximately 0.066. Meanwhile MCUF evenly distributes the sessions of C_1 to the three links without considering the session distribution of C_2. Thus $U_{1,1}(t)$, $U_{1,2}(t)$ and $U_{1,3}(t)$ become 0.244 (=[the bandwidth (0.733) required for C_1] / 3). Consequently the amounts of (normalized) input traffic that MCUF tries to allocate to l_1, l_2 and l_3 become 0.377 (=0.244+0.133), 0.3 (=0.244+0.066) and 0.3 respectively. Thus l_1 is overloaded and fully utilized. $U_1(t)$ becomes 0.377 (but limited by the physical bandwidth of l_1, 0.33). But l_2 and l_3 are underutilized.

In time interval [500, 700], we can see how fairly the algorithms allocate the total bandwidth to two service classes. The total bandwidths allocated by all algorithms are nearly same as the physical bandwidth supplied by the switch. But in terms of the fairness on the allocated bandwidth to each class, MLUF and SLDF show poor performances. The bandwidths allocated to C_1 and C_2 by MLUF and SLDF are fluctuated around the reserved bandwidths. Meanwhile SCDF and MCUF stably allocate the bandwidths nearly equal to the reserved amounts.

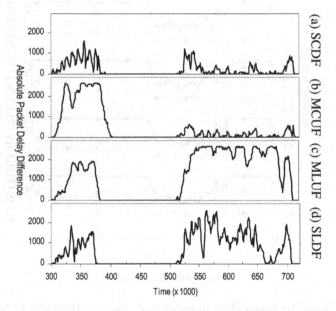

Fig. 5. Absolute packet delay differences of session in C_1

To measure how fairly the algorithms balance packet delays of sessions, we choose absolute packet delay difference between sessions as the metric. Fig. 5 shows the maximum absolute delay difference between $C_{1,1}$, $C_{1,2}$ and $C_{1,3}$ in time interval [300, 700]. That is, the figure shows $max\{|D_{1,1}-D_{1,2}|, |D_{1,1}-D_{1,3}|, |D_{1,2}-D_{1,3}|\}$ at each time t in the interval. The delay difference under SCDF is relatively well-balanced and small in the entire interval. Meanwhile MCUF produces the smallest delay in time interval [500, 700] but a very large delay in [300, 400]. The traffic of $C_{1,1}$ overloaded by MCUF is much more than l_1 handles (as mentioned above) in [300, 400]. Thus the time for $C_{1,1}$ packets to stay in queue gets longer and packet delay in l_1 reaches to the maximum for a packet to get through the queue. Meanwhile the traffic loads of $C_{1,2}$ and $C_{1,3}$ are smaller than that of $C_{1,1}$. Thus packet delays in l_2 and l_3 become smaller. Consequently the delay difference in the interval becomes large. The traffic load of $C_{1,1}$ allocated by MLUF is much larger than those of $C_{1,2}$ and

$C_{1,3}$ in [500, 700]. The load difference makes packet delay difference large in the interval. SLDF allocates traffic load very unstably in the entire interval (especially in [500, 700]). Thus the packet delay difference by SLDF varies very widely.

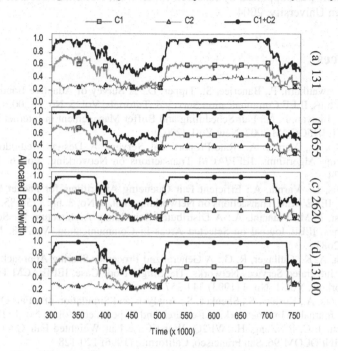

Fig. 6. Bandwidths allocated by SCDF with different queue sizes

Fig. 6 illustrates the allocated bandwidths by SCDF when the length of queue varies from 131(1ms packet delay at a Giga-bit speed network), 655(5ms), 2620(20ms) to 13100(100ms) packets large. As the queue length increases, link utilization also increases. That is, there exists a clear trade-off between queue length and link utilization. It means that an appropriately long queue is required to keep a certain QoS level. If queue length is shorter than 131 packets, link utilization will become less stable but the overall shape looks like one the switch has a longer queue. When queue length becomes larger than 13100 packets, nearly no significant change in link utilization is observed.

4 Conclusion

We proposed several link distribution algorithms for switch with multiple output links. The simulation shows that SCDF shows the best performance in terms of link utilization and packet delay differences between sessions in a service class. The performance of SCDF is pretty close to that of PFQ algorithm (SGLK) running in switch with a single link. SCDF improves link utilization and reduces packet delay difference by minimizing the overload assigned to each output link.

Acknowledgement

This research was supported by research funds from National Research Lab program and Chosun University, 2004.

References

1. Siripongwutikorn, P., Banerjee, S., Tipper, D.: A Survey of Adaptive Bandwidth Control Algorithms, IEEE Communication Survey & Tutorials, Vol. 5, No. 1 (2003) 14-26
2. Ni, N., Bhuyan, L. N.: Fair Scheduling and Buffer Management in Internet Routers, Proc. of IEEE INFOCOM 2002, New York (2002)
3. Stiliadis, D., Varma, A.: Rate-Proportional Servers: A Design Methodology for Fair Queueing Algorithms, IEEE/ACM Transactions on Networking, Vol. 6, No. 2 (1998), 164-174
4. Stiliadis, D., Varma, A.: Efficient Fair Queueing Algorithms for Packet Switched Networks, IEEE/ACM Transactions on Networking, Vol. 6, No. 2, pp. 175-185, Apr. 1998.
5. Chiussi, F. M., Francini, A.: A Distributed Scheduling Architecture for Scalable Packet Switches, IEEE Journal on Selected Areas in Communication, Vol. 18, No. 12 (2000) 2665-2683
6. Parekh, A. K., Gallager, R. G.: A Generalized Processor Sharing Approach to Flow Control in Integrated Services Networks: The Single-Node Case, IEEE/ACM Transactions on Networking, Vol. 1, No. 3 (1993) 344-357
7. Demmers, A., Keshav, S., Shenker, S.: Analysis and Simulation of a Fair Queueing Algorithm, Journal of Internetworking Research and Experience, Vol. 1, No. 1 (1990) 3-26
8. Bennett, J. C. R., Zang, H.: WF2Q: Worst-Case Fair Weighted Fair Queueing, Proc. of IEEE INFOCOM'96, San Francisco, California, (1996) 120-128
9. Bennett, J. C. R., Zang, H.: Hierarchical Packet Fair Queueing Algorithms, IEEE/ACM Transactions on Networking, Vol. 5, No. 5 (1997) 675-689
10. Stallings, W., Operating Systems: Internals and Design Principles, 4th ED., Prentice Hall (2001) 408-411

An Efficient Bridging Support Mechanism Using the Cache Table in the RPR-Based Metro Ethernet*

Ji-Young Huh[1], Dong-Hun Lee[1], Jae-Hwoon Lee[1,**],
Hyeong-Sup Lee[2], and Sang-Hyun Ahn[3]

[1] Dept. of Information and Communication Engineering, Dongguk University,
26, 3 Pil-dong, Chung-gu Seoul, 100-715 Korea
jaehwoon@dongguk.edu
[2] ETRI, 161 Gajeong-dong, Yuseong-gu, Daejeon, 305-350 Korea
[3] University of Seoul, Seoul 130-743 Korea

Abstract. RPR (Resilient Packet Ring) is a protocol proposed for the efficient delivery of frames over the metro network and is required to provide the bridging functionality for the interconnection of other external LANs. To do this, the node with the RPR capability should broadcast frames received from external networks to the entire RPR network. Since the RPR is proposed to operate in a metro network, most of RPR nodes are not hosts but bridges. Therefore broadcasting of all frames from external networks may significantly deteriorate the performance of the RPR network. In this paper, we propose a mechanism in which an RPR bridge node can efficiently support the bridging functionality by using a cache table with the information on external nodes.

1 Introduction

Internet consists of private and public networks and a public network is comprised of a backbone network, metro networks and access networks. Up until now, the capacity of backbone and access networks has been increased drastically. However, since a metro network uses SONET/SDH based on the synchronous TDM, the bandwidth utilization of a metro network is relatively low and it is difficult to increase the capacity. In order to resolve this problem, the metro Ethernet that allows high-speed transmission and easy configuration has been proposed and RPR (Resilient Packet Ring) is the protocol proposed for this purpose [1-4].

Since RPR is proposed for the use in the metro network, it has to provide the bridging functionality for the interoperation among LANs. For the efficient usage of link capacity, RPR provides the destination stripping mechanism in which destinations remove the unicast frames from the ring. However some bridge nodes may not have the information on some external nodes, so those nodes

---

* This work was supported by ETRI and University IT Research Center project.
** Corresponding author.

D.-K. Baik (Ed.): AsiaSim 2004, LNAI 3398, pp. 635–642, 2005.
© Springer-Verlag Berlin Heidelberg 2005

may not perform the bridging capability properly. In order to provide the full bridging functionality, RPR nodes receiving frames from external networks are required to broadcast those frames on the entire RPR network [5]. However this may deteriorate the performance of the RPR network due to the broadcast of all unicast frames from external networks. Since most RPR nodes are not hosts but bridges interconnecting LANs, the RPR bridging mechanism can result in the inefficient use of link capacity due to broadcasting a large number of frames generated from external networks.

In this paper, to enhance the performance of the RPR network, we propose a cache based bridging support mechanism which can efficiently deliver frames from external nodes by maintaining a cache with the information on external nodes from which frames are generated to the RPR network.

This paper is organized as follows; in section 2, the operation and the problem of the original RPR bridging mechanism are described. In section 3, we describe the proposed cache-based bridging support mechanism. The structure of the NS-2 simulator used for the performance analysis of the proposed scheme is provided in section 4. In section 5, we compare the performance of the original RPR and the proposed cache-based bridging support mechanism by using the simulator described in section 4. And section 6 concludes this paper.

2   RPR Bridging Mechanism

First, the basic operation of the MAC bridge is briefly described, and then the bridging mechanism defined in the RPR standard and its problem are described.

2.1   MAC Bridge

A MAC bridge can provide the LAN interconnection capability and utilizes bandwidth efficiently by having the filtering function [6]. Upon receiving a frame, a bridge looks up for the source address of the frame in its FDB (Filtering Database) and, if the address is not in its FDB or has been changed, the bridge adds the address into its FDB or updates the corresponding entry in its FDB, respectively. Also the bridge checks whether the destination address of the received frame is registered in its FDB or not. If the destination is on the same port as that of the source, it just discards (filters) the frame. Otherwise, it relays the frame to the port for the destination as specified in its FDB. However, if the destination address is not specified in its FDB or is the multicast/broadcast address, the bridge floods the frame to all other ports except that from the source. This operation is carried out by the relay layer which is the next higher layer of the MAC layer.

2.2   Bridging Support Mechanism

The RPR network consists of two rings, but the relay layer considers the network as a broadcast network with no loop and allocates one port of a MAC bridge to

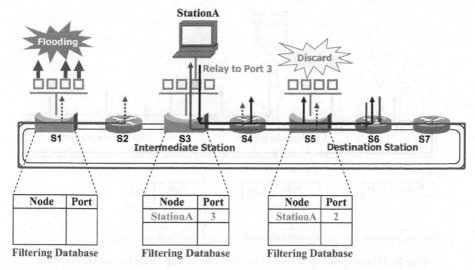

Fig. 1. Frame delivery scenario in the case of frame discard at the destination.

the RPR network. Thus, the RPR MAC layer has to determine which ring to use for the delivery of frames forwarded from the relay layer.

RPR uses the destination stripping mechanism, where a unicast frame is discarded at its destination for the efficient use of bandwidth. However if a unicast frame from an external network is discarded at the destination, some bridges may not perform the bridging function properly.

Figure 1 shows the problem of discarding a frame from an external network at the destination in an RPR network with bridges S1, S3 and S5. If Station A in an external network sends a frame to S6, bridge S3 receives the frame. The relay layer of S3 stores the source, i.e. Station A, information in its FDB, and forwards the frame to its RPR MAC layer. The RPR MAC layer determines which ring to use to forward the received frame. The outer ring is chosen and used for the delivery of the frame to S6 in Figure 1. The frame passes by S4 and S5 and then arrives at S6 where it is discarded. S5 which is on this delivery path can store the information on Station A in its FDB. However, S1 which is not on the path can not.

In this case, in order for S6 to send an acknowledgement (ACK) for the receipt of the frame, S6 has to broadcast the ACK on the entire RPR network. For the delivery of the ACK, a bridge forwards it from the RPR MAC layer to the relay layer and perform the function of delivering or discarding it to other networks based on the information in its FDB. That is, since S3 and S5 have the information on Station A, S3 can forward the ACK to its port 3 and S5 discards the ACK because the destination port of the ACK is its port 2 which is connected to the RPR network from which the ACK has come. However, frames sent from S6 to Station A are flooded to other networks connected to S1 because S1 does not have the information on Station A. In this case, S1 does not perform the bridging function properly, which incurs lots of overhead.

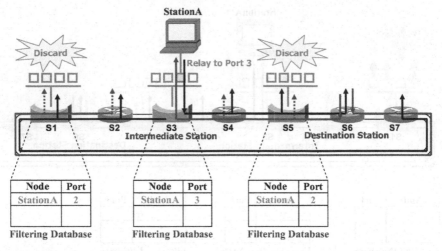

Fig. 2. Frame delivery scenario in the case of the bridging interoperation.

In order to resolve this problem, RPR standard broadcasts frames from external networks as shown in figure 2 so that bridges can store the information on sources residing in external networks [5]. In figure 2, if Station A delivers a frame to S6, S3 stores the information on Station A in its FDB and forwards the frame to the RPR MAC layer. If the frame is from an external network, the RPR MAC layer broadcasts it by using both of its rings so that all bridges in the RPR network can store the information on Station A in their FDBs. Hence, even when S6 sends a frame to Station A, bridges in the RPR network can perform the proper bridging function without flooding the frame.

However, since RPR is proposed for the metro Ethernet, most of the nodes in an RPR network are not hosts, but bridges connecting to external networks. Also most of the frames in an RPR network are generated not from inside, but from external networks. Therefore, broadcasting all frames from external networks may cause lots of overhead.

3   Cache-Based Bridging Support Mechanism

In this paper, we propose an efficient bridging support mechanism which uses the cache table. The operation of the proposed mechanism is shown in figure 3.

In figure 3, each of S1, S3 and S5 which are bridges in the RPR network maintains a cache which can store the information on external nodes. At the first time, when Station A sends a frame to S6, bridge S3 which connects an external network with the RPR network receives the frame. Since the information on the source node, i.e., Station A, of the frame is not in S3's FDB, S3 stores the information in its FDB and passes the frame to the RPR MAC layer. The RPR MAC layer checks whether the address of Station A is in its cache table, and, if it is not, the RPR MAC layer stores the address of Station A in its cache table and broadcasts the frame to the RPR network. By broadcasting the first frame

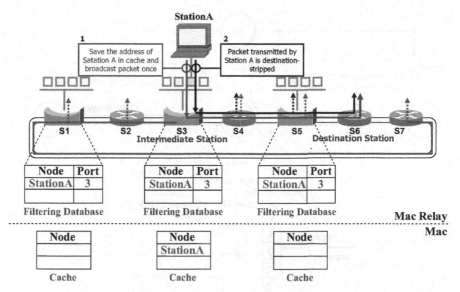

Fig. 3. Operation of the cache-based bridging support mechanism.

from Station A, all bridges in the RPR network can store the information on Station A in their own FDBs.

Once all bridges in the RPR network store the information on Station A in their FDBs, they do not need to broadcast frames received from Station A unless the lifetime of the corresponding entry is expired or the information in the entry is changed. So when S3 having the information on Station A in its cache table receives frames from Station A, it does not have to broadcast them. Therefore those frames can be discarded at their destinations, which can result in a significant reduction of broadcast frames. Furthermore, by setting the lifetime of cache table entries to some smaller value than that of FDB entries of the relay layer, all bridges in the RPR network can keep on maintaining the information on Station A.

4   Structure of the Simulator

We implemented a simulator by extending NS-2 [7], in order to analyze the proposed cache-based bridging support mechanism. In this simulator, RPR_Client and RPR802_17 are implemented between so that nodes in the RPR network can perform the RPR functionality. RPR_Client has the ring selection capability of RPR and the topology database maintenance function. RPR802_17 includes the functions which should be executed by a ring such as the fairness algorithm, the topology discovery and the frame transmitting/receiving capabilities. And RPR_Client of a node acting as a bridge in the RPR network is added with the functions such as the cache table maintenance and the bridging mechanism using the cache table. Figure 4 shows the structure of the simulator implemented for the analysis of the performance of the proposed cache-based bridging support mechanism.

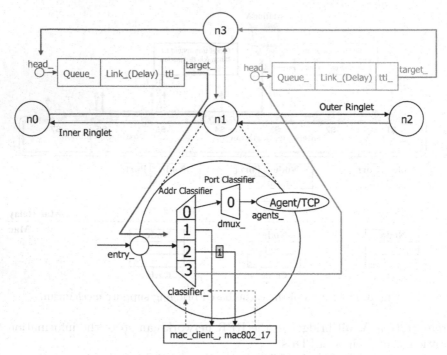

Fig. 4. Structure of the simulator using NS-2.

In figure 4, n1 is a bridge in the RPR network and n3 a host residing in an external network. If n1 received a frame from n3, the frame is delivered to RPR_Client of n1. RPR_Client of n1 uses its topology database to decide whether the frame is from an external node or not. Since the frame is from outside of the RPR network, RPR_Client of n1 searches for the source address of the frame in its cache table. If the source address is in the cache table, n1 directly delivers the frame to the destination by using the ring chosen by the ring selection capability and the destination discards the frame. Otherwise, n1 stores the source address in its cache table and broadcasts the frame to all RPR nodes by using both rings. If the cache table is full, it just broadcasts the frame to all RPR nodes without storing the source address in the cache table. Also, a timer is maintained for each cache table entry and, if the timer for an entry is expired, the entry is discarded. If a frame whose source address is in the cache table is received, the corresponding timer is reset.

5   Performance Evaluation

In this section, the performance comparison between the bridging support mechanism of RPR and the proposed cache-based bridging support mechanism is provided. The network model for simulation is shown in figure 5.

In figure 5, the data rate of each link in the RPR network is 622 Mbps and the link propagation delay 0.1 msec. The data rate of the link connecting S1

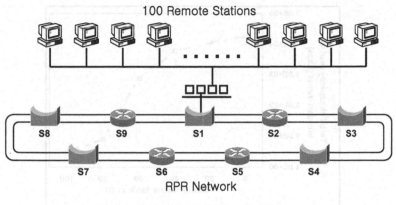

Fig. 5. Network Model for Simulation.

which is a bridge in the RPR network and the external network is 100 Mbps and the propagation delay of that link is 1 msec. In this simulation, 100 external nodes connected to S1 send frames to a node in the RPR network at the rate of 0.5 Mbps.

Figure 6 and 7 show the number of broadcast frames and the total link capacity used by broadcast frames with varying the size of the cache table at S1, respectively. With the size of the cache table being zero, S1 broadcasts all frames from the external network as the original RPR does. As the size of the cache table increases up to the number of external nodes, the amount of information stored in the cache table increases, hence the number of broadcast frames and the amount of total link capacity used by broadcast frames are reduced.

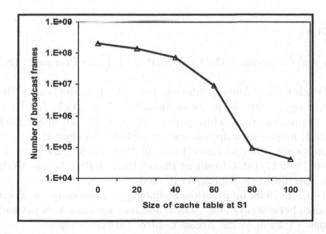

Fig. 6. Number of broadcast frames vs. size of cache table at S1.

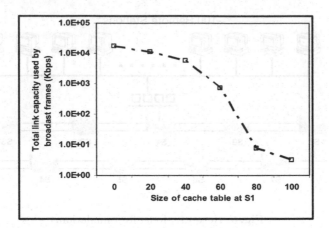

Fig. 7. Total link capacity used by broadcast frames vs. size of cache table at S1.

6   Conclusions

RPR is a protocol proposed for the metro Ethernet and provides the bridging functionality that allows the RPR network to be interconnected with other LANs. However RPR has the problem of broadcasting all frames from external networks to the entire RPR network.

In this paper, we proposed an efficient cache-based bridging support mechanism which stores the information on external nodes in the cache table. The performance of the proposed mechanism has been compared with the bridging function of RPR throughout the simulation. According to the simulation result, we can see that the number of broadcast frames reduces significantly as the size of the cache table increases in the proposed mechanism.

References

1. D. Tsiang and G. Suwala, "The Cisco SRP MAC Layer Protocol", IETF RFC 2892, 2000.
2. Resilient Packet Ring Alliance white paper, "An Introduction to Resilient Packet Ring Technology", 2001, http://www.rpralliance.org/articles/ACF16.pdf.
3. APPIAN Communications white paper, "IEEE 802.17 Resilient Packet Ring Networks", 2001; http://www.appiancom.com/RPR_Strategy.pdf.
4. Cisco white paper, "Spatial Reuse Protocol Technology", 2000.
5. IEEE Draft P802.17/D2.3, Resilient Packet Ring (RPR) Access Method and Physical Layer Specifications, September 6, 2003.
6. IEEE Std 802.1D-1998, Information Technology – Telecommunications and information exchange between systems – Local and metropolitan area networks – Common speifications – Part 3: Media Access Control (MAC) Bridges.
7. LBL, Xerox PARC, UCB, USC/ISI, VINIT Project, The Network Simulator ns-2, www.isi.edu/nsnam/ns.

Simulating Cyber-intrusion
Using Ordered UML Model-Based Scenarios

Eung Ki Park[1], Joo Beom Yun[1], and Hoh Peter In[2,*]

[1] National Security Research Institute, 62-1 Hwa-Am-Dong, Yu-Seong-Gu
Daejeon, 305-718, Republic of Korea
{ekpark,netair}@etri.re.kr
[2] Dept.of Computer Science and Engineering, Korea University, Seoul, 136-701
Republic of Korea
hoh_in@korea.ac.kr

Abstract. Network security simulator is required for the study on the cyber intrusion and defense as cyber terrors have been increasingly popular. Until now, network security simulation aims at estimation of a small-size network or performance analysis of information protection systems. However, a systematic way to develop network security simulation is needed. In this paper, an ordered UML-based scenario model and its network security simulator are proposed. A network security scenario is presented as a case study of our proposed simulation model.

1 Introduction

Network security simulation is widely used to understand cyber attack and defense mechanisms, their impact analysis, and network patterns and behaviors because it is difficult to study them by implementing such dangerous attacks in the real world. At the early stage of network security simulation, Mostow, et al. proposed the Internet Attack Simulator (IAS) to simulate three scenarios [4]. Effectiveness of password sniffing and firewall systems were simulated in [5]. Smith and Bhattac Harya simulated a firewall system to analyze the performance of a small size network by changing the architecture and configuration of the firewall system. Simulation was also used to estimate the network performance by changing the network topology [2,3]. However, these simulation techniques have limitation to represent the status and configuration of the real-world network, and express realistic security scenarios. For example, the existing simulators could not handle ordered scenarios in efficient ways to represent realistic attack and defense mechanisms. In addition, there is no systematic way to develop simulation model from informal network security requirements.

This paper proposes an ordered scenario-based network security simulation model to overcome the limitation of existing network security simulators. The Unified Modeling Language (UML) is used to develop the simulation model from informal network security requirements in a systematic way. Our proposed network security simulator aims at the following goals: First, the components of the network security

---

* Hoh Peter In is the corresponding author.

D.-K. Baik (Ed.): AsiaSim 2004, LNAI 3398, pp. 643–651, 2005.

simulation model could be reusable. Once the components of the network security simulation model are defined in a domain, they can be reusable for other similar domains or the domains with similar scenarios with minimum reconfiguration or parameter modification. Second, the simulation model can simulate various scenarios of cyber intrusion and defense mechanisms. Third, it needs an easy-to-use feature. For example, it provides a convenient graphical user interface to edit cyber intrusion and defense scenarios and observe their results clearly.

This paper is organized as follows: the context of the work is presented in Section 2. The overview of our proposed network security simulation model is described in Section 3. A case study of distributed denial of service is explored in Section 4. The conclusion and future work are presented in Section 5.

2 The Context: Object-Oriented Network Security Simulator

An object-oriented network security simulation model was proposed by Donald Welch et al.[5]. The following object classes were presented with their attributes and behaviors:

- *Node Address*: an object which is used to identify network nodes such as a workstation, a server, and a router. It has attributes such as processor speed, memory capacity, and provided services. Its behavior represents node actions.
- *Connection*: a physical link among the nodes. Examples of its attributes are a speed, information of node connection, and reliability.
- *Interaction*: an exchange of information among the nodes. "Interaction" shown in [5] represents only an important event, not a packet level.
- *Infotron*: the smallest piece of interesting information. This can be a whole set of databases or a particular database

An example of the object-oriented network security simulation models with a scenario of password sniffing is described in Figure 1. The left side of Figure 1 represents the scenario of password sniffing. The right side describes an overview of executing the simulation of the scenario. This object-oriented simulation model has several good features such as easy-to-use, inheritance, information hiding, and reusability. The object classes can be reused for other scenarios.

However, it is difficult to model the sequence of events because cyber attack scenarios are represented into the graph shown in the left side of Fiture 1. In addition, the "interaction" in the right side is not intuitively understood and does not represent its behaviors in the detailed. For example, "Sniff" behavior does not have the detailed information. Therefore, it is needed to describe the sequence of events and detailed interaction. For example, the orders (or steps) of "sniffing" are: 1) connect to a user computer, 2) execute a sniffing program, 3) sniffing all network packets, and 4) acquiring user's credit card information.

In this paper, a new network security simulation model is proposed to overcome these limitations and develop the ordered scenario-based simulation model using the UML modeling techniques. The new model and its support tool are described in Section 3.

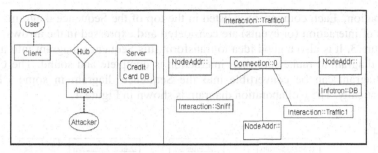

Fig. 1. Donald Welch, et al.'s Simulation Model

3 Our Proposed Work: Modeling Ordered UML-Based Scenarios

Our proposed network security simulation model adapts the attack tree [8] with ordered information and begins with informal requirements description. The informal description is transformed into several UML diagrams, and is finally driven into our proposed ordered scenario simulation model.

As the first step, a Use Case diagram is used to understand who players are and how they interact with. An example of the Use Case diagram is shown Figure 2. For example, suppose that a hacker wants to sniff a user's credit card information. Thus, the hacker needs to connect the user's computer, execute a sniff program and begin sniffing all packets from the user's computer. A benefit of using the Use Case diagrams is to enable model developers to have insights of the interaction between players and system components or between the components and the components.

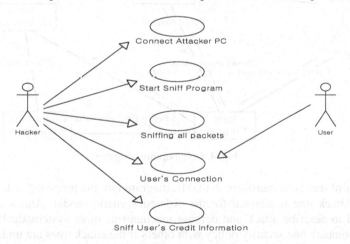

Fig. 2. Use case diagram

The second step is to elaborate the Use Case diagrams and develop Sequence diagrams as shown in Figure 3. Among the Use Case diagram as shown in Figure 2, for example, the model developer identifies important model components such as a hacker, an attacker system, the sniff program, the user connection, and the credit card information,. Each component is listed in the top of the Sequence diagram. Then, the orders of interaction (or events) are considered and expressed in the arrows, as shown in Figure 3. It is also a good idea to transform this Sequence diagram into a Collaboration diagram to make sure their interaction is complete and sound. The Collaboration diagram can be convertible into the Sequence diagram in some UML tools automatically. The Collaboration diagram is shown in Figure 4.

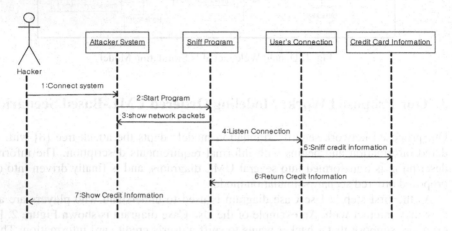

Fig. 3. Sequence diagram

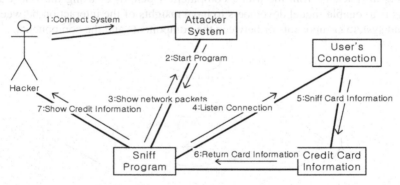

Fig. 4. Collaboration diagram

The third step is to transform the UML diagram into the proposed ordered scenario model. Attack tree is adapted for the ordered scenario model. Attack tree [8] was proposed to describe attack and defense mechanisms more systematically. It is also easy to compare one security policy with others if the attack trees are understood. The attack tree model consists of a root node and subnodes. The root node is the final target of intrusion, and subnodes are preceding nodes to reach the root node. It is convenient to understand the hierarchical structure of subnodes. An example of an attack tree is shown in Figure 5. However, it is difficult to understand the order of each subnodes. It is needed to present the order of subnodes when a cyber attack scenario is simulated.

Our proposed ordered scenario model is developed by adding numbers which represents execution orders into the Attack Tree. An example of ordered scenario model is presented in Figure 6. Since a sequence number is assigned to each node, it is easy to understand the order of simulation execution and map each step into a position of the attack tree.

Two-type nodes are defined: "Actor node" and "Operation node." "Actor Node" is a terminal node of the tree and executes events. A node name is described as an event name. "Operation Node" has one or more subnodes, and determines the order of subnode execution. It has a logical operation semantics: "AND" and "OR". Operation nodes are used to represent the relationships between nodes and to group actor nodes into a category having similar event characteristics. Furthermore, the "Operation node" can represent the order with Sequential_AND, Sequential_OR, Parallel_AND, and Parallel_OR.

Sequential_AND represents that all lower nodes should be executed sequentially and then the sibling nodes can be executed. Sequential_OR represents that the sibling nodes can be executed if one of lower nodes is executed sequentially. Parallel_AND and Parallel_OR are similar with Sequential_AND and Sequential_OR accordingly, except the subnodes are executed in parallel.

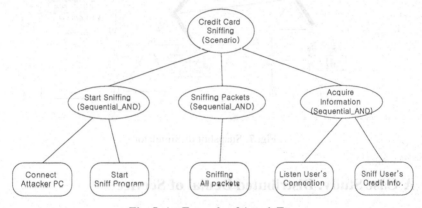

Fig. 5. An Example of Attack Tree

1 Sniffing Scenario (Scenario)
   1.1 Sniffing Start (SequentialAND)
      1.1.1 Connecting Attacker PC (Actor)
      1.1.2 Start Sniffing Program (Actor)
   1.2 Sniffing Network Packet (SequentialAND)
      1.2.1 Sniff all packets passing through a Hub (Actor)
   1.3 Acquiring a credit card information (SequentialAND)
      1.3.1 Listen User's Connection (Actor)
      1.3.2 Sniff a credit card information (Actor)

Fig. 6. Ordered Scenario Model

The snapshot of our proposed network security simulator is shown in Figure 7. Network topology and configuration of network architecture are modeled intuitively using GUI. A simulation scenario of sniffing credit card information (upper right) is executed on the network. The thick lines represent the communication path of the important credit information between nodes. Our proposed simulator shows graphically the simulation process and expresses the nodes more than 100,000. The simulation results are shown in a graph (below right). The objects shown in Figure 7 were implemented by reusing and adapting the classes of the SSFNet simulator.

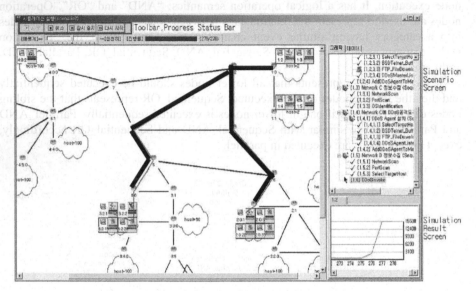

Fig. 7. Snapshot of Simulator

4   A Case Study: Distributed Denial of Service

In this section, a scenario of distributed denial of service (DDoS) attacks is simulated to test the effectiveness of our proposed simulator. The assumption and experimental setting is presented in Section 4.1, simulation scenario in Section 4.2, and the simulation results in Section 4.3.

4.1   Simulation Setting

A network topology to simulate DDoS attacks is shown in Figure 8. The host 3:1 starts the ICMP Flooding Attack, one of the DDoS, while two HTTP clients (host 2:1 and 2:2) communicate HTTP messages from and to a HTTP server (host 1:1). Experiment is set up to observe and measure the number of ICMP packets per second to the HTTP server in our proposed simulator to compare with real-world experimental results.

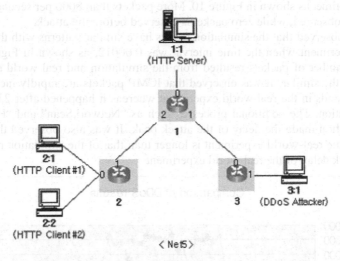

Fig. 8. Network Environment

The host used in this experiment is a Pentium III 700MHz PC with a 128MB memory. Its operating system is RedHat Linux 6.0. The network bandwidth is 10Mbps. The HTTP server is the Apache web server, and the HTTP clients are SPECweb99. The tool of the DDoS is TFN2K.

4.2 Simulation Scenario: DDoS

In the experiment a DDoS attack from the host 3:1 to the host 1:1 is performed using TFN2K tool. However, the execution is performed as shown in Figure 9 because our simulator has to pass over a "NetworkScan" phase.

```
1. Scenario
   1.1 SequentialAND (SequentialAND)
       1.1.1 NetworkScan
       1.1.2 PortScan
       1.1.3 DDoSAgentListen
       1.1.4 DDoSMasterListen
       1.1.5 AddDDoSAgentToMaster
       1.1.6 DDoSInvoker
```

Fig. 9. DDoS Scenario

4.3 Verification of Simulation Results

Only ICMP packets were observed, and TCP and UDP packets are ignored. The packet number of inputs to a target system is measured in this experiment. The number of ICMP packets is highly increased from 12 seconds to 42 seconds, ICMP Flooding attack time, as shown in Figure 10. More packets than 8000 per second during the attack are observed, while zero packet is observed before the attack.

It was observed that the simulation results have similar patterns with those of real-world experiment when the time interval was 0.00012, as shown in Figure 10. The average number of packets resulted from the simulation and real-world experiments were also the similar. It was observed that ICMP packets are rapidly increased after the 12 seconds in the real-world experiment whereas it happened after 23 seconds in the simulation. The additional processes such as "NetworkScan" and "PortScan" in the simulation made the delay of the attack peak. It was also observed that the peak period of the real-world experiment is longer than that of the simulation result due to the network delay of the real-world experiment.

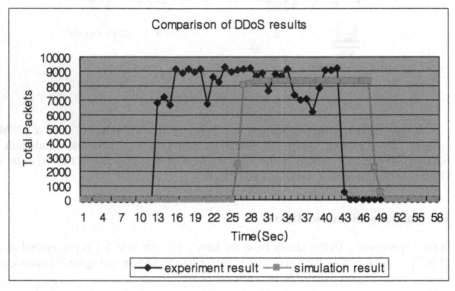

Fig. 10. The Comparison of the DDoS results

5   Conclusions

An ordered scenario-based network security simulation model and its simulator are proposed. The simulation model is driven from informal requirements description through UML diagrams. The simulator was also implemented based on SSF(Scalable Simulation Framework) and SSFNet [9]. The benefits of the simulator are: 1) to enables to simulate scenarios that require the orders of activities; 2) to have easy-to-use GUI to model the latest cyber terror scenarios such as DDoS and worm propagation; and 3) to reuse object classes into new scenarios. The simulation results of the DDoS attacks were compared with those of real-world experiment.

The future work is to apply the proposed simulator into various cyber attack scenarios and tune it to acquire more accurate results. It is also needed to apply it into the very large scale of attack and defense simulations. The real-world experimental data are necessary to validate the results. In addition, it is under consideration to add human components into the simulator to understand interaction between human and system components so that we can optimize all the resources including human and system components.

References

1. Smith, R and Bhattac Harya, "Firewall Placement In a Large Network Topology" in Proc. 6th IEEE workshop on Future Trends of Distributed Computing Systems, 1997.
2. Breslau, L., et al., "Advances in Network Simulation" Computer, Col. 33, No.5, May 2000.
3. Optimum Network Performance, OPNET Modeler, http://www.opnet.com/products/modeler/ home.html, March 2001.
4. John R. Mostow, John D. Roberts, John Bott, "Integration of an Internet Attack Simulator in an HLA Environment", Proc. IEEE workshop on Information Assurance and Security, West Point, NY, June 2000.
5. Donald Welch, Greg Conti, Jack Marin, "A framework for an Information Warfare Simulation", Proc. IEEE workshop on Information Assurance and Security, June 2001.
6. "SSF Simulator Implementation", http://www.ssfnet.org/ssfImplementations.html.
7. "Domain Modeling Language(DML)", http://www.ssfnet.org/homePage.html.
8. Schneier, B., "Attack Tree Secrets and Lies", pp.318-333, John Wiley and Sons, New York, 2000.
9. "Scalable Simulation Framework", http://www.ssfnet.org/homePage.html.

Automatic Global Matching of Temporal Chest MDCT Scans for Computer-Aided Diagnosis

Helen Hong[1], Jeongjin Lee[2], Yeni Yim[2], and Yeong Gil Shin[2]

[1] School of Electrical Engineering and Computer Science BK21: Information Technology,
Seoul National University, San 56-1 Shinlim-dong Kwanak-gu, Seoul 151-742, Korea
hlhong@cse.snu.ac.kr
[2] School of Electrical Engineering and Computer Science, Seoul National University
{jjlee,shine,yshin}@cglab.snu.ac.kr

Abstract. We propose a fast and robust global matching technique for detecting temporal changes of pulmonary nodules. For the registration of a pair of CT scans, a proper geometrical transformation is found through the following steps. First, an automatic segmentation is used for identifying lung surfaces in chest MDCT scans. Second, optimal cube registration is performed for the initial gross registration. Third, for allowing fast and robust convergence on the optimal value, a 3D distance map is generated by the narrow band distance propagation. Finally, the distance measure between surface boundary points is repeatedly evaluated by the selective distance measure to align lung surfaces. Experimental results show that the computation time and robustness of our registration method is very promising compared with conventional methods. Our method can be used for investigating temporal changes such as pulmonary infiltration, tumor masses, or pleural effusions.

1 Introduction

Chest MDCT (Multi-detector Computed Tomography) is widely used to diagnose pulmonary metastasis of oncology patients and evaluate disease progression and regression during treatment [1]. In clinical practice, radiologists often compare current chest MDCT with previous one of the same patient to detect temporal changes such as newly developed pulmonary infiltration, tumor masses, or pleural effusions. However, it is often very difficult to identify subtle temporal changes particularly in lesions that involve overlap with anatomic structures such as ribs, vessels, heart and diaphragm. Even established radiologists can miss important changes over time when they observe serial chest MDCT images. Therefore, the automatic matching of corresponding region in temporal chest MDCT scans would be very useful for radiologists to improve nodule detection and follow-up.

Several methods have been suggested for the automated matching of temporal lung CT images. In Betke et al. [2-4], anatomical landmarks such as the sternum, vertebrae, and tracheal centroids are used for initial global registration. Then the initial surface alignment is refined step by step by an iterative closest point (ICP) process. Most part of the computation time for the ICP process is to find the point correspondences of lung surfaces obtained from two time interval CT scans. Hong et al. [5] proposed an efficient multilevel method for surface registration to cope with the problem of Betke [2]. The multilevel method first reduces the original number of points

D.-K. Baik (Ed.): AsiaSim 2004, LNAI 3398, pp. 652–662, 2005.

and aligns them using an ICP algorithm. In addition, they proposed a midpoint approach to define the point correspondence instead of using the point with the smallest Euclidean distance as in the original ICP algorithm. However the midpoint approach has a tradeoff between accuracy and efficiency, because additional processing time is needed to find the second closest point and to compute the midpoint. Mullaly et al. [6] developed multi-criterion nodule segmentation and registration that improve the identification of corresponding nodules in temporal chest CT scans. The method requires additional nodule segmentation process and measures for multi-criterion. Gurcan et al. [7] developed an automated global matching of temporal thoracic helical CT scans. The method uses three-dimensional anatomical information such as the ribs without requiring any anatomical landmark identification or organ segmentation. But it is difficult to align correctly since the method uses only limited information obtained by Maximum Intensity Projection (MIP) images of two time-interval CT scans.

Current approaches still need some progress in computational efficiency and accuracy for investigating changes of lung nodules in temporal chest CT scans. In this paper, we propose a fast and robust global matching technique to speed-up the computation time and to increase robustness. Our global matching method for aligning a pair of CT images has four main steps. First, automatic segmentation method is used to identify lungs in chest CT images. Second, optimal cube registration is performed to correct gross translational mismatch. This initial registration does not require any anatomical landmarks. Third, a 3D distance map is generated by narrow band distance propagation, which derives fast and robust convergence on the optimum value. Fourth, the distance measure between surface boundary points is evaluated repeatedly by the selective distance measure. Then the final geometrical transformations are applied to align lung surfaces in the current CT scans with lung surfaces in the previous CT scans. Experimental results show that our method is faster than the chamfer matching-based registration and more robust in the sense that the algorithm always convergence to an optimum value.

The organization of the paper is as follows. In Section 2, we discuss how to extract the lungs from other organs in chest CT images and how to correct the gross translational mismatch. Then we propose a narrow band distance propagation and selective distance measure to find exact geometrical relationship in two time interval images. In Section 3, experimental results show how the method rapidly and robustly aligns lung surfaces of current and previous CT scans. In addition, the improved lung nodule correspondence of our method is discussed. This paper is concluded with a brief discussion of the results in Section 4.

2 Automatic Global Matching

For the registration of the current CT scan, called target volume, with the previous scan, called template volume, we apply the pipeline shown in Fig. 1 to two time interval CT images. At first, lung surfaces are automatically segmented from each volume and saved as binary volumes. In the second step, initial alignment is performed using the optimal cube registration for correcting the gross translational mismatch. In the third step, initial alignment is repeatedly refined by the selective distance measure and optimization process. In order to find exact geometrical relationship between two time interval volumes, the target volume is moved during the iterative alignment procedure. Interpolating the target volume at grid positions of the template volume is required for each iteration depending on the transformation. After global matching, corresponding lung surfaces and nodules of template and target volume are displayed by data mixing and volume rendering. The pulmonary nodules are detected manually in this implementation.

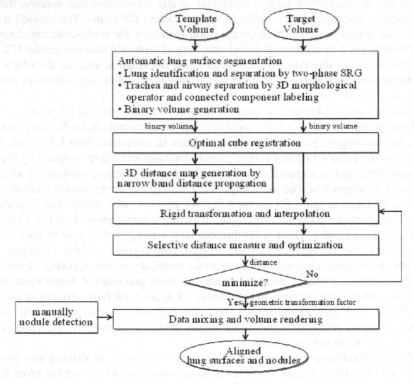

Fig. 1. The pipeline of automatic global matching

In our method, we have two assumptions as follows: 1) Each CT scan is acquired at the maximal inspiration. 2) The entire lung regions are included in each CT scan. Based on this assumption, we found that rigid transformation is sufficient for the registration of temporal chest CT scans. Thus we use rigid transformation – three translations and three rotations about the x-, y-, z-axis.

2.1 Automatic Lung Segmentation

A precursor to the whole process for global matching is a lung segmentation. Our automatic segmentation method consists of two main steps: an extraction step to identify the lungs, and a separation step to delineate the lungs from the trachea and large airways.

In the lung extraction step, the voxels corresponding to lung tissue is separated from the voxels corresponding to the surrounding anatomy by the two-phase seeded region growing (SRG) with automatic seed point selection. After applying the two-phase SRG, the voxels corresponding to air surrounding the body, the lungs, and other low-density regions within the volume are removed. Then three-dimensional connected component labeling is used to identify the lung voxels. Small, disconnected regions are discarded if the region volume is too small. To identify the lungs, we retain the only two largest components in the volume, with the additional constraint that each component must be larger than a predetermined minimum volume. The high-density vessels in the lungs are labeled as body voxels during the two-phase SRG process. As a result, the three-dimensional lung regions contain unwanted interior cavities. Thus hole filling is used to fill the lung regions and eliminate the interior cavities. In the separation step, the trachea and large airways are identified and separated from the lungs. The trachea and left and right mainstem bronchi are identified and discarded by repeated three-dimensional opening and connected component labeling.

2.2 Optimal Cube Registration

According to the imaging protocol and the patient's respiration and posture, the position of lung surfaces between template and target volume can be quite different. For the efficient registration of such volumes, an initial gross correction method is usually applied. Several landmark-based registrations have been used for the initial gross correction. To achieve the initial alignment of lung surfaces, these landmark-based registrations require the detection of landmarks and point-to-point registration of corresponding landmarks. These additional processes much degrade the performance of the whole system.

To minimize the computation time and maximize the effectiveness for initial registration, we propose a simple method of global alignment using the circumscribed boundary of lung surfaces. An optimal cube of bounding volume which includes left and right lung surfaces is generated as shown in Fig. 2. For initial registration of two volumes, we align centers of optimal cubes automatically.

Our optimal cube registration dramatically reduces the processing time since initial alignment is performed without any anatomical landmark detection. In addition, our method leads to robust convergence to the optimal value since the search space is limited near an optimal value.

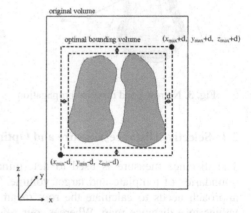

Fig. 2. The generation of an optimal cube

2.3 Narrow Band Distance Propagation

In a surface registration algorithm, the calculation of distance from a surface boundary to a certain point can be done using a preprocessed distance map based on chamfer matching. Chamfer matching reduces the generation time of a distance map by an approximated distance transformation instead of a Euclidean distance transformation. However, the computation time of distance is still expensive by two-step distance transformation of forward and backward mask. In particular, the generation of a 3D distance map of whole volume dataset is unnecessary when the initial alignment almost corrects the gross translational mismatch. From this observation, we propose the narrow band distance propagation for the efficient generation of a distance map.

For generating a 3D distance map, we approximate the global distance computation with repeated propagation of local distances within a small neighborhood. To approximate Euclidean distances, we consider 26-neighbor relations for a 1-distance propagation as seen in Eq.(1). The positive distance value of the 3D distance map tells how far it is apart from a surface boundary point. The narrow band distance propagation shown in Fig. 3 is applied to surface boundary points only in the template volume.

$$DP(i) = \min\{DP(i-1)+, DP(i-1)\} \tag{1}$$

Fig. 4 shows the result of a 3D distance map using the narrow band distance propagation. The generation time can be considerably reduced since pixels need to be propagated only in the direction of increasing distances to the maximum neighborhood. In addition, we do not need the backward propagation, which reduces the size of required neighborhoods.

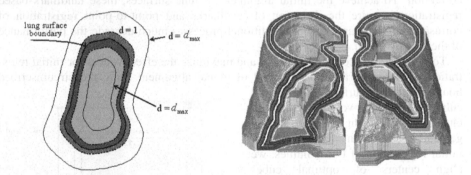

Fig. 3. Narrow band distance propagation   **Fig. 4.** The result of 3D distance map

2.4 Selective Distance Measure and Optimization

The distance measure is used to determine the degree of resemblance of surface boundaries of template and target volume. To get the distance measure, the current approach needs to calculate the root mean square of distance differences of whole values in a distance map. Whereas, our selective distance measure(SDM) only uses distance values near to the surface boundary that can be easily found in the already generated a 3D distance map. Since distance values in the selected regions are used, we can reduce the computation time for the distance measure and optimization.

As can be seen in Eq. (2), the distance value $D_{\text{target}}(i)$ of target volume is subtracted from the distance value $D_{\text{template}}(i)$ of the 3D distance map of template volume. We assume that $D_{\text{target}}(i)$ are all set to 0. N_C is the total number of surface boundary points in target volume. Then SDM reaches minimum when surface boundary points of template and target volumes are aligned correctly.

$$SDM = \frac{1}{N_C} \sum_{i=0}^{N_C-1} \left| D_{\text{template}}(i) - D_{\text{target}}(i) \right| \qquad (2)$$

We use the Powell's method to evaluate SDM. Since the search space of our distance measure is limited to the surrounding lung surface boundaries, we do not need to use a more powerful optimization algorithm such as simulated annealing.

3 Experimental Results

All our implementation and test were performed on an Intel Pentium IV PC containing 2.4 GHz CPU and 1.0 GB of main memory. Our registration method has been applied to five pairs of successive MDCT scans whose properties are described in Table 1. The performance of our method is evaluated with the aspects of visual inspection, accuracy and robustness.

Table 1. Image conditions of experimental datasets

(mm)

Case #		Image size	Slice #	Pixel size	Slice thickness	FOV	Nodule #
1	Template	512 x 512	358	0.64 x 0.64	2.0	328 x 328	1
	Target	512 x 512	316	0.66 x 0.66	2.0	336 x 336	1
2	Template	512 x 512	300	0.57 x 0.57	2.0	292 x 292	1
	Target	512 x 512	330	0.55 x 0.55	2.0	284 x 284	1
3	Template	512 x 512	407	0.62 x 0.62	2.0	317 x 317	1
	Target	512 x 512	454	0.64 x 0.64	2.0	326 x 326	1
4	Template	512 x 512	446	0.55 x 0.55	2.0	281 x 281	2
	Target	512 x 512	379	0.54 x 0.54	2.0	279 x 279	2
5	Template	512 x 512	301	0.60 x 0.60	2.0	308 x 308	8
	Target	512 x 512	311	0.51 x 0.51	2.0	262 x 262	8

Fig. 5 shows the results of global matching of patient 1. Fig. 5(b) shows the effectiveness of the optimal cube for initial registration. The positional difference between lung surfaces of template and target volumes shown in Fig. 5(a) is much reduced as shown in Fig. 5(b) by the optimal cube registration. This initial registration is further refined by lung surface registration until lung surfaces of template and target volumes are aligned exactly like a Fig. 5(c).

Fig. 6 shows nodule correspondences for patients 1 and 5 with nodules in target volume (blue sphere) and nodules transformed from template volume into target volume (red sphere). The lung nodule alignment results of five patients are reported in Fig. 7 on a per-center point basis using the root mean squared (RMS) error between corresponding nodules of template and target volumes. The reduction of initial RMS error is obtained with the optimal cube registration and subsequent iterative surface registration.

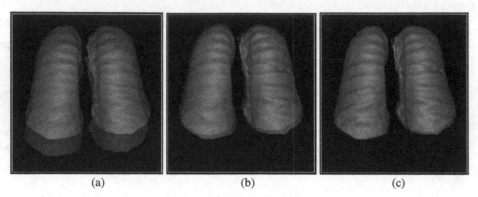

(a)   (b)   (c)

Fig. 5. Results of automatic global matching of lung surfaces in patient 1 (a) initial position (b) after initial registration (c) after lung surface registration

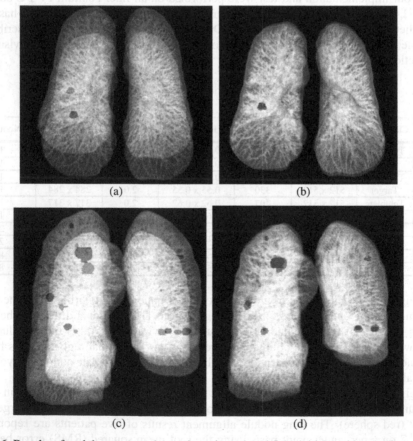

(a)   (b)

(c)   (d)

Fig. 6. Results of nodule correspondences in patients 1 and 5 (a) initial position (patient 1) (b) after global matching (patient 1) (c) initial position (patient 5) (d) after global matching (patient 5)

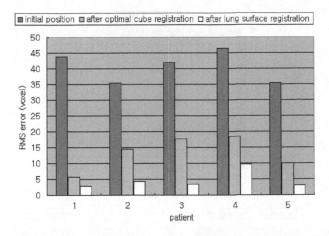

Fig. 7. Accuracy evaluation of corresponding nodules using the average RMS error per patient

Fig. 8 shows how we can reduce the error measure, the sum of squared distance difference (SSD), using the optimal cube registration and subsequent iterative surface registration. After the first iteration, the SSD of our method is significantly reduced by optimal cube registration compared to chamfer matching-based surface registration. Moreover in early iterations the SSD of our method (Method 2) is much smaller than that of chamfer matching-based surface registration (Method 1).

The total processing time is summarized in Table 2 where execution time is measured for the generation of a distance map, the distance measure and optimization processes.

Table 2. Total processing time

(sec)

Case #		3D distance map generation	Distance measure and optimization	Total processing time
1	Method 1	46	120	166
	Method 2	5	72	77
2	Method 1	41	122	163
	Method 2	7	78	85
3	Method 1	54	196	250
	Method 2	10	154	164
4	Method 1	42	123	165
	Method 2	8	92	100
5	Method 1	46	169	215
	Method 2	8	125	133

The robustness of the selective distance measure (SDM) criterion has been evaluated by comparing SDM measure traces (represented by square dot line) with chamfer distance measure (CDM) measure traces (represented by diamond dot line). As shown in Fig. 9, the changes of SDM measure are smooth near to the minimal position, but CDM measure is changed rapidly. This means that SDM measure is more likely to converge to an optimum value.

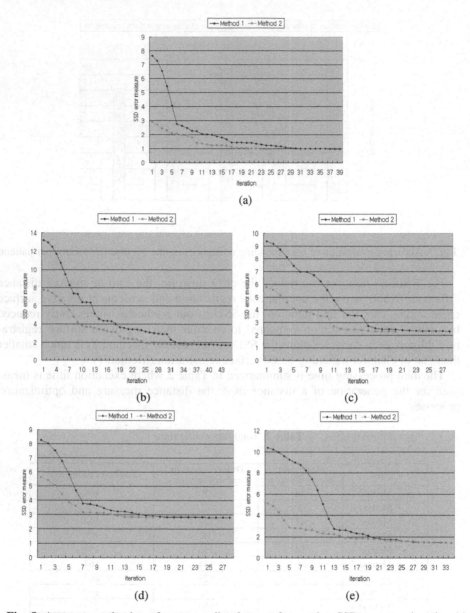

Fig. 8. Accuracy evaluation of corresponding lung surfaces using SSD error per iteration (a) patient 1 (b) patient 2 (c) patient 3 (d) patient 4 (e) patient 5

4   Conclusion

We have developed a fast and robust global matching technique for effectively detecting the correspondence of lung surfaces and nodules in two time interval CT scans. Using the optimal cube registration, the initial gross registration can be done much

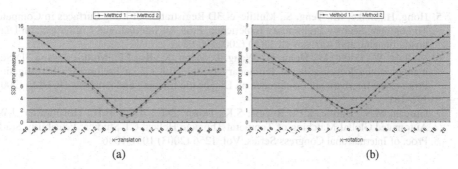

Fig. 9. Comparison of our proposed method and chamfer distance-based surface registration (a) the error in x-translation (b) the error in x-rotation

fast and effectively without any detection of anatomical landmarks. Selective distance measure using a 3D distance map generated by the narrow band distance propagation allows rapid and robust convergence on the optimal value. Five pairs of temporal chest CT scans have been used for the performance evaluation with the aspects of visual inspection, accuracy and robustness. In early iterations, the SSD of our method is much smaller than that of chamfer matching-based registration by using optimal cube registration and lung surface registration. Experimental results also show that SDM measure has more chance of converging to an optimum value than CDM measure. Our method can be successfully used for investigating temporal changes such as pulmonary infiltration, tumor masses, or pleural effusions.

Acknowledgements

The authors are grateful to Prof. Kyung Won Lee from the Seoul National University Hospital of Bundang, Korea for providing lung MDCT datasets shown in this paper and giving advice unsparingly to our research. This work was supported in part by the Korea Research Foundation under the Brain Korea 21 Project. The ICT at Seoul National University provides research facilities for this study.

References

1. Yankelevitz, D.F., Reeves, A.P., Kostis, W.J., Binsheng, Z., henschke, C.I., Small Pulmonary Nodules: Volumetrically Determined Growth Rates Based on CT Evaluation, Radiology, Vol. 217 (2000) 251-256.
2. Betke, M., Hong, H., Ko, J.P., Automatic 3D Registration of Lung Surfaces in Computed Tomography Scans, Proc. Of Medical Image Computing and Computer-Assisted Intervention (MICCAI) (2001) 725-733.
3. Betke, M., Hong, H., Ko, J.P., Automatic 3D Registration of Lung Surfaces in Computed Tomography Scans, CS Technical Report 2001-004, Boston University.
4. Betke, M., Hong, H., Thomas, D., Prince, C., Ko, J.P., Landmark Detection in the Chest and Registration of Lung Surfaces with an Application to Nodule Registration, Medical Image Analysis, Vol. 7 (2004) 265-281.

5. Hong, H., Betke, M., Teng, S., Multilevel 3D Registration of Lung Surfaces in Computed Tomography Scans – Preliminary Experience, Proc. Of International Conference on Diagnostic Imaging and Analysis (ICDIA) (2002) 90-95.
6. Mullaly, W., Betke, M., Hong, H., Wang, J., Mann, K., Ko, J.P., Multi-criterion 3D Segmentation and Registration of Pulmonary Nodules on CT: a Preliminary Investigation, Proc. of the International Conference on Diagnostic Imaging and Analysis (ICDIA) (2002) 176-181.
7. Gurcan, M.N., Hardie, R.C., Rogers, S.K., Dozer, D.E., Allen, B.H., Hoffmeister, J.W., Automated Global Matching of Temporal Thoracic Helical CT Studies: Feasibility Study, Proc. of International Congress Series, Vol. 1256 (2003) 1031-1036.

High Quality Volume Rendering
for Large Medical Datasets Using GPUs

Taek-Hee Lee[1], Young J. Kim[2,*], and Juno Chang[3]

[1] Seoul National University, Seoul Korea
thlee@cglab.snu.ac.kr
[2] Ewha Womans University, Seoul Korea
kimy@ewha.ac.kr
[3] Division of Media Technology, Sangmyung University, Seoul Korea
jchang@smu.ac.kr

Abstract. In this paper, we present efficient, high-quality volume rendering techniques for large volume datasets using graphics processing units (GPUs). We employ the 3D texture mapping capability commonly available in modern GPUs as a core rendering engine and take advantage of combinations of HW-supported occlusion queries, stencil tests and programmable shaders to accelerate the whole rendering process. As a preprocessing step, we subdivide the entire volume dataset into a union of subvolumes of a uniform size. For each subvolume, we also create a *filtered visible subvolumes* (FVS). The FVS is defined as a set of subvolumes that contain the visible voxels. Before executing an interactive rendering loop, using FVS, we find the *boundary subvolumes* that are closest to the bounding planes enclosing the entire volume data, and pre-fetch them from main memory to texture memory as they are likely to be rendered regardless of the change of a viewpoint. Then, by rendering the boundary subvolumes onto stencil buffer, we create an initial occlusion map. At runtime, as we render each subvolume, the occlusion map is updated accordingly. Moreover, using the occlusion map, we issue a series of HW-supported occlusion queries to cull away occluded subvolumes and also perform an early ray termination based on the stencil test. We have implemented the volume rendering algorithm and, for a large volume data of $512 \times 512 \times 1024$ dimensions, we achieve real-time performance (i.e., $2 \sim 3$ FPS) on a Pentium IV 2.8 GHz PC equipped with ATI 9800Pro graphics card with 256MB video memory and 256MB AGP memory without any loss of image quality.

1 Introduction

The direct volume rendering[1] is a technique to render a volume dataset, usually stored at 3D regular grids, without explicitly constructing its surface representation. Thanks to the development of various volume rendering techniques, volume rendering has been extensively used in many areas such as medical imaging [1, 2], CAD [3], scientific imaging [4, 5], etc. At the same time, the need for rendering large volume datasets is

---

* Corresponding author.
[1] Throughout the paper, by simply volume rendering, we mean the direct volume rendering.

D.-K. Baik (Ed.): AsiaSim 2004, LNAI 3398, pp. 663–674, 2005.
© Springer-Verlag Berlin Heidelberg 2005

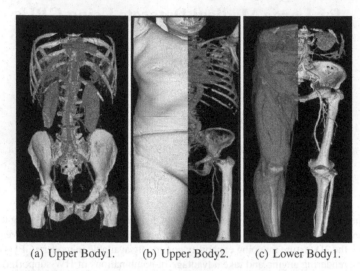

(a) Upper Body1.   (b) Upper Body2.   (c) Lower Body1.

Fig. 1. High Quality Volume Rendering of CT-Scanned Human Bodies. The dimensions of the volume datasets in (a), (b) and (c) are $512 \times 512 \times 917 \times 2$ Bytes, $512 \times 512 \times 967 \times 2$ Bytes, and $512 \times 512 \times 1024 \times 2$ Bytes, respectively. These images are volume-rendered onto a 512×512 window without a loss of image quality.

ever increasing because recent volume acquisition devices can generate volume datasets at a very high resolution. For example, in medical imaging, a CT Scanner can generate up to 1500 slices of 2D images, each having a resolution of 512×512.

In the meantime, the recent advancement in graphics processing units (GPUs) has enabled a direct volume rendering at interactive rates. A typical way of performing GPU-based volume rendering is the use of the 3D texture mapping capability widely supported in commodity GPUs. However, commodity GPUs have limited texture memory resources[2] which, in turn, has prevented a straightforward method to render large volume datasets. In the literature, most of earlier work has addressed this issue by compressing the original dataset and uncompressing it in texture memory with some loss of image quality. However, this approach can be problematic in such applications as medical imaging and scientific visualization. For example, in medical imaging, the visualization of critical regions of patients may be missed by data compression methods, and thus it can mislead medical doctors to diagnose erroneous results. Moreover, medical doctors want to see the visualization results up close through a viewing window of a high resolution (typically 512×512) all the time, and, as a result, a loss of image quality is highly noticeable and thus can lead to an incorrect diagnosis of patients [6].

Main Results. In this paper, we present a high quality volume rendering technique to render a large medical volume dataset of $512 \times 512 \times 1024$ dimensions at interactive rates. We employ 3D texture mapping as a core rendering engine and take advantage of combinations of HW-supported visibility tests such as occlusion queries, stencil tests and programmable shaders to accelerate the whole rendering process. All of the above

---

[2] A typical maximum texture memory size in GPUs is 512MB including AGP accelerated memory.

tests are performed using a novel occlusion representation based on stencil buffer and accelerated by GPUs. Moreover, using the *voxel counter table* and *filtered visible subvolumes*, we reduce the data traffic between CPU and GPU by eliminating the subvolumes that only have a small number of visible voxels to be rendered. As a result, we are able to achieve an interactive performance to render a large volume data of a $512 \times 512 \times 1024$ dimension without any loss of image quality.

Organization. The rest of the paper is organized as follows. In Sec. 2, we briefly review the earlier work on GPU-based large volume rendering. Sec 3 gives an overview of our algorithm, and present each step of the pipeline of our algorithm in detail. In Sec. 4, we analyze the performance of our algorithm, and in Sec. 5 we conclude the paper and also present our future work.

2 Prior Work

In this section, we briefly survey earlier work on GPU-based volume rendering and large volume rendering.

2.1 GPU-Based Volume Rendering

Most of GPU-based volume rendering methods rely on 3D texture mapping hardware. Cullip and Neumann [7] have been able to achieve realtime performance on rendering 256^3 volume data using GPUs. Meißner et al. [8] have presented a GPU-based volume rendering algorithm using a multipass rendering technique. Recently, using the programmability of GPUs, various optimization techniques such the early ray termination and empty space skipping can be effectively implemented on GPUs. [9–11]

2.2 Large Volume Rendering

In the literature, two different approaches have been developed to render large volume datasets; rendering with and without data reduction. Multi-resolution rendering and wavelet-based volume compression schemes are typical methods used in the data reduction approach [12, 13]. Chamberlin [14] has used a multi-resolution rendering scheme for volume rendering in 2D. LaMar et al. [15] have employed a similar approach based on Octree construction of volume data combined with 3D texture mapping. Nguyen et al. [16, 17] have presented a method to subdivide a volume into boxes of a uniform size and compress each of the boxes. Recently, J. Schneider et al. [18] have presented a relatively fast compression/decoding algorithm using GPUs. Relatively, there is not much work reported on rendering large volume without any data reduction. A major issue of rendering without data reduction is how to efficiently load necessary subvolume data into GPU memory.

3 Volume Rendering Algorithm

Our volume rendering algorithm is based on 3D texture mapping capability widely supported in modern GPUs. Classical 3D texture mapping-based algorithms such as [7, 19] render volume data using the following steps:

1. Slice the volume with planes parallel to the viewing plane.
2. Project the volume onto each of the slicing planes by trilinear interpolation.
3. Blend the planes in front-to-back order.

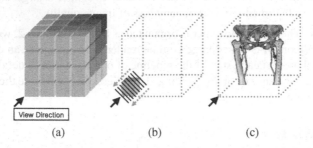

(a)   (b)   (c)

Fig. 2. Basic Volume Rendering Algorithm. (a) From a given view direction, each subvolume is traversed in front-to-back order. The darker subvolumes are traversed earlier than the lighter ones. (b) Inside a subvolume, slices are rendered in back-to-front order. (c) Blending all the slices contained in each subvolume generates a final result.

Our volume rendering algorithm also follows the basic mechanism of the conventional texturing-based algorithms. However, the main difference of our algorithm from others is that we uniformly subdivide the entire volume into a set of disjoint subvolumes, and render each subvolume in front-to-back order while rendering each slice contained in the subvolume in back-to-front order, as illustrated in Fig. 2.

One of the major challenges in GPU-based volume rendering algorithms is the limited size of texture memory imposed by current GPU architecture; Typically the maximum memory size of modern GPUs is 256MB [20]. We attempt to overcome this limitation by rendering only visible parts of volume data without a loss of image quality. In order to accomplish this objective, we perform various visibility tests on volume data such as empty space skipping, occlusion culling and early ray termination. Using these tests, we cull away invisible portion of the data, and, consequently, load only potentially visible data into texture memory. For different visibility tests, we adopt different testing units. For example, the empty space skipping and occlusion test are performed on a per-subvolume basis whereas the early ray termination is on a per-pixel basis.

The overall pipeline of our algorithm can be subdivided into four major stages (see Fig. 3):

1. **Voxel Counter Table Creation** (Sec. 3.1): For each subvolume, we construct a voxel counter table (VCT) that can quickly tell us the number of visible voxels. The table is also used to update *filtered visible subvolumes* that is explained below.
2. **Filtered Visible Subvolumes Update** (Sec. 3.2): We use *filtered visible subvolumes* (FVS) to efficiently identify whether a subvolume is empty or not. The FVS is defined as a set of subvolumes that contain a certain number of visible voxels to be rendered after OTF classification. This number depends on a pre-defined threshold value, and users can filter out empty subvolumes by changing the threshold value. Notice that FVS is independent from the viewpoint.

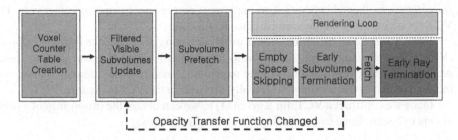

Fig. 3. Rendering Pipeline. We color-coded differently for different units of operations in the pipeline. The operation unit for red colored blocks is per-subvolume whereas the unit for blue colored blocks is per-pixel.

3. **Subvolume Prefetch** (Sec. 3.3): We select a subset of FVS that are closest to the six planes bounding the entire volume grid; we call the subset *boundary subvolumes*. Then, we fetch the boundary subvolumes from main memory into texture memory because they have a high probability to be visible from any viewpoint and occlude other subvolumes as well. We iterate the pre-fetching process until the total size of pre-fetched subvolumes reaches a predefined threshold, η. Here, η is a constant depending on the maximum texture memory size; Typically 50% of total texture memory. This process is performed independent of the change of a viewpoint.

4. **Rendering Loop** (Sec. 3.4): At runtime, we traverse each subvolume in front-to-back order (see Fig. 2-(a)), performs a series of visibility tests to determine whether the subvolume is actually to be rendered, and fetch it if it needs to be rendered and does not reside in texture memory. Only the portion of a subvolume that passes all the visibility tests is finally rendered. The visibility tests used in the algorithm includes the following:

   (a) *Empty Space Skipping:* If the currently traversed subvolume does not belong to FVS, we consider it empty and simply skip it.

   (b) *Early Subvolume Termination:* We do not need to consider rendering the subvolumes that are completely occluded by already rendered subvolumes. We conservatively check for the occlusion of subvolumes by applying HW-supported occlusion query to the axis aligned bounding box (AABB) of the subvolume.

   (c) *Early Ray Termination:* For partially visible subvolumes, we fetch them into texture memory first if necessary. Then, we render only visible pixels of the subvolume by using a stencil-based occlusion map.

If an OTF changes, we reiterate the algorithm from step 2 (FVS update). Each step of the above rendering pipeline is explained in detail below.

3.1 Voxel Counter Table

For each subvolume, we construct a voxel counter table (VCT) that provides us with the number of visible voxels after OTF classification. VCT is used to create *filtered visible subvolumes*, explained in Sec. 3.2, to implement empty space skipping effectively. The VCT for a given subvolume is constructed as follows:

First, we create an one-dimensional array, consisting of a sequence of *voxel counters* (VC), whose element index is mapped to the range of a certain voxel density. Then, we visit all the voxels contained in the subvolume, evaluate their density values, and increment their corresponding VC. Finally, we add up the VCs to create a VCT.

Once we construct a VCT, for a given OTF, we can obtain the proportion of visible voxels in a subvolume by computing R_1 and R_2 as follows:

$$R_1 = \frac{\text{VCT}[OTF_{MAX}]}{\text{VCT}[VCT_{MAX}]} \quad \text{and} \quad R_2 = \frac{\text{VCT}[OTF_{MIN}]}{\text{VCT}[VCT_{MAX}]} \tag{1}$$

Here, OTF_{MAX} and OTF_{MIN} denote the maximum and minimum density values of an OTF, respectively, and VCT_{MAX} denotes the maximum VCT entry. R_1 in Eq. 1 becomes zero when the density values of all the voxels are greater than OTF_{MAX}. Similarly, R_2 in Eq. 1 becomes one when all the density values are less than OTF_{MIN}. As a result, if either R_1 is zero or R_2 is one, we conclude that the given subvolume is empty.

3.2 Filtered Visible Subvolumes Update

When we render a large volume dataset containing lots of empty spaces, we need to determine which subvolume corresponds to an empty space as a result of OTF classification. Moreover, there can be a non-empty subvolume but contains only a small number of voxels to be rendered. We filter out these *almost-empty* subvolumes using the VCT along with user-defined two threshold values, α and β. We call the remaining subvolumes surviving the filtering operation filtered visible subvolumes (FVS).

We compute view-independent visibility information for each subvolume and construct FVS as follows:

1. Traverse all the subvolumes with user-provided minimum (OTF_{MIN}) and maximum (OTF_{MAX}) values of OTF.
2. For the currently visited subvolume, if R_1 in Eq. 1 is greater than α or R_2 is less than β, add the subvolume to the FVS.
3. Since FVS depends on OTF, we should update FVS whenever OTF changes.

FVS will be used for an empty space test during the rendering loop. We simply reference FVS to determine whether a subvolume is empty or not.

3.3 Subvolume Prefetch

In general, subvolumes are fetched on demand during the rendering loop and replaced based on a LRU replacement scheme whenever necessary. However, this method has no consideration about the change of a viewpoint. In this case, every time we change the viewpoint, we need to release some subvolumes from texture memory and fetch newly visible subvolumes. This can cause a serious performance degradation when the texture memory is fully utilized all the time; i.e., thrashing. In order to resolve the issue,

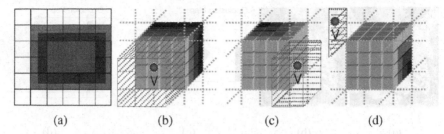

(a)   (b)   (c)   (d)

Fig. 4. Boundary Subvolumes and Traversal Order. (a) shows a simple example of boundary subvolumes(blue colored rectangles). (b), (c) and (d) show a subvolume traversal order using Voronoi Cells. Lightly shaded subvolumes are traversed before darker ones. (b) shows when a view point (V) belongs to the Voronoi cell of the face (F) of bounding planes. Similarly, (c) and (d) correspond to the cases where V belongs to the Voronoi cell of an edge (E) and a vertex (V), respectively.

we pre-fetch the subvolumes that have a high probability to be rendered regardless of the change of a viewpoint and occlude other subvolumes. The *boundary subvolumes* satisfy the above two conditions. The boundary subvolumes are defined as a subset of FVS that are closest to the boundary planes enclosing the original volume data. An example of boundary subvolumes is illustrated in Fig. 4-(a). We fetch these boundary subvolumes prior to the rendering loop and allocate them to texture memory. The pre-fetched subvolumes are not replaced by other subvolumes during the rendering loop.

3.4 Rendering Loop

When a large volume dataset is rendered, many parts (i.e., subvolume) can be occluded by others. Moreover, there may also exist many empty parts depending on a given OTF. By avoiding unnecessary rendering of occluded and empty parts, we can greatly accelerate the overall rendering performance. In order to address these issues, we provide different levels of visibility tests such as per-subvolume level empty test, per-subvolume level occlusion test and per-pixel level skipping, all of which are performed based on occlusion map and GPU-accelerated. In our algorithm, only the subvolumes and their contained voxels that pass all of the three visibility tests are considered to be rendered.

   Since our per-subvolume visibility tests require information about earlier rendering results, subvolumes must be sorted; i.e., subvolumes that are rendered earlier must have a shorter distance from the viewpoint than rendered later. But sorting every subvolume at each frame can cause a severe performance loss. In order to address this issue, we pre-calculate a set of rendering orders depending on a location of viewpoints as follows. Since there are 6 faces (F), 8 edges (E), and 8 vertices (V) comprising the boundary planes of volume dataset, one can have 22 *Voronoi cells* that partition the external space of the volume dataset (Fig. 4-(b)). As a result, one can have 22 different locations of viewpoints having different rendering orders. We pre-calculate the Voronoi cells. At runtime, we determine the rendering order based on which Voronoi cell contains the viewpoint, perform three visibility tests, and finally render the portion of a subvolume that pass the tests. A detailed description of the visibility tests are as follows.

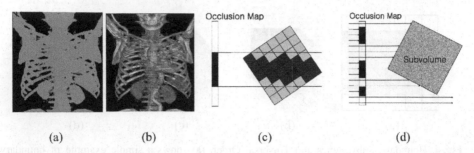

(a)   (b)   (c)   (d)

Fig. 5. (a) shows an example of an occlusion map for rendering a human upper body. (b) illustrates a rendering result using (a) as an OM. (c) shows the early subvolume termination result using occlusion query. Each square denotes the AABB of each subvolume. If an AABB is completely occluded by earlier rendering results, stored at stencil buffer (occlusion map), using HW-supported occlusion query, its underlying subvolume skips rendering. Otherwise, in (d), per-pixel visibility test is performed. (d) describes the stencil test process. Pixel is not rendered when the corresponding stencil bit is set to a predefined value.

Emtpy Space Skipping. The goal of empty space skipping is to identify subvolumes whose contained voxels have all zero opacity values as a result of OTF classification and to cull away these subvolumes. In our algorithm, for a given OTF, we approximate the number of voxels with a non-zero opacity value for each subvolume by using the FVS. More specifically, for each subvolume, we check whether it belongs to the FVS, and if not, we do not render it.

Early Subvolume Termination. Since we subdivide a large volume dataset into many chunks of disjoint subvolumes, some of the subvolumes may be completely occluded by others from a certain viewpoint. In this case, we do not want to pass these subvolumes through the rest of the rendering pipeline which heavily uses relatively costly per-pixel operations such as 3D texture mapping and stencil tests. Instead, we use a HW-supported occlusion query [20] to conservatively check whether the bounding box of a subvolume is completely occluded by the current occlusion map. More specifically, we render the axis aligned bounding box (AABB) of a subvolume onto the current occlusion map, and use the occlusion query to count how many pixels are actually rendered. If the pixel count is zero, we do not consider the subvolume for further volume rendering; also see Fig. 5-(c).

Early Ray Termination. When subvolumes are rendered, the rendering speed can be improved by excluding the fragments already occluded by previous rendering. Since slices are rendered in back-to-front order (see Fig. 2-(b)), their occlusion information cannot be obtained during the rendering of each subvolume. Instead, the information comes from the rendering of previous subvolumes. After each subvolume is rendered, an occlusion map is constructed using stencil buffer. During the back-to-front rendering process, if the accumulated opacity is greater than a threshold value, the corresponding stencil bit is set to one for each pixel. Otherwise the bit remains as zero. Since the subvolumes are rendered in front-to-back order, the occlusion map can be also used in the rendering of next subvolumes, in order to exclude hidden fragments. As shown in Fig. 5-(d), this process is implemented using simple stencil operations supported by GPUs.

Name	Resolution	# of Slices	Size of Voxel(BYTE)	Data Size(MB)
Upper Body1	512 x 512	917	2	458
Upper Body2	512 x 512	967	2	483
Lower Body1	512 x 512	1024	2	512

Fig. 6. Benchmark Data Set. The rendering results are shown in Fig. 1.

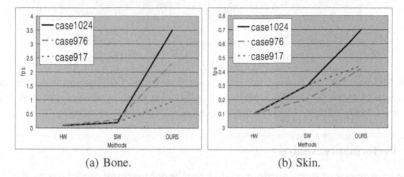

        (a) Bone.   (b) Skin.

Fig. 7. Performance Comparison with Other Rendering Methods. HW, SW and OURS denote the performance of hardware volume rendering without any acceleration techniques mentioned in this paper, a highly optimized software ray casting method, and our volume rendering method, respectively.

4   Results and Analysis

In this section, we explain the implementation issues of our volume rendering algorithm and analyze its performance.

4.1   Implementation Issues

Since 3D texture mapping supported by the current GPU architecture allows a texture size only in powers of two, we need to split the entire volume into textures in powers of two. For example, for a volume data of $512 \times 512 \times 700 \times 2$ bytes, we subdivide it into 704 unit subvolumes, each having $64 \times 64 \times 64 \times 2$ bytes. However, an excessive subdivision of subvolumes can cause a context switching overhead during 3D texturing. In order to address this issue, we create fourteen 3D textures of a $256 \times 256 \times 256$ size in texture memory, and fetch a $64 \times 64 \times 64$ unit subvolume into one of the fourteen textures when needed. This technique can reduce the number of context switches because only fourteen context switches are required in the worst case.

We have implemented our volume rendering algorithm using Visual C, Pixel Shader 2.0 and DirectX 9.0, and measure its performance on a 2.4 GHz Pentium4 PC equipped with ATI RADEON 9800 PRO. The benchmarking volume models we have used in our experiments are shown in Fig. 6, having data sizes of 458MB, 483MB and 512MB, respectively. We also have used $2048 \times 2048 \times 4$ bytes (16MB) of pre-integrated dependent textures for OTF classification.

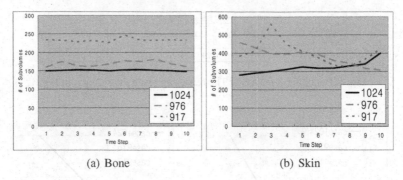

(a) Bone   (b) Skin

Fig. 8. The Number of Rendered Subvolumes (a) Bone areas. Different color denotes different resolutions of data sets. Many subvolumes are culled away because of empty space skipping. (b) Skin areas. Many subvolumes are culled away because of the occlusion test.

4.2   Performance Analysis

The performance of our algorithm on different benchmarking models as well as visibility test results are shown in Fig. 7 and 8. The rendering performance of our algorithm is 2 FPS on average for 512MB volume data, rendered onto a 512×512 screen window. In Fig. 7-(a) and (b), we compare the performance (OURS) of our algorithm with other methods like software-based direct volume rendering (SW) [21] and hardware volume rendering without any visibility tests (HW). The results show our algorithms consistently performs better than others. For example, in Fig. 7-(a) (bone data), our algorithm shows particularly high frame rates because of the empty space test. In particular, our approach (OURS) out-performs the software-based approach (SW) by a factor of fifteen. Moreover, the number of subvolumes that are actually rendered are only 15~30% of the entire volume dataset. As a result, our algorithm can fetch all visible subvolumes into texture memory. In Fig. 7-(b) (skin data) OURS out-performs SW by a factor of two. In this case, the early subvolume termination based on occlusion query and early ray termination based on pixel skipping play an important role. In this case, as shown in Fig. 8-(b), 40% (204 MB) of the entire volume dataset is rendered.

5   Conclusions

We have presented a HW-accelerated volume rendering method for large datasets that exceed the typical texture memory size. These datasets are often used in practical medical applications. By taking advantage of HW-supported occlusion query, stencil tests, programmable shaders, we can render large volume datasets at interactive rates. However, our algorithm is not suitable for a out-of-core volume data that does not fit into the size of main memory because it is too slow to fetch the volume data from hard disk. Another limitation is the data transfer overhead between CPU and GPU. We expect that new bus technology such as PCI-Express can greatly alleviate such a overhead in the near future. Finally, we will like to look for other possibilities where we can apply our rendering algorithms, for example, time-dependent (4D) volume rendering and volume rendering with wavelet-based data compression schemes.

Acknowledgements

The authors thank Yeong-Gil Shin and Dongho Kim for providing insightful advice and discussions on the research. This research was sponsored in part by the INIFINITT Co. in Seoul, Korea and the grant no. R08-2004-000-10406-0 from the Korean Ministry of Science and Technology.

References

1. E. Steen and B. Olstad: Volume rendering of 3D medical ultrasound data using direct feature mapping. IEEE Transactions on Medical Imaging 13 (1994) 517–525
2. Elizabeth Bullitt and Stephen R. Aylward: Volume Rendering of Segmented Image Objects. IEEE Transactions on Medical Imaging 21 (2002)
3. K. Tsuchiyal and S. Katase and J. Hachiyal and Y. Shiokawa: Volume-Rendered 3D Display Of MR Angiograms in the Diagnosis of Cerebral Arteriovenous Malformations. Acta Radiologica 44 (2003) 675
4. M. Magnor and G. Kindlmann and C. Hansen and N. Duric: Constrained inverse volume rendering for planetary nebulae. Proceeding of IEEE Visualization 2004 (2004)
5. J.M. Kniss and C. Hansen and M. Grenier and T. Robinson: Volume Rendering Multivariate Data to Visualize Meteorological Simulations: A Case Study. Proceedings of Eurographics - IEEE TCVGSymposiumon Visualization (2002)
6. Andreas Pommert and Karl Heinz Hohne: Evaluation of Image Quality in Medical VolumeVisualization: The State of the Art. In: Medical Image Computing and Computer-Assisted Intervention. (2002) Proc. MICCAI 2002 Part II Lecture Notes in Computer Science 2489 Springer–Verlag Berlin 2002 598–605
7. T. J. Cullip and U. Neumann: Accelerating Volume Reconstruction with 3D Texture Mapping Hardware. Technical Report TR93-027, Department of Computer Science at the University of North Carolina, Chapel Hill (1993)
8. M. Meißner and U. Hoffmann and W. Straßer: Enabling Classification and Shading for 3D Texture Mapping based Volume Rendering using OpenGL and Extensions. In: IEEE Visualization '99 Proc. (1999)
9. Klaus Engel and Martin Kraus and Thomas Ertl: High-Quality Pre-Integrated Volume Rendering Using Hardware-Accelerated Pixel Shading. In: 2001 SIGGRAPH / Eurographics Workshop on Graphics Hardware, ACM SIGGRAPH / Eurographics / ACM Press (2001)
10. Jens Krueger and Ruediger Westermann: Acceleration Techniques for GPU-based Volume Rendering. In: Proceedings IEEE Visualization 2003. (2003)
11. Wei Li and Klaus Mueller and Arie Kaufman: Empty Space Skipping and Occlusion Clipping for Texture-based Volume Rendering. IEEE Visualization (2003)
12. P. Cignoni and C. Montani and E. Puppo and and R. Scopigno: Multiresolution Representation and Visualization of Volume Data. IEEE Transactions on Visualization and Computer Graphics 3 (1997) 352–369
13. Peter Schroder and Wim Sweldens and Michael Cohen and Tony DeRose and David Salesin: Wavelets in Computer Graphics. In: SIGGRAPH 1996 Course. (1996)
14. Bradford Chamberlain and Tony DeRose and Dani Lischinski and David Salesin and John Snyder: Fast rendering of complex environments using a spatial hierarchy. In: Graphics Interface. (1996) 132–141
15. Eric LaMar and Bernd Hamann and Kenneth I. Joy: Multiresolution Techniques for Interactive Texture-based Volume Visualization. In: IEEE Visualization '99. (1999) 355–362

16. Eric J. Stollnitz and Tony D. DeRose and David H. Salesin: Wavelets for Computer Graphics: Theory and Applications. Morgan Kaufmann Publishers, Inc. (1996)
17. Ky Giang Nguyen and Dietmar Saupe: Rapid High Quality Compression of Volume Data for Visualization. Computer Graphics Forum **20** (2001)
18. J. Schneider and R. Westermann: Compression Domain Volume Rendering. In: IEEE Visualization. (2003) 39
19. B. Cabral and N. Cam and J. Foran: Accelerated Volume Rendering and Tomographic Reconstruction Using Texture Mapping Hardware. In: Workshop on Volume Visualization. (1994) 91–98
20. Guennadi Riguer: Performance Optimization Techniques for ATI Graphics Hardware with DirectX 9.0. Technical report, ATI Technologies Inc. (2002)
21. Marc Levoy: Efficient ray tracing of volume data. ACM Transactions on Graphics (TOG) **9** (1990) 245–261

Visualization of Tooth for 3-D Simulation

Hoon Heo, M. Julius Hossain, Jeongheon Lee, and Oksam Chae

Department of Computer Engineering, Kyung Hee University,
1 Seochun-ri, Kiheung-eup, Yongin-si, Kyunggi-do, 449-701, Korea
hhoon@naver.com, mdjulius@yahoo.com,
opendori@paran.com oschae@khu.ac.kr

Abstract. In this paper, we present a 3D tooth visualization system for the simulation of endodontic, orthodontic, and other dental treatments. Detail 3D model of individual tooth in mandible can be simulated and it can be rotated virtually around any axis to determine the best approach of surgery. To obtain higher accuracy of the reconstructed 3D model, a new segmentation method based on adaptive optimal thresholding and genetic snake is applied on the 2D CT images to separate individual tooth region as well as to extract the tooth contour. Surface reconstruction phase utilizes the contour data and tile the vertices of the contours of successive slices of CT images to form the triangular surface mesh of tooth. The paper, therefore, reports on experimentation with novel interactive techniques designed to work with the available geometric representation to teeth.

1 Introduction

Dentistry, compared with other medical fields, has not received a great deal of attention from computer visualization researchers. While existing medical imaging makes use of technologies such as magnetic resonance imaging (MRI) and computerized tomography (CT) to produce computer-generated 3D models such as mandible or maxilla reconstruction and simulation of chewing process, very few advances are found on the reconstruction and visualization of individual tooth for the simulation of endodontic, orthodontic treatment. Our research investigates the adaptation of 3D reconstruction techniques of individual tooth along with mandible that will aid both practicing dentists and dental patients in illustration of dental structures and procedures. To make a success of the model, it should facilitate the simulation of preoperative planning. The function of simulation allow doctors to make a treatment plan for endodontic and orthodontic treatment, persuade a patient of costly operation, i.e. implant grafting, root canal surgery, as a result, the patient can rely on his proposals.

Since the simulation of tooth is performed on individual tooth, special attention should be paid to the segmentation of the teeth in CT images. There are lots of segmentation methods, which show their own limitations in the case of separating individual tooth region on the CT images[1,2,3]. In this paper, we present an adaptive optimal thresholding scheme, based on the fact that the shape and intensity of each tooth change gradually among CT image slices. It computes an optimal thresholding value[4] for every tooth region for a given slice to generate initial boundary of tooth for the genetic snake fitting. In the CT image tooth may have similar intensity profile

D.-K. Baik (Ed.): AsiaSim 2004, LNAI 3398, pp. 675–684, 2005.
© Springer-Verlag Berlin Heidelberg 2005

with neighboring alveolar bone. In this study, to overcome this problem, we present a new genetic snake, specially designed segmentation method by which an accurate contour of tooth can be obtained based on gradient forces. We have devised the fitness function of the algorithm to search the contour maximizing the sum of gradient magnitude while preventing it from being fitted to neighboring spurious edges. The fitting algorithm successfully isolates each tooth boundary from nearby teeth and alveolar bone. Extracted contours in every slice are numbered according to the tooth sequence. We use B-spline for the parametric function of snake and represent tooth contour as the B-spline. This curve not only allows the contour to be smooth but also prevents it from being trapped by spurious isolated edge points. With this curve we can generate more intermediate points, which assist the reconstruction process to realize accurate level-of-details 3D model of individual tooth. There are two types of methodological studies concerning the reconstruction of 3D model from set of contours; one is a surface based method[5,6] where two successive contours of a tooth of neighboring slices are connected along vertical direction by lines forming triangular facets and the other refers to a voxel based method[7,8] where reconstruction of tooth accompanies cubical voxel by defining voxel grid along with adjusting 2D regular grid over slices. In case of the latter method, the stacking of the voxel could lead to coarse quality of the model unless voxels are very small; while in the former one, generation of triangular facet allows the reconstructed model to be more descriptive.

In reconstruction we conserve the surface based approach and concentrate on realizing accurate level-of-details 3D model. For dentist, the 2D image of dental CT slice is not suitable for visualizing and understanding the actual structure of tooth since tooth is often surrounded by alveolar bone and has complex branched structure of roots. The proposed methods of segmentation and reconstruction eliminate these problems. The reconstructed models of individual tooth and mandible facilitate diagnoses, pre-operative planning and prosthesis design of various dental treatments.

2 Structure of Algorithm

Tooth segmentation should be worked upon individually since each tooth has its own shape and intensity profiles. The segmentation procedure starts by picking up the reference slice in which teeth and jaw have more disparity than that in other slices. The initial boundaries of each tooth and jaw in the reference slice are determined using the threshold values provided by user and passed to adjacent slices. Once tooth to be segmented is taken its turn following the determination of initial boundary, a tooth is segmented using adaptive optimal thresholding[4] automatically slice by slice. The boundary of tooth region presented by the thresholding step is used as an initial contour for the fitting process. The initial contour is much closer to the actual boundary than the contour of previous slice. This can lead to more accurate tooth boundary detection in the fitting process with small size of mask. This mask is usually required to reduce the effect of neighboring bone. In order to settle final contour of tooth, this initial contour is pushed to gradient features by the proposed genetic snake. We use B-spline for the parametric function of snake, as it is widely used[3,9], and represent tooth contour as the B-spline. For this, B-spline interpolation[10] is applied to the boundary of thresholded region before the fitting process. The curve can be localized on the maxima of the gradient by means of its inflation or deflation caused by an interaction between gradient information and proposed fitness function. Hence, we extract tooth contour by using the proposed fitting process, genetic snake, and mentioned small size of mask formed by adaptive optimal thresholding while overcoming the effect of spurious edges and hard tissues in its close vicinity. In reconstruction step, the facets of tooth surface are generated using the segmented contours. The controllability of the number of facets provided in this step allows the model to be rendered fast without loss of its quality.

3 Generation of Initial Contour

To determine initial contour for fitting, adaptive optimal thresholding works upon through two steps[4]. First, it defines the approximate tooth boundary using the temporary threshold values predicted from the information passed by neighboring slices. Then, it generates the histograms for the inside region and outside region of a tooth. From these two histograms, we compute optimal thresholding value, which generates the initial contour. This choice of the value is based on the fact that intensity profile of inner region of tooth changes gradually between two slices, though shape of tooth may change significantly. After finding out final tooth contour by the fitting, predicted threshold value is obtained and passed to the next slice. Since final contour can also present its inside and outside histograms, optimal threshold can also be computed again, as a predicted threshold value for the next slice, based on the two histograms.

4 Tooth Segmentation Using Genetic Snake

After the initial contour of tooth is obtained by adaptive optimal thresholding, it is represented as B-spline so as to be fitted to gradient features with genetic snake for final contour of tooth. The design of fitness function in genetic algorithm is important since it measures the accuracy of fitness of the possible contour to the object boundary in the current slice. The fitness value is a basis for determining the termination of the evolution process and selecting elite chromosomes in mating pool generation. In the existing active contour models, the energy function like fitness function consists of the internal forces controlling the smoothness of the contour and the external force used for representing the object boundary information in the image [9,11]. One drawback of this representation is that it requires the determination of the weight values balancing these two components, which is a very tedious task for the user. Brigger et. al[3] suggest a simple energy function having only the external force computed based on the gradient magnitude on the contour. They eliminate the internal force term by substituting stiffening parameter for it and rely on the B-spline representation of a contour, which the smoothness constraints implicitly built into.

However, in the image data such as dental CT image slices, those fitting function often generates the contour fitted to the boundary of nearby object as shown in Figure 1. They also require proper value of stiffening parameter in order to prevent the curve from being twisted. To alleviate these problems, Liu et. al[2] suggest the energy function that sums up the distance between each contour point and the nearest point of object boundary. It, however, requires the accurate object boundary detection to evaluate the energy function and yet does not solve the malfitting problems shown in Figure 1.

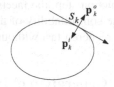

(a) Fitting to neighbor   (b) twisted contour

Fig. 1. Malfitting to neighboring object boundary

Fig. 2. Definition of inner and outer regions

In this paper, we propose an energy function (or fitness function) to solve those problems under the condition that the relative intensity between the inside and outside of an object contour is maintained throughout the contour. Under this condition, if a contour is expanded out to other object boundary, the relative intensity between the two sides of a contour is reversed. The fitness function is designed by taking the advantages of this property of the data. It subtracts magnitude of gradient from fitness value being accumulated to penalize part of contour fitted to the neighboring object. And it also prevents the contour from being twisted during fitting process by penalizing the portion of contour twisting. To compute the fitness value for a possible solution (or chromosome), we first generate the contour points like snaxels[11] on the B-spline curve and trace the contour as shown in Figure 2. At the k'th contour point $r(s_k)$ of B-spline, compute a normal vector $n(s_k)$. And, compute the inner region and outer region pixel location p_k^i and p_k^o respectively by using the following equation.

$$p_k^o = r(s_k) + n(s_k) \quad , \quad p_k^i = r(s_k) - n(s_k) \tag{1}$$

Thereafter, the fitness value can be determined based on gradient magnitude information at each contour point by using the following equation.

$$f(\mathbf{x}) = \sum_{k=0}^{M-1}(grad_k - \alpha_k) \quad , \quad grad_k = \begin{cases} |\ \nabla I(r(s_k))\ | & I(p_k^i) - I(p_k^o) > 0 \\ -|\ \nabla I(r(s_k))\ | & I(p_k^i) - I(p_k^o) \le 0 \end{cases} \tag{2}$$

$$\alpha_k = \begin{cases} C & r(s_k) = r(s_j) \\ 0 & r(s_k) \ne r(s_j) \end{cases} , \ \forall\, j \in \{0,1,...,M-1\} \wedge j \ne k \ , \tag{3}$$

where $I(p_k^i)$ and $I(p_k^o)$ are grayscale values of the inside and the outside of the k'th contour point respectively. Twist penalty, C is set to 10 empirically. The proposed fitness function deducts the gradient magnitude values of the contour pixel from the accumulated value when it lies on the neighboring object boundary or it overlaps with any other, to avoid the malfitting shown in Figure 1.

5   Reconstruction of Tooth

The reconstruction problem can be broken into three sub problems: the correspondence problem (which contours should be connected by the surface?), the tiling problem (how should the contours be connected?), and the branching problem (what do we do when there are branches in the surface?). We present our system for surface reconstruction from sets of contours, which efficiently solves each of these sub problems. In the first step, we need to decide how detail 3D model of tooth will be represented based on continuous curve segmented out. In order to control the quality of model, we compute the number of sample points for one curve, as follows.

$$sampling\ points = total\ contour\ points * user\ specified\ ratio. \tag{4}$$

Optimal number of sampling points per contour has been researched on according to the case of application and found as $50/2\pi$ [12]. Since the contour of tooth is represented as B-spline, wrinkles in the contour are not required to be filtered out in our research. The contours of neighboring two slices are denoted by C_k and C_{k+1}, consisting of $m+1$ and $j+1$ sample points respectively, where

$$C_k = \{V_{k,0}, V_{k,1}, V_{k,2}, \cdots, V_{k,m}\} \ , \ C_{k+1} = \{V_{k+1,0}, V_{k+1,1}, V_{k+1,2}, \cdots, V_{k+1,j}\}. \tag{5}$$

The reconstruction process can be summarized as follows:

1. For every point in C_k, find out the nearest point among the points of C_{k+1}. Connection is made by straight lines for every point in C_k with its corresponding nearest points.
2. Repeat step 3 for $0 \le n \le m-1$ where n is even.
3. Let $V_{k+1,i}$ and $V_{k+1,i+t}$ be the nearest points from $V_{k,n}$ and $V_{k,n+1}$ respectively. If t=1 i.e. points are neighbor, then $V_{k+1,i}$ is connected to $V_{k,n+1}$ or $V_{k+1,i+1}$ is connected to $V_{k,n}$ forming triangular facet. If t>1 then, let P_1, P_2, ..., P_t be t points between $V_{k+1,i}$ and $V_{k+1,i+t}$. Let d_i be the maximum value between the distance from P_i to $V_{k+1,i}$ and the distance from P_i to $V_{k+1,i+t}$. Let $s \in \{1, 2, ..., t\}$ and we select P_s where $P_s \in \{P_1, P_2, ..., P_t\}$ and d_s is the minimum value of the set $\{d_1, d_2, ..., d_t\}$. The selected point P_s is connected to both $V_{k,n}$ and $V_{k,n+1}$ to generate triangular facet with better shape. All the points from $V_{k+1,i}$ to P_s are connected to $V_{k,n}$ and all the points from P_s to $V_{k+1,i+t}$ are connected to $V_{k,n+1}$ to form rest of the triangular facets.
4. Repeat step 3 for the two nearest points of $V_{k,0}$ and $V_{k,m}$.

When numbers of contours for a particular tooth in neighboring slices are not equal, we construct a composite contour from the set of contours that defines the branches[13]. The composite contour is treated as a single contour and tiled with the corresponding pre branch contour from an adjacent section. Composite contour is constructed by introducing a new node midway between the closest nodes on the branches that connects the adjacent contours. The Z coordinate of this new node is the average of the Z coordinates of the two contour levels involved. The nodes of the branches and the new nodes are numbered again such that they can be considered as being one loop. This means that the new nodes and their immediate neighbors are numbered twice as shown in Figure 3. This composite contour is then tiled with the single contour from the adjacent section.

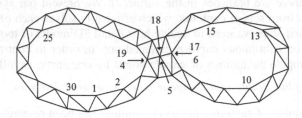

Fig. 3. Solving branching problem inserting new vertices and renumbering

Our method for generating triangular facet allows us to reconstruct 3D tooth model with low complexity of computation. Modeling surface of jaw can be carried out in an analogous manner. Figure 4 shows wireframe models of tooth and jaw generated with proposed method.

   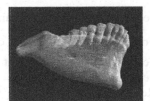

(a) 3D reconstruction of tooth   (b) 3D reconstruction of mandible

Fig. 4. Wireframe models of tooth and mandible

6 Experimental Results

We applied the proposed algorithm to two CT sets of mandible, which were acquired using a conventional CT system, where slice thicknesses were 0.67 and 1mm respectively. Visual C++ and 3D graphics library OpenGL were used for environment tools.

Table 1 lists parts of numerical results of the segmentation algorithm. N is the number of slices over which each tooth spans. FPE (False Positive Error) is the percent of area reported as a tooth by the algorithm, but not by manual segmentation. FNE (False Negative Error) is the percent of area reported by manual segmentation, but not by the algorithm. Similarity and dissimilarity indices[14,15], which show the amounts of agreement and disagreement between the area of the algorithm and manual segmentation, are computed by:

$$S_{agr} = 2\frac{A_{man} \cap A_{alg}}{A_{man} + A_{alg}} , \quad S_{dis} = 2\frac{A_{man} \cup A_{alg} - A_{man} \cap A_{alg}}{A_{man} + A_{alg}} . \tag{6}$$

These indices are calculated for validation on N slices of each tooth. Averaged values are shown in Table 1 and we conclude that proposed method for segmentation isolates individual region of tooth successfully.

Table 1. Segmentation results for 8 teeth

Tooth	N	FPE[%]	FNE[%]	S_{agr}	S_{dis}
1	20	4.43	8.37	0.935	0.131
2	22	7.88	3.45	0.945	0.111
3	25	8.96	4.48	0.935	0.131
4	24	8.46	6.47	0.926	0.148
5	27	5.81	8.29	0.929	0.143
6	26	2.07	7.05	0.953	0.094
7	25	5.21	3.79	0.955	0.089
8	23	5.69	1.42	0.965	0.069

Figure 5a shows the initial contours generated by the proposed optimal threshold algorithm overlaid on the grayscale image. Figure 5b shows the contour extracted in the previous slice and passed to the current slice overlaid on the current grayscale image. Liu and Ma[2] use the contour detected in the previous image as an initial contour for the fitting process. In the figures, however, the initial contour generated by the proposed algorithm is much closer to the actual boundary. This leads to more accurate tooth boundary detection in the fitting process.

(a) The contour by the proposed algorithm   (b) The contour passed from previous slice

Fig. 5. The initial contours overlaid on the current slice

For comparison purpose, segmentation results of our algorithm and existing B-spline snake were shown in following figures. If too many control points represent single contour, it reduce smoothing effect on the curve and consequently generate twisted parts of contour as shown in Figure 6. Figure 7 shows part of test results using different set of slices, which have lower resolution. With the aid of relative exact initial contour, a window of 10x10 size suffices for each control point. As shown in Figure 7, individual tooth often appears with neighboring hard tissues such as teeth and alveolar bones, and the proposed algorithm produces better results than B-spline snake. Stiffening parameter is set to 2 for the following examples of B-spline snake.

Parts of segmentation results in slices are shown in Figure 8 and those of molar having more complicate shape are shown in Figure 9. The figures at most left side show the results of teeth initialization by applying proper threshold to each tooth interactively. As the segmentation is performed slice by slice, in contrast with the result of proposed method, mal-fitting error contained in results of existing method increases. Figure 10 shows the test result in terms of the similarity index, S_{agr} along with the slice number for a molar. The plots are made using the contours of the proposed segmentation algorithm, the proposed adaptive optimal thresholding, and B-spline snake. The left side of horizontal axis in the figure is corresponding to the slice containing the crown of tooth and the right side represents the slice containing the tooth root. As the slice number gets bigger, the amount of alveolar bone surrounding the tooth increases. Figure 10 shows that B-spline snake algorithm fails when the amount of bone increases while the proposed algorithm detects the accurate contours.

(a). By the proposed   (b). By B-spline   (a). By the proposed   (b). By B-spline
method   snake   method   snake

Fig. 6. Tooth contours extracted from CT   **Fig. 7.** Tooth contours extracted from CT
image (number of control points CP = 16)   image sequence (CP = 8)

(a). By the proposed method   (a). By the proposed
                                                 method

(b). By B-spline snake   (b). By B-spline snake

Fig. 8. Tooth contours extracted from CT image sequence   **Fig. 9.** Extracted contours
(CP = 16)   of molar (CP = 32)

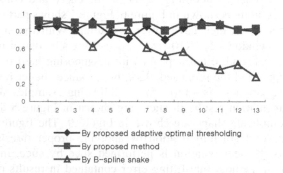

Fig. 10. The similarity index values at each slice for a given tooth

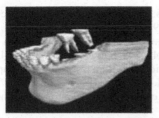
(a) Individual reconstruction of tooth

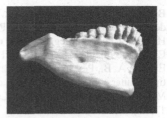
(b) Reconstruction of teeth and jaw

Fig. 11. Reconstruction results of teeth and jaw

Individual tooth forming the model of human mandible was reconstructed based on the segmentation results, as shown in Figure 11. Every tooth in the jaw can be manipulated from jaw for simulation of dental treatments, as shown in Figure 12.

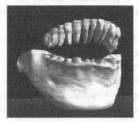
(a) Every tooth can be manipulated

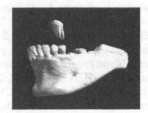
(b) Simulation of having tooth out

Fig. 12. Manipulation of tooth

7 Conclusions and Future Work

In this paper, a novel method is presented for the automated segmentation and surface reconstruction that utilizes a set of 2D CT images of human mandible to obtain the radiography of the jaw, an accurate mechanical model of the individual tooth and those of its root. These anatomical 3-D models can be used for facilitating diagnoses, pre-operative planning and prosthesis design. A significant reduction in operating time and blood loss during dental surgery can also be achieved through the application of these models. The model can be stored along with the patient data and retrieved on demand. The next phase of the research will include the analysis and simulation of the endodontics and orthodontic operations. The further research can also include the modification of the proposed method for reconstructing the 3D model from 2D CT images with low resolution and reconstructing other human organs.

References

1. Ryu, J. H., Kim, H. S., Lee, K. H.: Contour based algorithms for generating 3D medical models. Scanning Congress 2001: Numerization 3D session, Paris, France (2001)
2. Liu, S., Ma, W.: Seed-growing segmentation of 3-D surfaces from CT-contour data. Computer-Aided Design, Vol. 31 (1999) 517-536

3. Brigger, P., Hoeg, J., Unser, M.: B-Spline snakes: A flexible tool for parametric contour detection. IEEE Trans. on Image Processing, Vol. 9, No. 9 (2000) 1484-1496
4. Heo, H., Chae, O.: Segmentation of tooth in CT images for the 3D reconstruction of teeth. Electronic Imaging, Proc. of IS&T SPIE Conf., Vol. 5298. San Jose, USA (2004) 455-466
5. Ekoule, A. B., Peyrin, F., Odet, C. L.: A triangulation algorithm from arbitrary shaped multiple planar contour. ACM Trans. of Graphics, Vol. 10, No. 2 (1999) 182-199
6. Meyers, D., Skinner, S., Sloan, K.: Surfaces from contours. ACM Trans. of Graphics, Vol. 11, No. 3 (1992) 228-258
7. Shimabukuro, M. H., Minghim, R.: Visualization and reconstruction in dentistry. Information Visualization, Proc. of IEEE Conf. (1998) 25-31
8. Bors, A. G., Kechagias, L., Pitas, I.: Binary morphological shape-based interpolation applied to 3-D tooth reconstruction. IEEE Trans. on Medical Imaging, Vol. 21, No. 2 (2002) 100-108
9. Menet, S., Saint-Marc, P., Medioni, G.: Active contour models: overview, implementation and applications. Systems, Man and Cybernetics, IEEE Conf. (1990) 194-199
10. Farin, G.: Curves and Surfaces for CAGD. Academic Press, California (1997)
11. Ballerini, L., Bocchi, L.: Multiple genetic snakes for bone segmentation. EvoWorkshops, Vol. 2611 of LNCS., Springer (2003) 346-356
12. Seipel, S., Wagner, I., Koch, S., Schneider, W.: Three-dimensional visualization of the mandible: A new method for presenting the periodontal status and diseases. Computer Methods and Programs in Biomedicine, Vol. 46 (1995) 51-57
13. Christiansen, H. N., Sederberg, T. W.: Conversion of complex contour line definitions into polygonal element mosaics. ACM SIGGRAPH Computer Graphics, Proceedings of the 5th annual conference on Computer graphics and interactive techniques, Vol. 12 Issue. 3 (1978) 187-192
14. Zijdenbos, A. P., Dawant, B. M., Margolin, R. A., Palmer, A. C.: Morphometric analysis of white matter lesions in MR images: Method and validation. IEEE Trans. Medical Imaging, Vol. 13 (1994) 716-724
15. Klemencic, J., Valencic, V., Pecaric, N.: Deformable contour based algorithm for segmentation of the hippocampus from MRI. CAIP2001, Vol. 2124 of LNCS., Springer (2001) 298-308

Analysis of Facial Configuration
from Realistic 3D Face Reconstruction
of Young Korean Men

Seongah Chin[1] and Seongdong Kim[2]

[1] Division of Multimedia, Sungkyul University, Anyang-City, Korea
solideo@sungkyul.edu
[2] Department of Gameware, Kaywon School of Art and Design, Uiwang-City, Korea
sdkim@kaywon.ac.kr

Abstract. In this paper we analyze facial configuration for specifying young Korean men's 3D virtual face. The approach is performed by defining facial feature points based on muscle-based animation, sampling of the images taken, conducting alignment of the feature vertices on the reference mesh model to feature points on the sampling images and clustering on the feature vertices. Finally we analyze comparative metric, which draws the reasonable conclusion that each mesh model represents highly distinctive patterns.

1 Introduction

It is well known that humans communicate by intrinsically emotional expressions as well as their own languages. Various facial expressions can be thought of as conveying a bunch of information even more than a language itself. Indeed, facial modeling and animation techniques [1, 3–5] have been challenged as one of the fundamental issues and difficult topics caused by extremely complicated expressions of human faces. Modeling human faces for facial animation and recognition has been challenging research topics [5–9].

A considerable amount of research into facial animation has been widely presented to model and animate realistic facial expressions whereas there still exist shortcomings in applying facial patterns for specifying standard ethnic features of face model. This area, however, has been typically less challenged by the research society where they paid attention to the facial animation only for the specific model. Quite a different geometric feature and varying proportion of individual faces between components provide us the certain motivation to challenge the solution of facial expression reconstructing and facial recognition. Our approach begins with defining 3D facial feature points based on muscle-based animation parameter by Parke [1]. The feature points can be utilized facial animation as well as facial recognition. After collecting sample images from Korean young men it is necessary to align the feature vertices on the reference 3D mesh model to the feature points of the individual sample image. Finally Fuzzy c-mean clustering has been applied to gain proper clustering center vertices standing for Korean facial patterns. By experiments, we have shown acceptable results and reasonable conclusions.

D.-K. Baik (Ed.): AsiaSim 2004, LNAI 3398, pp. 685–693, 2005.

2   Facial Model Configuration

Facial model configuration plays a key role in deriving reasonable facial patterns. At first facial feature points should be introduced with respect to 3D vertices followed by collecting sample images from Korean young men. It is necessary to align the feature vertices on a reference 3D mesh model to the feature points of the individual sample image.

2.1   Facial Feature Points

Formulating of the facial features is thought of as the key aspect in the early stage in order to analyze facial patterns properly with respect to comparison metrics among various facial models. The ultimate goal is to derive facial patterns expressing characteristics of each 3D facial model. Accordingly facial feature points determining facial patterns should be defined in terms of usability and distinction between models. In the sequel facial features can be utilized in facial expressions or facial recognition. Intentionally our facial feature points note both key aspects carrying reusability such as muscle motion directions for facial animation, eye, nose, mouth and shape of the face for facial recognition.

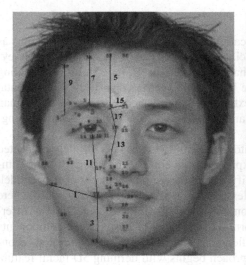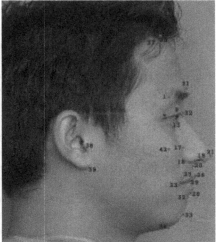

Fig. 1. 42 facial feature points and 9 muscle for one half of the face

Representing effective facial feature points is an essential prerequisite for deriving facial patterns as well as keeping usability. Facial patterns determined by 42 feature points (for one half of the face) are based on 9 muscle (one half of the face) based animation parameters defined by Parke [1] and FACS(Facial Action Coding System) [2] as shown in Figure 1. Figure 1 shows frontal (left) and side (right) view of the feature points.

2.2 Feature Point Alignment

Given identical background and illumination sample images have been gathered by taking facial pictures from a number of university students in twenties. In addition we have models remove accessory such as earrings, necklace and glasses as well to avoid even delicate effects. Models are demanded to move front hair back for the purpose of preventing them from hiding the shape of the face. Canon PowerShot G3 has been used to produce 1024×768 pixels positioning in the fixed distance 1.5m between the model and the lens of the camera. Sample image size should be regarded as same size as our 3D reference mesh model in terms of frontal view and side view.

Once completion of the gathering sample image, skew correction has been performed by obtaining the skew angle between one vertical straight line passing through the nose and one vertical straight line determined by two horizontal line on the image as shown in Figure 2. In addition the size of each sample image is normalized by 450pixel distance between the boundary of the front head and the bottom of the chin.

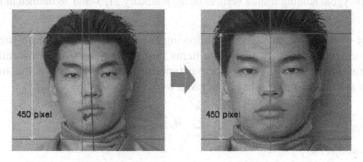

Fig. 2. A sample image and skew correction

How to align feature points on the sample images to feature vertices on the 3D reference mesh model? Thus 42 feature points defined in the previous section need to be marked on the sample images as accurately as possible. Now feature vertices on the 3D reference model associating with ones in the sample image should be adjusted accurately. First, our 3D reference mesh model has been manipulated into transparent model simply by switching opacity. We locate the sample images behind the 3D reference mesh model so that it is possible to align feature vertices to feature points in a sample image by moving feature vertices as shown in Figure 3. Coloring of the specific polygon enable us to avoid ambiguous decision when selecting vertices as shown in Figure 3. For calculating x, y coordinates front view has been used and side view for z coordinate respectively. Finally we are expected to collect transformed 3D mesh model with respect to feature vertices. In the sequel, those feature vertices have been clustered to derive facial patterns.

3 Feature Vertex Clustering

The Fuzzy C-Mean (FCM) algorithm found by Bezdek consists in the iteration of the given Eq. (1)(2). It is a commonly used clustering approach. It is a natural gener-

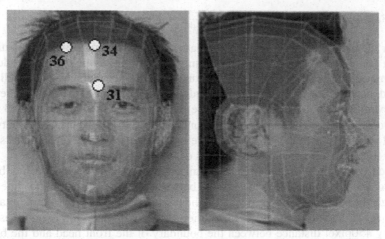

Fig. 3. Aligning feature vertices on the 3D reference model to ones on the sample image with colored polygons holding feature vertices on the boundary 31, 34and 36 defined in Fig. 1

alization of the K-means algorithm allowing for soft clusters based on fuzzy set theory. To classify a data set of N data items into C classes FCM can formulated as a minimization problem of the objective function J with respect to the membership function u and centroid c, where J is given by

$$J = \sum_{k=1}^{C} \sum_{i=1}^{N} u_{ki}^{m} d_{ki}^{2}$$

and it is subject to

$$u_{ki} \in [0,1], \sum_{k=1}^{C} u_{ki} = 1, 0 < \sum_{k=1}^{N} u_{ki} < N,$$
$$1 \le i \le N, 1 \le K \le C \tag{1}$$

Here m (>1) is control parameter of determining the amount of fuzziness of the clustering results. $d_{ki} = \|v_i - c_k\|$ is the Euclidean distance between the observed data v_i and the class centroid c_k and u_k is the membership value reflecting the degree of similarity between v_i and c_k . The objective function J is minimized when high membership values are assigned to feature vertices whose coordinates are close to the centroid of its particular class, and low membership vales are assigned to them when the vertices are far from the centriod. Taking the first derivative of J in Eq. (1) with respect to u_{kj} , c_k we can obtain the following necessary conditions to minimize the objective function J:

$$u_{ki} = \frac{(d_{ki})^{-2/(m-1)}}{\sum_{l=1}^{C}(d_{li})^{-2/(m-1)}}, c_k = \frac{\sum_{i=1}^{N} u_{ki}^{m} v_i}{\sum_{i=1}^{N} u_{ki}^{m}} \tag{2}$$
$$1 \le i \le N, 1 \le K \le C$$

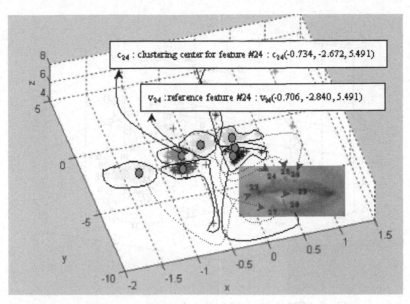

Fig. 4. FCM clusters: 7 feature clustering vertices of the mouth. The clustering center vertex is marked by the red solid circle along with its vertex coordinates. Blue dotted lines indicate corresponding feature points on the sample image

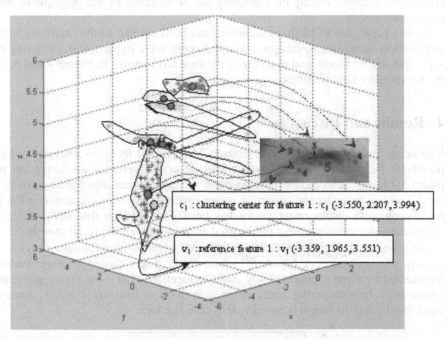

Fig. 5. FCM clusters: 6 feature clustering vertices of the eyebrow. The clustering center vertex is marked by the red solid circle along with its vertex coordinates. Blue dotted lines indicate corresponding feature points on the sample image

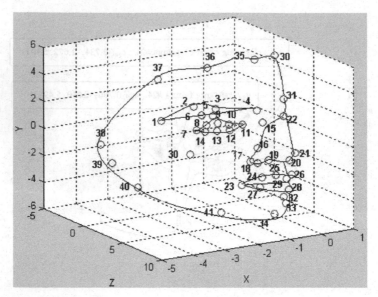

Fig. 6. FCM clusters: 42 feature clustering center vertices

After initialization of the centroids, u_{ki} and c_k are iteratively calculated until some stop criteria are reached. Finally the clustering can be obtained by the principle of maximum membership.

In this paper, the FCM fuzzy algorithm has been applied to the clustering of feature vertices using sample images of young Korean men. In particular, clustering data sets representing mouth and eyebrow of the face have shown in Figure 4 and 5. Figure 6 expresses 42 clustering center vertices.

4 Results and Discussion

Clustering center vertices acquired from FCM represent feature vertices of our proposed 3D facial mesh model. Given clustering center vertices, we derive our proposed 3D facial mesh model. Displacement vectors between the feature vertices on the 3D reference model and clustering center vertices have been calculated in Eq. (3). Each coordinate displacement vector has been computed so that new vertices so called clustering center vertices can be used as our proposed 3D facial models.

Let f_c be the clustering center vector such that $f_c=(c_1, c_2, c_3, \ldots c_N) = (c_{11}, c_{12}, c_{13}, c_{21}, c_{22}, c_{23}\ldots\ldots c_{N1}, c_{N2}, c_{N3}) \in R^{3N}$ where $N=42$. Each clustering center vertex is represented by three coordinates x, y and z sequentially. Let v_i be a reference feature vertex and c_i be a clustering center vertex such that $v_i=(x(u_1), y(u_1), z(u_1))$, $c_i=(x(u_2), y(u_2), z(u_2))$ then the length between v_i and c_i has the form,

$$l_i = \int_{u_1}^{u_2} \sqrt{(\tfrac{dx}{du})^2 + (\tfrac{dy}{du})^2 + (\tfrac{dz}{du})^2}\, du \qquad (3)$$

and c_i is the form,

$$c_i = (x(u_2), y(u_2), z(u_2))$$
$$= (x(u_1) + \frac{dx}{du}, y(u_1) + \frac{dy}{du}, z(u_1) + \frac{dz}{du})$$

Once our 3D facial model is obtained by adjusting the 3D reference model based on displacement vectors. Prior to addressing facial configuration among various models at first comparative facial metric should be introduced in terms of expressing reasonable comparisons. Thus we define 15 comparative metric totally standing for distinguishable configuration such as facial width, head height, head width, eyebrow width, eye height, eye width, glabella width, nose width, nose length, philtrum height, lip width, lip height, nose depth, lip depth, and eye depth as shown in Figure 7. Facial shape is also regarded. The aim of comparisons with four mesh models is to represent relative difference with respect to comparative metric as shown in Figure 7, Figure 8, and Table 1. The clustering mesh model tends to stand for young Korean men's 3D shape by the fact that mean model shown in the second column in Figure 8 may not lie on the sampling images since it does not regard clustering. Furthermore, reasonable comparisons have been acquired between Parke mesh model (the fourth column in Figure 8) since Parke mesh model may be one of western people. Before comparison, normalization has been conducted by adjusting size of mesh models. The clustering model so called young Korean men's standard is rounded shape comparing with Parke model. The clustering model expresses small eyes, short nose, broad front head and thick lip. The clustering model tends to have wide face among four facial models whereas Parke model shows narrow width. For glabella width, the clustering model has wide glabella over Parke model. Parke model expresses wide lips and longer nose over other three models. In Figure 9, mainly distinctive patterns are marked by dotted circles.

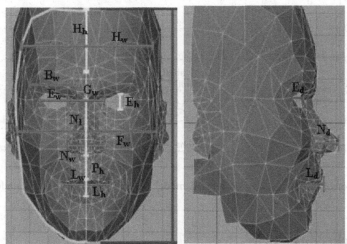

[Notation] F_w: facial width, H_h: head height, H_w: head width, B_w: eyebrow width, E_h: eye height, E_w: eye width, G_w: glabella width, N_w: nose width, N_l: nose length, P_h: philtrum height, L_w: lip width, L_h: lip height, N_d: nose depth, L_d: lip depth, E_d: eye depth

Fig. 7. Comparative metric for facial configuration

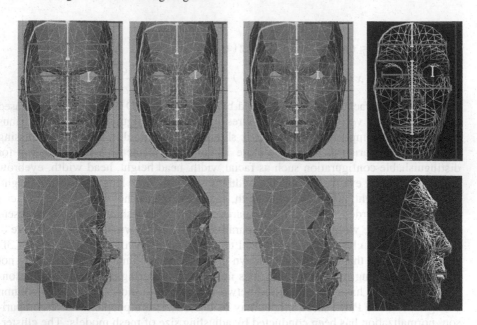

Fig. 8. Comparison of the facial mesh models: reference model, mean model, clustering model and Parke model from the left

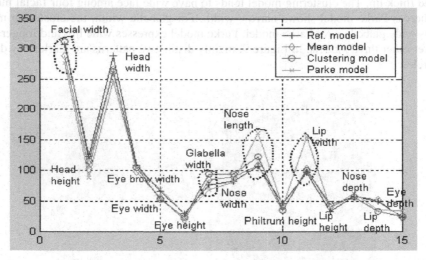

Fig. 9. Plot of facial patterns: Reference model, mean model, clustering model and Parke model

5 Conclusions

In this paper we propose facial configuration for specifying young Korean men's 3D virtual characters. The approach is derived by defining facial feature points based on muscle-based animation, sampling of the images taken, conducting alignment feature vertices on the reference mesh model to feature points on the sampling images and clustering on the feature vertices. Finally we define comparative metric, which draws the reasonable conclusion that each mesh model represents highly distinctive patterns.

Table 1. Comparison among models (in pixels)

Metric\Model	Ref. model	Mean model	Clustering model	Parke model
Facial Width	308	289	312	277
Head Height	122	102	116	88
Head Width	289	266	263	247
Eyebrow Width	108	98	105	104
Eye Height	66	55	53	70
Eye Width	30	20	23	30
Glabella Width	77	85	94	68
Nose Width	83	89	94	81
Nose Length	107	108	123	162
Philtrum Height	46	34	35	41
Lip Width	97	103	100	155
Lip Height	33	40	45	39
Nose Depth	58	58	57	62
Lip Depth	49	52	33	35
Eye Depth	23	22	25	34

References

1. Parke , F. I. , Waters, K: Computer Facial Animation. A K Peters. (1996)
2. Ekman, P., Friesen W.V: Manual for the Facial Action Coding System. Consulting Psychologist Press, Palo Alto. (1978)
3. Guenter, B Grimm, C., Wood, D., Malvar, H. Pighin F..:Making Faces. Proc. ACM SIGGRAPH98 Conf., (1998.)
4. Lee, W.S., Thalmann, N.:Fast head modeling for animation. Journal Image and Vision Computing, Volume 18, Number 4, Elsevier, 1 March. (2000) 355-364
5. Pantic, M., Rothkrantz, L.J.M.:Automatic Analysis of Facial Expressions: The State of the Art. IEEE Pattern Analysis and Machine Intelligence, Vol. 22. (2000) 1424-1445
6. Lee, Y, Terzopoulos, D., Waters, K.:Realistic modeling for facial animation. Proc. ACM SIGGRAPH95 Conf. (1995) 55-62,
7. Guenter,B., Grimm, C., Wood,D., Malvar,H., Pighin, F.: Making Faces. Proc. ACM SIGGRAPH98 Conf. (1998)
8. Kshirsagar, S., Garachery, S., Sannier, G., Thalmann, N.M.: Synthetic Faces:Analysis and Applications. International Journal of Images Systems & Technology. Vol. 13. (2003) 65-73
9. Hsu, R.L., Mottaleb, M-A., Jain, A.K.:Face Detection in Color Images. . IEEE Pattern Analysis and Machine Intelligence, Vol. 24. (2002) 696-706
10. Zhang, Y., Prakash, E.C, Sung, E.: Constructing a realistic face model of an individual for expression animation. International Journal of Information Technology, Vol.8. (2002)
11. Theodoridis, S., Koutroumbas, K:Pattern Recognition. Academic Press. (1999)

Finite Element Modeling
of Magneto-superelastic Behavior
# of Ferromagnetic Shape Memory Alloy Helical Springs

Jong-Bin Lee[1], Yutaka Toi[2], and Minoru Taya[3]

[1] Graduate School, University of Tokyo, Komaba, Meguro-ku,
Tokyo 153-8505, Japan
jongbin@iis.u-tokyo.ac.jp
[2] Institute of Industrial Science, University of Tokyo, Komaba, Meguro-ku,
Tokyo 153-8505, Japan
toi@iis.u-tokyo.ac.jp
[3] Center for Intelligent Materials and Systems, University of Washington, Seattle,
WA 98195-2600, USA
tayam@u.washington.edu

Abstract. The previous study by thefor the finite element analysis of superelastic, large deformation behaviors of one-dimensional shape memory alloy (SMA) devices is extended to the magneto-superelastic analysis of ferromagnetic shape memory alloy (FSMA) devices. The commercial code ANSYS/ Emag for magnetic analysis is combined with the SMA superelastic analysis program developed by the authors. The numerical results for a SMA (NiTi) beam are compared with the experimental results to illustrate the reliability of the superelastic analysis program. The magneto-superelastic analysis of FSMA helical springs is conducted to show the validity of the developed analysis system.

1 Introduction

Ferromagnetic shape memory alloy (abbreviated as FSMA) is under development and is expected as a new material of shape memory alloy (abbreviated as SMA) actuators. The conventional thermoelastic SMA exhibits large strain and also bears large stress. However their actuation speed is usually slow, since the actuation speed is limited by the heating/cooling rates of the SMA. On the other hand, FSMA is driven by applied magnetic field and hence can provide very fast actuation speed with reasonably large strain and stress capability. For this reason, FSMA actuator is attracting attention as a new device, for example, for a morphing aircraft, which is capable of changing the shape of wings like a bird to reduce drag and improve fuel consumption.

It is expected that a computational tool will be used more widely in the design of FSMA-based actuators. However, the computational method has not yet been established for the magneto-superelastic analysis of FSMA helical springs, which are standard components of FSMA actuators. In the present study, the finite element analysis is conducted for the magneto-superelastic behavior of FSMA helical springs. The commercial code ANSYS/ Emag and the superelastic analysis program developed by the authors [1] are used for the magnetic analysis and the superelastic analysis, respectively. In the program, Brinson's one-dimensional constitutive modeling [2],

D.-K. Baik (Ed.): AsiaSim 2004, LNAI 3398, pp. 694–703, 2005.
© Springer-Verlag Berlin Heidelberg 2005

which is relatively simple and phenomenological, is extended to consider the asymmetric tensile and compressive behavior as well as the torsional behavior. The incremental finite element formulation by the total Lagrangian approach is employed for the layered linear Timoshenko beam element equipped with the extended Brinson's constitutive equation. The present method is simple and efficient as it employs the beam element. The calculated results for the superelastic deformation analysis of a SMA (Ni-Ti-10%Cu) beam subjected to 4-point bending as well as the magneto-superelastic analysis of a FSMA helical spring subjected to magnetic forces generated by the hybrid magnet composed of a permanent and electro magnet are compared with the experimental results to show the validity of the present method.

2 Magneto-superelastic Analysis of FSMA Helical Springs

2.1 Magnetic Analysis

The commercial code ANSYS/Emag is used for the magnetic analysis. The method of difference scalar potential is employed with eight-noded hexahedron elements for magnetic solids and four-noded tetrahedron elements (SOLID96) for space. Six-noded trigonal prism elements (INFINI11) are used as infinite elements. SOURCE36 is used to model electric source. In ANSYS, magnetic forces are calculated in the air elements adjacent to the body of interest. Sufficiently fine and non-distorted mesh should be used to obtain accurate magnetic forces, however, it requires considerable meshing effort and computing time. Then the helical spring is simplified as an assembly of circular rings as shown in Fig. 1 with a rectangular cross-section as shown in Fig. 2 in order to avoid the difficulty in meshing for the three-dimensional space of complicated shape surrounding the surface of the helical spring.

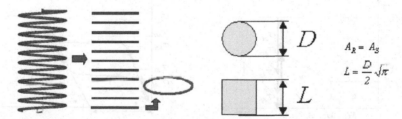

Fig. 1. Idealization of helical spring   **Fig. 2.** Cross-sections

Fig. 3 shows the correspondence between the models for the magnetic and the structural (superelastic) analysis (see also Fig. 9). In the superelastic analysis, the first turn of the helical spring is subjected to the calculated magnetic force on the first ring as shown in Fig. 3(a). The n-th turn of the helical spring is subjected to the calculated magnetic force on the n-th ring under the condition that all rings above the n-th ring are stacked on the yolk of the electromagnet as shown in Fig. 3(b). The stacking effect in the FSMA spring is important as the magnetic flux transmits the spring wire in self-contact. The calculated magnetic forces here give the upper bound values as they are calculated on the assumption that all rings above the ring of interest are stacked. Then, the calculated magnetic forces are reasonable under the strong magnetic flux, while this assumption can make some error under the weak magnetic flux.

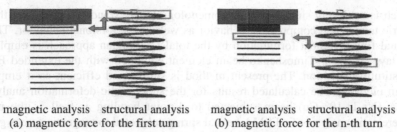

magnetic analysis   structural analysis   magnetic analysis   structural analysis
(a) magnetic force for the first turn   (b) magnetic force for the n-th turn

Fig. 3 Modeling for magnetic analysis and structural (superelastic) analysis

2.2 Superelastic Deformation Analysis

2.2.1 Constitutive Equation for Shape Memory Alloy

The mechanical property of SMA is schematically shown in Fig. 4 [2]. Fig. 4(a) and
Fig. 4(b) are the critical transformation stress versus and temperature and the superelastic stress-strain behavior respectively, in which the following symbols are used: T;
temperature, σ; stress, ε; strain, σ_f^{cr} and σ_s^{cr}; critical finishing and starting stress of
martensite transformation, C_M and C_A; slope for the relation between critical trans-
formation stress and temperature, M_f and M_s; critical finishing and starting temperature of martensite transformation, A_s and A_f; critical starting and finishing temperature of austenite transformation. The loading and unloading at the temperature
higher than A_f cause the superelastic stress-strain behavior as shown in Fig. 4(b).

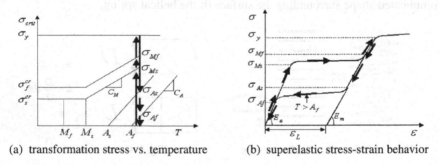

(a) transformation stress vs. temperature   (b) superelastic stress-strain behavior

Fig. 4 Mechanical property of shape memory alloy

The one-dimensional stress-strain relation is generally written as

$$\sigma - \sigma_0 = E(\varepsilon - \varepsilon_0) + \Omega(\xi_S - \xi_{S0}) + \theta(T - T_0) \tag{1}$$

where E; Young's modulus, Ω; transformation coefficient, ξ_S; stress-induced martensite volume fraction, θ; thermal elastic coefficient, T; temperature. The subscript
'0' indicates the initial values. Ω is expressed as

$$\Omega = -\varepsilon_L E \tag{2}$$

where ε_L is the maximum residual strain. Young's modulus E is a function of the martensite volume fraction ξ, which is given by

$$E = E_a + \xi(E_m - E_a) \tag{3}$$

where E_m and E_a are Young's modulus of austenite phase and martensite phase respectively. The total martensite volume fraction ξ is expressed as

$$\xi = \xi_S + \xi_T \tag{4}$$

where ξ_T is the temperature-induced martensite volume fraction. ξ, ξ_S and ξ_T are functions of the temperature T and the stress σ. To consider the difference between tensile and compressive behavior, von Mises equivalent stress σ_e in the evolution equations of ξ, ξ_S and ξ_T is replaced with Drucker-Prager equivalent stress σ_e^{DP} [3] defined as

$$\sigma_e^{DP} = \sigma_e + 3\beta p \tag{5}$$

where β is the material parameter and p is the hydrostatic pressure given by

$$p = \frac{1}{3}(\sigma_x + \sigma_y + \sigma_z) \tag{6}$$

In one-dimensional case, the equivalent stress in eq. (5) is expressed as

$$\sigma^{DP} = |\sigma| + \beta\sigma \tag{7}$$

Substituting eq. (7) into the evolution equations of ξ, ξ_S and ξ_T given by Brinson [2], the evolution equations for the transformation to martensite phase and austenite phase are expressed as follows:

(i) transformation to martensite phase
$T > M_s$ and $\sigma_s^{cr}(1+\beta) + C_M(1+\beta)(T - M_s) < \sigma^{DP} < \sigma_f^{cr}(1+\beta) + C_M(1+\beta)(T - M_s)$:

$$\xi_S = \frac{1 - \xi_{S0}}{2}\cos\left\{\frac{\pi}{\sigma_s^{cr}(1+\beta) - \sigma_f^{cr}(1+\beta)}\left[\sigma^{DP} - \sigma_f^{cr}(1+\beta) - C_M(1+\beta)(T - M_s)\right]\right\} + \frac{1 + \xi_{S0}}{2} \tag{8}$$

$$\xi_T = \xi_{T0} - \frac{\xi_{T0}}{1 - \xi_{S0}}(\xi_S - \xi_{S0}) \tag{9}$$

$T < M_s$ and $\sigma_s^{cr}(1+\beta) < \sigma^{DP} < \sigma_f^{cr}(1+\beta)$:

$$\xi_S = \frac{1 - \xi_{S0}}{2}\cos\left\{\frac{\pi}{\sigma_s^{cr}(1+\beta) - \sigma_f^{cr}(1+\beta)}\left[\sigma^{DP} - \sigma_f^{cr}(1+\beta)\right]\right\} + \frac{1 + \xi_{S0}}{2} \tag{10}$$

$$\xi_T = \xi_{T0} - \frac{\xi_{T0}}{1 - \xi_{S0}}(\xi_S - \xi_{S0}) + \Delta_{T\xi} \tag{11}$$

where $M_f < T < M_s$ and $T < T_0$:

$$\Delta_{T\xi} = \frac{1 - \xi_{T0}}{2}\{\cos[a_M(T - M_f)] + 1\} \tag{12}$$

Otherwise

$$\Delta_{T\xi} = 0 \tag{13}$$

(ii) transformation to austenite phase
$T > A_s$ and $C_A(1+\beta)(T - A_f) < f < C_A(1+\beta)(T - A_s)$:

$$\xi = \frac{\xi_0}{2}\left\{\cos\left[a_A\left(T - A_s - \frac{f}{C_A(1+\beta)}\right)\right]+1\right\} \tag{14}$$

$$\xi_S = \xi_{S0} - \frac{\xi_{S0}}{\xi_0}(\xi_0 - \xi) \tag{15}$$

$$\xi_T = \xi_{T0} - \frac{\xi_{T0}}{\xi_0}(\xi_0 - \xi) \tag{16}$$

where a_M and a_A are given by the following equations:

$$a_M = \frac{\pi}{M_s - M_f}, \quad a_A = \frac{\pi}{A_f - A_s} \tag{17}$$

It is assumed for simplicity that the superelastic shear deformation behavior is qualitatively similar to the normal deformation behavior and both are independent with each other [4]. In the evolution equations for the martensite volume fractions due to the shear stress, ξ_τ, $\xi_{S\tau}$ and $\xi_{T\tau}$ are used for the shear deformation. $\sqrt{3}|\tau|$ is employed instead of σ^{DP} in eq. (7). The shear stress-shear strain relation is expressed by the following equation:

$$\tau - \tau_0 = G(\gamma - \gamma_0) + \Omega_\tau(\xi_{S\tau} - \xi_{S\tau0}) \tag{18}$$

where G; shear modulus, Ω_τ; shear transformation constant, $\xi_{S\tau}$; shear stress-induced martensite volume fraction, T; temperature. The subscript '0' indicates the initial value. Ω_τ is expressed as follows:

$$\Omega_\tau = -\gamma_L G_\tau \tag{19}$$

where γ_L is the maximum residual strain. The shear modulus G is a function of the martensite volume fraction ξ_τ, which is given by

$$G_\tau = G_a + \xi_\tau(G_m - G_a) \tag{20}$$

where G_m and G_a are the elastic shear modulus of martensite phase and austenite phase respectively. The total martensite volume fraction ξ_τ is expressed as

$$\xi_\tau = \xi_{S\tau} + \xi_{T\tau} \tag{21}$$

where $\xi_{T\tau}$ is the temperature-induced martensite volume fraction. ξ_τ, $\xi_{S\tau}$ and $\xi_{T\tau}$ are functions of the temperature T and the shear stress τ.

$\sqrt{3}|\tau|$ is used as the equivalent stress to express the evolution equations of the martensite volume fractions due to shear, which are given by the following replacements in eqs. (8) to (17):

$$f \to \sqrt{3}|\tau|, \quad \beta = 0, \quad \xi \to \xi_\tau, \quad \xi_0 \to \xi_{\tau0}, \quad \xi_S \to \xi_{S\tau}, \quad \xi_{S0} \to \xi_{S\tau0},$$
$$\xi_T \to \xi_{T\tau}, \quad \xi_{T0} \to \xi_{T\tau0}, \quad \Delta_{T\xi} \to \Delta_{T\tau\xi} \tag{22}$$

2.2.2 Finite Element Formulation

Incremental Constitutive Equation

The layered linear Timoshenko beam element [5] as shown in Fig. 5 is used in the finite element analysis of SMA helical springs. The superelastic behavior is assumed for the normal stress (σ)-normal strain (ε) behavior associated with the axial and bending deformation as well as the shear stress (τ)-shear strain (γ) behavior associated with the torsional deformation.

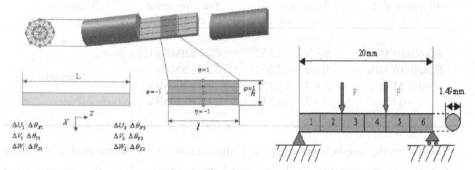

Fig. 5 Layered Timoshenko beam element   **Fig. 6** Ni-Ti-10%Cu alloy beam

The shear deformation associated with the bending deformation is assumed to be elastic and the shear strain energy due to bending is treated as a penalty term because the effect of bending is smaller than torsion in helical springs. The incremental stress-strain relation for the analysis of helical springs is written in the following form:

$$\{\Delta\sigma\} = [D_{se}](\{\Delta\varepsilon\} - \{\Delta\varepsilon_{se}\}) \tag{23}$$

where

$$\{\Delta\sigma\} = \begin{Bmatrix} \Delta\sigma \\ \Delta\tau_{xz} \\ \Delta\tau_{yz} \\ \Delta\tau \end{Bmatrix}, \quad [D_{se}] = \begin{bmatrix} E_{se} & 0 & 0 & 0 \\ 0 & G & 0 & 0 \\ 0 & 0 & G & 0 \\ 0 & 0 & 0 & G_{se} \end{bmatrix}, \quad \{\Delta\varepsilon\} = \begin{Bmatrix} \Delta\varepsilon \\ \Delta\gamma_{xz} \\ \Delta\gamma_{yz} \\ \Delta\gamma \end{Bmatrix}, \quad \{\Delta\varepsilon_{se}\} = \begin{Bmatrix} \Delta\varepsilon_{se} \\ 0 \\ 0 \\ \Delta\gamma_{se} \end{Bmatrix} \tag{24}$$

in which τ_{xz} and τ_{yz} (γ_{xz} and γ_{yz}) are the shear stresses (strains) due to bending. The final form of eq. (23) is given by Toi et al. [1].

Incremental Stiffness Equation

The effect of large deformation is taken into account by using the incremental theory by the total Lagrangian approach in which the nonlinear terms with respect to the displacement in the axial direction are neglected [1]. The following element stiffness equation in an incremental form is obtained by the finite element formulation based on the total Lagrangian approach [5, 6]:

$$([k_0]+[k_L]+[k_G])\{\Delta u\}=\{\Delta f\}+\{f_R\}+ \int_{V_e} [\overline{B}]^T [D]\{\Delta \varepsilon_\theta\} dV^{(0)} \tag{25}$$

where

$$[k_0]= \int_{V_e} [B_0]^T [D][B_0] dV^{(0)} \tag{26}$$

$$[k_L]= \int_{V_e} ([B_0]^T [D][B_L]+[B_L]^T [D][B_0]+[B_L]^T [D][B_L]) dV^{(0)} \tag{27}$$

$$[k_G]= \int_{V_e} [G]^T [S][G] dV^{(0)} \tag{28}$$

Table 1. Material constants of Ni-Ti-10%Cu alloy

Elastic moduli and β	Transformation temperatures	Transformation constants	Maximum residual strain
E_a=60x10³MPa E_m=20x10³MPa θ=0.55MPa/°C β=0.15	M_f= T_0−72.5°C M_s= T_0−52.5°C A_s= T_0−21.7°C A_f= T_0−14.5°C	C_M= 8.0MPa/°C C_A=13.8MPa/°C σ_s^{cr}=100MPa σ_f^{cr}=180MPa	ε_L=0.067

The following symbols are used: $[k_0]$; the incremental stiffness matrix, $[k_L]$; the initial displacement matrix, $[k_G]$; the initial stress matrix, $\{\Delta f\}$; the external force increment vector, $\{f_R\}$; the unbalanced force vector, $[D_{se}]$; the superelastic stress-strain matrix, $\{\Delta \varepsilon_{se}\}$; the superelastic initial strain vector, $[G]$; the gradient matrix, $[S]$; the initial stress matrix, V_e; the element volume.

3 Numerical Examples

3.1 Superelastic Analysis of SMA Beam

The numerical analysis for a SMA (Ni-Ti-10%Cu) beam subjected to 4-point bending has been conducted to illustrate the validity of the superelastic analysis program. Fig. 6 shows a simply supported Ni-Ti-10%Cu alloy beam subjected to 4-point bending. The material constants of Ni-Ti-10%Cu alloy have been determined, based on the material test results [3]. Fig. 7 shows the comparison between the experimental and the assumed stress-strain curves for Ni-Ti-10%Cu alloy. The assumed material constants are shown in Table 1. Fig. 8 shows the calculated load-central deflection curves assuming the symmetric ($\beta = 0$) and asymmetric ($\beta = 0.15$) tensile and compressive stress-strain behaviors. The calculated load-displacement curve by using Drucker-Prager equivalent stress ($\beta = 0.15$) has agreed much better with the experimental curve given by Auricchio and Taylor [3] than the result with von Mises equivalent stress ($\beta = 0$).

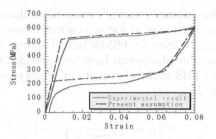

Fig. 7. Stress-strain relations

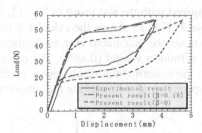

Fig. 8. Load-displacement curves

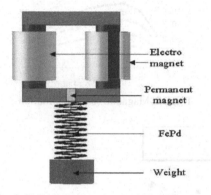

Fig. 9. FePd helical spring on hybrid magnet

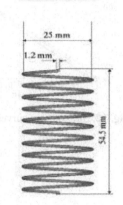

Fig. 10. Dimensions of FePd spring

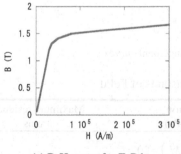

(a) B-H curve for FePd

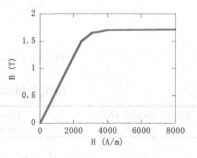

(b) B-H curve for yolk

Fig. 11. Permeability of FePd and yolk

3.2 Magneto-superelastic Analysis of FSMA Helical Spring

Fig. 9 shows a FePd helical spring with a weight subjected to a magnetic force by the permanent magnet (Niodume35, relative permeability : 1.17, coercive force : 835563 A/m) and the electro-magnet (798 turns, 0~1.0A). Fig. 10 shows the dimensions of the helical spring to be analyzed.

Fig. 11 shows B-H curves for FePd and yolk. The material constants of FePd have been determined, based on the tensile test result. The assumed stress-strain curve is compared with the test result in Fig. 12. The assumed material constants are shown in Table 2. Fig. 13 shows the calculated magnetic forces at various current levels. Fig. 14 shows the calculated current-displacement curves for FSMA helical springs with and without a weight (0.49N). The calculated displacements have corresponded well with the experimental results given by the CIMS at the University of Washington as an upper bound solution (see the subsection 2.1).

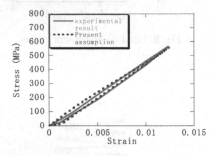

Fig. 12. Stress-strain curves for FePd   **Fig. 13.** Magnetic force distribution

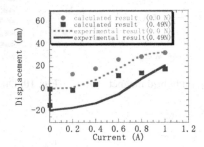

Fig. 14. Current-displacement curves

Table 2. Material constants of FePd

Elastic moduli (MPa) and β	Transformation constants (MPa)	Maximum residual strains
$E_m = 49000$ $E_a = 53000$ $\beta = 0.15$	$\sigma_{Ms} = \sigma_s^{cr} + C_M(T - M_s) = 20$ $\sigma_{Mf} = \sigma_f^{cr} + C_M(T - M_s) = 560$ $\sigma_{As} = C_A(T - A_s) = 18$ $\sigma_{Af} = C_A(T - A_f) = 2$	$\varepsilon_L = 0.001$ $\gamma_L = 0.001$

4 Concluding Remarks

The finite element analysis has been conducted for the magneto-superelastic behavior of FSMA helical springs in the present study, in which Brinson's constitutive modeling has been extended to consider the asymmetric tensile and compressive behavior as well as the torsional behavior. The numerical result for a SMA beam subjected to 4-point bending has agreed well with the experimental result. The magneto-superelastic behavior of a FSMA helical spring has been analyzed. The calculated current-upper bound displacement curves have corresponded well with the experimental results.

References

1. Toi Y, Lee JB, Taya M, Finite element analysis of superelastic, large deformation behavior of shape memory alloy helical springs, Computers & Structures, 82(20/21), 2004, 1685-1693.
2. Brinson LC, One-dimensional constitutive behavior of shape memory alloy: thermomechanical derivation with non-constant material functions and redefined martensite internal variable, Journal of Intelligent Material Systems and Structures, 4, 1993, 229-242.
3. Auricchio F, Taylor RL, Shape memory alloy: modeling and numerical simulations of the finite-strain superelastic behavior, Comput. Methods Appl. Mech. Engng., 143, 1997, 175-194.
4. Sun QP, Li ZQ, Phase transformation in superelastic NiTi polycrystalline micro-tubes under tension and torsion: from localization to homogeneous deformation, Int. J. Solids Structures 2002, 39, 3797-3809.
5. Bathe KJ, Finite Element Procedures, Prentice Hall, 1996.
6. Washizu K, Variational methods in elasticity and plasticity, 3rd Ed., Pergamon Press, 1982.

LSTAFF: System Software
for Large Block Flash Memory

Tae-Sun Chung[1], Dong-Joo Park[2], Yeonseung Ryu[1], and Sugwon Hong[1]

[1] Department of Computer Software, MyoungJi University,
Kyunggido 449-728, Korea
{tschung,ysryu,swhong}@mju.ac.kr
[2] College of Information Science, School of Computing, Soongsil University,
Seoul 156-743, Korea
djpark@computing.ssu.ac.kr

Abstract. Recently, flash memory is widely used in embedded applications since it has strong points: non-volatility, fast access speed, shock resistance, and low power consumption. However, due to its hardware characteristics, it requires a software layer called FTL (flash translation layer). We present a new FTL algorithm called LSTAFF (Large STAFF). LSTAFF is designed for large block flash memory. That is, LSTAFF is adjusted to flash memory with pages which are larger than operating system data sector sizes. We provide performance results based on our implementation of LSTAFF and previous FTL algorithms using a flash simulator.

1 Introduction

Flash memory has strong points: non-volatility, fast access speed, shock resistance, and low power consumption. Therefore, it is widely used in embedded applications, mobile devices, and so on. However, due to its hardware characteristics, it requires specific software operations in using it.

The basic hardware characteristics of flash memory is erase-before-write architectures [5]. That is, in order to update data on flash memory, if the physical location on flash memory was previously written, it has to be erased before the new data can be rewritten.

Thus, the system software called FTL (Flash Translation Layer) [1–3, 6, 8–10] is proposed. The proposed FTL algorithms are designed for small block flash memory in which the size of a physical page of flash memory is same to the size of data sector of operating system. Recently, however, major flash vendors have produced large block flash memory. In large block flash memory, the basic chunk for reading and writing is larger than operating system data sector sizes. That is, large block flash devices physically read and write data in chunks of 2K or 1K bytes. On the other hand, the data sector size of the most operating systems is 512 bytes.

Thus, the new FTL algorithm for large block flash memory should be addressed. We propose a high-speed FTL algorithm called LSTAFF (Large STAFF) for the large block flash memory system. LSTAFF is adapted to large block flash

D.-K. Baik (Ed.): AsiaSim 2004, LNAI 3398, pp. 704–712, 2005.

memory using STAFF [3] which is an FTL algorithm for small block flash memory.

This paper is organized as follows. Problem definition and previous work is described in Section 2. Section 3 shows our FTL algorithm and Section 4 presents performance results. Finally, Section 5 concludes.

2   Problem Definition and Previous Work

2.1   Problem Definition

In this paper, we define operations units in the large block flash memory system as follows.

Definition 1. *Sector is the smallest amount of data which is read or written by the file system.*

Definition 2. *Page is the smallest amount of data which is read or written by flash devices.*

Definition 3. *Block is the unit of the erase operation on flash memory. The size of block is some multiple of the size of page.*

Figure 1 shows the software architecture of a flash file system. We will consider the FTL layer in Figure 1. The File System layer issues a series of read and write commands with logical sector number to read from, and write data to, specific addresses of flash memory. The given logical sector number is converted to a real physical sector offset within a physical page of flash memory by some mapping algorithm provided by FTL layer.

Thus, the problem definition of FTL in the large block flash memory system is as follows. We assume that flash memory is composed of l physical pages which is composed of m physical sectors and the file system regards flash memory as n logical sectors. The number n is less than or equal to $l * m$.

Definition 4. *Flash memory is composed of blocks, each block is composed of pages, and each page is composed of sectors. Flash memory has the following characteristics: If the physical sector location within a physical page on flash*

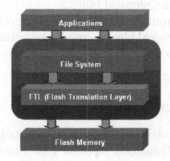

Fig. 1. Software architecture of flash memory system.

memory was previously written, it has to be erased in the unit of block before the new data can be rewritten in the same location. The FTL algorithm in large block flash memory is to produce the physical sector offset within a physical page in flash memory from the logical sector number given by the file system.

2.2 Previous FTL Algorithms

Since large block flash memory has recently been produced, there is no special work on FTL algorithm for large block flash memory. In small block case, we can classify the previous work into three categories: sector mapping [1], block mapping [2, 6, 9], and hybrid mapping [8, 10, 3].

Sector Mapping. First intuitive algorithm is sector mapping [1]. In sector mapping, if there are m logical sectors seen by the file system, the raw size of logical to physical mapping table is m.

If the physical sector location on flash memory was previously written, the FTL algorithm finds another sector that was not previously written. If it finds it, the FTL algorithm writes the data to the sector location and changes the mapping table. If it can not find it, a block should be erased, the corresponding sectors should be backed up, and the mapping table should be changed.

Block Mapping. Since the sector mapping algorithm requires the large size of mapping information, the block mapping FTL algorithm [2, 6, 9] is proposed. The basic idea is that the logical sector offset within the logical block corresponds to the physical sector offset within the physical block.

In block mapping method, if there are m logical blocks seen by the file system, the raw size of logical to physical mapping table is m.

Although the block mapping algorithm requires the small size of mapping information, if the file system issues write commands to the same sector frequently, the performance of the system is degraded since whole sectors in the block should be copied to another block.

Hybrid Mapping. Since the both sector and block mapping have some disadvantages, the hybrid technique [8, 10, 3] is proposed. The hybrid technique first uses the block mapping technique to find the physical block corresponding to the logical block, and then, the sector mapping techniques are used to find an available sector within the physical block.

When reading data from flash memory, FTL algorithm first finds the physical block number from the logical block number according to the mapping table, and then, by reading the logical sector numbers within the physical block, it can read the requested data.

3 LSTAFF (Large STAFF)

LSTAFF is our FTL algorithm for the large block flash memory system. LSTAFF applies STAFF (State Transition Applied Fast Flash Translation Layer) [3] in the large block flash memory system.

3.1   Review: STAFF

STAFF introduced the states of the block. A block in STAFF has the following states.

- F state: If a block is an F state block, the block is a free state. That is, the block is erased and is not written.
- O state: If a block is an O state block, the block is an old state. The old state means that the value of the block is not valid any more.
- M state: The M state block is a first state from a free block and is in place. That is, the logical sector offset within the logical block is identical to the physical sector offset within the physical block.
- S state: The S state block is created by the swap merging operation. The swap merging operation is occurred when a write operation is requested to the M state block which has no more space.
- N state: The N state block is converted from an M state block and is out of place.

We have constructed a state machine according the states defined above and various events occurred during FTL operations. The state machine is formally defined as follows. Here, we use the notation about automata in [7]. An automaton is denoted by a five-tuples $(Q, \sum, \delta, q_o, F)$, and the meanings of each tuple are as follows.

- Q is a finite set of states, namely $Q = \{F, O, M, S, N\}$.
- $\sum$ is a finite input alphabet, in our definition, it corresponds to the set of various events during FTL operations.
- δ is the transition function which maps $Q \times \sum$ to Q.
- q_0 is the start state, that is a free state.
- F is the set of all final states.

Figure 2 shows the block state machine. The initial block state is F state. When an F state block gets the first write request, the F state block is converted to the M state block. The M state block can be converted to two states of S and N according to specific events during FTL operations. The S and N state block is converted to O state block in the event e4 and e5, and the O state block is converted to the F state block in the event e6. The detailed description of the events is presented in [3].

3.2   Large Block Extension

Since a page is composed of more than one sector in large block flash memory, we should consider more complicated logical to physical mapping scheme. That is, we can classify the mapping technique in small block case into two kinds: sector and block mapping. In large block case, however, we should consider sector, page, and block level mapping.

To solve the FTL problem presented in Section 2.1, we first get logical page number, logical block number, and logical sector offset from the input logical

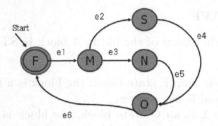

Fig. 2. Block state machine.

sector number. If the input logical sector (lsn) is given, the logical block number (lbn), logical page number (lpn), and logical sector offset (lso) is calculated as follows. Here, np is the number of pages in a block and ns is the number of sectors in a page within a block.

$$lbn = lsn/np \tag{1}$$
$$tmp = lsn\%np \tag{2}$$
$$lpn = tmp/ns \tag{3}$$
$$lso = tmp\%ns \tag{4}$$

Example 1. If a flash memory is composed of 1024 blocks, each block composed of 64 pages, and each page is composed of 4 sectors, the capacity of the flash memory is 128MB. If the logical sector number is 101, the lbn is 1, lpn is 9, and lso is 1.

To store mapping information on flash memory, we designed the page format of LSTAFF as in Figure 3.

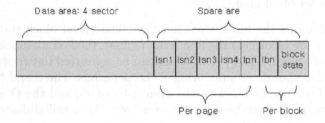

Fig. 3. Page format.

In Figure 3, we assume that a page in a block is composed of four sectors. As mentioned earlier, three levels of logical to physical mapping exits in the large block flash system: sector, page, and block mapping. Thus, logical sector numbers (lsn1, lsn2, lsn3, and lsn4), logical page number (lpn), and logical block number is stored in spare area which exits in each page and is usually used for storing meta information. Here, the lsns and lpn should be stored in each page and the logical block number may be stored in a page per block. From the mapping information, the logical to physical mapping table may be constructed by scanning the spare are of flash memory. Finally, the state of the block may be stored per block.

In LSTAFF, the F, S, O, and M state blocks do not require lsns and lpn since the logical page (sector) offsets are identical to the physical page (sector) offsets in those state blocks. The N state block only require lsns or lpn. There are some options in writing to the N state block as follows.

- Page Mapping: By using lpn area in Figure 3, we can map a logical page number to a physical page number. In this case, the logical sector numbers (lsn1, lsn2, lsn3, and lsn4) need not be written to flash memory.
- Sector Mapping: By using lsn area in Figure 3, we can map a logical sector number to a physical sector number. In this case, the logical page number (lpn) need not be written to flash memory.
- 1:n Mapping: In both page and sector mapping above, a logical page (sector) may be mapped to more than one physical page (sector).

Example 2. Figure 4 shows a write example of page and sector mapping. For the input lsn sequence (0,1,2,3,0,1,2,3,8,9,9), the upper side of Figure 4 shows the page mapping and the lower side of Figure 4 shows sector mapping. If we assume 1:1 page mapping, the data D8 in the upper side of Figure 4 should be copied to the location of physical page number 3. In reading from flash memory, most recently written data is valid data. For instance, in upper side of Figure 4, the valid data of lsn 9 is the one in the physical page number 3.

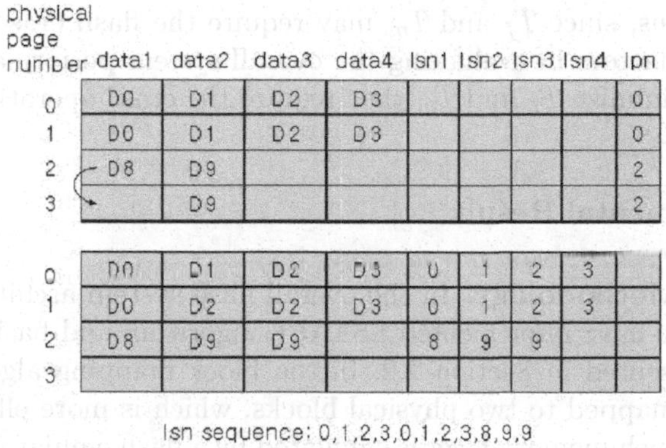

physical
page
number

	data1	data2	data3	data4	lsn1	lsn2	lsn3	lsn4	lpn
0	D0	D1	D2	D3					0
1	D0	D1	D2	D3					0
2	D8	D9							2
3		D9							2

	data1	data2	data3	data4	lsn1	lsn2	lsn3	lsn4	lpn
0	D0	D1	D2	D3	0	1	2	3	
1	D0	D2	D2	D3	0	1	2	3	
2	D8	D9	D9		8	9	9		
3									

lsn sequence: 0,1,2,3,0,1,2,3,8,9,9

Fig. 4. Map example.

4   Experimental Evaluation

4.1   Cost Estimation

The cost function for LSTAFF is similar to that of STAFF. However, since more than one sectors in a page can be read or written simultaneously in large block flash memory, the overall performance of LSTAFF is estimated to be better than that of STAFF. In LSTAFF, the read/write cost can be measured by the following equations:

$$C_{read} = p_M T_r + p_N k_1 T_r + p_S T_r \; (where \; p_M + p_N + p_S = 1) \qquad (5)$$

$$\begin{aligned} C_{write} \quad &= p_{first}[(T_f + T_w)] + (1 - p_{first})[p_{merge}\{T_m + \qquad (6)\\ &\quad p_{e_1} T_w + (1 - p_{e_1})(k_2 T_r + T_w)\} + \\ &\quad (1 - p_{merge})\{p_{e_2}(T_r + T_w) + (1 - p_{e_2})(k_3 T_r + T_w) + \\ &\quad T_r + p_{MN} T_w\}] \end{aligned}$$

where $1 \leq k_1, k_2, k_3 \leq n$. Here, n is the number of pages within a block. In the equation (5), p_M, p_N, and p_S are the probability that data is stored in the M, N, and S state block, respectively.

In the equation (6), p_{first} is the probability that the write command is the first write operation with the input logical block and p_{merge} is the probability that the write command requires the merging operation. p_{e_1} and p_{e_2} are the probability that input logical sector can be written to the in place location with merging and without merging operation, respectively. T_f is the cost for allocating a free block. It may require the merging and the erasing operation. T_m is the cost for the merging operation. Finally, p_{MN} is the probability that the write operation converts the M state block to the N state block. When the write operation converts the M state block to the N state block, a flash write operation is needed for marking states.

The cost function shows that the read and write operations to the N state block requires some more flash read operation than the M or S state block. However, in flash memory the read cost is very low compared to the write and erase cost. Thus, since T_f and T_m may require the flash erase operation, they are dominant factors in evaluating the overall system performance. LSTAFF is designed to minimize T_f and T_m that require the erase operation.

4.2   Experimental Result

Simulation Methodology. In the overall flash system architecture presented in Figure 1, we have implemented LSTAFF algorithm and basic block mapping algorithm presented in Section 2.2. In the block mapping algorithm, a logical block can be mapped to two physical blocks, which is more efficient algorithm. The physical flash memory layer is simulated by a flash emulator which has same characteristics as real flash memory.

We assume that the file system layer in Figure 1 is the FAT file system [4] which is widely used in embedded systems. Figure 5 shows the disk format of the FAT file system. It includes a boot sector, one or more file allocation tables, a root directory, and the volume files. Please refer [4] for more detailed description of the FAT file system. Here, we can see that logical spaces corresponding to the boot sector, file allocation tables, and the root directory are accessed more frequently than the volume files. We got access patterns that the FAT file system on Symbian operating system [11] issues to the block device driver when it gets a file write request. The access patterns are very similar to the real workload in embedded applications.

Boot Sector	File Allocation Table 1 (FAT 1)	File Allocation Table 2 (FAT 2)	Root Directory Entries	Files	

Fig. 5. FAT file system.

Result. Figure 6-(a) shows the total elapsed time. The x axis is the test count and the y axis is the total elapsed time in millisecond. At first, flash memory is empty, and flash memory is occupied as the iteration count increases. The result shows that LSTAFF has much better performance than block mapping.

Figure 6-(b) shows the erase count. The result is similar to the result of the total elapsed time. This is because the erase count is a dominant factor in the overall system performance. [5] says that the running time ratio of read (1 page), write (1 page), and erase (1 block) is 1:4:20 approximately. In addition, LSTAFF shows the consistent performance although flash memory is fully occupied.

Figure 7-(a) and Figure 7-(b) shows the read and write counts respectively. We can see that LSTAFF requires much smaller read and write operations. This is because that the merging operation occurs frequently in the block mapping method.

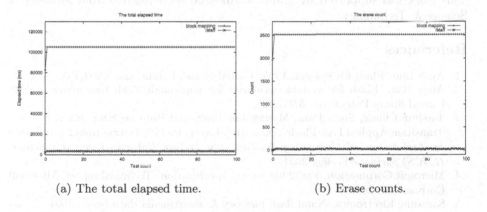

(a) The total elapsed time.   (b) Erase counts.

Fig. 6. The total elapsed time and erase counts.

5 Conclusion

In this paper, we propose a novel FTL algorithm called LSTAFF for large block flash memory. LSTAFF is designed to provide maximum performance in large block flash memory whose page size is larger than file system's data sector size. In LSTAFF which have same characteristics as STAFF, the state of the erase block is converted to the appropriate states according to the input patterns, which minimizes the erase operation. Additionally, we have provided some heuristics for storing to large block flash memory.

712   Tae-Sun Chung et al.

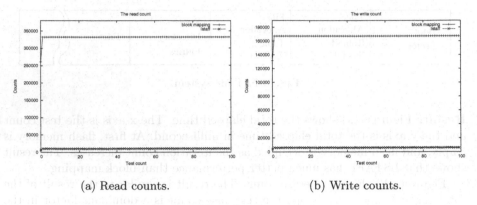

(a) Read counts.   (b) Write counts.

Fig. 7. The read and write counts.

Compared to the previous work, our cost function and experimental results show that LSTAFF has better performance.

Acknowledgment

This work was supported by grant No.R08-2004-000-10391-0 from Ministry of Science & Technology

References

1. Amir Ban. Flash file system, 1995. United States Patent, no. 5,404,485.
2. Amir Ban. Flash file system optimized for page-mode flash technologies, 1999. United States Patent, no. 5,937,425.
3. Tae-Sun Chung, Stein Park, Myung-Jun Jung, and Bumsoo Kim. STAFF: State Transition Applied Fast Flash Translation Layer. In *17th International Conference on Architecture of Computing Systems with Lecture Notes in Computer Science (LNCS) Springer-Verlag*, 2004.
4. Microsoft Corporation. Fat32 file system specification. Technical report, Microsoft Corporation, 2000.
5. Samsung Electronics. Nand flash memory & smartmedia data book, 2004.
6. Petro Estakhri and Berhanu Iman. Moving sequential sectors within a block of information in a flash memory mass storage architecture, 1999. United States Patent, no. 5,930,815.
7. John E. Hopcroft and Jeffrey D. Ullman. *Introduction to automata theory, languages, and computation*. Addison-Wesley Publishing Company, 1979.
8. Jesung Kim, Jong Min Kim, Sam H. Noh, Sang Lyul Min, and Yookun Cho. A space-efficient flash translation layer for compactflash systems. *IEEE Transactions on Consumer Electronics*, 48(2), 2002.
9. Takayuki Shinohara. Flash memory card with block memory address arrangement, 1999. United States Patent, no. 5,905,993.
10. Bum soo Kim and Gui young Lee. Method of driving remapping in flash memory and flash memory architecture suitable therefore, 2002. United States Patent, no. 6,381,176.
11. Symbian. http://www.symbian.com, 2003.

On the Repair of Memory Cells
with Spare Rows and Columns for Yield Improvement

Youngshin Han, Dongsik Park, and Chilgee Lee

School of Information and Communication Engineering, Sungkyunkwan University,
300 Chunchun-dong, Jangan-gu, Suwon, Kyunggi-do, 440-746, S. Korea
yshan@ece.skku.ac.kr

Abstract. As the density of memory chips increases, the probability of having defective components is also increased. It would be impossible for us to repair a memory device if it has numerous defects that cannot be dealt with properly. However, in case of a small number of defects, it is desirable to reuse a defective die (standard unit measuring a device on a wafer) after repair rather than to discard it, because reuse is an essential element for memory device manufactures to cut costs effectively. To perform the reuse, laser-repair process and redundancy analysis for setting an accurate target in the laser-repair process is needed. In this paper new approach for the repair of large random access memory (RAM) devices in which redundant rows and columns are added as spares. So, cost reduction was attempted by reducing time in carrying out a new type of redundancy analysis after simulating each defect.

1 Introduction

Manufacturing of large density memories has become a reality [1]. This has been possible by integration techniques such as very large scale integration (VLSI) and wafer scale integration (WSI). Redundancy has been extensively used for manufacturing memory chips and to provide repair of these devices in the presence of faulty cells. This type of memory has been referred as Redundant Random Access memory (RRAM). Redundancy consists of spare cells arranged into spare rows and columns. These rows and columns are used to replace those rows and columns in which faulty cells lie. This process, commonly referred as memory repair, has considerably increased the yield of these devices, and hence, their cost effectiveness [2–5]. A fabrication method using redundancy is called laser-repair. This method plays a significant role in cutting costs by increasing the yield of faulty devices. Until now, many experiments have been conducted to repair RRAM in the form of wafers. Repair of the algorithm consists of gathering addresses of rows and columns that can be repaired with fault counters on every row and column consisting of a memory device. The fault-driven comprehensive redundancy algorithm [6] depends on user defined preferences to achieve an optimal repair solution in every possible collation. In efficient spare allocation in reconfigurable arrays [7], two algorithm methods for RRAM repair were suggested. The first algorithm uses the branch-and-bound approach which screens subjects in the initial stage of fabrication. The second one uses a heuristic criterion. The proposed CRA (Correlation Repair Algorithm) simulation, beyond the idea of the conventional redundancy analysis algorithm, aims at reducing the time

D.-K. Baik (Ed.): AsiaSim 2004, LNAI 3398, pp. 713–720, 2005.

spent in the process and strengthening cost competitiveness by performing redundancy analysis after simulating each case of defect. In this paper, cost reduction was achieved by reducing the time of unnecessary tests in carrying out a new type of redundancy analysis, after simulating each defect.

2 The Conventional Redundancy Analysis Algorithm

An RRAM usually consists of a rectangular memory die of M rows and N columns. Redundancy consists of R_A rows and C_A columns.

The fault-driven algorithm (FDA)
In fault-driven, repair solutions are generated according to user defined preferences. Repair is implemented using a two-stage analysis: forced-repair and sparse-repair. The first is forced-repair analysis chooses a certain row or column that will be substituted by spare cells located in the same direction as the defect. The second is the sparse-repair analysis that determines the repair method of remaining defects after the first phase by using spare cells that were not used in forced-repair analysis. In the fault-driven algorithm, number of records is generated in repairing the defects [6].

The fault line covering approach algorithm (FLCA)
The following features must be assumed in the repair process.

- There are always enough spare rows and columns such that every parent record in the tree can have two descendant records.
- The total number of faulty cells after forced-repair is T_F. The total number of single faulty cells is S_F. The complexity of the repair process is analyzed after forced-repair.
- The deletion and optimization processes in filtering multiple copies of a record are not considered.

R_A represents the number of spare cells in a row, C_A represents the number of spare cells in a column, the maximum value in the fault counters be denoted as M_F.

In the fault-driven algorithm, $2^{(T_F+1)} - 1$ numbers of records are generated in repairing the defects. In the fault line covering approach algorithm, $2^{(\frac{T_F-S_F}{min}+1)} - 1$ number of records, less than the number of records in the fault-driven algorithm, are generated [9].

Consider the fault pattern shown in Fig. 1, $R_A=C_A=3$. This pattern is the same as considered in [6]. Fig. 2 (A)-(C) shows the iterations required by the repair-algorithm using FLCA.

The Effective Coefficient Algorithm (ECA)
The Effective Coefficient Algorithm [9] is referred as the effective coefficient, because it establishes a relationship between a faulty line and all other lines to repair faulty cells. The coefficient is referred as effective, because it does not only consider the fault counter of the faulty line, but also the complements. The effective coefficient of $R_i(C_i)$ is denoted as $F(R_i)$ $(F(C_i))$.

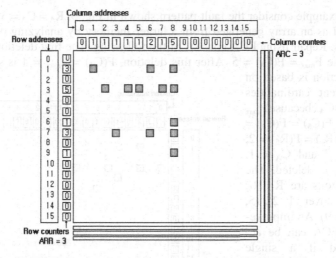

Fig. 1. An example of the occurrence of a defect (FLCA)

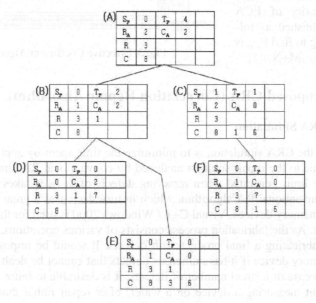

Fig. 2. FLCA algorithm

$F(R_i)$ is given as

$$F(R_i) = N_{F0} + N_{F1} - N_{F2}.$$

N_{F0} is the number of faulty cells on R_i. N_{F1} is the number of faulty cells on of R_i which have no column complement faulty cell. N_{F2} is the number of faulty cells on R_i which have only a single column complement for each cell and the column have no row faulty cell. $F(C_i)$ can be equivalently defined by interchanging row with column in the definitions of N_{F0}, N_{F1} and N_{F2}.

As an example consider the fault pattern shown in Fig. 3, $R_A = C_A = 3$. This can be considered as an array generated from a larger memory by analyzing only the lines with faults. Repair using ECA consists of the following. The first deletion is given by C_8, because $F_{max} = F(C_8) = 5$. After this deletion, $F(C_4) = F_{max} = 4$ is selected. The third selection is based on the different cardinalities of R_A and C_A, because F_{max} $= F(C_2) = F(C_6) = F(C_9) =$ $F(R_3) = F(R_4) = F(R_5) = 2$, but $R_A = 3$ and $C_A = 1$. Hence, R_3 is deleted. The next deletions are R_1, R_5, and C_9 to cover (1, 2), (5, 6), and (4, 9). An improvement to ECA can be accomplished if a single faulty cell counter and a totality counter are used. The complexity of ECA can be established as follows. Sorting to find F_{max} is $O((M+N)\log(M+N))[5]$.

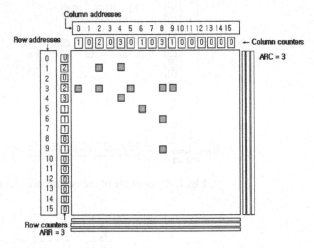

Fig. 3. Effective Coefficient Algorithm

3 The Proposed CRA (Correlation Repair Algorithm)

3.1 The CRA Simulation

The goal of the CRA simulation is to minimize the time spent by applying a correlation technique to the repair solution analyzed by conventional RA simulation. Using the correlate limit generated when repairing defects of devices takes less time than using the conventional RA algorithm, which in turn decreases a great deal of cost in fabricating memory devices. Visual C++ / Windows 2000 is used for the experimental environment. As the fabrication process consists of various operations, there is a possibility of fabricating a final product with defects. It would be impossible for us to repair a memory device if it has numerous defects that cannot be dealt with properly. However, in case of a small number of defects, it is desirable to reuse a defective die (standard unit measuring a device on a wafer) after repair rather than to discard it, because reuse is an essential element for memory device manufactures to cut costs effectively. With this, Laser repair process is needed, and from Laser repair process, to establish exact target, RA became an important factor. Therefore from our research, we've applied CRA (Correlation Repair Algorithm) in our research.

3.2 Experiment

- Sample pattern: 10,000 memory arrays
- Device size (n×n):16×16, 32×32, 64×64, 128×128, 256×256, 1024×1024 rectangular array

- Line redundancy number: ARR (Available redundant rows) = ARC (Available redundant column)
- Input factor of CRA: random fault pattern

Table 1. Simulation condition of CRA

Array Size	Number of Defective Cells	Number of ARR	Number of ARC
32 x 32	13	2	2
64 x 64	13	4	4
128 x 128	28	4	4
256 x 256	30	5	5
512 x 512	45	10	10
1024 x 1024	60	20	20

3.3 CRA (Correlation Repair Algorithm) Define

Correlation coefficient of pattern X, Y is as following.

$COR\ (X,Y)$

$$= COV\ (Z_X, Z_Y) = E\left[\left(\frac{X - \mu_X}{\sigma_X}\right)\left(\frac{Y - \mu_Y}{\sigma_Y}\right)\right]$$

$$= \frac{COV\ (X,Y)}{\sqrt{V(X)\,V(Y)}} = \frac{E(XY) - E(X)E(Y)}{\sqrt{V(X)\,V(Y)}}$$

$$\mu_X = E(X),\quad \sigma_X^2 = V(X),\quad Z_X = \frac{(X - \mu_X)}{\sigma_X},\quad Z_Y = \frac{(Y - \mu_Y)}{\sigma_Y}$$

$$COV(X,Y) = E[(X - \mu_X)(Y - \mu_Y)] = E(XY) - E(X)E(Y)$$

$$V(X) = E[(X - \mu_X)^2] = E(X^2) - \{E(X)\}^2$$

Improvement plan of speed with massive DB
Improvement plan of speed with massive DB using correlation algorithm, correlation coefficient of input pattern X and registered pattern Y1, Y2, YN is as following.

$COR(X,Y)$

$$= \frac{E[(X - \mu_X)(Y - \mu_Y)]}{\sqrt{E[(X - \mu_X)^2]\,E[(Y - \mu_Y)^2]}}$$

Available Preprocessing Term
μ_Y
$E[(Y - \mu_Y)^2]$

$$= \frac{E(XY) - E(X)E(Y)}{\sqrt{[E(X^2) - \{E(X)\}^2][E(Y^2) - \{E(Y)\}^2]}}$$

$$= \frac{E(XY) - \mu_X \mu_Y}{\sqrt{[E(X^2) - \mu_X^2][E(Y^2) - \mu_Y^2]}}$$

Available Preprocessing Term
μ_Y
$[E(Y^2) - \mu_Y^2]$

Calculated time $E[(X - \mu_X)^2]$ is faster than $E(X^2) - \mu_X^2$, and even if $Y - \mu_Y$ is float(4byte), Y is unsigned char (1byte). Therefore whole memory size decreases.

3.4 Main Flowchart

This Fig.5. shows the correlate's flow in detail. We extract the correlation value by obtaining the average, covariance, and dispersion by retrieves each fail type store in the database.

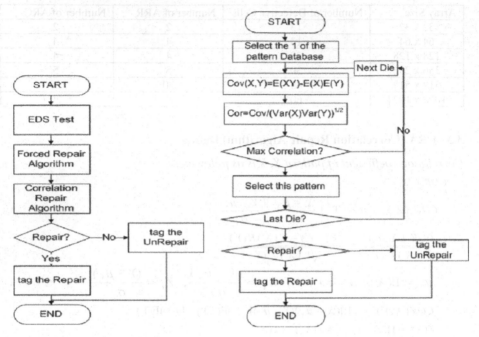

Fig. 4. CRA process simulator flow   **Fig. 5.** Correlate's flow in details

3.5 Result of the Simulation

Process of CRA Simulation

- Store information of all the possibilities for repair with spare cells in the database
- Load files containing fail bit map information obtained from the previous test.
- Begin the simulation by starting the correlate process for each fail type stored in the database and the fail type of files containing fail bit map information.

Fig.6 shows the CRA simulation result of saved pattern in data base and defect type pattern with 32 X 32 sized devices. Fig.7. shows the run time by square memory array size. The vertical line represents the run time, and the horizontal line represents array size. The conventional RA Algorithm has weak points since it needs many records and exponential has complexity. To produce the best output with RA Algorithm, the fault pattern should be analyzed several times. But with CRA, the fault pattern is already stored in the database and selects the best fitting pattern. Therefore this can improve on the performance of the conventional algorithm.

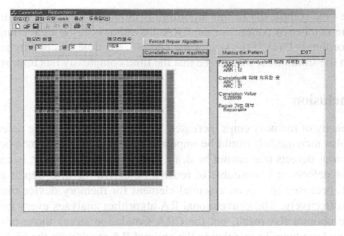

Fig. 6. CRA simulation result

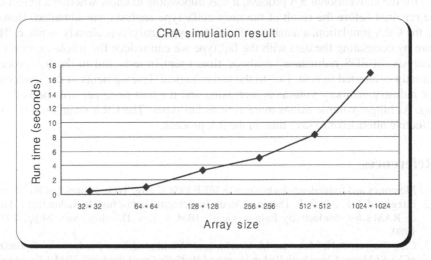

Fig. 7. The run time of square memory array size

Table 2. An evaluation of performance of the conventional RA algorithm versus the proposed CRA algorithm

	RA Algorithm	CRA
Size of ARR = N Size of ARC = K	O((N+K)log(N+K))	O(1)
Device size = I	O(1)	O(I)
Pattern number = J	O(1)	O(J)

3.6 Performance Evaluation

In the conventional RA Algorithm, as the number fault cells increases, the time of performance increases. But, with the proposed in this paper, the only object which can affect the time complexity of this algorithm is number of patterns and the size of the cell. If the number of patterns in the CRA algorithm is J, the time complexity of the CRA algorithm is O(J), if the size of the cell is I, the time complexity is O(I). Therefore, with the simulation, we can see that the time complexity has improved.

4   Conclusion

As the density of memory chips increases, the probability of having defective components is also increased. It would be impossible for us to repair a memory device if it has numerous defects that cannot be dealt with properly. However, in case of a small number of defects, it is desirable to reuse a defective die after repair rather than to discard it, because reuse is an essential element for memory device manufactures to cut costs effectively. The conventional RA algorithm analyzes every fail type to calculate the optimal RA result, but the CRA simulation stores those fail types in the database and use them in calculating the optimal RA result with the highest similarity. As for the conventional RA process, it was impossible to know whether a defect could be repaired before the result of the main cell's type analysis was obtained. However, in the CRA simulation, a database of each fail type analysis is already in place. Therefore by correlating the data with the fail type we can reduce the whole process of the analysis. In EDS redundancy analysis, time spent in tests and in the RA process is directly connected to cost. Due to the technological developments in the semiconductor industry, memory volume is increasing and the unit price per volume is decreasing. As bigger volume means more various fail types, The CRA simulation will be an effective alternative to save time in the RA process.

References

1. Memories and redundancy techniques in IEEE ISSCC Dig. Tech. Papers, pp. 80-87, 1981.
2. Fitzgerald, B.F. and E. P. Thoma, "Circuit Implementation of fusible Redundant Addresses of RAMS for Productivity Enhancement", IBM J. Res. Develop., vol. 24,pp. 291-298, 1980.
3. C.H. Stapper, A.N. McLaren, M. Dreckmann, "Yield model for Productivity Optimization of VLSI Meory Chips with Redundancy and Partially Good Product", IBM J.Res. Develop. Vol. 24, No. 3, May 1980.
4. Evans, R.C., "Testing repairable RAMs and Mostly good Memories", Proc. Int. Test Conf., pp. 49-55, 1981.
5. M. Tarr, D. Boudreau, R. Murphy, "Defect analysis system speeds Test and repair of redundant memories", Electronics, pp.175-179, January 12, 1984.
6. J. R. Day, "A fault-driven comprehensive redundancy algorithm for repair of dynamic RAMs" IEEE Design & Test, vol. 2, no. 3, pp.33-44, 1985
7. S-Y. Kuo and W. K. Fuchs, "Efficient spare allocation in reconfigurable arrays, IEEE Design & Test, vol. 4, pp. 24-31, 1987.
8. C-L. Wey and F.Lombardi, "On the repair of redundant RAM's" IEEE Trans. ComputerAided Design & Test, vol. 2, no.3, pp33-44, 1985.
9. Wei Kang Huang, Yi Nan Shen, Fabtrizio Lombardi, "New Approaches for the Repairs of Memories with Redundancy by Row/ Column Deletion for Yield Enhancement", Transactions on Computer-Aided Design, Vol. 9, no. 3, march 1990.
10. Averill M. Law, W.David Kelton, "Simulation Modeling and Analysis", Third Edition, Chap4. pp. 235-260 , 2000.

Effective Digital IO Pin Modeling Methodology
Based on IBIS Model

Won-Ok Kwon and Kyoung Park

Internet Server Department, Electronics and Telecommunication Research Institute,
161 Gajeong-dong, Yuseong-gu, Daejeon, 305-350, South Korea
{happy,kyoung}@etri.re.kr

Abstract. IBIS (I/O buffer information specification) model is widely used in
signal integrity analysis of on-board high-speed digital systems. IBIS model is
converted equivalent SPICE behavioral model when used board-level simulations. It is important to represent accurately output buffer's switching characteristics converting IBIS model to SPICE behavioral model. This paper proposes a
new modeling algorithm to represent output buffer's switching characteristics
in IBIS model. The accuracy of the proposed algorithm has verified through
SPICE simulation with other behavioral models.

1 Introduction

The higher integration density of digital circuits and clock speed, the more effective
simulation that analyzes signal integrity (SI) on PCB is needed. Such simulation is
needed electrical model that describes driver and receiver's behavior of integrated
circuit (IC). For this purpose, IBIS model is made and it is used to analyze SI on
PCB.

Recently, IBIS model becomes standard to express input/output behavioral characteristics of IC providing current versus voltage (DC IV), rising and falling voltage
versus time (VT) and packaging information of I/O pin in type of table information.
IBIS model has many advantages compared to SPICE model. It can protect proprietary information (IP) about both the circuit design and the underlying fabrication
process and run much faster than SPICE model. At the same time no accuracy is
sacrificed. So IBIS model has become widely used among EDA vendors, semiconductor vendors and system designers, and it becomes an international standard.

But widely used SPICE-based EDA tools do not support IBIS model in electronic
analysis on chips. Therefore the conversion of IBIS model into equivalent SPICE
model becomes significant to many applications. The conversion algorithm of IBIS
model into SPICE behavioral model makes it possible to analysis SI on PCB fast and
efficiently.

In this paper, a new method is proposed to convert IBIS model into SPICE behavioral model. This new algorithm can represent the dynamic feature of output buffer
by extracting accurate switching time coefficient (STC) from pullup and pulldown
transistor at IBIS output behavioral model.

Generally STC can be extracted accurately from IBIS model which has two-pair
VT table [1]. But this algorithm cannot be applied to general IBIS model which has
one-pair VT table. On the other hand, some papers show general STC extraction

D.-K. Baik (Ed.): AsiaSim 2004, LNAI 3398, pp. 721–730, 2005.
© Springer-Verlag Berlin Heidelberg 2005

algorithm from one-pair VT table [2]. But it has a defect of low accuracy in compared with STC extracted two-pair VT table. A new algorithm proposed this paper is made of linear and transient model using only one-pair VT table. The proposed model accuracy is verified through SPICE simulation under test load condition with other models.

2   SPICE Behavioral Model Based on IBIS Model

2.1   Modeling Issue at IBIS Model

IBIS output buffer behavioral model widely divides three elements as figure 1. At the first part, pullup and pulldown transistor have static DC IV table and dynamic rising and falling VT table. As the second element, power clamp and ground clamp diodes have only DC IV table. And pad capacitor and packaging lumped RLC are just passive component.

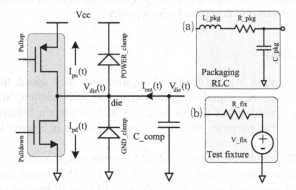

Fig. 1. IBIS output buffer behavioral model. (a) Package RLC, (b) VT table measurement load

DC IV curve of pullup and pulldown which has no influences on each transistor is in the steady-state value. That is, DC IV curve is recorded at the pullup and pulldown table each when output buffer is logic high and low state. DC IV curve of clamping diode is recorded at power clamp table and ground clamp table individually.

When the output IBIS model is translated into the analog SPICE behavioral model, DC IV table can be represented by G-table which is voltage controlled current source (VCCS) of SPICE. Packaging lumped RLC and pad capacitor can be used by SPICE directly. In IBIS model, all static characteristics can be translated into SPICE without much trouble. But the main problem is how to generate a suitable large signal model for transient simulation based on IBIS dynamic information, VT table.

Figure 1(b) shows load condition when VT table measure. In this circuit, die node voltage, Vdie(t) is measured by timescale under the condition of eliminating packaging component and adding specific load, R_fix and V_fix. Most IBIS models provide ramp or VT table information. Optionally over version 2.1 IBIS model provides two-pair VT table information under two specific load conditions. Two-pair VT table's model is generated under two different load conditions that is Vcc and GND at V_fix value in figure 1(b). The other hand one-pair VT table's model is generated from only one load condition; V_fix is set to Vcc or GND.

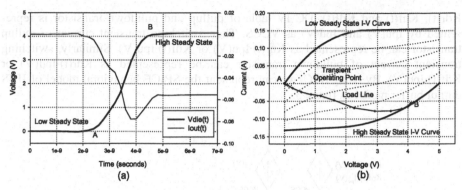

Fig. 2. Transient current using load-line method. (a) Vdie(t), Iout(t), (b) Pullup and pulldown transient IV characteristic curves and load-line

Figure 2 shows transient current Iout(t) using load-line method in figure 1. When Vdie(t) value changes from low steady state to high steady state, transient current Iout(t) changes as figure 2(a). IV characteristic curve of pullup and pulldown transistor is represented as shown figure 2 (b). The above bold-line is low steady state IV curve of pulldown transistor and the blow bold-line is high steady state IV curve of pullup transistor. There are transient IV curves between them. We can solve the transient operating point in rising period by using load-line method as figure 2(b) using transient current.

But this method has critical problem because we don't know exact transient IV curves during transient time. STC is closely related to transient IV curve. STC is the time coefficient of pullup and pulldown transistor during switching time.

2.2 SPICE Behavioral Model

When a high-to-low state transition takes place, transient current has transit process from high steady-state to low steady-state gradually.[2] During transition, initial high steady-state current is correspond to pullup DC IV curve and end of low steady-state current is correspond to pulldown DC IV curve. That is, pullup transistor gradually transits from turn-off state to turn-on and pulldown transistor gradually transits from turn-on state to turn-off. On the contrary, during a low-to-high state transition, it will change from low steady-state to high steady-state passing through transition state. That is, at low-to-high transition, pullup transistor's STC increases from '0'to '1'and pulldown transistor's STC decreases from '1'to '0'. The '0' value of STC means turn-off state of transistor and '1' means fully turn-on state.

As the result of paper [1], there is much difference at transient simulation result between purely based on a DC behavioral model which don't represent switching information of buffer and SPICE model simulation. STC extracted from VT table represents dynamic characteristics of pullup and pulldown transistor. So it is very important to represent STC effect accurately when making SPICE behavioral model from IBIS.

We define STC of pullup transistor and pulldown transistor as Ku(t) and Kd(t) respectively. When considering rising and falling case, STC can be defined as Kur(t),

Kdr(t), Kuf(t) and Kdf(t). DC IV table of pullup and pulldown transistor is represented by Ipu(V) and Ipd(V) as VCCS. Then switching transient current of pullup transistor can be represented as Kur(t)·Ipu(V) or Kuf(t)·Ipu(V). Similarly, switching transient current of pulldown transistor can be represented as Kdr(t)·Ipd(V) or Kdf(t)·Ipd(V). By using STC, we can represent the SPICE behavioral model of IBIS output buffer model as figure 3.

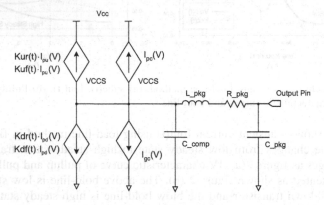

Fig. 3. SPICE behavioral model of IBIS output buffer using STC

2.3 Modeling Using Switching Time Coefficient

Let us move on the extraction methods of four STC that is multiplied to the pullup and pulldown transistor's DC IV table. By using circuit equation in figure 1(b), transient current Iout(t) can be expressed as formula (1).

$$I_{out}(t) = -\left(C_{comp} \frac{d}{dt} V_{die}(t) + \frac{V_{die}(t) - V_{fix}}{R_{fix}} \right) \qquad (1)$$

And Iout(t) is equal to formula (2) in figure 3.

$$I_{out}(t) = Kux(t) \times I_{pu}(V_{die}) + Kdx(t) \times I_{pd}(V_{die}) + I_{pc}(V_{die}) + I_{gc}(V_{die}) \qquad (2)$$

Two unknown variables, Kux(t) and Kdx(t) can be solved at simultaneous equation of formula (1) and (2) under two different VT table condition. So if IBIS model has two different VT table, it can be solved the Kux(t) and Kdx(t) using above equation. In this paper, we call it two-pair method.

But it is impossible to solve simultaneous equation if only one-pair VT table is given. Actually many semiconductor vendors still provide only one-pair VT table IBIS model. To solve these equations, we can simply think about steady state equation as formula (3). It represents linearized switching process using steady state of STC. This is the initial and final condition during switching. Such STC extraction method is called linear method.

$$Kux(t) + Kdx(t) = 1 \qquad (3)$$

But linear method has disadvantage of low accuracy because it can't represent practical switching process. Through the real examples we can see the linear method is far from the real STC.

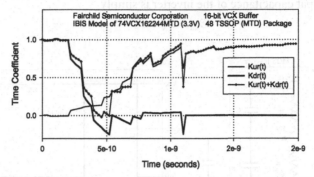

Fig. 4. Switching time coefficient extracted from two-pair method: Fairchild 74VCX162244 IBIS model

Figure 4 shows STC extracted from 74VCX162244 IBIS model by using two-pair method. This model provided Fairchild Semiconductor has two-pair VT tables. As shown the figure, the sum of Kur(t) and Kdr(t) is not 1 during switching time. Only when transistor be in initial and steady state, sum of Kur(t) and Kdr(t) close to 1.

2.4 Proposed Algorithm

To reflect STC's effect at IBIS output model given one-pair VT table, we do modeling about the switching behavioral of push-pull CMOS inverter. Figure 5 shows linear model of a CMOS buffer using pullup and pulldown DC IV tables in IBIS model. As above said linear model is not reflected STC during transient time. So we make new model to represent CMOS switching characteristics.

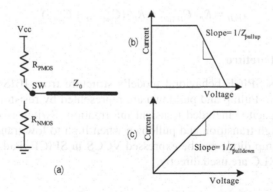

Fig. 5. (a) Linear model of a COMS buffer, (b) Pullup linear impedance, (c) Pulldown linear impedance

Figure 6(a) shows CMOS Inverter model reflected switching characteristic. Cp and Cn capacitor are parallelly connected to linear model. By these register and capacitor we can reflect STC characteristic. Figure 6(b) shows CMOS inverter's input and output waveform. Although the model is shown with both switches open, in practice one of the switches is closed, keeping the output connected to Vcc or ground.[10] The effective output capacitance of the inverter is simply

$$C_{out} = C_N + C_P \tag{4}$$

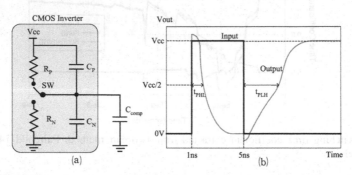

Fig. 6. (a) CMOS Inverter model reflected switching characteristic, (b) Inverter switching state changing

In figure t_{PHL} and t_{PLH} are the propagation delay of inverter and they can be get rising and falling VT table given one-pair IBIS model. The capture point of them is set to when input voltage is same with output voltage that is both pullup and pulldown is in the saturation region. High to low propagation delay, t_{PHL} is product of linear pulldown register and effective falling CMOS output capacitor, C_{out_HL} as formula (5). The same way low to high propagation delay, t_{PHL} is product of linear pullup register and effective rising CMOS output capacitor, C_{out_LH} as formula (6). C_{out_HL} and C_{out_LH} can be solved because all other values are known.

$$t_{PHL} = R_N \cdot C_{LOAD1} = R_N \cdot (C_{out\_HL} + C_{comp}) \tag{5}$$

$$t_{PLH} = R_P \cdot C_{LOAD2} = R_P \cdot (C_{out\_LH} + C_{comp}) \tag{6}$$

2.5 Output Structure

Figure 7 shows SPICE behavioral model's structure from IBIS model applied proposed algorithm. Pullup and pulldown are represented by resistor included DC information and capacitor included transient information. Switch is connected to pullup when low to high transition and pulldown when high to low transition. Power clamp and ground clamp diode can be expressed VCCS in SPICE. Pad capacitor and packaging lumped RLC are used directly.

3 SPICE Simulation

The accuracy of SPICE behavioral model can be verified as figure 8. At first generate IBIS model with two-pair VT table from SPICE model using SPICE to IBIS conversion utility[8]. Then make three different output SPICE behavioral models from IBIS model using two-pair method, linear method and proposed method in this paper. In case of input SPICE behavioral model, all the models are the same. Finally we make test topology then simulate and compare the result each other. The barometer of accuracy is how close to the SPICE result. We can expect two-pair model is the most accurate model because of its accurate STC. The main interest will be the accuracy of proposed model using only one-pair VT table.

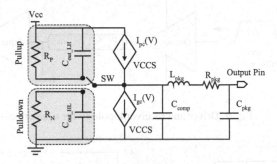

Fig. 7. IBIS output SPICE behavioral model

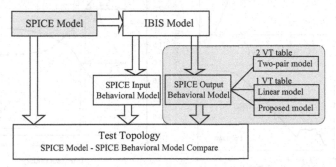

Fig. 8. The verification procedure of SPICE behavioral model generated from IBIS model

This paper performed simulation using 74AC244SC model provided Fairchild Semiconductor Corporation. It is used to memory address driver and clock driver. It is provided to customer that SPICE model and IBIS model with two-pair VT table.

Figure 9 shows simulation topology which is general board inter-connection. Output buffer will be placed with three different behavioral models and SPICE model. The transmission line's impedance is set to 50ohm and its delay is set to 1nS. We will force pulse waveform to output buffer and then measure the wave in front of input buffer.

Figure 10 shows the process of extraction modeling coefficients from 74AC244SC IBIS model using proposed algorithm. Figure 10(a) and 10(b) illustrate how to solve linear resistor of pullup and pulldown. Figure 10(c) shows provided two-pair VT table and we get the transient factor t_{PHL} and t_{PLH} from falling and rising VT table each. We use only VT1 for proposed model. The solved values are as blow. Pad capacitor, C_{comp} value is 5pF. The solved values put into output structure as figure 7 for simulation.

$$R_P = 9\Omega, R_N = 50\Omega, t_{PLH} = 0.5nS, t_{PHL} = 0.4nS, C_{out\_LH} = 50pF, C_{out\_HL} = 3pF$$

We also solve the STC for two-pair model from IBIS model and make SPICE behavioral output buffer as figure 3. In case of linear model, we can make output buffer directly without solve STC.

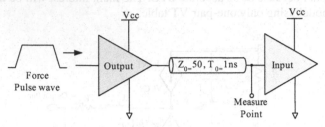

Fig. 9. Driver and receiver topology for simulation

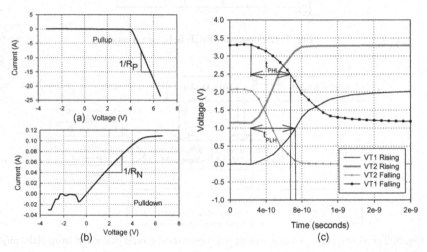

Fig. 10. Extraction modeling coefficient from (a) Pullup IV table, (b) Pulldown IV table, (c) VT table

Figure 11 and 12 show measured rising and falling waveform. The bold-line is SPICE model wave and the dot-line is proposed model wave. The dot-line having asterisk is two-pair model and straight thin line is linear model. As we expected, linear model is far from SPICE model. The performance of two-pair model is fairly good because it is similar to SPICE model. The proposed model also shows good performance under consideration of using one-pair VT table. Although it has some delay compare with two-pair model, it shows much better performance than linear model.

4   Conclusion

This paper proposed a new algorithm making SPICE behavioral model from IBIS model. This algorithm is made to compensate for linear model's inaccuracy given

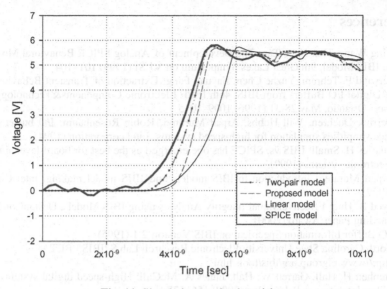

Fig. 11. Simulation result when rising

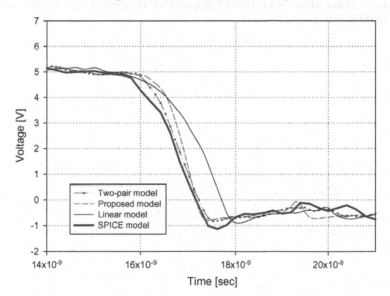

Fig. 12. Simulation result when falling

IBIS model with only one-pair VT table. The main idea of this algorithm is to represent steady state by resistor and transient state by capacitor. The register value is get from DC pullup and pulldown table and capacitor value is get from rising and falling VT table in IBIS model. We do simulation proposed model with two-pair model and linear model. Although proposed model was made by one-pair VT table, it was as accurate as two-pair model in result. Therefore this algorithm can replace linear model converting IBIS model to SPICE behavioral model given one-pair VT table.

References

1. Ying Wang, Han Ngee Tan: The Development of Analog SPICE Behavioral Model Based on IBIS Model, 9th Great Lakes Symposium on VLSI (1999) 101–104
2. Peivand F. Tehrani, Yuzhe Chen, Jiayuan Fang: Extraction of Transient Behavioral Model of Digital I/O Buffers from IBIS, 46th IEEE Electronic Components&Technology Conference, Orlando, May 28-31 (1996) 1009–1015
3. Derrick Duehren, Will Hobbs, Arpad Muranyi, Robin Rosenbaum: I/O buffer modeling spec simplifies simulation for high-speed systems, Intel corporation (1994)
4. Charies H. Small: IBIS vs. SPICE has one emerged as the best for board-level simulation?, Electronic systems (1999)
5. Arpad Muranyi: Introduction to IBIS models and IBIS model making, Intel Corporation (1999)
6. Syed B. Huq: Effective Signal Integrity Analysis using IBIS Models, DesignCon2000 Outstanding Paper Award (2000)
7. I/O Buffer Information Specification IBIS Version 2.1 (1995)
8. North Carolina State University Electronic Research Lab S2IBIS3 BETA: http://www.eigroup.org/ibis/tools.htm
9. Stephen H. Hall, Garrett W. Hall, James A. McCall: High-speed digital system design. A Wiley-Interscience Publication (2000) 156–177
10. R.Jacob Baker, Harry W.Li, David E.Boyce: CMOS circuit design, Layout, and Simulation. IEEE Press (1998) 201–211

Author Index

Lecture Notes in Artificial Intelligence (LNAI)